Partial Differential Equations of Mathematical Physics and Integral Equations

Ronald B. Guenther

and

John W. Lee

Department of Mathematics
Oregon State University

DOVER PUBLICATIONS, INC.

New York

Copyright

Bibliographical Note

This Dover edition, first published in 1996, is an unabridged, corrected and enlarged republication of the work first published by Prentice-Hall, Englewood Cliffs, N.J., 1988. For this edition the authors have corrected a number of typographical errors and provided a new section, ''Solutions and Hints to Selected Problems.''

Library of Congress Cataloging-in-Publication Data

Guenther, Ronald B.
Partial differential equations of mathematical physics and integral equations / Ronald B. Guenther and John W. Lee.
p. cm.
Originally published: Englewood Cliffs, N.J. : Prentice-Hall, c1988.
Includes index.
ISBN 0-486-68889-5 (pbk.)
1. Differential equations, Partial. 2. Integral equations. 3. Mathematical physics. I. Lee, John W., 1942- . II. Title.
QC20.7.D5G83 1995
530.1'55353—dc20 95-35578
CIP

Manufactured in the United States of America
Dover Publications, Inc., 31 East 2nd Street, Mineola, N.Y. 11501

Contents

Preface

During the past several years, we have had the opportunity to give advanced undergraduate and beginning graduate-level courses dealing with those areas of partial differential and integral equations that arise frequently in physical applications. Our students were future mathematicians, physical scientists, engineers, oceanographers, and others. They were good, highly motivated students; however, their backgrounds varied widely and we tried to design our courses to fill, simultaneously, their diverse needs. The mathematicians as a rule were strong mathematically but lacked physical intuition and expertise; yet they wanted to know how to use their mathematical training to set up and solve nontrivial concrete problems. The nonmathematicians, on the other hand, were accustomed to physical reasoning and it was their desire to learn modern mathematical techniques in a setting compatible with their interests so that they could set up and analyze rigorously problems arising in their disciplines. At the same time, the nonmathematicians usually found abstract presentations beyond their stage of mathematical development. Indeed, one of their main reaons for taking a course such as this was to develop their mathematical ability to the point where they could take advantage of modern mathematical techniques to solve problems in their own fields. The present book is our attempt to strike a balance between the needs of both groups. On the one hand, the mathematics used is rigorous but not overwhelming. On the other hand, we model nontrivial physical situations with care and emphasize the lively feedback between a beginning model, physical experiments, mathematical predictions, and the subsequent refinement and reevaluation of the physical model itself.

The minimal mathematical prerequisites for this text are a course in advanced calculus or familiarity in dealing with mathematical concepts as presented in a good course in mathematics for engineers, and a basic exposure to elementary matrix methods. Most of the mathematical prerequisites are covered in Chapter 3 on Fourier series. This prerequisite material can be learned here for the first time; however, a previous course in advanced calculus would provide the reader with a better point of departure. Although we derive virtually all our physical models in detail and from first principles, a modest general background in physical reasoning is needed. A standard one-year university physics course will provide adequate background.

We have developed the mathematical theory in close connection with physical problems. In fact, we begin in Chapter 1 with a discussion of various physical problems and equations that play a central role in applications. In this setting, a number of results which are obtained in later chapters appear natural and intuitively obvious. The following chapters take up the theory of partial differential equations, including detailed discussions of uniqueness, existence, and continuous dependence questions, as well as techniques for constructing solutions. The mathematical treatment is interspersed with physical discussions which are used to motivate and guide the development. On the other hand, mathematical rigor is not sacrificed. Nowhere is physical reasoning used as a "proof". Engineers and scientists are not well served by substituting plausibility arguments for solid proofs and by leaving out technical and mathematical details from hypotheses and proofs of theorems. Indeed, one of the principal tests instilling confidence in the soundness of a proposed mathematical model is mathematical proof that the model predicts physically reasonable results. Engineers and scientists need the mathematical tools and training required to develop and understand such proofs. For their part, mathematicians are not well served by believing that scientists and engineers will formulate applied problems, which mathematicians will then solve. The mathematician must have a reasonable understanding of the physical situation and the underlying assumptions before he or she can attack the problem intelligently. Difficult physical problems require the combined expertise of mathematicians, scientists, and engineers. To engage productively in such interactions, the mathematician must become familiar with rigorous physical thinking. Although both the mathematical and physical treatments are solid, we have not tried to achieve maximum generality in the physical models or mathematical results given here. Instead, we have developed the models and supporting mathematics only in enough generality to provide a natural setting for the problem at hand. Sometimes the modeling is easy and the mathematics is elementary, and sometimes both are difficult. It is important for students to see both situations. In particular, a student is not well served by thinking that either a problem has a simple, elegant solution or that nothing can be done. Hard problems usually require hard work.

The fields of partial differential equations and integral equations are vast. The early development was intimately linked with applications from science and technology, and it is this internal relationship that we try to exploit. This book is designed to be an introduction to and motivation for further study in these areas. It makes no attempt to be encyclopedic. Almost any book in the field can be used to supplement ours and, by consulting other texts, one will arrive at a broader and deeper understanding of the topics presented here. We highly recommend such outside reading and, to facilitate further study, have included a bibliographic section which can be used to find alternative approaches, deeper and more abstract results, additional topics, more varied applications, and sources for continued study.

Some general remarks about the overall organization of the text are in order. Chapters 2 through 6 deal with problems in one spatial dimension. Chapter 7 is a detailed introduction to the theory of integral equations; then Chapters 8 through 12 treat problems in more spatial variables. Each chapter begins with a discussion of problems that can be treated by elementary means, such as separation of variables or integral transforms, and lead to explicit, analytical representations of solutions. These sections can serve as an introduction to the basic ideas and techniques of partial differential equations and can be given to advanced juniors and to seniors. Chapter 7 is an introduction to the Fredholm theory of integral equations of the second kind, as well as some related topics. The treatment stands on its own, but taken in conjunction with the preceding chapters, it helps clarify and extend the discussion of Fourier series expansions in the method of separation of variables and, moreover, it prepares the way for the existence theorems of potential theory. We have also given this material at the senior level. Chapters 8 through 12 deal with elliptic, hyperbolic, and parabolic problems in more spatial dimensions. The emphasis is on three spatial dimensions, with the case of two spatial dimensions usually left for the problems. Higher-dimensional cases are left to the problems when they are straightforward extensions of the three-dimensional case. Our final chapter, Chapter 12, is devoted to a sampling of additional applications of the methods developed earlier. Each section gives a thumbnail sketch of some interesting and important topic and prepares the way for further study.

The following more detailed suggestions indicate how this text may be used to support courses at different mathematical levels. As we mentioned above, a basic introduction to partial differential equations in one spatial variable is given in Chapters 1 through 6. The text is written to allow flexibility in the amount of rigor used. For example, the method of characteristics is developed and illustrated in one section and proofs are given in another. Thus, the instructor may choose to skip the proofs as he or she sees fit. Similarly, the basic facts about Fourier series and Fourier integrals are discussed and illustrated before detailed proofs are given. Once again, the instructor can adjust the depth of coverage to best serve the students. In any event, all the proofs will be available for the interested student even if they are not covered in class. In an introductory course, the material in one spatial dimension can be augmented with some problems in more spatial variables by including topics selected from Sections 8-1, 8-2, 9-1, 9-2, 10-1, 10-2, 10-3, 10-4, 11-1, and 11-2. The material mentioned above can be covered in a one-semester or two-quarter course. In a year-long sequence, additional topics from Chapters 7 through 12 can be covered. For the most part, this additional material is at a higher mathematical level and we normally give it to graduate students. Whatever course is given, it is likely that several topics will not be covered. Thus, we hope that the text will serve as an inviting source for further study by the student and as a useful future reference.

Finally, we wish to thank our students who saw this material in its unpublished form and whose critical questions forced us to clarify and rework much of the material presented here. Thanks are also due to the editorial staff at Prentice-Hall. They arranged for several very helpful reviews of the original manuscript, and encouraged and supported us throughout the writing of this text.

R.B.G / J.W.L.

1

Elementary Modeling

1-1. Introduction

Whenever a physical situation is described in mathematical terms, certain idealizations must be made to make the problem amenable to mathematical treatment. Care must be taken not to oversimplify because then significant physical properties of the system may be lost. In the end the success or failure of the model is governed to a large extent by how close to reality the idealized model is.

The type of model developed depends in part on the sort of information sought. For example, an entire physical system is often modeled by a variable u which gives the state of the system at time t. Then physical laws are used to derive an equation for u. This approach typically leads to an ordinary differential equation or a system of such equations if u is a vector variable. A typical example of such an approach is in the study of currents in electric circuits. If $u = u(t)$ is the current at a certain time in one of the circuits, a differential equation is derived for it. However, if one asks what the current is at a certain point x in the circuit at the time t, the ordinary differential equation gives no information. Indeed, it is assumed implicitly in the derivation of the differential equation that u is independent of the spatial dimension x. In order to obtain such information a new model and equation involving both spatial and time dependence must be derived. If it is realistic to think of the medium as a continuum, then the attempt to obtain specific information about the system at some point in space and at some time leads in general to a partial differential equation, an integral equation, or to systems of such equations, which may even have to be combined with systems of ordinary differential equations. These equations are typically accompanied by initial and/or boundary conditions dictated by the physical situation.

We begin our study with a derivation of several important partial differential equations. The physical assumptions are stated clearly so that in the event corrections or generalizations must be made, it will be clear where the model must be modified. The solution to these equations and their properties can often best be understood from knowledge of the underlying physical phenomenon itself. So these derivations serve both as a motivation for the study of the partial differential equations that arise and as an aid in understanding them. The derivations also make clear what boundary conditions to pose and the physical meaning of these conditions.

1-2. Small Vibrations of an Elastic String

Let us consider a string tautly stretched between two posts. To set it in motion, we can strike it, pluck it, bow it, and so on. The problem is to study the resulting vibrations. Let us assume that the string has length L and is fastened to posts at each end. We set the origin of the coordinates at one post and lay out an x axis along the line connecting the two posts so that the second post is located at the position $x = L$.

We must now make some assumptions about the string itself and the general nature of vibrations that will occur. Then the mathematical ramifications of these assumptions must be explored. First, we think of the string as a continuum. This means that we will be modeling the motion of the string as we perceive it with our senses. We do not think of the string as being made up of billions of tiny molecules somehow held together by molecular forces. From a mathematical standpoint this means that it is possible to introduce a continuous density function, ρ, such that the integral of ρ over any segment of the string gives the mass of string in that segment. In addition, the following assumptions will be made:

1. The string is perfectly elastic, that is, the string is so flexible that it does not resist deformations and the tension in the string acts tangentially.
2. The vibrations are so small that if we focus our attention on a point x in the equilibrium position, it moves approximately on a line which is perpendicular to the line connecting the posts located at the positions $x = 0$ and $x = L$.

We construct a u axis perpendicular to the x axis at $x = 0$, and assume that the equilibrium position of the string is the horizontal segment $0 \leq x \leq L$. After a certain time t has elapsed, a point whose position in the equilibrium position was x will have a new location which will be denoted by $u(x, t)$. Then for fixed time t, the function $u = u(x, t)$ will give the shape of the string at that time. Now let $\rho_0(x)$ denote the density of the string in the equilibrium position and $\rho(x, t)$ the density at time t. As the string stretches, ρ and ρ_0 will be somewhat different. If we focus our attention on an arbitrary interval between $x = x_1$ and $x = x_2$ along the string, we find that the mass m in this interval satisfies

$$\int_{x_1}^{x_2} \rho_0(x)\, dx = m = \int_{x_1}^{x_2} \rho(x, t)[1 + u_x^2(x, t)]^{1/2}\, dx.$$

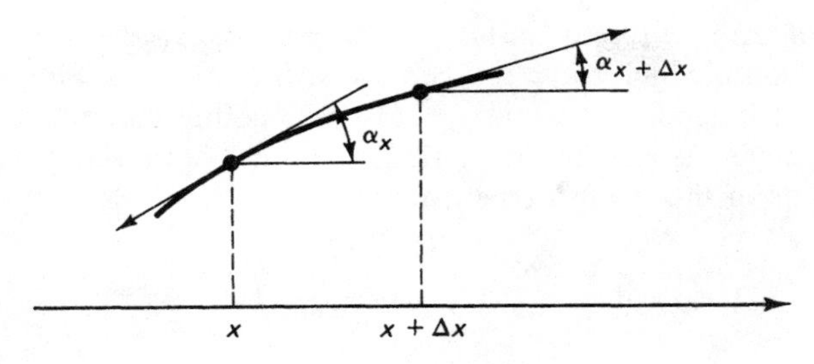

Figure 1-1

Since the interval $[x_1, x_2]$ is arbitrary and the integrands are continuous, we conclude that

$$\rho_0 = \rho(1 + u_x^2)^{1/2}. \tag{2-1}$$

This equation simply expresses the conservation of mass. Bear in mind here that the string moves perpendicular to its equilibrium position.

Next consider an element of the string over an arbitrary interval $[x, x + \Delta x]$. We determine the forces acting on this small piece of string and obtain an expression for the balance of linear momentum. Denote by T_x the tension at the point x and consider Fig. 1-1.

The first observation to make is that since only vertical motion takes place, the forces in the horizontal direction have to balance, which in view of the diagram implies that

$$T_{x+\Delta x} \cos \alpha_{x+\Delta x} - T_x \cos \alpha_x = 0$$

for all Δx, or upon dividing by Δx and letting $\Delta x \to 0$,

$$\frac{\partial}{\partial x} T_x \cos \alpha_x = 0,$$

whence

$$T_x \cos \alpha_x = \tau, \tag{2-2}$$

where τ is either a constant or possibly a function of t. Note that $\tau > 0$ because it is the magnitude of the horizontal component of the tension.

The vertical motion is determined by the fact that the time rate of change of linear momentum is equal to the sum of the forces acting in the vertical direction. The momentum of the small element is given, using (2-1), by

$$\int \rho u_t \, ds = \int_x^{x+\Delta x} \rho(1 + u_x^2)^{1/2} u_t \, dx = \int_x^{x+\Delta x} \rho_0 u_t \, dx$$

and its time rate of change is

$$\frac{d}{dt} \int_x^{x+\Delta x} \rho_0 u_t \, dx = \int_x^{x+\Delta x} \rho_0 u_{tt} \, dx.$$

We distinguish two kinds of forces acting on the segment of the string, those due to the tension, which serve to keep the string taut and whose horizontal components were balanced previously, and those acting along the whole length of the string, such as the weight. From (2-2) and our diagram we find the net force acting at the ends of the string element to be

$$\begin{aligned} T_{x+\Delta x} \sin \alpha_{x+\Delta x} - T_x \sin \alpha_x &= \tau\left(\frac{\sin \alpha_{x+\Delta x}}{\cos \alpha_{x+\Delta x}} - \frac{\sin \alpha_x}{\cos \alpha_x}\right) \\ &= \tau(\tan \alpha_{x+\Delta x} - \tan \alpha_x) \\ &= \tau[u_x(x + \Delta x, t) - u_x(x, t)]. \end{aligned}$$

Next, the weight of the string acting downward is

$$-\int \rho g \, ds = -\int_x^{x+\Delta x} \rho g(1 + u_x^2)^{1/2} \, dx = -\int_x^{x+\Delta x} \rho_0 g \, dx.$$

If an external load is acting on the string (e.g., a violin string is bowed), we model the resultant force by

$$\int \rho f \, ds = \int_x^{x+\Delta x} \rho_0 f(x, t) \, dx,$$

where $f(x, t)$ is a force density which describes the action of the external force. This force density $f(x, t)$ must be given. Finally, it is possible that the medium where the vibrations are taking place will impede the motion. For example, the string may be immersed in water. The form of the law of friction must be given or at least determined empirically. Here we assume a linear law of friction of the form

$$-\int \rho k u_t \, ds = -\int_x^{x+\Delta x} k\rho(1 + u_x^2)^{1/2} u_t \, dx = -\int_x^{x+\Delta x} \rho_0 k u_t \, dx.$$

Now an application of Newton's second law yields

$$\int_x^{x+\Delta x} \rho_0 u_{tt} \, dx = \tau[u_x(x + \Delta x, t) - u_x(x, t)] - \int_x^{x+\Delta x} k\rho_0 u_t \, dx + \int_x^{x+\Delta x} \rho_0(f - g) \, dx.$$

Divide by Δx and let Δx tend to zero to obtain the equation

$$\rho_0 u_{tt} = \tau u_{xx} - k\rho_0 u_t + \rho_0(f - g)$$

or after dividing by ρ_0, setting $c^2 = \tau/\rho_0$ and $F = f - g$,

$$u_{tt} + k u_t = c^2 u_{xx} + F. \tag{2-3}$$

Equation (2-3) describes the vibrations of our string once it is set into motion. The assumption of "smallness" had led us to a single linear equation in u.

The equation (2-3) is often referred to as the *damped one-dimensional wave equation* because of the term ku_t. If the friction is negligible, we can take k to be zero and obtain the so-called inhomogeneous wave equation

(2-4) $$u_{tt} = c^2 u_{xx} + F.$$

If there are no external forces and the weight of the string is negligible, F can realistically be taken equal to zero, and we obtain

(2-5) $$u_{tt} = c^2 u_{xx},$$

the *one-dimensional wave equation.* Since u has the units of length, u_{tt} has the units of acceleration and u_{xx} the units of 1 over length, we see that c has the units of velocity.

PROBLEMS

1. Derive an equation governing the small, horizontal vibrations of an elastic chain hanging under its own weight. Neglect frictional forces and other external forces.
2. Suppose that the chain in Prob. 1 has a mass m hanging on the end. Derive a set of equations governing the motion of the combined system.
3. Equation (2-3) governs the "transverse" vibrations of an elastic string; that is, the vibrations are taking place perpendicular to the direction of the disturbance. Now derive an equation governing the longitudinal vibrations of an elastic bar with constant cross-sectional area A. Assume that cross sections of the bar remain planar while vibrating. Let $x + u(x, t)$ be the location of the cross section at time t which was at position x in equilibrium. Assume that the density ρ (mass/unit volume) is constant. The bar is set in motion by pulling or pressing the end. The forces at the ends of an element are restoring forces. Assume the validity of Hooke's law; that is, the force on the cross section at x is

$$T = AEu_x,$$

where E is a constant. Show that the vibrations satisfy

$$\rho u_{tt} = Eu_{xx} \quad \text{or} \quad u_{tt} = c^2 u_{xx} \quad \text{with} \quad c^2 = \frac{E}{\rho}.$$

4. Derive a two-dimensional wave equation governing the small transverse vibrations of a thin elastic membrane. Keep careful track of your assumptions. Use rectangular coordinates and assume that the membrane occupies a portion of the xy plane when in equilibrium.

1-3. Heat Conduction

Consider the problem of finding the temperature in a solid. Since temperature is related to heat, which is a form of energy, it seems reasonable to use the law of conservation of energy to determine the temperature. To use this law, it must be formulated in a precise way. Let V be an arbitrary volume, that is, a bounded, open, connected set, contained in the interior of the solid. We postulate the law of conservation of thermal energy in the following form.

The rate of change of thermal energy with respect to time in V is equal to the net flow of energy across the surface of V plus the rate at which heat is generated within V.

Having set down this general law, we must next translate the words into mathematical expressions. First, let u denote the temperature at the position $x = (x_1, x_2, x_3)$ and at the time t. We assume that the solid is at rest and that it is rigid so that the only energy present is thermal energy and the density $\rho = \rho(x)$ is independent of the time t and temperature u. Let e denote the specific internal energy of the solid, that is, e is the energy per unit mass. Then the amount of thermal energy in V is given by

$$\int_V \rho e\, dx, \qquad dx = dx_1\, dx_2\, dx_3.$$

The time rate of change of thermal energy in V is, therefore,

$$\frac{d}{dt}\int_V \rho e\, dx = \int_V \rho e_t\, dx. \tag{3-1}$$

Next, let B denote the boundary of V. Let $q = (q_1, q_2, q_3)$ denote the heat flux vector. Then $q \cdot n$ represents the flow of heat per unit cross-sectional area per unit time crossing a surface element with exterior unit normal $n = (n_1, n_2, n_3)$. Thus,

$$-\int_B q \cdot n\, dS \tag{3-2}$$

is the amount of heat per unit time flowing in across the boundary B of V. Here dS represents the element of surface area, and n is the unit exterior normal to B at x. The minus sign reflects the fact that if more heat flows out of the region than in, the energy in V decreases. Finally, if there are any sources or sinks in the medium, some information regarding these sources must be provided. We denote the rate at which heat is produced per unit volume by f so that the source term becomes

$$\int_V f\, dx. \tag{3-3}$$

f is assumed to be known. The law of conservation of energy now takes the form

$$\int_V \rho e_t\, dx = -\int_B q \cdot n\, dS + \int_V f\, dx$$

from (3-1)–(3-3). Apply the Gauss divergence theorem to the integral over B to obtain

$$\int_V (\rho e_t + \operatorname{div} q - f)\, dx = 0. \tag{3-4}$$

Finally, use the fact that this equality holds for arbitrary V to conclude that

$$\rho e_t = -\operatorname{div} q + f. \tag{3-5}$$

Here we have tacitly assumed the continuity of the integrand in (3-4). This is the basic form of our heat conduction law. The variables e and q are all unknown and additional information of an empirical nature is needed to determine an equation for the temperature u. First, for many materials, over fairly wide but not too large temperature ranges, the function $e = e(u)$ depends nearly linearly on u, so that

$$e_t = cu_t. \tag{3-6}$$

Here c, called the specific heat, is assumed to be constant. Next we relate the temperature u and heat flux q. Common experience and experimental measurements show that heat flows from regions of high temperature to regions of low temperature. Also, the rate of heat flow is small or large according as temperature changes between neighboring regions are small or large. To describe these quantitative properties of heat flow, we postulate a linear relationship between the rate of heat flow and the rate of temperature change. Let x be a point in the heat-conducting medium and n a unit vector specifying a direction at x. The rate of heat flow at x in the direction n is $q \cdot n$ and the rate of change of temperature is $\partial u/\partial n = \nabla u \cdot n$, the directional derivative of temperature. Since $q \cdot n > 0$ requires $\nabla u \cdot n < 0$, and vice versa, our linear relation takes the form $q \cdot n = -k \nabla u \cdot n$, with $k > 0$. Since n specifies any direction from x, this is equivalent to the assumption

$$q = -k \nabla u, \tag{3-7}$$

which is *Fourier's law*. We assume at present that $k > 0$ is a constant.

Insert (3-6) and (3-7) into (3-5) to get

$$u_t = a \Delta u + F, \tag{3-8}$$

where $a = k/c\rho > 0$ and $F = f/c\rho$. The quantity k is referred to as the *thermal conductivity* and a is sometimes called the *thermal diffusivity*. The operator

$$\Delta = \operatorname{div} \nabla = \nabla^2 = \frac{\partial^2}{\partial x_1^2} + \frac{\partial^2}{\partial x_2^2} + \frac{\partial^2}{\partial x_3^2},$$

or more generally

$$\Delta = \sum_{j=1}^{n} \frac{\partial^2}{\partial x_j^2},$$

is called the *Laplace operator*. If u is independent of the time t, the equation for u in (3-8) becomes

$$\Delta u = -\frac{F}{a} \tag{3-9}$$

and is called the *Poisson equation* or the inhomogeneous Laplace equation. If F vanishes identically, (3-9) becomes

$$\Delta u = 0 \tag{3-10}$$

and is simply called the *Laplace equation*.

PROBLEMS

1. Consider one-dimensional heat flow. Take for a model a thin, laterally insulated rod with constant cross-sectional area A. Start from scratch and derive the one-dimensional heat equation $u_t = au_{xx} + F$, where a and F are as given following (3-8).
2. If the left end of the rod in Prob. 1 is held at temperature zero, then $u(0, t) = 0$ for $t \geq 0$. What boundary condition at $x = L$ expresses the fact that this end is insulated (i.e., no heat crosses the end $x = L$)?
3. The specific heat c and thermal conductivity k in Fourier's law (3-7) become functions of position x when the heat-conducting medium is inhomogeneous. Find the analogue of (3-8) for such a medium.
4. Suppose that the left end of the rod in Prob. 1 is in contact with a medium, perhaps air, held at a constant temperature T_0. *Newton's law of cooling*, confirmed by experiment, states that heat flow across the end $x = 0$ will be proportional to the temperature difference $u(0, t) - T_0$. What boundary condition does this law give for u at $x = 0$? Suppose that a similar medium at the end $x = L$ is held at temperature T_L. Find the boundary condition at $x = L$.
5. Heat flows in a region D in space that is bounded by a surface B. Within D heat flow is governed by (3-8). The surface B is surrounded by a medium held at the constant temperature T_∞. Use Newton's law of cooling to find an equation (boundary condition) giving the heat flow across B.
6. Suppose that the region D in Prob. 5 has an insulated boundary. What is the boundary condition? Explain briefly how this boundary condition can be regarded as a limiting case of the situation in Prob. 5.

1-4. Diffusion–Dispersion Phenomena

Diffusion problems lead to partial differential equations rather similar to those of heat conduction. Suppose that C represents the concentration of a substance, that is, the mass per unit volume, which is dissolving into a moving liquid or gas. The amount of the substance in a given volume V is then

$$\int_V C\, dx.$$

The law of conservation of mass states that the time rate of change of mass in V is equal to the rate at which mass flows into V plus the rate at which mass is produced due to sources in V. We shall assume that there are no internal sources. Let $q = (q_1, q_2, q_3)$ denote the mass flux vector such that $q \cdot n$ gives the mass per unit area per unit time crossing a surface element with unit normal n. Then if $n = (n_1, n_2, n_3)$ is the unit exterior normal to the boundary B of V, we find that

$$\frac{d}{dt}\int_V C\, dx = \int_V C_t\, dx = -\int_B q \cdot n\, dS.$$

As in the case of heat flow, we apply the Gauss divergence theorem to the boundary integral, convert it to a volume integral, and use the fact that V is arbitrary to obtain

$$C_t = -\text{div}\, q \tag{4-1}$$

as the equation describing the conservation of mass. The vector q is related to C by the empirical relationship

$$q = -D\,\nabla C + Cv, \tag{4-2}$$

which is called *Fick's law of diffusion.* In (4-2), $v = (v_1, v_2, v_3)$ represents the velocity of the fluid and is assumed known. The quantity $D > 0$ is called the *diffusion coefficient.* In general it depends on C but we shall assume that it is constant here. Inserting (4-2) into (4-1) yields

$$C_t = D\,\Delta C - \text{div}\,(Cv) \tag{4-3}$$

as the basic equation governing diffusion–dispersion phenomena. If the velocity of the medium is negligible, the diffusion equation is the same as (3-8) with $F = 0$, while if D is negligibly small, (4-3) becomes the first-order partial differential equation

$$C_t + v \cdot \nabla C + C\,\text{div}\, v = 0. \tag{4-4}$$

PROBLEMS

1. Suppose that a substance is dissolving into a liquid which is at rest in a container. Assume further that no mass enters or escapes the container. On the basis of (4-2) with $v = 0$, give a physical argument showing that

$$\lim_{t \to \infty} C(x, t) = C_0,$$

where C_0 is a constant. Argue that

$$C_0 = \frac{1}{|V|} \int_V C(x, 0)\, dx,$$

where V is the set making up the container and $|V|$ is its volume.

2. Give a physical motivation for the validity of Fick's law. Model your argument along the lines of Fourier's law.
3. What form does (4-3) have when the diffusion coefficient D depends on C and internal sources are present?

1-5. Saturated Flows Underground

Consider a fluid such as water flowing underground. If one cuts out a small volume V, both soil particles and fluid particles will be present, so that only a fraction of the space underground is actually available to the fluid. Denote this fraction by ϕ, which is called the *porosity*, and in general varies with position, temperature, and pressure. However, in this model we shall assume that porosity depends only on the position $x = (x_1, x_2, x_3)$. If all the volume that is available to the fluid is actually filled by it, the flow is said to be saturated.

If one cuts out an arbitrary volume in a saturated flow, the mass of fluid in the volume is given by

$$\int_V \rho\phi \, dx,$$

where ρ is the density of the fluid, so that the rate of change in the mass is

$$\frac{d}{dt}\int_V \rho\phi \, dx = \int_V \phi\rho_t \, dx. \tag{5-1}$$

This rate must equal the net rate at which mass flows over the boundary B of V plus the rate at which mass is produced by sources in V, namely

$$-\int_B \rho q \cdot n \, dS + \int_V f \, dx,$$

where $q = (q_1, q_2, q_3)$ is the *volumetric flow rate*, the rate at which a volume of fluid flows over a cross-sectional area per unit area per unit time. Conservation of mass gives

$$\int_V \phi\rho_t \, dx = -\int_V \operatorname{div}(\rho q) \, dx + \int_V f \, dx,$$

where the Gauss divergence theorem was used to transform the boundary integral. As usual, since V is an arbitrary volume in the flow region, this conservation of mass equation can be expressed as the partial differential equation

$$\phi\rho_t = -\operatorname{div}(\rho q) + f. \tag{5-2}$$

In the case of slow flows, empirical evidence shows that

$$q = -\frac{k}{\mu}(\nabla p + \rho\vec{g}), \tag{5-3}$$

which is called *Darcy's law*. Here p is the pressure, ρ the density of the fluid, $\vec{g}$ the vector giving the acceleration due to gravity, $k > 0$ is called the permeability of the medium, and μ is the viscosity of the fluid. Combining (5-2) and (5-3) yields

$$\phi\rho_t = \operatorname{div}\left[\frac{k\rho}{\mu}\left(\nabla p + \rho\vec{g}\right)\right] + f \tag{5-4}$$

as the equation describing the flow of fluids underground.

If the fluid is incompressible, as in the case of water, then ρ is a constant so that ρ_t vanishes. If, in addition, k and μ are constant, we arrive at the equation

$$\frac{k\rho}{\mu}\Delta p + f = 0. \tag{5-5}$$

which must be solved for the pressure p to determine the state of the system at each point.

PROBLEMS

1. In the case of a gas, gravitational effects may be neglected. Assume a relationship between the density ρ and the pressure p of the form

$$\frac{p}{p_0} = \left(\frac{\rho}{\rho_0}\right)^{\lambda}$$

where p_0 and ρ_0 are constant "mean" pressure and density values. Assume that both k and μ are constant and derive a nonlinear partial differential equation governing the flow of a gas underground.

2. Contrast Fourier's law, Fick's law, and Darcy's law.

1-6. Telegrapher's System

In this section we use average values of variables rather than integrals over a segment x to $x + \Delta x$ to evaluate pertinent physical quantities. This is another, somewhat more informal approach used to derive differential equations for a physical model. Compare with Prob. 5.

Consider the flow of electricity in a transmission line, such as a coaxial cable. Figure 1-2 shows a small segment of such a cable. We wish to determine the voltage $v(x, t)$ across the cable and the current $i(x, t)$ in the inner wire at position x and time t. (An equal but opposite current flows through the outer cable.) The electrical properties of the cable are described by R, its resistance per unit length; L, its inductance per unit length; C, its capacitance per unit length, and G, its conductance per unit length. We use Kirchhoff's laws to derive equations for the voltage and current. First, Kirchhoff's loop law, the sum of the potential drops

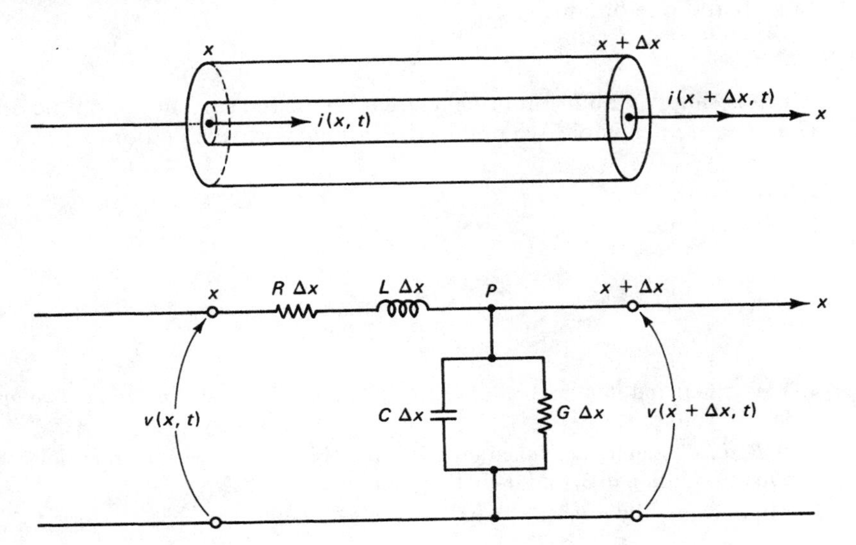

Figure 1-2

around a closed circuit is zero, leads to

(6-1) $$v(x, t) - (R\,\Delta x)\tilde{i} - (L\,\Delta x)\tilde{i}_t - v(x + \Delta x, t) = 0,$$

when applied to the loop in Fig. 1-2. Here $\tilde{i}$ is the average value for the nearly constant current in the segment from x to $x + \Delta x$. Second, by conservation of charge, Kirchhoff's current law, the sum of the currents entering P in Fig. 1-2 must be zero. The current $i(x, t)$ enters P while currents $i(x + \Delta x, t)$, $(C\,\Delta x)\tilde{v}_t$, and $(G\,\Delta x)\tilde{v}$ leave P, where $\tilde{v}$ is the average value of the voltage in the cable from x to $x + \Delta x$. Thus,

(6-2) $$i(x, t) - i(x + \Delta x, t) - (C\,\Delta x)\tilde{v}_t - (G\,\Delta x)\tilde{v} = 0.$$

Divide by Δx in (6-1) and (6-2) and let Δx tend to zero to obtain the *telegrapher's system*

(6-3) $$i_x + Cv_t + Gv = 0,$$

(6-4) $$v_x + Li_t + Ri = 0.$$

If i and v are twice continuously differentiable, and if R, L, C, and G are constants, this system can be simplified. In fact, differentiate the first equation with respect to x and the second with respect to t, multiply the second equation by C, and subtract to obtain

$$i_{xx} + Gv_x - LCi_{tt} - CRi_t = 0.$$

Now insert the expression for v_x from (6-4) to obtain

(6-5) $$i_{xx} = LCi_{tt} + (RC + GL)i_t + RGi.$$

Equation (6-5) is referred to as the *telegrapher's equation.* If the leakage to the ground is negligible and the self-induction is small, then G and L may be set equal to zero and one obtains

$$i_{xx} = RCi_t,$$

which is simply a diffusion or heat equation for the current. If, on the other hand, G and R are negligible, we obtain the ordinary wave equation

$$i_{xx} = LCi_{tt}$$

for the current.

PROBLEMS

1. Using (6-3) and (6-4) and assuming C, G, L, and R to be constant, derive an equation for the voltage v.
2. If R, L, C, and G are functions of x, that is, the wire is inhomogenous, find single equations, using (6-3) and (6-4), that i and v must satisfy.
3. Multiply (6-3) by v, (6-4) by i, and add to get

$$-(iv)_x = Ri^2 + Gv^2 + \tfrac{1}{2}(Li^2 + Cv^2)_t.$$

Integrate from x_1 to x_2 with $x_1 < x_2$, solve for $(iv)(x_2)$, and give a physical interpretation of the resulting equation.

4. The telegrapher's equation has the form

$$u_{xx} = LCu_{tt} + (RC + GL)u_t + RGu.$$

Given the physical situation, it is desirable that solutions propagate with undistorted form along the telegraph line. A signal input at $x = 0$ should arrive at $x = b$ in recognizable form. For solutions to the wave equation (2-5) this is true, but it need not be true for solutions to the telegrapher's equation. However, if the cable is built correctly, the signal will propagate in a damped but distortionless way:

(a) Let $u = v(x, t)e^{\alpha t}$ and choose α so that the differential equation for v has no v_t term.

(b) Now, find the relation between R, L, C, and G so that the equation in part (a) reduces to $v_{xx} = LCv_{tt}$.

(c) Conclude that the signal arriving at $x = b$ will be a damped version of the signal sent at $x = 0$.

5. Derive the telegrapher's system by the use of integrals (as for the vibrating string) rather than mean values.

-7. Flow of Ideal Gases

As a final example of a physical problem that leads to a system of partial differential equations, we consider the flow of a gas in a large container. Let ρ denote the density of the gas, p the pressure, and $v = (v_1, v_2, v_3)$ the velocity of the gas at the position $x = (x_1, x_2, x_3)$ at the time t. If we again let V denote an arbitrary volume in the gas, the rate of change of mass in V must equal the net rate at which mass flows out over the boundary B of V plus the rate at which mass is introduced or taken out of the system in V. We assume that there are no internal sources or sinks, so that the latter rate is zero. Let $q = (q_1, q_2, q_3)$ denote the mass flux, that is, $q \cdot n$ is the rate at which mass is transported across a unit cross-sectional area with unit normal n per unit time. Since $q = \rho v$, the equation describing the conservation of mass in V is given by

$$\frac{d}{dt}\int_V \rho \, dx = \int_V \rho_t \, dx = -\int_B \rho v \cdot n \, dS.$$

Upon applying the Gauss divergence theorem and using the fact that V is arbitrary, we have

$$\rho_t + \operatorname{div}(\rho v) = 0. \tag{7-1}$$

Next, we need to obtain equations describing the velocity v. In our discussion of heat conduction and diffusion, we used an empirical law relating $q = \rho v$ to the temperature or the concentration gradients. These empirical laws essentially took the place of equations of motion. In the case of gases, such empirical laws do not exist and we must turn to the equations of motion themselves, which are simply expressions of Newton's second law. Thus in this case we will obtain a system of partial differential equations just as we did in the derivation of the equations governing the flow of electricity in a wire.

We consider an arbitrary system of particles making up the volume V_0 at time $t = 0$. As time progresses, the particles will move and occupy a volume V_t at the time t. The momentum of this system is

$$\int_{V_t} \rho v \, dx,$$

and Newton's second law states that the rate of change of momentum is equal to the sum of the forces acting on V_t. As in the case of the vibrating string, two types of forces will be distinguished, body forces, which act on the particles in V_t as a whole, and surface forces, which act on the surface B_t which bounds V_t. We assume that the only body forces acting are those due to gravity. If we take the positive direction for x_3 to be in the upward direction, the body forces are given by

$$-\int_{V_t} \rho \vec{g} \, dx,$$

where $\vec{g} = (0, 0, g)$, $g > 0$, is the acceleration due to gravity. We assume that the surface forces per unit area are generated by a function p, the pressure, acting normal to the surface B_t so that the surface forces are given by

$$-\int_{B_t} pn \, dS = -\int_{V_t} \nabla p \, dx.$$

Newton's second law now yields

$$\frac{d}{dt} \int_{V_t} \rho v \, dx = -\int_{V_t} \nabla p \, dx - \int_{V_t} \rho \vec{g} \, dx. \tag{7-2}$$

The final step consists in differentiating the volume integral whose domain depends on the time t. In Prob. 3 you are asked to prove that

$$\frac{d}{dt} \int_{V_t} f(x, t) \, dx = \int_{V_t} f_t(x, t) \, dx + \int_{B_t} f(x, t) v \cdot n \, dS.$$

Apply this result to the ith component of the momentum and make use of (7-1) to get

$$\begin{aligned} \frac{d}{dt} \int_{V_t} \rho v_i \, dx &= \int_{V_t} (\rho v_i)_t \, dx + \int_{B_t} \rho v_i v \cdot n \, dS \\ &= \int_{V_t} (\rho v_{it} + \rho_t v_i) \, dx + \int_{V_t} \operatorname{div} (\rho v_i v) \, dx \\ &= \int_{V_t} \rho \left[\frac{\partial}{\partial t} v_i + (v \cdot \nabla) v_i \right] dx. \end{aligned}$$

We denote by $(v \cdot \nabla)v$ the vector whose ith component is given by

$$\sum_{j=1}^{3} v_j \frac{\partial}{\partial x_j} v_i$$

so that (7-2) takes the form

$$\int_{V_t} \rho[v_t + (v \cdot \nabla)v]\, dx = -\int_{V_t} \nabla p\, dx - \int_{V_t} \rho\vec{g}\, dx.$$

Since V_t was arbitrary, we conclude that

$$\rho[v_t + (v \cdot \nabla)v] = -\nabla p - \rho\vec{g}. \tag{7-3}$$

The system (7-1), (7-3) represents a nonlinear system of four equations in the five unknowns p, ρ, and $v = (v_1, v_2, v_3)$. Evidently, some further information is needed about the gas. However, notice that this system is very general and describes any gas that satisfies the rather mild assumptions made in our derivation. To apply the system to a particular gas, some additional information about the gas itself is needed. That is, we require an equation of state relating certain of the variables. This equation is determined empirically. One such determination relates pressure and density through the equation

$$\frac{p}{p_0} = \left(\frac{\rho}{\rho_0}\right)^{\lambda}. \tag{7-4}$$

Now (7-1), (7-3), and (7-4) give a system of five equations in five unknowns.

The system (7-1), (7-3), (7-4) is very difficult to treat, mainly because of the nonlinearities. An important special case occurs when the velocity v is "small" and the pressure p and the density ρ do not differ significantly from the mean pressure p_0 and density ρ_0. Specifically, we assume that $\rho/\rho_0 = 1 + u$, where u is "small," meaning that u^2 and higher powers of u can be neglected. Furthermore, v is so small that products of components of v and u or components of v and derivatives of components of v may be neglected. By the binomial theorem (7-4) has the form

$$\left(\frac{\rho}{\rho_0}\right)^{\lambda} = (1 + u)^{\lambda} = 1 + \lambda u. \tag{7-5}$$

Making use of the assumptions above and (7-5) allows us to rewrite (7-1) and (7-3) in the form

$$u_t + \operatorname{div} v = 0 \tag{7-6}$$

and

$$\rho_0 v_t = -p_0\lambda\, \nabla u - \rho_0(1 + u)\vec{g}, \tag{7-7}$$

or in the event that gravitational effects can be neglected,

$$v_t = -c^2\, \nabla u, \qquad c^2 = \frac{\lambda p_0}{\rho_0}. \tag{7-8}$$

Assuming sufficient differentiability on the functions v and u, we can eliminate v from (7-6) and (7-8) to obtain

$$u_{tt} = c^2\, \Delta u. \tag{7-9}$$

PROBLEMS

1. Rewrite (7-9) in terms of a differential equation for p.
2. If gravitational effects cannot be neglected in (7-7), eliminate v from the system (7-6) and (7-7) and find the differential equation that u satisfies.
3. Prove that

$$\frac{d}{dt}\int_{V_t} f(x, t)\,dx = \int_{V_t} f_t(x, t)\,dx + \int_{B_t} f(x, t)v \cdot n\,dS.$$

Hint. Change coordinates by $x = \phi(a, t)$ so that V_t goes into a fixed volume V_0, differentiate with respect to t, use the divergence theorem and transform back. Detailed calculations are involved.

1-8. Well-Posed and Ill-Posed Problems

Return to the small vibrations of a string. The shape $u(x, t)$ of the string satisfies

$$u_{tt} + ku_t = c^2 u_{xx} + F(x, t), \tag{8-1}$$

by (2-3). This equation was derived principally from Newton's second law and describes how the string moves as time passes. It does not tell us the initial state of the string, which clearly affects its subsequent motion. Thus, we need to specify the initial position and initial velocity of each particle on the string before we can determine the motion. That is, we must prescribe

$$u(x, 0) = f(x) \quad \text{and} \quad u_t(x, 0) = g(x) \tag{8-2}$$

at each point x on the string, where $f(x)$ is the initial shape of the string and $g(x)$ is its initial velocity profile. Finally, recall that the ends of the string were pinned. These boundary conditions require that

$$u(0, t) = 0 \quad \text{and} \quad u(L, t) = 0 \tag{8-3}$$

for all times $t \geq 0$. At this point, physical intuition suggests that we have a *well-posed problem.* That is, the initial, boundary value problem (8-1)–(8-3) should have a unique solution that varies continuously with the given inhomogeneous data. That is, small changes in the data should cause only small changes in the solution. It is important to show mathematically that the problem (8-1)–(8-3) is well-posed, as the physical situation suggests. Indeed, if the model were not well-posed, we should go back to the drawing board and seek a new mathematical description for the vibrating string. We shall return to this matter in Chap. 4.

For the moment, we simply observe that well-posedness is a very desirable attribute for any physical problem and its mathematical model. In practice, the initial and/or boundary data are measured and so small errors occur. Very often the mathematical model must be solved numerically and truncation and round-off errors come into play. If the problem is well-posed, these unavoidable small errors produce only slight errors in the computed solution, and, hence, useful results are obtained.

There are important, but very delicate problems which are not well posed, so-called *ill-posed problems.* Suppose, for example, that we measure the concentration $f(x)$ of an impurity in a large lake with negligible currents at a certain time, say $t = 0$, and that we would like to determine the concentration $C(x, t)$ for all previous times. This could conceivably allow us to determine the source of the impurity. Applying (4-3) with $v = 0$, we arrive at the model

$$\text{(8-4)} \qquad \begin{cases} C_t = a\,\Delta C, & t < 0, \quad x \text{ in the lake,} \\ C(x, 0) = f(x), \end{cases}$$

where we have set the diffusion coefficient $D = a$ for convenience. If we set

$$u(x, t) = C(x, -t) \quad \text{for} \quad t > 0,$$

then (8-4) becomes

$$\text{(8-5)} \qquad \begin{cases} u_t = -a\,\Delta u, & t > 0, \quad x \text{ in the lake,} \\ u(x, 0) = f(x). \end{cases}$$

The differential equation in (8-5) is called the *backward heat equation.* [Compare with (3-8).]

On physical grounds, we expect (8-5) to be a nearly impossible problem. To see this, take several identical glasses of water. Now, add a drop of red dye to each glass. Use the same amount of dye in each glass, but insert the dye at different points. After some time has passed, the dye will have diffused throughout each glass and the resulting dye distributions will appear indistinguishable. If we measure the concentration at such a time, there is no way to tell how things started out. In other words, very different initial conditions lead to virtually the same final state. Thus, solving the diffusion equation backward in time, equivalently solving the backward heat equation, seems suspect at best.

We can see mathematically that (8-5) is ill-posed. For simplicity, we work in one space dimension, consider our large lake to be $-\infty < x < \infty$, assume the measured concentration $f(x) \equiv 1$, and set $a = 1$. Then (8-5) becomes

$$\text{(8-6)} \qquad \begin{cases} u_t = -u_{xx}, & t > 0, \quad -\infty < x < \infty, \\ u(x, 0) = 1, & -\infty < x < \infty. \end{cases}$$

Of course, $u(x, t) \equiv 1$ is an obvious solution. On the other hand,

$$u_n(x, t) = 1 + \frac{1}{n} e^{n^2 t} \sin(nx)$$

solves

$$\text{(8-7)} \qquad \begin{cases} u_t = -u_{xx}, & t > 0, \quad -\infty < x < \infty, \\ u(x, 0) = 1 + \dfrac{1}{n} \sin(nx), \end{cases}$$

for $n = 1, 2, \ldots$. For n large enough, the initial condition in (8-7) differs from that in (8-6) by an arbitrarily small amount. Nonetheless, the solutions $u_n(x, t)$ and

$u(x, t)$ differ by $n^{-1}e^{n^2t} \sin(nx)$, which grows rapidly as n increases. Thus, the solution to (8-6) does not depend continuously on the initial data and the problem is ill-posed.

PROBLEMS

1. Consider the laterally insulated rod in Prob. 1 of Sec. 1-3. The rod extends from 0 to L along the x axis. On physical grounds, argue which initial and/or boundary conditions can be added to the heat equation $u_t = au_{xx} + F(x, t)$ to get a well-posed problem. *Hint.* There are several choices for boundary conditions (see Probs. 2 and 4 of Sec. 1-3).
2. Show that

$$\begin{aligned} &u_t = -u_{xx}, \quad t > 0, \quad 0 < x < \pi, \\ &u(0, t) = u(L, t) = 0, \quad t \geq 0, \\ &u(x, 0) = f(x), \quad 0 \leq x \leq L, \end{aligned}$$

   is not well-posed. *Hint.* Check that

$$v_n(x, t) = \frac{1}{n} e^{n^2t} \sin\left(\frac{n\pi x}{L}\right)$$

   solves $v_t = -v_{xx}$, $v(0, t) = v(\pi, t) = 0$, $v(x, 0) = (1/n) \sin(nx)$. What problem does $u(x, t) + v_n(x, t)$ solve?
3. Show that

$$\begin{aligned} &u_{xx} + u_{yy} = 0, \quad y > 0, \quad -\infty < x < \infty, \\ &u(x, 0) = f(x), \quad u_y(x, 0) = g(x), \quad -\infty < x < \infty, \end{aligned}$$

   is ill-posed. *Hint.* Consider the analogous problem satisfied by $u_n(x, y) = (1/n)(\cos nx)(\cosh ny)$, and note that $u_n(0, 1) = (\cosh n)/n \to \infty$ as $n \to \infty$.
4. Show that (8-5) is ill-posed in two and three space dimensions. *Hint.* Consider

$$u_n(x, t) = \frac{1}{n} e^{\alpha n^2 t} \sin n(x_1 + x_2 + x_3)$$

   and determine α appropriately.

2

Partial Differential Equations of the First Order

2-1. The Method of Characteristics for Quasilinear Equations

The partial differential equation

$$a(x, y, u)u_x + b(x, y, u)u_y = c(x, y, u) \tag{1-1}$$

is called *quasilinear* because it is linear in its highest derivatives. Purely geometric reasoning indicates how to solve such an equation. In this section our reasoning will be somewhat intuitive as we develop the solution procedure known as the method of characteristics. However, once this method has been set forth, it will be easy to justify rigorously. This is done in Sec. 2-2. For the moment we concentrate on the geometry and some illuminating examples.

Observe that a solution

$$u = u(x, y) \tag{1-2}$$

to (1-1) represents a surface in the three-dimensional xyu space and that the problem of solving (1-1) amounts to determining a surface whose normal $(u_x, u_y, -1)$ is constrained by (1-1). Since a surface can be built up from curves lying in it, we begin to construct a solution surface by seeking some curves that must lie in it. Let (x_0, y_0, u_0) be a point on a solution surface and $(x(\tau), y(\tau), u(\tau))$ be a curve that passes through (x_0, y_0, u_0) when $\tau = 0$. We seek conditions on this curve which force it to remain in the solution surface for $\tau \neq 0$. If the curve does remain in the solution surface, its tangent vector $(x'(0), y'(0), u'(0))$ at $\tau = 0$ must lie in the tangent plane to the solution surface. Now a normal to the surface at the point (x_0, y_0, u_0) is $(p, q, -1)$, where $p = u_x(x_0, y_0)$, $q = u_y(x_0, y_0)$. From (1-1) this normal must satisfy the equation

$$c(x_0, y_0, u_0) = a(x_0, y_0, u_0)p + b(x_0, y_0, u_0)q. \tag{1-3}$$

Since we really do not know a solution surface through (x_0, y_0, u_0), we do not know p and q in (1-3); hence, we regard (1-3) as a constraint on the normal $(p, q, -1)$ of the tangent plane

$$u - u_0 = p(x - x_0) + q(y - y_0) \tag{1-4}$$

to the solution surface at (x_0, y_0, u_0). Then (1-3) and (1-4) define a one-parameter family of planes that pass through (x_0, y_0, u_0), and among these planes is the tangent plane to a solution surface. We can guarantee that the vector $(x'(0), y'(0), u'(0))$ will lie in the tangent plane to a solution surface if we require it to lie in *all* the planes of the one-parameter family (1-4). We will show now that this one-parameter family of planes has the line

$$\frac{x - x_0}{a(x_0, y_0, u_0)} = \frac{y - y_0}{b(x_0, y_0, u_0)} = \frac{u - u_0}{c(x_0, y_0, u_0)}$$

in common. Given this, the vector $(x'(0), y'(0), u'(0))$ will be tangent to a solution surface through (x_0, y_0, u_0) provided that it is a scalar multiple of the vector $(a(x_0, y_0, u_0), b(x_0, y_0, u_0), c(x_0, y_0, u_0))$. Taking the constant of proportionality to be 1 (which amounts to a possible reparametrization of the curve) yields

$$x'(0) = a(x_0, y_0, u_0), \qquad y'(0) = b(x_0, y_0, u_0), \qquad u'(0) = c(x_0, y_0, u_0).$$

Since this argument can be repeated for every point on our curve, we arrive at the following system of ordinary differential equations, which should force the curve to remain in a solution surface:

$$\begin{cases} x'(\tau) = a(x(\tau), y(\tau), u(\tau)) \\ y'(\tau) = b(x(\tau), y(\tau), u(\tau)) \\ u'(\tau) = c(x(\tau), y(\tau), u(\tau)). \end{cases} \tag{1-5}$$

Solutions to this system are called *characteristic curves* for the quasilinear differential equation (1-1). The projections of the characteristic curves onto the xy plane, $(x(\tau), y(\tau))$, are traditionally referred to as the *characteristics*.

To see why the line, whose symmetric equations are given above, is common to the set of planes (1-4), we shall determine the envelope of this family. Differentiate (1-3) and (1-4) with respect to p to get

$$\frac{dq}{dp} = -\frac{a(x_0, y_0, u_0)}{b(x_0, y_0, u_0)} = -\frac{x - x_0}{y - y_0}.$$

Consequently,

$$\frac{x - x_0}{a(x_0, y_0, u_0)} = \frac{y - y_0}{b(x_0, y_0, u_0)}.$$

Now from (1-4), this equality, and (1-3),

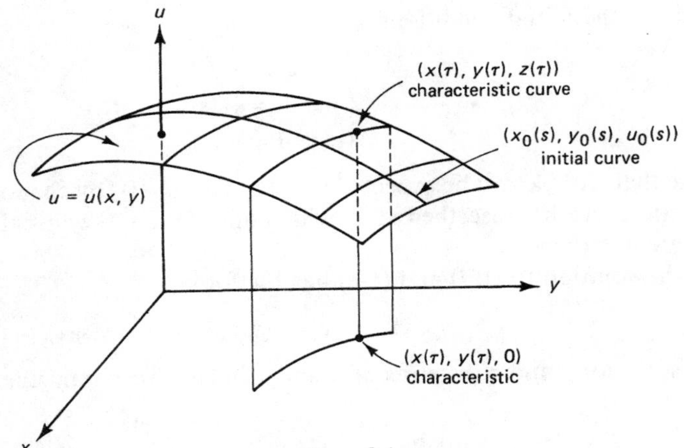

Figure 2-1

$$
\begin{aligned}
u - u_0 &= p(x - x_0) + q\,\frac{b(x_0, y_0, u_0)}{a(x_0, y_0, u_0)}(x - x_0) \\
&= \frac{pa(x_0, y_0, u_0) + qb(x_0, y_0, u_0)}{a(x_0, y_0, u_0)}(x - x_0) \\
&= \frac{c(x_0, y_0, u_0)}{a(x_0, y_0, u_0)}(x - x_0).
\end{aligned}
$$

Thus,

$$
\frac{u - u_0}{c(x_0, y_0, u_0)} = \frac{x - x_0}{a(x_0, y_0, u_0)}.
$$

Combining this ratio with the previous one yields the symmetric equations for the line enveloped by the one-parameter family of planes (1-4).

The characteristic system (1-5) can be used to construct solution surfaces to (1-1) which contain a given curve, the "initial data" for the solution surface. For example, if the curve

$$
(x_0(s), y_0(s), u_0(s)) \tag{1-6}
$$

is to lie in the solution surface, we could solve the system (1-5) subject to the initial condition for $\tau = 0$ that the solution pass through a point on the given curve (see Fig. 2-1). Therefore, to construct the solution surface one solves the system of ordinary differential equations in τ, and depending on s as parameter,

$$
\begin{cases}
\dfrac{dx}{d\tau} = a(x, y, u) \\[2ex]
\dfrac{dy}{d\tau} = b(x, y, u) \\[2ex]
\dfrac{du}{d\tau} = c(x, y, u)
\end{cases} \tag{1-7}
$$

subject to the initial conditions

$$\text{(1-8)} \qquad \begin{cases} x(s, 0) = x_0(s) \\ y(s, 0) = y_0(s) \\ u(s, 0) = u_0(s). \end{cases}$$

Notice that trouble can be expected if $(x_0(s), y_0(s), u_0(s))$ by chance specifies a characteristic curve because then (1-7) will simply follow the initial data and fail to generate a surface.

The solution to (1-7) and (1-8) has the form

$$\text{(1-9)} \qquad x = x(s, \tau), \qquad y = y(s, \tau), \qquad u = u(s, \tau).$$

Next solve for s and τ in terms of x and y in (1-9) to obtain the functions

$$s = s(x, y), \qquad \tau = \tau(x, y)$$

and insert these expressions into the function u in (1-9) to obtain, finally,

$$\text{(1-10)} \qquad u = u(s(x, y), \tau(x, y)) = u(x, y)$$

as the desired solution.

Example 1. Solve

$$xu_x + (x + y)u_y = u + 1,$$
$$u(x, 0) = x^2.$$

The characteristic equations are

$$x' = x, \qquad y' = x + y, \qquad u' = u + 1.$$

The initial curve can be parametrized in the form

$$x_0(s) = s, \qquad y_0(s) = 0, \qquad u_0(s) = s^2.$$

Integrating the equation for x yields

$$x = se^\tau.$$

For y we find

$$y' - y = x = se^\tau$$

and upon integration

$$y = s\tau e^\tau.$$

Finally, we integrate the differential equation for u to obtain

$$u = (s^2 + 1)e^\tau - 1.$$

Next, solve for s and τ in terms of x and y and insert these expressions into the formula for u to find after a little simplification

$$u = x^2 \exp\left(-\frac{y}{x}\right) + \exp\left(\frac{y}{x}\right) - 1.$$

It is easily checked that for $x \neq 0$, the function u satisfies all the conditions of the problem.

Example 2. Solve

$$xu_x + yu_y = u + 1, \qquad u(x, x) = x^2.$$

The characteristic equations are

$$x' = x, \qquad y' = y, \qquad u' = u + 1$$

and a natural parameterization of the initial curve is

$$x_0(s) = s, \qquad y_0(s) = s, \qquad u_0(s) = s^2.$$

We obtain upon integrating the system of characteristic equations

$$x = se^\tau, \qquad y = se^\tau, \qquad u = s^2e^\tau + e^\tau - 1.$$

Unfortunately, this system cannot be solved for s and τ in terms of x and y. On the other hand, if the data is given on a different curve, say

$$u(x, x^2) = x^2,$$

and the initial curve is parameterized by

$$x_0(s) = s, \qquad y_0(s) = s^2, \qquad u_0(s) = s^2,$$

we obtain

$$x = se^\tau, \qquad y = s^2e^\tau, \qquad u = s^2e^\tau + e^\tau - 1.$$

Solve for s and τ in terms of x and y and insert the result into the expression for u to find

$$u = y - 1 + \frac{x^2}{y}.$$

These examples point up two important difficulties. First, the solutions are not necessarily valid for all values of x and y, only for values in certain sets. Second, data cannot be given arbitrarily on any curve. In the case of Ex. 2, we had first prescribed data on a characteristic so that as already noted above there was no way for the differential equation to obtain values off the initial curve and so construct a surface. This points up the fact that we need a theorem justifying our procedure and giving precise information about when this solution procedure can actually be carried out.

We remark, finally, that in the physical applications, one usually takes $y = t$, where t is the time.

PROBLEMS

1. Generalize the solution procedure described above to the problem

$$\sum_{i=1}^{n} a_i(x, u)u_{x_i} = c(x, u),$$

where $x = (x_1, \ldots, x_n)$ and the initial data are given in the form

$$x = x^\circ(s), \qquad u = u^\circ(s),$$

with $s = (s_1, \ldots, s_{n-1})$.

2. Generalize the solution procedure described above to the problem

$$u_t + \sum_{i=1}^{n} a_i(x, u)u_{x_i} = c(x, t, u), \qquad x = (x_1, \ldots, x_n), \qquad u(x, 0) = f(x),$$

where $f(x)$ is given.

3. Solve

$$u_t + au_x = 0, \qquad a \text{ is a constant,}$$
$$u(x, 0) = f(x).$$

For a special case, take $a = 1$ and for $0 \leq x \leq 1$, let $f(x) = 2x^3 - 3x^2 + 1$, $f(x) = 0$ for $x > 1$, and for negative values of x, take $f(x) = f(-x)$. For various values of t, sketch the solution u as a function of x.

4. Solve

$$xu_x + u_y = 1, \qquad u(x, 0) = e^x.$$

5. Solve

$$xu_x + (y^2 + 1)u_y = u, \qquad u(x, 0) = e^x$$

6. Suppose that $v > 0$ is a constant and that $f(t) \geq 0$ with $f(t) \not\equiv 0$ but with $f(0) = 0$. Solve

$$u_t + vu_x = 0 \quad \text{in} \quad x > 0, \qquad t > 0,$$
$$u(x, 0) = 0 \quad \text{in} \quad x \geq 0$$
$$u(0, t) = f(t) \quad \text{for} \quad t \geq 0.$$

Interpret the solution physically.

2-2. Justification of the Method of Characteristics

To carry out the solution procedure given above, we must put hypotheses on the differential equation which allow us to solve the characteristic system, and additionally we must place conditions on the initial curve which allows us to solve for the s and τ in terms of x and y. The implicit function theorem will guarantee that s and τ can be expressed in terms of x and y if we require that the Jacobian

$$\frac{\partial(x, y)}{\partial(\tau, s)} = \begin{vmatrix} \dfrac{\partial x}{\partial \tau} & \dfrac{\partial x}{\partial s} \\ \dfrac{\partial y}{\partial \tau} & \dfrac{\partial y}{\partial s} \end{vmatrix}$$

be nonzero. Since we are dealing with continuous functions, it is enough to assume that the Jacobian be nonzero initially, since then it will be nonzero for values of τ sufficiently close to zero and we will be able to solve for s and τ in a neighborhood of the initial curve. Therefore, we assume that

$$\begin{vmatrix} \dfrac{\partial x}{\partial \tau} & \dfrac{\partial x}{\partial s} \\ \dfrac{\partial y}{\partial \tau} & \dfrac{\partial y}{\partial s} \end{vmatrix}_{\tau=0} = \begin{vmatrix} a(x_0(s),\ y_0(s),\ u_0(s)) & x_0'(s) \\ b(x_0(s),\ y_0(s),\ u_0(s)) & y_0'(s) \end{vmatrix} \neq 0$$

for the relevant values of s. It is obvious from this discussion that the initial curve should be differentiable. Thus, we are led to formulate the following result.

Theorem 2-1. Suppose that the functions $a(x, y, u)$, $b(x, y, u)$, and $c(x, y, u)$ are continuously differentiable in a domain (open, connected set) of the xyu-space. Suppose further that the curve $(x_0(s), y_0(s), u_0(s))$ lies inside this domain and is continuously differentiable there. Assume finally that

$$\begin{vmatrix} a(x_0(s),\ y_0(s),\ u_0(s)) & x_0'(s) \\ b(x_0(s),\ y_0(s),\ u_0(s)) & y_0'(s) \end{vmatrix} \neq 0.$$

Then there exists a unique continuously differentiable solution to the problem

$$\text{(2-1)} \qquad \begin{cases} a(x, y, u)u_x + b(x, y, u)u_y = c(x, y, u) \\ u(x_0(s), y_0(s)) = u_0(s) \end{cases}$$

in some neighborhood of the given initial curve.

Proof. The existence proof amounts to a check of our solution procedure. Construct the unique solution to the system

$$x' = a(x, y, u), \qquad y' = b(x, y, u), \qquad u' = c(x, y, u)$$

satisfying the initial conditions

$$x(s, 0) = x_0(s), \qquad y(s, 0) = y_0(s), \qquad u(s, 0) = u_0(s).$$

Our assumptions guarantee unique (local) solutions $x = x(s, \tau)$, $y = y(s, \tau)$, and $u = u(s, \tau)$ which are continuously differentiable functions of each argument. Also by hypothesis and the implicit function theorem, we can solve the system

$$x = x(s, \tau), \qquad y = y(s, \tau)$$

for s and τ in terms of x and y to obtain

$$s = s(x, y), \qquad \tau = \tau(x, y)$$

with

$$s = s(x_0(s), y_0(s)), \qquad 0 = \tau(x_0(s), y_0(s)).$$

Consequently, we can define u as a function of x and y by

$$u = u(x, y) = u(s(x, y), \tau(x, y)).$$

The initial conditions are satisfied because

$$u(x_0(s), y_0(s)) = u(s, 0) = u_0(s).$$

u also satisfies the differential equation, for by the chain rule,

$$u_x = u_s s_x + u_\tau \tau_x$$
$$u_y = u_s s_y + u_\tau \tau_y,$$

and consequently

$$au_x + bu_y = (as_x + bs_y)u_s + (a\tau_x + b\tau_y)u_\tau.$$

Now from the characteristic system

$$0 = \frac{\partial}{\partial \tau} s = s_x x' + s_y y' = as_x + bs_y$$

and

$$1 = \frac{\partial}{\partial \tau} \tau = \tau_x x' + \tau_y y' = a\tau_x + b\tau_y,$$

so that

$$au_x + bu_y = \frac{\partial}{\partial \tau} u = c.$$

The uniqueness of the solutions follows from the uniqueness theory for ordinary differential equations. Suppose that $u(x, y)$ is any solution to the problem (2-1). We construct a curve on the surface by constructing for a fixed s the functions $x(\tau)$, $y(\tau)$ obtained by solving the set of equations

$$x' = a(x, y, u(x, y)), \qquad y' = b(x, y, u(x, y))$$

subject to the initial condition

$$x(0) = x_0, \qquad y(0) = y_0.$$

Lefine

$$u(\tau) = u(x(\tau), y(\tau))$$

and observe that $u(\tau)$ satisfies

$$u'(\tau) = u_x x' + u_y u' = au_x + bu_y = c$$

and

$$u(0) = u(x(0), y(0)) = u(x_0, y_0) = u_0$$

so that the curve $(x(\tau), y(\tau), u(\tau))$ satisfies the characteristic equations and initially passes through a point (x_0, y_0, u_0) on the surface. The uniqueness now follows from the uniqueness theorem for ordinary differential equations.

PROBLEMS

1. Extend Th. 2-1 to the problem

$$\sum_{i=1}^{n} a_i(x, u)u_{x_i} = c(x, u),$$

where $x = (x_1, \ldots, x_n)$ is a point in n dimensions, such that

$$u(x^\circ(s)) = u^\circ(s)$$

with $s = (s_1, \ldots, s_{n-1})$.

2. Extend Th. 2-1 to the problem

$$u_t + \sum_{i=1}^{n} a_i(x, u)u_{x_i} = c(x, u),$$

where $x = (x_1, \ldots, x_n)$, such that u satisfies the initial condition

$$u(x, 0) = f(x).$$

2-3. Immiscible Displacement

One of the most important technical applications of the theory of first-order partial differential equations occurs in treating the immiscible displacement of one fluid by another in a porous medium. The assumption that the fluids are immiscible means that the fluids preserve their identity and do not react with each other chemically nor does one dissolve in the other. We shall assume that two fluids are flowing underground and shall distinguish between the two fluids by means of subscripts. Recall that the porosity ϕ tells how much pore space is available to the fluids. Since there are two fluids, they cannot both be taking up all the available pore space. Instead, one fluid will be taking up a certain fraction of the available pore space, call it S_1, and the second fluid will be taking up another fraction, S_2, of the available pore space. We shall assume that the medium is completely saturated so that

$$S_1 + S_2 = 1.$$

Since the fluids are immiscible, the mass of each fluid must be conserved:

$$(\rho_1 \phi S_1)_t + \operatorname{div}(\rho_1 q_1) = 0, \tag{3-1}$$

$$(\rho_2 \phi S_2)_t + \operatorname{div}(\rho_2 q_2) = 0. \tag{3-2}$$

The variables ρ_j, ϕ, S_j, and q_j, $j = 1, 2$, in (3-1) and (3-2) denote, respectively, the density, the porosity, the *saturation*, and the volumetric flow rates of the fluids in question. A modified form of Darcy's law holds for each fluid separately; it has been found empirically that

$$q_1 = -k\frac{k_1}{\mu_1}(\nabla p_1 + \rho_1 \vec{g}) \tag{3-3}$$

and

$$q_2 = -k\frac{k_2}{\mu_2}(\nabla p_2 + \rho_2\vec{g}). \tag{3-4}$$

p_1 and p_2 are the pressures for the fluids in question and $\vec{g}$ is the acceleration due to gravity. k is the permeability, which depends on the medium and is the same for both fluids. μ_1 and μ_2 are the viscosities of the fluids. Since, however, two fluids are flowing in the same medium, each will impede the flow of the other, so that the permeability, which is a measure of how well a given medium will pass a fluid, must be modified in the presence of another fluid. The functions k_1 and k_2, called relative permeabilities, depend on the saturation. In our case, since there are only two fluids flowing and the total flow is saturated, we can set $S = S_1$ so that $S_2 = 1 - S$ and k_1 and k_2 can be regarded as functions of the single variable S.

There is another effect that must be considered when more than one fluid is present. We know from experience that when a block of soil is placed on water but not submersed, the soil will take up the water. In this case the two fluids are air and water and one says that due to capillarity effects the soil has absorbed or sucked up water. Capillarity effects depend on the saturation of the fluids, and consequently one determines empirically a function, $p_c(S)$, the capillary pressure, by measuring the difference

$$p_1(S) - p_2(S) = p_c(S),$$

so that $p_c(S)$ is a known function of S.

We now make the following additional assumptions on the flow. We assume that k, ϕ, ρ_1, ρ_2, μ_1, and μ_2 are all constant and that gravitational effects are negligible. $k_1(S)$ and $k_2(S)$ are given functions of the unknown function, the fraction of pore space occupied by fluid 1. We finally assume that the capillary pressure may be taken equal to zero so that $p_1 = p_2 \equiv p$. Under these conditions the equations of mass conservation or continuity (3-1) and (3-2) take the form

$$\phi S_t + \operatorname{div} q_1 = 0, \tag{3-5}$$

$$-\phi S_t + \operatorname{div} q_2 = 0 \tag{3-6}$$

and the Darcy equations (3-3) and (3-4) become

$$q_1 = -k\frac{k_1}{\mu_1}\nabla p, \tag{3-7}$$

$$q_2 = -k\frac{k_2}{\mu_2}\nabla p. \tag{3-8}$$

Let $q = q_1 + q_2$. Adding (3-5) and (3-6) yields

$$\operatorname{div} q = 0.$$

Adding (3-7) and (3-8) gives

$$q = -k\left(\frac{k_1}{\mu_1} + \frac{k_2}{\mu_2}\right)\nabla p$$

or

$$\nabla p = -\left[k\left(\frac{k_1}{\mu_1} + \frac{k_2}{\mu_2}\right)\right]^{-1} q.$$

The divergence of this expression is

$$\operatorname{div} \nabla p = \Delta p = -q \cdot \nabla\left[k\left(\frac{k_1}{\mu_1} + \frac{k_2}{\mu_2}\right)\right]^{-1},$$

because $\operatorname{div} q = 0$. Next,

$$\begin{aligned}\phi S_t = -\operatorname{div} q_1 &= -k\,\nabla p \cdot \nabla \frac{k_1}{\mu_1} - k\frac{k_1}{\mu_1}\Delta p \\ &= \left(\frac{k_1}{\mu_1} + \frac{k_2}{\mu_2}\right)^{-1} q \cdot \nabla \frac{k_1}{\mu_1} + \frac{k_1}{\mu_1} q \cdot \nabla\left(\frac{k_1}{\mu_1} + \frac{k_2}{\mu_2}\right)^{-1} \\ &= q \cdot \nabla \frac{k_1/\mu_1}{k_1/\mu_1 + k_2/\mu_2}.\end{aligned}$$

Typically, the relative permeabilities k_1 and k_2 have the form shown in Fig. 2-2. Let

$$H(S) = \frac{k_1/\mu_1}{k_1/\mu_1 + k_2/\mu_2}$$

and $H'(S) = h(S)$. Then S satisfies the differential equation

$$\phi S_t = h(S) q \cdot \nabla S, \tag{3-9}$$

called the *Buckley–Leverett equation*. In the event that q is known, this equation can be solved by the method of characteristics to obtain the saturation S. Very often q can be determined independently. In particular, if the flow is one-dimensional, q can often be taken to be a known constant and S can be determined from (3-9).

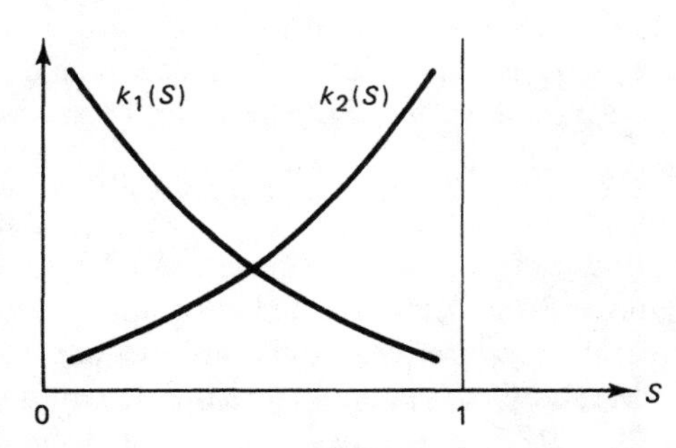

Figure 2-2

PROBLEMS

1. Suppose that $k_1/\mu_1 = 1 - S^2$, $k_2/\mu_2 = S^2$, $\phi = \frac{1}{2}$, and the flow can be considered to be one-dimensional with $q = \frac{1}{2}$. Solve the resulting differential equation given that $S(x, 0) = f(x)$.
2. Suppose in three space dimensions that $k_1/\mu_1 = 1 - S$ and $k_2/\mu_2 = S$, $\phi = \frac{1}{2}$, and

$$q = (\tfrac{1}{2}e^{-x_1}, \tfrac{1}{2}e^{-x_2}, \tfrac{1}{2}e^{-x_3}).$$

Solve the resulting differential equation (3-9) subject to the condition that $S(x, 0) = f(x)$.

2-4. The Method of Characteristics for Nonlinear Partial Differential Equations of the First Order

We turn to the problem of solving the general nonlinear equation

$$F(x, y, u, p, q) = 0, \tag{4-1}$$

where $p = u_x$ and $q = u_y$. The argument for setting up the characteristic equations parallels that of Sec. 2-1, but is complicated by the fact that p and q are no longer linearly related. We begin with the geometric analysis that leads to the characteristic system.

To guarantee that (4-1) is actually a partial differential equation, we assume that $F_p^2 + F_q^2 \neq 0$ in the region of (x, y, u, p, q) space under consideration. For definiteness, take $F_q \neq 0$.

Just as in Sec. 2-1, a solution $u = u(x, y)$ to (4-1) defines a surface in xyu space. Let (x_0, y_0, u_0) be a point lying in the surface and consider a curve $(x(\tau), y(\tau), u(\tau))$ which passes through the point (x_0, y_0, u_0) when $\tau = 0$. We seek conditions on this curve which will guarantee that it lies in the solution surface. As a first step we require that the tangent vector $(x'(0), y'(0), u'(0))$ lie in the tangent plane to the solution surface $u = u(x, y)$ at the point (x_0, y_0, u_0). Just as for the quasilinear equation the tangent plane has the equation

$$u - u_0 = p(x - x_0) + q(y - y_0), \tag{4-2}$$

where $p = u_x(x_0, y_0)$ and $q = u_y(x_0, y_0)$. However, we do not know $p = u_x(x_0, y_0)$ and $q = u_y(x_0, y_0)$ because the solution surface $u = u(x, y)$ is unknown. We know from the differential equation (4-1) only that p and q are constrained by the equation

$$F(x_0, y_0, u_0, p, q) = 0. \tag{4-3}$$

Thus, (4-2) and (4-3) determine a one-parameter family of planes, which includes the tangent plane to a solution surface through (x_0, y_0, u_0). In the quasilinear case, this family of planes envelopes a unique line through (x_0, y_0, u_0) which determines the characteristic system. Here the situation is more complicated. The family of planes (4-2), (4-3) envelopes a cone, called the *Monge cone* (see Fig. 2-3). One generator of this cone is tangent to the solution surface and can be used to determine the characteristics.

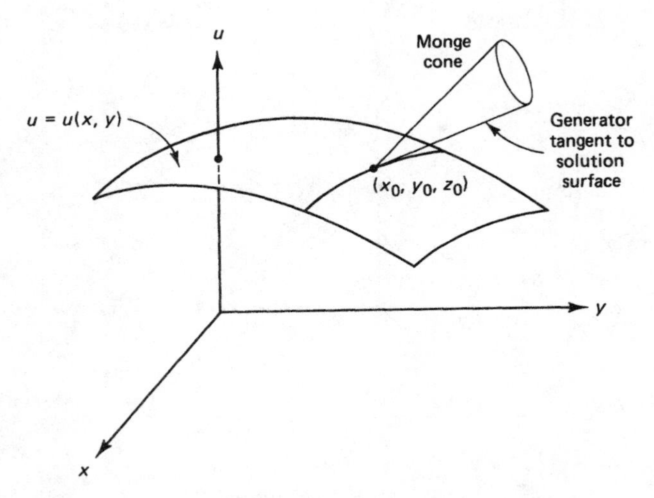

Figure 2-3

To determine the envelope of the family of planes (4-2), (4-3), differentiate (4-2) with respect to p to obtain

$$0 = x - x_0 + q'(p)(y - y_0),$$

where we have used the condition $F_q \neq 0$, which allows us to determine $q = q(p)$ from (4-3). Also from (4-3), $q' = -F_p/F_q$. Thus,

$$\frac{x - x_0}{F_p} = \frac{y - y_0}{F_q}.$$

Inserting this into (4-2) and simplifying leads to

$$\frac{x - x_0}{F_p} = \frac{y - y_0}{F_q} = \frac{u - u_0}{pF_p + qF_q},$$

which is one of the lines that generates the Monge cone. Different generators are determined by different choices of p and q which satisfy (4-3). Since one of these generators lies in the tangent plane to a solution surface through (x_0, y_0, u_0), we first require that our curve $(x(\tau), y(\tau), u(\tau))$ satisfy

$$x'(0) = F_p, \qquad y'(0) = F_q, \qquad u'(0) = pF_p + qF_q, \tag{4-4}$$

where F_p and F_q are evaluated at (x_0, y_0, u_0, p, q). In contrast to the quasilinear case, these ordinary differential equations do not determine the characteristic curves because the unknowns p and q occur explicitly in the equations. Evidently, we must supplement (4-4) by differential equations for p and q which guarantee that $(p, q, -1)$ specifies a normal to the solution surface $u = u(x, y)$ as τ varies. Thus, for each τ we must determine five quantities

$$(x(\tau), y(\tau), u(\tau), p(\tau), q(\tau)), \tag{4-5}$$

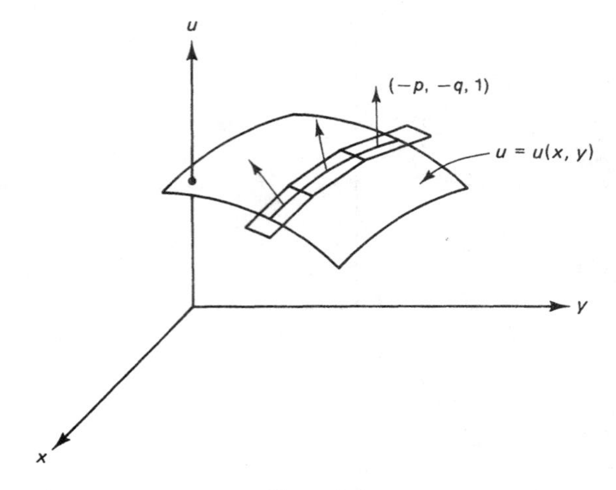

Figure 2-4

which determine both the point through which our curve passes and a tangent plane to this curve which is also tangent to a solution surface. This situation is illustrated in Fig. 2-4. For obvious reasons, such a set of "small" planes is called a strip and the five quantities (4-5) are called its elements.

To determine differential equations for p and q to supplement (4-4), first differentiate (4-1) with respect to x and use $p_y = q_x$ to obtain

$$F_x + F_u p + F_p p_x + F_q p_y = 0.$$

Consequently, if $(x(\tau), y(\tau), u(\tau))$ lies in the solution surface $u = u(x, y)$ and (4-4) holds, we must have

$$x'(0)p_x + y'(0)p_y = -(F_x + pF_u).$$

On the other hand, we want to determine $p(\tau)$ so that $p(\tau) = p(x(\tau), y(\tau))$. Consequently, the expression on the left of the preceding equation is just $p'(0)$. That is,

$$p'(0) = -(F_x + p(0)F_u), \tag{4-6}$$

and similarly,

$$q'(0) = -(F_y + q(0)F_u). \tag{4-7}$$

Since this argument can be repeated at each point on the curve, we conclude from (4-4), (4-6), and (4-7) that

$$\begin{cases} x'(\tau) = F_p(x, y, u, p, q) \\ y'(\tau) = F_q(x, y, u, p, q) \\ u'(\tau) = pF_p(x, y, u, p, q) + qF_q(x, y, u, p, q) \\ p'(\tau) = -(F_x(x, y, u, p, q) + pF_u(x, y, u, p, q)) \\ q'(\tau) = -(F_y(x, y, u, p, q) + qF_u(x, y, u, p, q)) \end{cases} \tag{4-8}$$

for the determination of the characteristic strip.

Since (4-8) implies that

$$\frac{d}{d\tau} F(x(\tau), y(\tau), u(\tau), p(\tau), q(\tau)) = 0,$$

we see that if we have a single value $(x_0, y_0, u_0, p_0, q_0)$ with $F(x_0, y_0, u_0, p_0, q_0) = 0$, then the curve determined by (4-8) with these initial values satisfies

$$F(x, y, u, p, q) = 0.$$

To determine an initial strip once initial values (x_0, y_0, u_0) are given, we must solve the equation

$$F(x_0, y_0, u_0, p_0, q_0) = 0$$

together with another condition which will allow us to "move" the solution curve on the surface in such a way that an integral surface will be swept out. We suppose, therefore, that $(x_0(s), y_0(s), u_0(s))$ is a curve lying in the surface. It serves as initial data for (4-1). Since we seek a solution surface $u = u(x, y)$ through this curve, we must have

$$u_0(s) = u(x_0(s), y_0(s))$$

and hence

$$u_0'(s) = u_x x_0'(s) + u_y y_0'(s).$$

But here $u_x(x_0(s), y_0(s))$ and $u_y(x_0(s), y_0(s))$ are the initial directions p_0 and q_0 which determine the normal to the solution surface and must also satisfy $F(x_0, y_0, u_0, p_0, q_0) = 0$. Consequently, given an initial curve $(x_0(s), y_0(s), u_0(s))$ the remaining components $p_0(s), q_0(s)$ of the initial data for (4-8) are obtained by solving the system

$$\text{(4-9)} \qquad \begin{cases} F(x_0, y_0, u_0, p_0, q_0) = 0, \\ p_0 x_0' + q_0 y_0' = u_0'. \end{cases}$$

Now we can summarize our solution procedure: To determine a solution surface $u = u(x, y)$ which contains the initial curve $(x_0(s), y_0(s), u_0(s))$, solve the system of ordinary differential equations (4-8) for $x = x(s, \tau)$, $y = y(s, \tau)$, ... subject to the initial conditions, depending on the parameter s,

$$\text{(4-10)} \qquad \begin{cases} x(s, 0) = x_0(s), & y(s, 0) = y_0(s), \quad u(s, 0) = u_0(s), \\ p(s, 0) = p_0(s), & q(s, 0) = q_0(s), \end{cases}$$

where $x_0(s)$, $y_0(s)$, and $u_0(s)$ are given and $p_0(s)$ and $q_0(s)$ are determined by solving the system (4-9). In this way, functions $x = x(s, \tau)$, $y = y(s, \tau)$ are obtained, which are then solved for s and τ in terms of x and y and the result is inserted into the expression $u = u(s, \tau)$ to obtain a solution.

Example. Solve $u = p^2 - 3q^2$, $u(x, 0) = x^2$. Let $x_0(s) = s$, $y_0(s) = 0$, $u_0(s) = s^2$. $p_0(s)$ and $q_0(s)$ are solutions to the system

$$s^2 = p_0^2 - 3q_0^2, \qquad 2s = p_0,$$

so that $s^2 = 4s^2 - 3q_0^2$ or $q_0^2 = s^2$. $q_0(s)$ is, of course, not uniquely determined. It is $\pm s$. Let us take $q_0(s) = s$. Then taking $F = p^2 - 3q^2 - u$, we find

$$F_p = 2p, \qquad F_q = -6q, \qquad F_u = -1, \qquad F_x = F_y = 0,$$

so that the characteristic equations are

$$x' = 2p, \qquad y' = -6q, \qquad u' = 2p^2 - 6q^2 = 2(p^2 - 3q^2) = 2u,$$
$$p' = p, \qquad q' = q.$$

We find first that $p = 2se^{\tau}$ and $q = se^{\tau}$, so that after inserting these values into the characteristic equations for x and y, we find immediately

$$x = 4s(e^{\tau} - 1) + s, \qquad y = -6s(e^{\tau} - 1), \qquad u = s^2 e^{2\tau}.$$

Solve for s and τ in terms of x and y, insert these values into the expression for u and simplify to obtain for the solution

$$u = \left(x + \frac{y}{2}\right)^2.$$

Had we chosen $q_0(s) = -s$, we would have found a diffferent solution in precisely the same way, namely

$$u = \left(x - \frac{y}{2}\right)^2.$$

PROBLEMS

1. Solve $pq = 1$, $u(x, 0) = \log x$.
2. Solve $u = p^2 + q^2$, $u(x, 0) = x^2 + 1$.
3. Solve $u = p^2 + q^2$ for u which passes through the curve

$$x_0(s) = \cos s, \qquad y_0(s) = \sin s, \qquad u_0(s) = 1.$$

4. Solve $q = -p^2$, $u(x, 0) = x$.
5. Solve $p^2 - q^2 = x^2 - y$, $u(x, 0) = x$.
6. Solve $yq - xp = 2xyu$, $u(x, x) = x$.
7. Let $x = (x_1, \ldots, x_n)$, $p = (p_1, \ldots, p_n)$, where $u = u(x)$ and $p_j = \partial u/\partial x_j$, $j = 1, \ldots, n$. Consider the problem of solving

$$F(x, u, p) = 0.$$

Argue that the characteristic system is

$$\frac{d}{d\tau} x_j = F_{p_j}(x, u, p), \qquad j = 1, \ldots, n,$$
$$\frac{d}{d\tau} u = \sum_{j=1}^{n} p_j F_{p_j}(x, u, p),$$
$$\frac{d}{d\tau} p_j = -(F_{x_j}(x, u, p) + p_j F_u(x, u, p)), \qquad j = 1, \ldots, n.$$

If u is given on an initial surface, $u = u^\circ(s)$ when $x = x^\circ(s)$, where $s = (s_1, \ldots, s_{n-1})$, argue further that the initial conditions for p are given as solutions to the system

$$F(x^\circ(s), u^\circ(s), p^\circ(s)) = 0,$$

$$\frac{\partial}{\partial s_j} u^\circ(s) = \sum_{k=1}^{n} p_k^\circ(s) \frac{\partial x_k^\circ(s)}{\partial s_j}, \qquad j = 1, \ldots, n-1.$$

8. Consider the partial differential equation

$$c^2(p_1^2 + p_2^2 + p_3^2) = 1, \qquad c \text{ a constant,}$$

which arises in geometrical optics. Suppose that on the sphere of radius r, $u = u^\circ = \alpha$ is a constant. Determine u. The characteristics are called rays and the surfaces $u = \text{const}$ are called the wavefronts. The equation itself is called the equation of the *eiconal*.

2-5. Existence and Uniqueness for Solutions of Nonlinear Equations of the First Order

We now take up the question of the existence of a solution to

$$F(x, y, u, p, q) = 0, \tag{5-1}$$

where $p = u_x$ and $q = u_y$, given that

$$u = u_0(s) \quad \text{when} \quad x = x_0(s) \quad \text{and} \quad y = y_0(s). \tag{5-2}$$

That is, we seek a solution surface $u = u(x, y)$ which passes through the initial curve $(x_0(s), y_0(s), u_0(s))$ and is defined in some neighborhood of this curve.

Our approach is to put sufficient conditions on $F(x, y, u, p, q)$ and the initial curve $(x_0(s), y_0(s), u_0(s))$ which guarantee that the computations described in Sec. 2-4 can be carried out and furnish a solution. Our discussion will make it clear just when a solution is unique.

First we must solve the system

$$\begin{cases} x' = F_p, \quad y' = F_q, \quad u' = pF_p + qF_q, \\ p' = -(F_x + pF_u), \quad q' = -(F_y + qF_q), \end{cases} \tag{5-3}$$

where the prime indicates differentiation with respect to τ, subject to the initial conditions that x, y, u, p, and q reduce to

$$(x_0(s), y_0(s), u_0(s), p_0(s), q_0(s)) \tag{5-4}$$

when $\tau = 0$. Recall that $p_0(s)$ and $q_0(s)$ are determined by solving the system

$$\begin{cases} F(x_0(s), y_0(s), u_0(s), p_0(s), q_0(s)) = 0, \\ x_0'(s)p_0(s) + y_0'(s)q_0(s) = u_0'(s), \end{cases} \tag{5-5}$$

for p_0 and q_0. We assume that this algebraic system can be solved for $p_0(s)$ and $q_0(s)$. At this point we could assert that the characteristic initial value problem (5-3), (5-4) has a solution

$$(x(s, \tau), y(s, \tau), u(s, \tau), p(s, \tau), q(s, \tau)) \tag{5-6}$$

provided that F_x, F_y, F_u, F_p, and F_q are continuous in a domain containing the initial data (5-4). However, for later purposes we will need to assume that the solution (5-6) has continuous second-order partial derivatives. This follows from standard results about ordinary differential equations if we assume that F has continuous second-order partials with respect to all its arguments and that (5-4) has a continuous second derivative. Given this, the initial value problem (5-3), (5-4) has a unique solution (5-6) which has continuous second-order partial derivatives.

To determine the solution to (5-1), (5-2) from (5-6) we must solve for s and τ in terms of x and y in a neighborhood of the initial curve. In view of the implicit function theorem and the continuity of the partial derivatives in (5-6), this will be possible if we require that

$$\frac{\partial(x, y)}{\partial(s, \tau)} = \begin{vmatrix} x_s & x_\tau \\ y_s & y_\tau \end{vmatrix}_{\tau=0} = \begin{vmatrix} x_0' & F_p(x_0, y_0, u_0, p_0, q_0) \\ y_0' & F_q(x_0, y_0, u_0, p_0, q_0) \end{vmatrix} \neq 0. \tag{5-7}$$

Solving for (s, τ) in terms of (x, y), we obtain unique, twice-differentiable functions $s = s(x, y)$, $\tau = \tau(x, y)$ which are then inserted into the expressions for (u, p, q) to obtain twice-differentiable functions

$$\begin{aligned} u &= u(s(x, y), \tau(x, y)) = u(x, y) \\ p &= p(s(x, y), \tau(x, y)) = p(x, y) \\ q &= q(s(x, y), \tau(x, y)) = q(x, y). \end{aligned}$$

Finally, we must show that $u(x, y)$ solves the partial differential equation (5-1) and initial data (5-2). The latter condition is evidently satisfied by our construction. Next from (5-3), for each fixed s,

$$\frac{\partial}{\partial \tau} F(x, y, u, p, q) = F_x x_\tau + F_y y_\tau + F_u u_\tau + F_p p_\tau + F_q q_\tau = 0.$$

Thus, upon integration,

$$F(x, y, u, p, q) = F(x_0(s), y_0(s), u_0(s), p_0(s), q_0(s)) = 0$$

by (5-5). Consequently,

$$F(x, y, u, p, q) = 0,$$

where $u(x, y)$, $p(x, y)$, and $q(x, y)$ are the functions defined following (5-7). This equation becomes (5-1) if we can prove that $p(x, y) = u_x(x, y)$ and $q(x, y) = u_y(x, y)$. Thus, it remains to show that $p = u_x$ and $q = u_y$.

Obviously,

$$u_s = u_x x_s + u_y y_s, \qquad u_\tau = u_x x_\tau + u_y y_\tau,$$

which we can regard as a system of linear equations satisfied by u_x and u_y. This system has a unique solution because its determinant is nonzero by (5-7). Consequently, if we can show that p and q also satisfy this same system,

$$\begin{cases} u_s = p x_s + q y_s, \\ u_\tau = p x_\tau + q y_\tau, \end{cases} \tag{5-8}$$

then it follows that $p = u_x$ and $q = u_y$ as desired. Thus, we need only establish (5-8). The second equation in (5-8) is evident from the characteristic system (5-3):

$$u_\tau = pF_p + qF_q = px_\tau + qy_\tau.$$

Now use of the second equation in (5-8), the characteristic system, and the fact that $F \equiv 0$ in (s, τ) yields

$$\begin{aligned}
\frac{\partial}{\partial \tau}(u_s - px_s - qy_s) &= \frac{\partial}{\partial \tau}(u_s - px_s - qy_s) - \frac{\partial}{\partial s}(u_\tau - px_\tau - qy_\tau) \\
&= -p_\tau x_s - q_\tau y_s + p_s x_\tau + q_s y_\tau \\
&= x_s(F_x + pF_u) + y_s(F_y + qF_u) + p_s F_p + q_s F_q \\
&= \frac{\partial}{\partial s} F(x, y, u, p, q) - u_s F_u + x_s p F_u + y_s q F_u \\
&= -(u_s - px_s - qy_s)F_u.
\end{aligned}$$

This is a simple ordinary differential equation in τ for $\Delta(\tau) = u_s - px_s - qy_s$, with s regarded as a parameter. When $\tau = 0$,

$$\Delta(0) = u_s - px_s - qy_s = u_0'(s) - p_0(s)x_0'(s) - q_0(s)y_0'(s) = 0$$

by (5-5). Integration gives

$$\Delta(\tau) = \Delta(0) \exp\left(-\int_0^\tau F_u \, d\tau\right) = 0,$$

which validates the first equation in (5-8).

We summarize this discussion as follows.

Theorem 5-1. Consider the partial differential equation

$$F(x, y, u, p, q) = 0, \qquad p = u_x, \qquad q = u_y. \tag{5-9}$$

Suppose that F is twice continuously differentiable with respect to each of its arguments in a domain D of the five-dimensional space (x, y, u, p, q). Suppose that $F_p^2 + F_q^2 \neq 0$ in D. Consider further a twice continuously differentiable curve Γ parameterized by s:

$$\Gamma : x = x_0(s), \quad y = y_0(s), \quad u = u_0(s). \tag{5-10}$$

Assume finally that the system

$$\begin{cases} F(x_0, y_0, u_0, p_0, q_0) = 0 \\ x_0' p_0 + y_0' q_0 = u_0' \end{cases} \tag{5-11}$$

is solvable for (p_0, q_0) yielding a twice-continuously differentiable initial strip $(x_0(s), y_0(s), u_0(s), p_0(s), q_0(s))$ lying in D such that

$$\begin{vmatrix} x_0' & F_p(x_0, y_0, u_0, p_0, q_0) \\ y_0' & F_q(x_0, y_0, u_0, p_0, q_0) \end{vmatrix} \neq 0. \tag{5-12}$$

Then there exists a solution $u = u(x, y)$ to the problem (5-9) and (5-10) in some neighborhood of Γ and the solution u is twice continuously differentiable.

In general, solutions to (5-9), (5-10) are not unique (see the example from Sec. 2-4). However, our analysis above shows that we have uniqueness at each stage of the solution procedure once $p_0(s)$ and $q_0(s)$ are determined from (5-11). This establishes (i) and (ii) in the following theorem.

Theorem 5-2. Let F and Γ satisfy the hypotheses of Th. 5-1.

(i) If the system (5-11) is uniquely solvable for p_0 and q_0 and (5-12) holds along the resulting strip, then (5-9), (5-10) is uniquely solvable.

(ii) If the system (5-11) possesses multiple solutions (p_0, q_0), then each choice of (p_0, q_0) determines a strip. If (5-12) holds along each strip, then (5-9), (5-10) will possess a unique solution corresponding to each such strip.

(iii) If the system (5-11) does not have a solution, then the problem (5-9), (5-10) is not solvable.

Statement (iii) follows from the observation that if $u(x, y)$ solves (5-9), (5-10), then

$$x_0(s),\ y_0(s),\ u_0(s),\ p_0(s) = u_x(x_0(s), y_0(s)),$$
$$q_0(s) = u_y(x_0(s), y_0(s))$$

solves (5-11).

Example. Solve $p - q^2 = 0$, $u(x, 0) = f(x)$. Parameterize the initial curve by

$$x_0(s) = s, \qquad y_0(s) = 0, \qquad u_0(s) = f(s).$$

Equation (5-11) takes the form

$$p_0 - q_0^2 = 0, \qquad p_0 = f'(s).$$

If $f'(s) < 0$, no real solution exists. If $f'(s) > 0$, then

$$q_0 = \pm\sqrt{p_0} = \pm\sqrt{f'(s)},$$

(5-12) holds because

$$\begin{vmatrix} 1 & 1 \\ 0 & \pm 2\sqrt{f'(s)} \end{vmatrix} = \pm 2\sqrt{f'(s)} \neq 0,$$

and there are two solutions. If there are values of s for which $f'(s)$ vanishes, the foregoing theory tells us nothing but we may be able to solve the problem in specific cases. On the other hand, suppose that the initial data consist of $u(0, y) = g(y)$, so that $x_0(s) = 0$, $y_0(s) = s$, $u_0(s) = g(s)$. The system (5-11) is now

$$p_0 - q_0^2 = 0, \qquad q_0 = g'(s),$$

which is uniquely solvable. Also, (5-12) becomes

$$\begin{vmatrix} 0 & 1 \\ 1 & -2g'(s) \end{vmatrix} = -1 \neq 0.$$

Thus, the new initial value problem has a unique solution.

An important special case of (5-9) arises for the initial value problem

$$q = G(x, y, u, p), \qquad p = u_x, \qquad q = u_y, \tag{5-13}$$

$$u(x, 0) = f(x). \tag{5-14}$$

We can parameterize this initial curve by

$$x_0(s) = s, \qquad y_0(s) = 0, \qquad u_0(s) = f(s)$$

and (5-11) has the form

$$q_0(s) = G(s, 0, f(s), p_0(s)), \qquad p_0(s) = f'(s),$$

which determines $p_0(s)$ and $q_0(s)$ uniquely. Furthermore, (5-12) is

$$\begin{vmatrix} 1 & -G_p \\ 0 & 1 \end{vmatrix} = +1 \neq 0.$$

An application of Th. 5-2 yields our next result.

Theorem 5-3. **Suppose that f is three times continuously differentiable and G is twice continuously differentiable. Then in a neighborhood of the initial curve $u = f(x)$, $y = 0$, there exists a unique solution $u = u(x, y)$ to the problem (5-13), (5-14).**

PROBLEMS

1. Let $x = (x_1, \ldots, x_n)$, $p = (p_1, \ldots, p_n)$ with $p_j = \partial u/\partial x_j$, $j = 1, \ldots, n$. Prove the analogues of Ths. 5-1, 5-2, and 5-3 for the partial differential equation

$$F(x, u, p) = 0$$

given that $u = u^\circ(s)$ when $x = x^\circ(s)$, where $s = (s_1, \ldots, s_{n-1})$.

2. Theorems 5-1, 5-2, and 5-3 are all local in nature, that is, they refer to the existence and uniqueness in the neighborhood of the given initial data. This restriction is very real, as the following example will show.

Solve

$$u_t + uu_x = 0, \qquad -\infty < x < \infty, \qquad t > 0,$$
$$u(x, 0) = f(x), \qquad -\infty < x < \infty.$$

The variable y has been replaced by t, which is customary in physical applications.

(a) Show that the solution is given implicitly by

$$u = f(x - tu).$$

If $f'(x) \geq 0$, show that there is a function $u = u(x, t)$ satisfying this implicit equation for u. If $f'(x) \leq -\alpha$, $\alpha > 0$ a constant, show that after a finite time the solution no longer exists.

(b) More explicitly suppose that

$$f(x) = \begin{cases} 1, & x \leq 0 \\ 1 - 3x^2 + 2x^3, & 0 < x < 1, \\ 0, & x \geq 1. \end{cases}$$

Take for the initial parameterization $x_0(s) = s$, $t_0(s) = 0$, $u_0(s) = f(s)$. Construct the curves $x = x(s, \tau)$, $t = t(s, \tau)$ and graph for fixed values of s, say s equal to $0, \frac{1}{2}, 1$, the curves $x = x(s, \tau)$, $t = t(s, \tau)$ for $\tau > 0$ and interpret the results. In certain physical applications it is nevertheless necessary to seek solutions for all time. Severe difficulties arise which we shall postpone here and take up in Chap. 12.

3. Suppose that $F_p^2 + F_q^2 \neq 0$ holds because $F_p \neq 0$ and $F_q \neq 0$. Show that a solution $u(x, y)$ in Th. 5-1 possesses continuous third-order partial derivatives.

2-6. Classification of Second-Order Equations

As a final application of the theory of first-order partial differential equations, we consider the problem of simplifying and classifying partial differential equations of the second order in two independent variables, x and y. Specifically, consider

$$a(x, y)u_{xx} + 2b(x, y)u_{xy} + c(x, y)u_{yy} = f(x, y, u, u_x, u_y), \tag{6-1}$$

where a, b, and c are twice continuously differentiable functions of x and y. Several equations of this kind were derived in Chap. 1. If $a = 1$, $b = 0$, $c = -1$ and y represents time, we get the wave equation

$$u_{xx} - u_{yy} = f \tag{i}$$

with the units chosen so that the speed of wave propagation is 1 [see (2-3) of Chap. 1]. If $a = 1$, $b = 0$, $c = 0$, and $f = u_y - F(x, y)$, the one-dimensional heat equation

$$u_{xx} = u_y - F(x, y) \tag{ii}$$

with thermal diffusivity normalized to 1 results [see (3-8) of Chap. 1]. Finally, if $a = 1$, $b = 0$, $c = 1$, and $f = F(x, y)$, where x and y are now space variables, then

$$u_{xx} + u_{yy} = f \tag{iii}$$

describes the two-dimensional Poisson equation for steady-state heat flow.

We will show that these important special cases are typical. That is, the second-order partial differential equation (6-1) can be reduced, by a suitable change of coordinates, to a simpler equation whose second-order terms agree with (i), (ii), or (iii) in certain domains in the xy plane. This reduction is of evident theoretical interest. It also has practical significance because it makes it clear, by comparison with the appropriate wave, heat, or Poisson equation, just what boundary and/or initial data are appropriate for the given partial differential equation.

The wave equation (i) can also be put in the alternative form

(i)′ $$u_{xy} = f$$

by means of the simple change of variables $\xi = x - y, \eta = x + y$, followed by a relabeling of the independent variables.

To achieve one of the standard or canonical forms (i)′, (ii), or (iii), we make a suitable twice continuously differentiable change of variables

(6-2) $$\alpha = \phi(x, y), \qquad \beta = \psi(x, y).$$

Since we want the transformed equation to be equivalent to the original equation, we insist that the Jacobian

$$\frac{\partial(\phi, \psi)}{\partial(x, y)} = \begin{vmatrix} \phi_x & \phi_y \\ \psi_x & \psi_y \end{vmatrix} \neq 0,$$

which guarantees (at least locally) that the inverse transformation

(6-3) $$x = \Phi(\alpha, \beta), \qquad y = \Psi(\alpha, \beta)$$

to (6-2) exists and is twice continuously differentiable. In terms of the new variables

$$u = u(\Phi(\alpha, \beta), \Psi(\alpha, \beta)) \equiv u(\alpha, \beta).$$

Use of the chain rule gives

$$u_x = u_\alpha \phi_x + u_\beta \psi_x, \qquad u_y = u_\alpha \phi_y + u_\beta \psi_y$$

and

$$\begin{aligned}
u_{xx} &= u_{\alpha\alpha}\phi_x^2 + 2u_{\alpha\beta}\phi_x\psi_x + u_{\beta\beta}\psi_x^2 + u_\alpha\phi_{xx} + u_\beta\psi_{xx} \\
u_{yy} &= u_{\alpha\alpha}\phi_y^2 + 2u_{\alpha\beta}\phi_y\psi_y + u_{\beta\beta}\psi_y^2 + u_\alpha\phi_{yy} + u_\beta\psi_{yy} \\
u_{xy} &= u_{\alpha\alpha}\phi_x\phi_y + u_{\alpha\beta}(\phi_x\psi_y + \phi_y\psi_x) + u_{\beta\beta}\psi_x\psi_y + u_\alpha\phi_{xy} + u_\beta\psi_{xy}.
\end{aligned}$$

Consequently,

$$\begin{aligned}
au_{xx} + 2bu_{xy} + cu_{yy} = {} & (a\phi_x^2 + 2b\phi_x\phi_y + c\phi_y^2)u_{\alpha\alpha} \\
& + 2(a\phi_x\psi_x + b\phi_x\psi_y + b\phi_y\psi_x + c\phi_y\psi_y)u_{\alpha\beta} \\
& + (a\psi_x^2 + 2b\psi_x\psi_y + c\psi_y^2)u_{\beta\beta} + (a\phi_{xx} + 2b\phi_{xy} + c\phi_{yy})u_\alpha \\
& + (a\psi_{xx} + 2b\psi_{xy} + c\psi_{yy})u_\beta.
\end{aligned}$$

Thus, the transformed equation (6-1) has the form

(6-1)′ $$A(\alpha, \beta)u_{\alpha\alpha} + 2B(\alpha, \beta)u_{\alpha\beta} + C(\alpha, \beta)u_{\beta\beta} = F_1(\alpha, \beta, u, u_\alpha, u_\beta),$$

with coefficients

$$\begin{aligned}
A &= a\phi_x^2 + 2b\phi_x\phi_y + c\phi_y^2 \\
C &= a\psi_x^2 + 2b\psi_x\psi_y + c\psi_y^2 \\
B &= a\phi_x\psi_x + b(\phi_x\psi_y + \phi_y\psi_x) + c\phi_y\psi_y,
\end{aligned}$$

which can be expressed in terms of α and β via (6-3).

Now we specialize our change of coordinates (6-2) to obtain one of the standard forms (i)′, (ii), or (iii). If $a(x, y) = c(x, y) = 0$ in the region of interest for (6-1), then (6-1) can be put into the normal form (i)′ simply by division by $2b(x, y)$, which we assume nonzero. Thus, we may suppose that one of $a(x, y)$ or $c(x, y)$ is nonzero. Since x and y appear symmetrically in (6-1), we may further assume without loss of generality that $a(x, y) \neq 0$. In view of the expressions for A and C, we can achieve $A = C = 0$ in (6-1)′ if ϕ and ψ are solutions to

$$av_x^2 + 2bv_xv_y + cv_y^2 = 0. \tag{6-4}$$

This quadratic, first-order partial differential equation factors as

$$a\left(v_x - \frac{-b + \sqrt{b^2 - ac}}{a}\, v_y\right)\left(v_x - \frac{-b - \sqrt{b^2 - ac}}{a}\, v_y\right) = 0.$$

It is now apparent that there are three cases to consider, depending on the sign of the discriminant $b^2 - ac$ of (6-1).

Case 1. $b^2 - ac > 0$ in the region of interest. In this case we call (6-1) a *hyperbolic equation*, in analogy to the theory of plane quadratic curves. Then (6-4) has solutions ϕ and ψ determined by the quasilinear first-order equations

$$\phi_x - \frac{-b + \sqrt{b^2 - ac}}{a}\,\phi_y = 0, \qquad \psi_x - \frac{-b - \sqrt{b^2 - ac}}{a}\,\psi_y = 0. \tag{6-5}$$

If we can find solutions to these equations which satisfy the Jacobian condition following (6-2), then $\alpha = \phi(x, y)$ and $\beta = \psi(x, y)$ will define a change of variables for which (6-1)′ becomes

$$2B(\alpha, \beta)u_{\alpha\beta} = F_1(\alpha, \beta, u, u_\alpha, u_\beta). \tag{6-6}$$

Now from (6-5),

$$\frac{\partial(\phi, \psi)}{\partial(x, y)} = \begin{vmatrix} \phi_x & \phi_y \\ \psi_x & \psi_y \end{vmatrix} = \frac{2}{a}\sqrt{b^2 - ac}\,\phi_y\psi_y.$$

So the required Jacobian condition holds provided that $\phi_y, \psi_y \neq 0$. To construct such a ϕ, solve the characteristic system

$$\frac{dx}{d\tau} = 1, \qquad \frac{dy}{d\tau} = -\frac{-b + \sqrt{b^2 - ac}}{a}, \qquad \frac{d\phi}{d\tau} = 0$$

with initial curve taken to be, for example,

$$x_0(s) = x_0, \qquad y_0(s) = y_0 + s, \qquad \phi_0(s) = s,$$

where the point (x_0, y_0) is chosen so that the curve $(x_0(s), y_0(s))$ lies within the region of the xy plane where $b^2 - ac \geq 0$. (It is convenient to allow the possibility that the initial curve lies on the boundary of the region where $b^2 - ac > 0$.) The Jacobian condition in Th. 2-1, namely,

$$\begin{vmatrix} 1 & 0 \\ \dfrac{b - \sqrt{b^2 - ac}}{a} & 1 \end{vmatrix} = 1 \neq 0$$

holds. Therefore, the characteristic initial value problem leads to a unique solution $\phi(x, y)$. Furthermore, since $d\phi/d\tau = 0$, $\phi(s, \tau) = s$ and consequently $1 = \phi_s = \phi_x x_s + \phi_y y_s$. Now if $\phi_y = 0$, then by (6-5), $\phi_x = 0$, which contradicts the last equation. Thus, $\phi_y \neq 0$. In the same way, we infer that $\psi_y \neq 0$, where $\psi(x, y)$ is the solution obtained from the characteristic initial value problem

$$\frac{dx}{d\tau} = 1, \qquad \frac{dy}{d\tau} = \frac{b + \sqrt{b^2 - ac}}{a}, \qquad \frac{d\psi}{d\tau} = 0,$$

$$x_0(s) = x_0, \qquad y_0(s) = y_0 + s, \qquad \psi_0(s) = s.$$

In summary, the change of variables $\alpha = \phi(x, y)$, $\beta = \psi(x, y)$ defined in this way has a nonzero Jacobian and reduces (6-1)′ to (6-6). Finally, a routine calculation yields

$$B = 2\left[\frac{(ac - b^2)}{a}\right]\phi_y\psi_y \neq 0.$$

So (6-6) reduces to the normal form

$$u_{\alpha\beta} = F(\alpha, \beta, u, u_\alpha, u_\beta)$$

where $F = F_1/2B$. Thus in case 1, the partial differential equation can be reduced to the normal form (i)′, equivalently (i).

Notice that the reduction to normal form is a "local" result. The reduction is possible in a region in the xy plane where (6-1) is hyperbolic. Also, the special initial curve used above can be replaced by any other curve $(x_0(s), y_0(s), \phi_0(s))$ with $\phi_0(s) = s$ and $(x_0(s), y_0(s))$ specifying a curve in the xy plane in the region where (6-1) is defined and for which the determinant condition in Th. 2-1 holds. The curves $\phi(x, y) = \text{const}$ and $\psi(x, y) = \text{const}$ are called the *characteristics* of (6-1). Therefore, in the region where (6-1) is hyperbolic, the equation has two sets of characteristics.

Example. Consider the partial differential equation

$$yu_{xx} - xu_{yy} = 0$$

in the (open) first quadrant $x > 0$, $y > 0$. Here $a(x, y) = y$, $c(x, y) = -x$, $b(x, y) = 0$, and $b^2 - ac = xy > 0$, so that the equation is hyperbolic. We construct a solution ϕ to the partial differential equation

$$\phi_x - \sqrt{\frac{x}{y}}\,\phi_y = 0$$

from the characteristic initial value problem

$$\frac{dx}{d\tau} = 1, \qquad \frac{dy}{d\tau} = -\sqrt{\frac{x}{y}}, \qquad \frac{d\phi}{d\tau} = 0$$

$$x_0(s) = 0, \qquad y_0(s) = s, \qquad \phi_0(s) = s$$

for $s > 0$, so $(x_0(s), y_0(s))$ is the positive y axis. Solving yields

$$x = \tau, \qquad y^{3/2} + \tau^{3/2} = s^{3/2}, \qquad \phi = s,$$

whence

$$\alpha = \phi(x, y) = (x^{3/2} + y^{3/2})^{2/3}.$$

Similarly,

$$\beta = \psi(x, y) = (y^{3/2} - x^{3/2})^{2/3}.$$

So the desired transformation of coordinates is

$$\alpha = (x^{3/2} + y^{3/2})^{2/3}, \qquad \beta = (y^{3/2} - x^{3/2})^{2/3}$$

for $x, y > 0$. Notice that the second solution $\psi(x, y)$ to $\psi_x + \sqrt{x/y}\,\psi_y = 0$ only extends off the positive y axis into the triangular region $y > x > 0$. The transformation $\alpha = \phi(x, y)$, $\beta = \psi(x, y)$ maps this region one-to-one, onto the triangular region $\alpha > \beta > 0$ in the $\alpha\beta$ plane. The inverse transformation is

$$x = \left(\frac{\alpha^{3/2} - \beta^{3/2}}{2}\right)^{2/3}, \qquad y = \left(\frac{\alpha^{3/2} + \beta^{3/2}}{2}\right)^{2/3},$$

so that the differential equation, expressed in the new coordinates is

$$u_{\alpha\beta} = -\frac{1}{2(\alpha^3 - \beta^3)}(\beta^2 u_\alpha - \alpha^2 u_\beta).$$

This same normal form is obtained for the lower triangular region $x > y > 0$ because the transformation $\alpha = \phi(x, y)$, $\beta = \psi(x, y)$ also maps this triangle one-to-one, onto the same triangular region $\alpha > \beta > 0$ in the $\alpha\beta$ plane. To reduce the original equation to normal form in a neighborhood of a point (x_0, x_0) with $x_0 > 0$, other initial curves must be used for the characteristic system; for example, $x_0(s) = x_0$, $y_0(s) = s$, $\phi_0(s) = s$.

Case 2. $b^2 - ac \equiv 0$ in the region of interest. In this event (6-1) is called *parabolic*. If $a \equiv 0$, then $b \equiv 0$ and we assume that $c \neq 0$ in the region of interest. In this situation (6-1) is already in the normal form (ii) with x and y interchanged. So we can assume that $a \neq 0$ in the region and (6-4) factors as

$$a\left(v_x + b\frac{v_y}{a}\right)^2 = 0.$$

We can use the single quasilinear equation

$$v_x + b\frac{v_y}{a} = 0$$

to determine a solution $\alpha = \phi(x, y)$ as in case 1. Then $A = 0$. Whatever choice we make for $\psi(x, y)$, we find that $B = 0$ because $b^2 = ac$; indeed,

$$\begin{aligned} B &= a\phi_x\psi_x + b(\phi_x\psi_y + \phi_y\psi_x) + c\phi_y\psi_y = \psi_x(a\phi_x + b\phi_y) + \psi_y(b\phi_x + c\phi_y) \\ &= \psi_y(b\phi_x + c\phi_y) = \frac{1}{a}\psi_y(ab\phi_x + ac\phi_y) = \frac{b}{a}\psi_y(a\phi_x + b\phi_y) = 0. \end{aligned}$$

Consequently, choose $\beta = \psi(x, y)$ to be any convenient, twice-continuously differentiable function which is linearly independent of ϕ to obtain the normal or canonical form

$$u_{\beta\beta} = F(\alpha, \beta, u, u_\alpha, u_\beta)$$

for a parabolic equation.

Case 3. $b^2 - ac < 0$ in the region of interest. In this situation (6-1) is called *elliptic*. Examination of the characteristic initial value problem shows that there are no real characteristics. This substantially complicates the analysis leading to the canonical form (iii) under the differentiability assumptions that we have made. However, in many important cases, the coefficients $a(x, y)$, $b(x, y)$, $c(x, y)$ are analytic functions of x and y (i.e., they possess convergent power series expansions in x and y) and in this case the reduction to normal form proceeds much as for the hyperbolic case, but calculating with complex variables α and β. One arrives at an equation of the form

$$u_{\alpha\beta} = G(\alpha, \beta, u, u_\alpha, u_\beta).$$

To eliminate the complex variables, set

$$\alpha = \xi + i\eta \quad \text{and} \quad \beta = \xi - i\eta$$

to obtain the normal or canonical form for elliptic equations

$$u_{\xi\xi} + u_{\eta\eta} = F(\xi, \eta, u, u_\xi, u_\eta).$$

The terminology elliptic, hyperbolic, and parabolic is also used for equations in several variables. In this case of two variables, the equation (6-1) is elliptic if $b^2 - ac < 0$; in other words, if the matrix of the coefficients

$$\begin{bmatrix} a & b \\ b & c \end{bmatrix}$$

is positive definite, where we have taken $a > 0$ in the region under consideration. More generally, let $x = (x_1, \ldots, x_n)$ and $p = (p_1, \ldots, p_n)$, with $p_j = u_{x_j}, j = 1, \ldots, n$. The partial differential equation

$$Lu = \sum_{i,j=1}^{n} a_{ij} u_{x_i x_j} = F(x, u, p)$$

is *elliptic* if the matrix (a_{ij}) of its coefficients is positive definite. Here the a_{ij} may depend on the independent variables x, the solution u, and the first derivatives of the solution p. The partial differential equation

$$u_t = Lu + F(x, t, u, p)$$

is *parabolic* if L is elliptic, where the a_{ij} may depend on the variables (x, t, u, p). Finally, the partial differential equation

$$u_{tt} = Lu + F(x, t, u, u_t, p)$$

is *hyperbolic* if L is elliptic and the a_{ij} are allowed to depend on the variables (x, t, u, u_t, p). Notice that in the case where the coefficients a_{ij} depend on the solution u and its first derivatives, the classification may depend on the solution.

PROBLEMS

1. Show that the linear partial differential equation

$$a(x, y)u_{xx} + 2b(x, y)u_{xy} + c(x, y)u_{yy} + d(x, y)u_x + e(x, y)u_y + f(x, y)u = g(x, y)$$

remains linear after reduction to canonical form.

2. Consider the partial differential equation with constant coefficients

$$au_{xx} + 2bu_{xy} + cu_{yy} + du_x + eu_y + fu = g(x, y).$$

Show that the canonical forms are

$$\begin{aligned}
&\text{Hyperbolic:} && u_{\alpha\alpha} - u_{\beta\beta} + \bar{d}u_\alpha + \bar{e}u_\beta + \bar{f}u = \bar{g}(\alpha, \beta)\\
&\text{Elliptic:} && u_{\alpha\alpha} + u_{\beta\beta} + \bar{d}u_\alpha + \bar{e}u_\beta + \bar{f}u = \bar{g}(\alpha, \beta)\\
&\text{Parabolic:} && u_{\beta\beta} + \bar{d}u_\alpha + \bar{e}u_\beta + \bar{f}u = \bar{g}(\alpha, \beta),
\end{aligned}$$

where $\bar{d}$, $\bar{e}$, $\bar{f}$ are constants.

3. Make a substitution of the form $u = v \exp(\lambda\alpha + \mu\beta)$ with suitable values λ and μ, to bring the canonical forms in Prob. 2 into the form

$$\begin{aligned}
&\text{Hyperbolic:} && v_{\alpha\alpha} - v_{\beta\beta} + \tilde{f}v = \tilde{g}(\alpha, \beta)\\
&\text{Elliptic:} && v_{\alpha\alpha} + v_{\beta\beta} + \tilde{f}v = \tilde{g}(\alpha, \beta)\\
&\text{Parabolic:} && v_{\beta\beta} + \tilde{d}v_\alpha = \tilde{g}(\alpha, \beta),
\end{aligned}$$

where $\tilde{f}$ and $\tilde{d}$ are constants.

4. Classify the following partial differential equations.
(a) $x^2u_{xy} - yu_{yy} + u_x - 4u = 0$.
(b) $xyu_{xx} + 4u_{xy} - (x^2 + y^2)u_{yy} - u = 0$.
Note that in different parts of the xy plane, the partial differential equation can be of a different type.

5. Reduce the following partial differential equations to canonical form.
(a) $c^2u_{xx} - u_{yy} = 0$.
(b) $2u_{xx} + u_{xy} + yu_{yy} = 0$ in $y > 1$, so the equation is elliptic.
(c) $x^2u_{xx} - 2xyu_{xy} + y^2u_{yy} = 0$.
(d) $xu_{xx} - 4u_{xy} = 0$ in $x > 0$.

3

Elements of Fourier Series and Integrals

3-1. Introduction

In Sec. 4-1 we use the method of separation of variables to find a series solution to the initial, boundary value problem for a vibrating string. A key step in this procedure will require us to expand both the initial position and initial velocity profile of the string into a trigonometric series on $0 \leq x \leq L$, where L is the length of the string.

To get some idea about what should happen, suppose that $f(x)$ is a function that can be expanded in a trigonometric series, say,

$$f(x) = A + \sum_{n=1}^{\infty} \left[a_n \cos\left(\frac{n\pi x}{L}\right) + b_n \sin\left(\frac{n\pi x}{L}\right) \right] \tag{1-1}$$

for $-L \leq x \leq L$. Evidently, the coefficients A, a_n, and b_n must be determined by the function $f(x)$. The question is how. Some formal calculations will indicate what to expect: Integrate both sides of (1-1) from $-L$ to L, integrating term by term on the right, to find

$$A = \frac{1}{2L} \int_{-L}^{L} f(x)\, dx.$$

The coefficients a_n and b_n can be calculated in a similar way using the *orthogonality relations*

$$\int_{-L}^{L} \sin\left(\frac{m\pi x}{L}\right) \sin\left(\frac{n\pi x}{L}\right) dx = \int_{-L}^{L} \cos\left(\frac{m\pi x}{L}\right) \cos\left(\frac{n\pi x}{L}\right) dx = 0 \quad \text{for} \quad m \neq n,$$

$$\int_{-L}^{L} \sin\left(\frac{m\pi x}{L}\right) \cos\left(\frac{n\pi x}{L}\right) dx = 0 \quad \text{for} \quad \text{all } m, n,$$

and the identities

$$\int_{-L}^{L} \sin^2\left(\frac{n\pi x}{L}\right) dx = \int_{-L}^{L} \cos^2\left(\frac{n\pi x}{L}\right) dx = L \quad \text{for} \quad n \geq 1.$$

Multiply (1-1) by $\cos(m\pi x/L)$ and integrate from $-L$ to L to find

$$a_m = \frac{1}{L}\int_{-L}^{L} f(x)\cos\left(\frac{m\pi x}{L}\right) dx$$

for $m = 1, 2, 3, \ldots$. Similarly,

$$b_m = \frac{1}{L}\int_{-L}^{L} f(x)\sin\left(\frac{m\pi x}{L}\right) dx$$

for $m = 1, 2, 3, \ldots$. The coefficients a_m are originally defined only for $m = 1, 2, 3, \ldots$. However, it is convenient to use the formula above with $m = 0$ to define a_0 because then $A = a_0/2$.

These formal calculations suggest: If a function $f(x)$ has a trigonometric series expansion,

$$f(x) = \frac{1}{2}a_0 + \sum_{n=1}^{\infty}\left[a_n\cos\left(\frac{n\pi x}{L}\right) + b_n\sin\left(\frac{n\pi x}{L}\right)\right] \tag{1-2}$$

for $-L \leq x \leq L$, then

$$a_n = \frac{1}{L}\int_{-L}^{L} f(x)\cos\left(\frac{n\pi x}{L}\right) dx \quad \text{for} \quad n = 0, 1, 2, \ldots, \tag{1-3}$$

and

$$b_n = \frac{1}{L}\int_{-L}^{L} f(x)\sin\left(\frac{n\pi x}{L}\right) dx \quad \text{for} \quad n = 1, 2, 3, \ldots. \tag{1-4}$$

The trigonometric series

$$\frac{1}{2}a_0 + \sum_{n=1}^{\infty}\left[a_n\cos\left(\frac{n\pi x}{L}\right) + b_n\sin\left(\frac{n\pi x}{L}\right)\right]$$

is called the *Fourier series* of the function $f(x)$ defined on $-L \leq x \leq L$ if the coefficients a_n and b_n are given by (1-3) and (1-4). The numbers a_n and b_n are called the *Fourier coefficients* of $f(x)$. In this language, our basic problem is to determine reasonable conditions on a function $f(x)$ so that it will be the sum of its Fourier series. That is, if a_n and b_n are given by (1-3) and (1-4), what reasonable conditions on $f(x)$ lead to (1-2)? You might find the formal discussion above so convincing that you would expect (1-2) to hold for any reasonable function $f(x)$, say $f(x)$ continuous. However, there are continuous functions $f(x)$ for which (1-2) fails to hold! Thus, the possibility of expanding a given function in a Fourier series as in (1-2) is a more subtle question than first meets the eye.

In view of the preceding remarks, it is worthwhile to take a closer look at (1-2). We will describe three handy convergence criteria which guarantee that

(1-2) holds. The proofs are given in Sec. 3-3. The statements of the convergence tests may involve a few unfamiliar terms, such as $2L$ periodic, piecewise smooth, and uniformly convergent. They will be explained below. Here is our first convergence result:

Theorem 1-1. If the Fourier series of a $2L$ periodic, continuous function $f(x)$ converges uniformly over a period, then this Fourier series has sum $f(x)$:

$$f(x) = \frac{1}{2} a_0 + \sum_{n=1}^{\infty} \left[a_n \cos\left(\frac{n\pi x}{L}\right) + b_n \sin\left(\frac{n\pi x}{L}\right) \right].$$

An often used consequence of this result is

Corollary 1-1. Let $f(x)$ be $2L$ periodic and continuous. If the series of magnitudes of the Fourier coefficients $\sum |a_k|$ and $\sum |b_k|$ are convergent, then the Fourier series of $f(x)$ converges absolutely and uniformly to $f(x)$.

Our next convergence test is called *Dirichlet's theorem*:

Theorem 1-2. Let f be $2L$ periodic and piecewise smooth. Then the Fourier series of f converges for each x in $(-\infty, \infty)$ to

$$\frac{f(x-) + f(x+)}{2}.$$

Consequently, if f is continuous at the point x, its Fourier series converges to $f(x)$. If, in addition, f is continuous for all x, its Fourier series converges absolutely and uniformly to f on $(-\infty, \infty)$.

When $f(x)$ is $2L$ periodic, continuous, and piecewise smooth, Dirichlet's theorem asserts that its Fourier series converges uniformly on $(-\infty, \infty)$. However, it is often necessary to expand discontinuous functions in Fourier series. Here is an extension of Dirichlet's theorem that covers this case:

Theorem 1-3. Let $f(x)$ be $2L$ periodic and piecewise smooth. Then the Fourier series of $f(x)$ converges to $[f(x+) + f(x-)]/2$ for each x in $(-\infty, \infty)$. Moreover, the convergence is absolute and uniform on any closed subinterval that contains no points of discontinuity of $f(x)$.

We now elaborate on the terminology used in these convergence theorems. A function $f(x)$ defined on $(-\infty, \infty)$ is *$2L$ periodic* if $f(x + 2L) = f(x)$ for all x. Geometrically speaking, the graph of $f(x)$ repeats its shape in each interval of length $2L$. We call $2L$ the *period* of the function $f(x)$. A function $f(x)$ is *piecewise continuous on an interval* $[a, b]$ if it is continuous except possibly at a finite number of points in $[a, b]$, where it has jump discontinuities. The function may or may not be defined at the jump discontinuities. A function $f(x)$ has a *jump discontinuity* at c if both one-sided limits of $f(x)$ exist and are finite at $x = c$, in which case we write

$$\lim_{\substack{x \to c \\ x < c}} f(x) = f(c-) \quad \text{and} \quad \lim_{\substack{x \to c \\ x > c}} f(x) = f(c+).$$

Finally, we say that a function $f(x)$ defined for all x is *piecewise continuous* if it is piecewise continuous on every finite interval $[a, b]$ of the real line. In other words, $f(x)$ is continuous except possibly for jump discontinuities, of which there are at most a finite number in any bounded interval.

A function $f(x)$ is *piecewise smooth on* $[a, b]$ if both $f(x)$ and $f'(x)$ are piecewise continuous on $[a, b]$. As expected, $f(x)$ is piecewise smooth on $-\infty < x < \infty$, if it is piecewise smooth on each finite subinterval of the real line.

Finally, we turn to uniform convergence. This fundamental notion plays a central role in both the practical and theoretical analysis of Fourier series, and other series and sequences. For clarity, we first discuss uniform convergence for sequences of functions. A sequence of functions $\{g_n(x)\}$ is said to *converge uniformly on a set* S to a function $g(x)$ if given $\varepsilon > 0$ there corresponds a number N such that

$$(1\text{-}5) \qquad n > N \quad \text{implies that} \quad |g_n(x) - g(x)| < \varepsilon \quad \text{for} \quad \text{all } x \quad \text{in} \quad S.$$

Often S is an interval of real numbers, in which case uniform convergence can be visualized as in Fig. 3-1.

Call the region between the graphs of $y = g(x) + \varepsilon$ and $y = g(x) - \varepsilon$ for x in S the ε tube about the graph of $y = g(x)$. Then the analytical condition (1-5) defining uniform convergence simply means that the graph of $y = g_n(x)$ lies in the ε tube about the graph of $y = g(x)$ for all x in S. The word "uniform" refers to the fact that the graphs of $y = g_n(x)$ and $y = g(x)$ are within the tolerance ε of each other for *all* x in S as soon as n is chosen large enough.

Uniform convergence of an infinite series is now easy to understand: An infinite series $\sum_{k=0}^{\infty} h_k(x)$ *converges uniformly on a set* S if its sequence of partial sums $\{s_n(x)\}$, with

$$s_n(x) = h_0(x) + h_1(x) + \cdots + h_n(x),$$

converges uniformly on S.

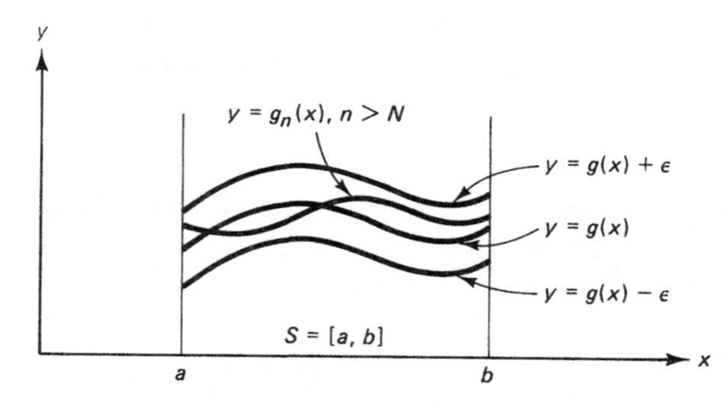

Figure 3-1

In the case of the Fourier series of $f(x)$, the nth partial sum is

$$S_n(x) = \frac{1}{2} a_0 + \sum_{k=1}^{n} \left[a_k \cos\left(\frac{k\pi x}{L}\right) + b_k \sin\left(\frac{k\pi x}{L}\right) \right].$$

Thus, the statement of uniform convergence in Dirichlet's theorem means that given any $\varepsilon > 0$, no matter how small, there is an index N such that

$$n > N \quad \text{implies that} \quad |f(x) - S_n(x)| < \varepsilon \quad \text{for} \quad \text{all } x.$$

Put informally,

$$f(x) \approx S_n(x) \quad \text{for} \quad \text{all } x,$$

and we can make this approximation as accurate as we like by choosing n large enough.

Recall from our remarks at the beginning of this section that the method of separation of variables also leads to the need for trigonometric series expansions for functions $f(x)$ originally defined only on the interval $0 \le x \le L$. Such expansions follow easily from corresponding results for functions defined on $-L \le x \le L$, and make use of even and odd extensions of a given function (see Probs. 7–10).

While studying Fourier series and Fourier series expansions, it is convenient to suppose that $L = \pi$. If L is not π to begin with, we simply make a change of scale to achieve $L = \pi$ (see Prob. 15). Then the Fourier series of $f(x)$, defined on $-\pi \le x \le \pi$, takes the simpler form

$$\frac{1}{2} a_0 + \sum_{n=1}^{\infty} [a_n \cos(nx) + b_n \sin(nx)]$$

with

$$a_n = \frac{1}{\pi} \int_{-\pi}^{\pi} f(x) \cos(nx)\, dx, \qquad n = 0, 1, 2, \ldots$$

and

$$b_n = \frac{1}{\pi} \int_{-\pi}^{\pi} f(x) \sin(nx)\, dx, \qquad n = 1, 2, 3, \ldots.$$

If $f(x)$ is the sum of its Fourier series, then

$$f(x) = \frac{1}{2} a_0 + \sum_{n=1}^{\infty} [a_n \cos(nx) + b_n \sin(nx)] \tag{1-6}$$

for $-\pi \le x \le \pi$. Each term on the right of (1-6) is 2π periodic. Consequently, the series on the right actually converges for all real x, not just x in $[-\pi, \pi]$, and is 2π periodic. In this situation it is convenient to extend the domain of $f(x)$ to $-\infty < x < \infty$ by the requirement that $f(x + 2\pi) = f(x)$ for all x. This *2π periodic extension* of f is still denoted by $f(x)$. Evidently, (1-6) holds for all x for this extended version of $f(x)$.

PROBLEMS

1. Let $f(x)$ be specified by
   (a) $f(x) = e^x$, $-2 \le x < 2$.
   (b) $f(x) = 4$, $-2 \le x < 0$ and $f(x) = x^2$, $0 \le x < 2$.
   Sketch the graph of the 4 periodic extension of $f(x)$. Find the Fourier series of $f(x)$ and its sum. Also, determine where the series is uniformly convergent.
2. If you have a programmable calculator or microcomputer, write a simple program for calculating $S_n(x)$, the nth partial sum of the Fourier series of $f(x)$ for $-L \le x \le L$. For the two functions in Prob. 1, sketch graphs of $S_n(x)$ for, say, $n = 5, 10, 100$.
3. Let $f(x) = x \sin(x)$ for $-\pi \le x \le \pi$. Sketch the graph of the 2π periodic extension of $f(x)$. Find the Fourier series of $f(x)$ and its sum for each x. Also, determine where the series is uniformly convergent.
4. Find the Fourier series of the 2π periodic square wave with $f(x) = 1$ for $0 < x < \pi$ and $f(x) = -1$ for $-\pi < x < 0$. When does the series converge to the square wave, and when is the convergence uniform?
5. Verify the orthogonality relations for integrating products of sines and/or cosines.
6. Multiply equation (1–1) by $\cos(m\pi x/L)$ and carry out the term-by-term integration to obtain

$$a_m = \frac{1}{L}\int_{-L}^{L} f(x)\cos\left(\frac{m\pi x}{L}\right) dx.$$

   Confirm the corresponding formula for b_m.
7. A function $g(x)$ defined on $-L \le x \le L$ is *odd* if $g(-x) = -g(x)$. [Notice that all odd powers of x are odd functions and that the functions $\sin(n\pi x/L)$ are odd.] Show that the Fourier series of an odd function reduces to a sine series. That is, $a_n = 0$ for all n and

$$b_n = \frac{1}{L}\int_{-L}^{L} g(x)\sin\left(\frac{n\pi x}{L}\right) dx = \frac{2}{L}\int_{0}^{L} g(x)\sin\left(\frac{n\pi x}{L}\right) dx.$$

8. A function $h(x)$ defined on $-L \le x \le L$ is *even* if $h(-x) = h(x)$. Show that the Fourier series of an even function reduces to a cosine series. That is, $b_n = 0$ for all n and

$$a_n = \frac{1}{L}\int_{-L}^{L} h(x)\cos\left(\frac{n\pi x}{L}\right) dx = \frac{2}{L}\int_{0}^{L} h(x)\cos\left(\frac{n\pi x}{L}\right) dx.$$

9. Let $f(x)$ be defined for $0 \le x \le L$. The odd function defined by $g(x) = f(x)$ for $0 \le x \le L$ and by $g(x) = -f(-x)$ for $-L \le x < 0$ is called the *odd extension* of f to the interval $[-L, L]$.
   (a) Show that the Fourier series for $g(x)$ reduces to

$$\sum_{n=1}^{\infty} b_n \sin\left(\frac{n\pi x}{L}\right) \quad \text{with} \quad b_n = \frac{2}{L}\int_{0}^{L} f(x)\sin\left(\frac{n\pi x}{L}\right) dx.$$

   This series is also called the *Fourier sine series* of $f(x)$ on $0 \le x \le L$.
   (b) Sketch the odd extension of $f(x) = x$, $0 \le x \le \pi$, to $[-\pi, \pi]$. Now sketch the 2π periodic extension of this odd extension, which is called the *2π periodic, odd extension of $f(x)$*.

(c) Find the Fourier sine series for $f(x) = x$, $0 \le x \le \pi$.
(d) When does the Fourier sine series converge to $f(x)$? When is the convergence uniform?

10. Let $f(x)$ be defined for $0 \le x \le L$. The even function defined by $h(x) = f(x)$ for $0 \le x \le L$ and by $h(x) = f(-x)$ for $-L \le x \le 0$ is called the *even extension* for $f(x)$ to $[-L, L]$.
(a) Show that the Fourier series of $h(x)$ reduces to

$$\frac{1}{2}a_0 + \sum_{n=1}^{\infty} a_n \cos\left(\frac{n\pi x}{L}\right) \quad \text{with} \quad a_n = \frac{2}{L}\int_0^L f(x)\cos\left(\frac{n\pi x}{L}\right) dx.$$

This series is also called the *Fourier cosine series* of $f(x)$ on $0 \le x \le L$.
(b) Sketch the even extension of $f(x) = x$, $0 \le x \le \pi$, to $[-\pi, \pi]$. Now sketch the 2π periodic extension of this even extension, which is called the *2π periodic, even extension of $f(x)$*.
(c) Find the Fourier cosine series for $f(x) = x$, $0 \le x \le \pi$.
(d) When does the Fourier cosine series converge to $f(x)$? When is the convergence uniform?
(e) What is the Fourier series of $f(x) = |x|$ on $-\pi \le x \le \pi$?

11. Let $f(x) = x$ for $0 \le x \le \pi$. Take a careful look at the Fourier sine and cosine series of $f(x)$. Which expansion would you prefer for numerical purposes? Evaluate the cosine series at $x = 0$ to get

$$\frac{\pi^2}{8} = \frac{1}{1^2} + \frac{1}{3^2} + \frac{1}{5^2} + \frac{1}{7^2} + \cdots.$$

Then evaluate the sine series at $x = \pi/2$ to find

$$\frac{\pi}{4} = 1 - \frac{1}{3} + \frac{1}{5} - \frac{1}{7} + - \cdots.$$

12. Assume that at time $t = 0$ the temperature in a rod, whose ends are in zero-degree temperature baths, is given by

$$T(x) = x(L - x), \qquad 0 \le x \le L.$$

To find the temperature distribution in the rod for $t > 0$ by separation of variables, $T(x)$ must be expanded in a Fourier sine series. Find this series and discuss its convergence properties.

13. Show that the formation of Fourier series is a linear process. That is, if $a_n(f)$ and $b_n(f)$ are the Fourier coefficients of an integrable function f defined on $[-\pi, \pi]$, then $a_n(cf) = ca_n(f)$ for any constant c, $a_n(f + g) = a_n(f) + a_n(g)$, and corresponding formulas hold for the Fourier sine coefficients.

14. Show that $\sin(nx)$ and $\cos(nx)$ for $-\pi \le x \le \pi$ are their own Fourier series. Then deduce from Prob. 13 that any trigonometric polynomial

$$\frac{1}{2}\alpha_0 + \sum_{k=1}^{n} [\alpha_k \cos(kx) + \beta_k \sin(kx)]$$

is its own Fourier series.

15. Given $f(x)$ defined for $-L \leq x \leq L$, set $t = \pi x/L$ and $g(t) = f(Lt/\pi)$ for $-\pi \leq t \leq \pi$. Show that

$$f(x) = \frac{1}{2} a_0 + \sum_{n=1}^{\infty} \left[a_n \cos\left(\frac{n\pi x}{L}\right) + b_n \sin\left(\frac{n\pi x}{L}\right) \right]$$

for $-L \leq x \leq L$, with a_n and b_n given by (1–3) and (1–4) is equivalent to

$$g(t) = \frac{1}{2} a_0 + \sum_{n=1}^{\infty} [a_n \cos(nt) + b_n \sin(nt)]$$

for $-\pi \leq t \leq \pi$ and where

$$a_n = \frac{1}{\pi} \int_{-\pi}^{\pi} g(t) \cos(nt)\, dt \quad \text{and} \quad b_n = \frac{1}{\pi} \int_{-\pi}^{\pi} g(t) \sin(nt)\, dt.$$

This equivalence shows that we can assume for simplicity that $L = \pi$ during our study of Fourier series expansions.

16. Use the Euler identities,

$$e^{\pm inx} = \cos(nx) \pm i \sin(nx)$$

to verify that

$$\frac{1}{2} a_0 + \sum_{n=1}^{\infty} [a_n \cos(nx) + b_n \sin(nx)] = \sum_{n=-\infty}^{\infty} c_n e^{inx},$$

where $b_0 = 0$ and

$$c_n = \tfrac{1}{2}(a_n - ib_n), \qquad c_{-n} = \bar{c}_n \quad \text{for} \quad n \geq 0.$$

Then show that

$$c_n = \frac{1}{2\pi} \int_{-\pi}^{\pi} f(x) e^{-inx}\, dx,$$

for $n = 0, \pm 1, \pm 2, \ldots$. Thus, the Fourier series of $f(x)$ can be expressed in the complex form $\sum_{n=-\infty}^{\infty} c_n e^{inx}$ with c_n given as above.

3-2. Least-Squares and Uniform Approximation

In Sec. 3-1 we remarked that Fourier series arise in solving a variety of initial and/or boundary value problems of mathematical physics by the method of separation of variables. Now we present another approach to Fourier series, which sheds more light on their importance.

A basic problem for both pure and applied mathematics is: Given a complicated function $f(x)$ find a simple, easily computable function $T(x)$ which approximates $f(x)$ as accurately as desired. For us the manner in which $T(x)$ should approximate $f(x)$ depends on the underlying physical situation. For example, energy methods require the approximation of a given function $f(x)$ by $T(x)$ in the sense that

$$\left[\int [f(x) - T(x)]^2\, dx \right]^{1/2}$$

is small, where the integral is taken over some relevant interval. This is called *least-squares*, or L_2, *approximation* and plays a prominent role in applied mathematics and statistics. On the other hand, if $T(x)$ is required to produce a table of values for $f(x)$, we want

$$|f(x) - T(x)|$$

small for *all* x in an appropriate interval. This is called *uniform approximation.* In this section we consider both modes of approximation where $T(x)$ is a trigonometric polynomial. A *trigonometric polynomial* is a function of the form

$$T_n(x) = \frac{1}{2}\alpha_0 + \sum_{k=1}^{n} [\alpha_k \cos(kx) + \beta_k \sin(kx)], \tag{2-1}$$

where α_k and β_k are certain constants. $T_n(x)$ has *degree n* if $|\alpha_n| + |\beta_n| > 0$.

We begin by asking how we can best approximate, in the least-squares sense, a given function $f(x)$ by a trigonometric polynomial of degree at most n. That is, we seek to minimize the integral

$$\int_{-\pi}^{\pi} [f(x) - T_n(x)]^2 \, dx \tag{2-2}$$

by proper choice of a trigonometric polynomial $T_n(x)$ of degree at most n. The value of the integral (2-2) depends quadratically on the $2n + 1$ variables α_0, $\alpha_1, \ldots, \alpha_n, \beta_1, \ldots, \beta_n$, and consequently we can find its minimum by completing the square. First, note that

$$\int_{-\pi}^{\pi} [f(x) - T_n(x)]^2 \, dx = \int_{-\pi}^{\pi} f(x)^2 \, dx - 2\int_{-\pi}^{\pi} f(x)T_n(x) \, dx + \int_{-\pi}^{\pi} T_n(x)^2 \, dx.$$

A simple calculation using (2-1) gives

$$\int_{-\pi}^{\pi} f(x)T_n(x) \, dx = \pi\left[\frac{1}{2}\alpha_0 a_0 + \sum_{k=1}^{n} (\alpha_k a_k + \beta_k b_k)\right],$$

where a_k and b_k are the Fourier coefficients of $f(x)$. A slightly longer, but routine calculation (see Prob. 4) using (2-1) yields

$$\int_{-\pi}^{\pi} T_n(x)^2 \, dx = \pi\left[\frac{1}{2}\alpha_0^2 + \sum_{k=1}^{n} (\alpha_k^2 + \beta_k^2)\right].$$

Combine these results to obtain

$$\int_{-\pi}^{\pi} [f(x) - T_n(x)]^2 \, dx = \int_{-\pi}^{\pi} f(x)^2 \, dx + \frac{1}{2}\pi(\alpha_0^2 - 2\alpha_0 a_0)$$
$$+ \pi \sum_{k=1}^{n} (\alpha_k^2 - 2\alpha_k a_k + \beta_k^2 - 2\beta_k b_k).$$

Finally, complete the square to find

$$\int_{-\pi}^{\pi} [f(x) - T_n(x)]^2\, dx = \int_{-\pi}^{\pi} f(x)^2\, dx + \frac{\pi}{2}(\alpha_0 - a_0)^2 + \pi \sum_{k=1}^{n} [(\alpha_k - a_k)^2 + (\beta_k - b_k)^2] - \frac{\pi}{2} a_0^2 - \pi \sum_{k=1}^{n} a_k^2 + b_k^2. \tag{2-3}$$

This expansion shows that (2-2) assumes its minimum value precisely when $\alpha_k = a_k$ and $\beta_k = b_k$, the Fourier coefficients of $f(x)$. That is, the trigonometric polynomial of degree at most n which minimizes (2-2) is

$$S_n(x) = \frac{1}{2} a_0 + \sum_{k=1}^{n} [a_k \cos(kx) + b_k \sin(kx)],$$

which is just the partial sum (with $n + 1$ terms) of the Fourier series of $f(x)$. We will call $S_n(x)$ the nth *partial sum of the Fourier series*. Thus, we have established

Theorem 2-1. Let $f(x)$ be square integrable on $[-\pi, \pi]$. Then the nth partial sum of its Fourier series provides the best least-squares approximation to $f(x)$ among all trigonometric polynomials of degree at most n. That is,

$$\int_{-\pi}^{\pi} [f(x) - S_n(x)]^2\, dx \le \int_{-\pi}^{\pi} [f(x) - T_n(x)]^2\, dx$$

for $T_n(x)$ any trigonometric polynomial of degree at most n. Equality holds only if $T_n(x) = S_n(x)$.

In Th. 2-1, $f(x)$ *square integrable* means that

$$\int_{-\pi}^{\pi} |f(x)|^2\, dx < \infty.$$

This assumption assures that all the integrals involved in the calculations above exist.

When $T_n(x) = S_n(x)$, (2-3) reduces to

$$\int_{-\pi}^{\pi} [f(x) - S_n(x)]^2\, dx = \int_{-\pi}^{\pi} f(x)^2\, dx - \pi\left[\frac{1}{2}a_0^2 + \sum_{k=1}^{n} (a_k^2 + b_k^2)\right], \tag{2-4}$$

which yields, since the left member of (2-4) is nonnegative,

$$\frac{1}{2} a_0^2 + \sum_{k=1}^{n} (a_k^2 + b_k^2) \le \frac{1}{\pi} \int_{-\pi}^{\pi} f(x)^2\, dx$$

for all $n \ge 1$. Let n tend to infinity to get *Bessel's inequality*,

$$\frac{1}{2} a_0^2 + \sum_{k=1}^{\infty} (a_k^2 + b_k^2) \le \frac{1}{\pi} \int_{-\pi}^{\pi} f(x)^2\, dx. \tag{2-5}$$

We turn now to the question of uniform approximation by trigonometric polynomials. Given a function $f(x)$ and a measure of error $\varepsilon > 0$, we wish to find a corresponding trigonometric polynomial $T(x)$ such that $|f(x) - T(x)| < \varepsilon$ for all x. If such an approximation is possible for arbitrarily small positive ε, then $f(x)$ must be 2π periodic and continuous (see Prob. 10). The *Weierstrass approximation theorem* guarantees that such approximations can be found:

Theorem 2-2. Let $\varepsilon > 0$ be arbitrary and $f(x)$ be a 2π periodic continuous function. Then there is a trigonometric polynomial $T(x)$ such that $|f(x) - T(x)| < \varepsilon$ for all x.

The proof of this fundamental result of real and applied analysis is deferred until the end of this section. Here is an important application of the Weierstrass theorem.

Theorem 2-3. If $f(x)$ is 2π periodic and continuous, then the Fourier series of $f(x)$ converges to $f(x)$ in the least squares, or L_2, sense. That is,

$$\lim_{n\to\infty} \left[\int_{-\pi}^{\pi} [f(x) - S_n(x)]^2\, dx\right]^{1/2} = 0,$$

where $S_n(x)$ is the nth partial sum of the Fourier series of $f(x)$.

By the Weierstrass approximation theorem, given $\varepsilon > 0$ there is a trigonometric polynomial $T(x)$ such that $|f(x) - T(x)| < \varepsilon/(2\pi)^{1/2}$ for all x. Let N be the degree of $T(x)$. By Th. 2-1, $S_n(x)$ is the best least-squares approximation to $f(x)$; consequently, for $n > N$, $T(x)$ is a trigonometric polynomial of degree at most n, and we conclude that

$$\int_{-\pi}^{\pi} [f(x) - S_n(x)]^2\, dx \leq \int_{-\pi}^{\pi} [f(x) - T(x)]^2\, dx < \varepsilon^2,$$

$$\left[\int_{-\pi}^{\pi} [f(x) - S_n(x)]^2\, dx\right]^{1/2} \leq \varepsilon.$$

Thus, the expression on the left has limit zero, as claimed, and Th. 2-3 is proved.

Corollary 2-1 (*Parseval's Relation*). If $f(x)$ is 2π periodic and continuous, then

$$\frac{1}{\pi}\int_{-\pi}^{\pi} f(x)^2\, dx = \frac{1}{2}a_0^2 + \sum_{k=1}^{\infty} (a_k^2 + b_k^2),$$

where a_k and b_k are the Fourier coefficients of $f(x)$.

This equality follows directly from Th. 2-3 and (2-4), upon letting n tend to infinity. Here is an important consequence of Parseval's relation.

Theorem 2-4. If the Fourier coefficients of a 2π periodic, continuous function are all zero, then the function is identically zero. Consequently, if two

2π periodic, continuous functions have identical Fourier coefficients, they are equal.

This follows directly from Parseval's relation, which reduces to

$$\frac{1}{\pi}\int_{-\pi}^{\pi} f(x)^2\,dx = 0,$$

and which in turn implies that the continuous function $f(x)$ is identically zero. If $f(x)$ and $g(x)$ are continuous and have the same Fourier coefficients, then $f(x) - g(x)$ has all zero Fourier coefficients. Thus, $f(x) - g(x) = 0$ for all x by what we just proved.

Theorem 2-4 and two basic facts about uniform convergence allow us to establish the often used convergence criterion for Fourier series given in Th. 1-1. The two pertinent facts from real analysis which we use without proof are:

Proposition 2-1. The uniform limit of a sequence of continuous functions is continuous.

Proposition 2-2. Let $\{g_n(x)\}$ converge uniformly on the finite interval $a \leq x \leq b$. Then

$$\lim_{n\to\infty}\int_a^b g_n(x)\,dx = \int_a^b \lim_{n\to\infty} g_n(x)\,dx.$$

Passing to the limit under the integral sign can be tricky. Prop. 2-2 gives a general, sufficient condition for this. But, see Prob. 12.

Often we shall deal with uniformly convergent series rather than sequences. Since uniform convergence of a series means uniform convergence of its sequence of partial sums, Props. 2-1 and 2-2 can be expressed in terms of series:

Proposition 2-1′. Let $g_n(x)$ be continuous on a set S and $\sum_0^\infty g_n(x)$ be uniformly convergent on S. Then $\sum_0^\infty g_n(x)$ is continuous on S.

Proposition 2-2′. Let $g_n(x)$ be continuous on the finite interval $a \leq x \leq b$, and let $\sum_0^\infty g_n(x)$ be uniformly convergent there. Then

$$\sum_0^\infty \int_a^b g_n(x)\,dx = \int_a^b \sum_0^\infty g_n(x)\,dx.$$

That is, a uniformly convergent series can be integrated term by term on a closed bounded interval.

Proposition 2-2′ has a useful consequence about term-by-term differentiation of series, which will be important for us later when we take a careful look at the method of separation of variables.

Proposition 2-3. Let $g_n(x)$ be continuously differentiable on an interval I and assume the following:
(i) $\sum_0^\infty g_n(a)$ converges at some point a in I;
(ii) $\sum_0^\infty g'_n(x)$ is uniformly convergent on I.
Then $\sum_0^\infty g_n(x)$ is differentiable on I and

$$\left[\sum_0^\infty g_n(x)\right]' = \sum_0^\infty g'_n(x).$$

To confirm this, let x be any point in I, and apply Prop. 2-2′ to get

$$\int_a^x \sum_0^\infty g'_n(t)\,dt = \sum_0^\infty \int_a^x g'_n(t)\,dt = \sum_0^\infty [g_n(x) - g_n(a)].$$

Add the convergent series $\sum_0^\infty g_n(a) = A$ to each side to conclude that

$$\sum_0^\infty g_n(x) = A + \int_a^x \sum_0^\infty g'_n(t)\,dt.$$

Since the integrand on the right is continuous (Prop. 2-1′), the fundamental theorem of calculus asserts that $\sum_0^\infty g_n(x)$ is differentiable with derivative $\sum_0^\infty g'_n(x)$, as asserted.

To apply these propositions, we must be able to recognize uniformly convergent series. Here is a handy test, called the *Weierstrass M-test*:

Theorem 2-5. Suppose that there are constants M_n such that $|g_n(x)| \le M_n$ for all x in a set S and $\sum M_n < \infty$. Then the series $\sum_0^\infty g_n(x)$ is absolutely and uniformly convergent on S.

Since $\sum_0^\infty |g_n(x)| \le \sum_0^\infty M_n < \infty$, the series $\sum_0^\infty g_n(x)$ is absolutely convergent. Let $s_n(x) = g_0(x) + \cdots + g_n(x)$. Then

$$s_n(x) \to s(x) = \sum_0^\infty g_k(x) \quad \text{and} \quad |s_n(x) - s(x)| = \left|\sum_{n+1}^\infty g_k(x)\right| \le \sum_{n+1}^\infty M_k$$

for all x in S. Since the numerical series $\sum_0^\infty M_k$ converges, its tail series $\sum_{n+1}^\infty M_k$ can be made less than any positive number ε by choosing n suitably large. Therefore, $|s_n(x) - s(x)| < \varepsilon$ for all x in S and all n sufficiently large. That is, $s_n(x)$ converges uniformly to $s(x)$ on S, which means that $\sum_0^\infty g_n(x)$ is uniformly convergent on S.

With Props. 2-1′ and 2-2′ at our disposal, it is an easy matter to prove Th. 1-1. Recall the situation: We are given a 2π periodic (remember that we can always assume that $L = \pi$), continuous function $f(x)$ whose Fourier series is known to converge uniformly, and we wish to show that the Fourier series has sum $f(x)$. To verify this, let $g(x)$ be the sum of the Fourier series of $f(x)$,

$$g(x) = \frac{1}{2}a_0 + \sum_{k=1}^\infty [a_k \cos(kx) + b_k \sin(kx)].$$

Clearly, $g(x)$ is 2π periodic, and it is continuous by Prop. 2-1′ because the series is uniformly convergent. Also, since the Fourier series is uniformly convergent, so are the series obtained from it by multiplication by $\cos(nx)$ or $\sin(nx)$ (see Prob. 13). Performing each of these multiplications and integrating term by term (Prop. 2-2′) reveals that the Fourier coefficients of $g(x)$ are a_n and b_n. That is, both $f(x)$ and $g(x)$ have the same Fourier coefficients. Since both of these functions are continuous and 2π periodic, Th. 2-4 reveals that they are equal. Thus, $f(x) = g(x)$ is the sum of its Fourier series and Th. 1-1 is established.

Corollary 1-1 is a simple consequence of the M-test (see Prob. 9).

As promised, we close this section with a proof of the Weierstrass approximation theorem. The method of proof is important in itself. The expression $c_n \cos^{2n}(u/2)$ in (2-6) below is called a "reproducing kernel" or "approximate δ function." Fix $\varepsilon > 0$ and let x be in $(-\infty, \infty)$. We will show that

$$T(x) = c_n \int_{-\pi}^{\pi} \cos^{2n}\left(\frac{u}{2}\right) f(x+u)\, du \tag{2-6}$$

is a trigonometric polynomial of degree at most n which satisfies $|f(x) - T(x)| < \varepsilon$ for n fixed large enough. Here c_n is chosen so that

$$c_n \int_{-\pi}^{\pi} \cos^{2n}\left(\frac{u}{2}\right) du = 1. \tag{2-7}$$

The cosine integral can be evaluated explicitly (Prob. 14) and we find that

$$c_n = \frac{2 \cdot 4 \cdots 2n}{2\pi \cdot 1 \cdot 3 \cdots (2n-1)}.$$

This formula gives the estimate

$$c_n = \frac{2n}{2\pi} \cdot \frac{2n-2}{2n-1} \cdots \frac{4}{5} \cdot \frac{2}{3} < n.$$

From (2-7),

$$c_n \int_{-\pi}^{\pi} \cos^{2n}\left(\frac{u}{2}\right) f(x)\, du = f(x)$$

for each x. Consequently, for any δ with $0 < \delta < \pi$,

$$\begin{aligned} |T(x) - f(x)| &= \left| c_n \int_{-\pi}^{\pi} \cos^{2n}\left(\frac{u}{2}\right) [f(x+u) - f(x)]\, du \right| \\ &\le c_n \int_{-\delta}^{\delta} \cos^{2n}\left(\frac{u}{2}\right) |f(x+u) - f(x)|\, du \\ &\quad + c_n \left(\int_{-\pi}^{-\delta} + \int_{\delta}^{\pi} \right) \cos^{2n}\left(\frac{u}{2}\right) |f(x+u) - f(x)|\, du \\ &= I_1 + I_2, \end{aligned} \tag{2-8}$$

where I_1 and $I_2 \ge 0$ are the summands on the right of the last inequality. We estimate I_1 and I_2 separately. Since the function $f(x)$ is continuous, and therefore uniformly continuous on $[-\pi, \pi]$, we can fix $\delta > 0$ so that

$$|f(x + u) - f(x)| < \frac{\varepsilon}{2} \quad \text{for} \quad |u| < \delta,$$

and for all x. Then

$$I_1 \leq \frac{\varepsilon}{2} c_n \int_{-\delta}^{\delta} \cos^{2n}\left(\frac{u}{2}\right) du < \frac{\varepsilon}{2}, \tag{2-9}$$

from (2-7). Also, the continuity of $f(x)$ guarantees that there is a constant M such that $|f(x)| \leq M$ on $[-\pi, \pi]$, and this same bound holds for all x because of periodicity. Thus,

$$I_2 \leq 2Mc_n\left(\int_{-\pi}^{-\delta} + \int_{\delta}^{\pi}\right) \cos^{2n}\left(\frac{u}{2}\right) du < 4Mn \int_{\delta}^{\pi} \cos^{2n}\left(\frac{u}{2}\right) du$$

because $c_n < n$. Now, for $\delta \leq |u| \leq \pi$ we have

$$\cos^{2n}\left(\frac{u}{2}\right) \leq \cos^{2n}\left(\frac{\delta}{2}\right) = \eta^{2n},$$

where $0 < \eta < 1$. Therefore, our estimate for I_2 yields $I_2 \leq 4M\pi n\eta^{2n}$. Since $n\eta^{2n}$ tends to zero as n tends to infinity, we can fix n so large that $I_2 \leq 4M\pi n\eta^{2n} < \varepsilon/2$. This estimate, (2-8), and (2-9) yield

$$|T(x) - f(x)| < \frac{\varepsilon}{2} + \frac{\varepsilon}{2} = \varepsilon$$

for all x.

It remains to prove that $T(x)$ is in fact a trigonometric polynomial. This follows easily from the change of variables $t = x + u$ in (2-6), which gives

$$T(x) = \int_{x-\pi}^{x+\pi} \cos^{2n}\left(\frac{t - x}{2}\right) f(t)\, dt = \int_{-\pi}^{\pi} \cos^{2n}\left(\frac{t - x}{2}\right) f(t)\, dt, \tag{2-10}$$

by 2π periodicity. Use of elementary trigonometric identities shows that this last integral evaluates as a trigonometric polynomial of degree at most n (see Prob. 15). Thus the Weierstrass approximation theorem is established.

PROBLEMS

1. Show that $|\sin(x)|$ is the sum of its Fourier series and that this series is uniformly convergent. Set $x = \pi/2$ to get an interesting formula for π. What does Parseval's relation state in this case?
2. Let (a) $f(x) = x$ and (b) $f(x) = x^2$ for $0 \leq x \leq \pi$. Write out Parseval's relation for both the even and odd, 2π periodic extensions of $f(x)$.
3. Show that the Fourier sine series of

$$f(x) = \begin{cases} (1 - s)x, & x \leq s, \\ (1 - x)s, & s < x, \end{cases}$$

where $0 \le x \le 1$ and s is regarded as a parameter, with $0 \le s \le 1$, is

$$\frac{2}{\pi^2} \sum_{n=1}^{\infty} \frac{\sin(n\pi s)\sin(n\pi x)}{n^2}.$$

Does this series converge to $f(x)$? Is the convergence uniform? Write out Parseval's relation; interesting formulas result for $s = \frac{1}{4}$ and $\frac{1}{2}$.

4. Verify the expansion

$$\int_{-\pi}^{\pi} T_n(x)^2\, dx = \pi\left(\frac{1}{2}\alpha_0^2 + \sum_{k-1}^{n} \alpha_k^2 + \beta_k^2\right)$$

Hint. Write $T_n(x)^2 = T_n(x)T_n(x)$, use different indices of summation in (2-1) for each factor on the right, and expand out.

5. Let $f(x)$ be 2π periodic and continuous. Show that the minimum least-squares error

$$\left\{\int_{-\pi}^{\pi} [f(x) - S_n(x)]^2\, dx\right\}^{1/2} = \left(\pi \sum_{k=n+1}^{\infty} [a_k^2 + b_k^2]\right)^{1/2}.$$

6. Let a_n, b_n, c_n, and d_n be the Fourier coefficients of functions f and g, respectively. Write out Parseval's relation for the function $f + g$. Apply Parseval's relation for f and g separately to deduce that

$$\frac{1}{2} a_0 c_0 + \sum_{n=1}^{\infty} a_n c_n + b_n d_n = \frac{1}{\pi} \int_{-\pi}^{\pi} f(x)g(x)\, dx,$$

which is also called Parseval's relation.

7. Let $f(x)$ be 2π periodic, continuous, and differentiable. Then the Fourier series of $f'(x)$ is obtained from the Fourier series of $f(x)$ by term-by-term differentiation. *Hint.* Integrate by parts the Fourier coefficients of f' to relate them to the Fourier coefficients of f. Now differentiate the Fourier series of f and see what you get.

8. Here is the analogue of Prob. 7 for integration of Fourier series. Use the Schwarz inequality to show that

$$\left|\int_0^x [f(t) - S_n(t)]\, dt\right| \le \left\{\int_0^x [f(t) - S_n(t)]^2\, dt\right\}^{1/2} \left(\int_0^x 1^2\, dt\right)^{1/2}.$$

Conclude that the Fourier series of a 2π periodic, continuous function $f(x)$ when integrated term by term from 0 to x equals to $\int_0^x f(t)\, dt$. Note that this equation holds even though the Fourier series of $f(x)$ need not converge to $f(x)$.

9. Prove Cor. 1-1. *Hint.* Use the M-test to show that the Fourier series is uniformly convergent.

10. Let $f(x)$ be a real-valued function defined for all x and such that: Given any $\varepsilon > 0$ there corresponds a trigonometric polynomial $T(x)$ with $|f(x) - T(x)| < \varepsilon$ for all x. Show that $f(x)$ is 2π periodic and continuous. *Hint.* Show that $f(x)$ is the uniform limit of trigonometric polynomials.

11. Given a 2π periodic continuous function $f(x)$, show that there is a sequence of trigonometric polynomials $T_1(x), T_2(x), \ldots, T_n(x), \ldots$ which converge uniformly to $f(x)$ on $-\infty < x < \infty$. *Hint.* Apply the Weierstrass approximation theorem with $\varepsilon = 1, \frac{1}{2}, \frac{1}{3}, \ldots, 1/n, \ldots$.

12. Limits cannot always be taken under the integral sign as in Prop. 2-2: Let $g_n(x) = n^2x^n(1-x)$ for $0 \le x \le 1$. Show that $g_n(x)$ has limit 0 as n tends to infinity, but

$$\int_0^1 g_n(x)\,dx = \frac{n^2}{(n+1)(n+2)} \to 1$$

as n becomes infinite.

13. Let $\sum_0^\infty g_k(x)$ converge uniformly on $[a, b]$. If $h(x)$ is a bounded function on $[a, b]$, then $\sum_0^\infty h(x)g_k(x)$ is uniformly convergent on $[a, b]$.

14. Show that $I_{2n} = \displaystyle\int_{-\pi}^{\pi} \cos^{2n}\left(\frac{u}{2}\right) du = 2\pi \frac{1 \cdot 3 \cdot 5 \cdots (2n-1)}{2 \cdot 4 \cdot 6 \cdots (2n)}$.

Hint. Write $\cos^{2n}(u/2) = \cos^{2n-1}(u/2)\cos(u/2)$ and integrate by parts to get $I_{2n} = (2n-1)I_{2n-2} - (2n-1)I_{2n}$.

15. Use the following steps to verify that $T(x)$ in (2-10) is a trigonometric polynomial of degree at most n.

(a) Use the addition formulas for the cosine to show that

$$2\cos(k\alpha)\cos\alpha = \cos(k+1)\alpha + \cos(k-1)\alpha$$

for $k = 1, 2, \ldots$. Then show by mathematical induction that

$$\cos^n \alpha = \sum_{k=0}^{n} c_k \cos(k\alpha)$$

for certain constants c_k.

(b) Use part (a) and a half-angle formula to deduce that

$$\cos^{2n}\left(\frac{\alpha}{2}\right) = \sum_{k=0}^{n} d_k \cos(k\alpha)$$

for certain constants d_k.

(c) Substitute part (b) in (2-10) to verify that $T(x)$ is a trigonometric polynomial of degree at most n

16. The Weierstrass approximation theorem for ordinary polynomials states:

> Let $\varepsilon > 0$ be given and $f(x)$ be continuous on $[a, b]$.
> Then there is a polynomial $P(x)$ such that
> $|f(x) - P(x)| < \varepsilon$ for all x in $[a, b]$.

Deduce this result from the trigonometric case as follows.

(a) Show that the theorem holds for any interval $[a, b]$ if it holds for the special interval $[0, 1]$. *Hint.* Consider the change of variables $x = a + t(b-a)$.

(b) Given $g(t)$ continuous on $[0, 1]$, extend g to a continuous function $G(t)$ on $[-\pi, \pi]$ so that G is linear on $[-\pi, 0]$ and $[1, \pi]$ with $G(-\pi) = G(\pi) = 0$. Now extend G to be 2π periodic.

(c) Apply the trigonometric approximation theorem to G to produce a trigonometric polynomial $T(t)$ such that

$$|G(t) - T(t)| < \frac{\varepsilon}{2} \quad \text{on} \quad [-\pi, \pi].$$

Conclude that $|g(t) - T(t)| < \varepsilon/2$ on $[0, 1]$.

(d) Now use the standard Taylor series expansions for $\sin(z)$ and $\cos(z)$ to approximate each term in $T(t)$ for t in $[0, 1]$ by an ordinary polynomial.

3-3. Convergence Theorems for Fourier Series

In Sec. 3-2 we established Th. 1-1: A 2π periodic, continuous function is the sum of its Fourier series, provided that the series converges uniformly. This is a very handy convergence criterion; however, it does not cover all cases of practical importance. In this section we present some additional convergence tests, and also prove both Dirichlet's theorem and Th. 1-3.

The so-called *Riemann–Lebesgue lemma* will play a key role in our reasoning here and in our analysis of Fourier transforms later:

Theorem 3-1. Let $g(x)$ be continuous except possibly at a finite number of points in an interval $[a, b]$ and assume that

$$\int_a^b |g(x)|\, dx < \infty.$$

Then

$$\lim_{\lambda \to \pm\infty} \int_a^b g(x) \cos(\lambda x)\, dx = \lim_{\lambda \to \pm\infty} \int_a^b g(x) \sin(\lambda x)\, dx = 0.$$

Proof. We may assume that $\lambda > 0$ because the case for $\lambda \to -\infty$ follows from the case $\lambda \to \infty$ and the identities $\sin(-\alpha) = -\sin(\alpha)$ and $\cos(-\alpha) = \cos(\alpha)$. Also, we may assume that $a + \pi/\lambda < b$ because only large values of λ are of interest.

For the moment, assume that $g(x)$ is continuous on $[a, b]$ and set

$$I = \int_a^b g(x) \sin(\lambda x)\, dx.$$

The change of variables $x = t + \pi/\lambda$ leads to

$$I = -\int_{a-\pi/\lambda}^{b-\pi/\lambda} g\left(t + \frac{\pi}{\lambda}\right) \sin(\lambda t)\, dt.$$

Add and rearrange to find that

$$\begin{aligned} \text{(3-1)} \qquad 2I &= \int_a^b g(x) \sin(\lambda x)\, dx - \int_{a-\pi/\lambda}^{b-\pi/\lambda} g\left(t + \frac{\pi}{\lambda}\right) \sin(\lambda t)\, dt \\ &= \int_{b-\pi/\lambda}^{b} g(x) \sin(\lambda x)\, dx \\ &\quad + \int_a^{b-\pi/\lambda} \left[g(x) - g\left(x + \frac{\pi}{\lambda}\right)\right] \sin(\lambda x)\, dx \\ &\quad - \int_{a-\pi/\lambda}^{a} g\left(x + \frac{\pi}{\lambda}\right) \sin(\lambda x)\, dx. \end{aligned}$$

The first integral on the right of (3-1) can be estimated by

$$\left|\int_{b-\pi/\lambda}^{b} g(x) \sin(\lambda x)\, dx\right| \le \frac{\pi M}{\lambda},$$

where M is the maximum of $|g(x)|$ on $[a, b]$. The third satisfies the same bound.

With these estimates we infer from (3-1) that

$$2|I| \leq \frac{2\pi M}{\lambda} + (b - a) \max_{a \leq x \leq (b - \pi/\lambda)} \left| g(x) - g\left(x + \frac{\pi}{\lambda}\right) \right|.$$

Since the continuous function $g(x)$ is uniformly continuous on $[a, b]$, the maximum above tends to 0 as λ tends to ∞ and we conclude that

$$\lim_{\lambda \to \infty} \int_a^b g(x) \sin(\lambda x)\, dx = 0.$$

The proof just given applies without change when $\sin(\lambda x)$ is replaced by $\sin(\lambda x + c)$, for any constant c. Thus,

$$\lim_{\lambda \to \infty} \int_a^b g(x) \sin(\lambda x + c)\, dx = 0.$$

The choice $c = \pi/2$ yields the statement for the cosine integral in Th. 3-1. Thus, Th. 3-1 holds for $g(x)$ continuous on $[a, b]$.

The extension to the more general case is easy: If $g(x)$ has discontinuities at a finite set of points, then

$$\int_a^b g(x) \sin(\lambda x)\, dx$$

may be expressed as a finite sum of integrals of the form

$$\int_c^d g(x) \sin(\lambda x)\, dx,$$

where c and d are consecutive points of discontinuity of $g(x)$. If each integral from c to d has limit 0 as λ tends to ∞, the same is true for the integral from a to b. Therefore, it suffices to prove Th. 3-1 when $g(x)$ is continuous except possibly at a and b and

$$\int_a^b |g(x)|\, dx < \infty.$$

Since this improper integral converges, given any $\varepsilon > 0$ we can find a $\delta > 0$ so that

$$\int_a^{a+\delta} |g(x)|\, dx < \frac{\varepsilon}{3} \quad \text{and} \quad \int_{b-\delta}^b |g(x)|\, dx < \frac{\varepsilon}{3}.$$

Then

$$\int_a^b g(x) \sin(\lambda x)\, dx = \int_a^{a+\delta} g(x) \sin(\lambda x)\, dx + \int_{a+\delta}^{b-\delta} g(x) \sin(\lambda x)\, dx$$
$$+ \int_{b-\delta}^b g(x) \sin(\lambda x)\, dx,$$

and the result follows immediately.

Before we can apply the Riemann–Lebesgue lemma, we need to express the partial sums of a Fourier series in a more convenient form. For the moment assume that $f(x)$ is 2π periodic and *absolutely integrable* on $[-\pi, \pi]$, meaning that

$$\int_{-\pi}^{\pi} |f(x)|\, dx < \infty.$$

Then the Fourier coefficients,

$$a_k = \frac{1}{\pi}\int_{-\pi}^{\pi} f(t)\cos(kt)\, dt \quad \text{and} \quad b_k = \frac{1}{\pi}\int_{-\pi}^{\pi} f(t)\sin(kt)\, dt$$

of $f(x)$ exist, and it turns out that there is a simple, closed form for the partial sums of the Fourier series:

$$\begin{aligned} S_n(x) &= \frac{1}{2}a_0 + \sum_{k=1}^{n} [a_k \cos(kx) + b_k \sin(kx)] \\ &= \frac{1}{\pi}\int_{-\pi}^{\pi}\left[\frac{1}{2} + \sum_{k=1}^{n} \cos(kt)\cos(kx) + \sin(kt)\sin(kx)\right] f(t)\, dt \\ &= \frac{1}{\pi}\int_{-\pi}^{\pi}\left[\frac{1}{2} + \sum_{k=1}^{n} \cos k(t-x)\right] f(t)\, dt. \end{aligned}$$

Now, the addition formula for the sine gives

$$2\cos(k\theta)\sin\tfrac{1}{2}\theta = \sin(k\theta + \tfrac{1}{2}\theta) - \sin(k\theta - \tfrac{1}{2}\theta).$$

Therefore,

$$2\sin\frac{1}{2}\theta \sum_{k=1}^{n}\cos(k\theta) = \sin(n\theta + \tfrac{1}{2}\theta) - \sin\tfrac{1}{2}\theta,$$

because the sum is telescoping, and we obtain the identity

$$\frac{1}{2} + \sum_{k=1}^{n}\cos(k\theta) = \frac{\sin(n+\frac{1}{2})\theta}{2\sin\frac{1}{2}\theta} \tag{3-2}$$

An alternative derivation of this result is given in Prob. 3. With this identity, the formula for $S_n(x)$ above simplifies to

$$S_n(x) = \frac{1}{\pi}\int_{-\pi}^{\pi} \frac{\sin(n+\frac{1}{2})(t-x)}{2\sin\frac{1}{2}(t-x)} f(t)\, dt. \tag{3-3}$$

The expression

$$\frac{\sin(n+\frac{1}{2})(t-x)}{2\sin\frac{1}{2}(t-x)}$$

is called the *Dirichlet kernel.* Since the integrand in (3-3) is 2π periodic, the change of variables $u = t - x$ gives

$$S_n(x) = \frac{1}{\pi}\int_{-\pi}^{\pi} \frac{\sin(n+\frac{1}{2})u}{2\sin\frac{1}{2}u} f(u+x)\, du. \tag{3-4}$$

In the special case $f(x) = 1$ for all x, we have $S_n(x) = 1$ by direct calculation (see Prob. 14 of Sec. 3-1), so

$$1 = \frac{1}{\pi}\int_{-\pi}^{\pi} \frac{\sin(n + \frac{1}{2})u}{2 \sin \frac{1}{2}u}\, du. \tag{3-5}$$

Multiplying this identity by $f(x)$ and subtracting from (3-4) yields

$$S_n(x) - f(x) = \frac{1}{\pi}\int_{-\pi}^{\pi} \frac{\sin(n + \frac{1}{2})u}{2 \sin \frac{1}{2}u}\,[f(x + u) - f(x)]\, du. \tag{3-6}$$

Thus, in order to show that the partial sums of the Fourier series converge to $f(x)$, we must find appropriate conditions on f which allow us to conclude that the integral in (3-6) has limit 0 as n tends to infinity. For example, we have

Theorem 3-2. Let f be 2π periodic, continuous except possibly for a finite number of points, and absolutely integrable. The Fourier series of f converges to $f(x_0)$ at each point x_0 for which

$$\int_{-\varepsilon}^{\varepsilon} \left|\frac{f(x_0 + u) - f(x_0)}{u}\right| du < \infty$$

for some $\varepsilon > 0$.

Remark. The existence of this integral is called a *Dini condition.* It holds for all Lipschitz functions and Hölder continuous functions (see Prob. 2). Notice that the Dini condition is a "local" condition. It shows that the convergence of the Fourier series at x_0 depends only on the behavior of $f(x)$ for x near x_0. This fact is known as Riemann's localization principle.

The proof of Th. 3-2 follows easily from (3-6),

$$S_n(x_0) - f(x_0) = \frac{1}{\pi}\int_{-\pi}^{\pi} \sin\left(n + \frac{1}{2}\right)u \left[\frac{u/2}{\sin(u/2)}\, \frac{f(x_0 + u) - f(x_0)}{u}\right] du. \tag{3-7}$$

The function $h(u) = (u/2)/\sin(u/2)$, with $h(0) = 1$, is continuous on $[-\pi, \pi]$. Consequently, the function in brackets in (3-7) will be absolutely integrable exactly when

$$\int_{-\varepsilon}^{\varepsilon} \left|\frac{f(x_0 + u) - f(x_0)}{u}\right| du < \infty$$

for some $\varepsilon > 0$. Since such an ε exists by hypothesis, the Riemann–Lebesgue lemma and (3-7) give $S_n(x_0) - f(x_0) \to 0$ as $n \to \infty$. That is, the Fourier series converges at x_0 to $f(x_0)$.

Corollary 3-1. If f is piecewise smooth and x_0 is a point of continuity of f, then the Fourier series converges to $f(x_0)$.

Since f is piecewise continuous on $[-\pi, \pi]$, the function

$$g(u) = \frac{f(x_0 + u) - f(x_0)}{u}$$

will be piecewise continuous on $[-\pi, \pi]$ provided that it has at worst a jump discontinuity at $u = 0$. However, since f is piecewise smooth and continuous at x_0, $f(x)$ is continuous for all x near x_0 and differentiable for $x \neq x_0$ and near x_0. The mean value theorem implies that

$$g(u) = f'(v) \quad \text{for} \quad \text{some } v \text{ between } x_0 \text{ and } x_0 + u$$

and hence,

$$g(u) \to f'(x_0+) \quad \text{as } u \text{ decreases to } 0,$$

and similarly,

$$g(u) \to f'(x_0-) \quad \text{as } u \text{ increases to } 0.$$

So $g(u)$ is piecewise continuous on $[-\pi, \pi]$. Since a piecewise continuous function clearly is bounded,

$$\int_{-\pi}^{\pi} \left| \frac{f(x_0 + u) - f(x_0)}{u} \right| du = \int_{-\pi}^{\pi} |g(u)|\, du < \infty,$$

and by Th. 3-2 the Fourier series of f converges to $f(x_0)$.

These arguments can be modified easily to show that the Fourier series of a piecewise smooth f converges to $[f(x_0+) + f(x_0-)]/2$ at each point x_0 in $(-\infty, \infty)$, which is the first conclusion of Dirichlet's theorem: Since the integrand in (3-5) is even, we have

$$\frac{1}{2} = \frac{1}{\pi} \int_0^{\pi} \frac{\sin (n + \frac{1}{2})u}{2 \sin \frac{1}{2}u}\, du,$$

and hence

$$\frac{f(x_0+) + f(x_0-)}{2} = \frac{1}{\pi} \int_0^{\pi} \frac{\sin (n + \frac{1}{2})u}{2 \sin \frac{1}{2}u} [f(x_0+) + f(x_0-)]\, du,$$

Similarly, express the integral in (3-4) as a sum of two integrals, one from $-\pi$ to 0 and the other from 0 to π, and then change variables from u to $-u$ in the former integral to find

$$S_n(x_0) = \frac{1}{\pi} \int_0^{\pi} \frac{\sin (n + \frac{1}{2})u}{2 \sin \frac{1}{2}u} [f(x_0 + u) + f(x_0 - u)]\, du.$$

Consequently,

$$\text{(3-8)} \quad S_n(x_0) - \frac{f(x_0+) + f(x_0-)}{2}$$
$$= \frac{1}{\pi} \int_0^{\pi} \frac{\sin (n + \frac{1}{2})u}{2 \sin \frac{1}{2}u} [f(x_0 + u) + f(x_0 - u) - f(x_0+) - f(x_0-)]\, du.$$

Notice that the integral on the right side of (3-8) can be expressed as

$$\frac{1}{\pi} \int_0^{\pi} \sin \left(n + \frac{1}{2}\right) u \frac{u}{2 \sin \frac{1}{2} u} \left[\frac{f(x_0 + u) - f(x_0+)}{u} + \frac{f(x_0 - u) - f(x_0-)}{u} \right] du.$$

Just as above, the Riemann–Lebesgue lemma implies that

(3-9) $$S_n(x_0) \to \frac{f(x_0+) + f(x_0-)}{2}$$

provided that the term in brackets above is absolutely integrable. To confirm this, apply the mean value theorem to see that

$$\frac{f(x_0 + u) - f(x_0+)}{u} \to f'(x_0+), \qquad \frac{f(x_0 - u) - f(x_0-)}{u} \to f'(x_0-),$$

as u decreases to 0. Therefore, the term in brackets is piecewise continuous on $[0, \pi]$, and so is absolutely integrable. Thus, (3-9) holds and we have proved all of Dirichlet's theorem except for the statement about absolute and uniform convergence.

To complete the proof of Dirichlet's theorem, assume that f is 2π periodic, continuous, and piecewise smooth. We must show that the Fourier series of $f(x)$ converges absolutely and uniformly to f. In view of Cor. 1-1, it suffices to show that the series

$$\sum |a_n| < \infty \quad \text{and} \quad \sum |b_n| < \infty,$$

where a_n and b_n are the Fourier coefficients of f. Let

$$x_0 = -\pi < x_1 \cdots < x_m < x_{m+1} = \pi$$

be the points at which $f'(x)$ has jump discontinuities. Then

$$a_n = \frac{1}{\pi}\int_{-\pi}^{\pi} f(x)\cos(nx)\,dx = \frac{1}{\pi}\sum_{k=0}^{m}\int_{x_k}^{x_{k+1}} f(x)\cos(nx)\,dx$$

$$= \sum_{k=0}^{m}\frac{1}{\pi}\left[f(x)\frac{\sin(nx)}{n}\bigg|_{x_k}^{x_{k+1}} - \int_{x_k}^{x_{k+1}} f'(x)\frac{\sin(nx)}{n}\,dx\right].$$

Since $f(x)$ is continuous, the boundary terms in the integration by parts cancel in pairs at each x_k for $k = 1, 2, \ldots, m$, and we find

$$a_n = -\frac{1}{n}\cdot\frac{1}{\pi}\int_{-\pi}^{\pi} f'(x)\sin(nx)\,dx = -\frac{1}{n}\beta_n,$$

where β_n is the nth Fourier sine coefficient of f'. Similarly, since f is 2π periodic, $b_n = (1/n)\alpha_n$, where α_n is the Fourier cosine coefficient of f'. Since $2rs \le r^2 + s^2$, we obtain

$$\sum |a_n| = \sum \frac{1}{n}|\beta_n| \le \frac{1}{2}\left(\sum n^{-2} + \sum \beta_n^2\right) < \infty,$$

in view of Bessel's inequality. Similarly, $\sum |b_n| < \infty$, and Dirichlet's theorem is proven.

When $f(x)$ is 2π periodic, continuous, and piecewise smooth, Dirichlet's theorem asserts that its Fourier series converges uniformly on $(-\infty, \infty)$. However, it is often necessary to expand discontinuous functions in Fourier series. A simple, but basic example is the 2π periodic function $w(x)$ which satisfies $w(x) = x$ for $-\pi < x < \pi$ and then is extended periodically, $w(x) = w(x + 2\pi)$, for all x. We have not defined $w(x)$ for x an odd multiple of π and it is not essential to do so; however, the specification $w((2k + 1)\pi) = 0$ is natural from the point of view of Dirichlet's theorem because the Fourier series of w converges to $w(x)$ at each $x \neq (2k + 1)\pi$ and converges to 0 for $x = (2k + 1)\pi$. The Fourier series of $w(x)$ cannot converge uniformly on any interval that contains a jump discontinuity of $w(x)$ by Prop. 2-1′. The most we could hope for is that the convergence is uniform on each closed subinterval which contains no discontinuity points of $w(x)$. We shall confirm this uniform convergence directly. Then, as a simple consequence, we shall establish the corresponding result for any piecewise smooth function. That is, we shall prove Th. 1-3.

The Fourier series of $w(x)$ is readily calculated (Prob. 9 of Sec. 3-1) and by Dirichlet's theorem,

$$w(x) = 2 \sum_{k=1}^{\infty} (-1)^{k+1} \frac{\sin(kx)}{k}$$

for x not an odd multiple of π. We show that this series converges uniformly to $w(x)$ on any closed interval that does not contain any odd multiple of π. In view of the 2π periodicity of $w(x)$, it suffices to establish this uniform convergence for $|x| < \pi - \delta$ for all small $\delta > 0$. For this purpose we use summation by parts:

$$\sum_{k=1}^{N} a_k b_k = \sum_{k=1}^{N-1} A_k(b_k - b_{k+1}) + A_N b_N$$

where $A_k = a_1 + \cdots + a_k$ (see Prob. 5). Let

$$a_k = a_k(x) = 2(-1)^{k+1} \sin(kx) \quad \text{and} \quad b_k = \frac{1}{k}.$$

Then the summation-by-parts formula expresses the Nth partial sum of the Fourier series of $w(x)$ as

$$S_N(x) = \sum_{k=1}^{N-1} A_k(x) \frac{1}{k^2 + k} + \frac{A_N(x)}{N}, \tag{3-10}$$

with

$$A_k(x) = 2 \sum_{j=1}^{k} (-1)^{j+1} \sin(jx).$$

To sum this expression for $A_k(x)$, we use the trigonometric identity

$$2 \sin \tfrac{1}{2}x \sin(jx) = \cos(jx - \tfrac{1}{2}x) - \cos(jx + \tfrac{1}{2}x),$$

which yields upon summation

$$2 \sin \frac{1}{2} x \sum_{j=1}^{k} \sin(jx) = \cos \frac{1}{2} x - \cos\left(k + \frac{1}{2}\right)x.$$

Replace x by $x + \pi$ and simplify to get

$$2 \cos \frac{1}{2} x \sum_{j=1}^{k} (-1)^j \sin (jx) = \cos \frac{1}{2} (x + \pi) - \cos \left(k + \frac{1}{2} \right)(x + \pi),$$

or

$$A_k(x) = -\frac{\cos \frac{1}{2}(x + \pi) - \cos (k + \frac{1}{2})(x + \pi)}{\cos \frac{1}{2}x}.$$

Thus, for $|x| < \pi - \delta$,

$$|A_k(x)| \leq \frac{2}{\cos \frac{1}{2}(\pi - \delta)} = \frac{2}{\sin \frac{1}{2}\delta},$$

a constant independent of x. Consequently, by the Weierstrass M-test the series

$$\sum_{k=1}^{\infty} \frac{A_k(x)}{k^2 + k}$$

is absolutely and uniformly convergent on $|x| < \pi - \delta$. Similarly, the estimate

$$|A_N(x)| \leq \frac{2}{\sin \frac{1}{2}\delta}$$

shows that the last term in (3-10), $A_N(x)/N$, converges uniformly to 0 on $|x| < \pi - \delta$. Therefore, $S_N(x)$ is absolutely and uniformly convergent to $w(x)$ on $|x| < \pi - \delta$.

With this special case in hand, we are able to prove Th. 1-3 by an appropriate application of Dirichlet's theorem: We can assume that $L = \pi$ in Th. 1-3. Also, since f is 2π periodic, it suffices to prove the uniform convergence of the Fourier series for each closed interval $[a, b]$, with $0 \leq a < b \leq 2\pi$, which contains no jumps of $f(x)$. Let $x_1 < \cdots < x_r$ be the points of discontinuity of $f(x)$ on $0 \leq x < 2\pi$ and denote the jump in $f(x)$ at the point x_k by

$$a_k = f(x_k+) - f(x_k-)$$

for $k = 1, 2, \ldots r$. Since the special function $w(x)$ has jump -2π at each odd multiple of π, the function

$$w_k(x) = \frac{1}{2\pi} w(x - x_k + \pi)$$

has jump -1 at $x = x_k + 2\pi j$ for any integer j. Therefore,

$$g(x) = f(x) + \sum_{k=1}^{r} a_k w_k(x)$$

is continuous for all x because the jumps of f are canceled by equal and opposite jumps of the functions w_k. Plainly, $g(x)$ is piecewise smooth. The Fourier series of $g(x)$ converges uniformly to g on $(-\infty, \infty)$ by Dirichet's theorem. Moreover, if $[a, b]$ contains no jumps of $f(x)$, it contains no jumps of $w_k(x)$ for $k = 1, 2, \ldots,$ r. By what we have just established for the function $w(x)$, the Fourier series of $w_k(x)$ converges uniformly to $w_k(x)$ on $[a, b]$ (see also Prob. 6). Since the Fourier series of $f(x)$ is the sum of the Fourier series of $g(x)$ and of $-a_1w_1(x), \ldots, -a_rw_r(x)$, the Fourier series of $f(x)$ is uniformly convergent on $[a, b]$. This completes the proof of Th. 1-3.

PROBLEMS

1. Use Th. 3-2 to discuss the convergence of the Fourier cosine series for (a) $f(x) = x^{1/2}$ with $0 \leq x \leq \pi$, and for (b) $f(x) = x^{-1/2}$. Why doesn't Dirichlet's theorem apply?

2. Let $f(x)$ satisfy

$$|f(x+u) - f(x)| < L|u|^{\alpha}$$

for some constants L and α. If $\alpha = 1$, we say that $f(x)$ is *Lipschitz continuous*, and if $0 < \alpha < 1$ that $f(x)$ is *Hölder continuous*. Show that if f is 2π periodic and Lipschitz or Hölder continuous, then its Fourier series converges to $f(x)$ for each x.

3. Use the formula for the sum of a finite geometric series to verify that

$$\frac{1}{2} + \sum_{k=1}^{n} e^{iku} = \frac{e^{i(n+1)u} - 1}{e^{iu} - 1} - \frac{1}{2}.$$

Separate this expression into its real and imaginary parts to obtain

$$\frac{1}{2} + \sum_{k=1}^{n} \cos(ku) = \frac{\sin(n+\frac{1}{2})u}{2\sin\frac{1}{2}u},$$

$$\sum_{k=1}^{n} \sin(ku) = \frac{\cos\frac{1}{2}u - \cos(n+\frac{1}{2})u}{2\sin\frac{1}{2}u}.$$

4. If $f(x)$ is continuous, 2π periodic, and piecewise smooth, we showed that

$$a_n = \frac{-\beta_k}{n} \quad \text{and} \quad b_n = \frac{\alpha_k}{n},$$

where a_n, b_n (resp., α_k, β_k) are the Fourier coefficients of f(resp., f').

(a) Show that there is a constant M such that

$$|a_n| \leq \frac{1}{n}M \quad \text{and} \quad |b_n| \leq \frac{1}{n}M.$$

That is,

$$|a_n| = O\left(\frac{1}{n}\right) \quad \text{and} \quad |b_n| = O\left(\frac{1}{n}\right).$$

(b) Suppose that $f(x)$ is 2π periodic, $f^{(r-1)}(x)$ is continuous, and $f^{(r)}$ is piecewise continuous for some $r \geq 1$. Show that

$$|a_n| = O\left(\frac{1}{n^r}\right) \quad \text{and} \quad |b_n| = O\left(\frac{1}{n^r}\right).$$

(c) Use part (b) to show: If f is 2π periodic and $f''(x)$ is continuous on $[-\pi, \pi]$, then the Fourier series of f converges to $f(x)$ for each x in $[-\pi, \pi]$.

5. Establish the summation-by-parts formula. *Hint.* Set $A_0 = 0$, write $a_kb_k = (A_k - A_{k-1})b_k$, and sum from $k = 1$ to $k = N$. Finally, reindex one sum.

6. In the proof of Th. 1-3, we implicitly assumed that the Fourier series of $w_j(x)$ could be obtained by replacing x by $x - x_j + \pi$ in the Fourier series for $w(x)$. Make this substitution. Then calculate the Fourier series of $w_j(x)$ directly and confirm our assumption.

3-4. Operational Aspects of Fourier Integrals and Transforms

Fourier series expansions apply to *periodic* functions. In a sense the periodicity restriction is mild because any function defined on an interval of finite length L can be extended periodically. On the other hand, a nonperiodic function defined on an infinite interval cannot be expanded in a Fourier series. Thus, Fourier series methods are restricted to physical problems that have finite extent. To apply "Fourier methods" to physical models with infinite extent ($L = \infty$), so-called Fourier integrals are needed. In this setting, Fourier integral representations replace Fourier series expansions. In this section we motivate and detail several of the most important aspects of Fourier integral representations and transforms. Careful proofs, which are a little tricky, are deferred until the next section.

We begin by showing how Fourier series expansions lead naturally to a Fourier integral representation: Consider a continuous function $f(x)$ defined for $-\infty < x < \infty$: Let $L > 0$ and set $f_L(x) = f(x)$ for $-L \leq x < L$. Extend $f_L(x)$ to be $2L$ periodic. For each fixed x,

$$f(x) = \lim_{L \to \infty} f_L(x). \tag{4-1}$$

If $f(x)$ is piecewise smooth on $-\infty < x < \infty$, then for $-L/2 < x < L/2$,

$$f_L(x) = \tfrac{1}{2}a_0 + \sum_{n=1}^{\infty}\left(a_n \cos\frac{n\pi x}{L} + b_n \sin\frac{n\pi x}{L}\right), \tag{4-2}$$

where

$$a_n = \frac{1}{L}\int_{-L}^{L} f(u)\cos\left(\frac{n\pi u}{L}\right) du,$$

and

$$b_n = \frac{1}{L}\int_{-L}^{L} f(u)\sin\left(\frac{n\pi u}{L}\right) du.$$

Our plan is to make a formal limit passage as $L \to \infty$ in (4-2). To be sure that the Fourier coefficients a_n and b_n are defined for all L, we assume that

$$\int_{-\infty}^{\infty} |f(x)|\, dx < \infty. \tag{4-3}$$

Insert the expressions for the Fourier coefficients a_n, b_n into (4-2) to obtain

$$f_L(x) = \frac{1}{2L}\int_{-L}^{L} f(u)\, du + \sum_{n=1}^{\infty} \frac{1}{L}\int_{-L}^{L} f(u)\cos\left(\frac{n\pi}{L}\right)(u - x)\, du.$$

We can interpret the series as an "infinite Riemann sum" for an improper integral over $0 \le v < \infty$. Indeed, let $v_n = n\pi/L$ and $\Delta v = v_{n+1} - v_n = \pi/L$, to find that

$$f_L(x) = \frac{1}{2L}\int_{-L}^{L} f(u)\,du + \frac{1}{\pi}\sum_{n=1}^{\infty} \Delta v \int_{-L}^{L} f(u)\cos v_n(u-x)\,du. \tag{4-4}$$

Now let $L \to \infty$ in (4-4) and proceed formally as follows: The left member of (4-4) has limit $f(x)$ by (4-1),

$$\frac{1}{2L}\int_{-L}^{L} f(u)\,du \to 0$$

by (4-3), and the "Riemann sum" in (4-4) has limit

$$\int_0^{\infty}\left[\int_{-\infty}^{\infty} f(u)\cos v(u-x)\right] du\,dv.$$

Thus, (4-4) yields

$$f(x) = \frac{1}{\pi}\int_0^{\infty}\left[\int_{-\infty}^{\infty} f(u)\cos v(u-x)\right] du\,dv. \tag{4-5}$$

This integral representation for a function $f(x)$ defined on $-\infty < x < \infty$ is the analogue of the Fourier series expansion for a function defined on a finite interval. The analogy is clearer if the term $\cos v(u - x)$ is expanded out (see Prob. 2). In the next section we establish the validity of (4-5) under hypotheses analogous to those used to obtain Fourier series expansions in Sec. 3-3. For the moment we just state two useful analogues of results given in Sec. 3-3 for Fourier series.

Theorem 4-1 (Fourier Integral Theorem). Let $f(x)$ be piecewise continuous and absolutely integrable on $(-\infty, \infty)$. Then

$$\frac{f(x+) + f(x-)}{2} = \frac{1}{\pi}\int_0^{\infty}\left[\int_{-\infty}^{\infty} f(u)\cos v(u-x)\,du\right] dv$$

at each point x for which there exists $\delta > 0$ with

$$\int_0^{\delta}\left|\frac{f(x+u) - f(x+)}{u}\right| du < \infty \tag{4-6}$$

and

$$\int_0^{\delta}\left|\frac{f(x-u) - f(x-)}{u}\right| du < \infty. \tag{4-6$'$}$$

Consequently, if f is also continuous at x,

$$f(x) = \frac{1}{\pi}\int_0^{\infty}\left[\int_{-\infty}^{\infty} f(u)\cos v(u-x)\,du\right] dv.$$

Corollary 4-1. If f is piecewise smooth on $(-\infty, \infty)$ and absolutely integrable, then

$$\frac{f(x+)+f(x-)}{2} = \frac{1}{\pi}\int_0^\infty \int_{-\infty}^\infty f(u)\cos v(u-x)\,du\,dv.$$

If, in addition, f is continuous at x,

$$f(x) = \frac{1}{\pi}\int_0^\infty \int_{-\infty}^\infty f(u)\cos v(u-x)\,du\,dv.$$

To verify Cor. 4-1, apply the mean value theorem to see that for $\delta > 0$ and small,

$$|f(x+u) - f(x+)| = |f'(z)u| \le M'u$$

for some point z in $(0, \delta)$, where M' bounds f' on $[0, \delta]$. Thus, the improper integral

$$\int_0^\delta \left|\frac{f(x+u) - f(x+)}{u}\right| du \le M'\delta < \infty,$$

and similarly the second integral in (4-6)$'$, is finite.

For most applications it is more convenient to express the conclusion of the Fourier integral theorem in terms of so-called Fourier transforms. These transforms are defined by improper integrals. Thus, we pause to review some standard terminology about such integrals.

For our purposes it suffices to restrict attention to integrands $g(x, y)$ with $a \le x \le b$ and $0 \le y < \infty$, which are *piecewise continuous in y*. This means that for each $Y > 0$, $g(x, y)$ is continuous on the rectangle of points (x, y) with $a \le x \le b, 0 \le y \le Y$, with the possible exception of the points on a finite number of lines $y =$ constant across which only jump discontinuities of $g(x, y)$ may occur. Then

$$\int_0^Y g(x, y)\,dy \text{ is defined and continuous at each } x \text{ in } [a, b],$$

and

$$\int_0^\infty g(x, y)\,dy = \lim_{Y\to\infty} \int_0^Y g(x, y)\,dy,$$

whenever this limit exists. The improper integral $\int_{-\infty}^\infty g(x, y)\,dy$ of a function defined on $a \le x \le b$, $-\infty < y < \infty$ and which is piecewise continuous in y is defined by

$$\int_{-\infty}^\infty g(x, y)\,dy = \lim_{\substack{Y\to\infty \\ Z\to\infty}} \int_{-Z}^Y g(x, y)\,dy,$$

whenever this limit exists. Notice that Y and Z become infinite independently. There are situations of great importance in mathematical physics where the limit above does not exist, but the symmetric limit $\lim_{Y\to\infty} \int_{-Y}^{Y} g(x, y)\, dy$ does exist. This limit is called a *Cauchy principal value.*

An improper integral of the form $\int_0^\infty g(x, y)\, dy$ is *uniformly convergent* for x in a set S if given $\varepsilon > 0$ there exists $Y(\varepsilon) > 0$ so that

$$\left| \int_M^\infty g(x, y)\, dy \right| < \varepsilon$$

for all $M \geq Y(\varepsilon)$ and all x in S. Briefly, the "tail integrals" can be made arbitrarily small for all x in S. Also,

$$\int_{-\infty}^{\infty} g(x, y)\, dy \quad \text{is uniformly convergent for } x \text{ in } S$$

$$\text{if both} \quad \int_{-\infty}^{0} g(x, y)\, dy \quad \text{and} \quad \int_0^\infty g(x, y)\, dy$$

are uniformly convergent on S.

Now we are prepared to express the conclusions of the Fourier integral theorem in terms of Fourier transforms. Assume that $f(x)$ is continuous, piecewise smooth, and absolutely integrable on $(-\infty, \infty)$. Then by the Fourier integral theorem

$$\begin{aligned} f(x) &= \frac{1}{\pi} \int_0^\infty \int_{-\infty}^{\infty} f(u) \cos v(u - x)\, du\, dv \\ &= \frac{1}{\pi} \lim_{M\to\infty} \int_0^M \int_{-\infty}^{\infty} f(u) \frac{e^{iv(u-x)} + e^{-iv(u-x)}}{2}\, du\, dv \\ &= \frac{1}{2\pi} \lim_{M\to\infty} \left[\int_0^M e^{-ivx} \int_{-\infty}^{\infty} f(u) e^{ivu}\, du\, dv \right. \\ &\qquad \left. + \int_{-M}^{0} e^{-ivx} \int_{-\infty}^{\infty} f(u) e^{ivu}\, du\, dv \right], \end{aligned}$$

after expressing the previous integral as a sum of two integrals and making the change of variables v into $-v$ in the second one. Thus,

$$f(x) = \frac{1}{2\pi} \lim_{M\to\infty} \int_{-M}^{M} e^{-ivx} \int_{-\infty}^{\infty} e^{ivu} f(u)\, du\, dv. \tag{4-7}$$

The *Fourier transform* of $f(x)$ is defined by

$$\hat{f}(\omega) = \int_{-\infty}^{\infty} e^{i\omega x} f(x)\, dx \tag{4-8}$$

whenever this integral exists. Evidently, the Fourier transform exists when $f(x)$ is absolutely integrable on $(-\infty, \infty)$. In view of (4-7), we can recover $f(x)$ from its transform by

$$f(x) = \frac{1}{2\pi} \lim_{M \to \infty} \int_{-M}^{M} e^{-i\omega x} \hat{f}(\omega)\, d\omega$$

or

(4-9)
$$f(x) = \frac{1}{2\pi} \int_{-\infty}^{\infty} e^{-i\omega x} \hat{f}(\omega)\, d\omega,$$

where the infinite integral is understood in the Cauchy principal value sense. In summary, we have

Theorem 4-2. Let $f(x)$ be a continuous, piecewise smooth, and absolutely integrable. Then

$$f(x) = \frac{1}{2\pi} \int_{-\infty}^{\infty} e^{-i\omega x} \hat{f}(\omega)\, d\omega.$$

In case $f(x)$ is just piecewise smooth and absolutely integrable, the reasoning used to establish Th. 4-2 leads to

(4-9)′
$$\frac{f(x+) + f(x-)}{2} = \frac{1}{2\pi} \int_{-\infty}^{\infty} e^{-i\omega x} \hat{f}(\omega)\, d\omega.$$

Fourier transforms are used in much the same way as Laplace transforms to solve a variety of problems involving ordinary or partial differential equations. In the case of ordinary differential equations, the transformed equation is algebraic. The transform of a partial differential equation is an ordinary differential equation. These "simpler" transformed problems are solved for $\hat{f}(\omega)$. Then $f(x)$ is recovered by inversion from (4-9). In practice this inversion is usually performed by reference to a table of Fourier transforms. For example, two excellent sets of tables are: H. Bateman, *Tables of Integral Transforms*, Vols. 1 and 2 (McGraw-Hill Book Company, New York, 1954) and F. Oberhettinger, *Tabellen zur Fourier Transformation* (Springer-Verlag, New York, 1957). Fourier transforms will be used to solve a variety of physical problems in subsequent chapters. For the moment we concentrate on establishing some basic properties of such transforms. Proposition 5-2 in the next section has as an immediate corollary,

Theorem 4-3. Let $f(x)$ be piecewise continuous and absolutely integrable. Then $\hat{f}(\omega)$ is continuous on $-\infty < \omega < \infty$.

Under the assumption that $f(x)$ is absolutely integrable, the Fourier transform (4-8) is absolutely and uniformly convergent. However, as noted above, the integral (4-9) may exist only as a Cauchy principal value (see Prob. 5). On the other hand, if $\hat{f}(\omega)$ is absolutely integrable on $(-\infty, \infty)$, the estimate $|e^{-i\omega x}\hat{f}(\omega)| \le |\hat{f}(\omega)|$ implies that (4-9) expresses $f(x)$ as an absolutely and uniformly convergent improper integral. Since absolutely and uniformly convergent integrals are easier to deal with than Cauchy principal value integrals, it is worthwhile to develop some simple and useful conditions that guarantee absolute and uniform convergence in (4-9).

To this end, assume that $f(x)$ and $f'(x)$ are continuous and absolutely integrable on $(-\infty, \infty)$. By the fundamental theorem of calculus,

$$f(x) = f(0) + \int_0^x f'(t)\,dt.$$

The integral on the right has a limit as x tends to $\pm\infty$ because we assumed that f' was absolutely integrable; hence, $f(x)$ has a limit as x becomes infinite and

$$\lim_{x\to\pm\infty} f(x) = \lim_{x\to\pm\infty} \left[f(0) + \int_0^x f'(t)\,dt \right].$$

Each limit must be 0 because $f(x)$ itself is integrable on $(-\infty, \infty)$. Now integrate by parts in (4-8) and use the fact that $f(x)$ has limit 0 at $\pm\infty$ to find

$$\hat{f}(\omega) = -\frac{1}{i\omega}\int_{-\infty}^{\infty} e^{i\omega x} f'(x)\,dx \quad \text{or} \quad \hat{f'}(\omega) = (-i\omega)\hat{f}(\omega).$$

Repeated application of this result leads to

Theorem 4-4. Let $f(x), f'(x), \ldots, f^{(n)}(x)$ be continuous and absolutely integrable on $(-\infty, \infty)$. Then $f, f', \ldots, f^{(n)}$ have Fourier transforms given by

$$[f^{(k)}]\hat{}\,(\omega) = (-i\omega)^k \hat{f}(\omega) \tag{4-10}$$

for $k = 1, 2, \ldots, n$.

Corollary 4-2. If $f(x)$, $f'(x)$, and $f''(x)$ are continuous and absolutely integrable on $(-\infty, \infty)$, then $\hat{f}(\omega)$ is absolutely integrable on $(-\infty, \infty)$, and the inversion formula (4-9) expresses $f(x)$ as an absolutely and uniformly convergent integral.

Indeed, from (4-10) for $\omega \neq 0$,

$$\hat{f}(\omega) = \frac{-1}{\omega^2}[f'']\hat{}\,(\omega),$$

$$|\hat{f}(\omega)| \leq \frac{1}{\omega^2}\int_{-\infty}^{\infty} |f''(x)|\,dx = \frac{M''}{\omega^2},$$

where M'' is a constant. This estimate and the continuity of $\hat{f}(\omega)$ imply that

$$\begin{aligned}\int_{-\infty}^{\infty} |\hat{f}(\omega)|\,d\omega &\leq \int_{-1}^{1} |\hat{f}(\omega)|\,d\omega + \left(\int_{-\infty}^{-1} + \int_{1}^{\infty}\right) M''\omega^{-2}\,d\omega \\ &= \int_{-1}^{1} |\hat{f}(\omega)|\,d\omega + 2M'' < \infty.\end{aligned}$$

Thus, $\hat{f}(\omega)$ is absolutely integrable on $(-\infty, \infty)$. Finally, the estimate

$$|e^{-i\omega x}\hat{f}(\omega)| \leq |\hat{f}(\omega)|$$

and the Weierstrass M-test (see Sec. 3-5) give the absolute and uniform convergence of the integral in (4-9).

Examples of Fourier Transforms

1. (*Laplace integrals*) Let $f(x) = e^{-a|x|}$ for fixed $a > 0$. Then

$$\hat{f}(\omega) = \int_{-\infty}^{\infty} e^{i\omega x} e^{-a|x|}\, dx$$
$$= \int_0^{\infty} e^{i\omega x} e^{-ax}\, dx + \int_{-\infty}^{0} e^{i\omega x} e^{ax}\, dx$$
$$= -\frac{1}{i\omega - a} + \frac{1}{i\omega + a} = \frac{2a}{\omega^2 + a^2}.$$

Invert to find

$$\frac{1}{2\pi} \int_{-\infty}^{\infty} e^{-i\omega x} \frac{2a}{\omega^2 + a^2}\, d\omega = e^{-a|x|},$$

which reduces to

$$\frac{a}{\pi} \int_{-\infty}^{\infty} \frac{\cos(\omega x)}{\omega^2 + a^2}\, d\omega = e^{-a|x|}.$$

2. $f(x) = e^{-ax^2}$ for $a > 0$. Here

$$\hat{f}(\omega) = \int_{-\infty}^{\infty} e^{i\omega x} e^{-ax^2}\, dx$$

cannot be evaluated by elementary quadratures. However, by Prop. 5-4 in the next section, we can differentiate under the integral sign to find

$$\hat{f}'(\omega) = \int_{-\infty}^{\infty} i e^{i\omega x} x e^{-ax^2}\, dx.$$

Then integration by parts gives

$$\hat{f}'(\omega) = \frac{-\omega}{2a} \int_{-\infty}^{\infty} e^{i\omega x} e^{-ax^2}\, dx = \frac{-\omega}{2a} \hat{f}(\omega).$$

This elementary first-order differential equation has the general solution

$$\hat{f}(\omega) = Ce^{-\omega^2/4a}.$$

Set $\omega = 0$ to find

$$C = \hat{f}(0) = \int_{-\infty}^{\infty} e^{-ax^2}\, dx = \frac{1}{\sqrt{a}} \int_{-\infty}^{\infty} e^{-u^2}\, du = \frac{\sqrt{\pi}}{\sqrt{a}}.$$

Thus, we obtain the transform

$$\hat{f}(\omega) = \frac{\sqrt{\pi}}{\sqrt{a}} e^{-\omega^2/4a}$$

and the inversion formula

$$e^{-ax^2} = \frac{1}{\sqrt{4a\pi}} \int_{-\infty}^{\infty} e^{-i\omega x - \omega^2/4a}\, d\omega,$$

which play important roles in diffusion processes and other problems (see Chap. 5).

If $f(x)$ is an odd (even) function, then its Fourier transform can also be expressed as a *Fourier sine* (*cosine*) *transform.* These transforms convert differentiations into multiplications in roughly the same manner as for Th. 4-4. The sine and cosine transforms are useful for problems modeled on semi-infinite intervals. Some of these results are explored in the problems here and in subsequent chapters.

All of the results presented above extend readily to functions of several independent variables. Let

$$x = (x_1, \ldots, x_n) \quad \text{and} \quad \omega = (\omega_1, \ldots, \omega_n)$$

be points in n-space. Then the Fourier transform of a function f defined on n-space is

$$\hat{f}(\omega) = \int_{R_n} e^{i\omega \cdot x} f(x)\, dx \tag{4-11}$$

whenever this integral exists, where the integral is over all of n-dimensional space,

$$x \cdot \omega = x_1\omega_1 + \cdots + x_n\omega_n$$

is the usual dot product in n-space, and $dx = dx_1 \cdots dx_n$ is the element of volume.

The Fourier inversion formula takes the form

$$f(x) = \left(\frac{1}{2\pi}\right)^n \int_{R_n} e^{-ix \cdot \omega} \hat{f}(\omega)\, d\omega. \tag{4-12}$$

This holds, in particular, if $f(x)$ and all its partial derivatives of orders equal to or less than $n + 1$ are continuous and absolutely integrable over R_n.

PROBLEMS

1. Let $f(x) = 0$ for $x < 0$, and $f(x) = e^{-x}$ for $x > 0$. Set $f(0) = \frac{1}{2}$. Show that

$$f(x) = \frac{1}{\pi} \int_0^{\infty} \frac{\cos(vx) + v \sin(vx)}{1 + v^2}\, dv = \begin{cases} 0, & x < 0, \\ \frac{1}{2}, & x = 0, \\ e^{-x}, & x > 0. \end{cases}$$

2. Assume that (4-5) holds and expand $\cos v(u - x)$ to obtain the Fourier integral representation

$$f(x) = \int_0^{\infty} [A(v) \cos(vx) + B(v) \sin(vx)]\, dv,$$

where

$$A(v) = \frac{1}{\pi} \int_{-\infty}^{\infty} f(u) \cos(vu)\, du$$

and

$$B(v) = \frac{1}{\pi} \int_{-\infty}^{\infty} f(u) \sin(vu)\, du.$$

This is the strict analogue of the Fourier series expansion for periodic functions.

3. Let $f(x)$ be continuous, piecewise smooth, and absolutely integrable.
   (a) If $f(x)$ is *even*, show that

$$f(x) = \int_0^{\infty} A(v) \cos(vx)\, dv,$$

   where

$$A(v) = \frac{2}{\pi} \int_0^{\infty} f(u) \cos(vu)\, du.$$

   (b) If $f(x)$ is *odd*, show that

$$f(x) = \int_0^{\infty} B(v) \sin(vx)\, dv,$$

   where

$$B(v) = \frac{2}{\pi} \int_0^{\infty} f(u) \sin(vu)\, du.$$

   (c) What are the conclusions in parts (a) and (b) if $f(x)$ is only piecewise smooth and absolutely integrable?

4. Let $f(x)$ be defined for $x \geq 0$. The *Fourier cosine* and *sine transform* of f are given by

$$C(f) = \frac{2}{\pi} \int_0^{\infty} f(x) \cos(\omega x)\, dx, \qquad S(f) = \frac{2}{\pi} \int_0^{\infty} f(x) \sin(\omega x)\, dx$$

provided that these improper integrals exist. Notice that these transforms are functions of ω and that Prob. 3 gives representations for $f(x)$ in terms of its cosine and sine transforms—just extend f to be even or odd on $(-\infty, \infty)$. Use integration by parts to verify that

$$C(f') = -\frac{2}{\pi} f(0) + \omega S(f), \qquad S(f') = -\omega C(f)$$

provided that f, f' are continuous, absolutely integrable, and $f(x) \to 0$ as $x \to \infty$. Use these results to express $C(f'')$ and $S(f'')$ in terms of $C(f)$ and $S(f)$, respectively. What must you assume about f, f', f''?

5. Let $f(x) = e^{-x}$ for $x > 0$, $f(x) = 0$ for $x < 0$, and $f(0) = \frac{1}{2}$. Show that $\hat{f}(\omega) = (1 - i\omega)^{-1} = (1 + i\omega)/(1 + \omega^2)$. Since f is piecewise smooth and absolutely integrable, (4-9)′ holds when $x = 0$. Check this directly. Notice that the integral when $x = 0$ in (4-9)′

must be interpreted as a Cauchy principal value: The improper integral involved does not converge.

6. Example 2 (Laplace integrals) shows that $e^{-a|x|}/2a$ has Fourier transform $(\omega^2 + a^2)^{-1}$. Use this to express the solution to

$$-y'' + a^2 y = g(x), \qquad -\infty < x < \infty, \qquad y(x) \to 0 \quad \text{as} \quad x \to \pm\infty,$$

as an integral involving $g(x)$. What must you assume about $g(x)$ for this solution formula to be valid?

7. In the context of multidimensional Fourier transforms, confirm that

$$\left(\frac{\partial f}{\partial x_j}\right)^{\wedge} = (-i\omega_j)\hat{f},$$

provided that f is continuous, absolutely integrable and $f(x) \to 0$ as $|x| \to \infty$.

8. In (4-11) some authors place the factor $(2\pi)^{-n/2}$ in front of the integral to define $\hat{f}$. Then $(2\pi)^{-n}$ in (4-12) is replaced by $(2\pi)^{-n/2}$. Use this modification in this problem. Let $f(x, y)$ be radially symmetric, that is, $f(x, y) = f(r)$ where $r = (x^2 + y^2)^{1/2}$. Introduce polar coordinates

$$x = r\cos(\theta), \qquad y = r\sin(\theta)$$

and

$$\omega_1 = \rho\cos(\alpha), \qquad \omega_2 = \rho\sin(\alpha)$$

in the expression for $\hat{f} = \hat{f}(\omega_1, \omega_2)$ in (4-11) to find that

$$\hat{f} = \hat{f}(\rho) = \int_0^\infty f(r) r \left[\frac{1}{2\pi}\int_0^{2\pi} \cos\left[\rho r\cos(\theta)\right] d\theta\right] dr.$$

The integral in brackets is equal to $J_0(\rho r)$, where $J_0(z)$ is the Bessel function of the first kind of order 0. So

$$\hat{f}(\rho) = \int_0^\infty J_0(\rho r) f(r) r\, dr.$$

Use the Fourier inversion formula to get

$$f(r) = \int_0^\infty J_0(r\rho)\hat{f}(\rho)\rho\, d\rho.$$

In this context, $\hat{f}(\rho)$ is called the *Hankel transform* of $f(r)$.

3-5. Fourier Integral and Transform Theorems

In this section we put our informal discussion of the Fourier integral theorem

$$f(x) = \frac{1}{\pi}\int_0^\infty \int_{-\infty}^\infty f(u)\cos v(u - x)\, du\, dv \tag{5-1}$$

and the basic properties of Fourier transforms on a firm mathematical foundation. The analysis rests on a careful study of the behavior of certain improper integrals.

In particular, we will need to interchange the order of integration in improper multiple integrals and to differentiate under improper integral signs. We begin with the relevant background material needed to justify these manipulations. The general results obtained will also play a key role in our analysis of heat conduction in an infinite rod in Chap. 5.

The *Weierstrass M-test* for improper integrals will play an important role:

Proposition 5-1. Suppose that

$$\int_{-\infty}^{\infty} g(x, y)\, dy$$

exists for each x in a set S and that there is a function $M(y)$ such that for x in S and $-\infty < y < \infty$,

$$|g(x, y)| \leq M(y) \quad \text{and} \quad \int_{-\infty}^{\infty} M(y)\, dy < \infty.$$

Then

$$\int_{-\infty}^{\infty} g(x, y)\, dy$$

is absolutely and uniformly convergent for x in S.

The proof is virtually the same as for the series version of the M-test (see Prob. 1). Proposition 5-1 has an evident analogue for $\int_0^\infty g(x, y)\, dy$.

Uniform convergence plays the same important role for improper integrals that it does for infinite series. For example, we have

Proposition 5-2. Let $g(x, y)$ be defined on $a \leq x \leq b$ and $-\infty < y < \infty$ and be piecewise continuous in y. Assume that $\int_{-\infty}^{\infty} g(x, y)\, dy$ is uniformly convergent for x in $[a, b]$. Then the integral defines a continuous function for x in $[a, b]$.

The proof is easy. For $n = 1, 2, \ldots,$

$$G_n(x) = \int_{-n}^{n} g(x, y)\, dy \text{ is continuous on } [a, b],$$

and

$$G(x) = \int_{-\infty}^{\infty} g(x, y)\, dy$$

is the uniform limit of the sequence of continuous functions $G_n(x)$. So $G(x)$ is continuous.

Proposition 5-3. Let $g(x, y)$ satisfy the hypotheses in Prop. 5-2. Then both of the following iterated integrals exist and

$$\int_a^b \left[\int_{-\infty}^{\infty} g(x, y)\, dy\right] dx = \int_{-\infty}^{\infty} \left[\int_a^b g(x, y)\, dx\right] dy.$$

Since

$$\int_{-\infty}^{\infty} g(x, y)\, dy = \int_{-\infty}^{0} g(x, y)\, dy + \int_0^{\infty} g(x, y)\, dy$$

it suffices to prove the corresponding proposition with only one limit of integration infinite, say,

$$\int_a^b \left[\int_0^{\infty} g(x, y)\, dy\right] dx = \int_0^{\infty} \left[\int_a^b g(x, y)\, dx\right] dy.$$

By Prop. 5-2,

$$\int_0^{\infty} g(x, y)\, dy \text{ is a continuous function on } [a, b]$$

and hence $\int_a^b \int_0^\infty g(x, y)\, dy\, dx$ exists. Also, given $\varepsilon > 0$, the uniform convergence in x guarantees that

$$\left|\int_M^{\infty} g(x, y)\, dy\right| < \varepsilon$$

for all x in $[a, b]$ and M sufficiently large. Since

$$\int_0^{\infty} g(x, y)\, dy = \int_0^{M} g(x, y)\, dy + \int_M^{\infty} g(x, y)\, dy,$$

use of the standard result on the interchange of order of integrals over a finite domain yields

$$\begin{aligned}\left|\int_a^b \int_0^{\infty} g(x, y)\, dy\, dx - \int_0^M \int_a^b g(x, y)\, dx\, dy\right| &= \left|\int_a^b \int_M^{\infty} g(x, y)\, dy\, dx\right| \\ &< \varepsilon(b - a)\end{aligned}$$

for all M sufficiently large. Since $\varepsilon > 0$ is arbitrary the previous inequality implies that there exists

$$\lim_{M\to\infty} \int_0^M \int_a^b g(x, y)\, dx\, dy = \int_a^b \int_0^{\infty} g(x, y)\, dy\, dx.$$

That is,

$$\int_0^{\infty} \int_a^b g(x, y)\, dx\, dy = \int_a^b \int_0^{\infty} g(x, y)\, dy\, dx.$$

The next result is a useful criterion for the interchange of differentiation and integration in improper integrals.

Proposition 5-4. Suppose that $g(x, y)$ and $g_x(x, y)$ are continuous for $-\infty \leq \alpha < x < \beta \leq \infty$ and $-\infty < y < \infty$. Then

$$\frac{d}{dx}\int_{-\infty}^{\infty} g(x, y)\, dy = \int_{-\infty}^{\infty} \frac{\partial g}{\partial x}(x, y)\, dy$$

for x in (α, β), provided that the integral on the right is uniformly convergent on each closed, bounded subinterval of (α, β) and the integral on the left is convergent.

To see this, fix a, b with $\alpha < a < b < \beta$. By Prop. 5-3, for each x in $[a, b]$,

$$\int_a^x \int_{-\infty}^{\infty} \frac{\partial g}{\partial x}(t, y)\, dy\, dt = \int_{-\infty}^{\infty} \int_a^x \frac{\partial g}{\partial x}(t, y)\, dt\, dy$$

$$= \int_{-\infty}^{\infty} [g(x, y) - g(a, y)]\, dy = \int_{-\infty}^{\infty} g(x, y)\, dy - A,$$

where A is a constant. Thus, for x in $[a, b]$,

$$\int_{-\infty}^{\infty} g(x, y)\, dy = A + \int_a^x \left[\int_{-\infty}^{\infty} \frac{\partial g}{\partial x}(t, y)\, dy\right] dt,$$

and the fundamental theorem of calculus gives

$$\frac{d}{dx}\int_{-\infty}^{\infty} g(x, y)\, dy = \int_{-\infty}^{\infty} \frac{\partial g}{\partial x}(x, y)\, dy,$$

as advertised. Since a, b are arbitrary with $\alpha < a$ and $b < \beta$, this equation holds for all x in (α, β).

Remark. Proposition 5-4 has evident counterparts when only one limit of integration is infinite.

To establish the Fourier integral representation (5-1), we need to know that

$$\int_0^{\infty} \frac{\sin y}{y}\, dy = \frac{\pi}{2}. \tag{5-2}$$

One way to confirm this uses Prop. 5-4 and the auxiliary integral

$$L(x) = \int_0^{\infty} e^{-xy} \frac{\sin y}{y}\, dy,$$

which converges uniformly for $x \geq 0$, as we will show presently. Consequently, $L(x)$ is continuous. Also, differentiating with respect to x under the integral sign gives the integral $-\int_0^{\infty} e^{-xy} \sin y\, dy$, which is uniformly convergent on each closed subinterval of $(0, \infty)$. This uniform convergence is immediate from the M-test (see Prob. 2). By Prop. 5-4, for $x > 0$,

$$L'(x) = -\int_0^{\infty} e^{-xy} \sin y\, dy = \frac{-1}{1 + x^2},$$

so

$$L(x) = -\arctan x + C.$$

Since $|\sin y| \le |y|$, as $x \to \infty$

$$|L(x)| = \left| \int_0^\infty e^{-xy} \frac{\sin y}{y} \, dy \right| \le \int_0^\infty e^{-xy} \, dy = \frac{1}{x} \to 0,$$

$\arctan x \to \pi/2$, and so $C = \pi/2$. Thus for $x > 0$,

$$L(x) = \frac{\pi}{2} - \arctan x.$$

Since $L(x)$ is continuous for $x \ge 0$, we conclude that

$$\int_0^\infty \frac{\sin y}{y} \, dy = L(0) = \lim_{x \to 0+} L(x) = \frac{\pi}{2}.$$

It remains to check that the improper integral defining $L(x)$ is uniformly convergent on $x \ge 0$: For each $M > 0$ let $n\pi$ be the *smallest* multiple of π that exceeds M. Then

$$\int_M^\infty e^{-xy} \frac{\sin y}{y} \, dy = \int_M^{n\pi} e^{-xy} \frac{\sin y}{y} \, dy + \sum_{k=n}^{\infty} \int_{k\pi}^{(k+1)\pi} e^{-xy} \frac{\sin y}{y} \, dy.$$

The series on the right has terms that alternate in sign and decrease to zero in magnitude. Such a series converges by the Leibniz alternating series test, and the magnitude of its sum is at most the magnitude of its first term. Therefore,

$$\begin{aligned} \left| \int_M^\infty e^{-xy} \frac{\sin y}{y} \, dy \right| &\le \int_M^{(n+1)\pi} e^{-xy} \frac{|\sin y|}{y} \, dy \\ &< \int_M^{(n+1)\pi} \frac{1}{M} \, dy = \frac{(n+1)\pi - M}{M} \le \frac{2\pi}{M} \end{aligned}$$

because $(n-1)\pi \le M < n\pi$. Since $2\pi/M \to 0$ as $M \to \infty$ independent of $x \ge 0$, the integral defining $L(x)$ is uniformly convergent for $x \ge 0$. This completes our verification of (5-2).

We note for future reference that the change of variables $y = cu$ yields

$$\int_0^\infty \frac{\sin cu}{u} \, du = \frac{\pi}{2} \qquad \text{for } c > 0. \tag{5-3}$$

Now we are prepared to establish (5-1), given reasonable assumptions on $f(x)$. Here is a useful analogue of some results given in Sec. 3-3 for Fourier series.

Theorem 5-1. (Fourier Integral Theorem). Let $f(x)$ be piecewise continuous and absolutely integrable on $(-\infty, \infty)$. Then

$$\frac{f(x+) + f(x-)}{2} = \frac{1}{\pi} \int_0^\infty \left[\int_{-\infty}^\infty f(u) \cos v(u - x) \, du \right] dv$$

at each point x for which there exists $\delta > 0$ with

$$\int_0^\delta \left| \frac{f(x+u) - f(x+)}{u} \right| du < \infty \tag{5-4}$$

and

$$\int_0^\delta \left| \frac{f(x-u) - f(x-)}{u} \right| du < \infty. \tag{5-4$'$}$$

Consequently, if f is also continuous at x,

$$f(x) = \frac{1}{\pi} \int_0^\infty \left[\int_{-\infty}^\infty f(u) \cos v(u - x)\, du \right] dv.$$

We must prove that

$$\frac{f(x+) + f(x-)}{2} = \lim_{M \to \infty} \frac{1}{\pi} \int_0^M \left[\int_{-\infty}^\infty f(u) \cos v(u - x)\, du \right] dv.$$

Without loss of generality we assume that $M > 1$. Since f is absolutely integrable and

$$|f(u) \cos v(u - x)| \le |f(u)|,$$

the M-test shows that the improper integral in brackets is uniformly convergent on $0 \le v < \infty$. Moreover, $f(u) \cos v(u - x)$ is piecewise continuous with respect to u on $-\infty < u < \infty$, $0 \le v \le M$, so Prop. 5-3 can be used to deduce that

$$\begin{aligned}
\int_0^M \int_{-\infty}^\infty f(u) \cos v(u - x)\, du\, dv &= \int_{-\infty}^\infty \int_0^M f(u) \cos v(u - x)\, dv\, du \\
&= \int_{-\infty}^\infty f(u) \frac{\sin M(u - x)}{u - x}\, du \\
&= \int_{-\infty}^\infty f(u + x) \frac{\sin (Mu)}{u}\, du.
\end{aligned}$$

Therefore, we must establish that

$$\frac{1}{\pi} \int_{-\infty}^\infty f(u + x) \frac{\sin (Mu)}{u}\, du \to \frac{f(x+) + f(x-)}{2} \tag{5-5}$$

as $M \to \infty$. For this purpose, we show that

$$\frac{1}{\pi} \int_0^\infty f(x + u) \frac{\sin (Mu)}{u}\, du \to \frac{f(x+)}{2} \tag{5-6}$$

and

$$\frac{1}{\pi} \int_0^\infty f(x - u) \frac{\sin (Mu)}{u}\, du \to \frac{f(x-)}{2}. \tag{5-7}$$

Indeed, given this, simply change variables from u to $-u$ in (5-7) and add the result to (5-6) to confirm (5-5).

Each of the last pair of limits has a similar verification. We consider only the first one. In view of (5-3),

$$\left|\frac{1}{\pi}\int_0^\infty f(x+u)\frac{\sin(Mu)}{u}\,du - \frac{f(x+)}{2}\right| = \left|\frac{1}{\pi}\int_0^\infty [f(x+u)-f(x+)]\frac{\sin(Mu)}{u}\,du\right|$$

$$\le \frac{1}{\pi}\int_0^\delta \left|\frac{f(x+u)-f(x+)}{u}\right| du + \frac{1}{\pi}\left|\int_\delta^b \frac{f(x+u)-f(x+)}{u}\sin(Mu)\,du\right|$$

$$+ \frac{1}{\pi}\int_b^\infty \left|\frac{f(x+u)}{u}\right| du + \frac{1}{\pi}|f(x+)|\left|\int_b^\infty \frac{\sin(Mu)}{u}\,du\right|$$

$$\le \frac{1}{\pi}\int_0^\delta \left|\frac{f(x+u)-f(x+)}{u}\right| du + \frac{1}{\pi}\left|\int_\delta^b \frac{f(x+u)-f(x+)}{u}\sin(Mu)\,du\right|$$

$$+ \frac{1}{\pi}\frac{1}{b}\int_{-\infty}^\infty |f(u)|\,du + \frac{1}{\pi}|f(x+)|\left|\int_{Mb}^\infty \frac{\sin u}{u}\,du\right|$$

$$= I_1 + I_2 + I_3 + I_4.$$

Let $\varepsilon > 0$. In view of (5-4), we can fix $\delta > 0$ small enough so that $I_1 < \varepsilon/4$. It is also evident (recall $M > 1$) that we can fix $b > \delta$ so that $I_3 < \varepsilon/4$ and $I_4 < \varepsilon/4$. With δ and b so fixed, the Riemann–Lebesgue lemma implies that $I_2 < \varepsilon/4$ as soon as M is sufficiently large. Combine these estimates to find

$$\left|\frac{1}{\pi}\int_0^\infty f(x+u)\frac{\sin(Mu)}{u}\,du - \frac{f(x+)}{2}\right| < \varepsilon$$

for all M sufficiently large. This proves (5-6), and as we noted already, (5-7) follows in the same fashion. Thus, the Fourier integral theorem is established.

PROBLEMS

1. Establish the Weierstrass M-test for improper integrals.
2. Show that $\int_0^\infty e^{-xy}\sin y\,dy$ is uniformly convergent on $[a, \infty)$ for each $a > 0$. *Hint.* Apply the M-test with $M(y) = e^{-ay}$.
3. Prove Th. 4-3 using Prop. 5-2.

4

Wave Propagation in One Spatial Dimension

4-1. General Wave Motion and d'Alembert's Solution

The undamped wave equation

(1-1) $$u_{tt} = c^2 u_{xx}$$

models a wide variety of phenomena. For example, we saw in Chap. 1 that u may be the transverse displacement of a vibrating string, the current or voltage in a long insulated wire with negligible resistance, or the velocity or variation in density of a gas under the linearized gas dynamics assumptions.

We begin our study of wave motion with some general observations about solutions to the undamped wave equation. The linear change of variables (Prob. 4).

$$\alpha = x + ct, \qquad \beta = x - ct$$

transforms (1-1) into the simpler form

(1-1)′ $$u_{\alpha\beta} = 0.$$

This form of the wave equation can be integrated twice to yield $u = A(\alpha) + B(\beta)$, where A and B are arbitrary functions. Thus, (1-1) has general solution

(1-2) $$u(x, t) = A(x + ct) + B(x - ct).$$

A routine check shows that (1-2) solves (1-1) for any choices of A and B, provided that these functions are twice differentiable.

It is important to understand the physical meaning of (1-2). Suppose first that $B \equiv 0$, so $u(x, t) = A(x + ct)$. Figure 4-1 shows how this wave propagates in time. It is easy to check that the graph of $u = A(x + ct)$ is obtained from the graph of $u = A(x)$ by translating each point on the wave form at time zero ct units to the left.

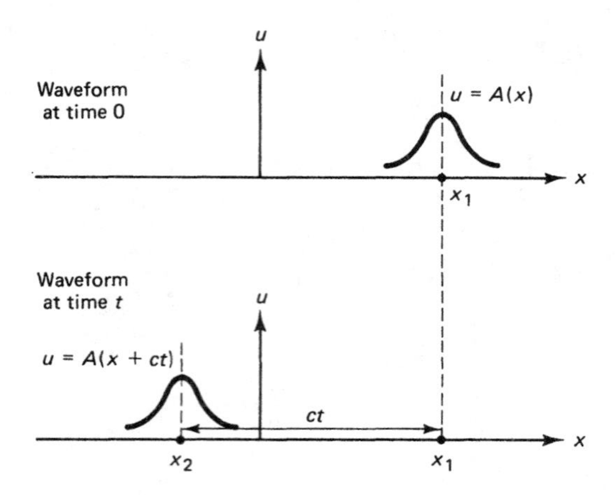

Figure 4-1

For example, the wave crest occurs at x_1 at time $t = 0$. At time t this crest will occur when $x + ct = x_1$ or $x = x_1 - ct$. The crest of the wave is located at $x_2 = x_1 - ct$ at time t, as shown in Figure 4-1, because $u(x_2, t) = A(x_2 + ct) = A(x_1)$. Thus, the solution $u = A(x + ct)$ is a wave traveling to the left with speed c. Similarly, the second term in (1-2) represents a wave that travels to the right at speed c. The general solution to the wave equation (1-1) is a superposition of two waves; one moves to the left, the other to the right, and both propagate at speed c. So c is the speed of wave propagation.

Return to Fig. 4-1. Notice that as time passes, the wave crest which is located at x_1 at time 0 occurs at the location x specified by $x + ct = x_1$ at time t. Also, each feature or special aspect of the wave form propagates along the lines $x + ct =$ constant. Similarly, the features of a wave $B(x - ct)$ propagate along the lines $x - ct =$ constant. The lines $x + ct =$ constant and $x - ct =$ constant are the *characteristics* of the wave equation (1-1).

Consider the vibrations of a very long string at points far from its ends. Since the vibrations propagate at a finite speed, we can ignore the manner in which the ends are supported, at least for some period of time. The simplest model for such wave motion is

$$(1\text{-}3)\qquad \begin{cases} u_{tt} = c^2 u_{xx}, & -\infty < x < \infty, \quad t > 0, \\ u(x, 0) = f(x), \quad u_t(x, 0) = g(x), & -\infty < x < \infty \end{cases}$$

where $f(x)$ and $g(x)$ specify the initial position and velocity. We can solve this initial value problem by adjusting the arbitrary functions in (1-2) to fit the data:

$$u(x, 0) = A(x) + B(x) = f(x), \qquad u_t(x, 0) = cA'(x) - cB'(x) = g(x).$$

The second equation yields

$$A(x) - B(x) = \frac{1}{c}\int_0^x g(s)\,ds + D,$$

where D is a constant of integration. Combine this with the first equation to find

$$A(x) = \frac{1}{2} f(x) + \frac{1}{2c}\int_0^x g(s)\,ds + \frac{D}{2}, \qquad B(x) = \frac{1}{2} f(x) - \frac{1}{2c}\int_0^x g(s)\,ds - \frac{D}{2}.$$

Finally, substitution into (1-2) gives

$$u(x, t) = \frac{1}{2}[f(x + ct) + f(x - ct)] + \frac{1}{2c}\int_{x-ct}^{x+ct} g(s)\,ds, \tag{1-4}$$

which is called *d'Alembert's solution* to the wave equation.

On physical grounds, we expect that (1-3) has a unique solution (the actual motion of the vibrating string), provided that reasonable data are given for the initial position and velocity. Moreover, small changes in these data should lead to small changes in the solution (motion of the string) over any finite time interval. We shall call a problem such a (1-3) *well-posed* if it has the following properties:

(i) If a solution exists, it is unique.
(ii) Under suitable restrictions on the data, there is a solution.
(iii) The solution depends continuously on the data.

Evidently, if (1-3) is a reasonable model for vibrations of a long string, it must be well-posed because experience shows that real strings have properties (i)–(iii). Let's check this. First, our reasoning leading to (1-4) shows that if (1-3) has a solution it must be given by (1-4). That is, (i) holds. Second, it is routine to check that (1-4) does solve (1-3) provided that f is twice differentiable and g is differentiable. So (ii) holds. Finally, let $u_1(x, t)$ and $u_2(x, t)$ be the solutions to (1-3) with f, g replaced, respectively, by f_1, g_1 and f_2, g_2. The data f_1, f_2 and g_1, g_2 will be close if $|f_1(z) - f_2(z)| \leq \varepsilon$, $|g_1(z) - g_2(z)| \leq \varepsilon$ for all z, where $\varepsilon > 0$ is small. Subtract the formulas for u_1 and u_2 given by (1-4) and simplify to obtain

$$|u_1(x, t) - u_2(x, t)| \leq \varepsilon + T\varepsilon = (1 + T)\varepsilon$$

for $-\infty < x < \infty$ and $0 \leq t \leq T$. Thus, u_1 and u_2 can be made as close together as desired over any finite time interval by choosing the initial data for these two solutions close enough. In other word, (iii) holds. In summary, we have

Theorem 1-1. Let $f(x)$ have a second derivative and $g(x)$ a first derivative for $-\infty < x < \infty$. Then the initial value problem (1-3) is well-posed. Its solution is given by (1-4).

The d'Alembert solution (1-4) displays another important feature of wave propagation. Figure 4-2 on page 92 shows a fixed point (x', t') and two characteristics emanating from it. The formula in (1-4) shows that the initial data $f(x)$ and $g(x)$, $-\infty < x < \infty$, only effect the solution (wave disturbance) at (x', t') through their values for x in the interval $[x' - ct', x' + ct']$, which is called the *domain of dependence* for the point (x', t'). This behavior is to be expected because the effects of the initial data propagate at the finite speed c. Thus, the only part of the initial data that can influence the solution at x' at time t' must be within ct' units of x'. This

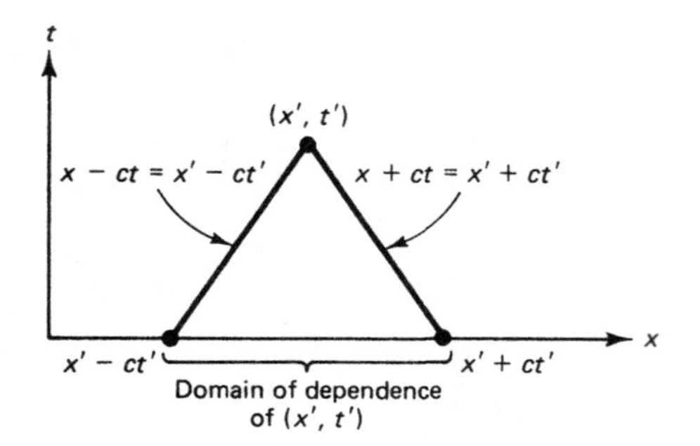

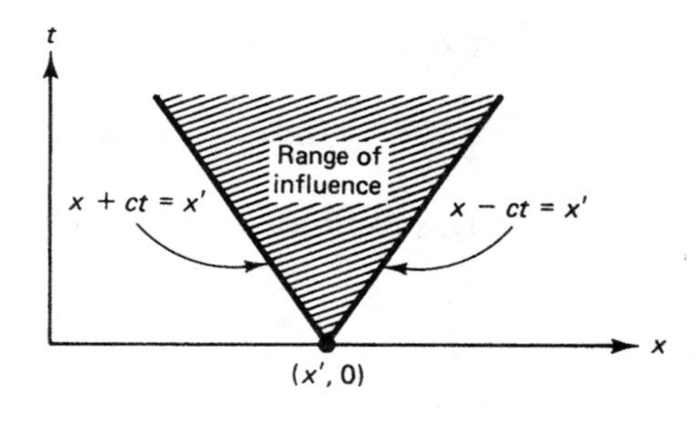

Figure 4-2

is precisely the data given for x in $[x' - ct', x' + ct']$. Similarly, the disturbance that originates at a fixed point x' at time $t = 0$ only affects points in the shaded region in Fig. 4-2, which is called the *range of influence of the point* $(x', 0)$.

In our discussion of the wave equation (1-1), we have implicitly assumed that a solution $u(x, t)$ should have second-order partial derivatives. This forced us to put smoothness restrictions on the initial data $f(x)$ and $g(x)$ in problem (1-3), and on the wave forms $A(\alpha)$ and $B(\beta)$ in (1-2). These are mathematical restrictions, not physical ones. For example, wave motions governed by (1-3) are often originated by triangular initial shapes, $f(x)$. The graph of $f(x)$ has a corner at each vertex of the triangle and so is not differentiable there. Thus, if we insist that $u(x, t)$ be twice differentiable, (1-4) is not a solution to (1-3). On the other hand, (1-4) does exhibit wavelike behavior and satisfies the initial data. So it should be the solution, properly interpreted. We need to extend our concept of a solution for the wave equation to include such nonsmooth initial data. This is done in Chap. 11. These generalized solutions to the wave equation will be limits of the smooth (or classical) solutions we have just considered.

PROBLEMS

1. Find the solution to (1-3) when $f(x) = x^2(1 - x)^2$ for $0 \le x \le 1$, $f(x) = 0$ elsewhere, and $g(x) \equiv 0$. Sketch the solution for $t = 0, 1, 2$. For simplicity take $c = 1$.
2. Repeat Prob. 1 for $f(x) \equiv 0$ and $g(x) = \sin x$.
3. Repeat Prob. 1 for $f(x) = x, 0 \le x \le 1$, $f(x) = 2 - x$, $1 \le x \le 2$, $f(x) = 0$ elsewhere, and $g(x) \equiv 0$.
4. Reduce (1-1) to canonical form to obtain (1-1)$'$.
5. Check that (1-2) solves (1-1) for A, B twice differentiable, and that (1-4) solves (1-3) for f twice differentiable and g differentiable.
6. Show that the solution u to

$$u_{tt} = c^2 u_{xx} + F(x, t), \qquad -\infty < x < \infty, \qquad t > 0,$$
$$u(x, 0) = f(x), \qquad u_t(x, 0) = g(x), \qquad -\infty < x < \infty$$

can be expressed as $u = v + w$, where v and w solve

$$v_{tt} = c^2 v_{xx} + F(x, t), \qquad -\infty < x < \infty, \qquad t > 0,$$
$$v(x, 0) = 0, \qquad v_t(x, 0) = 0, \qquad -\infty < x < \infty$$

and

$$w_{tt} = c^2 w_{xx}, \qquad -\infty < x < \infty, \qquad t > 0,$$
$$w(x, 0) = f(x), \qquad w_t(x, 0) = g(x), \qquad -\infty < x < \infty.$$

You can find w from (1-4). The next problem shows how to find v.

7. Consider the inhomogeneous wave equation $u_{tt} = c^2 u_{xx} + F(x, t)$ for $-\infty < x, t < \infty$ with $F(x, t) = 0$ for $t < 0$.

(a) Introduce the characteristic variables $\alpha = x + ct$, $\beta = x - ct$ and show that the differential equation becomes

$$u_{\alpha\beta} = \frac{-1}{4c^2} F\left(\frac{\alpha + \beta}{2}, \frac{\alpha - \beta}{2c}\right) \quad \text{for} \quad -\infty < \alpha, \beta < \infty.$$

(b) Fix a point (x', t') and let (α', β') be its image under the characteristic change of variables. Show that

$$u(\alpha', \beta') = \frac{1}{4c^2} \int_{\beta'}^{\infty} \int_{-\infty}^{\alpha'} F\left(\frac{\alpha + \beta}{2}, \frac{\alpha - \beta}{2c}\right) d\alpha \, d\beta$$

satisfies the transformed differential equation in part (a), assuming that F is suitably integrable.

(c) Make the change of variables $x = (\alpha + \beta)/2$, $t = (\alpha - \beta)/2c$ in the double integral in part (b) to find

$$u(x', t') = \frac{1}{2c} \int_0^{t'} \int_{x' - c(t' - t)}^{x' + c(t' - t)} F(x, t) \, dx \, dt.$$

Hint. Check that the corresponding regions of integration are as shown in Fig. 4-3 and that the Jacobian $\partial(\alpha, \beta)/\partial(x, t) = 2c$.

(d) You know that $u(x', t')$ in part (c) satisfies the inhomogeneous wave equation. Check also that $u(x', 0) = 0$ and $u_t(x', 0) = 0$. Thus, $u(x', t')$ solves the v-initial value problem

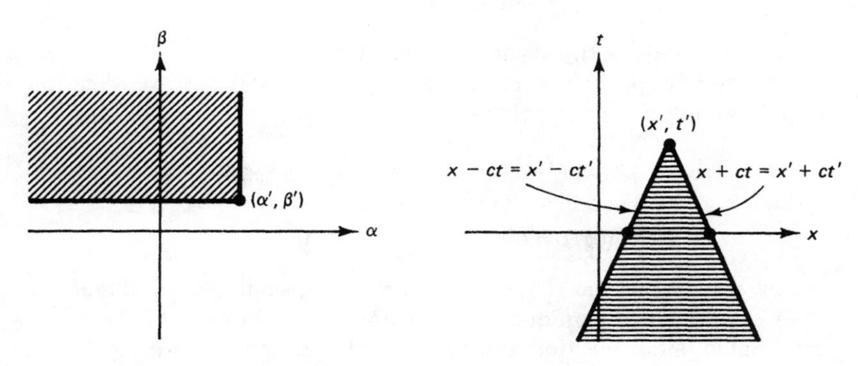

Figure 4-3

in Prob. 6. Observe that

$$u(x', t') = \frac{1}{2c} \iint_{T_{x't'}} F(x, t)\, dA,$$

where $T_{x't'}$ is the triangle shown in Fig. 4-2.

8. Use Probs. 6 and 7 to find d'Alembert's solution to the inhomogeneous initial value problem

$$\begin{aligned} &u_{tt} = c^2 u_{xx} + F(x, t), \qquad -\infty < x < \infty, \qquad t > 0, \\ &u(x, 0) = f(x), \qquad u_t(x, 0) = g(x), \qquad -\infty < x < \infty. \end{aligned}$$

Check that this problem is well-posed.

4-2. Homogeneous Initial, Boundary Value Problems

In Sec. 1-2 we derived the wave equation

$$u_{tt} = c^2 u_{xx} \tag{2-1}$$

for the displacement $u(x, t)$ of an elastic string undergoing "small" oscillations.

The motion of such a string clearly depends on more than the wave equation (2-1). We must also know the initial state of the string. In our derivation of the wave equation, we regarded the string as composed of tiny mass chunks and applied Newton's laws to each chunk. Thus, we need to specify the initial position and velocity for each chunk of mass. This is done by setting

$$\begin{aligned} u(x, 0) &= f(x), \qquad 0 \le x \le L, \\ u_t(x, 0) &= g(x), \qquad 0 \le x \le L, \end{aligned}$$

where $f(x)$ [resp., $g(x)$] is the initial position (resp., velocity profile) of the string whose length is L. Of course, for a violin or guitar string we also have the boundary conditions,

$$u(0, t) = u(L, t) = 0, \qquad t \ge 0,$$

because the ends of the string are pinned down.

In summary, we expect the motion of the string to be determined by the initial, boundary value problem

$$\begin{cases} u_{tt} = c^2 u_{xx}, \qquad 0 < x < L, \qquad t > 0, \\ u(x, 0) = f(x), \qquad u_t(x, 0) = g(x), \qquad 0 \le x \le L, \\ u(0, t) = u(L, t) = 0, \qquad t \ge 0 \end{cases} \tag{2-2}$$

for the wave equation. The physical reasoning leading to (2-2) makes it clear that (2-2) should have a unique solution (the actual motion of the string) provided reasonable initial position and velocity data are given. Moreover, small changes in these data should lead to small changes in the solution (motion of the string). That is, (2-2) must be well-posed.

As a first step toward confirming that (2-2) is well-posed, we will show that (2-2) has a solution for relatively smooth initial data f and g. We will discuss continuous dependence on the data and uniqueness in the next section.

In order to show that (2-2) does have a solution, we will construct it explicitly, using the method of *separation of variables.* Physically, this amounts to expressing the possibly very complicated motion of the string as a superposition of simple motions, called standing waves. A typical particle of the string, located at $(x, 0)$ when in equilibrium, moves vertically with an amplitude $X(x)$ depending on x and oscillates periodically in time. Let this periodic motion in time be given by $T(t)$. This suggests that such standing waves can be expressed in the form

$$u(x, t) = X(x)T(t). \tag{2-3}$$

In order for (2-3) to satisfy the wave equation, X and T must satisfy

$$X(x)T''(t) = c^2X''(x)T(t)$$

or

$$\frac{X''(x)}{X(x)} = \frac{T''(t)}{c^2T(t)}. \tag{2-4}$$

Imagine t fixed, say $t = 0$. Then the right side of (2-4) is a constant and hence $X''(x)/X(x)$ is constant as x varies from 0 to L. But this means that the left side of (2-4) is constant and hence the right side is, too. Let K be the constant value of the space and time ratios in (2-4). Then

$$\frac{X''(x)}{X(x)} = K \quad \text{and} \quad \frac{T''(x)}{c^2T(x)} = K$$

or

$$X'' - KX = 0 \quad \text{and} \quad T'' - c^2KT = 0. \tag{2-5}$$

We focus first on the differential equation for the displacements $X(x)$ along a standing wave. The pinned ends (boundary conditions) require that (2-3) vanish at $x = 0$ and $x = L$ for all t,

$$u(0, t) = X(0)T(t) = 0 \quad \text{and} \quad u(L, t) = X(L)T(t) = 0.$$

If $X(0) \neq 0$, the first condition gives $T(t) = 0$ for all t and (2-3) reduces to the zero solution to the wave equation. We cannot hope to construct the solution to (2-2) from these zero solutions (except in the trivial case of no motion, $f = g = 0$). Thus, we require that $X(0) = 0$. Of course, this simply means the standing wave has zero amplitude at its pinned left end. Also, we require that $X(L) = 0$. Therefore, the amplitudes along a standing wave must satisfy

$$\begin{cases} X''(x) - KX(x) = 0, & 0 \le x \le L, \\ X(0) = X(L) = 0. \end{cases} \tag{2-6}$$

No matter what choice we make for the constant K, the boundary value problem (2-6) has the so-called *trivial solution* $X(x) = 0$ for all x. This trivial solution leads to $u(x, t) = 0$ for all x and t in (2-3). Only nontrivial solutions to (2-6) will be useful for constructing solutions to (2-2). As we shall confirm momentarily, (2-6) has nontrivial solutions only for special choices of the constant K. These special values of K are called *eigenvalues* and the nontrivial solutions of (2-6) are called (corresponding) *eigenfunctions*. In this context, (2-6) is called an *eigenvalue problem*.

To gain some insight into the possible eigenvalues of (2-6), multiply the differential equation by X and integrate the first term by parts to obtain

$$\int_0^L [X'(x)^2 + KX(x)^2]\, dx = 0.$$

If $K \geq 0$ this equation implies that $X'(x) \equiv 0$. So $X(x)$ is a constant which must be zero because $X(0) = 0$. This shows that (2-6) can only have nontrivial solutions for $K < 0$. Set $K = -\lambda^2$, $\lambda > 0$. Then X must satisfy

$$X''(x) + \lambda^2 X(x) = 0, \qquad X(0) = X(L) = 0.$$

This differential equation has the general solution

$$X(x) = A \cos(\lambda x) + B \sin(\lambda x).$$

Since we require $X(0) = 0$, A must be zero and

$$X(x) = B \sin(\lambda x).$$

At $x = L$ we must have

$$X(L) = B \sin(\lambda L) = 0.$$

If $B = 0$, $X(x) \equiv 0$, but we can obtain a nontrivial solution $X(x)$ if we require instead that

$$\sin(\lambda L) = 0, \qquad \lambda L = n\pi$$

for $n = 1, 2, \ldots$.

In summary, (2-6) has nontrivial solutions exactly when $K = K_n = -\lambda_n^2$ where $\lambda_n = n\pi/L$ and $n = 1, 2, \ldots$. The corresponding nontrivial solutions are

$$X(x) = X_n(x) = B_n \sin(\lambda_n x) \quad \text{for } B_n \neq 0 \text{ any constant.}$$

In other words, the eigenvalues of (2-6) are $-\lambda_n^2$ and the corresponding eigenfunctions are the nonzero multiples of $\sin(\lambda_n x)$.

Insert $K_n = -\lambda_n^2$ into the equation for T in (2-5) and integrate to find

$$T(t) = T_n(t) = C_n \cos(\lambda_n ct) + D_n \sin(\lambda_n ct)$$

for arbitrary constants C_n and D_n. Thus, by (2-3), standing waves have the form

$$u_n(x, t) = X_n(x)T_n(t), \qquad n = 1, 2, \ldots$$

or

$$u_n(x, t) = [a_n \cos(\lambda_n ct) + b_n \sin(\lambda_n ct)] \sin \lambda_n x \tag{2-7}$$

where $\lambda_n = n\pi/L$ and $n = 1, 2, \ldots$. The constants $a_n = B_n C_n$ and $b_n = B_n D_n$ are arbitrary because B_n, C_n, and D_n are.

Each standing wave $u_n(x, t)$ satisfies the wave equation and boundary conditions in (2-2). Since the wave equation and boundary conditions are linear and homogeneous,

$$u(x, t) = \sum_{n=1}^{N} u_n(x, t)$$

also satisfies the wave equation and boundary conditions. What about the initial conditions? We want $u(x, 0) = f(x)$ and $u_t(x, 0) = g(x)$. We get

$$u(x, 0) = \sum_{n=1}^{N} a_n \sin\left(\frac{n\pi x}{L}\right), \qquad 0 \le x \le L,$$

$$u_t(x, 0) = \sum_{n=1}^{N} b_n \lambda_n c \sin\left(\frac{n\pi x}{L}\right), \qquad 0 \le x \le L,$$

from (2-7). Of course, these sums are probably not equal to $f(x)$ and $g(x)$, respectively, but if we allow infinite sums ($N = \infty$) our requirements become

$$\text{(2-8)} \qquad f(x) = \sum_{n=1}^{\infty} a_n \sin\left(\frac{n\pi x}{L}\right) \quad \text{and} \quad g(x) = \sum_{n=1}^{\infty} b_n \lambda_n c \sin\left(\frac{n\pi x}{L}\right)$$

for $0 \le x \le L$. Of course, this reminds us of the Fourier sine series expansions for $f(x)$ and $g(x)$. We would expect (2-8) to hold (for reasonable f and g) if we choose a_n and $b_n \lambda_n c$ to be the Fourier sine coefficients for the functions $f(x)$ and $g(x)$, respectively,

$$\text{(2-9)} \qquad a_n = \frac{2}{L} \int_0^L f(x) \sin\left(\frac{n\pi x}{L}\right) dx,$$

$$\text{(2-10)} \qquad b_n = \frac{2}{\lambda_n c L} \int_0^L g(x) \sin\left(\frac{n\pi x}{L}\right) dx,$$

where $\lambda_n = n\pi/L$.

This line of reasoning suggests that (2-2) has the solution

$$\text{(2-11)} \qquad u(x, t) = \sum_{n=1}^{\infty} u_n(x, t)$$

with a_n and b_n given by (2-9) and (2-10). At this point we refer to (2-11) as a *formal solution* to the initial, boundary value problem (2-2). We say formal because (2-11) was arrived at by formal calculations which seem reasonable but have not been fully justified from the mathematical point of view. It is clear that the series (2-11) satisfies the boundary conditions $u(0, t) = u(L, t) = 0$ because each $u_n(x, t)$ does. We do not know, however, that (2-11) satisfies the wave equation. To verify this from the fact that the standing waves $u_n(x, t)$ satisfy the wave equation, we must be able to differentiate (2-11) term by term twice with respect to both x and t. Term-by-term differentiation is not always possible, and so must be justified here to deduce that (2-11) satisfies the wave equation. Finally, to be certain that

(2-11) satisfies the initial conditions, we must guarantee that $f(x)$ and $g(x)$ have Fourier series expansions as in (2-8).

Let us show that (2-11) really does satisfy the wave equation by justifying the required term-by-term differentiations of the infinite series. Differentiating (2-11) term by term gives, provisionally,

$$u_x(x, t) = \sum_{n=1}^{\infty} [a_n\lambda_n \cos(\lambda_n ct) + b_n\lambda_n \sin(\lambda_n ct)] \cos(\lambda_n x),$$

$$u_{xx}(x, t) = \sum_{n=1}^{\infty} [-a_n\lambda_n^2 \cos(\lambda_n ct) - b_n\lambda_n^2 \sin(\lambda_n ct)] \sin(\lambda_n x).$$

By Prop. 2-3 of Chap. 3 these term-by-term differentiations are valid provided that the series obtained are uniformly convergent. For the series at hand, this uniform convergence follows from the M-test if we can find a constant C such that

$$|\lambda_n^2 a_n| \le \frac{C}{n^2} \quad \text{and} \quad |\lambda_n^2 b_n| \le \frac{C}{n^2}. \tag{2-12}$$

Calculation of u_t and u_{tt} term-by-term lead to similar series (Prob. 6) which are uniformly convergent if (2-12) holds. The estimates in (2-12) also justify the sine series expansions of $f(x)$ and $g(x)$ (see Prob. 8). Thus (2-11) with a_n and b_n given by (2-9) and (2-10) will solve (2-2) provided that (2-12) holds.

Finally, we determine conditions on the initial data $f(x)$ and $g(x)$ so that (2-12) holds. From the first initial condition in (2-2) we have $u(0, 0) = f(0)$. We also have $u(0, 0) = 0$ from the boundary data. So $f(0) = 0$ is a required compatibility condition. Similarly, $f(L) = 0$. Use this and integration by parts in (2-9) to find

$$a_n = \frac{2}{L\lambda_n} \int_0^L f'(x) \cos(\lambda_n x)\, dx.$$

Integrate by parts three more times and assume that $f''(0) = f''(L) = 0$ to obtain

$$a_n = \frac{2}{L\lambda_n^4} \int_0^L f^{(4)}(x) \sin(\lambda_n x)\, dx.$$

Since $\lambda_n = n\pi/L$ this leads to the estimate

$$|\lambda_n^2 a_n| \le \frac{C_1}{n^2} \tag{2-13}$$

for some constant C_1. In fact, we can take

$$C_1 = \frac{2L}{\pi^2} \int_0^L |f^{(4)}(x)|\, dx.$$

Similarly, we obtain (Prob. 7)

$$|\lambda_n^2 b_n| \le \frac{C_2}{n^2}, \tag{2-14}$$

where

$$C_2 = \frac{2L}{c\pi^2}\left[|g''(0)| + |g''(L)| + \int_0^L |g'''(x)|\,dx\right],$$

provided that $g(0) = g(L) = 0$, which expresses the natural requirement of zero velocity at the pinned ends. Therefore, (2-12) holds with C the maximum of C_1 and C_2, and we have proved Th. 2-1:

Theorem 2-1. Let $f(x)$ have a continuous fourth derivative on $0 \le x \le L$, $g(x)$ have a continuous third derivative there, $f(0) = f(L) = f''(0) = f''(L) = 0$, and $g(0) = g(L) = 0$. Then the initial boundary value problem (2-2) for the wave equation has a solution given by (2-9)–(2-11).

Our discussion of problem (2-2) has forced us to place heavier restrictions on the initial data than we would really like. They were forced by the mathematics used, not by the physics. From a physical standpoint, we need to assume the string is continuous and has fixed ends. So we would like to assume only that

$$f(x) \quad \text{is continuous on} \quad 0 \le x \le L \quad \text{and} \quad f(0) = f(L) = 0.$$

Although we may need to make some demands on the smoothness of f, we certainly do not want to assume differentiability at every point because we would like to study a string plucked with a rather sharp object.

For the initial velocity profile g, we would even like to allow discontinuities in order to model the vibrations of a string set into motion by the strike of a rigid hammer. Since the ends are pinned, they will not be set in motion. Thus, we would like to assume that, say,

$$g(x) \quad \text{is bounded on} \quad 0 \le x \le L \quad \text{and} \quad g(0) = g(L) = 0.$$

Under such weak assumptions on f and g one cannot find a solution $u(x, t)$ of the type we have been considering where u and its first- and second-order partials exist and are continuous. As in Sec. 4-1, we must extend the concept of a solution to the wave equation to include such realistic initial data. This will also be done in Chap. 11.

PROBLEMS

1. Solve formally by separation of variables

$$u_{tt} = c^2 u_{xx}, \qquad 0 < x < L, \qquad t > 0,$$
$$u(0, t) = u_x(L, t) = 0, \qquad t \ge 0,$$
$$u(x, 0) = f(x), \qquad u_t(x, 0) = g(x), \qquad 0 \le x \le L.$$

2. Solve formally by separation of variables

$$u_{tt} = c^2 u_{xx}, \qquad 0 < x < L, \qquad t > 0,$$
$$u_x(0, t) = u_x(L, t) = 0, \qquad t \ge 0,$$
$$u(x, 0) = f(x), \qquad u_t(x, 0) = g(x), \qquad 0 \le x \le L$$

3. For a string with damping, the initial, boundary value problem is

$$u_{tt} + ku_t = c^2 u_{xx}, \qquad 0 < x < L, \quad t > 0,$$
$$u(x, 0) = f(x), \qquad u_t(x, 0) = g(x), \qquad 0 \le x \le L,$$
$$u(0, t) = u(L, t) = 0, \qquad t \ge 0,$$

where $c^2 = \tau/\rho_0, k > 0$ is the damping coefficient, τ is the horizontal component of tension, and ρ_0 is the constant density of the string while in equilibrium. Find a formal solution.

4. Consider the telegrapher's system

$$\begin{cases} i_x + Cv_t + Gv = 0, \\ v_x + Li_t + Ri = 0, \end{cases} \qquad 0 < x < L, \qquad t > 0,$$
$$i(x, 0) = i_0(x), \qquad v(x, 0) = v_0(x), \qquad 0 \le x \le L,$$
$$i(0, t) = i(L, t) = 0, \qquad t \ge 0,$$

with C, G, L, and R constants. Find a formal solution. *Hint.* Let $i(x, t) = I(x)S(t)$ and $v(x, t) = V(x)T(t)$.

5. Consider a typical term $u_n(x, t)$ given by (2-7) of our series solution (2-11). Show that for each x, the particle on the string, located at x in equilibrium, vibrates with amplitude $\sqrt{a_n^2 + b_n^2}\sin(\lambda_n x)$ and moves periodically in time with period $T_n = 2\pi/c\lambda_n$ and frequency

$$\omega_n = \frac{1}{T_n} = \frac{n}{2L}\sqrt{\frac{\tau}{\rho}}.$$

Observe that the frequency of each standing wave can be increased by tightening the string, lowering its density, by shortening it, or by any combination of these. These facts are exploited in tuning musical instruments. Note, too, that $\omega_n = n\omega_1$. All frequencies are integer multiples of the fundamental frequency ω_1. A *node* of a vibrating string is a point that remains in the equilibrium position throughout the motion. What are the nodes of $u_n(x, t)$?

6. Calculate u_t and u_{tt} by differentiating (2-11) term by term. Check that these differentiations are legitimate provided that (2-12) holds.

7. Verify (2-14) with C_2 the constant given in the text.

8. Assume that the compatibility conditions $f(0) = f(L) = 0$ and $g(0) = g(L) = 0$ hold and that $f(x)$ and $g(x)$ have continuous second derivatives. Show that

$$|a_n| \le \frac{C}{n^2} \quad \text{and} \quad |b_n\lambda_n c| \le \frac{C}{n^2}$$

for a constant C. Then use an appropriate result from Chap. 3 to conclude that (the odd, $2L$ periodic extensions) of $f(x)$ and $g(x)$ have Fourier series expansions. This establishes (2-8).

9. Determine appropriate conditions on $f(x)$ and $g(x)$ so that the formal solution to Prob. 1 is assured to be a solution.

10. Determine appropriate conditions on $f(x)$ and $g(x)$ so that the formal solution to Prob. 3 is, in fact, a solution.

11. Find a formal solution to

$$\begin{aligned} &u_{tt} = c^2 u_{xx}, && 0 < x < L, && t > 0, \\ &u(0, t) = 0, && u(L, t) + u_x(L, t) = 0, && t \geq 0, \\ &u(x, 0) = f(x), && u_t(x, 0) = 0, && 0 \leq x \leq L. \end{aligned}$$

(a) First show that (2-6) is replaced by

$$X'' - KX = 0, \qquad X(0) = 0, \qquad X(L) + X'(L) = 0.$$

(b) Argue that $K = -\lambda^2$ for $\lambda > 0$
(c) Show that λ must satisfy $\tan(\lambda L) = -\lambda$ to give a nontrivial eigenfunction $X(x)$.
(d) Graph $y = \tan z$ and $y = -z/L$, where $z = \lambda L$, on the same graph to confirm that there is an infinite set of values $z_n > 0$ satisfying $\tan z = -z/L$. The eigenvalues are $-\lambda_n^2 = -(z_n/L)^2$.
(e) Approximate $z_1, \ldots, z_4$ to four places by Newton's method.
(f) Show that $z_n \approx (2n - 1)\pi/2$ for n large (like four for reasonable L).

4-3. Inhomogeneous Initial, Boundary Value Problems

The inhomogeneous wave equation

$$u_{tt} + ku_t = c^2 u_{xx} + F(x, t) \tag{3-1}$$

describes the vibrations of a damped string, with damping constant k, and $F(x, t)$ an external force (density) which drives the string. Recall from Sec. 1-2 that $c^2 = \tau/\rho_0$ where τ is the constant horizontal component of tension and ρ_0 is the constant linear mass density of the string in its equilibrium position. As in Sec. 4-2, we shall restrict our attention to solutions of (3-1) with continuous second-order partials.

An appropriate initial, boundary value problem for the damped wave equation is

$$\begin{cases} u_{tt} + ku_t = c^2 u_{xx} + F(x, t), & 0 < x < L, \quad t > 0, \\ u(x, 0) = f(x), & u_t(x, 0) = g(x), \quad 0 \leq x \leq L, \\ u(0, t) = r(t), & u(L, t) = s(t), \quad t \geq 0. \end{cases} \tag{3-2}$$

Evidently, (3-2) models the vibrations of a damped string with initial position $f(x)$, velocity profile $g(x)$, and with the location of its ends specified for all t by the functions $r(t)$ and $s(t)$. The initial and boundary conditions force the following compatibility conditions on the data:

$$f(0) = r(0), \qquad f(L) = s(0), \qquad g(0) = r'(0), \qquad g(L) = s'(0). \tag{3-3}$$

On physical grounds, we expect (3-2) to have a unique solution that varies continuously with the data F, f, g, r, and s. That is, we expect that problem (3-2) is well-posed. Of course this is the case, given reasonable assumptions on the data. We will not take the time here to prove in full detail that (3-2) is well-posed. However, we will do this for the special case of (3-2) treated in the preceding section where $F = r = s = 0$.

Some illuminating energy considerations show that (3-2) has at most one solution. Indeed, suppose that $u_1(x, t)$ and $u_2(x, t)$ both solve (3-2). Then their difference $u = u_1 - u_2$ satisfies the homogeneous, initial, boundary value problem

$$\text{(3-4)} \qquad \begin{cases} u_{tt} + ku_t = c^2 u_{xx}, & 0 < x < L, \quad t > 0, \\ u(x, 0) = 0, \quad u_t(x, 0) = 0, & 0 \le x \le L, \\ u(0, t) = 0, \quad u(L, t) = 0, & t \ge 0, \end{cases}$$

corresponding to (3-2). Evidently, $u \equiv 0$ solves (3-4). If we can show that this trivial solution is the only solution to (3-4), then $u_1 - u_2 \equiv 0$ or $u_1 \equiv u_2$, which shows that (3-2) has at most one solution. Suppose, then, that u solves (3-4) and express the damped wave equation in (3-4) as $\rho_0 u_{tt} + k\rho_0 u_t = \tau u_{xx}$. Multiply this equation by u_t and integrate to get

$$\int_0^L (\rho_0 u_{tt} u_t + k\rho_0 u_t^2)\, dx = \int_0^L \tau u_{xx} u_t \, dx.$$

Integrate by parts on the right and use the fact that u vanishes at $x = 0$ and $x = L$ for all t to obtain

$$\int_0^L \left(\frac{\rho_0}{2} \frac{\partial}{\partial t} u_t^2 + k\rho_0 u_t^2 \right) dx = -\int_0^L \frac{\tau}{2} \frac{\partial}{\partial t} u_x^2 \, dx$$

or

$$\text{(3-5)} \qquad \frac{d}{dt} \left(\int_0^L \frac{\rho_0}{2} u_t^2 \, dx + \int_0^L \frac{\tau}{2} u_x^2 \, dx \right) = -\int_0^L k\rho u_t^2 \, dx.$$

The first integral in (3-5) is the total kinetic energy of the string and the second is its stored or potential energy. Consequently, (3-5) says that the rate of change $E'(t)$ of the total energy $E(t)$ of the string is negative: $E'(t) \le 0$. Thus, total energy is decreasing,

$$\text{(3-6)} \qquad E(t) = \frac{1}{2} \int_0^L [\rho_0 u_t^2(x, t) + \tau u_x^2(x, t)]\, dx \le E(0) = 0$$

because $u_t(x, 0) = 0$ and $u_x(x, 0) = 0$ from the initial conditions in (3-4). It is clear from the integral in (3-6) that $E(t) \ge 0$, so we infer from (3-6) that $E(t) = 0$ for all $t \ge 0$. Since the integrand for the energy is nonnegative and continuous, $E(t) \equiv 0$ implies that $\rho_0 u_t^2 + \tau u_x^2 = 0$ and hence $u_t = 0$ and $u_x = 0$ for all $0 \le x \le L$, $t \ge 0$. That is, u is independent of x and t and hence is constant. However, when $t = 0$ the initial condition in (3-4) shows that $u(x, 0) = 0$, so the constant value of $u(x, t)$ is zero. We have shown that (3-4) has only the trivial solution and, as noted above, this proves uniqueness for (3-2).

Theorem 3-1. There exists at most one solution to the inhomogeneous, initial, boundary value problem (3-2).

Applying this result when $F = r = s = 0$, we see that our series solution (2-11) to (2-2) is the only solution to this problem. Thus, in order to confirm that the homogeneous, initial, boundary value problem (2-2) is well-posed, we need only prove that the solution (2-11) depends continuously on the initial data $f(x)$ and $g(x)$. We shall return to this point toward the end of this section.

For the moment, we return to the inhomogeneous wave equation. Separation of variables also can be used to solve such inhomogeneous differential equations. To see what is involved consider the motion of an undamped string initially at rest, with pinned ends, and driven by an external force $F(x, t)$:

$$\text{(3-7)} \qquad \begin{cases} u_{tt} - c^2 u_{xx} = F(x, t), & 0 < x < L, \quad t > 0, \\ u(x, 0) = 0, \quad u_t(x, 0) = 0, & 0 \le x \le L, \\ u(0, t) = 0, \quad u(L, t) = 0, & t \ge 0, \end{cases}$$

We seek a series solution to (3-7) similar to the series solution (2-11) to (2-2). Since the wave equation is now inhomogeneous, this suggests that we try variation of parameters. A glance at (2-7) and (2-11) reveals that the constants a_n, b_n occur in the time-dependent term of the solution. Replacing a_n and b_n by arbitrary functions of time, as variation of parameters suggests, amounts to replacing $u_n(x, t)$ by $u_n(t) \sin(\lambda_n x)$, where $u_n(t)$ is an arbitrary function of time. Thus, we shall seek a solution to (3-7) of the form

$$\text{(3-8)} \qquad u(x, t) = \sum_{n=1}^{\infty} u_n(t) \sin(\lambda_n x),$$

In order to match coefficients, and so determine the $u_n(t)$, after substituting into the inhomogeneous wave equation, we assume that $F(x, t)$ has a series expansion

$$\text{(3-9)} \qquad F(x, t) = \sum_{n=1}^{\infty} F_n(t) \sin(\lambda_n x), \qquad 0 \le x \le L.$$

A moment's reflection reveals that we are assuming that for each fixed t, $F(x, t)$ has a Fourier sine series expansion with

$$F_n(t) = \frac{2}{L} \int_0^L F(x, t) \sin(\lambda_n x)\, dx.$$

We assume that term-by-term differentiation is possible in (3-8), so that substitution into the inhomogeneous wave equation gives

$$\text{(3-10)} \qquad \sum_{n=1}^{\infty} [u_n''(t) + c^2 \lambda_n^2 u_n(t) - F_n(t)] \sin(\lambda_n x) = 0,$$

which will surely hold if we require that

$$\text{(3-11)} \qquad u_n''(t) + c^2 \lambda_n^2 u_n(t) = F_n(t).$$

Similarly, substituting (3-8) into the initial conditions leads to

$$\text{(3-12)} \qquad u_n(0) = u_n'(0) = 0.$$

The solution of the initial value problem (3-11)–(3-12) is

$$\text{(3-13)} \qquad u_n(t) = \frac{1}{c\lambda_n} \int_0^t F_n(\tau) \sin[c\lambda_n(t - \tau)]\, d\tau.$$

The function $u(x, t)$ defined by (3-8) with $u_n(t)$ given by (3-13) and $F_n(t)$ the Fourier coefficient given above yields a formal solution to (3-7). Of course, this formal solution is a solution given reasonable assumptions on the driving force $F(x, t)$. These mathematical points are explored in the problems. They are resolved much as for the initial, boundary value problem in (2-2). Here is the result.

> Theorem 3-2. Let $F(x, t)$ have continuous third-order partial derivatives for $0 \leq x \leq L$ and $t \geq 0$. Assume that $F(0, t) = F(L, t) = F_{xx}(0, t) = F_{xx}(L, t) = 0$ for all $t \geq 0$. Then the inhomogeneous, initial, boundary value problem (3-7) has a unique solution given by (3-8) and (3-13), where $\lambda_n = n\pi/L$.

In the problems we indicate how the solutions to problems of the type (2-2) and (3-7) can be used to construct the solution to the fully inhomogeneous problem (3-2).

We close this section by completing our discussion that (2-2) is well-posed. If $u_i(x, t)$ is the solution to (2-2) with f, g replaced by f_i and g_i for $i = 1, 2$, we must confirm that $u_1 - u_2$ can be made as small as desired by choosing f_1 "close" enough to f_2 and g_1 "close" enough to g_2. We assume that f_i and g_i satisfy the hypotheses of Th. 2-1 so we can be sure that the solutions u_i exist. Your first reaction is probably that f_1 "close" to f_2 should mean that $f_1(x) - f_2(x)$ is small for all x between 0 and L, and similarly for $g_1(x) - g_2(x)$. However, this idea of "close" simply is not good enough for this situation. We must require that f_1 be "close" to f_2 in the sense that $f_1(x) - f_2(x)$, $f'_1(x) - f'_2(x)$, and $f''_1(x) - f''_2(x)$ are simultaneously small. Correspondingly, we must require that $g_1(x) - g_2(x)$ and $g'_1(x) - g'_2(x)$ are both small for g_1 and g_2 to be "close."

In order to express these new and important ideas of closeness more precisely, we introduce some standard notation. Given a continuous function $h(x)$ defined for $0 \leq x \leq L$, we call

$$\|h\| = \max_{0 \leq x \leq L} |h(x)|$$

the (*maximum*) *norm* of h. In particular, if $h = f_1 - f_2$,

$$\|f_1 - f_2\| = \max_{0 \leq x \leq L} |f_1(x) - f_2(x)|,$$

and we regard $\|f_1 - f_2\|$ as a measure of the "distance" between the functions f_1 and f_2. Notice that $\|f_1 - f_2\|$ small means that the graphs of f_1 and f_2 are close together for all x between 0 and L.

In the context of our initial, boundary value problem (2-2) we shall call f_1 and f_2 "close" if

$$\|f_1 - f_2\|_2 = \max \{\|f_1 - f_2\|, \|f'_1 - f'_2\|, \|f''_1 - f''_2\|\}$$

is small. Geometrically, we regard f_1 as close to f_2 if the graphs of f_1 and f_2 are close at each point x in $[0, L]$, the slopes of these graphs are close at each x, and the curvatures of these graphs are close at each x. We regard g_1 as close to g_2 if

$$\|g_1 - g_2\|_1 = \max \{\|g_1 - g_2\|, \|g'_1 - g'_2\|\}$$

is small. Estimates very much like those leading to (2-13) and (2-14) yield

(3-14) $$|u_1(x, t) - u_2(x, t)| \le D \max \{\|f_1 - f_2\|_2, \|g_1 - g_2\|_1\}$$

where D is a constant (see Prob. 11). Thus, $u_1(x, t) - u_2(x, t)$ can be made as small as desired by choosing f_1 close enough to f_2 and g_1 close enough to g_2 in the sense just described. To sum up, we have established

Theorem 3-3. Let the initial data f and g in the initial, boundary value problem (2-2) satisfy the hypotheses of Th. 2-1. Then this problem is well-posed.

PROBLEMS

1. Consider the undamped, initial, boundary value problem (2-2). It is clear physically that energy is conserved. Establish this by showing that

$$E(t) = \frac{1}{2}\int_0^L [\rho_0 u_t^2 + \tau u_x^2]\, dx = E(0) = \frac{1}{2}\int_0^L [\rho_0 g(x)^2 + \tau f'(x)^2]\, dx.$$

2. Consider the damped, initial, boundary value problem (3-2) with pinned ends so $r = s = 0$. Show that a solution u satisfies the equation

$$\frac{d}{dt}\int_0^L \left[\frac{\rho_0}{2} u_t^2(x, t) + \frac{\tau}{2} u_x^2(x, t)\right] dx = \int_0^L \rho_0 F(x, t) u_t(x, t)\, dx - \alpha \int_0^L \frac{\rho_0}{2} u_t^2(x, t)\, dx,$$

$\alpha = 2k$. In words, the rate of change of energy in the string is equal to the rate at which work is being done on the string by external forces (or the rate at which power is being supplied to the string) minus the rate at which kinetic energy is being dissipated by friction.

3. Find a formal solution to

$$\begin{aligned} &u_{tt} = c^2 u_{xx} + F(x, t), && 0 < x < L, && t > 0,\\ &u(x, 0) = 0, \quad u_t(x, 0) = 0, && 0 \le x \le L,\\ &u(0, t) = 0, \quad u_x(L, t) = 0, && t \ge 0. \end{aligned}$$

4. Find a formal solution to

$$\begin{aligned} &u_{tt} = c^2 u_{xx} + F(x, t), && 0 < x < L, && t > 0,\\ &u(x, 0) = 0, \quad u_t(x, 0) = 0, && 0 \le x \le L,\\ &u_x(0, t) = 0, \quad u_x(L, t) = 0, && t \ge 0. \end{aligned}$$

5. Find a formal solution to the damped problem

$$\begin{aligned} &u_{tt} + k u_t = c^2 u_{xx} + F(x, t), && 0 < x < L, && t > 0,\\ &u(x, 0) = 0, \quad u_t(x, 0) = 0, && 0 \le x \le L,\\ &u(0, t) = 0, \quad u_x(L, t) = 0, && t \ge 0. \end{aligned}$$

6. This problem shows how the solution to the fully inhomogeneous problem (3-2) can be reduced to simpler problems of type (2-2) and (3-7).

(a) Let $U(x, t)$ be any convenient function that equals $r(t)$ when $x = 0$ and $s(t)$ when $x = L$. For example,

$$U(x, t) = \frac{L - x}{L} r(t) + \frac{x}{L} s(t).$$

Show that $u(x, t)$ solves (3-2) exactly when

$$w(x, t) = u(x, t) - U(x, t)$$

satisfies a problem of the form

$$\begin{aligned} &w_{tt} + kw_t = c^2 w_{xx} + F_1(x, t), \qquad 0 < x < L, \qquad t > 0, \\ &w(x, 0) = f_1(x), \qquad w_t(x, 0) = g_1(x), \qquad 0 \le x \le L, \\ &w(0, t) = 0, \qquad w(L, t) = 0, \qquad t \ge 0, \end{aligned}$$

and find F_1, f_1, and g_1.

(b) Show that $w(x, t) = w_1(x, t) + w_2(x, t)$ solves the problem for w in part (a) provided that w_1 and w_2 satisfy problems of types (3-7) and (2-2).

7. Find a formal solution to

$$\begin{aligned} &u_{tt} = c^2 u_{xx}, \qquad 0 < x < 1, \qquad t > 0, \\ &u(x, 0) = x + 1, \qquad u_t(x, 0) = x(1 - x), \qquad 0 \le x \le 1, \\ &u(0, t) = 1, \qquad u(1, t) = 2, \qquad t > 0. \end{aligned}$$

8. Consider the telegrapher's system

$$(*) \qquad i_x + Cv_t + Gv = 0, \qquad v_x + Li_t + Ri = 0$$

on $0 < x < L$, $t > 0$.

(a) Multiply the first equation by v, the second by i and add, and show that

$$(iv)_x + C\left(\frac{v^2}{2}\right)_t + L\left(\frac{i^2}{2}\right)_t + Gv^2 + Ri^2 = 0.$$

(b) Using part (a) show that there exists at most one solution to the system (*) satisfying the initial conditions

$$i(x, 0) = i_0(x), \qquad v(x, 0) = v_0(x), \qquad 0 < x < l,$$

and any one of the following sets of boundary conditions

$$\begin{aligned} &v(0, t) = \phi(t), \qquad v(l, t) = \psi(t), \qquad t > 0, \\ &i(0, t) = \phi(t), \qquad i(l, t) = \psi(t), \qquad t > 0, \\ &i(0, t) = \phi(t), \qquad v(l, t) = \psi(t), \qquad t > 0, \\ &v(0, t) = \phi(t), \qquad i(l, t) = \psi(t), \qquad t > 0. \end{aligned}$$

9. Solve formally the system (*) of Prob. 8 where

$$i(x, 0) = 0, \qquad v(x, 0) = A\frac{l - x}{l} + B\frac{x}{l}, \qquad 0 < x < l,$$

and

$$v(0, t) = A \sin(\omega t), \qquad v(l, t) = B \cos(\omega t), \qquad t > 0,$$

where A, B, and ω are positive constants. What is the behavior for large t? *Hint*. First reduce the boundary conditions to zero.

10. Prove Th. 3-2 using the following steps. Fix $T > 0$ and restrict x, t throughout to $0 \le x \le L$ and $0 \le t \le T$.

(a) Write out the expressions for u_x, u_{xx}, u_t, and u_{tt} assuming that term-by-term differentiation is permissible.

(b) Use (3-13) to confirm that

$$|\lambda_n^2 u_n(t)| \le T\lambda_n \frac{\|F_n\|}{c}, \qquad |u_n'(t)| \le T\|F_n\|, \qquad |u_n''(t)| \le |F_n(t)| + cT\lambda_n\|F_n\|,$$

where

$$\|F_n\| = \max_{0 \le t \le T} |F_n(t)|.$$

(c) Assume that $F(0, t) = F(L, t) = F_{xx}(0, t) = F_{xx}(L, t) = 0$ for all $t \ge 0$ and deduce that

$$|F_n(t)| \le \frac{2}{\lambda_n^3 L}\int_0^L |F_{xxx}(x, t)|\, dx.$$

(d) Use parts (b) and (c) to justify the differentiations in part (a) and the Fourier sine series expansion (3-9) for $0 \le x \le L$ and $0 \le t \le T$. Since T is arbitrary, this proves Th. 3-2.

11. Verify the estimate (3-14). *Hint*. Show that $u = u_1 - u_2$ solves the problem (2-2) with $f = f_1 - f_2$ and $g = g_1 - g_2$. Use the compatibility conditions and integration by parts to find $|\lambda_n b_n| \le E_1/n^2$, $|a_n| \le E_2/n^2$, where a_n and b_n are Fourier coefficients based on $f = f_1 - f_2$ and $g = g_1 - g_2$ and

$$E_1 = \frac{2L}{c\pi^2}\int_0^L |g'(x)|\, dx, \qquad E_2 = \frac{2L}{\pi^2}\int_0^L |f''(x)|\, dx.$$

Finally, use the series (2-11) for $u = u_1 - u_2$ to deduce the estimate (3-14).

4-4. Initial Value Problems via First-Order Systems

We have encountered first-order systems in Chap. 1. For example, the one-dimensional, linearized equations of gas dynamics (see Sec. 1-7) are

$$u_t + v_x = 0, \qquad v_t + c^2 u_x = 0, \tag{4-1}$$

where u is the variation in density and v is velocity. The flow of electricity in a long insulated wire with negligible resistance is given by the telegrapher's system

$$i_t + \left(\frac{1}{L}\right) v_x = 0, \qquad v_t + \left(\frac{1}{C}\right) i_x = 0, \tag{4-2}$$

where i is the current and v is the voltage.

If the functions in (4-1) and (4-2) are smooth enough, one of the unknown functions can be eliminated. Otherwise, the first order system must be treated directly, and accordingly we consider

$$U_t + AU_x = 0, \tag{4-3}$$

where $U = (u_1, u_2)$ is a vector-valued function of x and t and

$$A = \begin{bmatrix} 0 & a^2 \\ b^2 & 0 \end{bmatrix}, \qquad a, b > 0,$$

is a given 2×2 matrix. We begin by considering the case where the physical medium is infinite in extent, so that $-\infty < x < \infty$ and $t > 0$. In the case of the system (4-1), it is natural to specify the density and velocity of the gas initially, while for (4-2) we would specify the initial current and voltage. Thus, we augment (4-3) with prescribed initial data,

$$U(x, 0) = U^\circ(x) = (f_1(x), f_2(x)). \tag{4-4}$$

Our aim is to solve the first-order initial value problem (4-3)–(4-4), also called a Cauchy problem, for u. This can be done by uncoupling the system (4-3). That is, we seek a linear change of dependent variables $U = PV$ with P a 2×2 constant matrix and $V = (v_1, v_2)$ such that the first equation in the system for V only involves v_1 and the second equation only contains v_2. Substitute $U = PV$ into (4-3) to get

$$V_t + P^{-1}APV_x = 0. \tag{4-5}$$

For this system to be uncoupled, the matrix $\Lambda = P^{-1}AP$ must be diagonal. For this to happen the columns of P must be eigenvectors of the matrix A and the diagonal entries of Λ the corresponding eigenvalues (see Prob. 6). A routine calculation shows that A has eigenvalues $\lambda = \pm c$, where $c = ab$ and corresponding eigenvectors (a, b) and $(-a, b)$. Thus the required change of variables is

$$U = \begin{bmatrix} a & -a \\ b & b \end{bmatrix} V,$$

and the system (4-5) for V is

$$(v_1)_t + c(v_1)_x = 0, \qquad (v_2)_t - c(v_2)_x = 0. \tag{4-6}$$

Each of these equations can be solved by the method of characteristics, using the initial data

$$V(x, 0) = V^\circ(x) = P^{-1}U^\circ(x) = (F_1(x), F_2(x)) \tag{4-7}$$

where

$$F_1(x) = \frac{f_1(x)}{2a} + \frac{f_2(x)}{2b} \quad \text{and} \quad F_2(x) = -\frac{f_1(x)}{2a} + \frac{f_2(x)}{2b}. \tag{4-8}$$

We find that

$$v_1(x, t) = F_1(x - ct) \quad \text{and} \quad v_2(x, t) = F_2(x + ct). \tag{4-9}$$

Finally, use $U = PV$ and (4-8) to express (4-9) in terms of u, and so obtain

$$\text{(4-10)}\quad \begin{cases} u_1(x,t) = \dfrac{1}{2} f_1(x-ct) + \dfrac{1}{2}\dfrac{a}{b} f_2(x-ct) + \dfrac{1}{2} f_1(x+ct) - \dfrac{1}{2}\dfrac{a}{b} f_2(x+ct), \\ u_2(x,t) = \dfrac{1}{2}\dfrac{b}{a} f_1(x-ct) + \dfrac{1}{2} f_2(x-ct) - \dfrac{1}{2}\dfrac{b}{a} f_1(x+ct) + \dfrac{1}{2} f_2(x+ct), \end{cases}$$

as the solution to (4-3)–(4-4).

The reasoning leading to (4-10) assumed the initial value problem (4-3)–(4-4) had a solution. Hence, if there is a solution, it is unique and given by (4-10). It is clear that (4-10) satisfies the initial conditions (4-4) and a short calculation shows that it satisfies (4-3), provided that f_1 and f_2 are differentiable. So a solution does exist. The solution formula (4-10) also shows that small changes in the initial data will lead to small changes in the solution. More precisely, the solution depends continuously on the initial data. In summary, we have established:

Theorem 4-1. Let $f_1(x)$ and $f_2(x)$ be differentiable on $-\infty < x < \infty$. Then the initial value problem (4-3)–(4-4) is well-posed and its solution is given explicitly by (4-10).

Inhomogeneous, first-order systems can be solved by similar means. Consider

$$\text{(4-11)}\quad \begin{cases} U_t + AU_x = F(x,t), & -\infty < x < \infty, \quad t > 0, \\ U(x,0) = 0 & -\infty < x < \infty, \end{cases}$$

with A as in (4-3). The substitution $U = PV$, where P diagonalizes A, yields the system

$$\text{(4-12)}\quad \begin{aligned} V_t + \Lambda V_x &= G(x,t), & -\infty < x < \infty, \quad t > 0 \\ V(x,0) &= 0, & -\infty < x < \infty, \end{aligned}$$

where $G = P^{-1}F$. Solve by the method of characteristics to find

$$v_1(x,t) = \int_0^t g_1(x - c(t-\tau), \tau)\, d\tau,$$

$$v_2(x,t) = \int_0^t g_2(x + c(t-\tau), \tau)\, d\tau.$$

Transform back to the original variables to obtain

$$\text{(4-13)}\quad \begin{cases} u_1(x,t) = \displaystyle\int_0^t \frac{1}{2}\Big[f_1(x - c(t-\tau), \tau) + \frac{a}{b} f_2(x - c(t-\tau), \tau) \\ \qquad + f_1(x + c(t-\tau), \tau) - \dfrac{a}{b} f_2(x + c(t-\tau), \tau) \Big]\, d\tau, \\ u_2(x,t) = \displaystyle\int_0^t \frac{1}{2}\Big[\frac{b}{a} f_1(x - c(t-\tau), \tau) + f_2(x - c(t-\tau), \tau) \\ \qquad - \dfrac{b}{a} f_1(x + c(t-\tau), \tau) + f_2(x + c(t-\tau), \tau) \Big]\, d\tau. \end{cases}$$

As usual, we obtained (4-13) under the assumption that (4-11) had a solution. Thus, one should check that (4-13) does in fact solve (4-11). This point is left for the problems.

You may have noticed a striking similarity between the solutions (4-10) and (4-13). This leads to another approach to solving (4-11). Let $K(x, t - \tau, \tau)$ denote the integrand in (4-13). Then (4-13) can be expressed in the vector form

$$U(x, t) = \int_0^t K(x, t - \tau, \tau)\, d\tau, \tag{4-14}$$

where the function $K(x, t, \tau)$ satisfies, for each fixed τ,

$$\begin{cases} K_t + AK_x = 0, \\ K(x, 0, \tau) = F(x, \tau), \end{cases} \tag{4-15}$$

which is a homogeneous system of type (4-3)–(4-4). Conversely, if K is constructed to satisfy (4-15), then U defined by (4-14) solves (4-11). This procedure is called *Duhamel's principle.*

PROBLEMS

1. Consider

$$u_{tt} = c^2 u_{xx}$$
$$u(x, 0) = f(x), \qquad u_t(x, 0) = g(x)$$

Let $u_1 = u_x$, $u_2 = u_t$, and derive a first order system for (u_1, u_2). Solve by the methods of this section to get d'Alembert's solution.

2. Let $A = \begin{bmatrix} 4 & 1 \\ 3 & 2 \end{bmatrix}$ and solve

$$U_t + AU_x = 0, \qquad U(x, 0) = \begin{bmatrix} f(x) \\ 0 \end{bmatrix},$$

where $U = (u_1, u_2)$ and $f(x) = (1 - x^2)^2$ for $|x| \le 1$ and $f(x) = 0$ for $|x| \ge 1$. Discuss the behavior of the waves for $t > 0$.

3. Use the methods of this section to solve

$$u_{tt} = c^2 u_{xx} + F(x, t), \qquad -\infty < x < \infty, \qquad t > 0,$$
$$u(x, 0) = 0, \qquad u_t(x, 0) = 0, \qquad -\infty < x < \infty.$$

4. Let A be an $n \times n$ matrix with n linearly independent eigenvectors $P_1, \ldots, P_n$ and corresponding eigenvalues $\lambda_1, \ldots, \lambda_n$. Show how to solve the system

$$U_t + AU_x = 0, \qquad -\infty < x < \infty, \qquad t > 0,$$
$$U(x, 0) = U^\circ(x), \qquad -\infty < x < \infty,$$

where $U = (u_1, \ldots, u_n)$ and $U^\circ(x) = (f_1(x), \ldots, f_n(x))$ is given.

5. Derive a Duhamel's principle for

$$U_t + AU_x = F(x, t), \qquad -\infty < x < \infty, \qquad t > 0,$$
$$U(x, 0) = 0, \qquad -\infty < x < \infty,$$

where U and A are as in Prob. 4.

6. Suppose that

$$P^{-1}AP = \Lambda = \begin{bmatrix} \lambda_1 & 0 \\ 0 & \lambda_2 \end{bmatrix} \quad \text{and} \quad P = [P_1, P_2]$$

expresses P in terms of its columns. (For simplicity we have taken $n = 2$.) Show that $AP_i = \lambda_i P_i$ so that the P_i are eigenvectors of A. Conversely, if $AP_i = \lambda_i P_i$ and Λ is the diagonal matrix above, then $P^{-1}AP = \Lambda$.

7. Calculate the eigenvalues and vectors for $A = \begin{bmatrix} 0 & a^2 \\ b^2 & 0 \end{bmatrix}$ and so confirm the results stated in the text.

8. Use the method of characteristics to solve (4-6)–(4-7) and hence confirm (4-9).

9. Check that (4-10) solves (4-3)–(4-4).

10. Duhamel's principle could have been used to solve Prob. 3, after converting to a first-order system. How should Duhamel's principle be stated directly in terms of the second-order wave equation?

4-5 Semi-infinite Interval Problems

In this section we consider the first-order system

$$U_t + AU_x = 0, \qquad x > 0, \qquad t > 0, \tag{5-1}$$
$$U(x, 0) = U^\circ(x), \qquad x \geq 0, \tag{5-2}$$

together with a suitable boundary condition when $x = 0$. The notation of Sec. 4-4 is in force. So

$$U = (u_1, u_2), \qquad A = \begin{bmatrix} 0 & a^2 \\ b^2 & 0 \end{bmatrix}, \qquad U^\circ(x) = (f_1(x), f_2(x)),$$

and $a, b > 0$. As in Sec. 4-4, (5-1)–(5-2) may specify for example the flow of current and voltage in a semi-infinite wire or the motion of a gas in a semi-infinite tube. For the telegrapher's system we would expect to specify either the current or voltage at $x = 0$, but not both. Thus, appropriate boundary conditions are either

$$u_1(0, t) = h_1(t) \tag{5-3}$$

or

$$u_2(0, t) = h_2(t), \tag{5-3'}$$

where $h_1(t)$ and $h_2(t)$ are given for $t \geq 0$.

The system (5-1)–(5-2) can be solved by uncoupling just as in Sec. 4-4 to obtain

$$U(x, t) = \begin{bmatrix} aF_1(x - ct) - aF_2(x + ct) \\ bF_1(x - ct) + bF_2(x + ct) \end{bmatrix}, \tag{5-4}$$

where

$$F_1(x) = \frac{f_1(x)}{2a} + \frac{f_2(x)}{2b} \quad \text{and} \quad F_2(x) = -\frac{f_1(x)}{2a} + \frac{f_2(x)}{2b} \quad \text{for} \quad x \geq 0. \tag{5-5}$$

Now an important point emerges. The solution (5-4) to (5-1)–(5-2) is only defined when $x \geq ct$ because $F_1(z)$ is only defined for $z \geq 0$. We want to extend the domain of $F_1(z)$ to $z < 0$ so that (5-4) also solves (5-1)–(5-2) for $x < ct$ and so that either (5-3) or (5-3)′ holds. Suppose that we want (5-3) to hold. Then from (5-4) we want

$$h_1(t) = u_1(0, t) = aF_1(-ct) - aF_2(ct)$$

for $t \geq 0$. This will hold if we define $F_1(x)$ for negative arguments by

$$F_1(x) = \frac{1}{a} h_1\left(-\frac{x}{c}\right) + F_2(-x) \qquad \text{for} \quad x \leq 0. \tag{5-6}$$

In order for the initial data (5-2) to be compatible with (5-3) we must require that $f_1(0) = u_1(0, 0) = h_1(0)$. This is exactly the condition needed to assure that (5-5) and (5-6) give the same result when $x = 0$. Thus, $F_1(x)$ is defined and continuous on $-\infty < x < \infty$. Clearly, (5-4) will solve (5-1)–(5-3) provided that $F_1(x)$, $-\infty < x < \infty$, and $F_2(x)$, $0 < x < \infty$ are differentiable. This will be so if $f_1(x)$ and $f_2(x)$ are differentiable, except possibly at $x = 0$ for $F_1(x)$. Calculate the one-sided derivatives $F_1'(0+)$ and $F_1'(0-)$ from (5-5) and (5-6) to confirm that $F_1'(0)$ exists exactly when $h_1'(0) = -a^2 f_2'(0)$. This is also a natural compatibility condition: The first equation in the system (5-1) will also hold when $x = 0$, $t = 0$ precisely when $h_1'(0) = -a^2 f_2'(0)$. Thus, we have established:

> **Theorem 5-1.** Let $f_1(x)$, $f_2(x)$, and $h_1(t)$ be differentiable for $x \geq 0$ and $t \geq 0$. Assume that $f_1(0) = h_1(0)$ and $a^2 f_2'(0) = -h_1'(0)$. Then the initial, boundary value problem (5-1)–(5-3) is well-posed. Its solution is given by (5-4)–(5-6).

It is illuminating to interpret the solution (5-4)–(5-6) in terms of wave propagation. The general solution to the system (5-1) consists of waves traveling to the left and right at speed c. If we seek the solution at a point (x', t') with $x' - ct' > 0$, then the wave disturbances propagating from the boundary at $x = 0$ have moved a distance ct' and hence have not reached x'. As far as an observer located at x' at time t' is concerned, there is no boundary. The solution should be the same as for the doubly infinite string, and by (5-4)–(5-5) it is. At a point (x', t') with $x' - ct' \leq 0$, some waves generated at the boundary $x = 0$ or reflected from the boundary can reach x' at time t'. These effects are given precisely in (5-4) and (5-6).

The characteristic $x + ct =$ constant carries initial data to the left and $x - ct =$ constant carries it to the right to determine the solution at (x', t') with $x' \geq$

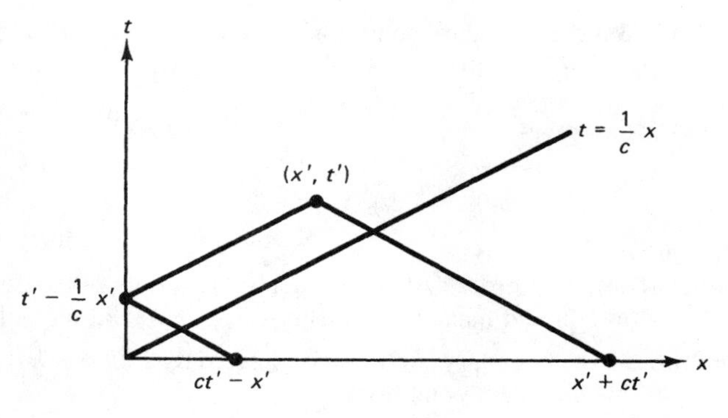

Figure 4-4

ct'. The solution $u(x', t')$ only involves the initial data at $x' - ct'$ and $x' + ct'$ [see (5-4) and (5-5)].

If $x' < ct'$, then the solution involves boundary effects as in Fig. 4-4. The solution involves the initial data on the backward (in time) extensions of the characteristics reflected at the boundary $x = 0$.

PROBLEMS

1. Solve the problem

$$U_t + \begin{bmatrix} 4 & 1 \\ 3 & 2 \end{bmatrix} U_x = 0, \qquad x > 0, \qquad t > 0,$$

$$U(x, 0) = \begin{bmatrix} f(x) \\ 0 \end{bmatrix}, \qquad x \geq 0,$$

$$U(0, t) = h(t), \qquad t \geq 0, \qquad \text{(where } h \text{ is a vector).}$$

What are the compatibility conditions?

2. Solve (5-1), (5-2), and (5-3)'. Specify the compatibility conditions and show the problem is well-posed.

3. Consider a semi-infinite vibrating string with its end pinned:

$$u_{tt} = c^2 u_{xx}, \qquad x > 0, \qquad t > 0,$$

$$u(x, 0) = f(x), \qquad u_t(x, 0) = g(x), \qquad x \geq 0,$$

$$u(0, t) = 0, \qquad t \geq 0.$$

Find an explicit solution to this problem. *Hint.* Show that $f(0) = g(0) = 0$ are compatibility conditions. Consider the motion of two doubly infinite strings, one with initial data $v(x, 0) = f(x)$, $v_t(x, 0) = g(x)$ for $x \geq 0$ and $v(x, 0) = v_t(x, 0) = 0$ for $x \leq 0$ and the other with initial data $w(x, 0) = w_t(x, 0) = 0$ for $x \geq 0$ and $w(x, 0) = -f(-x)$, $w_t(x, 0) = -g(-x)$ for $x \leq 0$. Draw a typical sketch of these initial data and argue on physical grounds that the two solutions are equal but opposite at $x = 0$. Hence, $u = v + w$ for $x \geq 0$, $t \geq 0$ should solve the problem and you know v and w.

4. Solve Prob. 3 if the boundary condition is replaced by $u_x(0, t) = 0$.

5. Find a d'Alembert-type solution to

$$u_{tt} = c^2 u_{xx}, \quad x > 0, \quad t > 0,$$
$$u(x, 0) = f(x), \quad u_t(x, 0) = g(x), \quad x \geq 0,$$
$$u(0, t) = h(t), \quad t \geq 0.$$

Do this in two ways.
(a) Convert to a first-order system and apply the results of this section.
(b) Start with (1-2) and make adjustments much as we did in deducing (5-6).

6. Solve Prob. 5 if the boundary condition is replaced by $u_x(0, t) = h(t)$. What is the physical meaning of this boundary condition?

7. Construct the solution to the inhomogeneous problem

$$U_t + AU_x = F(x, t), \quad 0 < x < \infty, \quad t > 0,$$
$$U(x, 0) = 0, \quad 0 \leq x < \infty,$$
$$u_1(0, t) = h_1(t), \quad t \geq 0.$$

What are the compatibility conditions?

8. Show, under certain conditions on the data, that the problem

$$u_{tt} = c^2 u_{xx} + f(x, t) \quad \text{in} \quad x > 0, \quad t > 0,$$
$$u(x, 0) = g(x), \quad u_t(x, 0) = h(x) \quad \text{in} \quad x \geq 0,$$
$$u(0, t) = \phi(t) \quad \text{in} \quad t \geq 0$$

is well-posed.

9. Let $u = (u_1, \ldots, u_n)$ and A be an $n \times n$ matrix of constants having n linearly independent eigenvectors. Consider the problem

$$u_t + Au_x = 0 \quad \text{in} \quad x > 0, \quad t > 0$$
$$u(x, 0) = 0 \quad \text{in} \quad x \geq 0.$$

What conditions at $x = 0$, $t > 0$ must one place on u so that the solution $u(x, t)$ can be determined for all x and t?

4-6. The Cauchy Problem for Hyperbolic Equations with Constant Coefficients

The telegrapher's equation and the damped wave equation are special cases of the equation

$$u_{tt} + au + bu_t + du_x = c^2 u_{xx} + F(x, t), \tag{6-1}$$

where a, b, $c > 0$ and d are constants. The initial value problem or the Cauchy problem requires finding a function $u = u(x, t)$ which satisfies (6-1) and the given initial conditions

$$u(x, 0) = f(x), \qquad u_t(x, 0) = g(x). \tag{6-2}$$

This equation can be reduced to canonical form by introducing the characteristic coordinates

$$\alpha = x + ct, \qquad \beta = x - ct \tag{6-3}$$

so that in terms of the $\alpha\beta$ coordinates, Eq. (6-1) becomes

$$u_{\alpha\beta} = \frac{bc + d}{4c^2} u_\alpha + \frac{-bc + d}{4c^2} u_\beta + \frac{a}{4c^2} u - \frac{1}{4c^2} F\left(\frac{\alpha+\beta}{2}, \frac{\alpha - \beta}{2c}\right). \tag{6-4}$$

This expression can be simplified by setting $u = v \exp(\lambda\alpha + \mu\beta)$ with

$$\lambda = \frac{-bc + d}{4c^2} \quad \text{and} \quad \mu = \frac{bc + d}{4c^2}.$$

Then (6-4) becomes

$$v_{\alpha\beta} = -kv + G(\alpha, \beta), \tag{6-5}$$

where

$$k = -\left(\frac{a}{4c^2} + \lambda\mu\right) \quad \text{and} \quad G(\alpha, \beta) = -\frac{1}{4c^2} e^{-(\lambda\alpha + \mu\beta)} F\left(\frac{\alpha + \beta}{2}, \frac{\alpha - \beta}{2c}\right).$$

The region $t > 0$ in the xt plane maps onto the region $\beta < \alpha$ in the $\alpha\beta$ plane. The x axis, where the initial data (6-2) are given, is mapped onto the line $\beta = \alpha$. So the "initial" data for (6-5) will be given on the line $\beta = \alpha$. Now, when $t = 0$, $x = \alpha = \beta$ and

$$v(\alpha, \alpha) = f(\alpha) \exp\left(-\frac{\alpha d}{2c^2}\right) \tag{6-6}$$

because $\lambda + \mu = d/2c^2$. Similarly, $u_t(x, 0) = c[u_\alpha(\alpha, \alpha) - u_\beta(\alpha, \alpha)]$, so the second initial condition becomes

$$v_\alpha(\alpha, \alpha) - v_\beta(\alpha, \alpha) = \left[\frac{b}{2c} f(\alpha) + \frac{1}{c} g(\alpha)\right] \exp\left(-\frac{\alpha d}{2c^2}\right). \tag{6-7}$$

Let $h(\alpha) = f(\alpha) \exp(-\alpha d/2c^2)$. Then $v(\alpha, \alpha) = h(\alpha)$ and $v_\alpha(\alpha, \alpha) + v_\beta(\alpha, \alpha) = h'(\alpha)$. Let $j(\alpha)$ be the known expression on the right of (6-7), and solve for $v_\alpha(\alpha, \alpha)$ and $v_\beta(\alpha, \alpha)$ to find

$$v_\alpha(\alpha, \alpha) = \frac{h'(\alpha) + j(\alpha)}{2} \equiv \phi(\alpha) \quad \text{and} \quad v_\beta(\alpha, \alpha) = \frac{h'(\alpha) - j(\alpha)}{2} \equiv \psi(\alpha).$$

Thus, the initial data for v are

$$v(\alpha, \alpha) = h(\alpha), \qquad v_\alpha(\alpha, \alpha) = \phi(\alpha), \qquad v_\beta(\alpha, \alpha) = \psi(\alpha), \tag{6-8}$$

where h, ϕ, and ψ satisfy the compatibility condition

$$h'(\alpha) = \phi(\alpha) + \psi(\alpha). \tag{6-9}$$

In this section we present two approaches for solving the initial value problem (6-5) and (6-8), and hence for solving (6-1) and (6-2). The first method, called

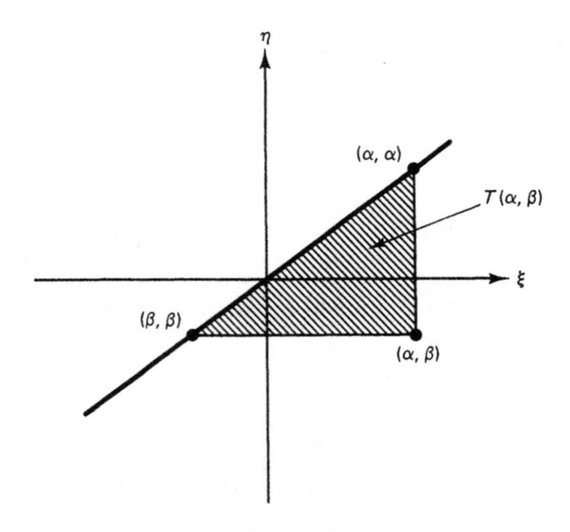

Figure 4-5

successive approximations, is a widely applicable procedure of great practical and theoretical value. In the context of differential equations, this method is applied after expressing the differential equation as an integral equation. We begin by integrating the mixed partial in (6-5) over the triangle $T(\alpha, \beta)$ in Fig. 4-5.

$$\begin{aligned}\iint_{T(\alpha,\beta)} v_{\xi\eta}(\xi, \eta)\, d\xi\, d\eta &= \int_\beta^\alpha \int_\eta^\alpha v_{\xi\eta}\, d\xi\, d\eta \\ &= \int_\beta^\alpha [v_\eta(\alpha, \eta) - v_\eta(\eta, \eta)]\, d\eta \\ &= h(\alpha) - v(\alpha, \beta) - \int_\beta^\alpha \psi(\eta)\, d\eta.\end{aligned}$$

Similarly, integrating in the opposite order gives

$$\iint_{T(\alpha,\beta)} v_{\xi\eta}\, d\xi\, d\eta = h(\beta) - v(\alpha, \beta) + \int_\beta^\alpha \phi(\xi)\, d\xi.$$

We choose to average these expressions to preserve the essential symmetry in α and β. Thus,

$$\iint_{T(\alpha,\beta)} v_{\xi\eta}\, d\xi\, d\eta = \frac{1}{2}[h(\alpha) + h(\beta)] - v(\alpha, \beta) + \frac{1}{2}\int_\beta^\alpha \phi(\xi)\, d\xi - \frac{1}{2}\int_\beta^\alpha \psi(\eta)\, d\eta.$$

Finally, use the differential equation (6-5) to obtain

$$v(\alpha, \beta) = k \iint_{T(\alpha,\beta)} v(\xi, \eta)\, d\xi\, d\eta + \Phi(\alpha, \beta) \tag{6-10}$$

where

$$\Phi(\alpha, \beta) = \frac{1}{2}[h(\alpha) + h(\beta)] + \frac{1}{2}\int_{\beta}^{\alpha} \phi(\xi)\, d\xi - \frac{1}{2}\int_{\beta}^{\alpha} \psi(\eta)\, d\eta - \iint_{T(\alpha,\beta)} G(\xi, \eta)\, d\xi\, d\eta$$

is a known function. To summarize, if $v(\alpha, \beta)$ satisfies (6-5) and (6-8), then it also solves (6-10). Conversely, it is routine to check that if $v(\alpha, \beta)$ is continuous and solves (6-10), then in fact $v(\alpha, \beta)$ has continuous second-order partials and solves (6-5) and (6-8) (see Prob. 6).

We shall use successive approximations to solve the integral equation (6-10), but first we use (6-10) to establish a basic uniqueness result.

Theorem 6-1. The initial value problem (6-1)–(6-2) has at most one solution.

If $u_1(x, t)$ and $u_2(x, t)$ both solve (6-1)–(6-2), then their difference $u = u_1 - u_2$ solves (6-1)–(6-2) with $F = f = g = 0$. It is easy to check that the corresponding problem (6-5), (6-8) for v has $G = h = \phi = \psi = 0$. Therefore, in this case, (6-10) is

$$v(\alpha, \beta) = k \iint_{T(\alpha,\beta)} v(\xi, \eta)\, d\xi\, d\eta. \tag{6-11}$$

Let $M = \max |v(\xi, \eta)|$ calculated over the triangle $T(\alpha, \beta)$. For any point (ξ, η) in $T(\alpha, \beta)$,

$$\begin{aligned} v(\xi, \eta) &= k \iint_{T(\xi,\eta)} v(r, s)\, dr\, ds, \\ |v(\xi, \eta)| &\le |k| \int_{\eta}^{\xi}\int_{s}^{\xi} M\, dr\, ds = \frac{M|k|(\xi - \eta)^2}{2}. \end{aligned} \tag{6-12}$$

Since this estimate holds for any point in $T(\alpha, \beta)$, (6-12) yields

$$\begin{aligned} |v(\xi, \eta)| &\le |k| \int_{\eta}^{\xi}\int_{s}^{\xi} |v(r, s)|\, dr\, ds \\ &\le |k| \int_{\eta}^{\xi}\int_{s}^{\xi} \frac{M|k|(r - s)^2}{2}\, dr\, ds = \frac{M|k|^2(\xi - \eta)^4}{4!} \end{aligned} \tag{6-13}$$

Again this estimate is valid for any point in $T(\alpha, \beta)$. Repeated substitution into (6-12) leads to

$$|v(\xi, \eta)| \le \frac{M|k|^n(\xi - \eta)^{2n}}{(2n)!} = \frac{M[\sqrt{|k|}(\xi - \eta)]^{2n}}{(2n)!}. \tag{6-14}$$

Since $z^m/m! \to 0$ as $m \to \infty$, (6-14) implies that $v(\xi, \eta) = 0$ for any point in $T(\alpha, \beta)$. In particular, $v(\alpha, \beta) = 0$ and thus

$$u_1(\alpha, \beta) - u_2(\alpha, \beta) = v(\alpha, \beta) \exp(\lambda\alpha + \mu\beta) = 0.$$

So $u_1 = u_2$ and uniqueness is established for the Cauchy problem (6-1)–(6-2).

Successive Approximations

Now, we return to (6-10) and the method of successive approximations. The idea is to guess at a solution to (6-10) and then allow the formula (6-10) to improve this guess. Let $v_0(\alpha, \beta)$ be the initial guess. Substitute it into the right side of (6-10). The result is a new function

$$v_1(\alpha, \beta) = k \iint\limits_{T(\alpha,\beta)} v_0(\xi, \eta)\, d\xi\, d\eta + \Phi(\alpha, \beta),$$

which is usually different from v_0. Of course, if $v_1 = v_0$, then $v_0(\alpha, \beta)$ solves (6-10). We now regard $v_1(\alpha, \beta)$ as a new "initial" guess, substitute it into the right side of (6-10), and calculate a new function $v_2(\alpha, \beta)$, and so on. In this way we generate a sequence $\{v_n(\alpha, \beta)\}$ which we hope converges to a solution to (6-10). Specifically, let

$$v_0(\alpha, \beta) \text{ be an arbitrary continuous function}$$

and define the sequence $\{v_n\}$ recursively by

$$v_{n+1}(\alpha, \beta) = k \iint\limits_{T(\alpha,\beta)} v_n(\xi, \eta)\, d\xi\, d\eta + \Phi(\alpha, \beta) \tag{6-15}$$

for $n = 0, 1, 2, \ldots$. It is quite reasonable to expect that the sequence $\{v_n\}$ may converge to a solution to (6-10). Indeed, suppose that $v_n \to v$ uniformly on $T(\alpha, \beta)$. Then we can let $n \to \infty$ in (6-15) and take the limit under the integral to obtain (6-10). Consequently, to solve (6-10) we need only prove that the sequence of successive approximations $\{v_n(\alpha, \beta)\}$ converges uniformly on $T(\alpha, \beta)$.

As our initial guess, we take $v_0(\alpha, \beta) \equiv 0$. Then

$$v_1(\alpha, \beta) = \Phi(\alpha, \beta),$$

$$v_2(\alpha, \beta) = \Phi(\alpha, \beta) + k \iint\limits_{T(\alpha,\beta)} \Phi(\xi, \eta)\, d\xi\, d\eta,$$

$$v_3(\alpha, \beta) = \Phi(\alpha, \beta) + k \iint\limits_{T(\alpha,\beta)} \Phi(\xi, \eta)\, d\xi\, d\eta + k^2 \iint\limits_{T(\alpha,\beta)} \left[\iint\limits_{T(\xi,\eta)} \Phi(\sigma, \tau)\, d\sigma\, d\tau \right] d\xi\, d\eta.$$

The last iterated integral is over the region of four-dimensional $(\xi, \eta, \sigma, \tau)$-space given by $\beta \le \eta \le \tau \le \sigma \le \xi \le \alpha$. Evaluating this four-dimensional integral in the opposite order gives

$$\int_\beta^\alpha \int_\tau^\alpha \int_\beta^\tau \int_\sigma^\alpha \Phi(\sigma, \tau)\, d\xi\, d\eta\, d\sigma\, d\tau = \iint\limits_{T(\alpha,\beta)} \Phi(\sigma, \tau)(\alpha - \sigma)(\tau - \beta)\, d\sigma\, d\tau.$$

Thus

$$v_3(\alpha, \beta) = \Phi(\alpha, \beta) + k \iint\limits_{T(\alpha,\beta)} \Phi(\xi, \eta)\, d\xi\, d\eta + k^2 \iint\limits_{T(\alpha,\beta)} (\alpha - \xi)(\eta - \beta)\Phi(\xi, \eta)\, d\xi\, d\eta.$$

Continuing in this way leads to

$$v_n(\alpha, \beta) = \Phi(\alpha, \beta) + k \iint_{T(\alpha,\beta)} \left[\sum_{j=0}^{n-2} k^j \frac{(\alpha - \xi)^j (\eta - \beta)^j}{(j!)^2} \right] \Phi(\xi, \eta) \, d\xi \, d\eta$$

for $n = 2, 3, \ldots$. Now

$$\sum_{j=0}^{\infty} \frac{z^j}{(j!)^2} = \sum_{j=0}^{\infty} \frac{(2\sqrt{z})^{2j}}{2^{2j}(j!)^2} = I_0(2\sqrt{z}),$$

where $I_0(z) = J_0(iz)$ is the modified Bessel function of order zero. This series has infinite radius of convergence and hence converges absolutely and uniformly for z in any bounded set. Thus, the sum in brackets converges uniformly to $I_0(2\sqrt{k(\alpha - \xi)(\eta - \beta)})$ and $v_n(\alpha, \beta)$ converges uniformly to

$$v(\alpha, \beta) = \Phi(\alpha, \beta) + k \iint_{T(\alpha,\beta)} I_0(2\sqrt{k(\alpha - \xi)(\eta - \beta)}) \Phi(\xi, \eta) \, d\xi \, d\eta, \tag{6-16}$$

which solves (6-10). We have established:

> Theorem 6-2. Let $F(x, t)$, $f(x)$, $f'(x)$, and $g(x)$ be continuous for $-\infty < x < \infty$, $t > 0$. Then the Cauchy problem (6-1)–(6-2) has a unique solution given by (6-16).

Since efficient methods for the evaluation or $I_0(z)$ are available, (6-16) is a very useful representation of the solution.

Riemann Function

Riemann suggested another way to solve (6-5) and (6-8), which is analogous to the manner in which systems of linear algebraic equations are solved. Let (α, β) be fixed and use (ξ, η) for the independent variables in (6-5) and (6-8) (see Fig. 4-5). Riemann's idea is to multiply (6-5) by a function $R = R(\alpha, \beta; \xi, \eta)$, integrate over the triangle $T(\alpha, \beta)$, and choose R so that integrals involving unknown values of v disappear. He also uses integration by parts to transfer derivatives on v over to R and thus solve for v. Here are the steps:

$$\begin{aligned}
\iint_{T(\alpha,\beta)} RG \, d\xi \, d\eta &= \iint_{T(\alpha,\beta)} R(v_{\xi\eta} + kv) \, d\xi \, d\eta \\
&= \iint_{T(\alpha,\beta)} [(Rv_\xi)_\eta - R_\eta v_\xi + kRv] \, d\xi \, d\eta \\
&= \iint_{T(\alpha,\beta)} [(Rv_\xi)_\eta - (R_\eta v)_\xi + R_{\eta\xi} v + kRv] \, d\xi \, d\eta \\
&= \iint_{T(\alpha,\beta)} [(Rv_\xi)_\eta - (R_\eta v)_\xi + (R_{\xi\eta} + kR)v] \, d\xi \, d\eta.
\end{aligned}$$

Since v is unknown in $T(\alpha, \beta)$, we get rid of the last term by requiring

$$R_{\xi\eta} + kR = 0 \quad \text{for} \quad \alpha > \xi \quad \text{and} \quad \eta > \beta.$$

Carrying out the integrations on the exact derivative terms with limits of integration read from Fig. 4-5 gives

$$\iint_{T(\alpha,\beta)} RG\, d\xi\, d\eta = -R(\alpha, \beta; \alpha, \beta)v(\alpha, \beta) + R(\alpha, \beta; \beta, \beta)h(\beta)$$

$$+ \int_\beta^\alpha R(\alpha, \beta; \xi, \xi)\phi(\xi)\, d\xi + \int_\beta^\alpha R_\xi(\alpha, \beta; \xi, \beta)v(\xi, \beta)\, d\xi$$

$$- \int_\beta^\alpha R_\eta(\alpha, \beta; \alpha, \eta)v(\alpha, \eta)\, d\eta + \int_\beta^\alpha R_\eta(\alpha, \beta; \eta, \eta)h(\eta)\, d\eta.$$

Since $v(\xi, \beta)$ and $v(\alpha, \eta)$ are unknown we require that

$$R_\xi(\alpha, \beta; \xi, \beta) = 0 \quad \text{for} \quad \alpha > \xi$$
$$R_\eta(\alpha, \beta; \alpha, \eta) = 0 \quad \text{for} \quad \eta > \beta,$$

and for simplicity we would like to have

$$R(\alpha, \beta; \alpha, \beta) = 1.$$

With these restrictions on R, we obtain

$$\text{(6-17)} \quad v(\alpha, \beta) = R(\alpha, \beta; \beta, \beta)h(\beta) + \int_\beta^\alpha R(\alpha, \beta; \xi, \xi)\phi(\xi)\, d\xi$$
$$+ \int_\beta^\alpha R_\eta(\alpha, \beta; \eta, \eta)h(\eta)\, d\eta - \iint_{T(\alpha,\beta)} R(\alpha, \beta; \xi, \eta)G(\xi, \eta)\, d\xi\, d\eta.$$

To summarize, (6-17) solves (6-5) and (6-8) provided that there is a function R such that

$$\text{(6-18)} \qquad \begin{cases} R_{\xi\eta}(\alpha, \beta; \xi, \eta) + kR(\alpha, \beta; \xi, \eta) = 0 & \text{for} \quad \alpha > \xi, \quad \eta > \beta, \\ R_\xi(\alpha, \beta; \xi, \beta) = 0 & \text{for} \quad \alpha > \xi, \\ R_\eta(\alpha, \beta; \alpha, \eta) = 0 & \text{for} \quad \eta > \beta, \\ R(\alpha, \beta; \alpha, \beta) = 1. \end{cases}$$

We can construct a solution to (6-18) by judicious guessing. Notice that $r = (\alpha - \xi)(\eta - \beta)$ will satisfy the conditions on R_ξ and R_η in (6-18). Any smooth function of r, say $R = \rho(r)$, will have the same property because

$$R_\xi = -\rho'(r)(\eta - \beta) = 0 \quad \text{when} \quad \eta = \beta$$

and

$$R_\eta = \rho'(r)(\alpha - \xi) = 0 \qquad \text{when} \quad \xi = \alpha.$$

Therefore, $R = \rho(r)$ will solve (6-18) provided that it satisfies the partial differential equation and the normalization condition. That is, we want

$$-r\rho''(r) - \rho'(r) + k\rho(r) = 0$$

and

$$\rho(0) = 1.$$

This problem has a series solution (by the method of Frobenius), namely

$$\rho(r) = \sum_{j=0}^{\infty} \frac{(kr)^j}{(j!)^2} = I_0(2\sqrt{kr}).$$

So

$$R(\alpha, \beta; \xi, \eta) = I_0(2\sqrt{k(\alpha - \xi)(\eta - \beta)}), \tag{6-19}$$

and is called the *Riemann function* for (6-5). The solution to (6-5) and (6-8) is given explicitly by

$$\begin{aligned} v(\alpha, \beta) = h(\beta) &+ \int_\beta^\alpha I_0(2\sqrt{k(\alpha - \xi)(\xi - \beta)})\phi(\xi)\, d\xi \\ &+ \int_\beta^\alpha I_1(2\sqrt{k(\alpha - \eta)(\eta - \beta)}) \frac{\alpha + \beta - 2\eta}{\sqrt{k(\alpha - \eta)(\eta - \beta)}} h(\eta)\, d\eta \\ &- \iint_{T(\alpha,\beta)} I_0(2\sqrt{k(\alpha - \xi)(\eta - \beta)})G(\xi, \eta)\, d\xi\, d\eta \end{aligned} \tag{6-20}$$

because $I_0'(z) = I_1(z)$.

PROBLEMS

1. Verify the calculations leading from (6-1) to (6-5).
2. Verify (6-6) and (6-7).
3. Solve explicitly the initial value problem

$$u_{tt} + ku_t = c^2 u_{xx}, \qquad -\infty < x < \infty, \qquad t > 0,$$
$$u(x, 0) = 0, \qquad u_t(x, 0) = f(x), \qquad -\infty < x < \infty.$$

Express your answer in terms of xt coordinates. Solve both by finding the Riemann function and by iteration.

4. Find the Riemann function for $v_{\alpha\beta} = 0$ and use it to solve the problem

$$u_{tt} = c^2 u_{xx} + F(x, t), \qquad -\infty < x < \infty, \qquad t > 0,$$
$$u(x, 0) = f(x), \qquad u_t(x, 0) = g(x), \qquad -\infty < x < \infty.$$

5. Consider the telegrapher's equation

$$i_{xx} = CLi_{tt} + (CR + GL)i_t + GRi$$

for the current $i(x, t)$ in $-\infty < x < \infty$, $t > 0$ and suppose that i satisfies the initial conditions

$$i(x, 0) = f(x), \qquad i_t(x, 0) = g(x), \qquad -\infty < x < \infty.$$

Show that as $G \to 0$ and $R \to 0$, the solution $i(x, t)$ tends to the solution of the problem

$$i_{xx} = CLi_{tt}, \qquad -\infty < x < \infty, \qquad t > 0,$$
$$i(x, 0) = f(x), \qquad i_t(x, 0) = g(x), \qquad -\infty < x < \infty.$$

which is what one would get by simply setting $G = 0$ and $R = 0$ in the telegrapher's equation.

6. $\Phi(\alpha, \beta)$ is continuous if $F(x, t)$, $f(x)$, $f'(x)$, and $g(x)$ are. Given this, show that a continuous solution of (6-10) has continuous second partials and solves (6-5) and (6-8).
7. Given the assumptions in Th. 6-2, show that problem (6-1)–(6-2) is well-posed over any finite time interval. *Hint.* Use (6-16).
8. Notice that the initial datum ψ does not appear explicitly in (6-17) or (6-20). It is involved implicitly through the compatibility condition (6-9). Find a more symmetric representation for $v(\alpha, \beta)$: In the calculations following Fig. 4-5, interchange the roles of ξ and η by writing

$$R(v_{\xi\eta} + kv) = (Rv_\eta)_\xi - (R_\xi v)_\eta + (R_{\xi\eta} + kR)v.$$

Then integrate to find a formula for $v(\alpha, \beta)$ analogous to (6-17). Average this formula and (6-17).

4-7. Nonlinear Hyperbolic Problems

In this section we extend the results of Sec. 4-6 to second-order hyperbolic equations with variable coefficients and/or nonlinear terms. The method of successive approximations will be used to construct solutions. In Sec. 4-6, this method lead to global solutions. That is, we were able to solve problem (6-1)–(6-2) for all $-\infty < x < \infty$ and $t > 0$. We can only expect local solutions for nonlinear problems, as you will see in a moment.

For simplicity we assume that the hyperbolic equation has already been reduced to normal form. So our initial value problem takes the form

$$u_{xy} = F(x, y, u, u_x, u_y) \tag{7-1}$$

with initial data given on a curve

$$\Gamma : y = \phi(x) \tag{7-2}$$

where $\phi(x)$ is continuously differentiable and either $\phi'(x) > 0$ or $\phi'(x) < 0$. We assume that u, u_x, and u_y are given on Γ:

$$u(x, \phi(x)) = f(x), \tag{7-3}$$

$$u_x(x, \phi(x)) = g(x) \quad \text{and} \quad u_y(x, \phi(x)) = h(x), \tag{7-4}$$

with f, f', g, and h continuous. Differentiate (7-3) to see that f, g, and h must satisfy the compatibility condition

$$f'(x) = g(x) + h(x)\phi'(x).$$

To be definite, we assume $\phi'(x) < 0$ and seek the solution to (7-1), (7-3), (7-4) in the portion of the xy plane given by $y > \phi(x)$ (i.e., above and to the right of Γ). The problem given in Fig. 4-6 shows that the solution may only exist near Γ. Its

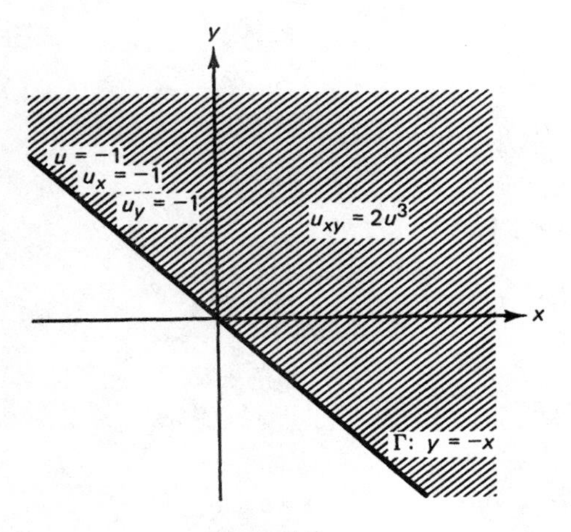

Figure 4-6

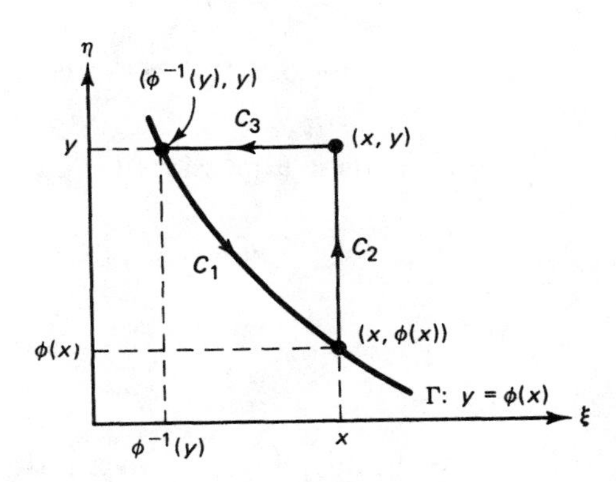

Figure 4-7

solution is

$$u(x, y) = (x + y - 1)^{-1}.$$

This solution extends off the initial curve Γ only as far as the line $x + y = 1$.

In order to show that the initial value problem for (7-1) has a solution, we first recast it as a system of integral equations and then solve this system by successive approximations. To this end, fix (x, y) and consider the triangular region $T(x, y)$ in Fig. 4-7. Assume that $u(\xi, \eta)$ solves (7-1), (7-3), (7-4), expressed in terms of the ξ, η variables. Integrate $u_{\xi\eta} = F(\xi, \eta, u(\xi, \eta), u_\xi, u_\eta)$ with respect to η along the vertical side of $T(x, y)$ to get

$$u_x(x, y) - u_x(x, \phi(x)) = \int_{\phi(x)}^{y} F(x, \eta, u(x, \eta), u_x, u_\eta)\, d\eta$$

or

$$u_x(x, y) = g(x) + \int_{\phi(x)}^{y} F(x, \eta, u(x, \eta), u_x(x, \eta), u_\eta(x, \eta))\, d\eta. \tag{7-5}$$

Likewise, integrate along the horizontal side of $T(x, y)$ to find

$$u_y(x, y) = h(\phi^{-1}(y)) + \int_{\phi^{-1}(y)}^{x} F(\xi, y, u(\xi, y), u_\xi(\xi, y), u_y(\xi, y))\, d\xi. \tag{7-6}$$

From (7-5) we obtain

$$\int_{\phi^{-1}(y)}^{x} u_\xi(\xi, y)\, d\xi = \int_{\phi^{-1}(y)}^{x} g(\xi)\, d\xi + \int_{\phi^{-1}(y)}^{x} \int_{\phi(\xi)}^{x} F(\xi, \eta, u(\xi, \eta), u_\xi, u_\eta)\, d\eta\, d\xi,$$

or since $u(\phi^{-1}(y), y) = f(\phi^{-1}(y))$,

$$u(x, y) = f(\phi^{-1}(y)) + \int_{\phi^{-1}(y)}^{x} g(\xi)\, d\xi + \iint_{T(x,y)} F(\xi, \eta, u, u_\xi, u_\eta)\, d\xi\, d\eta.$$

Similarly, from (7-6),

$$u(x, y) = f(x) + \int_{\phi(x)}^{y} h(\phi^{-1}(\eta))\, d\eta + \iint_{T(x,y)} F(\xi, \eta, u, u_\xi, u_\eta)\, d\xi\, d\eta.$$

Average these expressions for $u(x, y)$ to get

$$u(x, y) = \frac{1}{2}[f(x) + f(\phi^{-1}(y))] + \frac{1}{2}\int_{\phi^{-1}(y)}^{x} g(\xi)\, d\xi + \frac{1}{2}\int_{\phi(x)}^{y} h(\phi^{-1}(\eta))\, d\eta + \iint_{T(x,y)} F(\xi, \eta, u, u_\xi, u_\eta)\, d\xi\, d\eta. \tag{7-7}$$

Consequently, if $u(x, y)$ solves (7-1), (7-3), and (7-4) and if we set

$$p(x, y) = u_x(x, y), \qquad q(x, y) = u_y(x, y)$$

and

$$U(x, y) = \frac{1}{2}[f(x) + f(\phi^{-1}(y))] + \frac{1}{2}\int_{\phi^{-1}(y)}^{x} g(\xi)\, d\xi + \frac{1}{2}\int_{\phi(x)}^{y} h(\phi^{-1}(\eta))\, d\eta,$$

then, by (7-5)–(7-7), $u(x, y)$, $p(x, y)$, and $q(x, y)$ solve the system

$$\begin{cases} u(x, y) = U(x, y) + \displaystyle\iint_{T(x,y)} F(\xi, \eta, u(\xi, \eta), p(\xi, \eta), q(\xi, \eta))\, d\xi\, d\eta, \\ p(x, y) = g(x) + \displaystyle\int_{\phi(x)}^{y} F(x, \eta, u(x, \eta), p(x, \eta), q(x, \eta))\, d\eta, \\ q(x, y) = h(\phi^{-1}(y)) + \displaystyle\int_{\phi^{-1}(y)}^{x} F(\xi, y, u(\xi, y), p(\xi, y), q(\xi, y))\, d\xi. \end{cases} \tag{7-8}$$

Conversely, if $u(x, y)$, $p(x, y)$, and $q(x, y)$ are continuous and solve (7-8), then $p = u_x$, $q = u_y$ and u solves the Cauchy problem (7-1), (7-3), (7-4) (see Prob. 3).

We plan to solve (7-8) by successive approximations. Let

$$u^{(0)}(x, y) = U(x, y), \qquad p^{(0)}(x, y) = g(x), \quad \text{and} \quad q^{(0)}(x, y) = h(\phi^{-1}(y))$$

and determine $u^{(n)}$, $p^{(n)}$, $q^{(n)}$ by

$$\begin{cases} u^{(n+1)}(x, y) = U(x, y) + \displaystyle\iint_{T(x,y)} F(\xi, \eta, u^{(n)}(\xi, \eta), p^{(n)}(\xi, \eta) q^{(n)}(\xi, \eta))\, d\xi\, d\eta, \\ p^{(n+1)}(x, y) = g(x) + \displaystyle\int_{\phi(x)}^{y} F(x, \eta, u^{(n)}(x, \eta), p^{(n)}(x, \eta), q^{(n)}(x, \eta))\, d\eta, \\ q^{(n+1)}(x, y) = h(\phi^{-1}(y)) + \displaystyle\int_{\phi^{-1}(y)}^{x} F(\xi, y, u^{(n)}(\xi, y), p^{(n)}(\xi, y), q^{(n)}(\xi, y))\, d\xi. \end{cases} \tag{7-9}$$

Just as for (7-8), these equations imply that

$$\frac{\partial}{\partial x} u^{(n)} = p^{(n)} \quad \text{and} \quad \frac{\partial}{\partial y} u^{(n)} = q^{(n)}.$$

To show that the recursive calculations in (7-9) actually can be carried out and that $u^{(n)}, p^{(n)}, q^{(n)}$ converge to a solution of (7-8), we need to make several assumptions on the data and the nonlinear term F. Remember that we can only expect a local solution, so we only seek to solve (7-8) for (x, y) near Γ.

Suppose that Γ is given by $y = \phi(x)$ for $a \leq x \leq b$ with a, b finite. (If Γ is defined with a or b infinite, we apply the reasoning to follow on larger and larger finite intervals.) Assume that $F(x, y, u, p, q)$ is continuous on the portion of five-dimensional (x, y, u, p, q)-space given by (x, y) in D and $|u|, |p|, |q| \leq K$, where K is a fixed constant and

$$D = \{(x, y): a \leq x \leq b, y - \phi(x) \leq \rho, x - \phi^{-1}(y) \leq \rho\}$$

for some $\rho > 0$ and fixed. A glance at Fig. 4-7 shows that D consists of those points (x, y) whose horizontal and vertical distances from Γ are at most ρ. When ρ is small D consists of points near Γ. We suppose that F satisfies a Lipschitz condition on its domain. That is,

$$|F(x, y, u, p, q) - F(x, y, u', p', q')| \leq M[|u - u'| + |p - p'| + |q - q'|]$$

for (x, y) in D, $|u|, |u'|, |p|, |p'|, |q|, |q'| \leq K$ and some constant M. Since the continuous function F is bounded on its domain, we can assume M is so large that

$$|F(x, y, u, p, q)| \leq M$$

on the domain of F.

Now, to define $u^{(1)}$, $p^{(1)}$, and $q^{(1)}$ from (7-9), we need to confirm that the right sides of these formulas are defined when $n = 0$. This depends on the "size" of the initial data relative to the domain of F. To see this, let K' be a constant that bounds the continuous functions f, g, and h:

$$|f(x)|, |g(x)|, |h(x)| \leq K' \quad \text{for} \quad a \leq x \leq b.$$

Then from the expressions for $u^{(0)}$, $p^{(0)}$, and $q^{(0)}$ we have

$$|u^{(0)}(x, y)| \leq K' + \tfrac{1}{2}\rho K' + \tfrac{1}{2}\rho K' = (1 + \rho)K',$$
$$|p^{(0)}(x, y)| \leq K', \qquad |q^{(0)}(x, y)| \leq K'.$$

If we *assume* that $K' < K$, we can replace ρ by a smaller value if need be to achieve $(1 + \rho)K' \leq K$. With these restrictions on K', K and ρ, $|u^{(0)}|, |p^{(0)}|, |q^{(0)}| \leq K$ and the formulas in (7-9) define $u^{(1)}$, $p^{(1)}$, and $q^{(1)}$ for (x, y) in D. [In many applications $F(x, y, u, p, q)$ is defined for all u, p, q and we can take $K = K' + 1$ so that $K' < K$ is automatic and the restriction above on ρ simply means that (x, y) is "close" to Γ.] Furthermore, if

$$|u^{(n)}(x, y)|, |p^{(n)}(x, y)|, |q^{(n)}(x, y)| \leq K,$$

the formulas (7-9) yield

$$|u^{(n+1)}(x, y)| \le (1 + \rho)K' + \tfrac{1}{2}\rho^2 M$$
$$|p^{(n+1)}(x, y)|, |q^{(n+1)}(x, y)| \le K' + \rho M.$$

Consequently, if we choose ρ so small that

$$(1 + \rho)K' + \tfrac{1}{2}\rho^2 M \le K \quad \text{and} \quad K' + \rho M \le K,$$

we find that

$$|u^{(n+1)}(x, y)|, |p^{(n+1)}(x, y)|, |q^{(n+1)}(x, y)| \le K.$$

Thus, at each iteration step $(x, y, u^{(n)}, p^{(n)}, q^{(n)})$ is in the domain of F and (7-9) is a well-defined iteration scheme.

Finally, we shall show that $u^{(n)}(x, y)$, $p^{(n)}(x, y)$, $q^{(n)}(x, y)$ converge uniformly for (x, y) in D to functions $u(x, y)$, $p(x, y)$, $q(x, y)$. This allows us to let $n \to \infty$ under the integral signs in (7-9) and hence conclude that $u(x, y)$, $p(x, y)$, and $q(x, y)$ solve (7-8). This in turn proves that the original Cauchy problem has a solution. It is convenient to introduce the *maximum norm* on D,

$$\|w\| = \max |w(x, y)|,$$

where the maximum is computed for (x, y) in D. From (7-9) and the Lipschitz condition on F, we deduce that

$$u^{(n+1)}(x, y) - u^{(n)}(x, y) = \iint_{T(x,y)} [F(\xi, \eta, u^{(n)}, p^{(n)}, q^{(n)}) - F(\xi, \eta, u^{(n-1)}, p^{(n-1)}, q^{(n-1)})]\, d\xi\, d\eta,$$

$$\begin{aligned} |u^{(n+1)}(x, y) - u^{(n)}(x, y)| &\le \iint_{T(x,y)} M[|u^{(n)}(\xi, \eta) - u^{(n-1)}(\xi, \eta)| \\ &\qquad + |p^{(n)} - p^{(n-1)}| + |q^{(n)} - q^{(n-1)}|]\, d\xi\, d\eta \\ &\le \tfrac{1}{2}M\rho^2[\|u^{(n)} - u^{(n-1)}\| + \|p^{(n)} - p^{(n-1)}\| \\ &\qquad + \|q^{(n)} - q^{(n-1)}\|], \end{aligned}$$

$$\|u^{(n+1)} - u^{(n)}\| \le \tfrac{1}{2}M\rho^2 S_{n-1},$$

where

$$S_n = \|u^{(n+1)} - u^{(n)}\| + \|p^{(n+1)} - p^{(n)}\| + \|q^{(n+1)} - q^{(n)}\|.$$

Similarly,

$$\|p^{(n+1)} - p^{(n)}\| \le M\rho S_{n-1} \quad \text{and} \quad \|q^{(n+1)} - q^{(n)}\| \le M\rho S_{n-1}.$$

Add these three estimates to find $S_n \le (\tfrac{1}{2}\rho^2 + 2\rho)MS_{n-1}$ for $n \ge 1$. Once again we restrict ρ so that $\alpha = (\tfrac{1}{2}\rho^2 + 2\rho)M < 1$. Then

$$S_n \le \alpha S_{n-1} \le \alpha^2 S_{n-2} \le \cdots \le \alpha^n S_0.$$

Since $|\alpha| < 1$ the series

$$\sum_0^\infty S_n < \infty$$

and by definition

$$|u^{(n+1)}(x, y) - u^{(n)}(x, y)| \le S_n,$$
$$|p^{(n+1)}(x, y) - p^{(n)}(x, y)| \le S_n, \qquad |q^{(n+1)}(x, y) - q^{(n)}(x, y)| \le S_n.$$

The Weierstrass M-test guarantees the absolute and uniform convergence of the three series

$$u^{(o)}(x, y) + \sum_{m=0}^\infty (u^{(m+1)}(x, y) - u^{(m)}(x, y)),$$
$$p^{(o)}(x, y) + \sum_{m=0}^\infty (p^{(m+1)}(x, y) - p^{(m)}(x, y)),$$
$$q^{(o)}(x, y) + \sum_{m=0}^\infty (q^{(m+1)}(x, y) - q^{(m)}(x, y)).$$

The nth partial sums of these series are $u^{(n)}(x, y)$, $p^{(n)}(x, y)$, and $q^{(n)}(x, y)$. Thus, $u^{(n)}(x, y)$, $p^{(n)}(x, y)$, and $q^{(n)}(x, y)$ converge uniformly in D to the respective sums $u(x, y)$, $p(x, y)$, and $q(x, y)$ of the series above. As noted earlier, this proves that the Cauchy problem (7-1), (7-3), (7-4) has a solution in D. This solution is unique (Prob. 4) and depends continuously on the initial data (Prob. 5). In summary, we have established the following basic result.

Theorem 7-1. Let f, f', g, h be continuous and satisfy the compatibility condition $f'(x) = g(x) + h(x)\phi'(x)$. Assume that F is continuous and satisfies a Lipschitz condition near Γ and that $K' < K$ as specified above. Then the Cauchy problem (7-1), (7-3), (7-4) is well-posed for (x, y) in D.

PROBLEMS

1. Modify the initial value problem example involving $u_{xy} = 2u^3$ in Fig. 4-6 to show that the solution to a problem of type (7-1), (7-3), (7-4) may exist only on a region extending an arbitrarily short distance from the initial curve Γ.
2. Verify (7-6) and the expression for $u(x, y)$ preceding (7-7).
3. Check that if $u(x, y)$, $p(x, y)$, and $q(x, y)$ are continuous and satisfy (7-8), then $p = u_x$, $q = u_y$ and u solves the Cauchy problem (7-1), (7-3), and (7-4). *Hint.* Do not forget the compatibility condition.
4. Show that the problem (7-1), (7-3), (7-4) has at most one solution defined in D. *Hint.* If u_1 and u_2 are two solutions and p_1, p_2 and q_1, q_2 are their corresponding partials, then u_i, p_i, q_i solves (7-8) for $i = 1, 2$. Subtract the two systems and estimate the differences to get $S \le \alpha S$, where $S = \|u_1 - u_2\| + \|p_1 - p_2\| + \|q_1 - q_2\|$ and $\alpha = (\frac{1}{2}\rho^2 + 2\rho)M < 1$ is the constant from the text. Conclude that $S = 0$.

5. Show that the solution to (7-1), (7-3), (7-4) depends continuously on the initial data. *Hint*. Let f_i, g_i, h_i be two sets of initial data with $|f_1(x) - f_2(x)| < \varepsilon$, $|g_1(x) - g_2(x)| < \varepsilon$, and $|h_1(x) - h_2(x)| < \varepsilon$. Let

$$S = \|u_1 - u_2\| + \|p_1 - p_2\| + \|q_1 - q_2\|,$$

where u_i solves the Cauchy problem with data f_i, g_i, h_i, and p_i and q_i are the corresponding partials. Use (7-8) to deduce that $S \le (3\varepsilon + \rho\varepsilon) + \alpha S$ or $S \le (3 + \rho)\varepsilon/(1 - \alpha)$, where $\alpha < 1$ is the constant from the text.

6. Show that under certain assumptions on the given data, the problem

$$\begin{aligned} u_{xy} &= F(x, y, u, u_x, u_y), \qquad x > 0, \qquad y > 0, \\ u(x, 0) &= f(x), \qquad x > 0, \\ u(0, y) &= g(y), \qquad y > 0, \end{aligned}$$

is well-posed for (x, y) near $(0, 0)$ in the first quadrant. *Hint*. Integrate over the rectangle with vertices $(0, 0)$, $(0, x)$, (x, y), $(0, y)$ in the $\xi\eta$ plane.

7. Suppose that $\Gamma: x = \phi(y)$ is continuously differentiable, $\phi'(y) \ge 0$, and passes through the origin. Show that under certain assumptions on the given functions, that the problem

$$\begin{aligned} u_{xy} &= F(x, y, u, u_x, u_y), \qquad y > 0, \qquad x \text{ to the right of } \Gamma \\ u(\phi(y), y) &= f(y), \qquad y > 0, \\ u(x, 0) &= g(x), \qquad x > 0 \end{aligned}$$

has a unique solution near $(0, 0)$. This problem and Prob. 6 are called *Goursat problems*. *Hint*. Let (x, y) be fixed and integrate the differential equation with respect to ξ and η over the rectangle in Fig. 4-8.

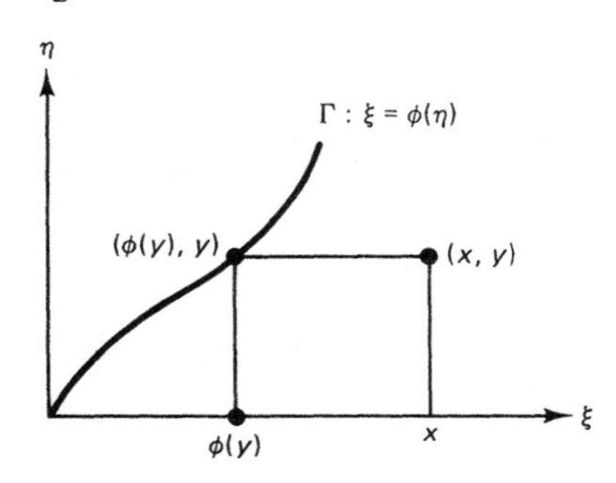

Figure 4-8

8. Assume that the differential equation in (7-1) is linear. That is, $F(x, y, u, p, q) = F(x, y) - a(x, y)p - b(x, y)q - c(x, y)u$. So the differential equation in (7-1) is

$$u_{xy} + a(x, y)u_x + b(x, y)u_y + c(x, y)u = F(x, y)$$

with a, b, c, and F continuous on $-\infty < x, y < \infty$. In this case, sho that Th. 7-1 can be strengthened to assert that the Cauchy problem has a unique solution defined for all (x, y) and this solution depends continuously on the data over any finite portion of

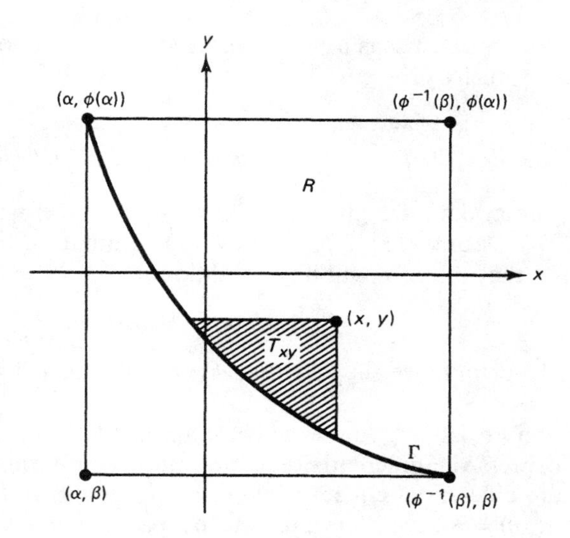

Figure 4-9

the xy plane. *Hint.* Proceed as in the nonlinear case to obtain

$$|u_{n+1}(x, y) - u_n(x, y)| \leq M^n Q^{n-1} S_0 \frac{(x - \alpha + y - \beta)^{n+1}}{(n+1)!}$$

$$|p_{n+1}(x, y) - p_n(x, y)| \leq M^n Q^{n-1} S_0 \frac{(x - \alpha + y - \beta)^n}{n!}$$

$$|q_{n+1}(x, y) - q_n(x, y)| \leq M^n Q^{n-1} S_0 \frac{(x - \alpha + y - \beta)^n}{n!},$$

where $Q = 2 + [\phi(\alpha) - \beta + \phi^{-1}(\beta) - \alpha]$. See Fig. 4-9.

9. Show that there exists a solution to the linear Goursat problem

$$u_{xy} + a(x, y)u_x + b(x, y)u_y + c(x, y)u = F(x, y), \qquad x > 0, \qquad y > 0,$$
$$u(x, 0) = f(x), \qquad x \geq 0,$$
$$u(0, y) = g(y), \qquad y \geq 0,$$

under certain conditions on the given functions which is valid for all $x \geq 0$, $y \geq 0$.

10. Solve

$$u_{tt} = c^2 u_{xx}, \qquad x > 0, \qquad t > 0,$$
$$u(x, 0) = f(x), \qquad u(x, x) = g(x), \qquad x > 0.$$

Determine $u_t(x, 0)$.

-8. Riemann's Method

Riemann's method is a technique for solving the Cauchy problem for linear, hyperbolic equations. This method was used for constant-coefficient equations in Sec. 4-6. Now, we consider problems with variable coefficients. For simplicity, we assume the problem has already been reduced to its normal form. Thus, we consider the Cauchy problem

$$\begin{cases} Lu \equiv u_{xy} + a(x, y)u_x + b(x, y)u_y + c(x, y)u = F(x, y), \\ u(x, \phi(x)) = f(x), \qquad u_x(x, \phi(x)) = g(x), \qquad u_y(x, \phi(x)) = h(x), \end{cases} \tag{8-1}$$

where the initial data are given on a curve $\Gamma: y = \phi(x)$ with either $\phi'(x) > 0$ or $\phi'(x) < 0$. We assume that a, b, c, and F are continuous, that f, f', g, h are continuous, and that the compatibility condition

$$f'(x) = g(x) + h(x)\phi'(x)$$

holds. To be definite we suppose that $\phi'(x) < 0$ and shall solve (8-1) to the right of Γ.

Riemann's reasoning runs exactly as in Sec. 4-6. Fix a point (x, y) to the right of Γ and express the differential equation in (8-1) in terms of the variables ξ, η. Multiply the differential equation by an as yet unknown function, the Riemann function $v(\xi, \eta) = R(x, y; \xi, \eta)$, integrate by parts, and place requirements on v which determine it and eliminate unknown terms. To prepare for the integration by parts, notice that

$$vu_{\xi\eta} = (vu_\xi)_\eta - v_\eta u_\xi = (vu_\xi)_\eta - (v_\eta u)_\xi + v_{\eta\xi}u$$

and symmetrically

$$vu_{\xi\eta} = (vu_\eta)_\xi - (v_\xi u)_\eta + v_{\xi\eta}u.$$

Average to get

$$vu_{\xi\eta} = \tfrac{1}{2}(vu_\eta - v_\eta u)_\xi + \tfrac{1}{2}(vu_\xi - v_\xi u)_\eta + uv_{\xi\eta}.$$

Consequently, a short calculation gives

$$vLu - uMv = [\tfrac{1}{2}(vu_\eta - v_\eta u) + auv]_\xi + [\tfrac{1}{2}(vu_\xi - v_\xi u) + buv]_\eta, \tag{8-2}$$

where

$$Mv = v_{\xi\eta} - (a(\xi, \eta)v)_\xi - (b(\xi, \eta)v)_\eta + c(\xi, \eta)v \tag{8-3}$$

is called the (formal) *adjoint* of the differential operator L.

The divergence theorem for a plane vector field $\vec{V} = A\vec{i} + B\vec{j}$ can be written as

$$\iint_T (A_\xi + B_\eta)\, d\xi\, d\eta = \int_C -B\, d\xi + A\, d\eta,$$

where the plane region T is bounded by the curve C, which is traversed counterclockwise. Apply this to the region $T(x, y)$ in Fig. 4-7 and vector field with

$$A = \tfrac{1}{2}(vu_\eta - v_\eta u) + auv \quad \text{and} \quad B = \tfrac{1}{2}(vu_\xi - v_\xi u) + buv \tag{8-4}$$

to find

$$(8\text{-}5)\qquad \iint_{T(x,y)} (vLu - uMv)\,d\xi\,d\eta = \int_{C_2} \left[\frac{1}{2}(vu_\eta - v_\eta u) + auv\right] d\eta - \int_{C_3} \left[\frac{1}{2}(vu_\xi - v_\xi u) + buv\right] d\xi + \int_{C_1} -B\,d\xi + A\,d\eta.$$

Since we plan to determine $v(\xi, \eta)$, the line integral along C_1 involves only known data on Γ. The line integral on C_2 can be expressed as

$$(8\text{-}6)\qquad \int_{C_2} \left[\frac{1}{2}(vu_\eta - v_\eta u) + auv\right] d\eta = \frac{1}{2} vu\Big|_R^P - \int_R^P u(v_\eta - av)\,d\eta$$

after an integration by parts on vu_η. Similarly,

$$(8\text{-}7)\qquad -\int_{C_3} \left[\frac{1}{2}(vu_\xi - v_\xi u) + buv\right] d\xi = \frac{1}{2} vu\Big|_Q^P - \int_Q^P u(v_\xi - bv)\,d\xi.$$

Since u is unknown inside $T(x, y)$ and on its vertical and horizontal sides, we require that

$$\begin{aligned} Mv = 0 &\quad \text{for} \quad \xi < x \quad \text{and} \quad \eta < y, \\ v_\eta - av = 0 &\quad \text{for} \quad \xi = x \quad \text{and} \quad \eta \le y, \\ v_\xi - bv = 0 &\quad \text{for} \quad \xi \le x \quad \text{and} \quad \eta = y. \end{aligned}$$

The last two conditions on the vertical and horizontal sides of $T(x, y)$ are met if we require that

$$v(x, \eta) = e^{\int_y^\eta a(x,s)\,ds} \quad \text{and} \quad v(\xi, y) = e^{\int_x^\xi b(s,y)\,ds}.$$

Notice that $v(P) = v(x, y) = 1$ with these choices. Now $Lu = F$, (8-5)–(8-7), and our requirements on v yield

$$(8\text{-}8)\qquad u(x, y) = \frac{v(Q)u(Q) + v(R)u(R)}{2} + \iint_{T(x,y)} vF\,d\xi\,d\eta + \int_{C_1} B\,d\xi - A\,d\eta.$$

In summary, we can represent the solution u to (8-1) by (8-8) provided that there is a function $v(\xi, \eta)$ such that for each fixed (x, y),

$$(8\text{-}9)\qquad \begin{cases} Mv = 0, & \xi < x \quad \text{and} \quad \eta < y, \\ v(x, \eta) = e^{\int_y^\eta a(x,s)\,ds}, & \eta \le y, \\ v(\xi, y) = e^{\int_x^\xi b(s,y)\,ds}, & \xi \le x. \end{cases}$$

The problem (8-9) is a linear Goursat problem and hence has a solution (see Probs. 7 and 9 of Sec. 4-7). So the *Riemann function* $v(\xi, \eta) = R(x, y; \xi, \eta)$ exists and (8-8) solves the Cauchy problem (8-1). Written out in more detail, (8-8) is

$$(8\text{-}10)\quad u(x, y) = \frac{1}{2}[R(x, y; x, \phi(x))f(x) + R(x, y; \phi^{-1}(y), y)f(\phi^{-1}(y))]$$
$$+ \iint_{T(x,y)} R(x, y; \xi, \eta)F(\xi, \eta)\, d\xi\, d\eta$$
$$+ \int_{\phi^{-1}(y)}^{x} \left\{\frac{1}{2}[R(x, y; \xi, \phi(\xi))g(\xi) - R_\xi(x, y; \xi, \phi(\xi))f(\xi)]\right.$$
$$\left. + b(\xi, \phi(\xi))R(x, y; \xi, \phi(\xi))f(\xi)\right\} d\xi$$
$$+ \int_{\phi(x)}^{y} \left\{\frac{1}{2}[R(x, y, \phi^{-1}(\eta), \eta)h(\phi^{-1}(\eta))\right.$$
$$- R_\eta(x, y; \phi^{-1}(\eta), \eta)f(\phi^{-1}(\eta))]$$
$$\left. + a(\phi^{-1}(\eta), \eta)R(x, y; \phi^{-1}(\eta), \eta)f(\phi^{-1}(\eta))\right\} d\eta.$$

Besides giving an explicit recipe for the solution to the Cauchy problem, this formula shows the following:

1. The solution depends linearly on $F(x, y)$.
2. The solution depends continuously on the initial data and F.
3. The domain of dependence for the point (x, y) is the part of Γ joining $(\phi^{-1}(y), y)$ and $(x, \phi(x))$.

Observe also that the Riemann function depends only on the differential operator L. So, once the Riemann function is known, the Cauchy problem can be solved for arbitrary initial data.

Example 1. Find the Riemann function for

$$Lu = u_{xy} - u_x.$$

Here $a = -1$, $b = c = 0$; therefore,

$$Mv = v_{xy} + v_x,$$

and we must solve the Goursat problem

$$\begin{cases} v_{\xi\eta} + v_\xi = 0, & \xi < x, \quad \eta < y, \\ v(x, \eta) = e^{-(\eta - y)}, & \eta \le y, \\ v(\xi, y) = 1, & \xi \le x. \end{cases}$$

The adjoint equation can be integrated with respect to ξ to yield

$$v_\eta + v = C(\eta),$$

where $C(\eta)$ is an arbitrary function of η. Another integration gives

$$v(\xi, \eta) = \alpha(\xi)e^{-\eta} + \beta(\eta)$$

for arbitrary functions $\alpha(\xi)$ and $\beta(\eta)$, which depend implicitly on the fixed point (x, y). That is, $\alpha(\xi) = \alpha(x, y; \xi)$ and $\beta(\eta) = \beta(x, y; \eta)$. The side conditions give

$$\alpha(x)e^{-\eta} + \beta(\eta) = e^{y-\eta}, \qquad \alpha(\xi)e^{-y} + \beta(y) = 1.$$

Set $\eta = y$ in the first equation and compare with the second one to infer that $\alpha(\xi) = \alpha(x)$. Then the first equation can be written as $\alpha(\xi)e^{-\eta} + \beta(\eta) = e^{y-\eta}$ or $v(\xi, \eta) = e^{y-\eta}$. Thus,

$$R(x, y; \xi, \eta) = e^{y-\eta}, \qquad \xi < x, \qquad \eta < y.$$

Example 2. Suppose that $\Gamma: y = -x$. Solve

$$\begin{cases} u_{xy} - yu_x = 1, & y > -x, \qquad -\infty < x < \infty \\ u(x, -x) = x, & u_x(x, -x) = x^2, \qquad u_y(x, -x) = x^2 - 1. \end{cases}$$

The Riemann function is given by

$$\begin{cases} v_{\xi\eta} + (\eta v)_\xi = 0, & \xi < x, \qquad \eta < y, \\ v(x, \eta) = e^{(y^2-\eta^2)/2}, & \eta \le y, \\ v(\xi, y) = 1, & \xi \le x. \end{cases}$$

Reason as in Example 1 to find

$$R(x, y; \xi, \eta) = e^{(y^2-\eta^2)/2}, \qquad \eta < y.$$

Since $y = \phi(x) = -x$ and $x = \phi^{-1}(y) = -y$, (8-10) yields

$$u(x, y) = \frac{1}{2}xe^{(y^2-x^2)/2} - \frac{1}{2}y + \frac{1}{2}\int_{-y}^{x} \xi^2 e^{(y^2-\xi^2)/2}\, d\xi$$
$$-\frac{1}{2}\int_{-x}^{y} e^{(y^2-\eta^2)/2}\, d\eta + \iint_{T(x,y)} e^{(y^2-\eta^2)/2}\, d\xi\, d\eta.$$

PROBLEMS

1. Show that the Riemann function for $Lu = u_{xy} + xu_x$ is

$$R(x, y; \xi, \eta) = \exp\,[\xi(\eta - y)]$$

and solve

$$Lu = 1, \qquad y > x, \qquad -\infty < x < \infty,$$
$$u(x, -x) = u_x(x, -x) = 0, \qquad -\infty < x < \infty.$$

2. Find the Riemann function for $Lu = u_{xy} + xu_x + yu_y + xyu$.

3. Find the Riemann function for $Lu = u_{xy} + a(y)u_x$.

4. Confirm the formula $R(x, y; \xi, \eta) = e^{(y^2-\eta^2)/2}$ given in Example 2.
5. Solve

$$u_{xy} - u_x = xy,$$
$$u(x, x^3) = \sin x, \qquad u_x(x, x^3) = \cos x, \qquad u_y(x, x^3) = 0$$

in the region above the curve $\Gamma: y = x^3$.
6. The solution formula (8-10) for the Cauchy problem (8-1) treats the data symmetrically. There are alternative representations, with fewer terms, which do not preserve this symmetry. You can find them by not averaging the two expressions for $vu_{\xi\eta}$ as we did to produce (8-2), but using them separately. Do this to find a representation for the solution which does not explicitly contain $h(x)$.
7. Write down the solution to (8-1) corresponding to zero initial data and a driving force $F(x, y)$. In this situation $u(x, y)$ may be regarded as the wave response at (x, y) to a system that is initially at rest and driven by an external excitation with intensity $F(x, y)$. Consider the configuration shown in Fig. 4-7. Imagine a point (x_0, y_0) inside the curvilinear triangle above Γ. Excite the system with pulses $F_n(x, y) \geq 0$ which are nonzero only in the disk of radius $1/n$ centered at (x_0, y_0) and have unit intensity:

$$\iint_{T(x,y)} F_n \, d\xi \, d\eta = 1.$$

Let $u_n(x, y)$ be the corresponding response of the system, so that u_n solves (8-1) with $f = g = h = 0$ and $F = F_n$. Show that $u_n(x, y) = R(x, y; \xi_n, \eta_n)$ for some point (ξ_n, η_n) within the disk of radius $1/n$ centered at (x_0, y_0). Conclude that

$$R(x, y; x_0, y_0) = \lim_{n \to \infty} u_n(x, y).$$

Use this to give a physical interpretation of the Riemann function.

4-9. Initial Value Problems for Hyperbolic Systems

In this section we study the Cauchy problem for first-order, quasilinear, hyperbolic systems:

$$\begin{cases} u_t + A(x, t, u)u_x = f(x, t, u) \\ u(x, 0) = \phi(x). \end{cases} \tag{9-1}$$

Here A is an $r \times r$ matrix and u, f, and ϕ are column vectors of length r. The theory we present is designed to cover the telegrapher's system for an inhomogeneous medium, and also represents a first step in analyzing the one-dimensional gas dynamics equations from Sec. 1-7. Equations (7-1) and (7-3) of Sec. 1-7 can be expressed as

$$\begin{bmatrix} \rho \\ v \end{bmatrix}_t + \begin{bmatrix} v & \rho \\ \dfrac{c^2}{\rho} & v \end{bmatrix} \begin{bmatrix} \rho \\ v \end{bmatrix}_x = \begin{bmatrix} 0 \\ 0 \end{bmatrix}, \tag{9-2}$$

where gravitational effects have been neglected and an equation of state of the form $p = p(\rho)$ has been assumed with $p'(\rho) > 0$. So p is a strictly increasing function of ρ, and we have set $c = c(\rho) = \sqrt{p'(\rho)}$. The eigenvalues of

$$A = \begin{bmatrix} v & \rho \\ \dfrac{c^2}{\rho} & v \end{bmatrix}$$

are $v \pm c$. They are real and distinct. We will call the system in (9-1) *hyperbolic* if the matrix $A(x, t, u)$ has *r real, distinct* eigenvalues $\lambda_i = \lambda_i(x, t, u)$, $i = 1, 2, \ldots, r$, in the region of (x, t, u)-space where we seek a solution. We also say that the matrix A is hyperbolic.

We use an iteration scheme to prove that (9-1) has a unique, smooth solution for small values of t. For large values of the time t, shocks or discontinuities in the solution may occur, and a deeper study becomes necessary. We postpone such considerations until Chap. 12.

We always assume that the data $A(x, t, u)$, $f(x, t, u)$, and $\phi(x)$ are smooth, say of continuity class $\mathscr{C}^k$ for some $k \geq 0$. This means that each of the components of these matrix or vector functions has continuous partials of order k at all points (x, t, u) in its domain. The iterative scheme we have in mind for solving (9-1) requires us to solve first the linear version of (9-1). That is, $A = A(x, t)$ does not depend on u, $f(x, t, u) = -B(x, t)u + f(x, t)$ and (9-1) specializes to

$$\text{(9-3)} \qquad \begin{cases} u_t + A(x, t)u_x + B(x, t)u = f(x, t), \\ u(x, 0) = \phi(x). \end{cases}$$

For each $T > 0$, assume for $-\infty < x < \infty$ and $0 \leq t \leq T$ that

(i) $A(x, t)$ is a class $\mathscr{C}^k$ for some $k \geq 2$ and is hyperbolic with eigenvalues $\lambda_i = \lambda_i(x, t)$;
(ii) $a_{ij}(x, t)$ are bounded functions for $i, j = 1, \ldots, r$.

Assumptions (i) and (ii) guarantee that the eigenvalues of $A(x, t)$ are also of class $\mathscr{C}^k$ and are bounded on the infinite strip. So there is a constant L, determined by the matrix $A(x, t)$ and T, such that

$$|\lambda_i(x, t)| \leq L$$

for $-\infty < x < \infty$, $0 \leq t \leq T$ (see Prob. 2).

We shall solve (9-3) by a successive approximation scheme. To prepare for this, we introduce some norms needed in the convergence analysis. Given fixed numbers X, T, and L, form a trapezoid Q as shown in Fig. 4-10.

If $h(x, t)$ is a continuous vector function defined on Q, set

$$\|h\|_Q = \max_{1 \leq i \leq r} \max_{(x,t) \in Q} |h_i(x, t)|.$$

If $H(x, t)$ is a continuous matrix function, let

$$\|H\|_Q = \max_{1 \leq i,j \leq r} \max_{(x,t) \in Q} |h_{ij}(x, t)|.$$

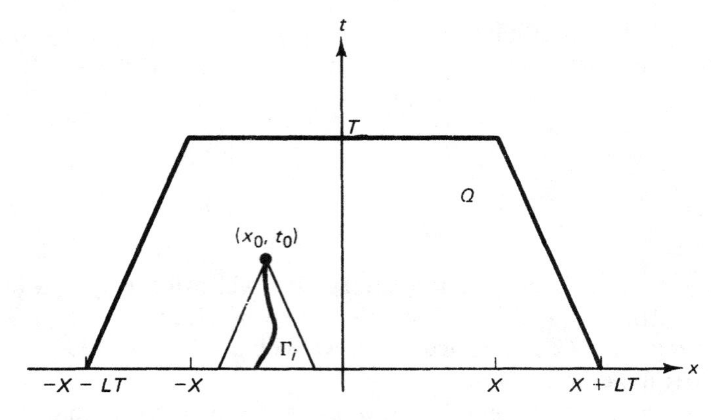

Figure 4-10

Finally, if h is bounded on the infinite strip $-\infty < x < \infty$, $0 \le t \le T$, set

$$\|h\| = \max_{1 \le i \le r} \sup_{(x,t)} |h_i(x, t)|,$$

where (x, t) varies over the strip.

The system (9-3) has *characteristic curves* Γ_i defined as the trajectories of the differential equation

$$\frac{dx_i}{dt} = \lambda_i(x_i, t)$$

for $i = 1, 2, \ldots, r$. For each i, there is a unique characteristic curve through each point (x_0, t_0) with $-\infty < x_0 < \infty$, $0 \le t_0 \le T$. This characteristic is defined for $0 \le t \le t_0$ because the slopes λ_i are uniformly bounded. Moreover, since $|\lambda_i(x, t)| \le L$, if (x_0, t_0) lies in Q, the characteristic curve extending backward in time from this point lies in the trapezoid Q for $0 \le t \le t_0$ (see Fig. 4-10).

It is convenient to reduce (9-3) to an equivalent linear system with A replaced by a diagonal matrix. Let $P = P(x, t)$ be the matrix whose columns are the eigenvectors of $A(x, t)$. Then

$$P^{-1}AP = \Lambda,$$

the diagonal matrix with entries $\lambda_1(x, t), \ldots, \lambda_r(x, t)$ on the main diagonal. Set $u = Pv$ in (9-3) to obtain the equivalent problem

$$\text{(9-4)} \qquad \begin{cases} v_t + \Lambda(x, t)v_x + C(x, t)v = g(x, t) \\ v(x, 0) = \psi(x), \end{cases}$$

where

$$C = P^{-1}(P_t + AP_x + BP), \qquad g = P^{-1}f, \quad \text{and} \quad \psi = P^{-1}\phi$$

for $-\infty < x < \infty$, $0 \le t \le T$. Notice that problems (9-3) and (9-4) have the same characteristics. We use these characteristics to find bounds on solutions to (9-4). Let $(x_i(t), t)$, $0 \le t \le t_0$, be the characteristic curve Γ_i joining the point (x_0, t_0) in

Q to the x axis. Define

$$v_i(t) = v_i(x_i(t), t),$$

where $v(x, t)$ is a solution to (9-4) which is defined at least on the trapezoid Q. Notice that

$$|v_i(0)| \le \alpha = \|\psi\|_Q, \qquad |g_i(x_i(t), t)| \le \beta = \|g\|_Q$$
$$|c_{ij}(x_i(t), t)| \le \gamma = \|C\|_Q$$

and

$$v_i'(t) = (v_i)_x \lambda_i + (v_i)_t.$$

So the system (9-4) yields

$$v_i'(t) = g_i - c_{ii} v_i - \sum_{j \ne i} c_{ij} v_j,$$
$$|v_i'| \le \beta + \gamma |v_i| + \gamma \sum_{j \ne i} |v_j|.$$

Now $|v_i(t)| = [v_i(t)^2]^{1/2}$ implies that $|v_i(t)|' \le |v_i'(t)|$ at any point where $v_i(t) \ne 0$. At such points,

$$|v_i|' - \gamma |v_i| \le \beta + \gamma \sum_{j \pm i} |v_j|,$$
$$(e^{-\gamma t} |v_i(t)|)' \le \beta e^{-\gamma t} + \gamma e^{-\gamma t} \sum_{j \ne i} |v_j(t)|.$$

Since $|v_i(0)| \le \alpha$ there is a time t_1 with $0 \le t_1 \le t_0$ such that $|v_i(t_1)| \le \alpha$ and $v_i(\tau) \ne 0$ on $t_1 \le \tau \le t_0$. [Here we assume that $v_i(t_0) = v_i(x_0, t_0) \ne 0$; otherwise, it is trivial to bound $|v_i(x_0, t_0)| = 0$.] Integrate the preceding inequality from t_1 to t, use $\beta e^{-\gamma t} \le \beta$, and simplify to get

$$e^{-\gamma t} |v_i(t)| \le \alpha + \beta T + \gamma \sum_{j \ne i} \int_{t_1}^{t} e^{-\gamma \tau} |v_j(\tau)| \, d\tau.$$

Let

$$s(t) = \sum_{i=1}^{r} |v_i(t)|$$

and sum the previous inequality to get

$$e^{-\gamma t} s(t) \le r(\alpha + \beta T) + r\gamma \int_{t_1}^{t} e^{-\gamma \tau} s(\tau) \, d\tau,$$

$$e^{-\gamma t} s(t) - r\gamma \int_{t_1}^{t} e^{-\gamma \tau} s(\tau) \, d\tau \le r(\alpha + \beta T), \tag{9-5}$$

$$\frac{d}{dt} \left[e^{-\gamma r t} \int_{t_1}^{t} e^{-\gamma \tau} s(\tau) \, d\tau \right] \le r(\alpha + \beta T) e^{-\gamma r t}.$$

Integrate from t_1 to t to find

$$\int_{t_1}^{t} e^{-\gamma \tau} s(\tau) \, d\tau \le \frac{(\alpha + \beta T)}{\gamma} e^{\gamma r t}.$$

Finally, substitute this estimate into (9-5) to obtain

$$s(t) \leq r(1 + e^{\gamma rt})e^{\gamma t}(\alpha + \beta T), \qquad s(t) \leq R'(\|\psi\|_Q + T\|g\|_Q),$$

where the constant

$$R' = r(1 + e^{\gamma rT})e^{\gamma T}$$

depends on T, on the trapezoid Q, and on the matrix C because $\gamma = \|C\|_Q$. Since $|v_i(x_0, t_0)| = |v_i(t_0)| \leq s(t_0)$ and (x_0, t_0) can be any point in Q, we conclude that

$$|v_i(x, t)| \leq R'(\|\psi\|_Q + T\|g\|_Q)$$

for any point (x, t) in Q. Thus

$$\|v\|_Q \leq R'(\|\psi\|_Q + T\|g\|_Q). \tag{9-6}$$

Since $P(x, t)$ and P^{-1} are bounded on Q, $u = Pv$, $\psi = P^{-1}\phi$, and $g = P^{-1}f$, this estimate implies the companion result

$$\|u\|_Q \leq R(\|\phi\|_Q + T\|f\|_Q), \tag{9-7}$$

where R depends on T, the region Q, and the matrices A and B. The reasoning leading to (9-6) only used the fact that $C(x, t)$ had bounded entries on Q. Thus, (9-7) holds with $A(x, t)$ of class $\mathscr{C}^1$, ϕ and f continuous, and $B(x, t)$ bounded on Q because $C = P^{-1}(P_t + AP_x + BP)$. Note for later purposes that if $T_1 \leq T$ and Q_1 is the trapezoid with horizontal top $t = T_1$ instead of $t = T$, then (9-6) and (9-7) hold with Q replaced by Q_1, T by T_1, and R the constant associated with the larger trapezoid Q.

We shall use (9-6) to show that the linear system (9-4), equivalently (9-3), has a solution on $-\infty < x < \infty, 0 \leq t \leq T$ for any $T > 0$. The bounds (9-6) and (9-7) imply uniqueness and continuous dependence for these problems. See Prob. 3 for the easy proof of the following result.

Theorem 9-1 Assume that $A(x, t)$ is of class $\mathscr{C}^1$, that f and ϕ are continuous, and that $B(x, t)$ is bounded on closed, bounded subsets of the infinite strip $-\infty < x < \infty, 0 \leq t \leq T$. Then (9-3) has at most one solution. This solution (if it exists) depends continuously on the data f and ϕ for (x, t) in any bounded region of the infinite strip.

To prove that (9-4), and hence (9-3), has a solution, consider the iteration scheme

$$\begin{cases} v^{(0)}(x, t) = 0 \\ v_t^{(n)} + \Lambda(x, t)v_x^{(n)} + C(x, t)v^{(n-1)} = g(x, t) \\ v^{(n)}(x, 0) = \psi(x) \end{cases} \tag{9-8}$$

for $n = 1, 2, \ldots$. Each component $v_i^{(n)}$ satisfies a first-order linear equation that can be solved by the method of characteristics (see Secs. 2-1 and 2-2). We will assume that $A(x, t)$, $B(x, t)$, $f(x, t)$, and $\phi(x)$ are of class $\mathscr{C}^2$. Then $v^{(n)}(x, t)$ will be of class $\mathscr{C}^2$ on $-\infty < \infty, 0 \leq t \leq T$. We will show that $v^{(n)}(x, t)$ converges to a solution to (9-4) for (x, t) in Q and t close enough to zero. To this end, consider

$$\sum_{n=1}^{\infty} \Delta^{(n)}(x, t), \qquad \sum_{n=1}^{\infty} \Delta_x^{(n)}(x, t), \quad \text{and} \quad \sum_{n=1}^{\infty} \Delta_t^{(n)}(x, t),$$

where $\Delta^{(n)}(x, t) = v^{(n)}(x, t) - v^{(n-1)}(x, t)$. We shall show that each of these series is uniformly convergent on the trapezoid Q_1 contained in Q with top $t = T_1 \leq T$ and $T_1 > 0$ chosen suitably small. Then

$$v(x, t) = \sum_{n=1}^{\infty} \Delta^{(n)}(x, t) = \lim_{n \to \infty} v^{(n)}(x, t)$$

has continuous first-order partials and solves (9-4) for (x, t) in Q_1.

Notice that $\Delta^{(n)}$ satisfies

$$\begin{cases} \Delta_t^{(n)} + \Lambda \Delta_x^{(n)} = -C(x, t)\Delta^{(n-1)} \\ \Delta^{(n)}(x, 0) = 0. \end{cases}$$

This is a problem of type (9-4); hence (9-6) yields

$$\|\Delta^{(n)}\|_Q \leq R'(0 + Tr\|C\|_Q\|\Delta^{(n-1)}\|_Q)$$

$$\|\Delta^{(n)}\|_Q \leq (rR'\|C\|_Q T)\|\Delta^{(n-1)}\|_Q. \tag{9-9}$$

Differentiate (9-8) with respect to x to infer that $\Delta_x^{(n)}$ satisfies

$$\begin{cases} (\Delta_x^{(n)})_t + \Lambda(\Delta_x^{(n)})_t + \Lambda_x(\Delta_x^{(n)}) = -C\Delta_x^{(n-1)} - C_x\Delta^{(n-1)} \\ \Delta_x^{(n)}(x, 0) = 0. \end{cases}$$

This is also a problem of type (9-4), so (9-6) yields

$$\|\Delta_x^{(n)}\|_Q \leq R'(r\|C\|_Q\|\Delta_x^{(n-1)}\|_Q + r\|C_x\|_Q\|\Delta^{(n-1)}\|_Q)T,$$

$$\|\Delta_x^{(n)}\|_Q \leq rR'(\|C\|_Q + \|C_x\|_Q)T(\|\Delta_x^{(n-1)}\|_Q + \|\Delta^{(n-1)}\|_Q). \tag{9-10}$$

Similarly,

$$\|\Delta_t^{(n)}\|_Q \leq rR'(\|C\|_Q + \|C_t\|_Q)T(\|\Delta_t^{(n-1)}\|_Q + \|\Delta^{(n-1)}\|_Q). \tag{9-11}$$

From (9-9)–(9-11) we conclude there is a constant $\tilde{R}$ depending on C and R' such that

$$\|\Delta^{(n)}\|_Q + \|\Delta_x^{(n)}\|_Q + \|\Delta_t^{(n)}\|_Q \leq \tilde{R}T(\|\Delta^{(n-1)}\|_Q + \|\Delta_x^{(n-1)}\|_Q + \|\Delta_t^{(n-1)}\|_Q).$$

As noted above, this estimate will also hold with Q replaced by the trapezoid Q_1 with top $t = T_1 \leq T$ and the constant $\tilde{R}$ determined by the larger trapezoid. We fix T_1 by

$$\tilde{R}T_1 = \tfrac{1}{2} \qquad \text{or} \qquad T_1 = \frac{1}{2\tilde{R}} \tag{9-12}$$

Then

$$M_n = \|\Delta^{(n)}\|_{Q_1} + \|\Delta_x^{(n)}\|_{Q_1} + \|\Delta_t^{(n)}\|_{Q_1}$$

satisfies

$$M_n \leq \tfrac{1}{2}M_{n-1} \leq (\tfrac{1}{2})^2 M_{n-2} \leq \cdots \leq (\tfrac{1}{2})^{n-1} M_1.$$

The Weierstrass M-test now implies the absolute and uniform convergence of the series

$$\sum_{n=1}^{\infty} \Delta^{(n)}(x, t)$$

and its term-by-term derivatives with respect to x and t on Q_1. Since term-by-term differentiation is justified by the uniform convergence,

$$v(x, t) = \sum_{n=1}^{\infty} \Delta^{(n)}(x, t) = \lim_{n \to \infty} v^{(n)}(x, t)$$

satisfies

$$v_x(x, t) = \sum_{n=1}^{\infty} \Delta_x^{(n)}(x, t) = \lim_{n \to \infty} v_x^{(n)}(x, t)$$

and

$$v_t(x, t) = \lim_{n \to \infty} v_t^{(n)}(x, t)$$

Thus, let $n \to \infty$ in (9-8) to see that $v(x, t)$ solves (9-4) in the trapezoid Q_1.

This iteration process can be repeated. We solve the differential equation in (9-8) with initial data $v(x, T_1)$. We obtain the same estimates (9-9)–(9-11) for the trapezoid $Q - Q_1$ and the same constant $\tilde{R}$. Thus (9-12) determines the same time interval T_1, and the iteration scheme in (9-8) extends the solution of (9-4) into $Q - Q_1$ for $T_1 \leq t \leq 2T_1$. After a finite number of steps the solution to (9-4) is extended to all of Q. Finally, since $X > 0$ is arbitrary, any fixed point (x, t) of the infinite strip $-\infty < x < \infty$, $0 \leq t \leq T$ is contained in the trapezoid Q for large X and we conclude:

Theorem 9-2. Let $A(x, t)$, $B(x, t)$, $f(x, t)$, $\phi(x)$ be of class $\mathscr{C}^2$. Then the Cauchy problem (9-3) has a unique solution defined on $-\infty < x < \infty$, $0 \leq t \leq T$.

We return to the quasilinear problem (9-1) and the following iteration procedure:

$$\text{(9-13)} \qquad \begin{cases} u^{(o)}(x, t) = \phi(x) \\ u_t^{(n)} + A(x, t, u^{(n-1)}(x, t))u_x^{(n)} = f(x, t, u^{(n-1)}(x, t)) \\ u^{(n)}(x, 0) = \phi(x). \end{cases}$$

As for the linear case, we shall show that this scheme converges to a solution $u(x, t)$ of (9-1) for (x, t) in a trapezoid with top $t = T_1 \leq T$ as in Fig. 4-10.

Assume the following:

(i)′ $A(x, t, u)$ is a class $\mathscr{C}^2$ and is hyperbolic with eigenvalues $\lambda_i = \lambda_i(x, t, u)$.

(ii)′ For each $T, U > 0$ the entries of $A(x, t, u)$ are bounded for $-\infty < x < \infty$, $0 \leq t \leq T$, $|u_i| \leq U$, and $i = 1, 2, \ldots, r$.

(iii)′ $f(x, t, u)$ is of class $\mathscr{C}^2$ and defined on $-\infty < x < \infty, 0 \le t \le T, |u_i| \le M$ for some constant M.

(iv)′ $\phi(x)$ is of class $\mathscr{C}^1$.

Let $U = M$ and fix $T, X > 0$. In view of (ii)′ there is a constant $L = L(M, X, T)$ such that

$$|\lambda_i(x, t, u)| \le L$$

for $-\infty < x < \infty$, $0 \le t \le T$, $|u_i| \le M$. Let Q be the trapezoid determined by T, X, and L as in Fig. 4-10. If $\|u_i^{(n-1)}(x, t)\|_Q \le M$, estimate (9-7) holds for $u^{(n)}$:

$$\text{(9-14)} \qquad \|u^{(n)}\|_Q \le R(\|\phi\|_Q + T\|f^{(n-1)}\|_Q),$$

where $f^{(n-1)} = f(x, t, u^{(n-1)}(x, t))$. We restrict our initial data ϕ so that $R\|\phi\| < M$. Then in the trapezoid Q_1 with top $t = T_1$, we have

$$\|u^{(1)}\|_{Q_1} \le R(\|\phi\|_Q + T_1 F) \le M$$

if $T_1 > 0$ is fixed small enough, where

$$F = \max_{1 \le i \le r} \max_{(x,t,u)} |f(x, t, u)|$$

with (x, t) in Q and $|u_i| \le M$. Since $|u_i^{(1)}(x, t)| \le M$ on Q_1, we can repeat the argument above to conclude that

$$|u_i^{(2)}(x, t)| \le M, \ |u_i^{(3)}(x, t)| \le M, \ldots .$$

Thus, the iteration scheme is well-defined for (x, t) in Q_1. As usual we show that

$$u(x, t) = \lim_{n \to \infty} u^{(n)}(x, t) = u^{(0)}(x, t) + \sum_{n=1}^{\infty} [u^{(n)}(x, t) - u^{(n-1)}(x, t)]$$

with the series uniformly convergent on a suitable trapezoid Q_2 contained in Q_1 and with top $t = T_2$. The difference

$$\Delta^{(n)}(x, t) = u^{(n)}(x, t) - u^{(n-1)}(x, t)$$

satisfies

$$\text{(9-15)} \qquad \begin{cases} \Delta_t^{(n)} + A(x, t, u^{(n-1)})\Delta_x^{(n)} = -[A(x, t, u^{(n-1)}) - A(x, t, u^{(n-2)})]u_x^{(n-1)} \\ \qquad\qquad\qquad\qquad + f(x, t, u^{(n-1)}) - f(x, t, u^{(n-2)}), \\ \Delta^{(n)}(x, 0) = 0. \end{cases}$$

This is a problem of type (9-3); hence, by (9-7) and an application of the mean value theorem to estimate the components on the right side of the system, we obtain

$$\text{(9-16)} \qquad \|\Delta^{(n)}\|_{Q_1} \le RT_1K_1\|\Delta^{(n-1)}\|_{Q_1},$$

where K_1 depends on the partials of A and f for (x, t) in Q and $|u_i| \le M$. [Note that our series solution to the linear problem assures that $u_x^{(n-1)}(x, t)$ is bounded on Q.]

Differentiate (9-13) with respect to x and t and form differences to arrive at analogues of (9-15) for $\Delta_x^{(n)}$ and $\Delta_t^{(n)}$ much as we did in the linear case. These analogues of (9-15) yield estimates

$$\|\Delta_x^{(n)}\|_{Q_1} \le RT_1K_2(\|\Delta_x^{(n-1)}\|_{Q_1} + \|\Delta^{(n-1)}\|_{Q_1}), \tag{9-17}$$

$$\|\Delta_t^{(n)}\|_{Q_1} \le RT_1K_3(\|\Delta_t^{(n-1)}\|_{Q_1} + \|\Delta^{(n-1)}\|_{Q_1}), \tag{9-18}$$

for constants K_2 and K_3. Adding (9-16)–(9-18) leads to the estimate

$$M_n \le RKT_1M_{n-1},$$

where

$$M_n = \|\Delta^{(n)}\|_{Q_1} + \|\Delta_x^{(n)}\|_{Q_1} + \|\Delta_t^{(n)}\|_{Q_1}.$$

Finally, replace T_1 by a smaller value T_2 so that $RKT_2 < \frac{1}{2}$, to get

$$M_n \le \tfrac{1}{2}M_{n-1} \le \cdots \le (\tfrac{1}{2})^{n-1}M_1.$$

As in the linear case, we conclude that the series for $u(x, t)$ above converges uniformly on the trapezoid Q_2 with top $t = T_2$ and that the series can be differentiated term by term with respect to both x and t. Let $n \to \infty$ in (9-13) to conclude that $u(x, t)$ solves (9-1) in Q_2. Uniqueness and continuous dependence follow from (9-7) (see Probs. 7 and 8).

Theorem 9-3. Assume that $A(x, t, u)$, $f(x, t, u)$, $\phi(x)$ satisfy (i)′–(iv)′ and that $R\|\phi\| < M$, where R is the constant in (9-14). Then (9-1) has a unique solution defined in a neighborhood of the initial line $t = 0$ in the xt-plane.

PROBLEMS

1. Verify that (7-1) and (7-3) in Sec. 1-7 can be written as the quasilinear system (9-2).
2. Assume that $A(x, t)$ satisfies (i) and (ii) and show that the eigenvalues $\lambda_i(x, t)$ are bounded for $-\infty < x < \infty$, $0 \le t \le T$. *Hint.* The eigenvalues are roots of the characteristic equation of A. Show that a polynomial in λ with coefficients which are bounded functions of (x, t) has bounded roots.
3. Prove Th. 9-1. *Hint.* For uniqueness, assume that u_1 and u_2 both solve (9-3). Find the problem solved by $\Delta = u_1 - u_2$ and apply (9-7).
4. Solve the initial value problem

$$\begin{aligned} &i_t + v_x + i = 0, && -\infty < x < \infty, && t > 0,\\ &v_t + i_x + v = 0, && -\infty < x < \infty, && t > 0,\\ &i(x, 0) = f(x), && v(x, 0) = g(x), && -\infty < x < \infty \end{aligned}$$

by iteration.

5. Consider the telegrapher's system (6-3)-(6-4) in Sec. 1-6, where the loss is not given by a linear law but instead by a nonlinear law so that the system has the form

$$i_x + Cv_t + G(v) = 0, \qquad v_x + Li_t + Ri = 0$$

in an infinite wire $-\infty < x < \infty$. Suppose initially that

$$i(x, 0) = 0, \qquad v(x, 0) = v_0(x),$$

and solve the resulting system for i and v by iteration.

6. If in Prob. 5, $G(v) = v^3$ and $v_0(x) = \exp(-x^2)$, evaluate the first few terms in the iterative procedure.

7. Show that (9-1) has at most one solution. *Hint.* Show that the difference $v = u_1 - u_2$ satisfies a problem of the form $v_t + A(x, t, u_1)v_x + B(x, t)v = 0$, $v(x, 0) = 0$, where $B(x, t)$ is constructed from the mean value theorem. Now apply (9-7).

8. Show continuous dependence for (9-1).

5

Parabolic Partial Differential Equations in One Space Dimension

5-1. Initial, Boundary Value Problems

Consider heat flow in a homogeneous rod of length L and insulated lateral surface. We shall assume that the heat flows only in the x direction and that the temperature u is constant over each cross section of the rod so that its value at the position of the cross section given by x and at the time t is $u(x, t)$. Finally, suppose that there are no internal heat sources. Then the heat equation (3-8) of Chap. 1 becomes

$$u_t = au_{xx} \quad \text{for} \quad t > 0 \quad \text{and} \quad 0 \leq x \leq L, \tag{1-1}$$

where $a = k/c\rho > 0$ is a constant, k is the thermal conductivity, c the specific heat, and ρ the density. (Recall that (1-1) also describes the concentration of a diffusing substance in the absence of convection.) Assume that the ends of the rod are held at a fixed temperature, say zero, and that the initial temperature distribution $f(x)$ for $0 \leq x \leq L$ is known. Then to find the subsequent temperature distribution in the rod, we must solve the heat equation (1-1) together with the boundary conditions

$$u(0, t) = 0, \qquad u(L, t) = 0 \quad \text{for} \quad t \geq 0, \tag{1-2}$$

and the initial condition

$$u(x, 0) = f(x) \quad \text{for} \quad 0 \leq x \leq L. \tag{1-3}$$

The initial, boundary value problem (1-1)–(1-3) can be solved by separation of variables in much the same way as we solved the vibrating string problem in Sec. 4-2: $u(x, t) = X(x)T(t)$ solves (1-1) if and only if $X'' - cX = 0$ and $T =$

$\alpha \exp(act)$, where c and α are constants. From (1-2) we conclude that $X(0) = X(L) = 0$ and consequently (see Prob. 2), $c < 0$. It is convenient to write $c = -\lambda^2$ with $\lambda \geq 0$ and consider again the eigenvalue problem

$$X'' + \lambda^2 X = 0, \qquad X(0) = 0, \qquad X(L) = 0.$$

Just as in the case of the vibrating string, the requirement that X not vanish identically forces $\lambda = \lambda_n = n\pi/L$ and $X = X_n(x) = \sin(\lambda_n x)$ up to a constant factor, for $n = 1, 2, \ldots$. Thus, (1-1) and (1-2) are satisfied by

$$u_n(x, t) = \exp(-a\lambda_n^2 t) \sin(\lambda_n x),$$

and also by any finite linear combination of these functions. More generally, the infinite series

$$u(x, t) = \sum_{n=1}^{\infty} \alpha_n \exp(-a\lambda_n^2 t) \sin(\lambda_n x) \tag{1-4}$$

will solve (1-1) and (1-2) provided that it converges and we can justify computing u_t and u_{xx} by term-by-term differentiation.

We side step this issue for the moment and turn to the initial temperature distribution in the rod. The series (1-4) will satisfy the initial condition (1-3) if

$$u(x, 0) = \sum_{n=1}^{\infty} \alpha_n \sin(\lambda_n x) = f(x), \qquad 0 \leq x \leq L.$$

For this equation to hold, we must have $f(0) = f(L) = 0$ because $\sin(\lambda_n x) = 0$ for $x = 0$ and $x = L$. Notice that the requirement $f(0) = f(L) = 0$ is a natural compatibility condition for the boundary conditions (1-2) and initial condition (1-3). This compatibility condition guarantees that the odd, $2L$ periodic extension of $f(x)$ is continuous. Consequently, by Th. 1-1 of Chap. 3, if

$$\alpha_n = \frac{2}{L} \int_0^L f(x) \sin(\lambda_n x)\, dx, \tag{1-5}$$

the Fourier (sine) series expansion

$$u(x, 0) = \sum_{n=1}^{\infty} \alpha_n \sin(\lambda_n x) = f(x), \qquad 0 \leq x \leq L,$$

will hold, provided that the series converges uniformly. Thus, the series (1-4) with α_n given by (1-5) is a formal solution to the initial, boundary value problem (1-1)–(1-3).

The term-by-term differentiations required to show that (1-4) satisfies the heat equation can be justified at each given point $(\tilde{x}, \tilde{t})$ with $\tilde{t} > 0$ and $0 \leq \tilde{x} \leq L$ by showing that the series (1-4) and the three series that result from term-by-term differentiation, once with respect to t and twice with respect to x, all converge uniformly in a region D which contains the point $(\tilde{x}, \tilde{t})$. Such a region is

$$D = \left\{(x, t) : t > \frac{\tilde{t}}{2} \quad \text{and} \quad 0 \leq x \leq L\right\}.$$

To check the uniform convergence, consider, as a typical case, the series obtained from (1-4) by differentiating term by term twice with respect to x,

$$-\sum_{n=1}^{\infty} \alpha_n \lambda_n^2 \exp(-a\lambda_n^2 t) \sin(\lambda_n x). \tag{1-6}$$

We assume that the initial temperature distribution $f(x)$ is bounded and integrable on $0 \le x \le L$: say, $|f(x)| \le M$. Then from (1-5),

$$|\alpha_n| \le 2M,$$

and the series (1-6) is dominated by the numerical series

$$\sum_{n=1}^{\infty} 2M\lambda_n^2 \exp\left(\frac{-a\lambda_n^2 \tilde{t}}{2}\right) \tag{1-7}$$

because $|\alpha_n \lambda_n^2 \exp(-a\lambda_n^2 t) \sin(\lambda_n x)| \le 2M\lambda_n^2 \exp(-a\lambda_n^2 \tilde{t}/2)$. The series (1-7) converges by the ratio test and the Weierstrass M-test guarantees the uniform convergence of (1-6) in D. Entirely similar reasoning applies to the other three series. Consequently, (1-4) satisfies the heat equation (1-1) for $t > 0$ and $0 \le x \le L$. Additionally, it trivially satisfies the boundary conditions (1-2) by choice of the λ_n.

To guarantee that the series (1-4) will also satisfy the initial condition (1-3), that is, that

$$\sum_{n=1}^{\infty} \alpha_n \sin(\lambda_n x) = f(x) \quad \text{for} \quad 0 \le x \le L \tag{1-8}$$

with α_n given by (1-5), further assumptions are needed about $f(x)$. From Chap. 3 on Fourier series, we know that (1-8) will hold if the series in (1-8) converges uniformly. The Weierstrass M-test assures this if $\sum_{n=1}^{\infty} |\alpha_n| < \infty$. The convergence of this numerical series depends on the smoothness of the initial temperature distribution $f(x)$. Recall that we have already assumed that $f(0) = f(L) = 0$, so that integrating twice by parts we get

$$\alpha_n = -\frac{2}{\lambda_n^2 L} \int_0^L f''(x) \sin(\lambda_n x)\, dx.$$

Hence, if $f''(x)$ is continuous on $0 \le x \le L$, then it is bounded say by M'', and

$$|\alpha_n| \le \frac{2}{\lambda_n^2 L} LM'' = \frac{2M''L^2}{\pi^2} \frac{1}{n^2}. \tag{1-9}$$

Consequently, $\sum |\alpha_n| < \infty$ and (1-8) holds with the Fourier sine series uniformly convergent to $f(x)$. This shows that (1-4) satisfies (1-3). The estimate (1-9) also implies that the series representation (1-4) for $u(x, t)$ is uniformly convergent for $t \ge 0$ and $0 \le x \le L$. Since $u(x, t)$ is the uniform limit of continuous functions, it is continuous. We summarize these results in

Theorem 1-1. If $f(x)$ is bounded and integrable, the series (1-4) yields a solution to (1-1) and (1-2) for $t > 0$ and $0 \le x \le L$. If, in addition, f has a continuous second derivative and $f(0) = f(L) = 0$, then $u(x, t)$ defined by (1-4)

is uniformly convergent and continuous for $t \geq 0$ and $0 \leq x \leq L$ and also satisfies the initial condition (1-3).

Suppose now that heat sources or sinks are present in the rod. Then [see (3-8) of Chap. 1] the initial, boundary value problem for the rod becomes

$$u_t = au_{xx} + F(x, t), \qquad 0 \leq x \leq L, \qquad t > 0, \tag{1-10}$$

$$u(0, t) = 0, \qquad u(L, t) = 0, \qquad t \geq 0, \tag{1-11}$$

$$u(x, 0) = f(x), \qquad 0 \leq x \leq L, \tag{1-12}$$

where $F(x, t)$ is a given source density. We seek a series solution to this new problem.

Since the differential equation (1-10) is inhomogeneous, it is natural to seek a solution using variation of parameters. That is, we let the constants in (1-4) vary and assume that (1-10) has a solution of the form

$$u(x, t) = \sum_{n=1}^{\infty} c_n(t) \exp(-a\lambda_n^2 t) \sin(\lambda_n x) \tag{1-13}$$

for $\lambda_n = n\pi/L$. In order to match coefficients after substituting this proposed solution into (1-10), we will need to require that the source term $F(x, t)$ have a similar series expansion

$$F(x, t) = \sum_{n=1}^{\infty} F_n(t) \exp(-a\lambda_n^2 t) \sin(\lambda_n x). \tag{1-14}$$

This is, for each t, a Fourier sine series for F in x and the quantity $F_n(t) \exp(-a\lambda_n^2 t)$ is the nth Fourier sine coefficient,

$$F_n(t) \exp(-a\lambda_n^2 t) = \frac{2}{L} \int_0^L F(x, t) \sin(\lambda_n x)\, dx, \tag{1-15}$$

provided that the series (1-14) is uniformly convergent in x (for each t). Using (1-14) and assuming that term-by-term differentiation is permissible in (1-13) for $t > 0$, we see that (1-13) solves the inhomogeneous heat equation (1-10) if

$$\sum_{n=1}^{\infty} [c_n'(t) - a\lambda_n^2 c_n(t) + a\lambda_n^2 c_n(t) - F_n(t)] \exp(-a\lambda_n^2 t) \sin(\lambda_n x) = 0$$

for $t > 0$ and $0 \leq x \leq L$. Thus, $c_n(t)$ must satisfy $c_n'(t) = F_n(t)$ for $n = 1, 2, \ldots$ or

$$c_n(t) - c_n(0) = \int_0^t F_n(\tau)\, d\tau. \tag{1-16}$$

Set $t = 0$ in (1-13) and enforce the initial condition

$$\sum_{n=1}^{\infty} c_n(0) \sin(\lambda_n x) = f(x), \qquad 0 \leq x \leq L, \tag{1-17}$$

to conclude that

$$c_n(0) = \frac{2}{L} \int_0^L f(x) \sin(\lambda_n x)\, dx = \alpha_n, \tag{1-18}$$

the Fourier coefficient from (1-5). So (1-16) can be written as

$$c_n(t) = \alpha_n + \int_0^t F_n(\tau)\,d\tau, \tag{1-16}'$$

and the initial, boundary value problem (1-10)–(1-12) has the formal solution

$$\begin{aligned} u(x, t) &= \sum_{n=1}^{\infty} \alpha_n \exp(-a\lambda_n^2 t) \sin(\lambda_n x) \\ &\quad + \sum_{n=1}^{\infty} \left[\int_0^t F_n(\tau)\,d\tau \right] \exp(-a\lambda_n^2 t) \sin(\lambda_n x). \end{aligned} \tag{1-13}'$$

To confirm that (1-13)′ is a solution to our problem, we need to justify the formal term-by-term differentiations leading to (1-13)′ and the Fourier expansions in (1-14) and (1-17). The first series in (1-13)′ is just (1-4) and (1-17) is just (1-8). Consequently, if we assume that $f(x)$ is bounded and integrable, the first series in (1-13)′ has the required term-by-term derivatives. Furthermore, the Fourier expansion (1-17) holds if $f(0) = f(L) = 0$ and f has a continuous second derivative on $0 \le x \le L$. Consider next the Fourier expansion (1-14). Since physically $F(x, t)$ is an internal source density, we can reasonably assume that $F(0, t) = F(L, t) = 0$. If, further, F has a continuous second derivative with respect to the space variable x, we find, upon integrating (1-15) twice by parts, that

$$F_n(t) \exp(-a\lambda_n^2 t) = -\frac{2}{L\lambda_n^2} \int_0^L F_{xx}(x, t) \sin(\lambda_n x)\,dx$$

and

$$|F_n(t) \exp(-a\lambda_n^2 t)| \le \frac{2}{\lambda_n^2} M''(t) = \frac{2M''(t)L^2}{\pi^2} \frac{1}{n^2}, \tag{1-19}$$

where $M''(t)$ is the maximum of $|F_{xx}(x, t)|$ as x varies from 0 to L. This estimate shows that

$$\sum_{n=1}^{\infty} |F_n(t) \exp(-a\lambda_n^2 t)| < \infty$$

for each t, and consequently, the Weierstrass M-test implies the uniform convergence in x (for each t) of the sine series (1-14) and so justifies this expansion.

Finally, we justify the term-by-term differentiations of the second series in (1-13)′ by showing that the resulting series are uniformly convergent. As a typical example, consider the series obtained by differentiating termwise once with respect to t,

$$\sum_{n=1}^{\infty} \left\{ F_n(t) \exp(-a\lambda_n^2 t) - a\lambda_n^2 \left[\int_0^t F_n(\tau)\,d\tau \right] \exp(-a\lambda_n^2 t) \right\} \sin(\lambda_n x). \tag{1-20}$$

Use (1-19) to estimate the terms in (1-20):

$$\begin{aligned} |\{\cdots\} \sin(\lambda_n x)| &\le \frac{2M''(t)}{\lambda_n^2} + a\lambda_n^2 \left[\int_0^t \frac{2M''(\tau)}{\lambda_n^2} \exp(a\lambda_n^2 \tau)\,d\tau \right] \exp(-a\lambda_n^2 t) \\ &\le \frac{2M''(t)}{\lambda_n^2} + 2a\tilde{M}''(t) \left[\int_0^t \exp(a\lambda_n^2 \tau)\,d\tau \right] \exp(-a\lambda_n^2 t), \end{aligned}$$

where $\tilde{M}''(t)$ is the maximum of $|F_{xx}(x, \tau)|$ over the rectangle $0 \le x \le L$ and $0 \le \tau \le t$. Clearly, $\tilde{M}''(t) \ge M''(t)$ and evaluation of the previous integral leads to

$$|\{\cdots\} \sin(\lambda_n x)| \le \frac{2\tilde{M}''(t)}{\lambda_n^2} + \frac{2\tilde{M}''(t)}{\lambda_n^2}(1 - e^{-a\lambda_n^2 t}) \le \frac{4\tilde{M}''(t)}{\lambda_n^2}.$$

Now the Weierstrass M-test implies the uniform convergence of (1-20) for $0 \le x \le L$ and t in any bounded time interval. The term-by-term differentiations with respect to x are verified in the same fashion. Thus, we have established,

Theorem 1-2. Suppose that $f(x)$ is bounded and integrable on $0 \le x \le L$ and $F(x, t)$ has a continuous second derivative with respect to x on $0 \le x \le L$, $t \ge 0$ with $F(0, t) = F(L, t) = 0$. Then the series (1-13) with $c_n(t)$ given by (1-16)′ is a solution to (1-10) and (1-11). If, in addition, $f(x)$ has a continuous second derivative and $f(0) = f(L) = 0$, then (1-13) also satisfies the initial condition (1-12).

A simple modification of the techniques used above to solve the initial, boundary value problems (1-1)–(1-3) and (1-10)–(1-12), which involve homogeneous boundary conditions, can be used to solve similar problems with inhomogeneous boundary data. Indeed, suppose that (1-11) is changed to

$$u(0, t) = g(t), \qquad u(L, t) = h(t), \qquad t \ge 0, \tag{1-11$'$}$$

where $g(t)$ and $h(t)$ are given differentiable functions that specify the temperature at the ends of the rod. To reduce to the case of homogeneous boundary conditions, we express the solution $u(x, t)$ to (1-10), (1-11)′ and (1-12) as

$$u(x, t) = v(x, t) + U(x, t), \tag{1-21}$$

where

$$U(x, t) = \frac{L - x}{L} g(t) + \frac{x}{L} h(t)$$

is chosen to satisfy the inhomogeneous boundary condition (1-11)′. It is easily verified that $u(x, t)$ given by (1-21) solves (1-10), (1-11)′, and (1-12) precisely when $v(x, t)$ satisfies

$$v_t = av_{xx} + F(x, t) - \frac{L - x}{L} g'(t) - \frac{x}{L} h'(t),$$

$$v(0, t) = 0, \qquad v(L, t) = 0,$$
$$v(x, 0) = f(x) - U(x, 0),$$

which is an initial, boundary value problem with homogeneous boundary conditions. Thus, v can be found as above and (1-21) yields the solution u to the corresponding inhomogeneous problem.

We pause to check that the solutions to (1-1)–(1-3) and (1-10)–(1-12) are physically reasonable. This check is essential in order to determine if the underlying model we have been working with corresponds to physical reality.

Consider the solution (1-4) to the initial, boundary value problem (1-1)–(1-3). We surely expect on physical grounds that this problem has a unique solution and that small changes in the given initial temperature $f(x)$ should lead to small changes in the solution. The question of uniqueness and continuous dependence on the data is treated completely in the next section using the maximum principle. But notice a strong suggestion of uniqueness in (1-4): The coefficients α_n are uniquely determined by the initial temperature distribution $f(x)$ (see also Prob. 11). Continuous dependence on the initial temperature distribution is also strongly suggested from direct estimates of the Fourier coefficients (Prob. 15).

Physical intuition and experience suggest that the solution to (1-1)–(1-3) should tend to zero as the time t tends to infinity. Furthermore, this should happen no matter what initial temperature distribution is given. The solution (1-4) clearly has these properties (Prob. 12). We would further expect that the internal energy $E(t)$ in the rod would tend to zero over time,

$$\lim_{t\to\infty} E(t) = \lim_{t\to\infty} c\rho \int_0^L u(x, t)\, dx = 0,$$

where c is the specific heat and ρ the density of the homogeneous rod. Again it is easy to check, using the uniform convergence of (1-4), that this limit relation holds. Suppose that the initial temperature $f(x) \geq 0$. Then we would expect that the temperature $u(x, t)$ would be nonnegative and the maximum temperature would be assumed initially; that is,

$$0 \leq u(x, t) \leq \max_{0\leq y\leq L} f(y) \tag{1-22}$$

for all $0 \leq x \leq L$ and $t \geq 0$. This result is difficult to verify directly from the series representation (1-4) for the temperature $u(x, t)$ and its verification will be deferred. In this context we would also expect the internal energy to decrease steadily to zero, not just tend to zero. That is, we expect

$$E'(t) = c\rho \frac{d}{dt} \int_0^L u(x, t)\, dx \leq 0 \tag{1-23}$$

for $t > 0$. This result is also difficult to obtain directly from (1-4); however, you can confirm it as suggested in Prob. 14.

If internal heat sources are present in the rod, we must deal with the problem (1-10)–(1-12) with $F(x, t) \not\equiv 0$. We expect that the effects of the initial temperature distribution $f(x)$ will become negligible over time and that only the continuing influence of the source term $F(x, t)$ will be felt. Our solution (1-13)′ clearly exhibits this behavior because its first term tends to zero as $t \to \infty$. Suppose that the source terms approach a time-independent state, $\lim_{t\to\infty} F(x, t) = F(x)$. Then the solution $u(x, t)$ should also "stabilize"; that is, $\lim_{t\to\infty} u(x, t) = u(x)$. Moreover, we would expect $u(x)$ to be a time-independent solution to (1-10), with $F(x, t)$ replaced by $F(x)$, and (1-11),

$$au''(x) + F(x) = 0, \qquad u(0) = u(L) = 0. \tag{1-24}$$

So we should have

$$\lim_{t\to\infty} u(x, t) = u(x), \tag{1-25}$$

where $u(x)$ is the solution to the boundary-value problem (1-24).

We give a formal verification of (1-25). The solution $u(x, t)$ to (1-10)–(1-12) is made up of two parts exhibited in (1-13)′. The first part, which contains the effects of the initial temperature, tends to zero as $t \to \infty$ and so can be ignored in confirming (1-25). The second part, which we denote by $w(x, t)$, contains the source terms:

$$w(x, t) = \sum_{n=1}^{\infty} \left[\int_0^t F_n(\tau)\, d\tau \right] \exp(-a\lambda_n^2 t) \sin(\lambda_n x).$$

To confirm (1-25), we must show that $w(x, t)$ tends to $u(x)$ as $t \to \infty$. Consider the limiting behavior of the coefficient of $\sin(\lambda_n x)$ in the preceding series. Use l'Hospital's rule and then (1-15) to deduce that

$$\begin{aligned}\lim_{t\to\infty} \frac{\int_0^t F_n(\tau)\, d\tau}{\exp(a\lambda_n^2 t)} &= \lim_{t\to\infty} \frac{F_n(t)}{a\lambda_n^2 \exp(a\lambda_n^2 t)} \\ &= \lim_{t\to\infty} \frac{1}{a\lambda_n^2} \frac{2}{L} \int_0^L F(x, t) \sin(\lambda_n x)\, dx = \frac{1}{a\lambda_n^2} F_n,\end{aligned}$$

where F_n is the nth Fourier sine coefficient for $F(x)$. Thus, a formal passage to the limit in the series for $w(x, t)$ yields

$$\lim_{t\to\infty} w(x, t) = \sum_{n=1}^{\infty} \frac{1}{a\lambda_n^2} F_n \sin(\lambda_n x). \tag{1-26}$$

However, using (1-24) and integrating twice by parts, we find that

$$\begin{aligned}F_n &= \frac{2}{L} \int_0^L F(x) \sin(\lambda_n x)\, dx \\ &= -\frac{2}{L} \int_0^L a u''(x) \sin(\lambda_n x)\, dx = a\lambda_n^2 \frac{2}{L} \int_0^L u(x) \sin(\lambda_n x)\, dx\end{aligned}$$

or

$$\frac{F_n}{a\lambda_n^2} = \frac{2}{L} \int_0^L u(x) \sin(\lambda_n x)\, dx.$$

Thus, the series on the right of (1-26) is just the Fourier sine series expansion of $u(x)$ on $0 \le x \le L$ and we obtain

$$\lim_{t\to\infty} u(x, t) = \lim_{t\to\infty} w(x, t) = u(x),$$

as expected.

Although certain mathematical details have been omitted, or deferred, it is clear that our series solutions (1-4) and (1-13) do exhibit the qualitative properties that intuition and experience tell us must hold. Thus, we can confidently use these heat conduction models to describe and predict temperature distributions for one-dimensional heat flow.

PROBLEMS

1. If the rod is insulated at both endpoints, no heat flows out of the rod at $x = 0$ and $x = L$; in other words, the heat flux in the direction going out of the rod must vanish. In the one-dimensional case, this means that $\vec{q} = -ku_x\vec{i}$ must vanish at $x = 0$ and $x = L$.

(a) Use separation of variables to solve the boundary value problem

$$u_t = au_{xx} \qquad 0 < x < L, \qquad t > 0,$$
$$u(x, 0) = f(x), \qquad 0 \le x \le L,$$
$$u_x(0, t) = u_x(L, t) = 0, \qquad t \ge 0,$$

governing the flow of heat in a rod with length L, where the ends are insulated.

(b) Physically, heat flows from regions of warmer to cooler temperatures. Give a physical argument for the statement that

$$\lim_{t\to\infty} u(x, t) = C,$$

where C is a constant; that is, after a long period of time, the temperature will equalize. Since energy must also be conserved, argue that

$$\int_0^L f(x)\,dx = \int_0^L u(x, t)\,dx = \lim_{t\to\infty} \int_0^L u(x, t)\,dx$$

must hold. Now derive this equality from the differential equation by integration. Finally, determine the constant C.

(c) Show that the solution obtained in part (a) actually has the properties required in part (b).

(d) Interpret the differential equation as governing a diffusion process and the solution u as a concentration. Reinterpret the arguments in part (b) for this case.

2. Consider separation of variables for (1-1)–(1-3). Explain on physical grounds why the separation constant c in $X'' - cX = 0$, $X(0) = X(L) = 0$ must be negative. Now prove this mathematically. (Consult Sec. 4-2.)

3. Solve formally the problem

$$u_t = au_{xx} + F(x, t), \qquad 0 < x < L, \qquad t > 0,$$
$$u_x(0, t) = u_x(L, t) = 0, \qquad t \ge 0,$$
$$u(x, 0) = f(x), \qquad 0 \le x \le L.$$

4. Continuing with Prob. 3, determine the behavior of $u(x, t)$ as t tends to infinity given that

$$\lim_{t\to\infty} F(x, t) = F(x).$$

You must be a little careful in this case since the solution $u(x)$ to the steady-state problem

$$au''(x) + F(x) = 0, \qquad 0 < x < L,$$
$$u'(0) = u'(L) = 0$$

is only determined up to a constant.

5. Consider the heat flow in a laterally insulated rod which is also insulated at $x = 0$, that is, $u_x(0, t) = 0$. Suppose that the uninsulated end is free so that heat can escape through radiation to the atmosphere, which will be assumed to be at the fixed temperature U. Use Newton's law of cooling (Prob. 4 in Sec. 1-3) to show that

$$q(L, t) = -ku_x(L, t) = \alpha(u(L, t) - U).$$

for some constant $\alpha > 0$. Then give a physical interpretation of the following problem and solve formally.

$$u_t = au_{xx}, \qquad 0 < x < L, \qquad t \geq 0,$$
$$u_x(0, t) = 0, \qquad u_x(L, t) + \mu u(L, t) = \mu U, \qquad t > 0,$$
$$u(x, 0) = f(x), \qquad 0 \leq x \leq L,$$

where $\mu = \alpha/k$. *Hint.* Let $w = u - U$ and solve first for w.

6. Physically, one would expect that $\lim_{t\to\infty} u(x, t) = U$, where u is the solution to Prob. 5. Prove that this is the case.

7. We showed in the text that problems with inhomogeneous boundary conditions can be reduced to equivalent problems with homogeneous boundary conditions. This reduction was also suggested in Prob. 5. Why is this reduction needed in the separation-of-variables solution procedure?

8. By integrating the differential equation in Prob. 5, show that

$$\frac{d}{dt}\int_0^L c\rho u(x, t)\, dx = -\alpha(u(L, t) - U)$$

and interpret this result physically. (Recall that $a = k/c\rho$.)

9. (a) Interpret the boundary value problem

$$u_t = au_{xx}, \qquad 0 < x < L, \qquad t > 0,$$
$$u(x, 0) = f(x), \qquad 0 \leq x \leq L,$$
$$u_x(0, t) - \mu u(0, t) = 0, \qquad u_x(L, t) + \mu u(L, t) = 0, \qquad t \geq 0,$$

where $\mu > 0$ is a constant, as a heat conduction model. What does the condition at $x = 0$ mean? Why should the sign be a minus at $x = 0$ and a plus at $x = L$?

(b) Find a formal solution.

10. Solve the boundary value problem

$$u_t = au_{xx}, \qquad 0 < x < L, \qquad t > 0,$$
$$u(x, 0) = f(x), \qquad 0 \leq x \leq L,$$
$$u(0, t) = 0, \qquad u_x(L, t) = 0, \qquad t \geq 0.$$

What does this problem mean in terms of a diffusion process, where u is the concentration of the substance? Determine both from a physical argument and from your solution the $\lim_{t\to\infty} u(x, t)$.

11. In our discussion of uniqueness of solutions to (1-1)–(1-3), we noted that the coefficients α_n in (1-4) are uniquely determined by the initial data $f(x)$. Why doesn't this prove that the initial, boundary value problem has a unique solution?

12. Verify, as asserted, that $u(x, t)$ given by (1-4) does tend to zero as $t \to \infty$. *Hint*. Notice that $\lambda_1 \geq \lambda_2 \geq \cdots \geq \lambda_n \geq \cdots$ and make some simple estimates.
13. Verify that the internal energy $E(t)$ of the rod described by (1-1)–(1-3) tends to zero as $t \to \infty$.
14. Suppose that $f(x) \geq 0$. We verified that (1-4) solves (1-1)–(1-3). Use the fact that $u(x, t)$ satisfies this initial, boundary value problem and the physically evident fact that $u(x, t) \geq 0$ to show that $E'(t) \leq 0$ as claimed in (1-23). *Hint*. Begin by using (1-1) and integration by parts in the integral in (1-23).
15. Let $u(x, t)$ be the solution to (1-1)–(1-3) and $\tilde{u}$ the solution to (1-1)–(1-3) with f replaced by $\tilde{f}$. Denote the respective Fourier coefficients by α_n and $\tilde{\alpha}_n$. Show $|\alpha_n - \tilde{\alpha}_n| \leq 2\|f'' - \tilde{f}''\|/\lambda_n^2$ and hence

$$\|u - \tilde{u}\| \leq 2\|f'' - \tilde{f}''\| \left(\sum \frac{1}{\lambda_n^2}\right)$$

where $\|h\| = \max |h(x)|$ for $0 \leq x \leq L$. Deduce continuous dependence.

5-2. The Maximum Principle and Consequences

In Sec. 5-1 we found solutions for the initial, boundary value problems (1-1)–(1-3) and (1-10)–(1-12). We also checked that these solutions exhibited several physically realistic properties. The complete verification of some of these properties, such as the fact that each initial, boundary value problem has a unique solution that depends continuously on the given initial and boundary data, was deferred until this section. These facts and many other important features of heat conduction problems follow from the so-called maximum or maximum–minimum principles given below in Th. 2-1 and its corollaries (Cors. 2-1 through 2-3).

Consider a laterally insulated rod whose ends are maintained at temperatures less than or equal to zero. Suppose further that the only internal activity in the rod arises from heat sinks and that initially the rod has a temperature distribution everywhere less than or equal to zero. Then the temperature $u(x, t)$ is governed by

$$\begin{aligned} &u_t - au_{xx} = F(x, t) \leq 0, \\ &u(0, t) = g(t) \leq 0, \qquad u(L, t) = h(t) \leq 0, \\ &u(x, 0) = f(x) \leq 0. \end{aligned}$$

It is physically apparent that $u(x, t) \leq 0$ for all x and t. This is the content of the maximum principle, Th. 2-1. We start with the essence of this principle:

> Proposition 2-1. Suppose that $w(x, t)$ satisfies the differential inequality $w_t - aw_{xx} < 0$ in $0 < x < L, 0 < t \leq T$, where $a \geq 0$. Then w cannot assume a local maximum value at any point in the rectangle $0 < x < L, 0 < t \leq T$.

To confirm this, assume to the contrary that $w(\tilde{x}, \tilde{t})$ is a local maximum for w with $0 < \tilde{x} < L, 0 < \tilde{t} \leq T$. Since $w(x, \tilde{t})$ has a local maximum at $\tilde{x}$ with $0 < \tilde{x} < L$, we have $w_{xx}(\tilde{x}, \tilde{t}) \leq 0$. Similarly, since $w(\tilde{x}, t)$ has a local maximum at $\tilde{t}$ with $0 < \tilde{t} \leq T$ we have $w_t(\tilde{x}, \tilde{t}) \geq 0$ (where strict inequality may occur if $\tilde{t} = T$). Consequently, $w_t(\tilde{x}, \tilde{t}) - aw_{xx}(\tilde{x}, \tilde{t}) \geq 0$, a contradiction.

Theorem 2-1. Let $a \geq 0$. Suppose that $u(x, t)$ is continuous on $0 \leq x \leq L$, $0 \leq t \leq T$ and satisfies

$$\begin{aligned}
&u_t - au_{xx} \leq 0 \quad \text{for} \quad 0 < x < L, \qquad 0 < t \leq T,\\
&u(0, t) \leq 0, \qquad u(L, t) \leq 0 \quad \text{on} \quad 0 \leq t \leq T,\\
&u(x, 0) \leq 0 \quad \text{on} \quad 0 \leq x \leq L.
\end{aligned}$$

Then $u(x, t) \leq 0$ for $0 \leq x \leq L$ and $0 \leq t \leq T$.

The proof is by contradiction. So assume to the contrary that u is positive at some point, say $u(\bar{x}, \bar{t}) > 0$. Form the auxiliary function

$$w(x, t) = u(x, t) - \varepsilon t,$$

where $\varepsilon > 0$ and $0 \leq x \leq L$, $0 \leq t \leq T$. For any choice of $\varepsilon > 0$,

$$\begin{aligned}
&w_t - aw_{xx} = u_t - au_{xx} - \varepsilon < 0,\\
&w(0, t) = u(0, t) - \varepsilon t \leq 0, \qquad w(L, t) = u(L, t) - \varepsilon t \leq 0,\\
&w(x, 0) = u(x, 0) \leq 0.
\end{aligned}$$

Proposition 2-1 and these boundary estimates show that for every $\varepsilon > 0$, $w(x, t)$ cannot have a positive maximum in the closed rectangle $0 \leq x \leq L$, $0 \leq t \leq T$. On the other hand, for small enough $\varepsilon > 0$ we have

$$w(\bar{x}, \bar{t}) = u(\bar{x}, \bar{t}) - \varepsilon \bar{t} > 0,$$

a contradiction that proves Th. 2-1.

Corollary 2-1. Let $a \geq 0$. Suppose that $u(x, t)$ is continuous on $0 \leq x \leq L$, $0 \leq t \leq T$ and satisfies

$$\begin{aligned}
&u_t - au_{xx} \geq 0 \quad \text{for} \quad 0 < x < L, \qquad 0 < t \leq T,\\
&u(0, t) \geq 0, \qquad u(L, t) \geq 0 \quad \text{on} \quad 0 \leq t \leq T,\\
&u(x, 0) \geq 0 \quad \text{on} \quad 0 \leq x \leq L.
\end{aligned}$$

Then $u(x, t) \geq 0$ for $0 \leq x \leq L$, $0 \leq t \leq T$.

To see this, simply notice that $v = -u$ satisfies the conditions in Th. 2-1.

Corollary 2-2. Let $a \geq 0$. Suppose that $u(x, t)$ is defined and continuous on $0 \leq x \leq L$, $0 \leq t \leq T$ and satisfies

$$u_t = au_{xx} \quad \text{in} \quad 0 < x < L, \qquad 0 < t \leq T.$$

Let

$$M = \max\left\{\max_{0 \leq x \leq L} u(x, 0),\ \max_{0 \leq t \leq T} u(0, t),\ \max_{0 \leq t \leq T} u(L, t)\right\}$$

$$m = \min\left\{\min_{0 \leq x \leq L} u(x, 0),\ \min_{0 \leq t \leq T} u(0, t),\ \min_{0 \leq t \leq T} u(L, t)\right\}.$$

Then

$$m \le u(x, t) \le M$$

for all $0 \le x \le L$, $0 \le t \le T$.

To get this, apply Th. 2-1 to $u(x, t) - M$ and $-u(x, t) + m$.

Corollary 2-3. Let $a \ge 0$. Suppose that $u(x, t)$ is defined and continuous on $0 \le x \le L$, $0 \le t \le T$ and satisfies

$$u_t - au_{xx} = F(x, t) \quad \text{for} \quad 0 < x < L, \qquad 0 < t \le T,$$

and that $F(x, t)$ is bounded by a constant N there. Let

$$M = \max \left\{ \max_{0 \le x \le L} |u(x, 0)|, \max_{0 \le t \le T} |u(0, t)|, \max_{0 \le t \le T} |u(L, t)| \right\}.$$

Then

$$|u(x, t)| \le M + TN \quad \text{for all} \quad 0 \le x \le L, \qquad 0 \le t \le T.$$

This corollary follows from Th. 2-1 applied to the functions $\pm u(x, t) - (M + tN)$.

We can now settle the unanswered questions from Sec. 5-1. First we computed one solution to each initial, boundary value problem, and it is the only one because of Th. 2-2.

Theorem 2-2. Let $a \ge 0$. There exists at most one function $u(x, t)$ which is continuous on $0 \le x \le L$, $t \ge 0$ and satisfies

$$\begin{aligned} &u_t - au_{xx} = F(x, t), \qquad 0 < x < L, \qquad t > 0, \\ &u(0, t) = g(t), \qquad u(L, t) = h(t), \qquad t \ge 0, \\ &u(x, 0) = f(x), \qquad 0 \le x \le L, \end{aligned}$$

where $f(x)$, $g(t)$, $h(t)$ are continuous on their domains of definition.

This is an easy consequence of either Cor. 2-2 or 2-3. If u_1 and u_2 are both solutions to the given initial, boundary value problem, then $u = u_1 - u_2$ satisfies $u_t - au_{xx} = 0$, $u(x, 0) = 0$, $u(0, t) = u(L, t) = 0$, and we infer that $u = 0$; that is, $u_1 = u_2$.

Corollary 2-2 allows us to verify (1-22) and (1-23). Recall that in these formulas $u(x, t)$ is the solution to (1-1)–(1-3) with $f(x) \ge 0$. Corollary 2-2 applies with $m = 0$ and $M = \max_{0 \le y \le L} f(y)$ to yield (1-22). In particular, $u(x, t) \ge 0$. We also know that $u(x, t)$ is given by the series expansion (1-4), which we have checked is termwise differentiable on $0 \le x \le L$. Since $u(x, t) \ge 0$ and $u(0, t) = u(L, t) = 0$, it follows immediately that $u_x(0, t) \ge 0$ and $u_x(L, t) \le 0$. Then the internal energy $E(t)$ must satisfy

$$\begin{aligned} E'(t) &= c\rho \int_0^L u_t(x, t)\, dx = c\rho a \int_0^L u_{xx}(x, t)\, dx \\ &= c\rho a[u_x(L, t) - u_x(0, t)] \le 0, \end{aligned}$$

which is (1-23). This confirms the physically evident fact that the thermal energy in the rod must decrease over time.

Finally, Cor. 2-3 implies that the solutions to our heat conduction problems vary continuously with the given data.

Theorem 2-3. Let $a \geq 0$. Suppose that $u_i(x, t)$ is continuous on $0 \leq x \leq L$, $0 \leq t \leq T$ and satisfies

$$\frac{\partial}{\partial t} u_i = a \frac{\partial^2}{\partial x^2} u_i + F_i(x, t), \qquad 0 < x < L, \qquad 0 < t \leq T,$$
$$u_i(0, t) = g_i(t), \qquad u_i(L, t) = h_i(t), \qquad 0 \leq t \leq T,$$
$$u_i(x, 0) = f_i(x), \qquad 0 \leq x \leq L,$$

for $i = 1, 2$. Suppose further that

$$|f_1(x) - f_2(x)| < \varepsilon, \qquad |g_1(t) - g_2(t)| < \varepsilon, \qquad |h_1(t) - h_2(t)| < \varepsilon,$$
$$|F_1(x, t) - F_2(x, t)| \leq \varepsilon$$

for $0 \leq x \leq L$, $0 \leq t \leq T$. Then

$$|u_1(x, t) - u_2(x, t)| \leq (1 + T)\varepsilon.$$

The maximum principle coupled with physical insight and judicious guessing provides a powerful means for approximating solutions to initial, boundary value problems. Consider, for example, the problem

$$u_t = u_{xx}, \qquad 0 < x < 1, \qquad t > 0$$
$$u(0, t) = 0, \qquad u(1, t) = 0, \qquad t \geq 0,$$
$$u(x, 0) = x(1 - x)e^x, \qquad 0 \leq x \leq 1.$$

Since the initial condition and boundary data are nonnegative, we infer that $u(x, t) \geq 0$. To bound u from above, notice that for $t > 0$ and small, $u(x, t)$ is nearly equal to its initial data. This suggests comparing $u(x, t)$ with

$$v(x, t) = \beta x(1 - x)e^{-\alpha t},$$

where we will try to adjust the parameters $\alpha, \beta > 0$ so that $u(x, t) \leq v(x, t)$. [The factor $e^{-\alpha t}$ is suggested by the terms in the series solution in Sec. 5-1. The factor $x(1 - x)$ is retained from the initial conditions so that v will satisfy the boundary data and e^x, which varies between 1 and e, is replaced by a constant β.] Now set $w = u - v$. Then

$$w_t - w_{xx} = \beta e^{-\alpha t}[\alpha x(1 - x) - 2] \leq 0$$

provided that $\alpha \leq 8$. Also, $w(0, t) = w(1, t) = 0$, and

$$w(x, 0) = x(1 - x)[e^x - \beta] \leq 0$$

provided that $\beta \geq e$. In particular, with the choices $\alpha = 8$ and $\beta = 3$, Th. 2-1 implies that $w \leq 0$ or $u \leq v$. Consequently,

$$0 \leq u(x, t) \leq 3x(1 - x)e^{-8t}.$$

This inequality is particularly valuable if we need to know how rapidly the effects of the initial conditions are damped out by diffusion. Already for $t = 1$ this inequality implies that $0 \le u(x, 1) \le 2.52 \times 10^{-4}$. For all practical purposes the rod has cooled down to temperature zero in one unit of time.

PROBLEMS

1. Show there exists at most one solution $u(x, t)$ which is once continuously differentiable with respect to x in $0 \le x \le L$, $0 \le t \le T$ and twice continuously differentiable with respect to x and once with respect to t in $0 < x < L$, $0 < t \le T$ and which satisfies

$$u_t = au_{xx} + F(x, t), \qquad 0 < x < L, \qquad 0 < t \le T,$$
$$u(x, 0) = f(x), \qquad 0 \le x \le L$$

and on the left either

$$\text{(i)} \quad u(0, t) = \phi(t)$$

or

$$\text{(ii)} \quad u_x(0, t) = \phi(t)$$

or

$$\text{(iii)} \quad u_x(0, t) - \alpha(t)u(0, t) = \phi(t), \qquad \alpha(t) \ge 0,$$

and on the right either

$$\text{(iv)} \quad u(L, t) = \psi(t)$$

or

$$\text{(v)} \quad u_x(L, t) = \psi(t)$$

or

$$\text{(vi)} \quad u_x(L, t) + \beta(t)u(L, t) = \psi(t), \qquad \beta(t) \ge 0,$$

where $\phi(t)$ and $\psi(t)$, and in the cases of (iii) and (vi), $\alpha(t)$ and $\beta(t)$ are given. *Hint.* Assume that there are two solutions, $u_1(x, t)$ and $u_2(x, t)$, and let $u = u_1 - u_2$. Observe that u satisfies the same type of problem with zero boundary and initial conditions and zero source term. Thus, one must show that u vanishes identically. For this let

$$I(t) = \frac{1}{2}\int_0^L u^2(x, t)\, dx.$$

Use the differential equation, integration by parts, and the boundary conditions to conclude that $I'(t) \le 0$.

2. For $T > 0$ let $Q_T = \{(x, t): 0 < x < L,\ 0 < t \le T\}$, be a rectangle with its bottom and two vertical sides removed. Let

$$\bar{Q}_T = \{(x, t): 0 \le x \le L, 0 \le t \le T\}.$$

Then $B_T = \bar{Q}_T - Q_T$ consists of the three missing sides of the original rectangle Q_T. Suppose that $w(x, t)$ satisfies the differential inequality

$$w_t - a(x, t)w_{xx} - b(x, t)w_x - c(x, t)w < 0 \quad \text{for} \quad (x, t) \in Q_T,$$

where $a(x, t) \ge 0$ and $c(x, t) \le 0$ in Q_T. Then $w(x, t)$ cannot achieve a positive local maximum in Q_T.

3. Let a, b, c be as in Prob. 2. Suppose that $u(x, t)$ is continuous on $\bar{Q}_T$ and satisfies

$$u_t - a(x, t)u_{xx} - b(x, t)u_x - c(x, t)u \le 0 \quad \text{on} \quad Q_T,$$
$$u(x, t) \le 0 \quad \text{on} \quad B_T.$$

Then $u(x, t) \le 0$ on $\bar{Q}_T$.

4. Let a, b, and c be as in Prob. 2. Suppose that $u(x, t)$ is continuous on $\bar{Q}_T$ and satisfies

$$u_t - a(x, t)u_{xx} - b(x, t)u_x - c(x, t)u = 0 \quad \text{in} \quad Q_T.$$

If u has a positive maximum, it is assumed on B_T, and if u has a negative minimum, it is assumed on B_T.

5. Let a, b, and c be as in Prob. 2. Suppose that $F(x, t)$ is defined in Q_T and bounded by the constant N there. Suppose that $u(x, t)$ is continuous on $\bar{Q}_T$, $M = \max_{B_T} |u(x, t)|$, and $u(x, t)$ satisfies

$$u_t - a(x, t)u_{xx} - b(x, t)u_x - c(x, t)u = F(x, t) \quad \text{in} \quad Q_T.$$

Then $|u(x, t)| \leq M + NT$ in Q_T.

6. Let $a(x, t) > 0$ and $c(x, t) \leq 0$ for all $0 < x < L$ and $t > 0$. Show that the initial, boundary value problem

$$\begin{aligned} &u_t - a(x, t)u_{xx} - b(x, t)u_x - c(x, t)u = F(x, t), \qquad 0 < x < L, \qquad t > 0, \\ &u(0, t) = \phi(t), \qquad u(L, t) = \psi(t), \qquad t \geq 0, \\ &u(x, 0) = f(x), \qquad 0 \leq x \leq L, \end{aligned}$$

has at most one solution which is continuous on $0 \leq x \leq L$ and $t \geq 0$. [Notice that if there is, in fact, a continuous solution, then the boundary and initial data must be compatible, that is, $\phi(0) = f(0)$, $\psi(0) = f(L)$.]

7. Suppose that $u(x, t)$ is the solution to

$$\begin{aligned} &u_t - u_{xx} = 1, \qquad 0 < x < 1, \qquad t > 0, \\ &u(0, t) = u(1, t) = 0, \qquad t \geq 0, \\ &u(x, 0) = 0, \qquad 0 \leq x \leq 1. \end{aligned}$$

Argue on physical grounds that there should be a steady state for this problem. Show that this steady-state solution is $u(x) = (x - x^2)/2$. Explain why it is physically plausible that $u(x, t) \leq (x - x^2)/2$, then derive the estimates

$$\frac{(x - x^2)(1 - e^{-8t})}{2} \leq u(x, t) \leq \frac{x - x^2}{2}.$$

How much time must elapse before these estimates permit the solution to be computed to within an error of 10^{-2}?

-3. The Green's Function

We have seen that the solution to the initial, boundary value problem

$$u_t = au_{xx}, \qquad 0 \leq x \leq L, \qquad t > 0, \tag{3-1}$$

$$u(x, 0) = f(x), \qquad 0 \leq x \leq L, \tag{3-2}$$

$$u(0, t) = u(L, t) = 0, \qquad t > 0, \tag{3-3}$$

is

$$u(x, t) = \sum_{n=1}^{\infty} c_n e^{-a\lambda_n^2 t} \sin(\lambda_n x), \qquad \lambda_n = \frac{n\pi}{L}, \tag{3-4}$$

where

$$c_n = \frac{2}{L}\int_0^L f(y) \sin(\lambda_n y)\, dy. \tag{3-5}$$

For this we assumed that the initial temperature $f(x)$ satisfied $f(0) = f(L) = 0$ and was twice continuously differentiable. For $t > 0$ insert (3-5) into (3-4) to obtain the following integral representation for the solution:

$$\begin{aligned} u(x, t) &= \sum_{n=1}^{\infty} \frac{2}{L} e^{-a\lambda_n^2 t}\left[\int_0^L f(y) \sin(\lambda_n y)\, dy\right] \sin(\lambda_n x) \\ &= \int_0^L \left[\frac{2}{L}\sum_{n=1}^{\infty} e^{-a\lambda_n^2 t} \sin(\lambda_n y) \sin(\lambda_n x)\right] f(y)\, dy. \end{aligned}$$

It is easy to check (as in Sec. 5-1) that the series in brackets converges uniformly together with all of its partial derivatives on $0 \le x \le L, t \ge t_0$ for any $t_0 > 0$. Consequently, the interchange of summation and integration above is valid for $0 \le x \le L$ and $t > 0$. Define

$$G(x, y, t) = \frac{2}{L}\sum_{n=1}^{\infty} e^{-a\lambda_n^2 t} \sin(\lambda_n y) \sin(\lambda_n x) \tag{3-6}$$

for $0 \le x, y \le L$ and $t > 0$. Then the solution (3-4) can be expressed as

$$u(x, t) = \int_0^L G(x, y, t) f(y)\, dy \tag{3-7}$$

for $0 \le x \le L$ and $t > 0$. The function $G(x, y, t)$ defined in (3-6) is called the *Green's function* for the heat equation. The uniform convergence noted above implies that the Green's function $G(x, y, t)$ is continuous on its domain.

In order to obtain the solution (3-4) or its reformulation (3-7) in terms of the Green's function, we imposed differentiability assumptions on the initial temperature distribution which have no physical significance. Now we can dispense with these smoothness assumptions by working directly with the Green's function, which we shall show has the following properties:

$$G_t(x, y, t) = aG_{xx}(x, y, t) = aG_{yy}(x, y, t) \tag{3-8}$$

for $0 \le x, y \le L, t > 0$;

$$G(0, y, t) = G(L, y, t) = 0 \quad \text{for} \quad 0 \le y \le L, \qquad t > 0; \tag{3-9}$$

$$\lim_{(x,t)\to(x_0,0+)} \int_0^L G(x, y, t) f(y)\, dy = f(x_0), \qquad 0 \le x_0 \le L \tag{3-10}$$

for every continuous function $f(x)$ on $0 \le x \le L$ with $f(0) = f(L) = 0$;

$$G(x, y, t) = G(y, x, t) \quad \text{for} \quad 0 \le x, y \le L, \qquad t > 0. \tag{3-11}$$

Equation (3-8) follows from (3-6) upon term-by-term differentiation in view of the uniform convergence already mentioned. So the Green's function satisfies the heat equation. By definition it also satisfies the boundary conditions (3-9) because $\lambda_n = n\pi/L$. The symmetry condition (3-11) is also clear. Property (3-10) has already been established under the added assumption that $f(x)$ has a continuous second derivative because then (3-7) holds and the results in Sec. 5-1 apply. We use the maximum principle to see that (3-10) still holds when $f(x)$ is merely continuous and $f(0) = f(L) = 0$. Given such initial data, choose a sequence of twice continuously differentiable functions $\{f_n(x)\}$ which converge uniformly to $f(x)$ and satisfy $f_n(0) = f_n(L) = 0$. Then given $\varepsilon > 0$ there is an index $N(\varepsilon)$ such that

$$n, m \geq N(\varepsilon) \quad \text{implies} \quad |f_n(x) - f_m(x)| < \varepsilon \quad \text{for} \quad 0 \leq x \leq L.$$

Let $u_n(x, t)$ be the solution to the heat equation with zero boundary values and initial temperature $f_n(x)$. This solution is given by (3-4) and (3-5), with f_n replacing f. For $t > 0$, it is also given by (3-7) with f_n replacing f. By the maximum principle

$$n, m \geq N(\varepsilon) \quad \text{implies that} \quad |u_n(x, t) - u_m(x, t)| < \varepsilon$$

for $0 \leq x \leq L$, $t \geq 0$, and consequently $\{u_n(x, t)\}$ converges uniformly to a function $u(x, t)$ on $0 \leq x \leq L$, $t \geq 0$. Evidently, $u(x, t)$ is continuous. Moreover, for each (x, t) with $t > 0$,

$$\begin{aligned} u(x, t) &= \lim_{n\to\infty} u_n(x, t) = \lim_{n\to\infty} \int_0^L G(x, y, t) f_n(y)\, dy \\ &= \int_0^L G(x, y, t) f(y)\, dy. \end{aligned}$$

Since $u(x, t)$ is continuous on $0 \leq x \leq L$, $t \geq 0$ and $u(x_0, 0) = f(x_0)$, Eq. (3-10) follows at once.

With these preparations, we easily obtain:

Theorem 3-1. Let the initial temperature $f(x)$ be continuous on $0 \leq x \leq L$ and satisfy $f(0) = f(L) = 0$. Then the initial, boundary value problem (3-1)–(3-3) has a unique, continuous solution defined on $0 \leq x \leq L$, $t \geq 0$. This solution is given by (3-7) for $t > 0$.

To confirm this, define $v(x, t)$ on $0 \leq x \leq L$, $t \geq 0$ by

$$v(x, t) = \begin{cases} \int_0^L G(x, y, t) f(y)\, dy, & 0 \leq x \leq L, \quad t > 0, \\ f(x), & 0 \leq x \leq L, \quad t = 0. \end{cases} \tag{3-12}$$

Then $v(x, t)$ satisfies the heat equation (3-1) by (3-8) and the uniform convergence already mentioned. Also, (3-2) holds trivially and (3-3) is satisfied because of (3-9). Finally, (3-10) shows that $v(x, t)$ is continuous on its domain. So $v(x, t)$ is the required solution. Of course, uniqueness follows from the maximum principle.

Consider a rod that is heated to an initial temperature distribution $f(x)$, $0 \leq x \leq L$, and whose ends are suddenly thrust into large ice baths at temperature zero. The resulting initial, boundary value problem for the rod is (3-1)–(3-3), but now we must admit the possibility that $f(0) \neq 0$ and/or $f(L) \neq 0$. What temperature distribution will result? We can no longer expect a continuous solution $u(x, t)$ to

(3-1)–(3-3). Indeed, assuming that $f(0) \neq 0$ and $f(L) \neq 0$, the initial, boundary value problem itself shows that $u(x, t)$ must be discontinuous at the points (0, 0) and (L, 0). This is the worst that can happen:

Theorem 3-2. Let the initial temperature $f(x)$ be continuous on $0 \leq x \leq L$. Then the initial, boundary value problem (3-1)–(3-3) has a solution that is defined on $0 \leq x \leq L$, $t \geq 0$ and continuous expect possibly at the points (0, 0) and (L, 0).

The solution is again given by (3-12), as we shall check. As above, $v(x, t)$ clearly satisfies (3-1)–(3-3). We must show that $v(x, t)$ is continuous except possibly at the indicated points. To this end let

$$l(x) = f(0) + \frac{f(L) - f(0)}{L} x$$

be the linear function that interpolates the values $f(0)$ and $f(L)$. We can express $v(x, t)$ as

$$v(x, t) = \begin{cases} \int_0^L G(x, y, t)[f(y) - l(y)]\, dy + \int_0^L G(x, y, t)l(y)\, dy, \\ \qquad\qquad 0 \leq x \leq L, \quad t > 0, \\ f(x), \quad 0 \leq x \leq L, \quad t = 0. \end{cases}$$

Let

$$w(x, t) = \int_0^L G(x, y, t)[f(y) - l(y)]\, dy.$$

Since $f(y) - l(y)$ vanishes at $y = 0$ and L, we infer from (3-10) that

$$w(x, t) \to f(x_0) - l(x_0) \quad \text{as} \quad (x, t) \to (x_0, 0+)$$

for $0 \leq x_0 \leq L$. Consequently, to show that $v(x, t)$ is continuous except possibly at (0, 0) and (L, 0), we must confirm that

$$\lim_{(x,t)\to(x_0,0+)} \int_0^L G(x, y, t)l(y)\, dy = l(x_0) \tag{3-13}$$

for $0 < x_0 < L$.

One justification of (3-13) is based on elementary properties of Fourier series. Since $t > 0$, the integral on the left of (3-13), which is the integral in (3-7) with $f(y) = l(y)$, can be expressed as (3-4), (3-5) with $f(y)$ replaced by $l(y)$:

$$\int_0^L G(x, y, t)l(y)\, dy = \sum_{n=1}^{\infty} c_n e^{-a\lambda_n^2 t} \sin(\lambda_n x), \tag{3-14}$$

where

$$c_n = \frac{2}{L} \int_0^L l(y) \sin(\lambda_n y)\, dy. \tag{3-15}$$

Now it is elementary that the Fourier sine series of $l(x)$ converges uniformly to $l(x)$ on any closed interval $[c, d]$ with $0 < c < d < L$ (see Th. 1-3 of Chap. 3). [In fact, c can be 0 when $l(0) = 0$ and d can be L when $l(L) = 0$.] Fix x_0 with $0 < x_0 < L$ and an interval $[c, d]$ with $0 < c < x_0 < d < L$. Then

$$l(x) = \sum_{n=1}^{\infty} c_n \sin(\lambda_n x), \qquad \text{uniformly on} \quad [c, d], \tag{3-16}$$

where c_n are given by (3-15).

For $t > 0$ and x in $[c, d]$, we have

$$\begin{aligned}
\int_0^L & G(x, y, t) l(y)\, dy - l(x_0) \\
&= \sum_{n=1}^{\infty} (e^{-a\lambda_n^2 t} - 1) c_n \sin(\lambda_n x) + l(x) - l(x_0) \\
&= \sum_{n=1}^{N} (e^{-a\lambda_n^2 t} - 1) c_n \sin(\lambda_n x) + \sum_{n=N+1}^{\infty} e^{-a\lambda_n^2 t} c_n \sin(\lambda_n x) \\
&\quad - \sum_{n=N+1}^{\infty} c_n \sin(\lambda_n x) + l(x) - l(x_0).
\end{aligned}$$

We will show that the second and third sums above can be made arbitrarily small independent of x in $[c, d]$ and $t > 0$ by choosing N large enough. Then we can let $t \to 0+$ and $x \to x_0$ to infer that (3-13) holds.

To fill in the missing steps, given $\varepsilon > 0$ use (3-16) to determine $N = N(\varepsilon)$ so that

$$M > N \quad \text{implies that} \quad \left| \sum_{n=M}^{\infty} c_n \sin(\lambda_n x) \right| \le \frac{\varepsilon}{2} \quad \text{for} \quad x \quad \text{in} \quad [c, d]. \tag{3-17}$$

This estimate with $M = N + 1$ takes care of the third sum above. It also implies that

$$\left| \sum_{n=N+1}^{\infty} e^{-a\lambda_n^2 t} c_n \sin(\lambda_n x) \right| \le \frac{\varepsilon}{2}, \tag{3-18}$$

which takes care of the second sum. To see this, let

$$a_n = e^{-a\lambda_n^2 t}, \qquad b_n = c_n \sin(\lambda_n x)$$

and

$$s_n = b_{N+1} + \cdots + b_n \quad \text{for} \quad n > N.$$

Then

$$\begin{aligned}
\sum_{n=N+1}^{P} a_n b_n &= a_{N+1} s_{N+1} + a_{N+2}(s_{N+2} - s_{N+1}) + \cdots + a_P(s_P - s_{P-1}) \\
&= (a_{N+1} - a_{N+2}) s_{N+1} + (a_{N+2} - a_{N+3}) s_{N+2} \\
&\quad + \cdots + (a_{P-1} - a_P) s_{P-1} + a_P s_P.
\end{aligned}$$

In our case $a_n - a_{n+1} \geq 0$ and the "tails" s_n are bounded by (3-17), so we obtain the estimate

$$\sum_{n=N+1}^{P} a_n b_n \leq (a_{N+1} - a_{N+2})\frac{\varepsilon}{2} + (a_{N+2} - a_{N+3})\frac{\varepsilon}{2} + \cdots + (a_{P-1} - a_P)\frac{\varepsilon}{2} + a_P\frac{\varepsilon}{2} = a_{N+1}\frac{\varepsilon}{2}.$$

Since $a_{N+1} < 1$ we have $\sum_{n=N+1}^{P} a_n b_n \leq \frac{\varepsilon}{2}$. Similarly, this sum is greater than $-\varepsilon/2$. Letting $P \to \infty$, we obtain (3-18). This completes the proof of Th. 3-2.

Notice that (3-13) and the reasoning above establishes the following modification of (3-10):

$$\text{(3-10)}' \qquad \lim_{(x,t)\to(x_0,0+)} \int_0^L G(x, y, t)f(y)\, dy = f(x_0), \qquad 0 < x_0 < L$$

for every continuous function $f(x)$ on $0 \leq x \leq L$.

Next consider a general one-dimensional heat flow problem:

$$\text{(3-19)} \qquad u_t = au_{xx} + F(x, t), \qquad 0 \leq x \leq L, \qquad t > 0,$$

$$\text{(3-20)} \qquad u(x, 0) = f(x), \qquad 0 \leq x \leq L,$$

$$\text{(3-21)} \qquad u(0, t) = \phi(t), \qquad u(L, t) = \psi(t), \qquad t \geq 0.$$

We assume that this problem has a solution $u(x, t)$ and proceed to derive a representation for it in terms of the Green's function (3-6). The identity

$$v(u_\tau - au_{yy}) + u(v_\tau + av_{yy}) = (uv)_\tau + a(uv_y - vu_y)_y$$

holds for any function v with the indicated derivatives. Set $v(y, \tau) = G(x, y, t - \tau)$, where G is the Green's function given by (3-6) and x and t are regarded as parameters with $0 \leq x, y \leq L$ and $\tau < t$. From (3-8),

$$v_\tau + av_{yy} = 0 \quad \text{for} \quad 0 \leq x, y \leq L, \qquad \tau < t.$$

Use of this equation in the previous identity leads to

$$\int_\varepsilon^{t-\varepsilon} \int_0^L G(x, y, t - \tau)F(y, \tau)\, dy\, d\tau$$

$$= \int_0^L G(x, y, \varepsilon)u(y, t - \varepsilon)\, dy - \int_0^L G(x, y, t - \varepsilon)u(y, \varepsilon)\, dy + a \int_\varepsilon^{t-\varepsilon} (uv_y - vu_y)\Big|_{y=0}^{y=L} d\tau.$$

Since $v = G(x, y, t - \tau)$ vanishes at $y = 0$ and L and (3-10)′ holds, we let $\varepsilon \to 0$ to obtain the representation

$$\text{(3-22)} \qquad u(x, t) = \int_0^L G(x, y, t)f(y)\, dy + a \int_0^t G_y(x, 0, t - \tau)\phi(\tau)\, d\tau - a \int_0^t G_y(x, L, t - \tau)\psi(\tau)\, d\tau + \int_0^t \int_0^L G(x, y, t - \tau)F(y, \tau)\, dy\, d\tau.$$

We obtained this solution formula under the assumption that (3-19)–(3-21) has a solution. This can now be confirmed by checking that (3-22) does satisfy this initial, boundary value problem. We defer this check until the next section, where another formula for the Green's function (3-6) is obtained. This alternative formula will make it easier to confirm that (3-22) solves the general heat conduction problem.

PROBLEMS

1. Show that there is at most one continuous function $G(x, y, t)$ such that (3-7) solves the heat conduction problem (3-1)–(3-3) for all smooth initial data $f(x)$ with $f(0) = f(L) = 0$. In short, (3-7) determines the Green's function uniquely. *Hint.* Assume that

$$u(x, t) = \int_0^L G(x, y, t)f(y)\,dy \quad \text{and} \quad u(x, t) = \int_0^L H(x, y, t)f(y)\,dy$$

both solve (3-1)–(3-3). Use the fact that the initial, boundary value problem has a unique solution and the arbitrariness of f to conclude that $G = H$.

2. Show that $G(x, y, t) \geq 0$. *Hint.* Assume the contrary and argue in a manner similar to Prob. 1.

3. Show that

$$\int_0^L G(x, y, t - t_1)G(y, z, t_1 - t_0)\,dy = G(x, z, t - t_0)$$

for $t > t_1 > t_0$. This result can be proven directly or it can be done rather "painlessly" by applying Prob. 1. That is, multiply the equality above by $f(z)\,dz$ and integrate from 0 to L. Consider the heat conduction problem solved by these functions.

4. Find a Green's function for the equation

$$\begin{aligned} &u_t = au_{xx} + bu_x + cu, \qquad 0 \leq x \leq L, \qquad t > 0,\\ &u(0, t) = u(L, t) = 0, \qquad t > 0,\\ &u(x, 0) = f(x), \qquad 0 \leq x \leq L. \end{aligned}$$

where $a > 0$, b, and c are constants.

5. Find a function like the Green's function for the equation

$$\begin{aligned} &u_t = au_{xx}, \qquad 0 \leq x \leq L, \qquad t > 0,\\ &u_x(0, t) = u_x(L, t) = 0, \qquad t > 0,\\ &u(x, 0) = f(x), \qquad 0 \leq x \leq L. \end{aligned}$$

Note. The function obtained is denoted by $N(x, y, t)$ and is called the *Neumann function.*

6. Derive a solution formula analogous to (3-22) for solutions to

$$\begin{aligned} &u_t = au_{xx}, \qquad 0 \leq x \leq L, \qquad t > 0,\\ &u_x(0, t) = \phi(t), \qquad u_x(L, t) = \psi(t), \qquad t > 0,\\ &u(x, 0) = f(x), \qquad 0 \leq x \leq L, \end{aligned}$$

using the Neumann function in place of the Green's function.

7. Look over the proof of Th. 3-2 and confirm that the initial, boundary value problem has a solution that is also continuous at (0, 0) when $f(0) = 0$ and is continuous at $(L, 0)$ when $f(L) = 0$.

5-4. Heat Flow in Infinite and Semi-Infinite Rods

The temperature $u(x, t)$ in an infinite rod, taken to be the entire x axis, is given by

$$u_t = au_{xx}, \qquad -\infty < x < \infty, \qquad t > 0, \tag{4-1}$$

$$u(x, 0) = f(x), \qquad -\infty < x < \infty, \tag{4-2}$$

where $f(x)$ is the given initial temperature distribution. Problem (4-1)–(4-2) also describes a diffusion process with $u(x, t)$ measuring the concentration of a diffusing substance, assuming that convection effects are negligible.

To solve this problem, we could first solve the finite interval problem as in Sec. 5-1 and then make a limit passage to obtain the solution to (4-1)–(4-2). This approach leads to a discussion analogous to that used in Chap. 3 to derive the Fourier integral formulas. Consequently, we solve (4-1) and (4-2) directly using Fourier integrals. The calculations will be strictly formal; therefore, the solution must be justified after it is obtained. Recall that a sufficiently well-behaved function $g(x)$ has Fourier transform

$$\hat{g}(\omega) = \int_{-\infty}^{\infty} e^{i\omega x}\, g(x)\, dx, \tag{4-3}$$

and the function $g(x)$ can be recovered from its transform via the inversion formula

$$g(x) = \frac{1}{2\pi}\int_{-\infty}^{\infty} e^{-i\omega x}\, \hat{g}(\omega)\, d\omega. \tag{4-4}$$

In order to transform the heat equation, we multiply (4-1) by $e^{i\omega x}$, integrate over $-\infty < x < \infty$, and then integrate by parts to find

$$\begin{aligned} 0 &= \int_{-\infty}^{\infty} e^{i\omega x}\,(u_t - au_{xx})\, dx = \frac{\partial}{\partial t}\int_{-\infty}^{\infty} e^{i\omega x} u(x, t)\, dx - a\int_{-\infty}^{\infty} e^{i\omega x} u_{xx}(x, t)\, dx \\ &= \hat{u}_t(\omega, t) + a\omega^2 \int_{-\infty}^{\infty} e^{i\omega x} u(x, t)\, dx = \hat{u}_t(\omega, t) + a\omega^2 \hat{u}(\omega, t). \end{aligned}$$

We have assumed that $u_x(x, t)$ and $u(x, t)$ tend to zero as x tends to $\pm\infty$ in the integration by parts. The initial condition (4-2) transforms into

$$\hat{u}(\omega, 0) = \hat{f}(\omega).$$

Thus, the partial differential equation (4-1) and initial data (4-2) transform into the simple initial value problem

$$\hat{u}_t = -a\omega^2 \hat{u}, \qquad \hat{u}(\omega, 0) = \hat{f}(\omega),$$

which has the solution

$$\hat{u}(\omega, t) = \hat{f}(\omega)e^{-a\omega^2 t}.$$

The solution $u(x, t)$ to the heat conduction problem is obtained from its Fourier transform by applying (4-4):

$$u(x, t) = \frac{1}{2\pi}\int_{-\infty}^{\infty} e^{-i\omega x}\hat{u}(\omega, t)\, d\omega = \frac{1}{2\pi}\int_{-\infty}^{\infty} e^{-i\omega x - a\omega^2 t}\hat{f}(\omega)\, d\omega$$

$$= \frac{1}{2\pi}\int_{-\infty}^{\infty}\left[\int_{-\infty}^{\infty} e^{-i\omega(x-y) - a\omega^2 t}\, d\omega\right] f(y)\, dy.$$

Refer to Ex. 2 of Sec. 3-4 on Fourier transforms to find

$$\int_{-\infty}^{\infty} e^{-i\omega\beta - \omega^2/4\alpha}\, d\omega = \sqrt{4\pi\alpha}\, e^{-\alpha\beta^2}, \qquad \alpha > 0.$$

Thus, with $\alpha = 1/4at$ and $\beta = x - y$, we obtain, for $t > 0$,

$$u(x, t) = \frac{1}{\sqrt{4\pi a t}}\int_{-\infty}^{\infty} e^{-(x-y)^2/4at} f(y)\, dy.$$

For convenience we write

$$u(x, t) = \int_{-\infty}^{\infty} k(x - y, at) f(y)\, dy, \tag{4-5}$$

where

$$k(x, t) = \frac{1}{\sqrt{4\pi t}}\, e^{-x^2/4t} \tag{4-6}$$

for $-\infty < x < \infty$ and $t > 0$. The function $k(x, t)$ is called the *fundamental solution* to the heat equation. It plays the same role for the infinite rod that the Green's function does for a rod with finite length.

Our derivation of the solution (4-5) is only formal and we will verify shortly that it really does solve the problem (4-1), (4-2). However, first we note that for numerical purposes it is usually more convenient to express (4-5) as

$$u(x, t) = \frac{1}{\sqrt{\pi}}\int_{-\infty}^{\infty} e^{-\xi^2} f(x - \sqrt{4at}\, \xi)\, d\xi, \tag{4-7}$$

obtained by using the change of variables $\xi = (x - y)/\sqrt{4at}$.

It is clear from (4-6) that (4-5) cannot give a solution to (4-1)–(4-2) for all continuous functions $f(x)$. Some restrictions must be placed on $f(x)$ to be sure that the integral will converge. For example, the integral (4-5) diverges when $f(x) = \exp(x^4)$. On the other hand, a realistic physical assumption is that the initial temperature distribution $f(x)$ is bounded. Then (4-5) gives a solution of (4-1), (4-2). We will establish an even stronger result shortly.

A routine calculation using (4-8) gives

$$\text{(4-8)}\qquad \begin{cases} k_x(x, t) = -\dfrac{2x}{4t}\, k(x, t), \\ k_t(x, t) = k_{xx}(x, t) = \dfrac{x^2 - 2t}{4t^2}\, k(x, t). \end{cases}$$

Thus, $k(x - y, at)$ satisfies (4-1), and (4-5) also will satisfy (4-1) if we can justify differentiation under the integral sign. We need the following technical estimates on the fundamental solution for this justification.

Lemma 4-1. Let $0 < t_0 < T$, $R > 0$ and $\gamma > 1$ be given. Then there is a constant N depending only on t_0, T, R and γ such that

$$0 \le k(x - y, at) \le N \exp\left(-\frac{y^2}{4\gamma aT}\right), \qquad |k_x(x - y, at)| \le N \exp\left(-\frac{y^2}{4\gamma aT}\right),$$

$$|k_t(x - y, at)| = |k_{xx}(x - y, at)| \le N \exp\left(-\frac{y^2}{4\gamma aT}\right)$$

for all $|x| \le R$ and $t_0 \le t \le T$.

The proof is a routine calculation.

$$k(x - y, at) = \frac{1}{\sqrt{4\pi at}} \exp\left[-\frac{(x - y)^2}{4at}\right] \le \frac{1}{\sqrt{4\pi at_0}} \exp\left[-\frac{(x - y)^2}{4aT}\right]$$

$$\le \frac{1}{\sqrt{4\pi at_0}} \exp\left(\frac{-x^2 + 2xy - y^2}{4aT}\right)$$

$$\le \frac{1}{\sqrt{4\pi at_0}} \exp\left(\frac{2R|y|}{4aT}\right) \exp\left\{-\frac{y^2[1 - (1/\gamma)]}{4aT}\right\} \exp\left(-\frac{y^2}{4\gamma aT}\right).$$

The coefficient of $\exp(-y^2/4\gamma aT)$ is continuous in y and has limit 0 as $|y| \to \infty$; consequently, this coefficient is bounded, which establishes the first inequality. Similar estimates follow for k_x, k_{xx}, and k_t using (4-8). The bound N in the lemma is the largest of the three bounds determined in this way.

Since

$$\int_{-\infty}^{\infty} \exp\left(-\frac{y^2}{4\gamma aT}\right) dy < \infty,$$

Lemma 4-1, the Weierstrass M-test, and Prop. 5-4 of Chap. 3 show that

$$\int_{-\infty}^{\infty} k(x - y, at) f(y)\, dy,$$

where $f(y)$ is any bounded, integrable function, can be differentiated under the integral sign once with respect to t and twice with respect to x for $|x| \le R$ and $0 < t_0 \le t \le T$. Let $t_0 \to 0$ and $R, T \to \infty$ to see that these derivatives can be taken under the integral sign for $-\infty < x < \infty$ and $t > 0$. Moreover, we have already seen that the fundamental solution satisfies

$$k_t(x, t) = k_{xx}(x, t) \quad \text{for} \quad -\infty < x < \infty, \qquad t > 0. \tag{4-9}$$

Consequently, (4-5) satisfies the heat equation (4-1), and we are well on our way to establishing the following results.

Theorem 4-1. Suppose that the initial temperature distribution is continuous and bounded on $-\infty < x < \infty$. Then the diffusion problem (4-1)–(4-2) has a solution which is continuous on $-\infty < x < \infty$, $t \geq 0$. For $t > 0$ such a solution is given by (4-5).

In fact, we shall show that (4-1)–(4-2) has the continuous solution

$$u(x, t) = \begin{cases} \displaystyle\int_{-\infty}^{\infty} k(x - y, at) f(y)\, dy, & -\infty < x < \infty, \qquad t > 0, \\ f(x), \quad -\infty < x < \infty, & t = 0. \end{cases} \tag{4-10}$$

We have just shown that (4-10) satisfies the heat equation (4-1). It obviously satisfies the initial condition (4-2). So only the continuity of $u(x, t)$ remains to be shown. That is, we must verify that

$$\lim_{(x,t)\to(x_0,0+)} \int_{-\infty}^{\infty} k(x - y, at) f(y)\, dy = f(x_0) \tag{4-11}$$

for $-\infty < x_0 < \infty$. The key to this is another basic property of the fundamental solution:

$$\int_{-\infty}^{\infty} k(x - y, at)\, dy = 1. \tag{4-12}$$

This follows easily from the change of variables $\xi = (x - y)/\sqrt{4at}$, which transforms the integral in (4-12) into

$$\frac{1}{\sqrt{\pi}} \int_{-\infty}^{\infty} \exp(-\xi^2)\, d\xi = 1,$$

a familiar result.

Now use (4-12) to get

$$\begin{aligned} &\int_{-\infty}^{\infty} k(x - y, at) f(y)\, dy - f(x_0) \\ &\quad = \int_{-\infty}^{\infty} k(x - y, at)[f(y) - f(x_0)]\, dy \\ &\quad = \int_{|y - x_0| < \delta} k(x - y, at)[f(y) - f(x_0)]\, dy \\ &\qquad + \int_{|y - x_0| \geq \delta} k(x - y, at)[f(y) - f(x_0)]\, dy \\ &\quad \equiv J_1 + J_2, \text{ respectively.} \end{aligned}$$

Given $\varepsilon > 0$, choose $\delta > 0$ so that $|f(y) - f(x_0)| < \varepsilon$ for $|y - x_0| < \delta$. Then, for all x and $t > 0$,

$$|J_1| \leq \int_{|y-x_0|<\delta} k(x - y, at)|f(y) - f(x_0)|\, dy$$

$$< \varepsilon \int_{|y-x_0|<\delta} k(x - y, at)\, dy < \varepsilon$$

by (4-12). Also, since $f(x)$ is bounded, $|f(x)| \leq M < \infty$ and

$$\begin{aligned}
|J_2| &\leq \int_{|y-x_0|\geq\delta} k(x - y, at)|f(y) - f(x_0)| \\
&\leq 2M \int_{|y-x_0|\geq\delta} k(x - y, at)\, dy \\
&= 2M \int_{-\infty}^{x_0-\delta} k(x - y, at)\, dy + 2M \int_{x_0+\delta}^{\infty} k(x - y, at)\, dy \\
&= \frac{2M}{\sqrt{\pi}} \left[\int_{-\infty}^{(x_0-\delta-x)/\sqrt{4at}} \exp(-\xi^2)\, d\xi + \int_{(x_0+\delta-x)/\sqrt{4at}}^{\infty} \exp(-\xi^2)\, d\xi \right]
\end{aligned}$$

with the change of variables $\xi = (y - x)/\sqrt{4at}$. Now for $|x - x_0| < \delta/2$ we obtain the estimate

$$|J_2| \leq \frac{2M}{\sqrt{\pi}} \left[\int_{-\infty}^{-\delta/2\sqrt{4at}} \exp(-\xi^2)\, d\xi + \int_{\delta/2\sqrt{4at}}^{\infty} \exp(-\xi^2)\, d\xi \right] < \varepsilon$$

provided that $t > 0$ is sufficiently close to 0. These estimates on J_1 and J_2 establish (4-11) and conclude the proof.

Theorem 4-1 leads to an important physical interpretation of the fundamental solution $k(x, t)$ in terms of a diffusion process. Assume initially that the entire mass of a substance is localized at the point x_0 and that the total mass is Q. There is not really a density function which can be used to describe the concentration of mass at a point, but we can idealize the situation somewhat. Let $\delta_n(x) \geq 0$ be a sequence of continuous functions which vanish for $|x - x_0| > 1/n$ and satisfy

$$\int_{-\infty}^{\infty} \delta_n(x)\, dx = 1$$

Then $Q\delta_n(x)$ is the concentration of a mass distribution nearly localized at x_0 and with total mass Q. Suppose that this mass distribution diffuses according to (4-1). Let $u_n(x, t)$ be the solution (concentration) determined by (4-1) and the initial data $Q\delta_n(x)$. Then for $t > 0$,

$$\begin{aligned}
u_n(x, t) &= \int_{-\infty}^{\infty} k(x - y, at) Q\delta_n(y)\, dy \\
&= \int_{x_0-1/n}^{x_0+1/n} k(x - y, at) Q\delta_n(y)\, dy.
\end{aligned}$$

According to the mean value theorem for integrals,

$$u_n(x, t) = k(x - y_n, at) \int_{x_0 - 1/n}^{x_0 + 1/n} Q\delta_n(y)\, dy = Qk(x - y_n, at)$$

for some y_n between $x_0 - 1/n$ and $x_0 + 1/n$. Let $n \to \infty$ to reach the idealized situation where all the mass is localized at x_0

$$\lim_{n \to \infty} u_n(x, t) = Qk(x - x_0, at).$$

Thus, $k(x - x_0, at)$ represents the concentration at x and t of a unit of mass concentrated at x_0 when $t = 0$. With this interpretation (4-12) expresses the conservation of mass, and (4-5) simply states that the mass distribution due to an arbitrary initial distribution $f(x)$ is the superposition of the distributions resulting from localized sources at x with intensity $f(x)$.

Some additional comments on the proof of Th. 4-1 are in order. Our reasoning shows that (4-5) solves the heat equation on $-\infty < x < \infty$ and $t > 0$ not only for bounded, continuous initial temperature distributions $f(x)$, but also for $f(x)$ bounded and piecewise continuous. For such functions the proof of Th. 4-1 shows that (4-11) still holds at each point x_0 where f is continuous. Consequently, for bounded, piecewise continuous initial temperature distributions, (4-1), (4-2) has a solution on $-\infty < x < \infty$, $t > 0$ which is continuous except at the points $(x_0, 0)$ corresponding to the points of discontinuity of $f(x)$. Such a solution is given by (4-10).

These observations help make it easy to solve heat conduction problems in semi-infinite rods. For example, consider

(4-13) $$u_t = au_{xx}, \qquad x > 0, \qquad t > 0,$$

(4-14) $$u(x, 0) = f(x), \qquad x > 0,$$

(4-15) $$u(0, t) = 0, \qquad t > 0.$$

This problem can be solved from scratch using the Fourier sine transform. However, the following physical considerations lead to the result more simply. Consider an infinite rod with the two initial temperature distributions indicated in Fig. 5-1, where the function $f(x)$ is that of (4-14). It is clear on physical grounds

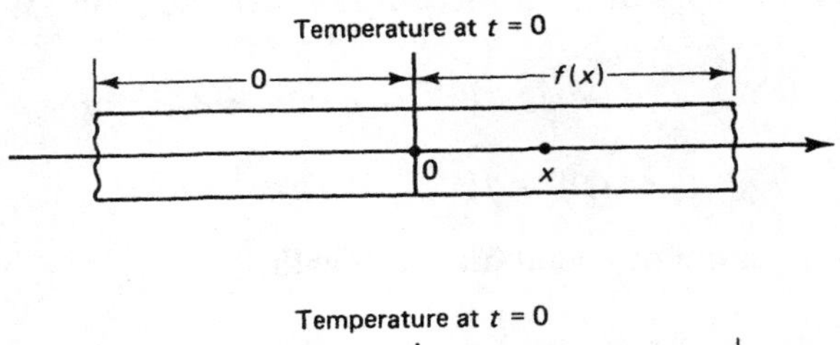

Figure 5-1

that under either initial temperature distribution the rod will have the same temperature at $x = 0$ at any time $t > 0$. Consequently, if $u^+(x, t)$, respectively, $u^-(x, t)$ is the subsequent temperature distribution for the first (resp., second) initial temperature distribution, then

$$u(x, t) = u^+(x, t) - u^-(x, t)$$

will satisfy (4-13)–(4-15). From our previous work

$$u^+(x, t) = \int_0^\infty k(x - y, at)f(y)\,dy$$

$$u^-(x, t) = \int_{-\infty}^0 k(x - y, at)f(-y)\,dy = \int_0^\infty k(x + y, at)f(y)\,dy.$$

Consequently,

$$u(x, t) = \int_0^\infty G(x, y, at)f(y)\,dy, \tag{4-16}$$

where

$$G(x, y, t) = k(x - y, t) - k(x + y, t) \tag{4-17}$$

is called the *Green's function for the semi-infinite interval* $0 < x < \infty$. Although our solution is based partly on physical reasoning, it is easy to check mathematically that (4-16) solves the semi-infinite problem.

By examining our proof of Th. 4-1 a little more carefully, we can show that (4-1) and (4-2) has a solution for continuous initial temperature distributions which grow quite rapidly as $|x| \to \infty$.

Theorem 4-2. Suppose that $f(x)$ is continuous on $-\infty < x < \infty$ and satisfies

$$|f(x)| \le M \exp(\mu x^2)$$

for given constants M and $\mu > 0$. Then the problem

$$u_t = au_{xx}, \qquad -\infty < x < \infty, \qquad 0 < t < \frac{1}{4a\mu},$$

$$u(x, 0) = f(x), \qquad -\infty < x < \infty,$$

has a continuous solution $u(x, t)$ with

$$u(x, t) = \int_{-\infty}^\infty k(x - y, at)f(y)\,dy \tag{4-18}$$

for $-\infty < x < \infty$, $0 < t < 1/4a\mu$.

As in Th. 4-1, we must justify differentiating under the integral sign for $-\infty < x < \infty$ and $0 < t < 1/4a\mu$. This is done by invoking Prop. 5-4 of Chap. 3 with the aid of the Weierstrass M-test and Lemma 4-1. For example, fix $0 < t_0 < T$ and $\gamma > 1$, $R > 0$. Use Lemma 4-1 to obtain a constant N such that

$$|k_{xx}(x - y, at)| \leq N \exp\left(-\frac{y^2}{4\gamma aT}\right)$$

for $|x| \leq R$ and $t_0 \leq t \leq T$. Then

$$|k_{xx}(x - y, at)f(y)| \leq NM \exp\left[-\left(\frac{1}{4\gamma aT} - \mu\right)y^2\right] \equiv M(y)$$

and

$$\int_{-\infty}^{\infty} M(y)\, dy < \infty \quad \text{provided that} \quad T < \frac{1}{4\gamma a\mu}.$$

Thus, u_{xx} can be calculated by differentiation under the integral sign for $|x| \leq R$ and $0 < t_0 \leq t \leq T$, with $T < 1/4\gamma a\mu$. Since R, t_0 and $\gamma > 1$ are arbitrary, we infer that this differentiation is valid for $-\infty < x < \infty$ and $0 < t < 1/4a\mu$. The same result holds for u_t. Consequently, (4-18) satisfies the heat equation on $-\infty < x < \infty$ and $0 < t < 1/4a\mu$. The fact that $u(x, t) \to f(x_0)$ as $(x, t) \to (x_0, 0+)$ is confirmed by a virtual repeat of the argument in Th. 4-1.

Notice that Th. 4-2 implies Th. 4-1. When $f(x)$ is bounded, say $|f(x)| \leq M$, we have $|f(x)| \leq M \exp(\mu x^2)$ for all $\mu > 0$. Consequently, we can let $\mu \to 0$ in $0 < t < 1/4a\mu$ to see that the solution (4-18) is defined for all $t > 0$. Similarly, if $|f(x)| \leq M \exp(\alpha|x|^p)$ for any $p < 2$, solution (4-18) exists for all $t > 0$.

We turn now to uniqueness questions. To establish that there is at most one solution to our infinite interval problem, we need to make assumptions on how solutions grow at infinity.

Theorem 4-3. There is at most one continuous solution to

$$u_t = au_{xx} + F(x, t), \qquad -\infty < x < \infty, \qquad 0 < t < \frac{1}{4a\mu},$$

$$u(x, 0) = f(x), \qquad -\infty < x < \infty,$$

$$|u(x, t)| \leq M \exp(\mu x^2), \qquad -\infty < x < \infty, \qquad 0 < t < \frac{1}{4a\mu},$$

where M and $\mu > 0$ are given constants.

We will first prove Th. 4-3 under the added assumption that $|u_x(x, t)| \leq M \cdot \exp(\mu x^2)$, that is, that the heat flux grows no faster than the bound on temperature. Let $u_1(x, t)$ and $u_2(x, t)$ both solve the heat conduction problem in Th. 4-3 and satisfy the indicated growth restrictions. Then $u = u_1 - u_2$ satisfies $u_t - au_{xx} = 0$, $u(x, 0) = 0$, and the growth estimates $|u|, |u_x| \leq 2M \exp(\mu x^2)$. Fix (x, t) with $-\infty < x < \infty$ and $0 < t < 1/4a\mu$ and define $v(y, \tau) = k(x - y, a(t - \tau))$. Then

$v_\tau + av_{yy} = 0$, so

$$0 = v(u_\tau - au_{yy}) + u(v_\tau + av_{yy}) = (uv)_\tau + a(uv_y - vu_y)_y.$$

Integrate over $-R < y < R$ and $0 < \tau < t - \varepsilon$ and use $u(x, 0) = 0$ to find, with $R > |x|$,

$$\int_{-R}^{R} k(x - y, a\varepsilon)u(y, t - \varepsilon)\, dy = a \int_0^{t-\varepsilon} (vu_y - uv_y)|_{y=-R}^{y=R}\, d\tau.$$

The growth assumptions on u and u_y imply that $(vu_y - uv_y)|_{y=-R}^{y=R} \to 0$ uniformly for $0 \le \tau \le t - \varepsilon$ as $R \to \infty$. Thus

$$\int_{-\infty}^{\infty} k(x - y, a\varepsilon)u(y, t - \varepsilon)\, dy = 0.$$

Letting $\varepsilon \to 0$ and using (4-11) finally gives $u(x, t) = 0$ for $-\infty < x < \infty, 0 < t < 1/4a\mu$ and the uniqueness is established in this case.

To obtain the uniqueness result as stated in Th. 4-3, we must drop the growth assumption on u_y. Then we cannot conclude that $(vu_y - uv_y)|_{y=-R}^{y=R} \to 0$ as $R \to \infty$ as above. To overcome this difficulty a technical trick is required. We introduce a "cutoff" function $\zeta_R(y)$ which in effect makes all functions of y vanish outside the finite interval $|y| < R + 1$. In fact, let $\zeta_R(y) = 1$ for $|y| \le R$, $\zeta_R(y) = 0$ for $|y| \ge R + 1$, $\zeta_R(y)$ be twice continuously differentiable and satisfy $0 \le \zeta_R(y) \le L$, $|\zeta_R'(y)| \le L$, and $|\zeta_R''(y)| \le L$, where L is a fixed constant independent of R. Now in the identity

$$(uw)_\tau + a[uw_y - u_y w]_y = w(u - au_{yy}) + u(w + aw_{yy}) \tag{4-19}$$

let $u(y, \tau)$ be as before and take $w(y, \tau) = \zeta_R(y)v(y, \tau)$ with $v(y, \tau) = k(x - y, a(t - \tau))$ to obtain

$$(u\zeta_R v)_\tau + a[u(\zeta_R v)_y - u_y\zeta_R v]_y = au(2\zeta_R' v_y + \zeta_R'' v).$$

Now integrate this identity over $-(R + 1) < y < (R + 1)$, where again $R > |x|$, $0 < \tau < t - \varepsilon$, and reason much as above using $\zeta_R(y) = 1$ for $|y| < R$ and $\zeta_R'(y) = \zeta_R''(y) = 0$ except on $-(R + 1) < y < -R$, $R < y < R + 1$ to conclude that $u(x, t) = 0$, as before. The details are left for the problems.

We conclude this section with a discussion of the heat conduction problem for the semi-infinite interval when the initial temperature is 0 and the end of the rod is maintained at temperature $\phi(t)$ for $t > 0$:

$$u_t = au_{xx}, \qquad x > 0, \qquad t > 0, \tag{4-20}$$

$$u(x, 0) = 0, \qquad x > 0, \tag{4-21}$$

$$u(0, t) = \phi(t), \qquad t > 0. \tag{4-22}$$

Assuming that a solution $u(x, t)$ exists, we can use (4-19) with $w(y, \tau) = G(x, y, a(t - \tau))$, the Green's function for the semi-infinite interval, and proceed as in the proof of Th. 4-3 to find the representation

$$u(x, t) = a \int_0^t G_y(x, 0, a(t - \tau))\phi(\tau)\, d\tau$$

or

$$u(x, t) = -2a \int_0^t k_x(x, a(t-\tau))\phi(\tau)\, d\tau, \tag{4-23}$$

for $x, t > 0$. Recall that $k_x(x, t) = (-x/2t)k(x, t)$, so

$$k_x(x, a(t-\tau)) = -\frac{x \exp(-x^2/4a(t-\tau))}{2a(t-\tau)\sqrt{4\pi a(t-\tau)}}.$$

From this equation it is easy to confirm that (4-23) satisfies (4-20) and (4-21). Finally, we need to show that (4-23) assumes its boundary values as $x \to 0+$. For this we need the fact that

$$\lim_{x \to 0+} -2a \int_0^t k_x(x, a(t-\tau))\, d\tau = 1. \tag{4-24}$$

Make the change of variables $\theta = x^2/4a(t-\tau)$ and use the expression for k_x above to find

$$-2a \int_0^t k_x(x, a(t-\tau))\, d\tau = \frac{1}{\sqrt{\pi}} \int_{x^2/4at}^{\infty} \theta^{-1/2} e^{-\theta}\, d\theta. \tag{4-25}$$

Let $x \to 0+$ to see that the limit is $(1/\sqrt{\pi})\Gamma(\frac{1}{2}) = 1$, as claimed.

With the help of this result we can conclude that

$$\lim_{x \to 0+} u(x, t) = \phi(t), \qquad t > 0,$$

for $\phi(t)$ any continuous function. Indeed, given $\varepsilon > 0$, choose $\delta > 0$ so that $|\phi(t) - \phi(\tau)| < \varepsilon$ for $0 < |t - \tau| < \delta$. Then

$$\begin{aligned} u(x, t) - \phi(t)\left[-2a \int_0^t k_x(x, a(t-\tau))\, d\tau\right] &= -2a \int_0^{t-\delta} k_x(x, a(t-\tau))[\phi(\tau) - \phi(t)]\, d\tau \\ &\quad - 2a \int_{t-\delta}^{t} k_x(x, a(t-\tau))[\phi(\tau) - \phi(t)]\, d\tau. \end{aligned}$$

The first integral on the right tends to zero as $x \to 0$. (See the expression for k_x above.) Since $k_x < 0$ the second integral is bounded by 2ε in view of (4-24) and (4-25). Thus sending $x \to 0+$ in the equation above and using (4-24), we find that $u(x, t) \to \phi(t)$ as $x \to 0+$. To summarize, we have:

Theorem 4-4. Suppose that $\phi(t)$ is continuous for $t > 0$. Then (4-23) solves (4-20)–(4-22). In addition, if $\phi(0) = 0$ and $\phi(t)$ is continuous for $t \geq 0$, there is a solution continuous for $x \geq 0, t \geq 0$. In the case when $\phi(0) = 0$ the continuous solution for $x > 0, t > 0$ is given by (4-23) for $x > 0, t > 0$; by 0 for $x > 0, t = 0$; and by $\phi(t)$ for $x = 0, t > 0$.

PROBLEMS

1. Show, using Fourier transforms, that the (formal) solution to

$$u_t = au_{xx} + F(x, t), \qquad -\infty < x < \infty, \qquad t > 0,$$
$$u(x, 0) = f(x), \qquad -\infty < x < \infty$$

is

$$u(x, t) = \int_{-\infty}^{\infty} k(x - y, at)f(y)\, dy + \int_0^t d\tau \int_{-\infty}^{\infty} k(x - y, a(t - \tau))F(y, \tau)\, dy.$$

2. Show that if $f(x)$ and $F(x, t)$ satisfy the growth conditions

$$|f(x)| \le M_1 \exp(\mu_1 x^2) \quad \text{and} \quad |F(x, t)| \le M_2 \exp(\mu_2 x^2),$$

then the solution given in Prob. 1 is valid for $-\infty < x < \infty$, $0 < t \le T < 1/4a\mu$, where $\mu = \max(\mu_1, \mu_2)$.

3. Suppose that $F(x, t) = 0$ for $t \ge t_0 > 0$ and that $\lim_{x \to \pm\infty} f(x) = 0$. Suppose further that $F(x, t)$ and $f(x)$ are continuous. Show that $\lim_{x \to \pm\infty} u(x, t) = 0$, where $u(x, t)$ is the solution of Prob. 1 and interpret this result physically. Is this result to be expected?

4. Show that a solution to

$$u_t = au_{xx} + F(x, t), \qquad x > 0, \qquad t > 0,$$
$$u(x, 0) = f(x), \qquad x > 0,$$
$$u(0, t) = \phi(t), \qquad t > 0,$$

is given formally by

$$u(x, t) = \int_0^{\infty} G(x, y, at)f(y)\, dy + a\int_0^t G_y(x, 0, a(t - \tau))\phi(\tau)\, d\tau$$
$$+ \int_0^t \int_0^{\infty} G(x, y, a(t - \tau))F(y, \tau)\, dy\, d\tau.$$

Hint. To obtain the representation, follow the same line of reasoning as was carried out in the proof of Th. 4-3 with v given by the function G in (4-17).

5. Check that the function $v(x, t) = k_x(x, at)$ solves the homogeneous problem

$$v_t = av_{xx}, \qquad -\infty < x < \infty, \qquad t > 0,$$
$$v(x, 0+) = 0, \qquad -\infty < x < \infty.$$

Show also that $v(x, t)$ becomes unbounded as $(x, t) \to (0, 0)$ along the curve $x^2 = 4at$. This example shows that for uniqueness to the initial value problem it is not sufficient to fix x and require only that $u(x, t) \to f(x)$ as $t \to 0^+$. This example also applies to Prob. 4.

6. Show that (4-16) does in fact solve the heat conduction problem (4-13)–(4-15) for the semi-infinite rod. *Hint.* Deduce this from facts about the infinite rod:

$$u(x, t) = u^+(x, t) - u^-(x, t)$$
$$= \int_{-\infty}^{\infty} k(x - y, at)f^+(y)\, dy - \int_{-\infty}^{\infty} k(x - y, at)f^-(y)\, dy,$$

where

$$f^+(x) = \begin{cases} f(x), & x \ge 0 \\ 0, & x < 0 \end{cases} \quad \text{and} \quad f^-(x) = \begin{cases} f(-x), & x \le 0 \\ 0, & x > 0. \end{cases}$$

7. Consider the heat conduction problem for a semi-infinite rod with insulated end:

$$u_t = au_{xx}, \qquad x > 0, \qquad t > 0,$$
$$u(x, 0) = f(x), \qquad x > 0,$$
$$u_x(0, t) = 0, \qquad t > 0.$$

Use physical reasoning analogous to that used to solve (4-13)–(4-15) to deduce that

$$u(x, t) = \int_0^\infty N(x, y, at)f(y)\,dy$$

solves this problem, where

$$N(x, y, t) = k(x - y, t) + k(x + y, t).$$

Once this solution formula is obtained, proceed as in Prob. 6 to prove that it is correct.

8. Show formally that a solution to

$$u_t = au_{xx} + F(x, t), \qquad x > 0, \qquad t > 0,$$
$$u(x, 0) = f(x), \qquad x > 0,$$
$$u_x(0, t) = \psi(t), \qquad t > 0$$

is given by

$$u(x, t) = \int_0^\infty N(x, y, at)f(y)\,dy - a\int_0^t N(x, 0, a(t - \tau))\psi(\tau)\,d\tau$$
$$+ \int_0^t\int_0^\infty N(x, y, a(t - \tau))F(y, \tau)\,dy\,d\tau.$$

9. Return to the setting in Th. 4-2. Modify the corresponding argument in Th. 4-1 to confirm that $u(x, t) \to f(x_0)$ as $(x, t) \to (x_0, 0+)$.

10. (a) Solve (4-13)–(4-15) using the Fourier sine transform. (b) Solve Prob. 7 using the Fourier cosine transform.

11. Solve formally by means of Fourier transforms the problem

$$u_t = a(t)u_{xx} + b(t)u_x + c(t)u, \qquad -\infty < x < \infty, \qquad t > 0,$$
$$u(x, 0) = f(x),$$

where $a(t)$, $b(t)$, $c(t)$ are continuous and defined on $t \geq 0$, $a(t) \geq a_0 > 0$, and $f(x)$ is continuous on $-\infty < x < \infty$.

12. Solve

$$u_t = u_{xx} + xu, \qquad -\infty < x < \infty, \qquad t > 0,$$
$$u(x, 0) = f(x)$$

by means of Fourier transforms. *Hint.* Show that the Fourier transform satisfies the first-order partial differential equation

$$\hat{u}_t + i\frac{\partial}{\partial\omega}\hat{u} = -\omega^2\hat{u}$$
$$\hat{u}(\omega, 0) = \hat{f}(\omega).$$

13. Solve

$$u_t = u_{xx} + xu_x$$
$$u(x, 0) = f(x)$$

by means of Fourier transforms.

14. Follow the steps indicated after (4-19) to prove Th. 4-3 without the growth assumption on u_x.

15. Return once again to the physical reasoning used to solve (4-13)–(4-15). Argue that the solution to

$$\begin{aligned} &u_t = au_{xx}, \quad 0 \le x \le L, \quad t > 0, \\ &u(x, 0) = f(x), \quad 0 \le x \le L, \\ &u(0, t) = u(L, t) = 0, \quad t > 0, \end{aligned}$$

is given by

$$u(x, t) = \int_{-\infty}^{\infty} k(x - y, at)F(y)\, dy,$$

where $F(x)$ is the odd, $2L$ periodic extension of the initial data $f(x)$. That is, $F(x) = f(x)$ for $0 \le x \le L$, $F(x) = -f(-x)$ for $-L \le x < 0$, $F(x + 2nL) = F(x)$. Now express this solution directly in terms of $f(x)$,

$$u(x, t) = \int_0^L \sum_{n=-\infty}^{\infty} [k(x - y - 2nL, at) - k(x + y - 2nL, at)]f(y)\, dy$$

and check that this is indeed the solution to the finite interval heat conduction problem. Thus, find a new representation for the Green's function for the finite interval of Sec. 5-1,

$$G(x, y, t) = \sum_{n=-\infty}^{\infty} [k(x - y - 2nL, t) - k(x + y - 2nL, t)].$$

16. Use the formula for $G(x, y, t)$ obtained in Prob. 15 to show that (3-22) is a valid representation for the solution to the general, finite-interval, heat conduction problem. Assume that $f(x)$, $\phi(t)$, and $\psi(t)$ are continuous and that $F(x, t)$ is continuously differentiable with respect to x.

17. Suppose that $\psi(t)$ is continuous on $0 \le t \le T$ and that $\psi(0) = 0$. Suppose further that $|\psi(t)| \le \varepsilon$ there. Let $u = u(x, t)$ be the solution to

$$\begin{aligned} &u_t = au_{xx}, \quad x > 0, \quad t > 0, \\ &u(x, 0) = 0, \quad x \ge 0, \\ &u_x(0, t) = \psi(t), \quad t > 0. \end{aligned}$$

Show that every bounded solution u to this problem is given by

$$u(x, t) = \int_0^t N(x, 0, a(t - \tau))\psi(\tau)\, d\tau.$$

Show that u satisfies an estimate of the form

$$|u(x, t)| \le MT^{1/2}\varepsilon,$$

where $M > 0$ is a constant. Use this result to show that bounded solutions to this problem depend continuously on the given data.

5. The Damped Wave Equation and the Heat Equation

In Chap. 1 we found that the differential equation

$$au_{tt} + bu_t = u_{xx} \tag{5-1}$$

with $a, b > 0$ describes the vibrations of a string subject to damping and also the conduction of electricity in a cable with negligible leakage. If $a = 1$ and $b = 0$ we have the standard wave equation, while if $a = 0$ and $b = 1$, (5-1) reduces to the heat or diffusion equation. In this section we examine the behavior of solutions to (5-1) in the limit as $a \to 0$.

Therefore, we consider the initial value problem

$$\varepsilon^2 u_{tt} + u_t = u_{xx}, \qquad -\infty < x < \infty, \qquad t > 0, \tag{5-2}$$

$$u(x, 0, \varepsilon) = f(x), \qquad -\infty < x < \infty, \tag{5-3}$$

$$u_t(x, 0, \varepsilon) = g(x), \qquad -\infty < x < \infty. \tag{5-4}$$

Under the assumptions that $f(x)$, $g(x)$ are bounded, $f(x)$ is twice continuously differentiable, and $g(x)$ is once continuously differentiable, we will show that the solution $u(x, t, \varepsilon)$ of the initial value problem for the damped wave equation (5-2)–(5-4) tends to a solution of the diffusion equation

$$u_t = u_{xx}, \qquad -\infty < x < \infty, \qquad t > 0, \tag{5-5}$$

$$u(x, 0) = f(x), \qquad -\infty < x < \infty, \tag{5-6}$$

as $\varepsilon \to 0$.

Solve (5-2)–(5-4) by Riemann's method to find (Prob. 1)

$$\begin{aligned} u(x, t, \varepsilon) = {} & \frac{1}{2}\left[f\left(x + \frac{t}{\varepsilon}\right) + f\left(x - \frac{t}{\varepsilon}\right)\right] \exp\left(-\frac{t}{2\varepsilon^2}\right) \\ & + \frac{1}{4\varepsilon}\int_{x-t/\varepsilon}^{x+t/\varepsilon} f(y) \exp\left(-\frac{t}{2\varepsilon^2}\right) I_0\left(\frac{tR}{2\varepsilon^2}\right) dy \\ & + \frac{1}{4\varepsilon}\int_{x-t/\varepsilon}^{x+t/\varepsilon} f(y) \exp\left(-\frac{t}{2\varepsilon^2}\right) \frac{1}{R} I_1\left(\frac{tR}{2\varepsilon^2}\right) dy \\ & + \frac{\varepsilon}{2}\int_{x-t/\varepsilon}^{x+t/\varepsilon} g(y) \exp\left(-\frac{t}{2\varepsilon^2}\right) I_0\left(\frac{tR}{2\varepsilon^2}\right) dy, \end{aligned} \tag{5-7}$$

where $R = \sqrt{1 - \varepsilon^2(x - y)^2/t^2}$ and

$$I_n(z) = \sum_{j=0}^{\infty} \frac{1}{j!(j+n)!}\left(\frac{z}{2}\right)^{2j+n}$$

for $n = 0, 1, \ldots$ is the modified Bessel function of the first kind of order n. Our goal is to show that $u(x, t, \varepsilon)$ given by (5-7) tends to a solution of (5-5) and (5-6) as $\varepsilon \to 0$. This requires some rather delicate estimates on the integrals involving $f(y)$ on the right of (5-7). These estimates show that each of these integrals has the limit

$$\frac{1}{2}\int_{-\infty}^{\infty} k(x-y, t)f(y)\,dy$$

as $\varepsilon \to 0$, where $k(x, t)$ is the fundamental solution to the heat equation. The other terms on the right of (5-7) are easily seen to have limit 0 as $\varepsilon \to 0$. So

$$\lim_{\varepsilon \to 0} u(x, t, \varepsilon) = \int_{-\infty}^{\infty} k(x-y, t)f(y)\,dy,$$

which is just the solution to the heat equation found in Sec. 5-4. To summarize, this will prove:

Theorem 5-1. Let $f(x)$ be twice continuously differentiable, $g(x)$ be continuously differentiable, and $f(x)$, $g(x)$ be bounded on $-\infty < x < \infty$. Then the solution $u(x, t, \varepsilon)$ to the damped wave equation (5-2)–(5-4) satisfies

$$\lim_{\varepsilon \to 0} u(x, t, \varepsilon) = \int_{-\infty}^{\infty} k(x-y, t)f(y)\,dy,$$

which is a solution to the heat conduction problem (5-5)–(5-6) for the infinite rod.

We now turn to the estimates that confirm Th. 5-1. Let F and G be bounds for $f(x)$ and $g(x)$:

$$|f(x)| \le F \quad \text{and} \quad |g(x)| \le G \quad \text{for} \quad -\infty < x < \infty.$$

We begin by establishing a kind of maximum principle for the solutions $u(x, t, \varepsilon)$. Consider the initial value problem (5-2)–(5-4) when $f(x) \equiv F$ and $g(x) \equiv G$. Since the initial data are independent of the space variable x, it is natural to expect that the solution is, too. That is, $u(x, t, \varepsilon)$ should be independent of x. With this observation, (5-2) reduces to an ordinary differential equation and the solution $U(x, t, \varepsilon)$ to (5-2)–(5-4) with constant initial data F and G is easily calculated to be

$$U(x, t, \varepsilon) = F + \varepsilon^2 G\left[1 - \exp\left(-\frac{t}{\varepsilon^2}\right)\right]. \tag{5-8}$$

Now the solution formula (5-7) shows that the initial value problem (5-2)–(5-4) has a nonnegative solution when the initial data are nonnegative. Apply this observation to the initial value problems solved by $U(x, t, \varepsilon) \pm u(x, t, \varepsilon)$ to infer that $|u(x, t, \varepsilon)| \le U(x, t, \varepsilon)$. Thus, we obtain

Theorem 5-2. Let $|f(x)| \le F$ and $|g(x)| \le G$. Then the solution $u(x, t, \varepsilon)$ to (5-2)–(5-4) satisfies

$$|u(x, t, \varepsilon)| \le F + \varepsilon^2 G\left[1 - \exp\left(-\frac{t}{\varepsilon^2}\right)\right].$$

Moreover, if $f(x), g(x) \geq 0$, then

$$0 \leq u(x, t, \varepsilon) \leq F + \varepsilon^2 G\left[1 - \exp\left(-\frac{t}{\varepsilon^2}\right)\right].$$

These estimates show that the solution $u(x, t, \varepsilon)$ is bounded in terms of its initial data and that the closer ε is to zero, the more nearly $u(x, t, \varepsilon)$ is bounded in terms of its initial values alone, just as for solutions to the heat equation. This suggests that as $\varepsilon \to 0$, $u(x, t, \varepsilon)$ may indeed tend to a solution of the heat equation, as we have claimed. Further evidence is provided by the constant initial data solution. Notice that $U(x, t, \varepsilon) \to F$ as $\varepsilon \to 0$ and $u(x, t) \equiv F$ solves the heat equation (5-5) with constant initial temperature $f(x) \equiv F$.

We return to estimating the terms in (5-7). Throughout the discussion, x and t are fixed with $-\infty < x < \infty$ and $t > 0$. First,

$$\left|\frac{1}{2}\left[f\left(x + \frac{t}{\varepsilon}\right) + f\left(x - \frac{t}{\varepsilon}\right)\right] \exp\left(-\frac{t}{2\varepsilon^2}\right)\right| \leq F \exp\left(-\frac{t}{2\varepsilon^2}\right)$$

and the right member obviously tends to 0 as $\varepsilon \to 0$. Next, (5-7) and Th. 5-2 yield

$$\left|\frac{\varepsilon}{2} \int_{x - t/\varepsilon}^{x + t/\varepsilon} g(y) \exp\left(-\frac{t}{2\varepsilon^2}\right) I_0\left(\frac{tR}{2\varepsilon^2}\right) dy\right| \leq G\varepsilon^2\left[1 - \exp\left(-\frac{t}{\varepsilon^2}\right)\right],$$

and again the right member tends to 0 as $\varepsilon \to 0$. The difficult estimates involve the two integrals involving $f(y)$ in (5-7) and we consider them next. For simplicity we merely sketch the reasoning needed to find their limits as $\varepsilon \to 0$.

Take the integral

$$\int_{x - t/\varepsilon}^{x + t/\varepsilon} \frac{1}{4\varepsilon} \exp\left(-\frac{t}{2\varepsilon^2}\right) f(y) I_0\left(\frac{tR}{2\varepsilon^2}\right) dy \tag{5-9}$$

with $R = [1 - \varepsilon^2(x - y)^2/t^2]^{1/2}$. Notice that $I_0(z)$ is an increasing function for $z \geq 0$, so one might expect that for small $\varepsilon > 0$ the major contribution to this integral should occur when y is close to x. This suggests that we look at the asymptotic behavior of $I_0(z)$ as $z \to \infty$. We find that (see Abramowitz and Stegun *Handbook of Mathematical Functions*, NBS, Washington, DC, 1972).

$$I_n(z) \sim \frac{e^z}{\sqrt{2\pi z}} \quad \text{as} \quad z \to \infty$$

where the $\sim$ sign is read "is asymptotic to" and is shorthand for the limit statement

$$\lim_{z \to \infty} \frac{I_n(z)}{e^z/\sqrt{2\pi z}} = 1.$$

We also need the elementary result

$$\sqrt{1 - z} \sim 1 - \frac{z}{2} \quad \text{as} \quad z \to 0.$$

Recall that x and t are fixed. For y varying in any bounded interval, the asymptotic results above give

$$
\begin{aligned}
\frac{1}{4\varepsilon}\exp\left(-\frac{t}{2\varepsilon^2}\right)I_0\left(\frac{tR}{2\varepsilon^2}\right) &\sim \frac{1}{4\varepsilon}\exp\left(-\frac{t}{2\varepsilon^2}\right)\exp\left(\frac{tR}{2\varepsilon^2}\right)\frac{1}{\sqrt{2\pi}}\sqrt{\frac{2\varepsilon^2}{tR}} \\
&\sim \frac{1}{4\sqrt{\pi}}\frac{1}{\sqrt{tR}}\exp\left(-\frac{t}{2\varepsilon^2}\right)\exp\left\{\frac{t}{2\varepsilon^2}\left[1-\frac{\varepsilon^2(x-y)^2}{2t^2}\right]\right\} \\
&\sim \frac{1}{4\sqrt{\pi t}}\exp\left[-\frac{(x-y)^2}{4t}\right] \\
&= \frac{1}{2}k(x-y,t)
\end{aligned}
$$

as $\varepsilon \to 0$. Now, let $\varepsilon \to 0$ to infer that the integral in (5-9) has limit

$$
\frac{1}{2}\int_{-\infty}^{\infty} k(x-y, at)f(y)\,dy.
$$

Similar reasoning leads to the same limit for the other integral which contains $f(y)$ in (5-7). Taken together, the preceeding estimates lead to the limit relation in Th. 5-1.

PROBLEMS

1. Use Riemann's method to derive the formula (5-7) for the solution to (5-2)–(5-4).
2. Let $f(x) = F$, $g(x) = G$ and solve (5-2)–(5-4) to get (5-8) assuming the solution is independent of x. By uniqueness there is no other solution.
3. Consider the initial value problem

$$
\begin{aligned}
&u_{tt} + \varepsilon u_t = u_{xx}, && -\infty < x < \infty, \quad t > 0, \\
&u(x, 0, \varepsilon) = f(x), && u_t(x, 0, \varepsilon) = g(x), \quad -\infty < x < \infty,
\end{aligned}
$$

and the initial value problem

$$
\begin{aligned}
&u_{tt} = u_{xx}, && -\infty < x < \infty, \quad t > 0, \\
&u(x, 0) = f(x), && u_t(x, 0) = g(x), \quad -\infty < x < \infty,
\end{aligned}
$$

and show that for fixed (x, t)

$$
\lim_{\varepsilon \to 0} u(x, t, \varepsilon) = u(x, t).
$$

4. Investigate whether the solutions to the boundary value problem

$$
\begin{aligned}
&\varepsilon^2 u_{tt} + u_t = u_{xx}, \quad 0 < x < L, \quad t > 0, \\
&u(0, t, \varepsilon) = u(L, t, \varepsilon) = 0, \quad t > 0, \\
&u(x, 0, \varepsilon) = f(x), \quad u_t(x, 0, \varepsilon) = g(x), \quad 0 < x < L,
\end{aligned}
$$

converge as $\varepsilon \to 0$ to the solution of

$$u_t = u_{xx}, \qquad 0 < x < L, \qquad t > 0,$$
$$u(0, t) = u(L, t) = 0, \qquad t > 0,$$
$$u(x, 0) = f(x), \qquad 0 < x < L.$$

5-6. Brownian Motion

The problems we have discussed so far have all concerned macroscopic phenomena. Partial differential equations also give important information about microscopic systems. However, in such cases, our basic standpoint and interpretation of the models must be modified. For example, consider a violin string. The wave equation with appropriate boundary and initial conditions determines the deflection $u(x, t)$ of a particle on the string relative to its equilibrium position at x. Consequently, we know the position and velocity of each particle that makes up the string. Suppose we want the same type of description for one mole of a gas. The state of the gas would be known precisely if we could determine the exact position and velocity of each molecule of gas at some instant in time. Indeed, given these initial data, we would simply solve the system of differential equations that governs the motion. This is not as simple as it sounds. First, there are 6.02×10^{23} equations of motion! It may be possible to solve them in principle, but in practice we are out of luck. A closer look shows that we are out of luck even in principle. How would we get our initial data? Basically, we would shine light on each molecule in the gas. We take the size of each molecule to be roughly 5.29×10^{-11} m, the radius of an atom. If we want the components of the position for each molecule to within 10%, we would need to use light with a wavelength of about 5.29×10^{-12} m or a frequency of $3 \times 10^8/5.29 \times 10^{-12} = 5.67 \times 10^{19}\ \text{s}^{-1}$. The energy imparted to each molecule is Planck's constant times this frequency or $(6.63 \times 10^{-34}\ \text{J}\cdot\text{s})(5.67 \times 10^{19}\ \text{s}^{-1}) = 3.76 \times 10^{-14}$ J. To get the total increase in internal energy for the mole of gas we multiply by the Avogadro–Loschmidt number, 6.02×10^{23}, to obtain 2.26×10^{10} J. The temperature of such a gas would increase to about $(2/3k)(3.76 \times 10^{-14}\ \text{J})$, where $k = 1.38 \times 10^{-23}$ J/deg is Boltzmann's constant. Consequently, starting from room temperature, say, the gas will have warmed up to about 1.82×10^9 °K (or degrees Celsius, since at this temperature they are about the same). This temperature is roughly that found inside the giant stars! To determine the initial data to within 10% we have radically altered the original object of study. We cannot even solve the problem in principle.

This example makes it clear that we cannot solve all problems in the physical sciences by decomposing a system into its constituent molecules (or atoms) and then writing a differential equation for each particle. Alternative models are needed. In previous chapters we have used the approach of continuum mechanics. Another useful class of models describes the state of a physical system in probabilistic terms. In this section we formulate a rather simple probabilistic model for Brownian motion.

Consider the movement of tiny particles suspended in a liquid. Such motion was first observed by Leeuwenhoek and then described in more detail by Brown in the 1820s. This so-called Brownian motion has proven to be of fundamental importance. Brownian motion is also observed in undissociated dilute solutions when a molecule of the dissolved substance is much larger than a molecule of solvent.

Microscopic particles suspended in a liquid or solute molecules in a solution are in constant motion, darting here and there in a completely irregular or random way. It is the "state" of such a system of particles or molecules that we wish to describe. The very random nature of the motion suggests a probabilistic description: Suppose for simplicity that the motion is one-dimensional. That is, in a time interval of length Δt a particle in the liquid will move a distance Δx, either to the right or to the left. We focus our attention on a particle located at position x when time $t = 0$. At time $t = n\,\Delta t$, this particle will be located somewhere between $x - n\,\Delta x$ and $x + n\,\Delta x$. It could be at one of these extremes, but more likely it is somewhere in between. Basically, our description of the system will answer the question: What is the chance that the particle will lie in a given interval, $a \le x \le b$, with $x - n\,\Delta x \le a < b \le x + n\,\Delta x$ at the time $t = n\,\Delta t$?

To answer this question and arrive at Einstein's model for Brownian motion, we need to make some assumptions about the motion. We have already made one simplifying assumption: Each particle will move either to the left or to the right a distance Δx in the time period Δt. Notice that Δx and Δt are not independent. The distance Δx a particle or molecule moves depends on the time increment Δt. Furthermore, different particles actually move slightly different distances. Thus, by Δx we mean an average displacement. More precisely, if there are n particles or molecules altogether, then

$$\Delta x = [(\Delta_1^2 + \cdots + \Delta_n^2)/n]^{1/2},$$

where Δ_j is the observed displacement of the particle or molecule labeled j. We further assume that the movement of a particle or molecule in each time interval Δt is not affected by its past history. Let p be the probability that a particle or molecule moves to the right and q the probability that it moves left. Then $p + q = 1$ because it must move one way or the other.

We cannot pinpoint a particle or molecule with perfect accuracy. Rather we ask: What is the probability that it will be located in the interval $[x - \frac{1}{2}\,\Delta x, x + \frac{1}{2}\,\Delta x]$ of length Δx at time t? We assume, as part of our model, that this question can be answered in terms of a probability density. Let $u = u(x, t)$ be the probability per unit length that a particle is located in the interval $[x - \frac{1}{2}\,\Delta x, x + \frac{1}{2}\,\Delta x]$ at time t. That is,

$$\Pr\{\text{a particle is in } [x - \tfrac{1}{2}\,\Delta x,\ x + \tfrac{1}{2}\,\Delta x] \text{ at time } t\} = u(x, t)\,\Delta x + o(\Delta x)$$

where $o(\Delta x)$ is a function such that $o(\Delta x)/\Delta x \to 0$ as $\Delta x \to 0$. Consequently,

$$\int_a^b u(x, t)\,dx$$

is the probability that a particle is located in the interval $[a, b]$ at time t.

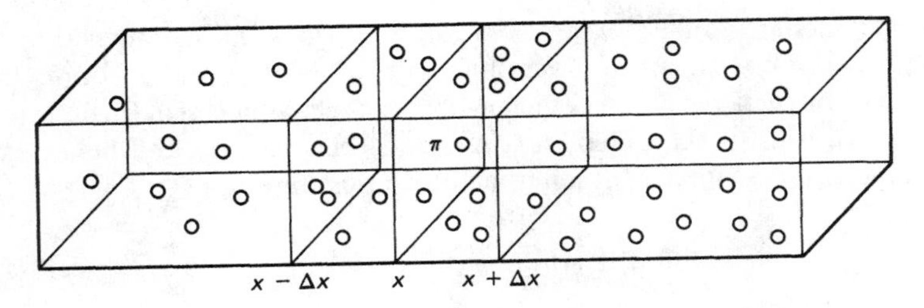

Figure 5-2

The probability density $u(x, t)$ describes the state of the system and we seek a differential equation that describes $u(x, t)$. To be specific, we concentrate on the Brownian motion of solute molecules and determine their flux across a plane perpendicular to the axis of motion at x (see Fig. 5-2). Suppose that there are n solute molecules, each with mass m. The solute molecules that can cross the plane π during the time interval from t to $t + \Delta t$ must be located in the liquid slabs between $x - \Delta x$ and x and between x and $x + \Delta x$ at time t. The mass of solute molecules in the slab between $x - \Delta x$ and x at time t is approximately

$$mnu(x - \tfrac{1}{2}\,\Delta x, t)\,\Delta x.$$

Thus, the mass of solute molecules crossing π from left to right is about

$$pmnu(x - \tfrac{1}{2}\,\Delta x, t)\,\Delta x.$$

Similarly, the mass of solute molecules crossing π from right to left at time t is approximately

$$qmnu(x + \tfrac{1}{2}\,\Delta x, t)\,\Delta x.$$

Consequently, the net mass flux across π in the positive x direction is about

$$p\,\frac{\Delta x}{\Delta t}\,mnu\left(x - \frac{1}{2}\,\Delta x, t\right) - q\,\frac{\Delta x}{\Delta t}\,mnu\left(x + \tfrac{1}{2}\,\Delta x, t\right).$$

This expression for the flux can be simplified using the estimates $u(x + \frac{1}{2}\,\Delta x, t) \approx u(x, t) + \frac{1}{2}\,\Delta x\,u_x(x, t)$. With this and the fact that $p + q = 1$, the flux is seen to be nearly

$$(p - q)\,\frac{\Delta x}{\Delta t}\,mnu(x, t) - \frac{1}{2}\,\frac{(\Delta x)^2}{\Delta t}\,mnu_x(x, t),$$

ignoring terms of order $(\Delta x)^3/\Delta t$. The concentration of solute molecules in a slab of length Δx centered at x is nearly

$$c(x, t) = \frac{mnu(x, t)\,\Delta x}{\Delta x} = mnu(x, t).$$

Therefore, the expression for the flux across π is

$$\left\{(p - q)\,\frac{\Delta x}{\Delta t}\right\} c(x, t) - \left[\frac{1}{2}\,\frac{(\Delta x)^2}{\Delta t}\right] c_x(x, t),$$

which is the familiar expression for flux when both convection and diffusion occur. The coefficient in braces has units of velocity and accounts for mass transfer by convection. The coefficient in brackets which scales the rate of change of concentration is the diffusion constant. For small Δt, and hence small Δx, these coefficients stabilize. Mathematically we postulate that

$$\lim\, (p - q)\frac{\Delta x}{\Delta t} = \lambda, \qquad \lim \frac{1}{2}\frac{(\Delta x)^2}{\Delta t} = \frac{1}{2} D > 0,$$

as Δt (and hence Δx) approaches zero. A limit passage in our previous expression for the flux gives

$$F = \lambda c(x, t) - \tfrac{1}{2} D c_x(x, t).$$

for the flux across π at time t. Applying conservation of mass to the small slab from x to $x + \Delta x$ and passing to the limit yields the differential equation

$$c_t = -\lambda c_x + \tfrac{1}{2} D c_{xx}$$

(see Prob. 1). Finally, using $c = mnu$, we obtain

(6-1) $$u_t = -\lambda u_x + \tfrac{1}{2} D u_{xx}$$

as the partial differential equation for the probability density $u(x, t)$. Here λ is the *drift constant* and D the *diffusion constant* for the process.

PROBLEMS

1. Carry out the conservation-of-mass reasoning that leads to (6-1). If necessary, review similar derivations in Chap. 1.
2. Suppose that the initial probability distribution for a system of particles is $f(x)$, where $f(x)$ is a continuous function, $0 \le f(x) \le 1$, and

$$\int_{-\infty}^{\infty} f(x)\, dx = 1.$$

Show that the solution to (6-1) satisfying $u(x, 0) = f(x)$ is given by

$$u(x, t) = \int_{-\infty}^{\infty} \frac{1}{\sqrt{2\pi D t}} \exp\left[\frac{-(x - \lambda t - y)^2}{2Dt}\right] f(y)\, dy.$$

Hint. Of course, you can check this formula by direct substitution. Alternatively, let $\xi = x - \lambda t$, $\tau = t$ and define $V(\xi, \tau) = u(x, t)$. Then $V_\tau = \frac{1}{2} D V_{\xi\xi}$, the heat equation. Use the familiar integral representation for V to get a formula for $u(x, t)$.

3. Show that the solution to Prob. 2 satisfies

$$\int_{-\infty}^{\infty} u(x, t)\, dx = 1 \quad \text{for all} \quad t > 0.$$

Hint. This can be done in two ways. First, you can simply integrate the expression for $u(x, t)$ with respect to x and interchange the order of integration. More painlessly, you can consider the expression

$$\frac{d}{dt}\int_{-\infty}^{\infty} u(x, t)\, dx,$$

make use of the differential equation, and make assumptions on how fast u_x and u_{xx} tend to zero as x tends to plus or minus infinity.

4. Suppose that it is known with absolute certainty that the particle is initially located at the point $x = 0$. Show that the subsequent probability distribution is given by

$$u(x, t) = \frac{1}{\sqrt{2\pi Dt}} \exp\left[-\frac{(x - \lambda t)^2}{2Dt}\right].$$

Hint. Since it is known that the particle is located at $x = 0$, we can take a sequence of density functions $\{f_n(x)\}$ such that $f_n(x) = 0$ for $|x| > 1/n$ and then let n tend to infinity in Prob. 2.

5. The expected value for a probability density is given by

$$\int_{-\infty}^{\infty} xu(x, t)\, dx.$$

Show that the expected value of the probability density $u(x, t)$ in Prob. 2 is

$$\lambda t + \int_{-\infty}^{\infty} xf(x)\, dx.$$

Now explain why λ is called the drift. *Hint.* Consider the expression

$$\frac{d}{dt}\int_{-\infty}^{\infty} xu(x, t)\, dx,$$

then make use of the differential equation and integrate by parts.

6. Show that if $u(x, t) = k(x - \lambda t, \frac{1}{2}Dt)$, the expected value is λt, so that if there is an equal probability that the particle move to the right or the left, the expected value is 0. In that case, for each t, $k(x, \frac{1}{2}Dt)$ is a Gaussian normal curve.

7. If $u(x, t)$ is the solution to Prob. 2 or 4, calculate the so-called second moment in both cases:

$$\int_{-\infty}^{\infty} x^2u(x, t)\, dx.$$

8. The variance is defined to be

$$V(u) = \int_{-\infty}^{\infty} x^2u(x, t)\, dx - \left[\int_{-\infty}^{\infty} xu(x, t)\, dx\right]^2.$$

Calculate $V(u)$ in the case that u is the solution to Probs. 2 and 4.

5-7. Periodic Boundary Conditions

There are diffusion and heat flow problems which are periodic in nature and for which no natural initial conditions are known. This occurs, for example, in the problem of determining the temperature near the surface of the earth and in certain blood flow problems.

Consider the problem of finding a continuous, bounded, periodic solution to

$$\text{(7-1)} \qquad \begin{cases} u_t = au_{xx}, & x > 0, \quad -\infty < t < \infty, \\ u(0, t) = \mu(t), & -\infty < t < \infty, \end{cases}$$

where μ is continuous and periodic with period T,

$$\text{(7-2)} \qquad \mu(t + T) = \mu(t) \quad \text{for all} \quad t.$$

Fourier used (7-1) and (7-2) to model the temperature beneath the surface of the earth. In this model, $x = 0$ is the surface of the earth, $x > 0$ inside the earth, and $\mu(t)$ represents the periodic, daily (monthly, yearly) variations in temperature over a portion of the earth. Fourier chose $\mu(t) = A \cos(\omega t)$. If reasonable experimental values of a, A, and ω are used, the solution (7-4) to (7-1) and (7-2) is quite realistic at moderate depths. At great depths, the model is invalid. For example, it ignores heating due to nuclear processes in the core of the earth.

If (7-1) and (7-2) has a solution u and t_0 is fixed arbitrarily, then u satisfies the initial, boundary value problem

$$\begin{aligned} &u_t = au_{xx}, \quad x > 0, \quad t > t_0, \\ &u(0, t) = \mu(t), \quad t > t_0, \\ &u(x, t_0) \quad \text{is given for} \quad x > 0. \end{aligned}$$

From (4-16) and (4-23) we infer that

$$\text{(7-3)} \qquad \begin{aligned} u(x, t) = &\int_0^\infty [k(x - y, a(t - t_0)) - k(x + y, a(t - t_0))]u(y, t_0)\, dy \\ &- 2a \int_{t_0}^t k_x(x, a(t - \tau))\mu(\tau)\, d\tau \end{aligned}$$

for $x > 0$ and $t > t_0$. In this expression, $u(y, t_0)$ really represents unknown initial data. Now, the effects of these initial data on the solution should be negligible when $t - t_0$ is large. We can guarantee that $t - t_0$ is large for all t by sending the reference time t_0 to $-\infty$. The time integral in (7-3) converges to

$$\int_{-\infty}^t k_x(x, a(t - \tau))\mu(\tau)\, d\tau$$

as $t_0 \to -\infty$, because μ is bounded. Let M be a bound on the periodic, continuous solution $u(x, t)$. Then

$$\begin{aligned} &\left| \int_0^\infty [k(x - y, a(t - t_0)) - k(x + y, a(t - t_0)]u(y, t_0)\, dy \right| \\ &\leq M \int_0^\infty [k(x - y, a(t - t_0)) - k(x + y, a(t - t_0))]\, dy \\ &= \frac{M}{\sqrt{\pi}} \left[\int_{-x/\sqrt{4a(t-t_0)}}^\infty e^{-\xi^2}\, d\xi - \int_{x/\sqrt{4a(t-t_0)}}^\infty e^{-\eta^2}\, d\eta \right] \to 0 \end{aligned}$$

as $t_0 \to -\infty$ for each fixed (x, t). Thus, we obtain

(7-4)
$$u(x, t) = -2a \int_{-\infty}^{t} k_x(x, a(t-\tau))\mu(\tau)\, d\tau$$

from (7-3) in the limit as $t_0 \to -\infty$. In review, if (7-1) and (7-2) has a bounded, continuous, periodic solution, it is given by (7-4). Thus, if a solution exists, it is unique. It is routine (see Prob. 1) to check that (7-4) is indeed a bounded, continuous solution to (7-1) and (7-2) and that u is periodic in time with period T.

Corresponding problems on finite intervals can be solved using Fourier series. Consider, for example, the temperature $u(x, t)$ in a homogeneous rod with insulated lateral surface and whose ends are maintained at given periodic temperature distributions. The boundary value problem for u is

(7-5)
$$\begin{cases} u_t = au_{xx}, & 0 < x < L, \quad -\infty < t < \infty, \\ u(0, t) = f(t), & u(L, t) = g(t), \quad -\infty < t < \infty, \end{cases}$$

where f and g are continuous and T periodic,

(7-6)
$$f(t + T) = f(t) \quad \text{and} \quad g(t + T) = g(t) \quad \text{for all} \quad t.$$

We proceed formally and assume that both f and g have Fourier series expansions. It is convenient to use the complex form of the Fourier series

$$f(t) = \sum_{n=-\infty}^{\infty} f_n \exp\left(\frac{2\pi int}{T}\right) \quad \text{and} \quad g(t) = \sum_{n=-\infty}^{\infty} g_n \exp\left(\frac{2\pi int}{T}\right)$$

where

$$f_n = \frac{1}{T} \int_{-T/2}^{T/2} f(t) \exp\left(-\frac{2\pi int}{T}\right)$$

and g_n is given by a corresponding formula. (Consult Prob. 16 of Sec. 3-1.) Note that $f_{-n} = \bar{f}_n$ and $g_{-n} = \bar{g}_n$ because f and g are real-valued. It is natural to expect that the solution $u(x, t)$ to (7-5) will be T periodic in time, so we set

(7-7)
$$u(x, t) = \sum_{n=-\infty}^{\infty} u_n(x) \exp\left(\frac{2\pi int}{T}\right).$$

Substitute this expression for u into the heat equation and differentiate term by term to get

$$\sum_{-\infty}^{\infty} u_n(x) \frac{2\pi in}{T} \exp\left(\frac{2\pi int}{T}\right) = \sum_{-\infty}^{\infty} au_n''(x) \exp\left(\frac{2\pi int}{T}\right).$$

Thus, the coefficients $u_n(x)$ in (7-7) must satisfy

(7-8)
$$\begin{cases} u_n''(x) = \dfrac{2\pi in}{aT} u_n(x), & 0 < x < L, \\ u_n(0) = f_n, \quad u_n(L) = g_n, \end{cases}$$

where the boundary conditions follow by requiring that (7-7) reduce to f when $x = 0$ and to g when $x = L$. Since the solution to (7-5) is real-valued, the Fourier coefficient $u_{-n}(x) = \overline{u_n(x)}$ for $n \geq 0$. Hence, it suffices to find $u_n(x)$ for $n = 0, 1, 2, \ldots$. If $n = 0$, we find from (7-8) that

$$u_0(x) = f_0 + \left(\frac{g_0 - f_0}{L}\right)x. \tag{7-9}$$

If $n \geq 1$ the general solution to the differential equation in (7-8) is easily found to be

$$u_n(x) = A_n e^{c_n(1+i)x} + B_n e^{-c_n(1+i)x} \tag{7-10}$$

where

$$c_n = \left(\frac{\pi n}{aT}\right)^{1/2}, \tag{7-11}$$

and the boundary conditions in (7-8) are satisfied when

$$\begin{cases} A_n = \dfrac{g_n - f_n \exp\left[-c_n(1+i)L\right]}{\exp\left[c_n(1+i)L\right] - \exp\left[-c_n(1+i)L\right]}, \\ B_n = \dfrac{f_n \exp\left[c_n(1+i)L\right] - g_n}{\exp\left[c_n(1+i)L\right] - \exp\left[-c_n(1+i)L\right]}. \end{cases} \tag{7-12}$$

The formal solution to (7-5) is (7-7) with $u_n(x)$ given by (7-9)–(7-12) for $n \geq 0$ and by $u_{-n}(x) = \overline{u_n(x)}$ for $n < 0$. This formal solution is a solution under reasonable assumptions on the periodic boundary data f and g (see Prob. 5).

PROBLEMS

1. (a) Use the fact that $k(x, t) = (1/\sqrt{4\pi t})e^{-x^2/4t}$ satisfies $k_t = k_{xx}$ to verify that $k_x(x, at)$ satisfies the heat equation with diffusion constant a. Then verify that (7-4) satisfies the heat equation in (7-1).
   (b) Show that (7-4) is T periodic in time. *Hint*. Use (7-4) with t replaced by $t + T$ and then make a change of variables in the integral.
   (c) Show that $u(x, t)$ given by (7-4) has limit $\mu(t)$ as $x \to 0+$. *Hint*. Review the reasoning following (4-23). The reasoning needed here is a little easier.
   (d) Show that $u(x, t)$ is bounded. In fact,

$$|u(x, t)| \leq \max_{0 \leq t \leq T} |\mu(t)|$$

   for $x > 0$, $t > 0$. Why should this result be expected?
2. In (7-1) and (7-2), choose $a = 1$ and $\mu(t) = \sin t$. Plot $u(x, t)$ as a function of t for $x = 1$, 10, and 50.
3. Solve (7-5) when $f(t) = 0$ and $g(t) = \sin(2\pi t/T)$.
4. In 1863, Ångström proposed the following method to determine the conductivity a of a metal. Make a rod of the metal with insulated lateral surface. Hold one end of the rod at a constant temperature and subject the other end to periodic temperature variations. After a certain time has passed, the effects of the initial data will be negligible and the temperature u will be determined by (7-5) with data like those in Prob. 3. Show how a can be determined from measurements of u taken inside the rod.

5. Determine reasonable conditions on f and g in (7-5) so that the formal solution (7-7) is a solution. *Hint.* Use (7-12) to show that

$$|A_n| \leq \frac{|g_n| + |f_n| \exp(-c_n L)}{\exp(c_n L) - \exp(-c_n L)}, \qquad |B_n| \leq \frac{|f_n| + |g_n| \exp(-c_n L)}{1 - \exp(-2c_n L)}.$$

These bounds allow you to estimate the decay rate for $|A_n|$ and $|B_n|$ in terms of the decay rates of $|f_n|$ and $|g_n|$. Now, determine what smoothness you need on f and g to justify the term-by-term differentiations used to obtain the formal solution.

6. Show that the boundedness assumption on $u(x, t)$ is necessary for uniqueness in (7-1) and (7-2). *Hint.* Show that $u(x, t) = e^{\alpha x + \beta t}$ solves $u_t = a u_{xx}$ if $\beta = a\alpha^2$. The exponential will be T periodic in the time t only if $\beta = \beta_n = 2\pi n i/T$ for $n = 0, 1, 2, \ldots$, in which case $\alpha_n = \pm c_n(1 + i)$ where c_n is given by (7-11). Then for *each fixed* $n = 1, 2, \ldots$

$$u_n(x, t) = \exp\left[\pm c_n(1 + i)x + \frac{2\pi n i t}{T}\right]$$

solves (7-1) with

$$\mu_n(t) = \exp\left(\frac{2\pi n i t}{T}\right).$$

Note that, for each n, one solution is unbounded as $x \to \infty$. Real-valued solutions can be gotten by taking the real parts of u_n and μ_n. What are they?

7. Another approach to measuring temperatures inside the earth follows from (7-5), where we take L very large. Here $x = 0$ is the surface of the earth and $x = L$ is some great depth; thus, $g(t)$ is unknown. However, show that $A_n \to 0$ and $B_n \to f_n$ as $L \to \infty$ so that the finite interval formal solution yields

$$u(x, t) = \sum_{n=-\infty}^{\infty} f_n e^{-c_n(1+i)x + 2\pi i n t/T}.$$

Use the relations $f_{-n} = \bar{f}_n$ to express this result in real form.

6

Higher-Order Equations in One Spatial Variable

6-1. The Vibrating Beam

Although second-order partial differential equations play a prominent role in modeling physical problems, they are by no means the only equations of importance. In this chapter we consider several problems that are modeled by higher-order equations. For example, the transverse vibrations of a beam are governed by a fourth-order differential equation.

For simplicity, we consider a homogeneous beam which has length L and rectangular cross sections, each with area A. Like a string, a beam is long compared to its cross-sectional area. However, in sharp contrast to a string, a beam is "stiff"; it strongly resists bending. We think of the unbent beam as composed of many longitudinal fibers each of length L. When the beam is deformed, some fibers will be compressed and others will be stretched. Somewhere between the top and bottom of the beam there will be a surface of fibers which maintain their original length. This surface is called the *neutral surface.*

We choose coordinates so that the xu plane is the longitudinal (axial) plane of symmetry for the beam. This plane and the neutral surface intersect in a curve called the *neutral axis* of the beam. We shall assume that all forces which act on the beam are constant on cross sections of the beam perpendicular to the x axis. Then the transverse vibrations of the beam can be described (see Fig. 6-1) by the shape of the neutral axis.

We suppose that the beam experiences small, transverse vibrations. This means that particles in the beam move essentially in the vertical direction only. In particular, a particle located at $(x, 0, 0)$ on the neutral axis of the unbent beam will be located at $(x, 0, u(x, t))$ at time t as the beam vibrates. Thus, $u(x, t)$ describes the deflection of the beam during vibration. For small deflections it has been verified experimentally that plane cross sections of the beam remain planar

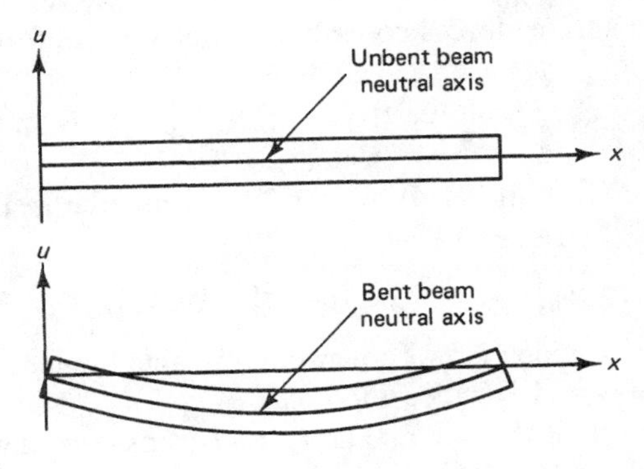

Figure 6-1

during bending. We assume this as part of our model. Also, during small vibrations the density ρ (mass per unit volume) of the beam remains virtually constant over time. Since we have chosen a homogeneous beam, ρ is also independent of position. Thus, we assume that the density ρ is constant. Finally, we assume that the small vibrations occur within the elastic limit of the material making up the beam. Thus, Hooke's law applies:

$$\text{(stress)} = E \cdot \text{(strain)}. \tag{1-1}$$

Here E is *Young's modulus*, a constant characteristic of the material making up the beam. For example, for aluminum $E \approx 10^7$ psi, for high-strength steel $E \approx 3 \times 10^7$ psi, and for wood $E \approx 1.5 \times 10^6$ psi. The stress is the force per unit area acting on a cross section of the beam. The strain is the corresponding relative elongation, the ratio of the change in length of a section of beam under stress to the length of that section when unstressed.

To find the differential equation of motion for the vibrating beam, consider a small chunk of the beam between the cross section at x and $x + \Delta x$. Since the vibrations are transverse, Newton's second law gives

$$\frac{d}{dt}\int_x^{x+\Delta x} A\rho u_t(\xi, t)\, d\xi = \sum \text{vertical forces}. \tag{1-2}$$

We distinguish two types of forces acting on the beam element. First, there are body forces that act on the element as a whole. These forces arise from the weight of the element, the external loading, and the resistance to vibration of the medium surrounding the beam. We assume a linear law of resistance and hence can express the body forces as

$$\int_x^{x+\Delta x} A[-\rho g - k u_t(\xi, t) + f(\xi, t)]\, d\xi, \tag{1-3}$$

where g is the acceleration due to gravity, $k > 0$ is a constant, and $f(x, t)$ is the force per unit volume due to loading. Second, we consider surface forces which act on the cross-sectional faces of the beam element. Let $F(x, t)$ be the vertical component of the surface forces due to the portion of the beam to the *right* of x. Then the surface forces due to the portion of the beam to the *left* of x is $-F(x, t)$

and hence the surface forces contribute a net vertical force of

$$F(x + \Delta x, t) - F(x, t) = \int_x^{x+\Delta x} F_x(\xi, t)\, d\xi. \tag{1-4}$$

Substitute (1-3) and (1-4) into (1-2) and use the fact that Δx is arbitrary to conclude that

$$A\rho u_{tt}(x, t) = F_x(x, t) + A[-ku_t(x, t) - \rho g + f(x, t)]. \tag{1-5}$$

The force $F(x, t)$, which is unknown in (1-5), acts tangentially to the cross section of the beam at x and is called the *shear*.

Next, we relate the shear, $F(x, t)$, and transverse deflection, $u(x, t)$. To this end, we introduce $M = M(x, t)$ the bending moment that arises in the cross section at x at time t due to the portion of the beam to the right of x. Let us consider the case of an upward (positive) shear force $F(x + \Delta x, t)$ and counterclockwise (positive) bending moment $M(x + \Delta x, t)$ acting on the cross section at $x + \Delta x$. The portion of the beam to the left of x will exert a downward shear force of magnitude $F(x, t)$ and a clockwise bending moment of magnitude $M(x, t)$. Equate counterclockwise and clockwise moments about the cross section at $x + \Delta x$ to find

$$M(x + \Delta x, t) + F(x, t)\, \Delta x = M(x, t);$$

hence, in the limit as Δx tends to zero,

$$\frac{\partial M}{\partial x} = -F. \tag{1-6}$$

See Prob. 6 concerning the minus sign in this equation. In the derivation of (1-6), we ignored the moments due to the various external forces considered above. Problem 7 shows that (1-6) still holds even when these forces are taken into account.

A final geometric argument relates the bending moment to the curvature of the neutral axis, whose equation is $u = u(x, t)$. Figure 6-2 shows the beam element. The bending is exaggerated for illustrative purposes. For small Δx the neutral axis given by $u = u(x, t)$ virtually conincides with the circle of radius R shown. A fiber z units above the neutral axis has length $(R - z)\, \Delta\theta$ and hence is compressed by the amount $z\, \Delta\theta$. When unstrained this fiber has length $\Delta s = \Delta x = R\, \Delta\theta$, the same as the part of the neutral axis in the element of beam shown in Fig. 6-2. Hence, the strain in the fiber is $z\, \Delta\theta/\Delta x = z/R$. Let h be the height and b the base of the rectangular cross sections of the beam. By Hooke's law the force on the small strip from z to $z + \Delta z$ across the face of the cross section at x is

$$E\left(\frac{z}{R}\right) b\, \Delta z.$$

The bending moment M about the axis perpendicular to the xu plane at $(x, u(x, t))$ is produced by these elastic forces. Thus, the magnitude of this moment is

$$|M| = \int_{-h/2}^{h/2} \frac{Ez^2 b}{R}\, dz = \frac{EI}{R}, \tag{1-7}$$

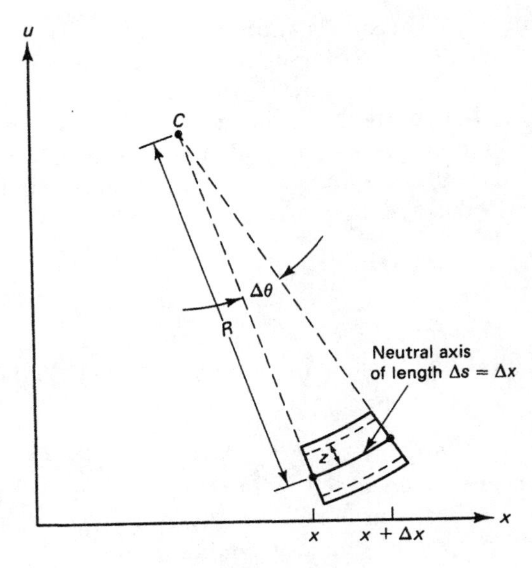

Figure 6-2

where $I = bh^3/12$ is the moment of inertia of the cross section at x relative to the axis just mentioned. In the limit as Δx tends to zero, the radius R of the circle in Fig. 6-2 approaches the radius of curvature of the neutral axis, which we also denoted by R. Since $1/R$ is the curvature of the neutral axis at x, we know from calculus that

$$\frac{1}{R} = \frac{|u_{xx}|}{(1 + u_x^2)^{3/2}} \approx |u_{xx}|,$$

under our small vibration assumption. Thus, from (1-7) and for small vibrations, we find that $|M| = EI|u_{xx}|$. By convention, M is positive if it compresses the top of the beam. In this case, the neutral line is concave up, $u_{xx} > 0$, and the equation above becomes $M = EIu_{xx}$. If M is negative, the neutral line is concave down, $u_{xx} < 0$, and we find $M = EIu_{xx}$ again. Thus,

$$M = EIu_{xx}(x, t), \tag{1-8}$$

and from (1-6) the shear can be expressed as

$$F = -EIu_{xxx}(x, t). \tag{1-9}$$

Insert (1-9) into (1-5) to obtain the differential equation

$$\rho u_{tt} + \frac{EI}{A} u_{xxxx} = -ku_t - \rho g + f(x, t) \tag{1-10}$$

for the transverse vibrations of a beam. If the frictional force can be neglected, the weight of the beam is negligible in comparison to its stiffness, and there is no loading, the beam satisfies

$$u_{tt} + a^2 u_{xxxx} = 0, \tag{1-11}$$

where $a^2 = EI/A\rho$.

Of course, the beam equation (1-10) or (1-11) must be solved subject to appropriate boundary and initial conditions. The initial conditions needed are obvious: the initial position, $u(x, 0)$, and initial velocity, $u_t(x, 0)$, must be given. The choice of boundary conditions depends on how the ends of the beam are supported. Suppose, for example, that the beam is set in concrete at the end $x = 0$. It is clear that the end at $x = 0$ cannot move and that the neutral axis is "flat" at $x = 0$. That is,

$$u(0, t) = 0 \quad \text{and} \quad u_x(0, t) = 0 \quad \text{for} \quad t \geq 0.$$

It this case, the beam is said to be *clamped* at $x = 0$. A beam may have an unconstrained or *free* end. If the beam end at $x = L$ is free, then there is no bending moment or shear on this end. In view of (1-8) and (1-9) the appropriate boundary conditions are

$$u_{xx}(L, t) = 0 \quad \text{and} \quad u_{xxx}(L, t) = 0 \quad \text{for} \quad t \geq 0.$$

A beam with one clamped and one free end is called a *cantilever beam*. Finally, a beam that just rests on a support, say at $x = 0$, is called *simply supported* at $x = 0$. The appropriate boundary conditions are

$$u(0, t) = 0 \quad \text{and} \quad u_{xx}(0, t) = 0 \quad \text{for} \quad t \geq 0$$

because the end at $x = 0$ does not move and the support does not provide a bending moment (although it does exert a shear).

PROBLEMS

1. Consider the case of a circular beam; that is, if the beam is cut perpendicular to its axis, the cross section is a circle. Derive a differential equation analogous to (6-10) which governs the transverse vibrations.
2. A *prismatic beam* has a straight axis and constant cross sections throughout its length. The rectangular beam in the text and the circular beam in Prob. 1 are examples. Find the beam equation in this case.
3. Consider a beam with rectangular cross sections which are tapered: When unstressed the intersections of the top and bottom of the beam with the xu-plane are given by $u = -mx + h$ and $u = mx - h$ for $0 \leq x \leq L$. Here the slope m satisfies $0 < m < h/L$ and $h > 0$. Find the beam equation in this case.
4. Consider a beam, such as the tapered beam in Prob. 3, whose cross sections vary with position along the axis of the beam. Assume also that the material composition of the beam varies along its axis. Derive the beam equation corresponding to (1-10).
5. Suppose that the beam is circular and has a hole in it, as in the case of a pipe, so that a cross section has the shape of an annulus. Obtain an equation similar to (1-10), which governs the transverse vibrations.
6. In Fig. 6-1 we take the u axis positive upward. In most engineering applications, loads on beams act downward. Hence, the u axis in Fig. 6-1 is usually taken positive downward, and accordingly shear forces are regarded as positive when they act downward. With these sign conventions and with counterclockwise moments still taken as positive, check that (1-6) becomes $\partial M/\partial x = F$. In words, the rate of change of bending moment equals the shear.

7. In the reasoning leading to (1-6) we ignored the external forces acting on the element. Show that (1-6) still holds even with these forces taken into account.

6-2. Initial, Boundary Value Problems for the Vibrating Beam

A typical problem in the engineering analysis of structures is to determine the vibrations of a stiff beam under various loadings. To be definite, we consider a beam of length L, as in Sec. 6-1, which is clamped at $x = 0$, simply supported at $x = L$, and initially is at rest with deflection $u(x, 0) = f(x)$. The initial, boundary value problem for this situation is

$$\text{(2-1)} \qquad \begin{cases} u_{tt} + a^2 u_{xxxx} = 0, & 0 < x < L, \quad t > 0, \\ u(x, 0) = f(x), \quad u_t(x, 0) = 0, & 0 \le x \le L, \quad t \ge 0, \\ u(0, t) = u_x(0, t) = 0, \quad u(L, t) = u_{xx}(L, t) = 0, & t \ge 0, \end{cases}$$

where $a^2 = EI/\rho A$.

Problem (2-1) can be solved by separation of variables. Subsitution of $u(x, t) = X(x)T(t)$ into the beam equation and boundary conditions leads to

$$\text{(2-2)} \qquad T'' + a^2 \lambda T = 0$$

and the eigenvalue problem

$$\text{(2-3)} \qquad \begin{cases} X'''' - \lambda X = 0 \\ X(0) = X'(0) = 0, \qquad X(L) = X''(L) = 0. \end{cases}$$

The eigenvalue problem (2-3) can have only positive eigenvalues: Suppose that (2-3) has a nontrivial solution X. Multiply the differential equation in (2-3) by $\bar{X}$, the complex conjugate of X, and integrate the fourth derivative term twice by parts to get

$$\int_0^L |X''|^2 \, dx = \lambda \int_0^L |X|^2 \, dx.$$

This equation shows that $\lambda \ge 0$. Furthermore, if $\lambda = 0$, we infer $X''(x) \equiv 0$, which implies that $X(x) \equiv 0$ because $X(0) = X'(0) = 0$. Thus, all the eigenvalues of (2-3) are positive and it is convenient to write $\lambda = \mu^4$ with $\mu > 0$. Then (2-3) becomes

$$\text{(2-4)} \qquad \begin{cases} X'''' - \mu^4 X = 0, \\ X(0) = X'(0) = 0, \qquad X(L) = X''(L) = 0. \end{cases}$$

It is elementary to check that the differential equation has the general solution

$$X(x) = A \cosh \mu x + B \sinh \mu x + C \cos \mu x + D \sin \mu x.$$

The boundary conditions at $x = 0$ imply that $C = -A$, $D = -B$, and hence

$$\text{(2-5)} \qquad X(x) = A(\cosh \mu x - \cos \mu x) + B(\sinh \mu x - \sin \mu x).$$

The boundary conditions at $x = L$ require

$$\text{(2-6)} \qquad \tan \mu L = \tanh \mu L$$

in order that (2-5) yields a nontrivial solution. You are asked to confirm these facts in Probs. 2 and 3. It is seen graphically that $\tan z = \tanh z$ has an infinite number of positive solutions $z_1, z_2, \ldots, z_n, \ldots$ (see Prob. 13). The positive solutions to (2-6) are therefore $\mu_1, \mu_2, \ldots$ where $\mu_n = z_n/L$ for $n = 1, 2, \ldots$. Notice (Prob. 13) that for large n, $\mu_n \approx (n + \frac{1}{4})\pi/L$. Thus, (2-4) has eigenvalues

$$\lambda_n = \mu_n^4, \qquad n = 1, 2, \ldots, \tag{2-7}$$

and corresponding eigenfunctions

$$X_n(x) = (\cosh \mu_n x - \cos \mu_n x) - \frac{\cosh \mu_n L - \cos \mu_n L}{\sinh \mu_n L - \sin \mu_n L} \times (\sinh \mu_n x - \sin \mu_n x), \tag{2-8}$$

which are determined up to a constant multiple. From (2-2) and (2-7) the time-dependent part of the separated solution is

$$T_n(t) = E_n \cos a\mu_n^2 t + F_n \sin a\mu_n^2 t, \tag{2-9}$$

where E_n and F_n are constants.

Thus, for $n = 1, 2, \ldots,$

$$u_n(x, t) = T_n(t)X_n(x),$$

with X_n and T_n given by (2-8) and (2-9), satisfy the beam equation and boundary conditions in (2-1). We hope to satisfy the initial conditions by superposing these solutions:

$$u(x, t) = \sum_{n=1}^{\infty} T_n(t)X_n(x). \tag{2-10}$$

In order to satisfy the initial conditions, we want

$$u(x, 0) = \sum_{n=1}^{\infty} T_n(0)X_n(x) = \sum_{n=1}^{\infty} E_n X_n(x) = f(x),$$

$$u_t(x, 0) = \sum_{n=1}^{\infty} T_n'(0)X_n(x) = \sum_{n=1}^{\infty} a\mu_n^2 F_n X_n(x) = 0.$$

Evidently, we should choose $F_n = 0$ for all n and the constants E_n so that

$$f(x) = \sum_{n=1}^{\infty} E_n X_n(x). \tag{2-11}$$

In our previous applications of separation of variables the X_n's were trigonometric functions and (2-11) a Fourier series expansion. Now the X_n are more complicated, but we still need (2-11) to hold for appropriate constants E_n, if separation of variables is to succeed.

In the Fourier series case, we determined the constants E_n by multiplying (2-11) by $X_m(x)$, integrating from 0 to L and using the orthogonality relation $\int_0^L X_n X_m\, dx = 0$ for $n \neq m$. This same relation holds for the more complicated functions X_n given by (2-8): Indeed, multiply $X_n^{(4)} - \lambda_n X_n = 0$ by X_m, multiply

$X_m^{(4)} - \lambda_m X_m = 0$ by X_n, and subtract to find

$$\int_0^L [X_n^{(4)} X_m - X_m^{(4)} X_n] \, dx = (\lambda_n - \lambda_m) \int_0^L X_n X_m \, dx.$$

Now integrate by parts twice on the left to get

$$(2\text{-}12) \quad (\lambda_n - \lambda_m) \int_0^L X_n X_m \, dx = [(X_n^{(3)} X_m - X_m^{(3)} X_n) + (X_n' X_m^{(2)} - X_m' X_n^{(2)})] \Big|_0^L .$$

Since X_n and X_m both satisfy the boundary conditions in (2-4) the right member of (2-12) vanishes. We conclude that

$$(2\text{-}13) \qquad \int_0^L X_n X_m \, dx = 0 \quad \text{for} \quad m \neq n.$$

With this result in hand, we proceed just as in the trigonometric case: Multiply (2-11) by $X_m(x)$, integrate from 0 to L, and simplify to find

$$(2\text{-}14) \qquad E_m = \frac{\int_0^L f(x) X_m(x) \, dx}{\int_0^L X_m(x)^2 \, dx},$$

assuming that term-by-term integration is permissible.

In summary, (2-10) provides a *formal* solution to the initial boundary value problem (2-1) if $F_n = 0$ and E_m is given by (2-14). This formal solution is of course a solution to (2-1) under reasonable restrictions on the initial data $f(x)$. We omit the mathematical details.

We certainly expect that (2-1) will have a unique solution which depends continuously on the initial deflection $f(x)$. To confirm these facts, we use energy considerations rather like those used for the vibrating string in Chap. 4. Consider a small element of the beam extending from x to $x + \Delta x$. The kinetic energy of this beam element is $(\rho A \, \Delta x) u_t^2/2$ and the entire beam has kinetic energy

$$(2\text{-}15) \qquad \frac{1}{2} \int_0^L \rho A u_t^2(x, t) \, dx,$$

where ρ is the constant density of the beam and A is its constant cross-sectional area. The bent beam also stores potential energy determined by the work done against the tensile forces. Recall that we have assumed that the beam vibrates within its elastic limits so Hooke's law applies. Refer to Fig. 6-2 and the accompanying notation. The fibers in the beam which are z to $z + \Delta z$ units from the neutral axis experience an elastic force F given by $F/b \, \Delta z = Ev/\Delta x$ when stretched or compressed a distance v. As the beam element in Fig. 6-2 deflects to the position shown, the tensile forces stretch or contract the fibers z units from the neutral axis by a total distance $z \, \Delta\theta$. Thus, the work done on the strip of fibers from z to $z + \Delta z$ across the face of the cross section at x is

$$\Delta W = \int_0^{z \, \Delta\theta} b \, \Delta z \, E \frac{v}{\Delta x} \, dv = Eb \frac{\Delta z}{\Delta x} \frac{z^2 (\Delta\theta)^2}{2}.$$

Since $z\,\Delta\theta/\Delta x = z/R = z|u_{xx}|$ under our assumption that vibrations are small, $\Delta W = (Eb/2)z^2u_{xx}^2\,\Delta x\,\Delta z$. The potential energy of the beam element is given by

$$\int_{-h/2}^{h/2} \frac{Eb}{2} u_{xx}^2\,\Delta x\, z^2\,dz = \frac{EI}{2} u_{xx}^2\,\Delta x,$$

where I is the moment of inertia of the cross section at x. Thus, the entire beam has potential energy

$$\frac{1}{2}\int_0^L EIu_{xx}^2(x, t)\,dx, \tag{2-16}$$

and total energy

$$E(t) = \frac{1}{2}\int_0^L [\rho A u_t^2(x, t) + EIu_{xx}^2(x, t)]\,dx. \tag{2-17}$$

In the beam equation in (2-1) external forces and internal damping have been ignored. Consequently, no power is supplied to the beam and the rate of change of energy in the system should be zero. That is, the total energy in the beam should be constant. To see that it is, compute

$$E'(t) = \int_0^L (\rho A u_t u_{tt} + EIu_{xx}u_{xxt})\,dx.$$

Integrate by parts twice on the second term and use the boundary conditions and beam equation to find

$$E'(t) = \int_0^L \rho A u_t(u_{tt} + a^2 u_{xxxx})\,dx = 0.$$

Consequently, the energy of the beam is constant; in particular,

$$E(t) = E(0) = \frac{1}{2}\int_0^L EIf''(x)^2\,dx \tag{2-18}$$

for the beam modeled by (2-1).

Now, suppose that u_1 and u_2 solve (2-1) and let $u = u_1 - u_2$, as usual. Then u satisfies (2-1) with zero initial data ($f(x) \equiv 0$). From (2-18), $E(t) \equiv 0$ and hence u_t and u_{xx} vanish for $0 \le x \le L$ and $t \ge 0$. Since $u_t = 0$, u depends only on x. Then $u_{xx} = 0$ implies that u is a linear function of x, $u(x, t) = Ax + B$. Finally, the boundary conditions in (2-1) imply that $A = B = 0$ and $u(x, t) \equiv 0$ or $u_1 = u_2$ for all x and t. Thus, (2-1) has at most one solution, as expected.

Continuous dependence of the solution $u(x, t)$ to (2-1) also follows from our energy considerations. Indeed, if u_1 and u_2 solve (2-1) with $f = f_1$ and f_2, respectively, then $u = u_1 - u_2$ solves (2-1) with $f = f_1 - f_2$. Thus, (2-18) yields

$$\int_0^L (\rho A u_t^2 + EIu_{xx}^2)\,dx = \int_0^L EI[f_1''(x) - f_2''(x)]^2\,dx. \tag{2-19}$$

From Taylor's theorem

$$u(x, t) = u(0, t) + u_x(0, t)x + \int_0^x (x - \xi)u_{xx}(\xi, t)\,d\xi.$$

The boundary conditions in (2-1) give $u(0, t) = u_x(0, t) = 0$. Therefore, an application of the Schwarz inequality to the integral above leads to the estimate

$$|u(x, t)| \le \left(\frac{L^3}{3}\right)^{1/2} \left(\int_0^L u_{xx}^2 \, dx\right)^{1/2}.$$

Then (2-19) gives

$$|u_1(x, t) - u_2(x, t)| \le \left(\frac{L^3}{3}\right)^{1/2} \left(\int_0^L [f_1''(x) - f_2''(x)]^2 \, dx\right)^{1/2}, \tag{2-20}$$

which shows that u_1 can be made as close to u_2 as we wish provided that we take f_1 close enough to f_2 in the sense of the norm $\|f_1 - f_2\|_2$ introduced in Sec. 4-3. In summary, we have proven the following result.

Theorem 2–1. The initial, boundary value problem (2-1) for the vibrating beam has at most one solution which (given existence) depends continuously on the initial data.

In Theorem 2-1 we assume, as in our calculations above, that a solution u has continuous partials u_{tt}, u_{xx}, u_{xt}, u_{xxt}, u_{xxx}, and u_{xxxx}.

PROBLEMS

1. Confirm the equation $\int_0^L |X''| \, dx = \lambda \int_0^L |X|^2 \, dx$ for a solution X of (2-3) and conclude that all the eigenvalues of (2-3) are positive.
2. Show that (2-4) has nontrivial solutions for X if and only if μ satisfies (2-6).
3. Check that the eigenfunctions of (2-4) are given by (2-8).
4. Use (2-8) to show that the eigenfunctions of (2-3) are bounded independent of n. That is, find a constant M such that $|X_n(x)| \le M$ for $0 \le x \le L$ and all n.
5. Use a root-finding method to find the first four positive roots z_1, z_2, z_3, and z_4 of $\tan z = \tanh z$ correct to four decimals. The corresponding eigenvalues of (2-3) are $\mu_n^4 = (z_n/L)^4$.
6. Take $L = 1$. Plot $X_n(x)$ for $n = 1, 2,$ and 3 using (2-8) and a programmable calculator.
7. Verify (2-12), (2-13), and (2-14).
8. Regard a tuning fork as a beam clamped at one end and free at the other. Strike the fork to start it vibrating. The initial, boundary value problem for this situation is

$$\begin{aligned} &u_{tt} + a^2 u_{xxxx} = 0, \qquad 0 < x < L, \qquad t > 0, \\ &u(x, 0) = 0, \qquad u_t(x, 0) = g(x), \qquad 0 \le x \le L, \\ &u(0, t) = u_x(0, t) = 0, \qquad u_{xx}(L, t) = u_{xxx}(L, t) = 0, \qquad t \ge 0. \end{aligned}$$

   (a) Show that nontrivial separated solutions $u(x, t) = X(x)T(t)$ are determined by the eigenvalue problem

$$\begin{aligned} &X'''' - \lambda X = 0, \\ &X(0) = X'(0) = 0, \qquad X''(L) = X'''(L) = 0, \end{aligned}$$

and differential equation

$$T'' + a^2\lambda T = 0.$$

(b) Show that the eigenvalues in part (a) are positive.
(c) Verify that the eigenvalues $\lambda = \mu^4$ are determined from the positive roots of the equation $1 + \cosh(\mu L)\cos(\mu L) = 0$. Calculate the first four roots correct to four decimals. Show that $L\mu_n \approx (n - \frac{1}{2})\pi$ for large n. Check this approximation for $n = 4$. It is even better for $n > 4$.
(d) Find the basic separated solutions $u_n(x, t) = X_n(x)T_n(t)$.
(e) Find the formal series solution to the initial, boundary value problem. Be sure to check that (2-13) holds in this case.
(f) You found in part (e) that the time dependent terms in the solution are $\sin(a\mu_n^2 t)$. This sinusoid has period $2\pi/a\mu_n^2$ and its frequency ω_n is the reciprocal of the period. If the tuning fork has fundamental frequency ω_1 cycles per second, show that the overtones have frequencies $\omega_n = (\mu_n^2/\mu_1^2)\omega_1$.

9. Formulate the initial, boundary value problem for a beam simply supported at both ends and subject to given initial data. Solve by separation of variables. Formulate conditions on the initial data which guarantee that the formal solution is a solution.

10. Show that for the beam of finite length L, there exists at most one solution u to

$$\begin{aligned} &u_{tt} + a^2 u_{xxxx} = 0, \qquad 0 < x < L, \qquad t > 0,\\ &u(x, 0) = f(x), \qquad 0 \le x \le L,\\ &u_t(x, 0) = g(x), \qquad 0 \le x \le L, \end{aligned}$$

and at the left, the beam is either clamped, or simply supported, or free, and at the right, it is either clamped, or simply supported, or free.

11. If one takes into account the tendency of the beam to rotate, one is led to the equation

(i) $$u_{tt} + a^2 u_{xxxx} = c u_{xxtt}, \qquad c > 0.$$

Suppose that the beam is clamped at the ends so that

(ii) $$u(0, t) = u_x(0, t) = u(L, t) = u_x(L, t) = 0, \qquad t > 0.$$

Show that u satisfies the energy relation

$$\frac{d}{dt}\int_0^L \left[\frac{1}{2}u_t^2(x, t) + \frac{a^2}{2}u_{xx}^2(x, t) + \frac{c}{2}u_{xt}^2(x, t)\right] dx = 0.$$

Use this result to show that there exists at most one function u satisfying (i) and (ii) and the initial conditions

(iii) $$u(x, 0) = f(x), \qquad u_t(x, 0) = g(x), \qquad 0 \le x \le L.$$

12. To solve the problem (i)–(iii) of Prob. 11 by separation of variables, one is led to the following eigenvalue problem for $X(x)$: Find values of λ for which there are nontrivial solutions $X(x)$ to

$$\begin{aligned} &X''''(x) - \lambda c X''(x) + \lambda X(x) = 0\\ &X(0) = X'(0) = X(L) = X'(L) = 0. \end{aligned}$$

(a) Verify this assertion and show that if $\lambda \ge 0$, $X(x)$ must vanish identically; there are no nonnegative eigenvalues.
(b) Suppose that λ is negative. Find an equation for the eigenvalues.

13. Graph tan z and tanh z on the same axes. Your graph should indicate that tanh $z <$ tan z on $0 < z < \pi/2$, so no eigenvalues arise from this interval. Confirm this inequality.

6-3. Vibrations of an Infinite Beam

We model a beam of infinite extent by

$$(3\text{-}1)\qquad \begin{cases} u_{tt} + a^2 u_{xxxx} = 0, & -\infty < x < \infty, \quad t > 0, \\ u(x, 0) = f(x), \quad u_t(x, 0) = g(x), & -\infty < x < \infty, \end{cases}$$

where $a = (EI/\rho A)^{1/2} > 0$, $f(x)$ specifies the initial deflection, and $g(x)$ the initial velocity. This initial value problem can be solved using Fourier transforms. Take the Fourier transform of (3-1) with respect to the spatial variable x to obtain

$$(3\text{-}2)\qquad \begin{aligned} &\hat{u}_{tt} + a^2\omega^4\hat{u} = 0, \\ &\hat{u}(\omega, 0) = \hat{f}(\omega), \qquad \hat{u}_t(\omega, 0) = \hat{g}(\omega), \end{aligned}$$

where, of course, $\hat{u} = \hat{u}(\omega, t)$ is the Fourier transform of $u(x, t)$. Now (3-2) is an ordinary differential equation in time which is easily solved together with the given initial data:

$$\hat{u}(\omega, t) = \hat{f}(\omega) \cos (a\omega^2 t) + \hat{g}(\omega) \frac{\sin (a\omega^2 t)}{a\omega^2}.$$

The solution $u(x, t)$ to (3-1) is obtained by computing the inverse transform

$$(3\text{-}3)\qquad u(x, t) = \frac{1}{2\pi} \int_{-\infty}^{\infty} e^{-i\omega x} \left[\hat{f}(\omega) \cos (a\omega^2 t) + \hat{g}(\omega) \frac{\sin (a\omega^2 t)}{a\omega^2} \right] d\omega.$$

At this point we must regard (3-3) as a formal solution to (3-1) because we have not justified the calculations. Let's check directly that (3-3) solves (3-1) under suitable restrictions on the initial data $f(x)$ and $g(x)$. From (3-3)

$$u(x, 0) = \frac{1}{2\pi} \int_{-\infty}^{\infty} e^{-i\omega x} \hat{f}(\omega)\, d\omega = f(x),$$

by Corollary 4-2 of Chap. 3, provided that we assume that $f^{(n)}(x)$ is continuous and absolutely integrable for $n = 0$, 1, and 2. Under the same assumptions on $g(x)$,

$$u_t(x, 0) = \frac{1}{2\pi} \int_{-\infty}^{\infty} e^{-i\omega x} \hat{g}(\omega)\, d\omega = g(x).$$

The differentiation under the integral sign in (3-3) needed to give this last equation is justified by Prop. 5-4 of Chap. 3 and the corollary mentioned above. Thus, (3-3) satisfies the initial conditions in (3-1).

We just found that (3-3) can be differentiated once under the integral sign with respect to t. The same reasoning allows a second such differentiation, so

$$(3\text{-}4)\qquad u_{tt} = \frac{-1}{2\pi} \int_{-\infty}^{\infty} e^{-i\omega x} [a^2\omega^4 \hat{f}(\omega) \cos (a\omega^2 t) + a\omega^2 \hat{g}(\omega) \sin (a\omega^2 t)]\, d\omega.$$

We can also use Prop. 5-4 of Chap. 3 to differentiate under the integral sign repeatedly with respect to x to get

$$(3\text{-}5) \qquad u_{xxxx} = \frac{1}{2\pi}\int_{-\infty}^{\infty} \omega^4 e^{-i\omega x}\left[\hat{f}(\omega)\cos(a^2\omega t) + \hat{g}(\omega)\frac{\sin(a\omega^2 t)}{a\omega^2}\right] d\omega,$$

provided that $\omega^4\hat{f}(\omega)$ and $\omega^2\hat{g}(\omega)$ are absolutely integrable. This integrability follows from Th. 4-4 of Chap. 3 if $f^{(n)}(x)$ is continuous and absolutely integrable for $n = 0, 1, \ldots, 6$, and $g^{(n)}(x)$ satisfies these hypotheses for $n = 0, 1, \ldots, 4$ (see Prob. 4). Now (3-4) and (3-5) give $u_{tt} + a^2u_{xxxx} = 0$. Thus, we have established the existence portion of the following result.

Theorem 3-1. Let $f^{(n)}(x)$ and $g^{(m)}(x)$ be continuous and absolutely integrable on $-\infty < x < \infty$ for $n = 0, 1, \ldots, 6$ and $m = 0, 1, \ldots, 4$. Then the initial value problem (3-1) for the infinite beam has a unique solution.

Uniqueness is treated in Prob. 3.

Although (3-3) is a useful representation of the solution to (3-1) for many purposes, it is also convenient to express the right side of (3-3) directly in terms of the given data $f(x)$ and $g(x)$. To do this, consider first the integral

$$\begin{aligned}
&\frac{1}{2\pi}\int_{-\infty}^{\infty} e^{-i\omega x}\hat{f}(\omega)\cos(a\omega^2 t)\, d\omega \\
&= \frac{1}{2\pi}\int_{-\infty}^{\infty} f(y)\int_{-\infty}^{\infty} e^{i\omega(y-x)}\cos(a\omega^2 t)\, d\omega\, dy \\
&= \frac{1}{2\pi}\cdot 2\cdot\frac{1}{4}\left(\frac{2\pi}{at}\right)^{1/2}\int_{-\infty}^{\infty}\left[\cos\frac{(y-x)^2}{4at} + \sin\frac{(y-x)^2}{4at}\right] f(y)\, dy,
\end{aligned}$$

upon reference to a table of Fourier transforms. Simplifying slightly, we find

$$(3\text{-}6) \quad \frac{1}{2\pi}\int_{-\infty}^{\infty} e^{-i\omega x}\hat{f}(\omega)\cos(a\omega^2 t)\, d\omega = \frac{1}{\sqrt{4\pi a t}}\int_{-\infty}^{\infty} f(y)\sin\left[\frac{(y-x)^2}{4at} + \frac{\pi}{4}\right] dy.$$

The second part of the integral in (3-3) takes the form

$$\begin{aligned}
(3\text{-}7) \qquad &\frac{1}{2\pi}\int_{-\infty}^{\infty} e^{-i\omega x}\hat{g}(\omega)\frac{\sin(a\omega^2 t)}{a\omega^2}\, d\omega \\
&= \frac{1}{2\pi}\int_{-\infty}^{\infty} g(y)\, dy\int_{-\infty}^{\infty} e^{+i\omega(y-x)}\frac{\sin(a\omega^2 t)}{a\omega^2}\, d\omega \\
&= \frac{1}{\pi a}\int_{-\infty}^{\infty}\left\{\frac{\pi(y-x)}{2}\left[S\left(\frac{(y-x)^2}{4at}\right) - C\left(\frac{(y-x)^2}{4at}\right)\right]\right. \\
&\qquad \left. + \sqrt{\pi a t}\,\sin\left(\frac{(y-x)^2}{4at} + \frac{\pi}{4}\right)\right\} g(y)\, dy,
\end{aligned}$$

from a table of Fourier transforms. Here the functions $C(z)$ and $S(z)$ are the Fresnel integrals defined by

$$C(z) = \frac{1}{\sqrt{2\pi}} \int_0^z s^{-1/2} \cos s \, ds \quad \text{and} \quad S(z) = \frac{1}{\sqrt{2\pi}} \int_0^z s^{-1/2} \sin s \, ds.$$

Define functions

$$K(x, t) = \frac{1}{\sqrt{4\pi a t}} \sin\left(\frac{x^2}{4at} + \frac{\pi}{4}\right)$$

and

$$L(x, t) = \frac{1}{\pi a}\left\{\frac{\pi x}{2}\left[S\left(\frac{x^2}{4at}\right) - C\left(\frac{x^2}{4at}\right)\right] + \sqrt{\pi a t}\,\sin\left(\frac{x^2}{4at} + \frac{\pi}{4}\right)\right\}.$$

Insert (3-6) and (3-7) into (3-3) to obtain

$$\text{(3-8)} \qquad u(x, t) = \int_{-\infty}^{\infty} [K(y - x, t)f(y) + L(y - x, t)g(y)] \, dy,$$

which expresses the solution to (3-1) directly in terms of the initial data.

Once the representation (3-8) for the solution to (3-1) has been found, its validity can be confirmed without the severe differentiability restrictions used in Th. 3-1. As (3-8) suggests, we simply need to assume that $f(x)$ and $g(x)$ decay sufficiently rapidly at infinity. We omit these details; however, Prob. 5 outlines a partial, direct verification that (3-8) solves (3-1).

The form of the function $L(x, t)$ is awkward at best. The functions $S(z)$ and $C(z)$ which occur in it arise from the factor $a\omega^2$ in the denominator of (3-7). These terms can be avoided if a suitable, second indefinite integral of $g(x)$ is known (see Prob. 2).

PROBLEMS

1. Solve explicitly

$$u_{tt} + a^2 u_{xxxx} = 0, \qquad -\infty < x < \infty, \qquad t > 0,$$
$$u(x, 0) = A \exp(-\alpha x^2), \qquad A > 0, \qquad \alpha > 0, \qquad -\infty < x < \infty,$$
$$u_t(x, 0) = 0, \qquad -\infty < x < \infty.$$

2. Suppose that a second indefinite integral is known for the given initial velocity $u_t(x, 0)$ in (3-1). Changing notation slightly, let $g(x)$ be this second indefinite integral. Then $g''(x)$ is the initial velocity and (3-1) can be expressed as

$$u_{tt} + a^2 u_{xxxx} = 0, \qquad -\infty < x < \infty, \qquad t > 0,$$
$$u(x, 0) = f(x), \qquad u_t(x, 0) = g''(x), \qquad -\infty < x < \infty.$$

Take Fourier transforms and obtain the representation

$$u(x, t) = \frac{1}{\sqrt{4\pi at}} \int_{-\infty}^{\infty} \left\{ f(y) \sin\left[\frac{(y-x)^2}{4at} + \frac{\pi}{4}\right] + \frac{1}{a} g(y) \sin\left[\frac{(y-x)^2}{4at} - \frac{\pi}{4}\right]\right\} dy.$$

3. Show that there is at most one solution to

$$u_{tt} + a^2 u_{xxxx} = F(x, t), \qquad -\infty < x < \infty, \qquad t > 0,$$
$$u(x, 0) = f(x), \qquad u_t(x, 0) = g(x), \qquad -\infty < x < \infty.$$

Hint. Consider the energy integral

$$E(t) = \frac{1}{2} \int_{-\infty}^{\infty} [\rho A u_t^2(x, t) + EI u_{xx}^2(x, t)]\, dx$$

and proceed as we did in Sec. 6-2 for the finite beam. What integrability assumptions must you make on a solution and its partial derivatives so that the calculations you perform are justified? For example, you will need to know that the energy integral is finite.

4. Assume that $h^{(k)}(x)$ is continuous and absolutely integrable for $k = 0, 1, \ldots, n$ and $n \geq 2$. Use Th. 4-4 of Chap. 3 to show $|\hat{h}(\omega)| \leq M/|\omega|^n$ for $|\omega| \geq 1$ and $M = \int_{-\infty}^{\infty} |h^{(n)}(x)|\, dx$. Conclude that $|\omega|^{n-2}\hat{h}(\omega)$ is absolutely integrable.
5. Consider (3-8) with the initial velocity $g(x) \equiv 0$ and $f(x)$ absolutely integrable.
   (a) Check that $K_{tt} + a^2 K_{xxxx} = 0$.
   (b) Justify differentiating under the integral sign in (3-8) for $-\infty < x < \infty$ and $t > 0$ to confirm that (3-8) satisfies the beam equation.
   (c) Show that

$$u(x, t) = \frac{1}{\sqrt{8\pi t}} \int_0^{\infty} [f(x+y) + f(x-y)] \left[\sin\left(\frac{y^2}{4t}\right) + \cos\left(\frac{y^2}{4t}\right)\right] dy.$$

   (d) Use the change of variables $v = y^2/4t$ in part (c) and $S(\infty) = C(\infty) = \frac{1}{2}$ to verify that $u(x, 0) = f(x)$.

6-4. The Internally Damped String

We used the wave equation

$$(4\text{-}1) \qquad u_{tt} = c^2 u_{xx} \quad \text{or} \quad \rho_0 u_{tt} = \tau u_{xx}$$

where $c^2 = \tau/\rho_0$ to model the small vibrations of an elastic string. Equation (4-1) describes the vibrations very well if no external forces act, external damping is negligible, and the time interval of interest is relatively short. However, (4-1) does predict that the string vibrates forever with undiminished amplitude and this is simply not the case. Of course, even in the absence of external damping, internal damping due to frictional effects causes the string to loose energy and the vibrations to stop. We seek to modify the wave equation (4-1) to take account of internal damping.

To this end, we reconsider our derivation in Sec. 1-2 of the wave equation for the vibrating string. Recall that ρ_0 is the constant density of the string when in its equilibrium position $0 \leq x \leq L$ along the x-axis, and that τ is the constant horizontal component of the tension. Since u_t is the velocity of a small particle of string as it executes its vertical vibrations, the string has kinetic energy

$$\frac{1}{2}\int_0^L \rho_0 u_t^2(x, t)\, dx. \tag{4-2}$$

The string also stores potential energy due to the work of the elastic forces. (Remember that external forces are regarded as negligible.) Since we assume small vibrations, the tension in the string is τ up to first-order terms in Δx. This force stretches the string a distance

$$\int_x^{x+\Delta x} [1 + u_x(\xi, t)^2]^{1/2}\, d\xi - \Delta x = [(1 + u_x^2)^{1/2} - 1]\,\Delta x = \tfrac{1}{2}u_x^2\,\Delta x,$$

up to first-order terms in Δx. Thus, the work done by the elastic forces on the segment from x to $x + \Delta x$ is

$$\tfrac{1}{2}\tau u_x^2\,\Delta x,$$

up to first-order terms. Sum and pass to the limit as Δx tends to 0 to find that the potential energy of the entire string is

$$\frac{1}{2}\int_0^L \tau u_x^2(x, t)\, dx. \tag{4-3}$$

The string has total energy

$$E(t) = \frac{1}{2}\int_0^L [\rho_0 u_t^2(x, t) + \tau u_x^2(x, t)]\, dx. \tag{4-4}$$

We assume, as usual, that the string has fixed ends, $u(0, t) = u(L, t) = 0$, so no energy is introduced at the ends of the string. If $u(x, t)$ satisfies (4-1) and the boundary conditions just mentioned, an easy calculation (Prob. 1) gives

$$E'(t) = \int_0^L u_t[\rho_0 u_{tt} - \tau u_{xx}]\, dx = 0. \tag{4-5}$$

Thus, energy is conserved, $E(t) = E(0)$ for $t \geq 0$.

In the case of external damping, the wave equation (4-1) is replaced by $\rho_0 u_{tt} + ku_t = \tau u_{xx}$, where $k > 0$ is the damping coefficient. Then (4-5) becomes

$$E'(t) = -\left(\frac{2k}{\rho_0}\right)\frac{1}{2}\int_0^L \rho_0 u_t^2\, dx \leq 0, \tag{4-6}$$

and energy is dissipated as expected. In words, (4-6) states that energy dissipates at a rate proportional to the kinetic energy. Thus

$$E'(t) < 0 \tag{4-7}$$

unless there is no motion.

Return now to the string without external damping. We seek a reasonable assumption that models internal energy dissipation and leads to the conclusion (4-7). The derivation of the undamped wave equation in Sec. 1-2 leads to

$$\rho_0 u_{tt} = V_x \tag{4-8}$$

where $V = V(x, t)$ is the vertical component of the tension. The small vibration assumptions were tantamount to

$$V(x, t) = \tau u_x(x, t) \tag{4-9}$$

where τ is the constant horizontal component of the tension. That is, the vertical forces acting at the ends of a string element were assumed proportional to the relative displacements u_x of the particles making up the string. Evidently, these particles are constantly rubbing against each other as the string vibrates. This rubbing converts some kinetic energy into heat and so decreases the tension in the string and damps the vibrations. The faster the vibrations (i.e., the faster the relative positions of the particles on the string change with time), the more heat is generated. Thus, the tension in the string depends not only on the relative displacements u_x as in (4-9), but also on the time rate of change of these relative displacements, u_{xt}. We therefore modify (4-9) to

$$V(x, t) = \tau u_x(x, t) + \varepsilon u_{xt}(x, t), \tag{4-10}$$

where $\varepsilon > 0$ is assumed constant. To reason that ε should be positive, write $u_{xt} = (u_x)_t$ and consider a point on the string with $u_x > 0$. If energy is steadily removed from the string, we expect the slope u_x of the wave form to decrease in time, so $(u_x)_t = u_{xt} < 0$. Since the tension will also decrease, we require $\varepsilon > 0$ in (4-10). Insert (4-10) into (4-8) to obtain the *wave equation with internal damping*

$$\rho_0 u_{tt} = \tau u_{xx} + \varepsilon u_{xxt} \tag{4-11}$$

or

$$u_{tt} = c^2 u_{xx} + a u_{xxt}, \tag{4-11$'$}$$

where $c^2 = \tau/\rho_0$ and $a = \varepsilon/\rho > 0$.

A number of properties of solutions to the wave equation with internal damping are developed in the problems. In particular, you are asked to confirm (4-7).

PROBLEMS

1. Verify (4-5). *Hint*. Differentiate (4-4), then integrate one term by parts, and use $u(0, t) = u(L, t) = 0$.
2. The initial, boundary value problem for a string with internal damping is

$$\begin{aligned} &\rho_0 u_{tt} = \tau u_{xx} + \varepsilon u_{xxt}, \qquad 0 < x < L, \qquad t > 0, \\ &u(x, 0) = f(x), \qquad u_t(x, 0) = g(x), \qquad 0 \le x \le L, \\ &u(0, t) = 0, \qquad u(L, t) = 0, \qquad t \ge 0. \end{aligned}$$

Show that a solution $u(x, t)$ satisfies the energy equation

$$E'(t) = -\varepsilon \int_0^L u_{xt}^2(x, t)\, dt,$$

where $E(t)$ is given by (4-4). Interpret this result physically; in particular, deduce (4-7).

3. Show that the initial, boundary value problem in Prob. 2 has at most one solution.
4. Use separation of variables in the internally damped wave equation (see Prob. 2) to find that separated solutions $u(x, t) = X(x)T(t)$ are determined by the eigenvalue problem

$$X'' + \lambda^2 X = 0, \qquad X(0) = X(L) = 0$$

just as for the undamped string. Conclude that $\lambda_n = n\pi/L$ for $n = 1, 2, \ldots$ and $X_n(x) = \sin(\lambda_n x)$. Show that the corresponding equation for $T_n(t)$ is $T_n'' + a\lambda_n^2 T_n' + c^2\lambda_n^2 T_n = 0$. Find $T_n(t)$. If $a = \varepsilon/\rho > 0$ is small, note that for small positive values of t, the solution $T_n(t)$ behaves much like the corresponding $T_n(t)$ obtained for the undamped wave equation.
5. Solve formally the initial, boundary value problem in Prob. 2. Show formally that $\lim_{t\to\infty} u(x, t) = 0$.
6. Find reasonable conditions on the initial data $f(x)$ and $g(x)$ so that the formal solution in Prob. 5 is a (rigorous) solution.
7. Extend the reasoning used in the derivation of the internally damped vibrating string to the case of an internally damped beam. Argue that the vibrations are governed by the equation

$$u_{tt} + a^2 u_{xxxx} + \varepsilon u_{xxxxt} = 0,$$

where $\varepsilon > 0$ is a "small" parameter.
8. Consider the initial, boundary value problem for an internally damped beam with clamped ends:

$$\text{(i)} \qquad u_{tt} + a^2 u_{xxxx} + \varepsilon u_{xxxxt} = 0, \qquad 0 < x < L, \qquad t > 0,$$

$$\text{(ii)} \qquad u(0, t) = u_x(0, t) = u(L, t) = u_x(L, t) = 0, \qquad t \geq 0,$$

$$\text{(iii)} \qquad u(x, 0) = f(x), \qquad u_t(x, 0) = g(x), \qquad 0 \leq x \leq L,$$

where $a^2 = EI/\rho A$. The energy $E(t)$ in the beam is given by (2-17). Show that (i) and (ii) give

$$E'(t) = -\rho A\varepsilon \int_0^L u_{xxt}^2(x, t)\, dx,$$

and use this result to show that there exists at most one solution to the initial, boundary value problem (i)–(iii).
9. Find a formal solution to the initial, boundary value problem in Prob. 8. Verify formally that $u(x, t) \to 0$ as $t \to \infty$.
10. In studying sound waves in a plasma, one is led to a system of the form

$$\phi_{xx} = e^{\phi} - \rho, \qquad \rho_t + (\rho v)_x = 0, \qquad v_t + v v_x = -\phi_x.$$

Here ϕ is the electrical potential, ρ is the ion density, v is the ion velocity, the problem is assumed to be one dimensional, and all the physical constants have been set equal to 1. Suppose that the disturbances around the solution $\rho = 1$, $v = 0$, $\phi = 0$ are small and that nonlinear effects can be neglected (in particular, $e^{\phi} = 1 + \phi$, etc.). Let $\rho = 1 + u$ and show that

$$\text{(i)} \qquad u_{ttxx} = u_{tt} - u_{xx}.$$

Show there exists at most one function u satisfying (i) and the boundary conditions.

$$\text{(ii)} \qquad u(0, t) = u(L, t) = 0, \qquad t \geq 0,$$

and the initial conditions

(iii) $$u(x, 0) = f(x), \qquad u_t(x, 0) = g(x), \qquad 0 \le x \le L.$$

11. Solve formally the initial, boundary value problem in Prob. 10. You will find, after setting $u(x, t) = X(x)T(t)$, that X will be determined as the solution to the eigenvalue problem

$$X''(x) + \lambda^2 X(x) = 0, \qquad X(0) = X(L) = 0,$$

and $T(t)$ will be given as the solution to

$$T''(t) + \frac{\lambda^2}{1 + \lambda^2} T(t) = 0.$$

Observe that the eigenvalues and the eigenfunctions are the same as in the case of the undamped vibrating string, but the oscillations described by $T(t)$ are different and in particular that the higher frequencies $(1/2\pi)\sqrt{1 + (1/\lambda^2)}$ are more important in this case.

6-5. Movement of Chemicals Underground

When solutes in the form of beneficial chemicals or pollutants enter the groundwater system, they can be carried far from their source and affect the livability of the surrounding environment. A mathematical description of such a process is based on a mass balance. We first discussed such models in Secs. 1-4 and 1-5. In this section we assume that the movement of chemicals is one-dimensional and take a closer look at their spread.

To be specific, assume that the underground flow is parallel to the x axis and consider a tube in the flow with constant cross-sectional area A. Let C be the concentration of the chemicals dissolved in the water, and ϕ be the porosity (assumed constant) of the soil through which the water moves. Then the mass of solute in the portion of the tube between x and $x + \Delta x$ is $A\phi \int_x^{x+\Delta x} C(x, t)\,dx$. The rate at which this mass is changing is $A\phi \int_x^{x+\Delta x} C_t(x, t)\,dx$. This expression is equal to the rate at which mass crosses the ends of the tube of length Δx minus the rate at which mass is being sorbed onto the soil. Thus, if we let $Q(x, t)$ be the rate per unit area per unit time that mass crosses a plane section through the flow at x at time t, and $N(x, t)$ be the sorbed chemical concentration in the soil, the mass balance can be expressed as

$$A\phi \int_x^{x+\Delta x} C_t(x, t)\,dx = -A[Q(x + \Delta x, t) - Q(x, t)] - A \int_x^{x+\Delta x} N_t(x, t)\,dx.$$

Divide by Δx and let it tend to zero to conclude that

(5-1) $$\phi C_t(x, t) = -Q_x(x, t) - N_t(x, t).$$

Now, we need equations relating Q and N to the concentration C. We relate Q and C by *Fick's law*,

(5-2) $$Q = -DC_x + \phi v C,$$

where $D > 0$ is called the diffusion constant and v is the velocity (which we assume constant) of the flow. This empirical law is the simplest mathematical expression which is consistent with the fact that mass flows from regions of high concentration toward regions of low concentration at a rate proportional to $|C_x|$, and that mass is transported by the bulk movement of the groundwater, which flows at speed v. Substitute (5-2) into (5-1) to obtain

$$C_t = KC_{xx} - vC_x - \frac{1}{\phi} N_t, \tag{5-3}$$

where $K = D/\phi$ and, as noted above, v is assumed constant.

The relationship between N and C is determined empirically. The simplest situation occurs when the concentration N of chemicals sorbed in the soil is proportional to the solute concentration C. That is,

$$N = \alpha C \tag{5-4}$$

for $\alpha > 0$, a constant. Insertion of (5-4) into (5-3) yields the standard *diffusion–dispersion equation*

$$C_t = K_0 C_{xx} - wC_x, \tag{5-5}$$

with constant coefficients $K_0 = K/(1 + \alpha/\phi)$ and $w = v/(1 + \alpha/\phi)$.

We turn from (5-4) to a more realistic assumption: Assume that there is a maximum concentration $\tilde{N}$ at which the soil becomes saturated and that the sorption rate is proportional to the difference $\tilde{N} - N$, so that

$$N_t = \alpha(\tilde{N} - N), \tag{5-6}$$

with $\alpha > 0$. If initially no chemical is present in the soil, $N(x, 0) = 0$ and we can integrate (5-6) to find $N(x, t) = \tilde{N}(1 - e^{-\alpha t})$, substitute this expression for N into (5-3) to obtain

$$C_t = KC_{xx} - vC_x - \gamma e^{-\alpha t}, \tag{5-7}$$

where $\gamma = \alpha\tilde{N}/\phi$.

Of course, the parameter α in (5-6), which we assumed constant, must depend on the concentration C. The simplest assumption taking this dependence into account is

$$N_t = \alpha C(\tilde{N} - N) \tag{5-8}$$

with $\alpha > 0$ a constant. Integration gives

$$N(x, t) = \tilde{N}\left\{1 - \exp\left[-\alpha \int_0^t C(x, \tau)\, d\tau\right]\right\}.$$

Use of this expression in (5-3) yields a highly nonlinear partial differential equation for C. However, if α is small and the concentrations C are not too large, we can use the approximation

$$N_t = \tilde{N}\alpha C \exp\left[-\alpha \int_0^t C(x, \tau)\, d\tau\right] \approx \tilde{N}\alpha C,$$

which neglects terms in $\alpha^2, \alpha^3, \ldots,$ in (5-3) to get

$$C_t = KC_{xx} - vC_x - \gamma C, \tag{5-9}$$

where $\gamma = \tilde{N}\alpha/\phi$.

The differential equations (5-5), (5-7), and (5-9) are all second-order equations of the sort we have treated in previous chapters. However, the interaction between chemicals and the soil underground can be extremely complicated, and (5-5), (5-7), and (5-9) are simply not always adequate descriptions of the physical situation.

Now, consider the following model. The sorption rate should increase as the concentration C increases. That is, N_t should be an increasing function of C. Similarly, as more and more chemical is sorbed, the ability of the soil to absorb the chemical will decrease. So, N_t will be a decreasing function of N. The simplest model with these characteristics is

$$N_t = \alpha C - \beta N, \tag{5-10}$$

with $\alpha, \beta > 0$ assumed constant. We can eliminate N from the system of partial differential equations (5-3) and (5-10): Substitute (5-10) into (5-3) and differentiate the result with respect to t. Then multiply (5-3) by β and add it to the result just obtained to find

$$C_{tt} + \left(\beta + \frac{\alpha}{\phi}\right)C_t = KC_{xxt} + \beta KC_{xx} - vC_{xt} - \beta vC_x. \tag{5-11}$$

If (5-11) is used to model the concentration C along an infinite tube $(-\infty < x < \infty)$, then Fourier transform solutions arise that are unpleasant to work with and from which it is hard to read off properties of the solution.

In the rest of this section, we consider a simpler version of (5-11) or the system (5-3), (5-10), which corresponds to the case when the groundwater is at rest, that is, $v = 0$. We wish to find the concentrations C and N in a tube of length L, $0 \le x \le L$ when C and N are known initially and the concentration of solute chemicals is reduced to zero at the ends of the tube. The initial, boundary value problem is

$$\begin{cases} C_t = KC_{xx} - \dfrac{1}{\phi}N_t, & 0 < x < L, \quad t > 0, \\ N_t = \alpha C - \beta N, & 0 < x < L, \quad t > 0, \\ C(x, 0) = f(x), \quad N(x, 0) = g(x), & 0 \le x \le L, \\ C(0, t) = 0, \quad C(L, t) = 0, & t > 0. \end{cases} \tag{5-12}$$

Theorem 5–1. There is at most one solution $C(x, t)$, $N(x, t)$ of the initial, boundary value problem (5-12) with $C, C_t, C_{tt}, C_x, C_{xt}, C_{xx}, C_{xxt}, N$ and N_t continuous on $0 \le x \le L$ and $t \ge 0$.

For the proof assume that there are two solutions (C_1, N_1) and (C_2, N_2) to (5-12). The differences $C = C_1 - C_2$ and $N = N_1 - N_2$ satisfy (5-12) with $f(x) = g(x) = 0$ for $0 \le x \le L$. Under the hypotheses in the theorem, C also satisfies the third-order equation in (5-11) with $v = 0$. Multiply this equation by C_t, integrate with respect to x from 0 to L, and use integration by parts to obtain

$$\frac{d}{dt}\int_0^L \frac{1}{2} C_t^2\,dx + \left(\beta + \frac{\alpha}{\phi}\right)\int_0^L C_t^2\,dx = -K\int_0^L C_{xt}^2\,dx - \frac{d}{dt}\int_0^L \frac{\beta K}{2} C_x^2\,dx$$

(see Prob. 3). Thus,

$$\frac{d}{dt}\int_0^L \left(\frac{1}{2} C_t^2 + \frac{\beta K}{2} C_x^2\right) dx \le 0.$$

Since $C(x, 0) = N(x, 0) = 0$ we find that $C_x(x, 0) = C_{xx}(x, 0) = 0$ and $N_t(x, 0) = 0$, from the second equation in (5-12). Then the first equation in (5-12) gives $C_t(x, 0) = 0$. Integrate the previous inequality from time zero to time t and use $C_t(x, 0) = C_x(x, 0) = 0$ to deduce

$$\frac{1}{2}\int_0^L [C_t^2(x, t) + K\beta C_x^2(x, t)]\,dx \le 0$$

for $t \ge 0$. Consequently, C_t and C_x must be identically zero because the integrand is clearly nonnegative and continuous. Thus, $C(x, t)$ is a constant. Since C vanishes when $t = 0$, the constant is zero and $C(x, t) \equiv 0$ or $C_1 = C_2$. With $C \equiv 0$ the first equation in (5-12) gives $N_t = 0$ and then the second equation yields $N \equiv 0$ or $N_1 = N_2$. Therefore, (5-12) has at most one solution as claimed.

The existence of a solution to (5-12) is shown most easily by applying separation of variables to (5-11) with $v = 0$. We must solve the differential equation

$$C_{tt} + aC_t = KC_{xxt} + K'C_{xx}, \qquad 0 < x < L, \qquad t > 0, \tag{5-13}$$

where $a = \beta + \alpha/\phi$ and $K' = \beta K$. [Compare (5-13) with (4-11)′ to see that (5-13) also describes wave motion with both internal and external damping.] From (5-12), C must satisfy the boundary conditions

$$C(0, t) = C(L, t) = 0, \qquad t \ge 0, \tag{5-14}$$

and the initial data

$$C(x, 0) = f(x), \qquad C_t(x, 0) = h(x), \qquad 0 \le x \le L, \tag{5-15}$$

where

$$h(x) = Kf''(x) - \frac{\alpha}{\phi} f(x) + \frac{\beta}{\phi} g(x).$$

Set $C(x, t) = X(x)T(t)$ in (5-13) to get

$$X(x)[T''(t) + aT'(t)] = X''(x)[KT'(t) + K'T(t)]. \tag{5-16}$$

This equation leads to a familar eigenvalue problem for $X(x)$, and we find

$$X_n(x) = \sin(\lambda_n x) \quad \text{and} \quad \lambda_n = \frac{n\pi}{L} \tag{5-17}$$

for $n = 1, 2, \ldots$ (see Prob. 4). The corresponding equation for $T_n(t)$ is

$$T_n''(t) + (a + \lambda_n^2 K)T_n'(t) + \lambda_n^2 K' T_n(t) = 0, \tag{5-18}$$

which has solution

(5-19) $$T_n(t) = e^{-r_n t}(a_n e^{s_n t} + b_n e^{-s_n t}),$$

where

(5-20) $$r_n = \frac{a + \lambda_n^2 K}{2} \quad \text{and} \quad s_n = \frac{\sqrt{(a + \lambda_n^2 K)^2 - 4\lambda_n^2 K'}}{2},$$

and a_n and b_n are arbitrary constants. Notice that s_n is real for n large, and for such n, $r_n > s_n$. In view of (5-17) and (5-19), the formal solution to (5-13)–(5-15) is

(5-21) $$C(x, t) = \sum_{n=1}^{\infty} e^{-r_n t}(a_n e^{s_n t} + b_n e^{-s_n t}) \sin(\lambda_n x),$$

where the coefficients a_n and b_n are chosen to satisfy,

$$C(x, 0) = f(x) = \sum_{n=1}^{\infty} (a_n + b_n) \sin(\lambda_n x),$$

and

$$C_t(x, 0) = h(x) = \sum_{n=1}^{\infty} [(s_n - r_n)a_n - (s_n + r_n)b_n] \sin(\lambda_n x).$$

Therefore, $a_n + b_n$ must be the nth Fourier sine coefficient for $f(x)$, and $[(s_n - r_n)a_n - (s_n + r_n)b_n]$ must be the nth Fourier sine coefficient for $h(x)$:

(5-22) $$a_n + b_n = \frac{2}{L}\int_0^L f(x) \sin(\lambda_n x)\, dx,$$

(5-23) $$(s_n - r_n)a_n - (s_n + r_n)b_n = \frac{2}{L}\int_0^L h(x) \sin(\lambda_n x)\, dx.$$

The solution to this system of equations for a_n and b_n is

(5-24) $$a_n = \frac{1}{Ls_n}\left[\int_0^L h(x) \sin(\lambda_n x)\, dx + (s_n + r_n)\int_0^L f(x) \sin(\lambda_n x)\, dx\right],$$

(5-25) $$b_n = \frac{1}{Ls_n}\left[-\int_0^L h(x) \sin(\lambda_n x)\, dx + (s_n - r_n)\int_0^L f(x) \sin(\lambda_n x)\, dx\right],$$

where r_n and s_n are given by (5-20).

Next, we determine conditions on the initial data which guarantee that the formal solution is a solution. In order to confirm that (5-21) satisfies (5-13) we need to justify the appropriate term-by-term differentiations. The portion of the series in (5-21) with coefficient b_n converges very rapidly due to the exponential decay of $\exp[-(r_n + s_n)t]$, because $r_n + s_n$ grows like λ_n^2. Assume that $h(x)$ and $f(x)$ are continuous. Then (5-24) and (5-25) imply that a_n and b_n are bounded independent of n, and the reasoning used in Sec. 5-1 for the heat equation shows that

(5-26) $$\sum b_n e^{-(r_n + s_n)t} \sin(\lambda_n x)$$

can be differentiated term by term any number of times with respect to x and/or t for $t > 0$ and $0 \le x \le L$. The other part of the series in (5-21),

$$\sum a_n e^{-(r_n - s_n)t} \sin(\lambda_n x), \tag{5-27}$$

is much less rapidly convergent (as are its term-by-term derivatives) because $r_n - s_n \approx K'/K$ for large n. Thus, the exponential term behaves like $\exp(-K't/K)$ and convergence will depend on how rapidly a_n tends to zero with increasing n. It is clear that series which result from (5-27) when (5-21) is differentiated term by term in (5-13) will all converge uniformly if the series

$$\sum \lambda_n^2 a_n e^{-(r_n - s_n)t} \sin(\lambda_n x) \tag{5-28}$$

corresponding to the C_{xx} term is uniformly convergent. (Recall that the factors $r_n - s_n$ which occur in C_t, C_{tt}, and C_{xxt} are approximately K'/K for large n.) We can assure uniform convergence in (5-28) if we require that $\lambda_n^2|a_n| \le A/\lambda_n^2$ for some constant A, because $\lambda_n = n\pi/L$. Now, (5-20) and (5-24) yield the estimate

$$|a_n| \le D\left[\frac{1}{\lambda_n^2}\left|\int_0^L h(x)\sin(\lambda_n x)\,dx\right| + \left|\int_0^L f(x)\sin(\lambda_n x)\,dx\right|\right] \tag{5-29}$$

for some constant D. If we assume that $h(0) = h(L) = 0$ and that $h(x)$ has a continuous second derivative, the first integral in (5-29) is bounded by a constant times $1/\lambda_n^2$, as we showed in Sec. 4-1. Similarly, if $f(0) = f(L) = f''(0) = f''(L) = 0$ and $f(x)$ has a continuous fourth derivative, then the second integral in (5-29) is bounded by a constant times $1/\lambda_n^4$. These observations and (5-29) reveal that $\lambda_n^2|a_n| \le A/\lambda_n^2$ for some constant A. Thus, the series in (5-28) is uniformly convergent, and the term-by-term differentiations needed to show that (5-21) satisfies (5-13) are justified.

Consider the Fourier sine series expansions following (5-21) and with coefficients determined by (5-22) and (5-23). The assumptions we have made on $f(x)$ and $h(x)$ allow us to estimate the Fourier coefficients in (5-22) and (5-23), as we just did in (5-29). Thus,

$$|a_n + b_n|, \quad |(s_n - r_n)a_n - (s_n + r_n)b_n| \le \frac{E}{\lambda_n^2}$$

for some constant E. Therefore, the Fourier sine series are uniformly convergent and since $f(0) = f(L) = h(0) = h(L) = 0$ these series converge to $f(x)$ and $h(x)$ as required (Th. 1-1 of Chap. 3).

Now we have established that $C(x, t)$ given by (5-21) solves (5-13)–(5-15). Finally, we obtain $N(x, t)$ by solving $N_t + \beta N = \alpha C$ with $N(x, 0) = g(x)$:

$$N(x, t) = g(x)e^{-\beta t} + \alpha\int_0^t C(x, \tau)e^{-\beta(t-\tau)}\,d\tau. \tag{5-30}$$

In summary, we have proved:

Theorem 5-2. Suppose that $g(x)$ has a continuous second derivative, $f(x)$ has a continuous fourth derivative, $f(0) = f(L) = f''(0) = f''(L) = 0$, and that $h(x) = Kf''(x) - (\alpha/\phi)f(x) + (\beta/\phi)g(x)$ satisfies $h(0) = h(L) = 0$. Then the initial, boundary value problem (5-12) has a unique solution $C(x, t)$ given by (5-21) and $N(x, t)$ given by (5-30).

PROBLEMS

1. Consider the following initial, boundary value problem for the standard diffusion–dispersion equation,

$$C_t = K_0 C_{xx} - wC_x, \qquad 0 < x < L, \qquad t > 0,$$
$$C(0, t) = C(L, t) = 0, \qquad t \geq 0,$$
$$C(x, 0) = f(x), \qquad 0 \leq x \leq L,$$

where K_0 and w are positive constants.

(a) What physical situation is described by this model?

(b) Show there exists at most one solution. *Hint.* Let C be the difference of two solutions,

$$I(t) = \frac{1}{2}\int_0^L C(x, t)^2\, dx,$$

and show that $I'(t) \leq 0$.

(c) Verify that nontrivial separated solutions $C(x, t) = X(x)T(t)$ are determined by the eigenvalue problem

$$K_0 X'' - wX' - \mu X = 0, \qquad X(0) = X(L) = 0.$$

Show that $\mu < 0$, say $\mu = -\lambda^2$ for $\lambda > 0$, and deduce that

$$\lambda_n^2 = \frac{w^2L^2 + 4\pi^2 n^2 K_0^2}{4K_0 L^2}, \qquad n = 1, 2, \ldots,$$

and

$$X_n(x) = e^{wx/2K_0} \sin\left(\frac{n\pi x}{L}\right), \qquad 0 \leq x \leq L,$$

up to a constant multiple.

(d) Find the corresponding $T_n(t)$ and the formal solution to the given problem.

2. Find reasonable conditions on the initial concentration in Prob. 1 which will imply that the formal solution is a solution.

3. Verify the steps in the uniqueness argument following (5-12) and leading to

$$\frac{d}{dt}\int_0^L \left(\frac{1}{2}C_t^2 + \frac{\beta K}{2}C_x^2\right) dx \leq 0.$$

4. Confirm (5-16)–(5-20). Check that $r_n + s_n \approx a + \lambda_n^2 K$ and that $r_n - s_n \approx K'/K$ for large n.

5. Use the arguments from Sec. 5-1 to carefully check the asserted term-by-term differentiability of the series in (5-26).

6. Use (5-20) and (5-24) to deduce (5-29).

7. In studying long waves, tides, solitary waves, and related phenomena, one is led to an equation (linearized version) of the form

$$u_t = \sigma u_{xxx}, \qquad \sigma > 0, \quad \text{a constant.}$$

This equation is referred to as the linearized *Korteweg–DeVries equation.* Use Fourier transforms to solve this differential equation in $-\infty < x < \infty$, $t > 0$, and subject to the initial condition $u(x, 0) = f(x)$. The solution will be of the form

$$u(x, t) = \int_{-\infty}^{\infty} k(x - y, \sigma t) f(y)\, dy$$

where

$$k(x, t) = \frac{1}{(3t)^{1/3}} Ai\left(-\frac{x}{(3t)^{1/3}}\right)$$

and $Ai(\lambda)$ is the *Airy function.*

8. Show there exists at most one solution to the boundary value problem

$$u_t = \sigma u_{xxx}, \qquad 0 < x < L, \qquad t > 0,$$
$$u(x, 0) = f(x), \qquad 0 \le x \le L,$$
$$u(0, t) = u_x(0, t) = 0, \qquad u(L, t) = 0, \qquad t \ge 0,$$

and solve it by means of separation of variables. *Hint.* For the uniqueness, consider the following integral for the difference of two solutions

$$I(t) = \frac{1}{2} \int_0^L u(x, t)^2\, dx$$

and show that $I'(t) \le 0$.

The next set of equations arise in the study of fluids flowing underground in a fractured or fissured medium. We shall assume that the fluid is incompressible. One then thinks of the medium as being made up of two continua superimposed on each other. The one continuum will be denoted by the subscript 1 and will refer to the fissured part of the medium and the subscript 2 will denote the second continuum and will refer to the porous part of the medium. In the case of one spatial dimension, one is led to the following system of partial differential equations for the pressures p_1 and p_2:

$$S_1 \frac{\partial p_1}{\partial t} - k_1 \frac{\partial^2 p_1}{\partial x^2} = -\alpha(p_1 - p_2)$$

$$S_2 \frac{\partial p_2}{\partial t} - k_2 \frac{\partial^2 p_2}{\partial x^2} = \alpha(p_1 - p_2),$$

where S_1, S_2, k_1, k_2, and α are positive constants.

9. Suppose, as is the case in many applications, that S_1 and k_2 can be considered to be small, that is, negligible in these equations. Show that p_2 satisfies an equation of the form

$$u_t = a u_{xx} + b u_{xxt},$$

where a and b are positive constants. This equation is referred to as *Barenblatt's equation.*

10. Suppose that S_1 is negligible but the other quantities are not. Show that p_2 satisfies an equation of the form

$$u_t = au_{xx} + bu_{xxt} - cu_{xxxx},$$

where a, b, and c are positive constants.

11. Suppose that none of the quantities are negligible. Show then that p_2 satisfies an equation of the form

$$u_{tt} + au_t = bu_{xx} + cu_{xxt} - du_{xxxx},$$

where a, b, c, and d are positive constants.

12. Using Fourier transforms, solve one of the equations in Probs. 9–11 in $-\infty < x < \infty$, $t > 0$ subject to the initial condition

$$u(x, 0) = f(x), \qquad -\infty < x < \infty.$$

7

Integral Equations, Green's Functions, and Eigenfunction Expansions

7-1. Introduction

Integral equations play an important role in many fundamental problems of mathematical physics. Questions of existence and uniqueness of solutions as well as eigenfunction representations for solutions are conveniently studied from the integral equations point of view. For example, we used the integral equation (6-10) in Sec. 4-6 to establish that the initial value problem (6-1) and (6-2) for linear hyperbolic equations has a unique solution. We also extended these existence and uniqueness results to Cauchy problems for nonlinear hyperbolic problems using the system of integral equations (7-8) in Sec. 4-7. In each case the integral equation was solved by the method of successive approximations.

Integral equations also arise naturally as reformulations of boundary value problems for differential equations. Such integral equations involve the Green's function or influence function of the boundary value problem. In Sec. 7-2 we study Green's functions for Sturm–Liouville boundary value problems.

The rest of this section is devoted to discussing several important problems related to the method of separation of variables and eigenfunction expansions. We have used separation of variables to solve the initial, boundary value problem for a homogeneous vibrating string in Sec. 4-2 and to solve the heat conduction problem for a homogeneous rod in Sec. 5-1.

As another example of separation of variables, consider the flow of heat in a heterogeneous rod of unit length which is imperfectly insulated so that it may lose heat through its lateral surface. If the rod is surrounded by a 0°C temperature bath, the partial differential equation that describes the heat flow is

$$\rho(x)c(x)\frac{\partial u}{\partial t} = \frac{\partial}{\partial x}\left(k(x)\frac{\partial u}{\partial x}\right) - l(x)u, \tag{1-1}$$

where $u(x, t)$ is the temperature at position x and time t, $\rho(x)$ is the density, $c(x)$ is the specific heat, $k(x) > 0$ is the thermal conductivity, and $l(x) \geq 0$ is a lateral radiation coefficient (see Prob. 1).

If we assume that Newton's law of cooling holds at the ends of the rod, the temperature $u(x, t)$ satisfies the boundary conditions

$$u(0, t) - h_0 u_x(0, t) = 0, \qquad u(1, t) + h_1 u_x(1, t) = 0, \tag{1-2}$$

for $t \geq 0$, where h_0 and h_1 are nonnegative constants. Finally, the temperature in the rod depends also on its initial temperature distribution, say

$$u(x, 0) = f(x) \tag{1-3}$$

for $0 \leq x \leq 1$.

It is natural to try to solve the initial, boundary value problem (1-1)–(1-3) by separation of variables. Substituting $u(x, t) = X(x)T(t)$ into (1-1) leads to

$$\frac{T'}{T} = K \quad \text{and} \quad \frac{(k(x)X')' - l(x)X}{\rho(x)c(x)X} = K,$$

where K is the separation constant. Thus, $T = e^{Kt}$, up to a constant multiple, and we expect that $K < 0$ on physical grounds (see Prob. 2). Set $K = -\lambda$ with $\lambda > 0$ and $r(x) = \rho(x)c(x) > 0$. Then $T(t) = e^{-\lambda t}$ and $u(x, t) = e^{-\lambda t}X(x)$ will satisfy the boundary conditions (1-2) provided that

$$X(0) - h_0 X'(0) = 0 \quad \text{and} \quad X(1) + h_1 X'(1) = 0.$$

Thus, $u(x, t) = e^{-\lambda t}X(x)$ will be a nontrivial solution of the heat equation (1-1) and boundary conditions (1-2) provided that X is a nontrivial solution of the *Sturm–Liouville eigenvalue problem*

$$\begin{cases} (k(x)X')' + (r(x)\lambda - l(x))X = 0, & 0 < x < 1, \\ X(0) - h_0 X'(0) = 0, & X(1) + h_1 X'(1) = 0, \end{cases} \tag{1-4}$$

where $h_0, h_1 \geq 0$, $k(x), r(x) > 0$, $l(x) \geq 0$ for $0 \leq x \leq 1$, and $k(x)$, $r(x)$, and $l(x)$ are continuous.

Notice that (1-4) reduces to the eigenvalue problem (2-6) of Chap. 4 for the vibrating string if $k = r = 1$, $l = h_0 = h_1 = 0$, and also to the eigenvalue problem in Chap. 5 for heat conduction in a homogeneous, laterally insulated rod. As in Chaps. 4 and 5, we call the values of λ for which (1-4) has nontrivial solutions *eigenvalues* and refer to the corresponding nontrivial solutions as *eigenfunctions*.

Our previous work makes it clear that (1-4) should have an infinite sequence of eigenvalues, say $\lambda_1, \lambda_2, \ldots, \lambda_n, \ldots$, and corresponding eigenfunctions, $X_n(x)$, for separation of variables to lead to a solution of (1-1)–(1-3). This leads us to a fundamental problem:

Basic Problem 1. Establish that the Sturm–Liouville problem (1-4) has an infinite sequence of eigenvalues and eigenfunctions.

Assume that (1-4) does have an infinite sequence of eigenvalues. Then $u_n(x, t) = e^{-\lambda_n t}X_n(x)$ is a nontrivial solution of the differential equation (1-1) and boundary conditions (1-2). Just as for the vibrating string or homogeneous, heat-conducting rod, we try to satisfy the initial condition (1-3) as well as (1-1) and (1-2) by superposing solutions. Thus, we set

$$u(x, t) = \sum_{n=1}^{\infty} c_n e^{-\lambda_n t} X_n(x) \tag{1-5}$$

and want to determine the coefficients c_n so that

$$\sum_{n=1}^{\infty} c_n X_n(x) = f(x), \qquad 0 \leq x \leq 1, \tag{1-6}$$

which means that $u(x, 0)$ satisfies the initial condition (1-3). We call (1-6) an *eigen-function expansion.* In Chaps. 4 and 5 with $L = 1$, $X_n(x) = \sin(n\pi x)$ and (1-6) is a Fourier sine series expansion. In summary, (1-5) is a formal solution to (1-1)–(1-3). To confirm that (1-5) is a solution, we must justify appropriate term-by-term differentiations in (1-5) and solve:

Basic Problem 2. Establish that any reasonably arbitrary function, $f(x)$, has an eigenfunction expansion, and determine the mode of convergence of this expansion.

Of course, part of this problem involves finding the coefficients c_n in (1-6). This issue is treated in the problems. The key is to discover the *orthogonality relations*

$$\int_0^1 X_n(x) X_m(x) r(x)\, dx = 0, \qquad n \neq m, \tag{1-7}$$

which lead to

$$c_n = \frac{\displaystyle\int_0^1 f(x) X_n(x) r(x)\, dx}{\displaystyle\int_0^1 X_n(x)^2 r(x)\, dx} \tag{1-8}$$

for $n = 1, 2, 3, \ldots$.

If heat is generated at a rate of $g(x, t)$ joules/kg·s in the rod in Fig. 7-1, the heat equation becomes

$$\rho(x)c(x)\frac{\partial u}{\partial t} = \frac{\partial}{\partial x}\left(k(x)\frac{\partial u}{\partial x}\right) - l(x)u + \rho(x)g(x, t) \tag{1-9}$$

and the boundary conditions remain the same. If $g(x, t) \to g(x)$ as $t \to \infty$, it is common experience that after a long time the temperature distribution in the rod approaches a *steady* (time independent) *state.* This state is described by a time-independent temperature distribution $u = u(x)$. Since $u_t = 0$, the steady-state temperature is determined by the *Sturm–Liouville boundary value problem*

$$\begin{cases} -(k(x)u')' + l(x)u = \rho(x)g(x), \\ u(0) - h_0 u'(0) = 0, \qquad u(1) + h_1 u'(1) = 0. \end{cases} \tag{1-10}$$

Basic Problem 3. Establish that this Sturm–Liouville problem has a solution.

It is often convenient to combine the study of the Sturm–Liouville eigenvalue and boundary value problems. Each is a special case of the *Sturm–Liouville problem*

$$\begin{cases} -(p(x)y')' + (q(x) - \lambda r(x))y = f(x), & 0 < x < 1, \\ y(0) - h_0 y'(0) = 0, & y(1) + h_1 y'(1) = 0. \end{cases} \tag{1-11}$$

Observe that the choice $f(x) \equiv 0$ leads to the eigenvalue problem (1-4) while the choice of a fixed λ (say, $\lambda = 0$) yields a problem of type (1-10).

PROBLEMS

1. Derive (1-1). Assume that Newton's law of cooling governs heat loss across the lateral surface of the rod.
2. Show that the separation constant K in the text must be negative. *Hint.* Review the argument used to show this for (2-6) in Chap. 4. It may help first to consider the case $h_0 = h_1 = 0$.
3. Establish (1-7). *Hint.* Multiply the differential equation satisfied by X_n (resp., X_m) by X_m (resp., X_n). Subtract these results and then integrate.
4. Assume that the series in (1-6) is uniformly convergent and confirm (1-8). *Hint.* Review the opening paragraphs in Sec. 3-1.
5. Consider the initial, boundary value problem for an inhomogeneous string with fixed ends:

$$u_{tt} = c^2(k(x)u_x)_x, \quad 0 < x < 1, \quad t > 0,$$
$$u(0, t) = 0, \quad u(1, t) = 0, \quad t \geq 0,$$
$$u(x, 0) = f(x), \quad u_t(x, 0) = g(x), \quad 0 \leq x \leq 1,$$

   with $k(x) > 0$.
   (a) Find the time-dependent part $T(t)$ of a separated solution $u(x, t) = X(x)T(t)$.
   (b) Formulate the Sturm–Liouville eigenvalue problem for $X(x)$. In the process, show that the separation constant K is negative and set $K = -\lambda$ for $\lambda > 0$.
   (c) Assume that the eigenvalue problem in part (b) has eigenvalues $\lambda_1, \lambda_2, \ldots$ and corresponding eigenfunctions $X_1(x), X_2(x), \ldots$. Find a formal solution. [You are not expected to find explicit formulas for the $X_n(x)$.]
6. Imagine that the heat conducting rod is bent into a ring (Fourier heat ring) with circumference 1. The corresponding initial value problem is (1-1), (1-3), and

$$u(0, t) = u(1, t), \quad u_x(0, t) = u_x(1, t)$$

   for $t > 0$.
   (a) Explain the physical meaning of these boundary conditions.
   (b) Discuss, paralleling the treatment in the text, how the temperature in the heat ring may be found using separation of variables.
7. Assume that (1-9), (1-2), (1-3) has a steady-state solution $u = u(x)$. Show that u satisfies a problem similar to (1-10).

7-2. Green's Functions for Sturm–Liouville Problems

Consider the Sturm–Liouville boundary value problem

$$\text{(2-1)}\qquad \begin{cases} -(p(x)y')' + q(x)y = f(x), & 0 < x < 1, \\ y(0) - h_0 y'(0) = 0, & y(1) + h_1 y'(1) = 0, \end{cases}$$

where $p(x) \neq 0$, $p'(x)$, $q(x)$, and $f(x)$ are continuous on $0 \leq x \leq 1$ and h_0, h_1 are constants. If $p(x) > 0$, $q(x) \geq 0$, and $h_0, h_1 \geq 0$ this Sturm–Liouville problem models the steady-state temperature in a rod [see (1-10)].

It is natural to seek a solution to (2-1) in the form

$$\text{(2-2)}\qquad y(x) = \int_0^1 g(x, s)f(s)\,ds,$$

where $g(x, s)$ is a continuous function called the *Green's* or *influence function* for the Sturm–Liouville problem (2-1). More precisely, a function $g(x, s)$ is called a Green's function for (2-1) if it is continuous on $0 \leq x, s \leq 1$ and (2-2) uniquely solves (2-1) for every continuous function $f(x)$. If (2-1) has a Green's function it is unique (see Prob. 2).

To see why the solution formula (2-2) is reasonable, consider the steady-state temperature in a rod. Then $f(x)$ represents a heat source intensity along the rod. In particular, consider a heat distribution $f = f_s(x)$ of unit intensity localized about the point s in the rod: That is, assume that $f_s(x) = 0$ for $|x - s| > \varepsilon$ where $\varepsilon > 0$ is small and

$$\int_{s-\varepsilon}^{s+\varepsilon} f_s(x)\,dx = 1$$

(see Fig. 7-1). Let $y(x) = g_\varepsilon(x, s)$ be the steady-state temperature induced by $f_s(x)$. That is, $g_\varepsilon(x, s)$ is the steady-state temperature at x produced by a unit heat source localized near the point s. Analytically speaking, $g_\varepsilon(x, s)$ is the solution of (2-1) when $f = f_s(x)$. As $\varepsilon \to 0$, it is plausible that the temperature distribution $g_\varepsilon(x, s)$ will converge to a limiting temperature distribution $g(x, s)$ corresponding to a heat

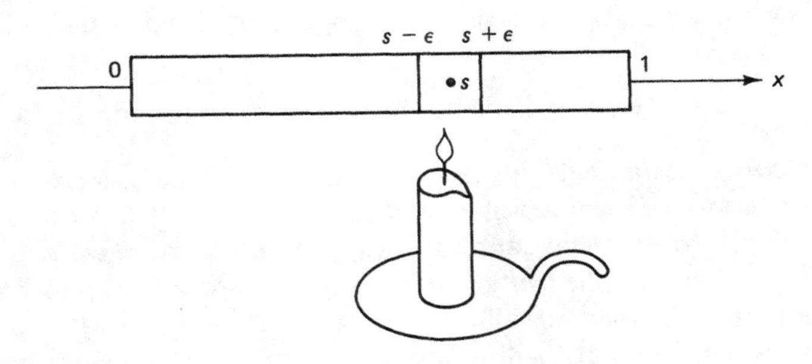

Figure 7-1

source of unit intensity applied at the point s. Thus, $g(x, s)$ is the temperature at x produced by a heat source of unit intensity applied at s.

Imagine that the rod is composed of a large number n of tiny chunks, and let s_k be a point in the kth chunk. Let $f(x)$ be a continuous heat source distribution in the rod. This distribution delivers an amount of heat nearly equal to $f(s_k)\,\Delta s$ to the chunk of rod containing s_k. Since the differential equation and boundary conditions in (2-1) are linear and homogeneous, the temperature at x caused by the heating near s_k is nearly $g(x, s_k)f(s_k)\,\Delta s$. Furthermore, since (2-1) is linear and homogeneous, the superposition principle holds (sums of solutions are solutions): The temperature at x due to the combined heat sources $f(s_k)\,\Delta s$ acting near s_k for $k = 1, 2, \ldots, n$ is nearly

$$\sum_{k=1}^{n} g(x, s_k)f(s_k)\,\Delta s.$$

Physical intuition suggests that as $n \to \infty$ and $\Delta s \to 0$ we obtain the exact temperature $y(x)$ induced by the heat distribution $f(x)$. That is,

$$y(x) = \lim_{n\to\infty} \sum_{k=1}^{n} g(x, s_k)f(s_k)\,\Delta s = \int_0^1 g(x, s)f(s)\,ds,$$

which is just (2-2).

This discussion suggests some fundamental properties of the Green's function $g(x, s)$. The solution $g_\varepsilon(x, s)$ to (2-1) with $f = f_s(x)$ satisfies

$$L[g_\varepsilon] = f_s(x) = 0 \quad \text{for} \quad |x - s| > \varepsilon,$$

where

$$Ly = -(py')' + qy \tag{2-3}$$

is called the *Sturm–Liouville differential operator*. Since $g(x, s)$ is the limit of $g_\varepsilon(x, s)$ as $\varepsilon \to 0$, the preceding equation for g_ε suggests that

(i)′ $\quad Lg = 0 \quad \text{for} \quad x \neq s,\quad$ where the differential operator L acts on the variable x.

Furthermore, $g(x, s)$ should satisfy the boundary conditions in (2-1) because each $g_\varepsilon(x, s)$ does:

(ii)′ $\quad g(x, s) \quad$ satisfies the boundary conditions in (2-1).

Now comes a delicate point. Each function $g_\varepsilon(x, s)$ has a continuous second derivative, by the meaning of a solution to (2-1). How smooth is the limit $g(x, s)$? The passage from a heat source of unit intensity localized *near* s to a heat source of unit intensity applied *at* s results in an abrupt infusion of heat energy at the single point s in the rod. The temperature at $x = s$ must respond accordingly. Intuition suggests that the limiting temperature should be continuous as x passes through s, but that the rate of change of temperature may well change abruptly in response to the infusion of heat at $x = s$.

The differential equation $L[g_\varepsilon] = f_s(x)$ can help us see what to expect. Since each heat source f_s has unit strength, we expect that the resultant temperatures $g_\varepsilon(x, s)$ should be bounded independent of ε, say $|g_\varepsilon(x, s)| \leq M$. Integrate $L[g_\varepsilon] = f_s(x)$ from $s - \varepsilon$ to $s + \varepsilon$ to find

$$\Big[-pg'_\varepsilon\Big]_{x=s-\varepsilon}^{x=s+\varepsilon} + \int_{s-\varepsilon}^{s+\varepsilon} q(x)g_\varepsilon(x, s)\,dx = 1,$$

where the prime indicates differentiation with respect to x. As $\varepsilon \to 0$ the integral tends to zero because it is bounded by $2\varepsilon M \max |q(x)|$, where the maximum is over $0 \leq x \leq 1$. Thus, we expect

$$\Big[-p(x)g'(x, s)\Big]_{x=s_-}^{x=s^+} = 1,$$

in the limit as $\varepsilon \to 0$. Since $p(x)$ is continuous at $x = s$, this can be expressed as:

(iii)′ The partial derivative of g with respect to x satisfies

$$g_x(s^+, s) - g_x(s^-, s) = \frac{-1}{p(s)}.$$

On physical grounds, we have shown that the Sturm–Liouville problem (2-1) should have a solution which is given by (2-2) in terms of the Green's function. Furthermore, our discussion suggests that properties (i)′–(iii)′ are characteristic of the Green's function. We turn now to the mathematical developments which confirm this physical analysis.

The solution formula (2-2) follows easily from the *Lagrange identity*

$$vLu - uLv = -\frac{d}{dx}\,[p(vu' - uv')], \tag{2-4}$$

which holds (Prob. 9) for u, v twice differentiable on $[0, 1]$, and Lagrange's method described below. For simplicity, we assume that $h_0 = h_1 = 0$ in (2-1). In this case, the boundary conditions become

$$y(0) = 0 \quad \text{and} \quad y(1) = 0, \tag{2-5}$$

and are called *Dirichlet boundary conditions.* The strategy of Lagrange's method will be clear from this special case, and the arithmetic will be a little easier.

Suppose that we can find a function $w \neq 0$ with $Lw = 0$. Apply the Lagrange identity with $v = w$ and $u = y$, the solution to (2-1) with $h_0 = h_1 = 0$, to get

$$wf = -\frac{d}{dx}\,p(wy' - w'y). \tag{2-6}$$

We can integrate this equation and obtain a *first-order* differential equation for y. Thus, we call w an *integrating factor* for $Ly = f$. Suppose that we can find integrating factors u and v such that

$$Lu = 0, \qquad u(0) = 0,$$
$$Lv = 0, \qquad v(1) = 0,$$

where each integrating factor satisfies one of the boundary conditions. Apply (2-6) with $w = u$ and $w = v$ and recall (2-5) to get

$$\int_0^x u(s)f(s)\,ds = -p(x)(uy' - u'y)(x) \tag{2-7}$$

and

$$\int_x^1 v(s)f(s)\,ds = p(x)(vy' - v'y)(x). \tag{2-8}$$

Now, y' can be eliminated from (2-7) and (2-8) by multiplying the first equation by $v(x)$, the second by $u(x)$, and adding the results:

$$v(x)\int_0^x u(s)f(s)\,ds + u(x)\int_x^1 v(s)f(s)\,ds = -p(x)[u(x)v'(x) - u'(x)v(x)]y(x). \tag{2-9}$$

The expression in brackets is just the Wronskian

$$W(x) = \begin{vmatrix} u(x) & v(x) \\ u'(x) & v'(x) \end{vmatrix}$$

of the two solutions u and v of $Lw = 0$. An elementary calculation (Prob. 11) shows that

$$p(x)W(x) = C, \tag{2-10}$$

a constant. Therefore, the coefficient of $y(x)$ in (2-9) is $-C$, and (2-9) yields a formula for the solution $y(x)$ to (2-1) with $h_0 = h_1 = 0$ provided that $C \neq 0$.

Suppose that $C \neq 0$ and let

$$v_0(x) = -\frac{u(x)}{C} \quad \text{and} \quad v_1(x) = v(x).$$

Clearly,

$$\begin{aligned} Lv_0 &= 0, & v_0(0) &= 0, \\ Lv_1 &= 0, & v_1(0) &= 0, \\ p(x)W(x) &= -1, \end{aligned}$$

where $W(x)$ is the Wronskian of v_0 and v_1, and (2-9) yields

$$y(x) = \int_0^x v_1(x)v_0(s)f(s)\,ds + \int_x^1 v_0(x)v_1(s)f(s)\,ds$$

or

$$y(x) = \int_0^1 g(x, s)f(s)\,ds,$$

where

$$g(x, s) = \begin{cases} v_1(x)v_0(s), & 0 \le s \le x \le 1, \\ v_0(x)v_1(s), & 0 \le x \le s \le 1. \end{cases} \tag{2-11}$$

In summary, the Sturm–Liouville problem (2-1) has solution given by (2-2), where $g(x, s)$ is given by (2-11) provided that we can find functions v_0 and v_1 satisfying

$$(2\text{-}12)\quad \begin{cases} Lv_0 = 0, & v_0 \quad \text{satisfies the boundary condition at } x = 0, \\ Lv_1 = 0, & v_1 \quad \text{satisfies the boundary condition at } x = 1, \\ pW = -1, & \end{cases}$$

where W is the Wronskian of v_0 and v_1.

In the analysis leading to (2-2) with $g(x, s)$ given by (2-11), we assumed that (2-1) had a solution. However, once functions satisfying (2-12) are known, it is a routine calculation to show that (2-2) with $g(x, s)$ given by (2-11) does solve (2-1) (see Prob. 8). Thus, whenever (2-12) holds the Sturm–Liouville problem (2-1) has a unique solution given by (2-2) and (2-11).

Notice that we can always find nontrivial functions v_0 and v_1 satisfying the first two conditions in (2-12). Simply take $v_i(x)$ to be the solution to the initial value problem $Lv = 0$, $v(i) = 0$, $v'(i) = 1$ for $i = 0, 1$. (Recall that we are taking $h_0 = h_1 = 0$.) Consequently, the existence of the Green's function hinges on obtaining the condition $pW = -1$ in (2-12). We are now in a position to determine exactly when (2-1) has a Green's function.

Theorem 2-1. The Sturm–Liouville problem (2-1) has a Green's function if and only if the corresponding homogeneous problem [with $f(x) \equiv 0$] has only the trivial solution, in which case the Green's function is given by (2-11) and (2-12).

For the proof, suppose first that (2-1) has a Green's function. Then (2-2) provides the unique solution to (2-1) for each continuous function $f(x)$. In particular, when $f(x) \equiv 0$, (2-2) yields the trivial solution $y(x) \equiv 0$.

Conversely, assume that the corresponding homogeneous equation has only the trivial solution. As above, we can find *nontrivial* solutions v_0 and v_1 to the first two conditions in (2-12). From (2-10), $p(x)W(x) = C$ a constant. If $C = 0$, v_0 and v_1 are linearly dependent (proportional) because their Wronskian vanishes. Since $v_0(0) = v_1(1) = 0$ and v_0 and v_1 are proportional, we also have $v_0(1) = v_1(0) = 0$. Then $v_0(x)$ [or $v_1(x)$] is a nontrivial solution of $Ly = 0$, $y(0) = y(1) = 0$. This contradicts our assumption that this homogeneous equation has only the trivial solution, $y \equiv 0$. We conclude that $C \neq 0$. Now we can replace v_0 by $-v_0/C$ and achieve $pW = -1$. Renaming $-v_0/C$ as v_0, we have found functions v_0 and v_1 satisfying (2-12). Consequently, (2-2) holds with $g(x, s)$ given by (2-11), and the proof is complete.

Recall properties (i)′–(iii)′ of the Green's function, which were deduced on physical grounds. We will formulate these properties more precisely and show that they characterize the Green's function.

Theorem 2-2. Assume that the homogeneous problem corresponding to (2-1) has only the trivial solution. Let $g(x, s)$ be the function defined by (2-11) and

(2-12). Then

(i) $g(x, s)$ is continuous on the square $0 \leq x, s \leq 1$ and has a continuous second derivative on each of the triangles $0 \leq s \leq x \leq 1$ and $0 \leq x \leq s \leq 1$. On each triangle $Lg = 0$, where L acts on functions of x;

(ii) $g(x, s)$ satisfies the boundary conditions at $x = 0$ and $x = 1$;

(iii) For each s with $0 < s < 1$,

$$g_x(s^+, s) - g_x(s^-, s) = \frac{-1}{p(s)}.$$

Conversely, properties (i)–(iii) uniquely determine a function such that (2-2) solves (2-1).

To confirm these assertions, first let $g(x, s)$ be defined by (2-11) and (2-12). It is clear from (2-12) that (i) and (ii) hold for $g(x, s)$. Also,

$$g_x(s^+, s) - g_x(s^-, s) = v_1'(s)v_0(s) - v_0'(s)v_1(s) = W(s) = -\frac{1}{p(s)}$$

by (2-12). Thus, $g(x, s)$ defined by (2-11) and (2-12) has the required properties.

Suppose that $\tilde{g}(x, s)$ also has properties (i)–(iii) and form $\Delta(x, s) = \tilde{g}(x, s) - g(x, s)$. Then $\Delta_x(s^+, s) - \Delta_x(s^-, s) = 0$ because (iii) holds for both g and $\tilde{g}$. Therefore, by (i), $z(x) = \Delta(x, s)$ is, for fixed s, continuous and has a continuous first derivative on $0 \leq x \leq 1$. Also by (i), $z'' = (-p'z' + qz)/p$ for $x \leq s$ and $x \geq s$. Since p, p', q, z, and z' are continuous at s, this equation shows that z'' is continuous on $0 \leq x \leq 1$. Then $Lz = 0$ and z satisfies the boundary conditions in (2-1) by (ii). Thus, $z(x) \equiv 0$ or $\tilde{g}(x, s) = g(x, s)$ because the homogeneous problem corresponding to (2-1) has only the trivial solution.

Example. Use the Green's function to represent the steady-state temperature in a homogeneous rod of unit length and with insulated lateral surface. Assume that the left end of the rod is at temperature zero, that the right end is insulated, and that heat generation along the rod is given by the source term $f(x)$.

From (1-10) the steady-state temperature $u = u(x)$ is determined by

$$-ku'' = f(x), \qquad u(0) = 0, \qquad u'(1) = 0,$$

where $k > 0$ is the constant thermal conductivity. The Green's function $g(x, s)$ is given by (2-11), where

$$\begin{aligned} -kv_0'' &= 0, & v_0(0) &= 0, \\ -kv_1'' &= 0, & v_1'(1) &= 0, \\ kW &= -1. \end{aligned}$$

Plainly, $v_0 = ax$ and $v_1 = b$ satisfy the first two conditions for any choice of constants a and b. The Wronskian requirement is $k(-ab) = -1$ or $ab = 1/k$.

Thus, we can choose $a = 1/k$ and $b = 1$ to get

$$g(x, s) = \begin{cases} \dfrac{s}{k}, & 0 \le s \le x \le 1, \\ \dfrac{x}{k}, & 0 \le x \le s \le 1 \end{cases}$$

from (2-11).

For future reference notice an important additional feature of the Green's function (2-11); it is *symmetric*,

$$g(x, s) = g(s, x)$$

for $0 \le x, s \le 1$. In the context of heat conduction this means that the temperature at x due to a unit heat source at s equals the temperature at s due to a unit heat source at x.

Theorem 2-3. The Green's function for the Sturm–Liouville problem (2-1) is symmetric, provided that it exists. The Green's function does exist if $p(x) > 0$, $q(x) \ge 0$, $h_0, h_1 \ge 0$, as in the case of the steady-state heat conduction model.

The asserted symmetry follows from (2-11) as we just observed. We need only establish existence, and for this we use Th. 2-1. Suppose that $z(x)$ is a solution to the homogeneous problem corresponding to (2-1):

$$-(pz')' + qz = 0,$$
$$z(0) - h_0 z'(0) = 0, \qquad z(1) + h_1 z'(1) = 0.$$

Multiply the differential equation by z, integrate by parts, and use the boundary conditions to find

$$h_1 p(1) z'(1)^2 + h_0 p(0) z'(0)^2 + \int_0^1 p(x) z'(x)^2 \, dx + \int_0^1 q(x) z(x)^2 \, dx = 0.$$

Since all four terms are nonnegative, we infer that each term is zero. In particular, the third term gives $z'(x) \equiv 0$ or $z(x) \equiv a$, a constant. Then the boundary condition at $x = 0$ gives $z(0) = 0$; hence, $z(x) \equiv 0$ and the Green's function exists.

Now, consider the general Sturm–Liouville problem (1-11), which can be expressed as

$$\text{(2-13)} \qquad \begin{cases} Ly - \lambda r(x) y = f(x), & 0 < x < 1, \\ y(0) - h_0 y'(0) = 0, & y(1) - h_1 y'(1) = 0. \end{cases}$$

If $\lambda = 0$, this problem reduces to (2-1), while if $f(x) \equiv 0$, the eigenvalue problem of Sturm–Liouville type (1-4) results. Assume that (2-1) has a Green's function, $g(x, s)$. Then any solution y of (2-13) satisfies

$$Ly = f(x) + \lambda r(x) y$$

and hence

$$y(x) = \int_0^1 g(x, s)[f(s) + \lambda r(s)y(s)]\, ds$$

or

$$y(x) = \lambda \int_0^1 g(x, s)r(s)y(s)\, ds + f_1(x), \tag{2-14}$$

where

$$f_1(x) = \int_0^1 g(x, s)f(s)\, ds.$$

Conversely, it is easy to verify that any continuous solution of (2-14) has in fact a continuous second derivative and solves the Sturm–Liouville problem (2-13). Thus, (2-13) and (2-14) are fully equivalent.

Equation (2-14) is called a *Fredholm integral equation of the second kind.* The function $g(x, s)r(s)$ multiplying the unknown function y is called its *kernel.* If $f(x) \equiv 0$ in (2-13) so that we have the Sturm–Liouville eigenvalue problem, then the equivalent integral equation is

$$y(x) = \lambda \int_0^1 g(x, s)r(s)y(s)\, ds \tag{2-15}$$

because $f_1(x) \equiv 0$. Thus, Basic Problems 1 and 2 of Sec. 7-1 can be rephrased in terms of the existence of eigenvalues and of eigenfunction expansions for the eigenvalue problem (2-15).

For later reference, we observe that (2-14) and (2-15) can be replaced by equivalent integral equations with symmetric kernels when $r(x) > 0$, as in the heat conduction problem of Sec. 7-1. Indeed, then (2-14) can be written

$$\sqrt{r(x)}y(x) = \lambda \int_0^1 g(x, s)\sqrt{r(x)r(s)}\,\sqrt{r(s)}y(s)\, ds + \sqrt{r(x)}f_1(x)$$

or

$$z(x) = \lambda \int_0^1 k(x, s)z(s)\, ds + f_2(x), \tag{2-16}$$

where

$$z(x) = \sqrt{r(x)}\,y(x), \qquad k(x, s) = g(x, s)\sqrt{r(x)r(s)}, \tag{2-17}$$

and $f_2(x) = r(x)f_1(x)$. Since $g(x, s)$ is symmetric, so is $k(x, s)$. Notice that (2-15) transforms into (2-16) with $f_2(x) \equiv 0$.

PROBLEMS

1. We assumed that $h_0 = h_1 = 0$ in the derivation of (2-11) and (2-12). Verify that these results still hold for nonzero choices of h_0 and h_1.

2. Show that the Green's function is unique, if it exists. *Hint*. If (2-2) and

$$y(x) = \int_0^1 h(x, s)f(s)\,ds$$

both uniquely solve (2-1) for any continuous f, then

$$\int_0^1 [g(x, s) - h(x, s)]f(s)\,ds = 0.$$

3. A flexible, homogeneous string pinned at $x = 0$ and $x = 1$ is stretched along the x-axis from $0 \le x \le 1$. A transverse load $f(x)$ causes the string to deflect slightly into the shape $y(x)$. Show that y satisfies

$$-Ty'' = f(x), \qquad y(0) = 0, \quad \text{and} \quad y(1) = 0,$$

where T is the constant horizontal component of tension in the string. Find the Green's function for this problem.

4. Interpret, for constant k and l,

$$-ky'' + ly = 0, \qquad y(0) - y'(0) = 0, \qquad y(1) = 0$$

in terms of steady-state heat conduction. Find the Green's function.

5. Suppose that $y'' - y = f(x)$. Find the Green's function for the boundary conditions $y(x) \to 0$ as $|x| \to \infty$. Give a physical interpretation of this problem.

6. Show that the steady-state temperature for a Fourier heat ring (see Prob. 6 of Sec. 7-1) is determined by the periodic Sturm–Liouville problem

$$-(k(x)y')' + l(x)y = f(x), \qquad 0 < x < 1,$$
$$y(0) = y(1), \qquad y'(0) = y'(1),$$

where $k(x) > 0$ and $l(x) \ge 0$. Discuss the existence and construction of the Green's function for this problem.

7. Find Green's function for the initial value problem

$$Ly = f(x), \qquad y(0) = 0, \qquad y'(0) = 0.$$

Show that the analogue of (2-2) has the form

$$y(x) = \int_0^x g(x, s)f(s)\,dx.$$

8. Check by direct substitution into (2-1) that (2-2) solves (2-1) with $g(x, s)$ given by (2-11) and (2-12).

9. Verify the Lagrange identity (2-4).

10. Verify that a continuous solution of (2-14) solves (2-13).

11. Confirm (2-10). *Hint*. Calculate $W'(x)$, use the differential equations satisfied by u and, v to eliminate second derivatives, and so obtain $W' = -p'W/p$.

12. To what extent do the results of this section solve Basic Problem 3 of Sec. 7-1?

7-3. Neumann Series

A *linear Fredholm integral equation of the second kind* is an equation of the form

$$y(x) = f(x) + \lambda \int_a^b k(x, s)y(s)\,ds. \tag{3-1}$$

Here $k(x, s)$ is the *kernel* of the integral equation, $f(x)$ is a given function, and λ is a real or complex parameter. The study of the Fredholm equation (3-1) proceeds along virtually the same lines regardless of the dimension of the underlying variables x and s. We always assume that x and s are real variables. Only minor changes in notation and reasoning are needed to handle integral equations when x and s are vector variables. These matters are left for the problem sets.

For the present, we assume that $-\infty < a < b < \infty$, $k(x, s)$ is continuous on the square $a \le x, s \le b$, and $f(x)$ is continuous on the interval $a \le x \le b$. These conditions hold, for example, when $k(x, s) = g(x, s)$, the Green's function for a Sturm–Liouville problem.

It is evident from (3-1) that any solution y must depend on both x and λ. We seek a power series solution in the form

$$y(x) = \sum_{n=0}^{\infty} \lambda^n y_n(x), \tag{3-2}$$

where the coefficients $y_n(x)$ are to be determined. Substitute (3-2) into (3-1) and integrate term by term to get

$$\sum_0^{\infty} \lambda^n y_n(x) = f(x) + \sum_0^{\infty} \lambda^{n+1} \int_a^b k(x, s) y_n(s)\, ds.$$

This equation will hold provided that the coefficients of like powers of λ on each side agree, which requires that

$$\begin{cases} y_0(x) = f(x), \\ y_{n+1}(x) = \displaystyle\int_a^b k(x, s) y_n(s)\, ds \end{cases} \tag{3-3}$$

for $n = 0, 1, 2, \ldots$. This recurrence formula uniquely determines the coefficients in (3-2), and (3-2) solves (3-1) provided that the term-by-term integration above can be justified.

We justify it by showing that the series (3-2) is uniformly convergent on $[a, b]$ provided that the parameter λ is not too large. To see this, let

$$M = \max |k(x, s)| \quad \text{and} \quad N = \max |f(x)|,$$

where the maximum is over $0 \le x, s \le b$. Then (3-3) yields the estimates

$$|y_0(x)| = |f(x)| \le N,$$

$$|y_1(x)| = \left| \int_a^b k(x, s) y_0(s)\, ds \right| \le MN(b - a),$$

$$|y_2(x)| = \left| \int_a^b k(x, s) y_1(s)\, ds \right| \le M[MN(b - a)](b - a) = NM^2(b - a)^2,$$

and, in general,

$$|y_n(x)| = \left| \int_a^b k(x, s) y_{n-1}(s)\, ds \right| \le NM^n(b - a)^n$$

for $n \geq 0$. Consequently,

$$|\lambda^n y_n(x)| \leq N[|\lambda| M(b-a)]^n \equiv M_n$$

for $n \geq 0$, and the Weierstrass M-test guarantees the absolute and uniform convergence of the series (3-2) provided that $\sum M_n < \infty$, which requires $|\lambda| M(b-a) < 1$.

In summary, the series (3-2) solves (3-1) provided that $|\lambda| < 1/M(b-a)$. Notice that the solution $y(x)$ given by (3-2) is continuous because the series is uniformly convergent and each function $\lambda^n y_n(x)$ is continuous. The series in (3-2) is called the *Neumann series.*

The Neumann series can be put into an alternative form as follows. We have

$$y_0(x) = f(x),$$

$$y_1(x) = \int_a^b k(x, s) f(s)\, ds,$$

$$y_2(x) = \int_a^b k(x, s) \left[\int_a^b k(s, t) f(t)\, dt \right] ds$$

or

$$y_2(x) = \int_a^b k_2(x, s) f(s)\, ds,$$

where $k_2(x, s) = \int_a^b k(x, t) k(t, s)\, dt$.

Quite generally, define recursively

$$\text{(3-4)} \qquad \begin{cases} k_1(x, s) = k(x, s), \\ k_n(x, s) = \displaystyle\int_a^b k(x, t) k_{n-1}(t, s)\, dt, \qquad n \geq 2, \end{cases}$$

where $k_n(x, s)$ is called the nth *iterated kernel* of $k(x, s)$. Then the reasoning above leads to

$$\text{(3-5)} \qquad y_n(x) = \int_a^b k_n(x, s) f(s)\, ds \quad \text{for} \quad n \geq 1,$$

which expresses the coefficient $y_n(x)$ directly in terms of the given data $k(x, s)$ and $f(x)$. Substitute (3-5) into (3-2) to find

$$y(x) = \sum_0^\infty \lambda^n y_n(x) = f(x) + \sum_1^\infty \lambda^n \int_a^b k_n(x, s) f(s)\, ds$$

$$= f(x) + \lambda \int_a^b \gamma(x, s; \lambda) f(s)\, ds,$$

where

$$\text{(3-6)} \qquad \gamma(x, s; \lambda) = \sum_{n=1}^\infty \lambda^{n-1} k_n(x, s).$$

You are asked to show in the problems that this series is absolutely and uniformly convergent on $a \leq x, s \leq b$ for $|\lambda| < 1/M(b-a)$. Consequently, the interchange of summation and integration performed above is valid. We have established all but the uniqueness assertion in Th. 3-1.

Theorem 3-1. Let $k(x, s)$ and $f(x)$ be continuous on $a \leq x, s \leq b$. Let $M = \max |k(x, s)|$ for $a \leq x, s \leq b$. If $|\lambda| < 1/M(b - a)$, the integral equation

$$y(x) = f(x) + \lambda \int_a^b k(x, s)y(s)\, ds$$

has the unique continuous solution

$$y(x) = f(x) + \lambda \int_a^b \gamma(x, s; \lambda)f(s)\, ds.$$

The function $\gamma(x, s; \lambda)$ defined in (3-6) is called the *resolvent kernel* because of the similarity in form of (3-1) and the solution formula above.

To prove the uniqueness of the solution to (3-1) for $|\lambda| < 1/M(b - a)$, assume that $u(x)$ and $v(x)$ are both continuous and solve (3-1). Then their difference $w(x) = u(x) - v(x)$ satisfies

$$w(x) = \lambda \int_a^b k(x, s)w(s)\, ds.$$

Consequently,

$$|w(x)| \leq |\lambda| M(b - a)\|w\|$$

where $\|w\| = \max |w(x)|$ for $a \leq x \leq b$. The previous inequality implies that

$$\|w\| \leq |\lambda| M(b - a)\|w\|, \qquad [1 - |\lambda| M(b - a)]\|w\| \leq 0, \qquad \|w\| \leq 0,$$

because the factor in brackets is positive. By definition, $\|w\| \geq 0$; so we infer that $\|w\| = 0$, which means that $w(x) = 0$ or $u(x) = v(x)$ for $a \leq x \leq b$. The solution is unique as claimed.

Let us take a fresh look at the Fredholm equation (3-1),

$$y(x) = f(x) + \lambda \int_a^b k(x, s)y(s)\, ds.$$

As above, assume that $k(x, s)$ and $f(x)$ are continuous for $a \leq x, s \leq b$. Since we seek a solution $y = y(x)$ to (3-1), we should concentrate more on this solution function and less on the variables x and s, which are difficult to ignore in (3-1) and all the analysis above. What we need is a better notation which focuses attention on the unknown function y and the given data—the kernel k and inhomogeneous term f. Fortunately, nothing really new is needed. We simply adapt the familiar functional notation of elementary calculus.

Remember that a function is just a rule for computing outputs from given inputs. Consider the integral term in (3-1), $\int_a^b k(x, s)y(s)\, ds$, in this context. This expression defines a function whose inputs are themselves functions. That is, given a continuous function y as input, the integral $\int_a^b k(x, s)y(s)\, ds$ defines a new output function. This output function depends on the input y and, of course, on the fixed kernel k. To keep this dependence clearly in mind, we denote the output function by Ky. Thus,

(3-7)
$$Ky = Ky(x) = \int_a^b k(x, s)y(s)\, ds.$$

Just as for the familiar functional notation of calculus, we regard y as the input (independent variable) for the function K and we denoted the output by Ky. It is convenient to write Ky rather than $K(y)$ in this context; otherwise, the middle term in (3-7) would have the cumbersome form $K(y)(x)$. Since the inputs and outputs of the function K are themselves functions, we have a language problem: K is a function whose domain and range consists of other functions. To avoid confusion we call K an *(integral) operator*; it operates or acts on y to produce Ky.

The domain of the operator K consists of all its allowed inputs. It is natural here to choose any continuous function y as an input. Thus, we choose $C[a, b]$, the space of all continuous functions on $[a, b]$, as the domain of K. The *range* of K is the set of all output functions Ky. Since $k(x, s)$ is continuous, (3-7) produces a continuous output from a continuous input function y. This means that the range of K is contained in $C[a, b]$. We write

$$K: C[a, b] \to C[a, b],$$

in accord with customary functional notation.

The *identity operator*

$$I\colon C[a, b] \to C[a, b]$$

is defined by $Iy = y$ for each y in $C[a, b]$. With this notation, the linear Fredholm integral equation (3-1) can be expressed as

$$y(x) = f(x) + \lambda Ky(x), \qquad y = f + \lambda Ky, \qquad y - \lambda Ky = f$$

or

(3-8)
$$(I - \lambda K)y = f.$$

Equation (3-8) simply expresses the Fredholm integral equation (3-1) in operator notation.

As you will see in subsequent sections, there is a natural, intimate relationship between integral equations and matrix equations. This fact as well as the form of (3-8) suggests the following approach for solving the Fredholm equation. The question "Does the Fredholm equation (3-8) have a solution?" can be phrased "Is f in the range of the operator $I - \lambda K$?" If so, and if $I - \lambda K$ has an inverse (operator), then the solution to (3-8) should be $y = (I - \lambda K)^{-1}f$. Notice that this is exactly the formula you would get if $I - \lambda K$ were an invertible matrix. But how do you find $(I - \lambda K)^{-1}$? One tempting approach is to pursue a formal analogy. You know that

$$(1 - t)^{-1} = \sum_{n=0}^{\infty} t^n \quad \text{for} \quad |t| < 1.$$

Perhaps

$$(I - \lambda K)^{-1} = \sum_{n=0}^{\infty} (\lambda K)^n = \sum_{n=0}^{\infty} \lambda^n K^n \quad \text{for} \quad |\lambda K| < 1.$$

If we can make sense of this, (3-8) has the solution

$$y = \left(\sum_{n=0}^{\infty} \lambda^n K^n \right) f = \sum_{n=0}^{\infty} \lambda^n K^n f, \tag{3-9}$$

provided that $|\lambda K| < 1$. This really works! The series is just the Neumann series, but written in operator form.

The operator point of view is very powerful and often suggests, as in the passage from (3-8) to (3-9), how a problem can be solved. Because of its importance, we devote the next section of this chapter to the basic properties of the operator calculus needed to justify the formal manipulations above.

PROBLEMS

1. Justify the calculations preceding (3-6) by showing that the series in (3-6) is absolutely and uniformly convergent on $[a, b]$ for $|\lambda| < 1/M(b - a)$.
2. Find the iterated kernels, the resolvent kernel, and solution of

$$y(x) = f(x) + \lambda \int_a^b e^{x-s} y(s)\, ds.$$

   In particular, find the solution when $f(x) = x^2$.
3. Find the iterated kernels, the resolvent kernel, and solution of

$$y(x) = f(x) + \int_0^1 xsy(s)\, ds.$$

   In particular, find the solution when $f(x) = \sin x$.
4. The *linear Volterra integral equation of the second kind* is

$$y(x) = f(x) + \lambda \int_a^x k(x, s)\, y(s)\, ds$$

   for x in $[a, b]$. The Volterra equation can be regarded as a Fredholm equation with kernel $m(x, s) = k(x, s)$ for $a \le s \le x \le b$ and $m(x, s) = 0$ for $a \le x \le s \le b$; however, this point of view is not always useful. Assume that $k(x, s)$ and $f(x)$ are continuous on $a \le s \le x \le b$ and $a \le x \le b$, respectively. Show that the Volterra equation has a unique continuous solution for all λ. *Hint.* Begin with a power series solution as in (3-2). Following the reasoning used in the Fredholm case, define appropriate iterated kernels and a resolvent kernel. Find a solution formula analogous to the one in Th. 3-1.
5. Let D be a closed, bounded set in the plane which has area denoted by $|D|$. Let $k(x, s)$ and $f(x)$ be continuous for x, s in D. So here $x = (x_1, x_2)$ and $s = (s_1, s_2)$ are vector variables. Set $ds = ds_1\, ds_2$. Show that the Fredholm integral equation

$$y(x) = f(x) + \lambda \int_D k(x, s) y(s)\, ds$$

   has a unique continuous solution defined in D provided that $|\lambda| < 1/M|D|$. Alternatively, solve this problem for D a closed, bounded set in n-space.
6. The method of *successive approximations* (see Sec. 4-6) or *iteration* can also be used to solve (3-1): Assume that $k(x, s)$ and $f(x)$ are continuous. Let $g(x)$, a continuous function,

be a guess at the solution to (3-1) and define

$$y_0(x) = g(x), \qquad y_{n+1}(x) = f(x) + \lambda \int_a^b k(x, s)y_n(s)\, ds,$$

for $n \geq 0$. Show that $y_n(x)$ converges uniformly on $a \leq x \leq b$ to a solution $y(x)$ of (3-1) when $|\lambda| < 1/M(b - a)$. *Hint.*

$$y_n(x) = y_0(x) + \sum_{j=1}^{n} z_j(x),$$

where $z_j(x) = y_j(x) - y_{j-1}(x)$. Show that the functions $z_n(x)$ satisfy a recurrence formula very much like (3-3). Use this to estimate $|z_n(x)|$.

7. Apply the method of successive approximations to show that the *nonlinear* integral equation

$$y(x) = f(x) + \lambda \int_a^b k(x, s, y(s))\, ds$$

has a unique continuous solution for small values of $|\lambda|$. Assume that $k(x, s, u)$ and $f(x)$ are continuous on $a \leq x,\ s \leq b$ and $-\infty < u < \infty$, and that $k(x, s, u)$ satisfies the Lipschitz condition

$$|k(x, s, u) - k(x, s, v)| \leq R|u - v|$$

for R a constant.

8. Show that the differential equation $\ddot{\theta} + a^2 \sin\theta = h(t)$, where $a^2 = g/L$ describes the angular displacement from the vertical of a simple pendulum driven by a tangential force $h(t)$. If at time $t = 0$ and $t = 1$, we wish the pendulum to be located at $\theta = 0$, the motion must satisfy

$$\ddot{\theta} + a^2 \sin\theta = h(t), \qquad \theta(0) = \theta(1) = 0.$$

Express this boundary value problem as an integral equation using the Green's function for $-\ddot{\theta} = f(t)$, $\theta(0) = \theta(1) = 0$ (see Prob. 3 of Sec. 7-2). Use Prob. 7 to discuss the solvability of this nonlinear boundary value problem.

7-4. The Linear Operator Point of View

The ideas developed in this section have wide-ranging application throughout applied mathematics. However, keep in mind our primary goal. We wish to justify the formal operator manipulations used to solve the Fredholm equation (3-8) by the operator form of the Neumann series given either by (3-9) or the equation preceding it:

$$(I - \lambda K)^{-1} = \sum_0^\infty \lambda^n K^n, \qquad y = \sum_0^\infty \lambda^n K^n f,$$

valid for small $|\lambda|$. So we must discuss the meaning of an inverse operator and what it means for a series of operators to converge. We begin by discussing convergence in the space of functions on which the operator K acts. This leads naturally to a meaning for convergence of operators.

In Sec. 7-3 the integral operator K given by (3-7) transformed continuous functions into other continuous functions; we expressed this by

$$K: C[a, b] \to C[a, b].$$

There are instances when it is more useful to regard the operator K as transforming square-integrable functions into square-integrable functions; this is expressed by

$$K: L_2(a, b) \to L_2(a, b),$$

where $L_2(a, b)$ is a standard notation for the square-integrable functions on the interval (a, b). In order to incorporate both these function spaces in the discussion to follow (as well as other spaces), we will let V stand for a space of functions. Also, T will stand for an operator that transforms functions in V into functions in V; of course, we write $T: V \to V$.

We refer to our function space V as a *linear space* because we always assume that the linear combination $\alpha f + \beta g$ is a function in V whenever f and g are functions in V and α and β are any numbers.

In addition to the linear structure just given, we will need to describe convergence of functions in V. This is needed, for example, to give meaning to the infinite series listed above. For this purpose, we assign a magnitude or *norm* $\|v\|$ to each function v in V, and require that

(i) $\|v\| \geq 0$ with equality only for $v = 0$.
(ii) $\|\alpha v\| = |\alpha| \, \|v\|$ for any scalar α.
(iii) $\|v + u\| \leq \|v\| + \|u\|$ for u, v in V.

Property (iii) is called the *triangle inequality*. A linear space V in which each element is assigned a norm is called a *normed linear space*. Convergence in a normed space is defined in the same way as in three space. We say a sequence $\{v_n\}$ of vectors in V *converges* to v in V and write

$$v_n \to v \quad \text{or} \quad \lim v_n = v$$

provided that $\|v_n - v\| \to 0$ as $n \to \infty$.

Suppose that $v_n \to v$. Then

$$\|v_n - v_m\| \leq \|v_n - v\| + \|v - v_m\| \to 0$$

as $n, m \to \infty$. So $\|v_n - v_m\| \to 0$ as $m, n \to \infty$. A sequence $\{v_n\}$ with this property is called a *Cauchy sequence*. Thus, we have shown

Lemma 4-1. Every convergent sequence is a Cauchy sequence.

Cauchy sequences are often obtained from both practical algorithms and theoretical arguments. These Cauchy sequences are constructed so that "in the limit" the problem in question is solved. So we want Cauchy sequences to converge. In other words, we want the converse of Lemma 4-1 to hold. Unfortunately, this need not be the case in general. However, it is the case for many important function spaces V. We call a function space V in which every Cauchy sequence converges (to a function in that space) *complete*. *A* complete, normed linear space is called a *Banach space*. Here are some important function spaces:

***The Continuous Functions on* $[a, b]$, $C[a, b]$**

The space $V = C[a, b]$ is a linear space because linear combinations of continuous functions are themselves continuous. $C[a, b]$ is a normed linear space with the norm of v in $C[a, b]$ given by

$$\|v\| = \max |v(x)|,$$

where the maximum is over $a \leq x \leq b$. This norm is called the *maximum norm.* Another important norm for the space of continuous functions is the *least-squares* or L_2 *norm,* given by

$$\|v\| = \sqrt{\int_a^b |v(x)|^2 \, dx.}$$

You will see shortly that the space $C[a, b]$ equipped with the maximum norm is a Banach space. When equipped with the least-squares norm it is not complete.

***The Bounded Functions on a Set* I, $B(I)$**

A function v is bounded on the set I if there is a constant M such that

$$|v(x)| \leq M \quad \text{for all} \quad x \quad \text{in} \quad I.$$

Evidently, linear combinations of bounded functions are bounded, so $B(I)$ is a linear space. The smallest number M that satisfies the inequality above is called the *least upper bound* of the function $|v(x)|$. It defines a norm on $B(I)$;

$$\|v\| = \operatorname*{lub}_{x \in I} |v(x)|,$$

where the abbreviation "lub" stands for "least upper bound." The terms "supremum" and "least upper bound" are synonymous. Thus, the norm of a bounded function v is also expressed as

$$\|v\| = \sup_{x \in I} |v(x)|.$$

The space $B(I)$ is complete; this is verified in Prop. 4-1. The convergence determined by the norm on $B(I)$ is simply *uniform convergence of functions on* I. To see this, simply observe that $\|v_n - v\| \to 0$, as $n \to \infty$ means that

$$\lim_{n \to \infty} \operatorname*{lub}_{x \in I} |v_n(x) - v(x)| = 0.$$

Thus, given any $\varepsilon > 0$ there is an index N_ε such that

$$n \geq N_\varepsilon \quad \text{implies that} \quad |v_n(x) - v(x)| < \varepsilon \quad \text{for all} \quad x \quad \text{in} \quad I,$$

which is uniform convergence of the sequence of functions $v_n(x)$ to the function $v(x)$ on the set I.

Now let $I = [a, b]$. Notice that a continuous function v on I is bounded and the norm can be expressed as

$$\|v\| = \max_{x \in I} |v(x)| = \operatorname*{lub}_{x \in I} |v(x)|.$$

Consequently, *convergence in the maximum norm is uniform convergence of continuous functions on the interval* $[a, b]$.

Proposition 4-1. $B(I)$ is a Banach space.

Proof. Let $\{v_n\}$ be a Cauchy sequence in $B(I)$. We must show that this sequence converges in $B(I)$. That is, we must find a function v in $B(I)$ for which $\|v_n - v\| \to 0$ as $n \to \infty$. Now for each x in I,

$$0 \le |v_n(x) - v_m(x)| \le \operatorname*{lub}_{x \in I} |v_n(x) - v_m(x)| = \|v_n - v_m\| \to 0 \quad \text{as} \quad m, n \to \infty.$$

Consequently, the sequence of numbers $\{v_n(x)\}$ is a Cauchy sequence. It is a fundamental property of the real (or complex) number system that Cauchy sequences converge. So the numerical sequence $\{v_n(x)\}$ has a limit; call it $v(x)$.

Next, we show that $v = v(x)$ is in $B(I)$ and that $\|v_n - v\| \to 0$ as $n \to \infty$, as expected. Let $\varepsilon > 0$. Then there is an index $N = N_\varepsilon$ so that

$$m, n \ge N \quad \text{implies that} \quad \|v_n - v_m\| \le \varepsilon$$

because $\{v_n\}$ is a Cauchy sequence in $B(I)$. Since

$$|v_n(x) - v_m(x)| \le \|v_n - v_m\|,$$

we have

$$m, n \ge N \quad \text{implies that} \quad |v_n(x) - v_m(x)| \le \varepsilon \quad \text{for all} \quad x \quad \text{in} \quad I.$$

Let $m \to \infty$ and use the fact that the numerical sequence $v_m(x) \to v(x)$ to infer that

$$\text{(4-1)} \qquad n \ge N \quad \text{implies that} \quad |v_n(x) - v(x)| \le \varepsilon \quad \text{for all} \quad x \quad \text{in} \quad I.$$

Apply (4-1) with $n = N$ to get

$$|v(x)| \le |v(x) - v_N(x)| + |v_N(x)| \le \varepsilon + \|v_N\|,$$

which shows that v is a bounded function. So v is in $B(I)$. Finally, (4-1) shows that ε is an upper bound for the function $|v_n(x) - v(x)|$ for $n \ge N$. Thus, its least upper bound $\|v_n - v\|$ is less than or equal to ε. That is,

$$n \ge N_\varepsilon \quad \text{implies that} \quad \|v_n - v\| \le \varepsilon,$$

which means that $\|v_n - v\| \to 0$ as $n \to \infty$.

Corollary 4-1. Let I be a closed bounded set in n-dimensional Euclidean space. Then $C(I)$, the continuous functions on I equipped with the maximum norm, is a Banach space.

Proof. Let $\{v_n\}$ be a Cauchy sequence in $C(I)$. Then $\{v_n\}$ is also a Cauchy sequence in $B(I)$ because continuous functions are bounded and

$$\|u\| = \max_{x \in I} |u(x)| = \operatorname*{lub}_{x \in I} |u(x)|,$$

for any continuous function u on I. So $\{v_n\}$ converges in $B(I)$ to a bounded function v in $B(I)$: $\|v_n - v\| \to 0$ as $n \to \infty$. But convergence in $B(I)$ is uniform convergence of functions on I, as we saw earlier and confirmed again in (4-1). Since the uniform limit of continuous functions is continuous, the function v is in fact in $C(I)$. Since $\|v_n - v\| \to 0$, the Cauchy sequence $\{v_n\}$ converges to v in $C(I)$.

The Lebesgue Square-Integrable Functions on a Set I, $L_2(I)$

The fact that $C[a, b]$ is not complete in the least-squares norm was one of the primary motivations leading to the Lebesgue theory of integration. We do not presuppose familiarity with this theory, but simply mention that the space $L_2(I)$ consists of functions $v(x)$ for which the Lebesgue integral $\int_I |v(x)|^2\, dx < \infty$ and that $L_2(I)$ is a complete normed space with

$$\|v\| = \sqrt{\int_I |v(x)|^2\, dx}.$$

Since we have a notion of convergence of sequences in a normed linear space V, we can define convergence of series via partial sums, just as for numerical series. We say that a series $\sum_{n=0}^{\infty} v_n$ of elements in V *converges* if its sequence of partial sums

$$s_0 = v_0, \quad s_1 = v_0 + v_1, \quad s_n = v_0 + v_1 + \cdots + v_n, \ldots,$$

converges. If $s_n \to s$ in V we call s the *sum of the series* and write $\sum_0^\infty v_n = s$.

Proposition 4-2. Let V be a Banach space. If the numerical series $\sum_0^\infty \|v_n\|$ converges, then the series $\sum_0^\infty v_n$ converges in V.

Proof. Let $s_n = v_0 + v_1 + \cdots + v_n$ as above. For $n \geq m$,

$$\|s_n - s_m\| = \|v_{m+1} + \cdots + v_n\| \leq \|v_{m+1}\| + \cdots + \|v_n\|,$$

which implies that

$$0 \leq \|s_n - s_m\| \leq \sum_{m+1}^{n} \|v_k\| \to 0$$

because the numerical series $\sum_0^\infty \|v_k\|$ converges. So $\{s_n\}$ is a Cauchy sequence in V and therefore converges, say to s in V. Then by definition $\sum_0^\infty v_n$ converges and its sum is s.

So far, we have introduced a new way of thinking about functions of a given class. We think of the functions (such as the set of continuous functions on $[a, b]$) as "points" or "vectors" in a function space (such as $C[a, b]$.) Now we describe how operators [such as the integral operator K given by (3-7)] act on function spaces.

Let V be a normed linear (function) space. A rule T that assigns to each function v in V another function Tv is called an *operator* (transformation or map) on V. We call T *continuous on* V if

$$v_n \to v \quad \text{implies that} \quad Tv_n \to Tv;$$

equivalently,

$$\|v_n - v\| \to 0 \quad \text{implies that} \quad \|Tv_n - Tv\| \to 0.$$

An operator $T: V \to V$ is *bounded* if there is a constant M such that

$$\|Tv\| \leq M\|v\| \quad \text{for all} \quad v \quad \text{in} \quad V,$$

and T is *linear* if

$$T(\alpha v + \beta u) = \alpha Tv + \beta Tu$$

for all scalars α, β and all u, v in V.

Example 1. Let $K: C[a, b] \to C[a, b]$ be the integral operator with continuous kernel $k(x, s)$ given in (3-7),

$$Kv = Kv(x) = \int_a^b k(x, s)v(s)\, ds.$$

Then K is linear, bounded, and continuous.

The operator K is linear because integration is a linear process:

$$\begin{aligned} K(\alpha v + \beta u) &= \int_a^b k(x, s)[\alpha v(s) + \beta u(s)]\, ds \\ &= \alpha \int_a^b k(x, s)v(s)\, ds + \beta \int_a^b k(x, s)u(s)\, ds \\ &= \alpha Kv + \beta Ku. \end{aligned}$$

If we use the maximum norm on $C[a, b]$ and note that $|k(x, s)| \leq M$, where M is the maximum of $|k(x, s)|$ on $a \leq x, s \leq b$, then

$$\begin{aligned} |Kv(x)| &= \left|\int_a^b k(x, s)v(s)\, ds\right| \leq \int_a^b M\|v\|\, ds, \\ |Kv(x)| &\leq M(b - a)\|v\|, \\ \|Kv\| &= \max_{a \leq x \leq b} |Kv(x)| \leq M(b - a)\|v\|. \end{aligned}$$

This shows that K is a bounded operator and establishes the useful estimate

$$\|Kv\| \leq M(b - a)\|v\|, \tag{4-2}$$

where M is the maximum of $|k(x, s)|$ on $a \leq x, s \leq b$. (In the problems you are asked to show that K is also bounded when $C[a, b]$ is equipped with the least-squares norm.) Finally, the boundedness of K implies that K is continuous since

$$\|Kv_n - Kv\| = \|K(v_n - v)\| \leq M(b-a)\|v_n - v\|.$$

Consequently, $Kv_n \to Kv$ whenever $v_n \to v$.

This last argument shows that any linear, bounded operator is automatically continuous, which proves half of

Proposition 4-3. Let $T: V \to V$ be linear. Then T is continuous if and only if T is bounded.

Proof. It remains to show that a continuous linear operator T is bounded. In particular, by continuity at 0, $Tv \to T0 = 0$ as $v \to 0$ or $\|Tv\| \to 0$ as $\|v\| \to 0$. Consequently, $\|Tv\|$ must remain smaller than or equal to 1 when $\|v\|$ is sufficiently small. That is, there is a number $\delta > 0$ such that

$$\|v\| \leq \delta \quad \text{implies that} \quad \|Tv\| \leq 1.$$

Now let u be any nonzero vector in V. Then the vector $v = \delta u/\|u\|$ has norm δ and hence

$$\left\|T\left(\frac{\delta u}{\|u\|}\right)\right\| \leq 1, \qquad \frac{\delta}{\|u\|}\|Tu\| \leq 1, \qquad \|Tu\| \leq \frac{1}{\delta}\|u\|.$$

This inequality clearly holds when $u = 0$, and so holds for all vectors in V. That is, T is a bounded operator.

Because of Prop. 4-3 the expressions "T is a continuous linear operator" and "T is a bounded linear operator" are synonymous. We let $\mathscr{L}(V)$ be the set of all bounded linear operators on V. $\mathscr{L}(V)$ is itself a linear space with addition and scalar multiplication given by

$$(T + S)(v) = Tv + Sv, \qquad (\alpha T)(v) = \alpha Tv.$$

Notice that these are the same rules used to define sums and scalar multiples for matrix operators.

Example 2. Let K and L be integral operators on $C[a, b]$ with kernels $k(x, s)$ and $l(x, s)$, respectively. Then the operator

$$(K + L)(v) = Kv + Lv = \int_a^b k(x, s)v(s)\,ds + \int_a^b l(x, s)v(s)\,ds$$

$$= \int_a^b [k(x, s) + l(x, s)]v(s)\,ds.$$

Thus, $K + L$ is the integral operator with kernel $m(x, s) = k(x, s) + l(x, s)$. Similarly, αK is the integral operator with kernel $\alpha k(x, s)$.

In addition to being a linear space, $\mathscr{L}(V)$ becomes a normed space if we define the norm of an operator T, $\|T\|$, to be the smallest value of M for which

$$\|Tv\| \leq M\|v\| \quad \text{for all} \quad v \text{ in } V.$$

Since $\|T\|$ is the smallest value of M for which this inequality holds, we have

(4-3) $$\|Tv\| \leq \|T\| \, \|v\| \quad \text{for all} \quad v \text{ in } V.$$

The number $\|T\|$ is called the *operator norm* of T. It satisfies the three defining properties (i)–(iii) of a norm (see Prob. 8).

Example 3. For our integral operator K, (4-2) shows that K is bounded and

$$\|K\| \leq M(b-a),$$

where M is the maximum of $|k(x, s)|$ over $a \leq x, s \leq b$.

Let T and S be operators on V. Then $TS\colon V \to V$ is the operator defined by

$$TS(v) = T(Sv) \quad \text{for all} \quad v \quad \text{in} \quad V.$$

Example 4. Let K and L be integral operators on $C[a, b]$ with kernels $k(x, s)$ and $l(x, s)$. Then the operator

$$\begin{aligned} KL(v) = K(Lv) &= \int_a^b k(x, s) Lv(s)\, ds \\ &= \int_a^b k(x, s) \int_a^b l(s, t) v(t)\, dt\, ds = \int_a^b m(x, s) v(s)\, ds, \end{aligned}$$

where

(4-4) $$m(x, s) = \int_a^b k(x, t) l(t, s)\, dt$$

is the kernel of the integral operator KL.

If T and S are bounded linear operators the operator TS defined above also has these properties:

$$\|TS(v)\| = \|T(Sv)\| \leq \|T\| \, \|Sv\| \leq \|T\| \, \|S\| \, \|v\|,$$

which shows that TS is bounded and that

(4-5) $$\|TS\| \leq \|T\| \, \|S\|.$$

To see that TS is linear, simply compute

$$\begin{aligned} TS(\alpha u + \beta v) &= T(S(\alpha u + \beta v)) = T(\alpha Su + \beta Sv) \\ &= \alpha T(Su) + \beta T(Sv) = \alpha TS(u) + \beta TS(v). \end{aligned}$$

Consequently, all the powers

$$T^2 = TT, \quad T^3 = TT^2, \quad \ldots, \quad T^{n+1} = TT^n, \quad \ldots$$

of a bounded operator are bounded and repeated use of (4-5) gives

(4-6) $$\|T^n\| \leq \|T\|^n.$$

For $n = 0$ we set $T^0 = I$, the identity operator, and of course $T^1 = T$.

Example 5. Let K be the integral operator on $C[a, b]$ with continuous kernel $k(x, s)$. Then for $n \geq 1$ the operator K^n is also an integral operator and its kernel is the nth iterated kernel $k_n(x, s)$ defined in (3-4).

To see this, just apply Ex. 4 with $L = K$ to see that K^2 is an integral operator with kernel

$$\int_a^b k(x, t)k(t, s)\,dt = k_2(x, s).$$

Now apply Ex. 4 with $L = K^2$ to conclude that $K^3 = KK^2$ is an integral operator with kernel $k_3(x, s)$, and so on.

Before proceeding further, remember our goal. We want to justify the expansions

$$(I - \lambda K)^{-1} = \sum_0^\infty \lambda^n K^n \quad \text{and} \quad y = \sum_0^\infty \lambda^n K^n f$$

obtained formally in Sec. 7-3. At this point we can make good sense of the terms in each series: $\lambda^n K^n$ is the integral operator on $C[a, b]$ with kernel $\lambda^n k_n(x, s)$, and $\lambda^n K^n f$ is the function $\int_a^b \lambda^n k_n(x, s) f(s)\,ds$. What remains is to discuss what convergence means in this operator context.

We have observed that the set of bounded linear operators $\mathscr{L}(V)$ is itself a normed linear space. Consequently, we already have a meaning for convergence of sequences and series of operators in $\mathscr{L}(V)$: The sequence $\{T_n\}$ of operators converges to the operator T if the operator norm $\|T_n - T\| \to 0$ as $n \to \infty$, and the series $\sum_{n=0}^\infty T_n$ converges if its sequence of partial sums $S_n = T_0 + T_1 + \cdots + T_n$ converges. If $\{S_n\}$ has for its limit the operator S, then $\sum_0^\infty T_n = S$. In the case of primary interest to us at the moment, we want to show that the operator series $\sum \lambda^n K^n$ converges in $C[a, b]$ (at least for small $|\lambda|$) and that its sum is the inverse of $I - \lambda K$. A crucial step for this is

Theorem 4-1. Let V be a Banach space. Then the normed linear space $\mathscr{L}(V)$ of bounded linear operators on V is also complete.

Proof. Let $\{T_n\}$ be a Cauchy sequence in $\mathscr{L}(V)$; so $\|T_n - T_m\| \to 0$ as $n \to \infty$. We must find a bounded linear operator T so that $\|T_n - T\| \to 0$ as $n \to \infty$. For any v in V,

$$\|T_n v - T_m v\| = \|(T_n - T_m)v\| \leq \|T_n - T_m\|\,\|v\| \to 0$$

as $m, n \to \infty$. So the sequence $\{T_n v\}$ is a Cauchy sequence in the complete space V. Thus $\{T_n v\}$ converges; call its limit Tv. So

$$\lim T_n v = Tv.$$

This assigns a vector Tv to each vector v in V. So T is an operator on V. In fact, T is linear:

$$\begin{aligned} T(\alpha v + \beta u) &= \lim T_n(\alpha v + \beta u) = \lim (\alpha T_n v + \beta T_n u) \\ &= \alpha \lim T_n v + \beta \lim T_n u = \alpha T v + \beta T u. \end{aligned}$$

The operator T is also bounded and $\|T_n - T\| \to 0$: Fix $\varepsilon > 0$. Then there is an index $N = N_\varepsilon$ so that

$$m, n \geq N \quad \text{implies that} \quad \|T_n - T_m\| \leq \varepsilon,$$

because $\{T_n\}$ is a Cauchy sequence. Then for any v in V,

$$m, n \geq N \quad \text{implies that} \quad \|T_n v - T_m v\| \leq \|T_n - T_m\| \, \|v\| \leq \varepsilon\|v\|$$

Let $m \to \infty$ to get

$$\text{(4-7)} \qquad n \geq N \quad \text{implies that} \quad \|T_n v - Tv\| \leq \varepsilon\|v\|.$$

For $n = N$ this gives

$$\|Tv\| \leq \|Tv - T_N v\| + \|T_N v\| \leq \varepsilon\|v\| + \|T_N\| \, \|v\|,$$
$$\|Tv\| \leq (\varepsilon + \|T_N\|)\|v\|,$$

which shows that T is bounded. Also, (4-7) shows that

$$n \geq N_\varepsilon \quad \text{implies that} \quad \|T_n - T\| \leq \varepsilon$$

by definition of the operator norm. The last assertion is exactly the statement that $\|T_n - T\| \to 0$ as $n \to \infty$. Thus, the Cauchy sequence $\{T_n\}$ converges to the operator T in the space $\mathscr{L}(V)$.

The next result states in effect that convergence of operators behaves as you would expect.

Lemma 4-2. Let T_n, T, S_n, S be bounded linear operators on a normed linear space V. Then

(i) $T_n \to T$ and $S_n \to S$ in $\mathscr{L}(V)$ implies that

$$\alpha T_n + \beta S_n \to \alpha T + \beta S, \qquad \alpha T_n \to \alpha T$$

for any scalars α and β.

(ii) $T_n \to T$ in $\mathscr{L}(V)$ implies that $T_n S \to TS$ and $ST_n \to ST$ in $\mathscr{L}(V)$.

(iii) $T_n \to T$ and $S_n \to S$ in $\mathscr{L}(V)$ implies that $T_n S_n \to TS$ and $S_n T_n \to ST$ in $\mathscr{L}(V)$.

The proof of this lemma is virtually the same as the case when the operator sequences are replaced by numerical sequences. For example, if $T_n \to T$, then

$$\|T_n S - TS\| = \|(T_n - T)S\| \leq \|T_n - T\| \, \|S\| \to 0$$

as $n \to \infty$, so $T_n S \to TS$, which proves half of (ii). The remaining checks are left for Prob. 4.

An operator T in $\mathscr{L}(V)$ is called *invertible* in $\mathscr{L}(V)$ if there is an operator S in $\mathscr{L}(V)$ such that $ST = I$ and $TS = I$, in which case S is called the *inverse* of T and we write $S = T^{-1}$. The next theorem contains the essence of our earlier Neumann series expansions.

Theorem 4-2. Let V be a Banach space, and T in $\mathscr{L}(V)$ have norm $\|T\| < 1$. Then the operator $I - T$ is invertible in $\mathscr{L}(V)$ and

$$(I - T)^{-1} = \sum_{n=0}^{\infty} T^n. \tag{4-8}$$

Proof. First $\mathscr{L}(V)$ is itself complete by Th. 4-1. Since $\|T\| < 1$ and $\|T^n\| \leq \|T\|^n$, we see that the numerical series

$$\sum \|T^n\| \leq \sum \|T\|^n = \frac{1}{1 - \|T\|} < \infty.$$

Then Prop. 4-2 [applied to the complete space $\mathscr{L}(V)$] implies that the series of operators $\sum T^n$ converges. In fact,

$$\sum_{n=0}^{\infty} T^n = S \quad \text{where} \quad S = \lim S_n \quad \text{and} \quad S_n = I + T + \cdots + T^n.$$

Now

$$(I + T + \cdots + T^n)(I - T) = I - T^{n+1} = (I - T)(I + T + \cdots + T^n)$$

or

$$S_n(I - T) = I - T^{n+1} = (I - T)S_n.$$

Since $\|T\| < 1$, $\|T^{n+1}\| \leq \|T\|^{n+1} \to 0$ as $n \to \infty$, which means that $T^{n+1} \to 0$, the zero operator, as $n \to \infty$. Also, $S_n \to S = \sum_0^\infty T^n$, so a limit passage in the equation above gives

$$S(I - T) = I = (I - T)S.$$

This shows that $I - T$ is invertible with

$$(I - T)^{-1} = S = \sum_0^{\infty} T^n.$$

The reasoning just given actually proves slightly more:

Theorem 4-2′. Let V be a Banach space and T in $\mathscr{L}(V)$ satisfy $\sum \|T^n\| < \infty$. Then $I - T$ is invertible in $\mathscr{L}(V)$ and

$$(I - T)^{-1} = \sum_{n=0}^{\infty} T^n.$$

Now we can set the discussion leading from (3-8) to (3-9) on a firm foundation. Recall that $k(x, s)$ is continuous on $a \leq x, s \leq b$ and that K: $C[a, b] \to C[a, b]$ is the integral operator

$$Ky = Ky(x) = \int_a^b k(x, s)y(s)\,ds.$$

We want to solve the Fredholm integral equation $(I - \lambda K)y = f$, where f is a continuous function on $[a, b]$. If we equip $C[a, b]$ with the maximum norm

$$\|y\| = \max_{a \le x \le b} |y(x)|,$$

then $C[a, b]$ is a complete normed space. Theorem 4-2 says that $I - \lambda K$ will have the inverse

$$(I - \lambda K)^{-1} = \sum_0^\infty \lambda^n K^n$$

provided that $\|\lambda K\| = |\lambda| \, \|K\| < 1$. Consequently, if $|\lambda| < 1/\|K\|$, the Fredholm integral equation $(I - \lambda K) = f$ has the unique solution

$$y = (I - \lambda K)^{-1} f \quad \text{or} \quad y = \left(\sum_{n=0}^\infty \lambda^n K^n \right) f \quad \text{or} \quad y = \sum_{n=0}^\infty \lambda^n K^n f.$$

Remember that convergence in $C[a, b]$ with the maximum norm is uniform convergence. So the series $\sum \lambda^n K^n f$ is uniformly convergent to y on $[a, b]$. Notice, too, that K^n is the integral operator with kernel $k_n(x, s)$ by Ex. 5. So, in the notation of Sec. 7-3, we can write

$$y(x) = \sum_0^\infty \lambda^n K^n f(x) = \sum_0^\infty \lambda^n y_n(x)$$

because

$$\int_a^b k_n(x, s) f(s) \, ds = y_n(x).$$

Thus, our operator form for the solution y is just the Neumann series expressed in a new notation.

We have just found that $(I - \lambda K)y = f$ has a unique solution for $|\lambda| < 1/\|K\|$, and this conclusion is actually a little better than our result in Sec. 7-3. There we showed that the Fredholm equation could be solved for $|\lambda| < 1/M(b - a)$, where $M = \max |k(x, s)|$ for $a \le x, s \le b$. But by Ex. 3, $\|K\| \le M(b - a)$, and typically, strict inequality holds here; so our new requirement on $|\lambda|$ is less restrictive. You are asked to show in Prob. 5 that

(4-9)
$$\|K\| = \max_{a \le x \le b} \int_a^b |k(x, s)| \, ds.$$

Next, for $|\lambda| < 1/\|K\|$ our solution formula for y can be expressed as

$$y = \sum_0^\infty \lambda^n K^n f = f + \sum_1^\infty \lambda^n K^n f = f + \lambda \left(\sum_1^\infty \lambda^{n-1} K^n \right) f.$$

The operator

(4-10)
$$\Gamma_\lambda = \sum_1^\infty \lambda^{n-1} K^n$$

is called the *resolvent operator*. It is an integral operator whose kernel is the resolvent kernel (3-6). With this notation, the Fredholm integral equation $y = f +$

λKy has unique solution $y = f + \lambda\Gamma_\lambda f$ for $|\lambda| < 1/\|K\|$. Compare this with the statement of Th. 3-1.

PROBLEMS

1. Check that properties (i)–(iii) hold for all the norms introduced in the spaces $C[a, b]$, $B(I)$, and $L_2(a, b)$. *Hint.* You will need the Schwarz inequality to confirm the triangle inequality for the least-squares norm.
2. Show that the integral operator in Ex. 1 is bounded when the least-squares norm is used on $C[a, b]$. *Hint.* Schwarz inequality.
3. For T in $\mathscr{L}(V)$ show that
$$\|T\| = \text{lub } \|Tv\|,$$
where the least upper bound is computed over either all v with $\|v\| \leq 1$ or over all v with $\|v\| = 1$. These are useful alternatives to the definition (4-3).
4. Complete the proof of Lemma 4-2.
5. Verify (4-9).
6. Assume $|\lambda| < 1/\|K\|$. Check that the resolvent operator Γ_λ is an integral operator with kernel $\gamma(x, s; \lambda)$, the resolvent kernel.
7. Verify $\Gamma_\lambda = K + K\Gamma_\lambda$, $\Gamma_\lambda = K + \Gamma_\lambda K$. Recall Lemma 4-2, which you may wish to express in series form.
8. Show that the operator norm, $\|T\|$, satisfies properties (i)–(iii) which define a norm.
9. Let $k(x, s)$ be continuous on the triangle $a \leq s \leq x \leq b$. Then
$$Ky = Ky(x) = \int_a^x k(x, s)y(s)\, ds$$
is called a *Volterra integral operator*. Regard K as an integral operator on $C[a, b]$. (Compare with Prob. 4 of Sec. 7-3.)
   (a) Show that K is a bounded linear operator on $C[a, b]$.
   (b) Show that K^n is a Volterra integral operator with kernel $k_n(x, s)$ defined by $k_1(x, s) = k(x, s)$ and
$$k_n(x, s) = \int_s^x k(x, t)k_{n-1}(t, s)\, dt.$$
   (c) Let M be the maximum of $|k(x, s)|$ over $a \leq s \leq x \leq b$. Use part (b) to show that
$$\|K^n y\| \leq \frac{M^n(b-a)^n}{n!}\|y\|,$$
where $\|y\|$ is the maximum norm on $C[a, b]$. Conclude that
$$\|K^n\| \leq \frac{M^n(b-a)^n}{n!}.$$
   (d) Deduce from Th. 4-2′ that the Volterra integral equation $(I - \lambda K)y = f$ has a unique continuous solution y for each f in $C[a, b]$ and any λ.
   (e) If the Volterra kernel $k(x, s)$ is continuous on $a \leq s \leq x < \infty$ and $f(x)$ is continuous on $a \leq x < \infty$, show that the Volterra equation $(I - \lambda K)y = f$ has a unique continuous solution $y = y(x)$ defined on $a \leq x < \infty$. *Hint.* Apply part (d) on $a \leq x \leq b$ and let $b \to \infty$.

7-5 Degenerate Kernels

The Fredholm integral equation

(5-1) $$y(x) = f(x) + \lambda \int_a^b k(x, s)y(s)\, ds$$

usually cannot be solved in closed form. Numerical methods are needed. This section is devoted to a brief introduction to one such method. Our aim is two-fold. First, we describe a practical method that can be used to generate numerical solutions. Second, we begin to establish a close link between integral equations and matrix equations. This link sheds considerable light on the general behavior of integral equations. We use it in Sec. 7-7 to establish the *Fredholm alternative*: Either (5-1) has a solution for each function $f(x)$ or the corresponding homogeneous equation

(5-2) $$y(x) = \lambda \int_a^b k(x, s)y(s)\, ds$$

has a nontrivial solution.

At present, we are interested in the approximation of solutions to (5-1) for *fixed* λ. Therefore, we assume that (5-1) has a unique solution, and as usual we suppose that $k(x, s)$ and $f(x)$ are continuous on $a \le x,\ s \le b$. In view of the Fredholm alternative, the λ we have fixed can be any value except one for which (5-2) has a nontrivial solution. These exceptional values, called eigenvalues, form either a finite set or an infinite sequence, as we shall prove in Sec. 7-8. Since λ is fixed, we can conveniently incorporate it into the kernel $k(x, s)$. That is, let $\tilde{k}(x, s) = \lambda k(x, s)$ and express (5-1) as

$$y(x) = f(x) + \int_a^b \tilde{k}(x, s)y(s)\, ds,$$

where we assume that 1 is not an eigenvalue of this problem. The net result is clearly this: For the present discussion we may and do assume that $\lambda = 1$ in (5-1) and that $\lambda = 1$ is not an eigenvalue for this problem.

With these agreements (5-1) is

(5-3) $$y = f + Ky$$

where K: $C[a, b] \to C[a, b]$ is the integral operator

$$Ky(x) = \int_a^b k(x, s)y(s)\, ds.$$

By the Fredholm alternative, (5-3) has a unique solution y for each choice of f and we want to approximate y. Many standard procedures for this first approximate the kernel $k(x, s)$, so the integral operator K, by a simpler kernel $k_n(x, s)$, which defines an approximate operator K_n. The approximation is made so that K_n converges to K in some appropriate sense and so that the approximate problem

(5-4) $$y_n = f + K_n y_n$$

is equivalent to a matrix equation. The matrix problem is solved numerically, which yields y_n, and we use y_n as our approximation to y: $y_n \approx y$. It is part of the theory of such approximation methods that $\lambda = 1$ is not an eigenvalue of (5-4) for "large" n, so this problem has a unique solution y_n, as our notation suggests. In other words, the matrix problem equivalent to (5-4) is nonsingular for large n.

In the discussion to follow, it is convenient both for practical applications and for developments in subsequent sections to allow our functions to assume complex values. Thus, bars over functions indicate complex conjugate values. In applications where only real-valued functions occur, the bars can be ignored.

A kernel $l(x, s)$ of the form

$$l(x, s) = \sum_{i=1}^{n} a_i(x)\overline{b_i(s)} \tag{5-5}$$

is called *degenerate*. (The reason for this terminology will become clear in Sec. 7-9.) We assume that $\{a_i(x)\}$ and $\{b_i(s)\}$ are linearly independent sets. The corresponding integral equation is

$$(I - L)y = f, \tag{5-6}$$

where

$$Ly(x) = \int_a^b l(x, s)y(s)\, ds.$$

We shall show that the integral equation (5-6) is equivalent to a matrix equation.

Suppose that $y = y(x)$ solves (5-6). Then

$$y(x) = f(x) + \int_a^b \sum_{j=1}^{n} a_j(x)\overline{b_j(s)}y(s)\, ds,$$

$$y(x) = f(x) + \sum_{j=1}^{n} \left[\int_a^b y(s)\overline{b_j(s)}\, ds\right] a_j(x).$$

Since the integrals on the right are specific numbers, an important fact emerges: If $y(x)$ solves (5-6), then $y(x)$ is a linear combination of $f(x)$ and the functions $a_1(x), \ldots, a_n(x)$. Once the coefficients of $a_1(x), \ldots, a_n(x)$ are known, the solution $y(x)$ is known. To express these coefficients conveniently, we use the inner product notation

$$\langle u, v\rangle = \int_a^b u(x)\overline{v(x)}\, dx.$$

Now, the formula for $y(x)$ can be expressed as

$$y(x) = f(x) + \sum_{j=1}^{n} \langle y, b_j\rangle a_j(x), \tag{5-7}$$

and we need to find the coefficients $\langle y, b_j\rangle$. To do this, multiply (5-7) by $\overline{b_i(x)}$ and integrate from a to b to obtain

$$\langle y, b_i\rangle = \langle f, b_i\rangle + \sum_{j=1}^{n} \langle a_j, b_i\rangle\langle y, b_j\rangle$$

for $i = 1, 2, \ldots, n$. So the vector $Y = [\langle y, b_1\rangle, \ldots, \langle y, b_n\rangle]^T$, where T means "transpose," satisfies the matrix equation

(5-8) $$Y = F + CY,$$

with

$$F = [\langle f, b_1\rangle, \ldots, \langle f, b_n\rangle]^T \equiv [F_1, \ldots, F_n]^T$$

and

(5-9) $$C = [c_{ij}]_{n\times n} = [\langle a_j, b_i\rangle]_{n\times n}.$$

If we set $Y_j = \langle y, b_j\rangle$, so that $Y = [Y_1, \ldots, Y_n]^T$, then (5-7)–(5-9) prove half of the equivalence:

(5-10) $$(I - L)y = f \leftrightarrow y = f + \sum_{j=1}^{n} Y_j a_j(x) \quad \text{and} \quad (I - C)Y = F.$$

For the converse, suppose that $(I - C)Y = F$ and y is given by the expression on the right side of (5-10). Then

$$\begin{aligned} y - Ly &= f(x) + \sum_{j=1}^{n} Y_j a_j(x) - \int_a^b \sum_{i=1}^{n} a_i(x)\overline{b_i(s)}\left[f(s) + \sum_{j=1}^{n} Y_j a_j(s)\right] ds \\ &= f(x) + \sum_{j=1}^{n} Y_j a_j(x) - \sum_{i=1}^{n} a_i(x)\left(F_i + \sum_{j=1}^{n} c_{ij} Y_j\right) \\ &= f(x) + \sum_{i=1}^{n} [Y_i - F_i - (CY)_i] a_i(x) = f(x) \end{aligned}$$

because $Y = F + CY$.

The relation (5-10) establishes the equivalence of an integral equation with a degenerate kernel and an associated matrix equation. Next, we present two common ways to approximate a given continuous kernel $k(x, s)$ by a degenerate kernel $k_n(x, s)$. In each case, we have $\|K_n - K\| \to 0$, where K_n and K are the corresponding integral operators.

Example 1. If $k(x, s)$ is continuous on $a \le x, s \le b$, then for each $n = 1, 2, 3, \ldots$ we can apply the Weierstrass approximation theorem to secure a polynomial

$$k_n(x, s) = \sum_{i,j=1}^{N_n} \alpha_{ij} x^i s^i$$

such that $|k(x, s) - k_n(x, s)| < 1/n$ for all $a \le x,\ s \le b$. The kernel $k_n(x, s)$ is degenerate; indeed,

$$k_n(x, s) = \sum_i x^i \left(\sum_j \alpha_{ij} s^j\right) = \sum_i a_i(x)\overline{b_i(s)},$$

with evident choices for $a_i(x)$ and $b_i(s)$. Furthermore, we have the operator norm estimate

$$\|K_n - K\| \le (b - a) \max_{a \le x, s \le b} |k_n(x, s) - k(x, s)| \le \frac{b - a}{n} \to 0$$

as $n \to \infty$, as you are asked to confirm in the problems.

Example 1 shows that an integral operator with a continuous kernel can be approximated arbitrarily closely by an integral operator with a degenerate kernel.

Example 2. A number of important kernels have the special form $k(x, s) = h(xs)$, where h has a power series expansion, say,

$$h(z) = \sum_0^\infty \alpha_i z^i.$$

A natural choice for a degenerate kernel approximation is

$$k_n(x, s) = \sum_0^n \alpha_i x^i s^i.$$

Then, for $a \le x, s \le b$,

$$|k(x, s) - k_n(x, s)| = \left|\sum_{n+1}^\infty \alpha_i x^i s^i\right| \le \sum_{n+1}^\infty |\alpha_i| r^{2i},$$

where $r = \max(|a|, |b|)$. If the radius of convergence of $\sum_0^\infty \alpha_i z^i$ is R and $r^2 < R$, then $\sum_{n+1}^\infty |\alpha_i| r^{2i} \to 0$ and, as in Ex. 1, $\|K_n - K\| \to 0$ as $n \to \infty$. Another important case, $k(x, s) = h(x - s)$, is considered in Prob. 2.

PROBLEMS

1. Confirm the operator norm estimate, $\|K_n - K\| \le (b - a)/n$, in Ex. 1.
2. Reasoning as in Ex. 2, discuss degenerate kernel approximation when $k(x, s) = h(x - s)$ and $h(z)$ has a power series expansion.
3. Discuss how Fourier series can be used in place of power series to provide degenerate kernel approximations when $k(x, s) = g(x - s)$ and $g(z)$ has a Fourier series expansion. Find reasonable conditions on g that guarantee operator norm convergence.
4. In our discussion of (5-10), we assume that all functions involved were continuous. Assume instead that

$$\int_a^b |a_i(x)|^2\, dx, \qquad \int_a^b |b_i(x)|^2\, dx, \qquad \int_a^b |f(x)|^2\, dx < \infty,$$

and check that our reasoning still leads to (5-10). In this setting we are interested in a solution y that is square integrable,

$$\int_a^b |y(x)|^2\, dx < \infty.$$

Be sure to explain where the square-integrability assumptions are used.

5. Solve explicitly.

(a) $y(x) = x + \dfrac{1}{2}\displaystyle\int_{-1}^{1} (x + s)y(s)\, ds.$

(b) $y(x) = x + \displaystyle\int_0^1 (1 + xs)y(s)\, ds.$

(c) $y(x) = 1 + \displaystyle\int_0^{2\pi} \sin(x - s)y(s)\, ds.$

7-6. Inner Product Spaces

In the next several sections we study the eigenvalue problem for a Fredholm integral equation. Much of this analysis rests on some key facts about orthogonality of vectors or functions. The notion of an inner product is central for all this work, and is a direct extension of the dot product for vectors in 2- and 3-space.

Just as in Sec. 7-4, we let V stand for a linear space (or vector space) whose elements we refer to as vectors. The vectors in V obey the same algebraic rules of addition, subtraction, and scalar multiplication as those for vectors in 2- and 3-space.

A linear space V is called an *inner product space* if to each pair of vectors u, v in V there is assigned a number $\langle u, v\rangle$, also denoted $u \cdot v$, called their *inner* (*dot* or *scalar*) *product* such that

(i) $\langle u, v\rangle = \overline{\langle v, u\rangle}$.
(ii) $\langle u + v, w\rangle = \langle u, w\rangle + \langle v, w\rangle$.
(iii) $\langle \alpha u, v\rangle = \alpha\langle u, v\rangle$ for all scalars α.
(iv) $\langle v, v\rangle \geq 0$ with equality only for $v = 0$.

Easy consequences of (i)–(iii) are

$$\langle u, v + w\rangle = \langle u, v\rangle + \langle u, w\rangle \quad \text{and} \quad \langle u, \alpha v\rangle = \bar{\alpha}\langle u, v\rangle.$$

By definition the *length* of a vector v in V is

$$\|v\| = +\sqrt{\langle v, v\rangle}.$$

As the notation indicates, $\|v\|$ is a norm on V (see Prob. 1). If V is complete with respect to this norm, we call V a *Hilbert space*.

We call an inner product space V *real* if $\langle u, v\rangle$ is always real-valued. If $\langle u, v\rangle$ assumes complex values, we call V a *complex* inner product space.

Example 1. *Real n-space*, R_n, is a real inner product space with

$$x \cdot y = \langle x, y\rangle = \sum_{i=1}^{n} x_i y_i,$$

where $x = (x_1, \ldots, x_n)$ and $y = (y_1, \ldots, y_n)$ are vectors with real components.

Example 2. *Complex n-space*, C_n, is a complex inner product space with

$$x \cdot y = \langle x, y\rangle = \sum_{i=1}^{n} x_i \bar{y}_i,$$

where the vectors x and y may have complex components.

Example 3. The space of *real-valued continuous functions*, $C[a, b]$, is a real inner product space with

$$f \cdot g = \langle f, g\rangle = \int_a^b f(x)g(x)\, dx.$$

Example 4. The space of *complex-valued continuous functions*, $C[a, b]$, is a complex inner product space with

$$f \cdot g = \langle f, g \rangle = \int_a^b f(x)\overline{g(x)}\, dx.$$

Example 5. $L_2(a, b)$, *the space of square-integrable functions on* (a, b), is a complex inner product space with

$$f \cdot g = \langle f, g \rangle = \int_a^b f(x)\overline{g(x)}\, dx.$$

We obtain a real inner product space if all functions and scalars are real.

The *Schwarz inequality*

$$|\langle u, v \rangle| \le \|u\|\,\|v\| \tag{6-1}$$

holds in any inner product space (see Prob. 3). This in turn implies that

$$\begin{aligned}
\|u + v\|^2 &= \langle u + v, u + v \rangle = u \cdot u + u \cdot v + v \cdot u + v \cdot v \\
&= \|u\|^2 + u \cdot v + \overline{u \cdot v} + \|v\|^2 \\
&= \|u\|^2 + 2 \operatorname{Re} u \cdot v + \|v\|^2 \\
&\le \|u\|^2 + 2|u \cdot v| + \|v\|^2 \\
&\le \|u\|^2 + 2\|u\|\,\|v\| + \|v\|^2 = (\|u\| + \|v\|)^2.
\end{aligned}$$

Therefore, the *triangle inequality*

$$\|u + v\| \le \|u\| + \|v\| \tag{6-2}$$

holds in any inner product space.

Any sequence of linearly independent vectors $v_1, v_2, \ldots$ in an inner product space V can be orthonormalized by the *Gram–Schmidt process*: Recall that a set of vectors is *orthogonal* if

$$s_1 \cdot s_2 = 0 \quad \text{for any} \quad s_1, s_2 \quad \text{in } S.$$

A vector is *normal* if its length is 1. An orthogonal set of normal vectors is called *orthonormal.* Now, let $v_1, v_2, \ldots$ be linearly independent and define

$$\begin{aligned}
w_1 &= v_1, \\
w_2 &= v_2 - \frac{v_2 \cdot w_1}{w_1 \cdot w_1} w_1, \\
w_3 &= v_3 - \frac{v_3 \cdot w_1}{w_1 \cdot w_1} w_1 - \frac{v_3 \cdot w_2}{w_2 \cdot w_2} w_2, \\
&\vdots
\end{aligned}$$

where w_n is v_n minus its components along all the previously constructed vectors $w_1, w_2, \ldots, w_{n-1}$. It is easy to check that the sequence $w_1, w_2, \ldots$ is orthogonal and that

$$u_1 = \frac{w_1}{\|w_1\|}, \qquad u_2 = \frac{w_2}{\|w_2\|}, \ldots$$

is an orthonormal set. Moreover, for each n,

$$\text{span}\{v_1, \ldots, v_n\} = \text{span}\{u_1, \ldots, u_n\}.$$

Recall that the *span* of a set of vectors is the set of all finite linear combinations of those vectors. We say that the orthonormal vectors $u_1, u_2, \ldots$ are obtained from $v_1, v_2, \ldots$ by the Gram–Schmidt process.

The next results extend familiar notions about (classical) Fourier series to inner product spaces. Let $v_1, v_2, \ldots, v_n, \ldots$ be an orthonormal sequence in an inner product space V. For each vector v in V, *Bessel's inequality*

$$\sum_{n=1}^{\infty} |\langle v, v_n\rangle|^2 \le \|v\|^2 \tag{6-3}$$

holds. To see this, let $c_k = \langle v, v_k\rangle$. Then

$$\begin{aligned}\left\|v - \sum_1^n c_k v_k\right\|^2 &= \left\langle v - \sum_1^n c_k v_k,\, v - \sum_1^n c_r v_r\right\rangle \\ &= \|v\|^2 - \sum_1^n \bar{c}_r\langle v, v_r\rangle - \sum_1^n c_k\langle v_k, v\rangle + \sum_1^n\sum_1^n c_k\bar{c}_r\langle v_k, v_r\rangle \\ &= \|v\|^2 - \sum_1^n \bar{c}_r c_r - \sum_1^n c_k\bar{c}_k + \sum_1^n c_s\bar{c}_s \\ &= \|v\|^2 - \sum_1^n |c_k|^2.\end{aligned}$$

Thus,

$$\left\|v - \sum_1^n c_k v_k\right\| = \|v\|^2 - \sum_1^n |c_k|^2. \tag{6-4}$$

This gives $\sum_1^n |c_k|^2 \le \|v\|^2$, and Bessel's inequality follows at once upon letting $n \to \infty$. Furthermore, (6-4) reveals that

$$v = \sum_{k=1}^{\infty} \langle v, v_k\rangle v_k \leftrightarrow \|v\|^2 = \sum_{k=1}^{\infty} |\langle v, v_k\rangle|^2.$$

The right member is called *Parseval's equation.* The series $\sum \langle v, v_k\rangle v_k$ is called the *Fourier series* of v relative to the orthonormal set $v_1, v_2, \ldots$, and $\langle v, v_k\rangle$ is the kth *Fourier coefficient* of v.

PROBLEMS

1. Check that $\|v\| = \sqrt{\langle v, v\rangle}$ is a norm for an inner product space V.
2. Verify that an inner product is *conjugate linear* in its second variable; that is, $\langle u, v + w\rangle = \langle u, v\rangle + \langle u, w\rangle$ and $\langle u, \alpha v\rangle = \bar{\alpha}\langle u, v\rangle$.
3. Prove the Schwarz inequality. *Hint.* Fix u, v and expand out $\|u + tv\|^2$ for t real. The result is a nonnegative, quadratic polynomial in t. The discriminant of this polynomial

is nonpositive. Conclude that $|\text{Re } u \cdot v| \le \|u\|\,\|v\|$. Apply this result with v replaced by $(u \cdot v)v$.

4. Confirm that the spaces in Exs. 1 to 5 are inner product spaces.

5. Check that the vectors $w_1, w_2, \ldots$ obtained in the Gram–Schmidt process are orthogonal. Why must the vectors $v_1, v_2, \ldots$ be linearly independent to apply the Gram–Schmidt process?

6. (a) Write out the Schwarz inequality for

$$u = (|x_1|, \ldots, |x_n|), \qquad v = (|y_1|, \ldots, |y_n|)$$

in n-space.

(b) Let $n \to \infty$ first on the right side of the inequality in part (a) and then on the left side to get the corresponding infinite series result.

7. Write out the Schwarz inequality for $|f(x)|$ and $|g(x)|$, where f and g are continuous or square-integrable functions.

8. Equip $C[0, 2\pi]$ with the usual inner product of Ex. 3 and consider only real-valued functions. Check that the sequence $v_1, v_2, \ldots$ given by

$$\frac{1}{\sqrt{2\pi}}, \quad \frac{1}{\sqrt{\pi}} \cos x, \quad \frac{1}{\sqrt{\pi}} \sin x, \quad \frac{1}{\sqrt{\pi}} \cos (2x), \quad \frac{1}{\sqrt{\pi}} \sin (2x), \quad \ldots$$

is orthonormal. If $f(x)$ is in $V = C[0, 2\pi]$ show that the series $\sum_1^\infty \langle f, v_k \rangle v_k$ is the usual Fourier series of $f(x)$.

9. Show that the sequence $\sin(\pi x), \sin(2\pi x), \sin(3\pi x), \ldots$ is an orthogonal system when the inner product

$$\langle u, v \rangle = \int_0^1 [u(x)v(x) + u'(x)v'(x)]\, dx$$

is used. Determine coefficients c_n so that $\{v_n = c_n \sin(n\pi x)\}$ is orthonormal. Find the Fourier coefficients of an f in $C^1[0, 1]$ relative to the orthonormal set $\{v_n\}$.

10. If $u_1, u_2, \ldots, u_n$ are orthonormal, prove that they are linearly independent. That is, show that $\sum \alpha_k u_k = 0$ implies that $\alpha_1 = \alpha_2 = \cdots = \alpha_n = 0$.

11. Let V be an inner product space and $T\colon V \to V$ a linear operator. A number λ such that

$$Tv = \lambda v \quad \text{for some} \quad v \ne 0$$

is called an *eigenvalue of* T, and v is called a corresponding *eigenvector*.

(a) Fix an eigenvalue λ. If u and v are eigenvectors corresponding to λ, show that $\alpha u + \beta v$ is also an eigenvector corresponding to λ for any scalars α and β. In other words, the set of eigenvectors belonging to λ is a subspace of V.

(b) Let $v_1, \ldots, v_n$ be linearly independent and satisfy $Tv_i = \lambda v_i$. Show that there are orthonormal vectors $u_1, \ldots, u_n$ such that $Tu_i = \lambda u_i$. That is, a linearly independent set of eigenvectors belonging to λ can be replaced by orthonormalized eigenvectors.

-7. The Fredholm Alternative

The most important use of the Fredholm alternative is to establish that certain boundary value problems have solutions. This is done by settling what is typically an easier problem, namely, showing that the given boundary value problem has at most one solution. Then existence of a solution is proven by converting the boundary value problem into an equivalent integral equation and invoking the

Fredholm alternative:

> Either the integral equation
>
> $$(I - \lambda K)y = f$$
>
> has a solution for each f or the corresponding homogeneous equation
>
> $$(I - \lambda K)y = 0$$
>
> has a nontrivial solution.

Another way to express this alternative is:

> Uniqueness is equivalent to existence for solutions of
>
> $$(I - \lambda K)y = f.$$

(See Prob. 16.)

The rest of this section is devoted to establishing the Fredholm alternative under various assumptions on the kernel $k(x, s)$ of the integral operator K. These assumptions cover the cases of principal physical interest.

We may assume that $\lambda = 1$ while establishing the Fredholm alternative. Indeed, we fix λ, set $\tilde{k}(x, s) = \lambda k(x, s)$, and prove the alternative for the kernel $\tilde{k}(x, s)$. This clearly amounts to proving the Fredholm alternative as stated for $\lambda = 1$. Thus, $\lambda = 1$ in what follows.

The Matrix Case

Let M be a real or complex matrix. Then either $MY = F$ has a solution for each column vector F or the corresponding homogeneous equation $MY = 0$ has a nontrivial solution.

This is, of course, the Fredholm alternative for matrices. Its proof is easy: If $\det(M) \neq 0$, the first possibility occurs, while if $\det(M) = 0$ the second possibility does.

Our approach will be to reduce the Fredholm alternative for integral equations to the matrix case.

Integral Operators with Degenerate Kernels

Let

$$l(x, s) = \sum_{i=1}^{n} a_i(x)\overline{b_i(s)} \tag{7-1}$$

be a degenerate kernel. In Sec. 7-5 we showed that the Fredholm equation

$$(I - L)y = f \tag{7-2}$$

is equivalent to the matrix problem

$$(I - C)Y = F, \tag{7-3}$$

where $C = [c_{ij}]_{n \times n}$, $F = [F_1, \ldots, F_n]^T$,

(7-4) $$c_{ij} = \langle a_j, b_i \rangle = \int_a^b a_j(x)\overline{b_i(s)}\, ds,$$

and

(7-5) $$F_i = \langle f, b_i \rangle = \int_a^b f(s)\overline{b_i(s)}\, ds,$$

in the sense that $y = y(x)$ solves (7-2) if and only if

(7-6) $$y = f(x) + \sum_{i=1}^{n} Y_i a_i(x)$$

where $Y = [Y_1, \ldots, Y_n]^T$ solves (7-3), in which case

(7-7) $$Y_i = \langle y, b_i \rangle, \qquad i = 1, 2, \ldots, n.$$

Our derivation of this equivalence assumed that all functions involved were continuous. However, the same result holds in the square-integrable case, when the functions are in $L_2(a, b)$ (see Prob. 4 of Sec. 7-5). Thus, we can regard L either as an integral operator on $C[a, b]$ or $L_2(a, b)$.

Suppose that the second possibility in the Fredholm alternative does not hold for (7-2). That is, suppose that $(I - L)y = 0$ implies that $y = 0$. The equivalent matrix problem is $(I - C)Y = 0$, because $f = 0$ implies that $F = 0$. Now, $(I - C)Y = 0$ has as its only solution $Y = 0$, by (7-7) and the fact that $y = 0$ is the only solution to $(I - L)y = 0$. The Fredholm alternative for matrices implies that $(I - C)Y = F$ has a solution for every F. In view of the equivalence of (7-2) and (7-3), we conclude that (7-2) has a solution for each f. Thus, the Fredholm alternative holds for integral equations with degenerate kernels.

Integral Operators with Continuous Kernels

Let $k(x, s)$ be continuous on $a \le x, s \le b$ and K: $C[a, b] \to C[a, b]$ be the integral operator

$$Ky = Ky(x) = \int_a^b k(x, s)y(s)\, ds.$$

As usual, $C[a, b]$ is the normed space of real- or complex-valued functions $y(x)$ on $[a, b]$ and $\|y\| = \max |y(x)|$ for x in $[a, b]$.

We can use the Weierstrass approximation theorem to obtain a polynomial $p(x, s)$ with

$$|k(x, s) - p(x, s)| < \frac{1}{2(b - a)}$$

for $a \le x, s \le b$. Let P: $C[a, b] \to C[a, b]$ be the integral operator with kernel $p(x, s)$ and R: $C[a, b] \to C[a, b]$ be the integral operator with kernel $r(x, s) = k(x, s) - p(x, s)$.

Now, we show that the Fredholm alternative holds for the integral operator $K: C[a, b] \to C[a, b]$. Notice that the reasoning which follows uses only four properties of $C[a, b]$, K, P, and R; namely,

(i) $C[a, b]$ is complete.
(ii) $K = P + R$.
(iii) P has a degenerate kernel.
(iv) R has operator norm, $\|R\| < 1$.

In the present context (i)–(iii) are evident and (iv) follows easily,

$$|Ry(x)| \leq \int_a^b |r(x, s)|\,|y(s)|\, ds \leq \frac{1}{2(b-a)} \|y\|(b-a) = \frac{1}{2}\|y\|,$$

$$\|Ry\| = \max_{a \leq x \leq b} |Ry(x)| \leq \tfrac{1}{2}\|y\|,$$

$$\|R\| \leq \tfrac{1}{2},$$

by the meaning of the operator norm.

Each of the following equations is equivalent (i.e., has the same solutions) because each step is reversible:

$$(7\text{-}8) \quad (I - K)y = f, \qquad (I - R)y - Py = f, \qquad y - (I - R)^{-1}Py = (I - R)^{-1}f,$$

or

$$(7\text{-}9) \qquad (I - L)y = \tilde{f},$$

where

$$(7\text{-}10) \qquad L = (I - R)^{-1}P \quad \text{and} \quad \tilde{f} = (I - R)^{-1}f.$$

We claim that $L: C[a, b] \to C[a, b]$ is an integral operator with a degenerate kernel. Assuming this fact, the Fredholm alternative for (7-8) follows easily: Suppose that the second possibility in the Fredholm alternative does not hold for the integral equation (7-8). Then $(I - K)y = 0$ implies that $y = 0$. The equivalence of (7-8) and (7-9) applied in the case $f = 0$, which yields $\tilde{f} = (I - R)^{-1}f = 0$, shows that $(I - L)y = 0$ has only the trivial solution $y = 0$. Since the Fredholm alternative holds for (7-9) this equation has a solution for every continuous function $\tilde{f}$. This equivalence of (7-8) and (7-9) yields a solution to (7-8) for every continuous f. Thus, the Fredholm alternative holds for the integral equation (7-8).

To complete our argument, we must confirm that the operator $L = (I - R)^{-1}P$ has a degenerate kernel. This is easy:

$$Ly = Ly(x) = (I - R)^{-1}Py(x) = (I - R)^{-1} \int_a^b \sum_{i=1}^n a_i(x)\overline{b_i(s)}y(s)\, ds$$

$$= \int_a^b \sum_{i=1}^n A_i(x)\overline{b_i(s)}y(s)\, ds,$$

where

$$p(x, s) = \sum_{i=1}^n a_i(x)\overline{b_i(s)} \quad \text{and} \quad A_i = (I - R)^{-1}a_i.$$

Thus, L has kernel

$$l(x, s) = \sum_{i=1}^{n} A_i(x)\overline{b_i(s)},$$

which is degenerate.

As we noted above, this reasoning only relies on properties (i)–(iv). Here is another situation where these properties hold.

Integral Operators with Square-Integrable Kernels

Let $k(x, s)$ be a square-integrable kernel,

$$\int_a^b \int_a^b |k(x, s)|^2 \, dx \, ds < \infty,$$

where $-\infty \leq a < b \leq +\infty$. Let $L_2(a, b)$ be the space of functions $y = y(x)$ which are square integrable on (a, b),

$$\int_a^b |y(x)|^2 \, dx < \infty.$$

$L_2(a, b)$ is a normed space with

$$\|y\| = \left[\int_a^b |y(x)|^2\right]^{1/2}.$$

For such functions y,

$$Ky = Ky(x) = \int_a^b k(x, s)y(s) \, ds \tag{7-11}$$

is also square integrable. Indeed by the Schwarz inequality,

$$|Ky(x)|^2 \leq \int_a^b |k(x, s)|^2 \, ds \int_a^b |y(s)|^2 \, ds,$$

$$\int_a^b |Ky(x)|^2 \, dx \leq \int_a^b \int_a^b |k(x, s)|^2 \, ds \, dx \int_a^b |y(s)|^2 \, ds < \infty. \tag{7-12}$$

So (7-11) defines a linear integral operator $K: L_2(a, b) \to L_2(a, b)$. Moreover, (7-12) shows that K is a bounded operator

$$\|Ky\| \leq \left(\int_a^b \int_a^b |k(x, s)|^2 \, ds \, dx\right)^{1/2} \|y\|,$$

$$\|K\| \leq \left(\int_a^b \int_a^b |k(x, s)|^2 \, ds \, dx\right)^{1/2}. \tag{7-13}$$

The space $L_2(a, b)$ is complete, provided that we regard the integrals as Lebesgue integrals. So property (i) holds. The elements of the Lebesgue theory of integration guarantee that there is a degenerate, square-integrable kernel $p(x, s)$

such that

$$\int_a^b \int_a^b |k(x, s) - p(x, s)|^2 \, dx \, ds < 1. \tag{7-14}$$

Readers familiar with the Lebesgue integral are asked to verify this in Prob. 3. Let P and R be the integral operators on $L_2(a, b)$ with kernels $p(x, s)$ and $r(x, s) = k(x, s) - p(x, s)$. By (7-13) and (7-14),

$$\|R\| \leq \left(\int_a^b \int_a^b |r(x, s)|^2 \, ds \, dx \right)^{1/2} < 1.$$

Thus, properties (i)–(iv) hold in the square-integrable setting. Consequently, the Fredholm alternative holds for integral operators on $L_2(a, b)$, which are defined by square-integrable kernels.

Integral Operators with Weakly Singular Kernels

The kernel

$$k(x, s) = \frac{m(x, s)}{|x - s|^r}, \tag{7-15}$$

where $0 < r < 1$ and $m(x, s)$ is continuous on $a \leq x, s \leq b$, is called *weakly singular.* The corresponding integral operator

$$Ky = Ky(x) = \int_a^b k(x, s) y(s) \, ds$$

is defined for any continuous function $y(x)$. That is, the improper integral converges. To see this, we use the comparison test: Let $M = \max |m(x, s)|$ for $a \leq x, s \leq b$. Since $0 < r < 1$,

$$\int_a^b |k(x, s)y(s)| \, ds \leq M\|y\| \int_a^b \frac{ds}{|x - s|^r}$$

$$= M\|y\| \frac{(x - a)^{1-r} + (b - x)^{1-r}}{1 - r} < \frac{2M\|y\|(b - a)^{1-r}}{1 - r} < \infty.$$

Moreover, Ky is a continuous function, so $K: C[a, b] \to C[a, b]$. To confirm this and to prepare the way for a degenerate kernel approximation, define a *continuous* kernel $k_n(x, s)$ by

$$k_n(x, s) = \begin{cases} k(x, s) & \text{when } |x - s| \geq \dfrac{1}{n}, \\[2ex] m(x, s)n^r & \text{when } |x - s| \leq \dfrac{1}{n}. \end{cases}$$

Notice that the two expressions for $k_n(x, s)$ match when $|x - s| = 1/n$. Let K_n:

$C[a, b] \to C[a, b]$ be the corresponding integral operator. For y in $C[a, b]$,

$$|K_n y(x) - Ky(x)| = \left| \int_{x-(1/n)}^{x+(1/n)} \left[m(x, s)n^r - \frac{m(x, s)}{|x - s|^r} \right] y(s)\, ds \right|$$

$$\leq M\|y\| \left[\int_{x-(1/n)}^{x+(1/n)} n^r\, ds + \int_{x-(1/n)}^{x+(1/n)} \frac{ds}{|x - s|^r} \right]$$

$$= M\|y\| \left[2n^{r-1} + \frac{2n^{r-1}}{1 - r} \right] = \frac{2M(2 - r)}{(1 - r)n^{1-r}} \|y\|.$$

Therefore,

$$|K_n y(x) - Ky(x)| \leq \frac{2M(2 - r)}{(1 - r)n^{1-r}} \|y\|. \tag{7-16}$$

Since $r < 1$ the right side of this inequality tends to zero independently of x in $[a, b]$. Thus, the continuous functions $K_n y$ converge uniformly to Ky on $[a, b]$; hence, Ky is continuous and K: $C[a, b] \to C[a, b]$, as we have said.

Taking the maximum over x in $[a, b]$ in (7-16) gives

$$\|K_n y - Ky\| \leq \frac{2M(2 - r)}{(1 - r)n^{1-r}} \|y\|.$$

$$\|K_n - K\| \leq \frac{2M(2 - r)}{(1 - r)n^{1-r}}. \tag{7-17}$$

Now the continuous kernel $k_n(x, s)$ has a polynomial approximation $p_n(x, s)$ such that $|k_n(x, s) - p_n(x, s)| \leq 1/n$ for $a \leq x, s \leq b$. The corresponding integral operator P_n has a degenerate kernel and $\|K_n - P_n\| \leq (b - a)/n$. Then

$$\|K - P_n\| \leq \|K - K_n\| + \|K_n - P_n\| \leq \frac{2M(2 - r)}{(1 - r)n^{1-r}} + \frac{b - a}{n} \to 0$$

as $n \to \infty$. Thus, we can fix n so that $\|K - P_n\| < 1$ and let $R_n = K - P_n$. Properties (i)–(iv) hold for these choices, and the Fredholm alternative is established for integral operators on $C[a, b]$ whose kernels are weakly singular.

The General Fredholm Alternative

The reasoning used above can be modified slightly to obtain a stronger form of the Fredholm alternative. Indeed, the Fredholm alternative for matrices can be strengthened as follows: Let $M = [m_{ij}]$ be an $n \times n$ real or complex matrix and $M^* = [m_{ij}^*]$ be its adjoint or transposed conjugate matrix, $m_{ij}^* = \bar{m}_{ji}$. As usual, $A \cdot B$ stands for the dot product of the vectors A and B. Consider the matrix problems

$$MY = F, \qquad M^*Z = G, \tag{7-18}$$

$$MY = 0, \qquad M^*Z = 0. \tag{7-19}$$

The following alternative holds:

> Either both equations in (7-18) have solutions for any column vectors F and G or the corresponding homogeneous equations in (7-19) both have nontrivial solutions. When the latter possibility occurs,
> (a) each equation in (7-19) has the same number of linearly independent solutions, say $Y_1, \ldots, Y_p$ and $Z_1, \ldots, Z_p$, respectively;
> (b) the equation $MY = F$ (resp., $M^*Z = G$) has a solution if and only if $F \cdot Z_j = 0$ (resp., $G \cdot Y_j = 0$) for $j = 1, 2, \ldots, p$.

Since $\det M^* = \overline{\det M}$, the first possibility in the general Fredholm alternative holds when $\det M \neq 0$. The second holds when $\det M = 0$. Conditions (a) and (b) express the familiar compatibility conditions that must be met for an inhomogeneous, singular matrix system to be solvable.

To extend this matrix alternative to integral operators, given an integral operator K, we must find an integral operator K^* which is related to K as M^* is related to M. In the matrix case, M^* comes up as the unique matrix such that $\langle MY, Z\rangle = \langle Y, M^*Z\rangle$ for all vectors Y and Z. To find the integral operator analogue, let K be the integral operator with continuous kernel $k(x, s)$ and calculate

$$\begin{aligned}\langle Ky, z\rangle &= \int_a^b Ky(x)\overline{z(x)}\,dx \\ &= \int_a^b \int_a^b k(x, s)y(s)\,ds\,\overline{z(x)}\,dx \\ &= \int_a^b y(s) \int_a^b k(x, s)\overline{z(x)}\,dx\,ds \\ &= \int_a^b y(s) \overline{\int_a^b \overline{k(x, s)}z(x)\,dx}\,ds = \langle y, K^*z\rangle.\end{aligned}$$

Here K^* is the integral operator defined by

$$K^*z(x) = \int_a^b k^*(x, s)z(s)\,ds,$$

where $k^*(x, s) = \overline{k(s, x)}$ is called the *adjoint* (or *transposed conjugate*) *kernel* of $k(x, s)$. Clearly, $k^*(x, s)$ is continuous, $K^*\colon C[a, b] \to C[a, b]$, and

$$\langle Ky, z\rangle = \langle y, K^*z\rangle \tag{7-20}$$

for all continuous functions y and z. The same reasoning applies for integral operators with square-integrable or weakly singular kernels. [Of course, in the square-integrable case, y and z are square integrable (see Prob. 6).]

Here is the integral equations analogue of the general Fredholm alternative for matrices. Consider the integral equations

$$(I - \lambda K)y = f, \qquad (I - \bar{\lambda}K^*)z = g, \tag{7-21}$$

$$(I - \lambda K)y = 0, \qquad (I - \bar{\lambda}K^*)z = 0. \tag{7-22}$$

The following alternative holds:

> Either both equations in (7-21) have a solution for any f and g or the corresponding homogeneous equations (7-22) both have nontrivial solutions. When the latter possibility occurs,
> (a) each equation in (7-22) has the same number of linearly independent solutions, say $y_1, \ldots, y_p$ and $z_1, \ldots, z_p$, respectively;
> (b) the equation $(I - \lambda K)y = f$ [resp., $(I - \bar{\lambda}K^*)z = g$] has a solution if and only if $\langle f, z_i \rangle = 0$ [resp., $\langle g, y_i \rangle = 0$] for $j = 1, 2, \ldots, p$.

This general Fredholm alternative is proven by reduction to the matrix case. No essentially new ideas are involved. We leave the details to Probs. 9–15, and outline the argument here. The first step is to prove the general alternative for degenerate kernels. This is done by reducing the degenerate kernel problems $(I - L)y = f$ and $(I - L^*)z = g$ to equivalent matrix problems of the form $(I - C)Y = F$ and $(I - C)^*Z = G$. Now, the alternative for matrices gives the corresponding result for integral operators with degenerate kernels. Next, given an integral operator K with a continuous, square-integrable, or weakly singular kernel, we can determine a splitting $K = P + R$ satisfying (i)–(iv). Automatically, $K^* = P^* + R^*$ satisfies (i)–(iv). These splittings are used to reduce the proof of the general Fredholm alternative to the degenerate kernel case.

In applications the basic form of the Fredholm alternative is used for Dirichlet problems and the generalized form is needed for Neumann problems.

PROBLEMS

1. Let D be a closed, bounded domain in n-space and $k(x, s)$ be a continuous function for x, s in D. Define $K: C(D) \to C(D)$ by

$$Ky(x) = \int_D k(x, s)y(s)\, ds,$$

where the integration is over the region D and the maximum norm is used in $C(D)$. Establish the basic Fredholm alternative. *Hint.* Verify (i)–(iv) for this setting.

2. Carry out the verification of the basic Fredholm alternative for (a) square-integrable and (b) weakly singular kernels defined on a domain in n-space. In the weakly singular case, $k(x, s) = m(x, s)/|x - s|^r$ where $r < n$, $m(x, s)$ is continuous, and $|x - s|$ is the distance between the vectors x and s. (In the weakly singular case, restrict yourself to $n = 2$ or 3 if you are not familiar with "polar coordinates" for $n > 3$.)
3. Verify (7-14) if you are familiar with Lebesgue integration.
4. Let $L_1(a, b)$ be the space of (Lebesgue) integrable functions on $-\infty < a < b < +\infty$. That is, y belongs to $L_1(a, b)$ if $\int_a^b |y(x)|\, dx < \infty$. Let $k(x, s)$ be a bounded, continuous kernel and $K: L_1(a, b) \to L_1(a, b)$ be the corresponding integral operator. Show that K is linear, bounded, and that the basic Fredholm alternative applies.
5. Apply the Fredholm alternative to a linear, Volterra integral operator $K: C[a, b] \to C[a, b]$ to show that $(I - \lambda K)y = f$ has a unique solution for every λ. *Hint.* Show $y = \lambda K y$ implies that $y = \lambda^n K^n y$ and $\|\lambda^n K^n\| \to 0$.

6. Verify (7-20) for a square-integrable or weakly singular kernel $k(x, s)$.

7. Let $V = C[a, b]$ or $L_2(a, b)$. Each integral operator K with a continuous, square-integrable, or weakly singular kernel has an adjoint operator K^* satisfying (7-20). Even if K is *not* an integral operator, we call an operator K^* satisfying (7-20) the adjoint of K. For example, the identity operator I has adjoint $I^* = I$. More precisely, we say that an operator T in $\mathscr{L}(V)$ has an *adjoint* if there is an operator S in $\mathscr{L}(V)$ such that $\langle Tu, v\rangle = \langle u, Sv\rangle$ for all u, v in V. Show that S is unique if it exists. This uniqueness justifies the notation $S = T^*$.

8. In the general context of Prob. 7, verify the following properties of adjoint operators:

$$(K_1 + K_2)^* = K_1^* + K_2^*, \quad (\lambda K)^* = \bar{\lambda} K^*, \quad (K_1 K_2)^* = K_2^* K_1^*.$$
$$(K^n)^* = (K^*)^n, \quad K^{**} = K.$$

Use the second property to verify that the general Fredholm alternative need only be proved for $\lambda = 1$.

The next several problems outline the proof of the general Fredholm alternative. Assume that $\lambda = 1$ (see Prob. 8).

9. Let $l(x, s) = \sum_{i=1}^{n} a_i(x)\overline{b_i(s)}$ and L be the corresponding integral operator. Recall that $(I - L)y = f$ is equivalent to $(I - C)Y = F$ in the sense that y solves the integral equation if and only if $y(x) = f(x) + \sum_1^n Y_j a_j(x)$, in which case $Y_j = \langle y, b_j\rangle$. Here $C = [c_{ij}]_{n\times n}$, $c_{ij} = \langle a_j, b_i\rangle$, $F = [F_i]^T$, and $F_i = \langle f, b_i\rangle$. Let L^* be the adjoint operator with kernel $l^*(x, s) = \overline{l(s, x)}$. Show that $(I - L^*)z = g$ is equivalent to the matrix equation $(I - C)^*Z = G$ in the sense that z solves the integral equation if and only if $z(x) = g(x) + \sum_1^n Z_j b_j(x)$, in which case $Z_j = \langle z, a_j\rangle$. Here $G = [G_i]^T$, $G_i = \langle g, a_i\rangle$.

10. In the context of Prob. 9 show that

$$\langle f, z\rangle = \langle f, g\rangle + \langle F, Z\rangle \quad \text{and} \quad \langle g, y\rangle = \langle g, f\rangle + \langle G, Y\rangle.$$

11. Confirm the general Fredholm alternative for an integral equation with a degenerate kernel.

12. For a kernel $k(x, s)$ which is continuous, square integrable, or weakly singular, show that the splitting $K = P + R$ which satisfies (i)–(iv) yields $K^* = P^* + R^*$ and that (i)–(iv) holds for these transposed operators.

13. Since $\|R\|$, $\|R^*\| < 1$ both operators $(I - R)^{-1}$ and $(I - R^*)^{-1}$ exist and are given by Neumann series. Show that $(I - R)^{-1}$ has an adjoint, namely, $(I - R^*)^{-1}$; that is,

$$[(I - R)^{-1}]^* = (I - R^*)^{-1}.$$

14. Use $K = P + R$ to show that $(I - K)y = f$ is equivalent to

$$[I - (I - R)^{-1}P]y = (I - R)^{-1}f,$$

while $(I - K^*)z = g$ is equivalent to

$$[I - P^*(I - R^*)^{-1}]\tilde{z} = g,$$

where $\tilde{z} = (I - R^*)z$. Now, use Prob. 13 to show

$$P^*(I - R^*)^{-1} = [(I - R)^{-1}P]^*.$$

Consequently, if $L = (I - R)^{-1}P$, then

$$\left.\begin{array}{l}(I - K)y = f \\ \text{is equivalent to} \\ (I - L)y = (I - R)^{-1}f\end{array}\right\} \quad \text{and} \quad \left\{\begin{array}{l}(I - K^*)z = g \\ \text{is equivalent to} \\ (I - L^*)\tilde{z} = g, \text{ where } \tilde{z} = (I - R^*)z.\end{array}\right.$$

Finally, use the general Fredholm alternative for the operator L with a degenerate kernel to deduce the alternative for K.

15. Spend a few minutes to convince yourself that the arguments needed to solve Probs. 6–14 hold when the underlying domain of the variables x and s is a set D in n-space, where D is closed and bounded for the continuous and weakly singular cases.

16. Verify that the following statement is equivalent to the Fredholm alternative: Uniqueness is equivalent to existence for solutions of $(I - \lambda K)y = f$.

-8. Eigenvalues and Eigenfunctions

Throughout this section, $k(x, s)$ is either continuous, square integrable, or weakly singular on $a \le x, s \le b$. (In the square-integrable case, the choices $a = -\infty$ and $b = +\infty$ are permitted.) Let K be the corresponding integral operator,

$$Ky(x) = \int_a^b k(x, s)y(s)\,ds.$$

As usual, we regard $K\colon C[a, b] \to C[a, b]$ when $k(x, s)$ is continuous or weakly singular, and regard $K\colon L_2(a, b) \to L_2(a, b)$ when $k(x, s)$ is square integrable. In all three cases, we use the complex inner product

$$\langle f, g \rangle = \int_a^b f(x)\overline{g(x)}\,dx.$$

Here f and g are assumed continuous when $K\colon C[a, b] \to C[a, b]$, and are assumed square integrable when $K\colon L_2(a, b) \to L_2(a, b)$. We make similar assumptions throughout this section. That is, when $K\colon C[a, b] \to C[a, b]$, all given function data are assumed continuous and solutions y are to be continuous. If $K\colon L_2(a, b) \to L_2(a, b)$, data and solutions are square integrable.

According to the Fredholm alternative, either $(I - \lambda K)y = f$ has a solution for all f or the corresponding homogeneous equation $(I - \lambda K)y = 0$ has a *nontrivial* solution. In the latter case, λ is called an *eigenvalue of the kernel* $k(x, s)$ and any nontrivial solution $y = y(x)$ is a *corresponding eigenfunction.*

Pursuing the analogy between matrices and integral operators, it is reasonable to expect that a given kernel will have eigenvalues. However, this expectation is not always realized. For example, a Volterra integral equation, which can be regarded as a Fredholm equation $(I - \lambda K)y = f$, where $k(x, s) = 0$ for $s > x$, always has a unique solution given by the Neumann series (see Prob. 4 of Sec. 7-3 or Prob. 9 of Sec. 7-4). In particular, the unique solution when $f = 0$ is clearly $y = 0$. A Volterra kernel has no eigenvalues. In contrast to this, we shall show that a *self-adjoint* kernel $k(x, s)$, that is, a kernel for which $k^*(x, s) = k(x, s)$, always has at least one eigenvalue, assuming the corresponding integral operator $K \ne 0$. Furthermore, a (nonzero) self-adjoint kernel actually has an infinite number of eigenvalues, unless it is degenerate.

Given that a kernel does have some eigenvalues and eigenfunctions, the following useful properties hold. The eigenvalues of $k(x, s)$ form an at most countable set, say $\lambda_1, \lambda_2, \ldots$. If this sequence has an infinite number of terms, then $|\lambda_n| \to \infty$

as $n \to \infty$. Consequently, in any finite portion of the complex λ plane there are at most a finite number of eigenvalues of $k(x, s)$. Each eigenvalue has at least one eigenfunction and at most a finite number of linearly independent ones. All these properties, which are proved below, follow from the fact, already used to prove the Fredholm alternative, that an integral equation with kernel $k(x, s)$ is equivalent to an integral equation with a degenerate kernel, which in turn reduces to a matrix problem.

Suppose that λ_0 is an eigenvalue of $k(x, s)$. Then the linear homogeneous equation $(I - \lambda_0 K)y = 0$ has, in addition to the trivial solution, nontrivial solutions. If $y_1, y_2, \ldots, y_p$ solve $(I - \lambda_0 K)y = 0$, so does the linear combination

$$\alpha_1 y_1 + \alpha_2 y_2 + \cdots + \alpha_p y_p$$

for any scalars $\alpha_1, \alpha_2, \ldots, \alpha_p$. That is, the set of solutions to $(I - \lambda_0 K)y = 0$ is a (linear) subspace of $C[a, b]$ or $L_2(a, b)$. This subspace is called the *eigenspace* of the eigenvalue λ_0. As noted above, λ_0 has at most a finite number of linearly independent eigenfunctions. The maximum number of linearly independent eigenfunctions belonging to λ_0 is the dimension of the eigenspace. This number is also called the *multiplicity* of the eigenvalue λ_0.

The next three theorems contain some elementary but very useful observations about eigenvalues and eigenfunctions.

Theorem 8-1. Let λ_0 be an eigenvalue of multiplicity m for the kernel $k(x, s)$. Then there are m orthonormal eigenfunctions which form a basis for the eigenspace of λ_0.

Proof. Let $y_1, \ldots, y_m$ be m linearly independent eigenfunctions corresponding to λ_0. By definition of multiplicity, these functions are a basis for the eigenspace of λ_0. Apply the Gram–Schmidt process to obtain orthonormal functions $u_1, \ldots, u_m$ from $y_1, \ldots, y_m$. Since each u_j is a linear combination of $y_1, \ldots, y_m$, each u_j is an eigenfunction corresponding to λ_0. Thus, $u_1, \ldots, u_m$ is the required orthonormal basis of eigenfunctions.

The fact that eigenfunctions corresponding to a given eigenvalue can be chosen to be orthonormal does not imply that eigenfunctions corresponding to distinct eigenvalues can be so chosen. However, this is possible when the kernel is self-adjoint.

Theorem 8-2. Let $k(x, s)$ be self-adjoint. Then the eigenvalues of $k(x, s)$ are all real and eigenfunctions corresponding to distinct eigenvalues are orthogonal.

Proof. Assume that λ is an eigenvalue of $k(x, s)$. Then $y = \lambda K y$ for some $y \neq 0$. Therefore, $\lambda \neq 0$ and

$$\langle y, y \rangle = \langle \lambda K y, y \rangle = \lambda \langle y, K y \rangle = \lambda \left\langle y, \frac{1}{\bar{\lambda}} y \right\rangle, \quad \langle y, y \rangle = \frac{\lambda}{\bar{\lambda}} \langle y, y \rangle,$$

$$(\bar{\lambda} - \lambda)\langle y, y \rangle = 0.$$

Since $\langle y, y \rangle = \|y\|^2 \neq 0$, we conclude that $\bar{\lambda} = \lambda$ and that λ is real.

Next, assume that $y = \lambda K y$, $z = \mu K z$ with $\lambda \neq \mu$ and eigenfunctions y and z. Since λ and μ are real,

$$\langle y, z \rangle = \langle \lambda K y, z \rangle = \lambda \langle y, Kz \rangle = \lambda \left\langle y, \frac{1}{\mu} z \right\rangle, \quad \langle y, z \rangle = \frac{\lambda}{\mu} \langle y, z \rangle,$$

$$(\mu - \lambda)\langle y, z \rangle = 0.$$

Since $\mu \neq \lambda$, we have $\langle y, z \rangle = 0$ and y and z are orthogonal.

A real-valued kernel $k(x, s)$ is called *symmetric* if $k(x, s) = k(s, x)$. Such kernels are self-adjoint. Recall that Sturm–Liouville boundary value problems can be recast as integral equations with symmetric kernels.

Theorem 8-3. Let $k(x, s)$ be a self-adjoint kernel. Let $\lambda_1, \lambda_2, \ldots$ be the eigenvalues of $k(x, s)$ where an eigenvalue of multiplicity m is repeated m times (is repeated to multiplicity) in this list. Then a corresponding set of orthonormal eigenfunctions $\phi_1(x), \phi_2(x), \ldots$ exists. Moreover, each $\phi_j(x)$ can be chosen real-valued when $k(x, s)$ is symmetric.

Proof. Each eigenvalue has an orthonormal basis for its eigenspace by Th. 8-1, while eigenfunctions belonging to distinct eigenvalues are automatically orthogonal by Th. 8-2. Thus, the entire collection of all orthonormal bases corresponding to the distinct eigenvalues of $k(x, s)$ comprises the required orthonormal sequence $\phi_1(x), \phi_2(x), \ldots$.

Assume now that $k(x, s)$ is symmetric. We must show that each eigenspace has a basis of real-valued eigenfunctions. Once this is established, the argument just given produces the required real-valued, orthonormal eigenfunctions.

Let λ be an eigenvalue and y a corresponding eigenfunction of the symmetric kernel $k(x, s)$: $y = \lambda K y$ with $y \neq 0$. We know λ is real, but y may be complex-valued, say, $y = u + iv$ with u, v real-valued. Since $k(x, s)$ is real-valued,

$$u + iv = \lambda K(u + iv) = \lambda K u + i \lambda K v$$

with Ku and Kv real-valued. Thus, $u = \lambda K u$ and $v = \lambda K v$. Either $u = 0$ or u is a real-valued eigenfunction corresponding to λ and similarly for v. Now, suppose that

$$y_1 = u_1 + iv_1, \ldots, y_m = u_m + iv_m$$

span the eigenspace for an eigenvalue λ of $k(x, s)$. If $\alpha_j = \beta_j + i\gamma_j$ are any complex scalars, then

$$\sum_1^m \alpha_j y_j = \sum_1^m (\beta_j u_j - \gamma_j v_j) + i \sum_1^m (\beta_j v_j + \gamma_j u_j).$$

There are at most m independent functions among $\{u_1, \ldots, u_m, v_1, \ldots, v_m\}$ because otherwise, λ would have multiplicity greater than m. On the other hand, if the previous list of functions contains fewer than m independent functions, then the expansion for $\sum_1^m \alpha_j y_j$ shows that the eigenspace of λ has a basis with fewer than m functions, contradicting the fact that the eigenspace has dimension m. Therefore, the list $\{u_1, \ldots, u_m, v_1, \ldots, v_m\}$ contains exactly m linearly independent, real-valued eigenfunctions. As noted earlier, this completes the proof of Th. 8-3.

We turn next to more subtle questions about the existence and distribution of the eigenvalues of a kernel $k(x, s)$. To this end, we review and extend slightly the reduction of a Fredholm equation to a matrix problem: Fix a positive number ρ. Reasoning as in Sec. 7-7, where $\rho = 1$, we can determine integral operators P and R, such that

(ii) $K = P + R$.
(iii) P has degenerate kernel $p(x, s)$.
(iv) R has operator norm $\|R\| < 1/\rho$.

Let λ be any real or complex number subject only to the restriction $|\lambda| \leq \rho$. Just as in passing from (7-8) to (7-9), we see that the equation

$$(I - \lambda K)y = f \tag{8-1}$$

is equivalent to

$$(I - L)y = (I - \lambda R)^{-1}f, \tag{8-2}$$

where

$$L = (I - \lambda R)^{-1}(\lambda P) = \lambda(I - \lambda R)^{-1}P.$$

Notice that $\|\lambda R\| = |\lambda| \, \|R\| \leq \rho \, \|R\| < 1$, so that $I - \lambda R$ is invertible. Furthermore (see Sec. 7-7), if

$$p(x, s) = \sum_{i=1}^{n} a_i(x)\overline{b_i(s)},$$

then L has kernel

$$l(x, s) = \sum_{i=1}^{n} A_i(x; \lambda)\overline{b_i(s)}, \tag{8-3}$$

where

$$A_i(x; \lambda) = \lambda(I - \lambda R)^{-1}a_i(x) = \sum_{m=0}^{\infty} \lambda^{m+1}R^m a_i(x) \tag{8-4}$$

is an analytic function of λ for $|\lambda| \leq \rho$, and

$$A_i(x; 0) = 0.$$

Again, as in Sec. 7-7, the integral equation (8-2) is equivalent to the matrix problem

$$(I - C(\lambda))Y = F(\lambda), \tag{8-5}$$

where

$$C(\lambda) = [c_{ij}(\lambda)]_{n \times n}, \qquad c_{ij}(\lambda) = \langle A_j, b_i \rangle, \tag{8-6}$$

$$F(\lambda) = [F_1(\lambda), \ldots, F_n(\lambda)]^T, \qquad F_j(\lambda) = \langle (I - \lambda R)^{-1}f, b_j \rangle, \tag{8-7}$$

in the sense that $y = y(x)$ solves (8-2) if and only if

$$y(x) = (I - \lambda R)^{-1} f(x) + \sum_{j=1}^{n} Y_j A_j(x; \lambda), \tag{8-8}$$

in which case $Y_j = \langle y, b_j \rangle$. Notice that $y = y(x)$ in (8-8) also depends on the parameter λ. To emphasize this we sometimes write $y = y(x; \lambda)$. We need two simple consequences of these reductions.

Lemma 8-1. The matrix $C(\lambda)$ and column vector $F(\lambda)$ in the system (8-5) are analytic in $|\lambda| \le \rho$. Also, $C(0) = 0$.

Lemma 8-2. Assume that $\det(I - C(\lambda)) \ne 0$ for $|\lambda| \le \rho$ and let $g(x)$ be continuous or square integrable, depending on the type of kernel $k(x, s)$. Then the function

$$h(\lambda) = \langle y, g \rangle$$

is analytic for $|\lambda| \le \rho$, where $y = y(x; \lambda)$ is the solution to $(I - \lambda K)y = f$.

To validate these lemmas, notice that for $|\lambda| \le \rho$,

$$\begin{aligned} c_{ij}(\lambda) = \langle A_j, b_i \rangle &= \int_a^b \sum_{m=0}^{\infty} \lambda^{m+1} R^m a_j(x) \overline{b_i(x)}\, dx \\ &= \sum_{m=0}^{\infty} \lambda^{m+1} \int_a^b R^m a_j(x) \overline{b_i(x)}\, dx, \end{aligned}$$

a convergent power series for $|\lambda| \le \rho$. The interchange of summation and integration is valid because the series in (8-4) converges uniformly in x in the $C[a, b]$ case and converges in the L_2 norm in the square-integrable case. Similarly, the components of $F(\lambda)$ are analytic in $|\lambda| \le \rho$. Evidently, $c_{ij}(0) = 0$, so $C(0) = 0$ and Lemma 8-1 holds.

Since $\det(I - C(\lambda)) \ne 0$ for $|\lambda| \le \rho$, the matrix equation (8-5) has a unique solution; hence, the integral equation (8-1) has a unique solution $y = y(x; \lambda)$ given by (8-8). Thus,

$$\begin{aligned} \langle y, g \rangle &= \left\langle \sum_{m=0}^{\infty} \lambda^m R^m f, g \right\rangle + \sum_{j=1}^{n} Y_j \left\langle A_j, g \right\rangle \\ &= \sum_{m=0}^{\infty} \lambda^m \left\langle R^m f, g \right\rangle + \sum_{j=1}^{n} Y_j \left\langle A_j, g \right\rangle. \end{aligned}$$

The interchange of summation and integration is permissible as above for $|\lambda| \le \rho$. That the inner product $\langle A_j, g \rangle$ is analytic in $|\lambda| \le \rho$ follows from (8-4). By Cramer's rule, the fact that $I - C(\lambda)$ and $F(\lambda)$ are analytic in $|\lambda| \le \rho$, and since $\det(I - C(\lambda)) \ne 0$, each component Y_j of Y is analytic in $|\lambda| \le \rho$. This establishes Lemma 8-2.

Theorem 8-4. The set of eigenvalues of $k(x, s)$ is at most countable and each eigenvalue has finite multiplicity. When the set of eigenvalues is infinite, say $\lambda_1, \lambda_2, \ldots, \lambda_n, \ldots$, we have $|\lambda_n| \to \infty$ as $n \to \infty$.

Proof. Fix a number $\rho > 0$. We apply (8-1)–(8-8) with $f = 0$ in (8-1). Then $F = 0$ in (8-5). Now, (8-5) will have a nontrivial solution Y and therefore (8-1) will have a nontrivial solution $y(x)$ given by (8-8) with $f = 0$ exactly when $\det(I - C(\lambda)) = 0$ for some λ with $|\lambda| \leq \rho$. (Notice that y is nontrivial because $Y_j = \langle y, b_j \rangle$.) It is a basic and elementary result of complex analysis that an analytic function in $|\lambda| \leq \rho$ which is not identically zero can vanish at most a finite number of times. In particular, the equation $\det(I - C(\lambda)) = 0$ has at most a finite number of zeros in $|\lambda| \leq \rho$; equivalently, the kernel $k(x, s)$ has at most a finite number of eigenvalues in $|\lambda| \leq \rho$. Let $\rho \to \infty$ to conclude that $k(x, s)$ has at most a countable set of eigenvalues, say $\lambda_1, \lambda_2, \ldots$, and if the set is infinite $|\lambda_n| \to \infty$. This proves all but the finite multiplicity statement.

Suppose that λ_0 is an eigenvalue of $k(x, s)$. Fix $\rho > 0$ so that $|\lambda_0| \leq \rho$. We apply (8-1)–(8-8) again, this time with $\lambda = \lambda_0$ and $f = 0$. Then (8-1) has a nontrivial solution. Suppose, in fact, that $y_1(x), \ldots, y_p(x)$ are linearly independent solutions to $(I - \lambda_0 K)y = 0$. Then $(I - C(\lambda_0))Y = 0$ has solutions

$$Y^{(i)} = [\langle y_i, b_1 \rangle, \ldots, \langle y_i, b_n \rangle]^T$$

for $i = 1, 2, \ldots, p$. Now, if $\sum_1^p \alpha_i Y^{(i)} = 0$, then (8-8) with $\lambda = \lambda_0$ and $f = 0$ yields $\sum_1^p \alpha_i y_i(x) \equiv 0$. Thus, $\alpha_1 = \alpha_2 = \cdots = \alpha_p = 0$ because $y_1(x), \ldots, y_p(x)$ are independent. Consequently, $Y^{(1)}, \ldots, Y^{(p)}$ are independent. Finally, each $Y^{(i)}$ is a vector in n-space, so $p \leq n$, and λ_0 has finite multiplicity.

An interesting alternative proof of the multiplicity result in Th. 8-4, based on orthogonality considerations, is given in Prob. 1. This proof requires $k(x, s)$ to be continuous or square integrable, but does not cover the case of a weakly singular kernel directly.

We come now to the matter of existence of eigenvalues. Our aim is to show that self-adjoint kernels have eigenvalues. Recall that $k(x, s)$ is self-adjoint if $k(x, s) = k^*(x, s)$; equivalently, $K = K^*$, where K^* is the adjoint operator of K.

Lemma 8-3. Let $k(x, s)$ be a self-adjoint kernel and K the corresponding integral operator. Then

$$\langle Kf, g \rangle = \langle f, Kg \rangle,$$

and more generally, K^n is also self-adjoint,

$$\langle K^n f, g \rangle = \langle f, K^n g \rangle$$

for all functions f and g in the domain of K.

Proof. Since $K = K^*$ the general relation $\langle Kf, g \rangle = \langle f, K^* g \rangle$ gives the first conclusion in the lemma. It is evident that K^2 is self-adjoint if K is:

$$\langle K^2 f, g \rangle = \langle K(Kf), g \rangle = \langle Kf, Kg \rangle = \langle f, K^2 g \rangle.$$

In the same way K^n is self-adjoint for $n = 3, 4, \ldots$ and $\langle K^n f, g \rangle = \langle f, K^n g \rangle$, as claimed.

In terms of the iterated kernels $k_n(x, s)$ the conclusions in Lemma 8-3 amount to the observation that $k_n(x, s)$ is self-adjoint whenever $k(x, s)$ is.

The existence of eigenvalues is a delicate matter. It turns out to be helpful to get at the eigenvalues of $k(x, s)$ indirectly through the eigenvalues of the second iterated kernel $k_2(x, s)$. It is easy to confirm (Prob. 6) that $k_2(x, s)$ is continuous, square integrable, or weakly singular whenever $k(x, s)$ is. Thus, all the foregoing results can be applied to $k_2(x, s)$ and its corresponding integral operator K^2. The next lemma shows that $k(x, s)$ has eigenvalues whenever $k_2(x, s)$ does.

Lemma 8-4. If μ is an eigenvalue of $k_2(x, s)$, then $\mu = \lambda^2$ for some eigenvalue λ of $k(x, s)$.

Proof. Suppose that $(I - \mu K^2)z = 0$ for some $z \neq 0$. Then

$$(I + \sqrt{\mu}\, K)(I - \sqrt{\mu}\, K)z = 0.$$

Either $(I - \sqrt{\mu}\, K)z = 0$, in which case $\lambda = \sqrt{\mu}$ is an eigenvalue of $k(x, s)$, or $(I - \sqrt{\mu}\, K)z = y \neq 0$, in which case

$$(I + \sqrt{\mu}\, K)y = 0$$

and $\lambda = -\sqrt{\mu}$ is an eigenvalue of $k(x, s)$. In either case, λ is an eigenvalue of $k(x, s)$ and $\lambda^2 = \mu$.

We need a final technical result:

Lemma 8-5. Let $k(x, s)$ be self-adjoint. Suppose that $k_2(x, s)$ has no eigenvalues in the disk $|\lambda| \leq \rho$, where $\rho > 0$ is fixed. Then

$$\langle Kz, Kz\rangle \leq \frac{1}{\rho}\langle z, z\rangle \tag{8-9}$$

for all functions z in the domain of K.

Proof. Fix a function $z \neq 0$ and define g by the equation

$$z - \rho K^2 z = g. \tag{8-10}$$

We claim that

$$\langle z, g\rangle \geq 0. \tag{8-11}$$

Assume this for the moment. Then (8-10) gives

$$\langle z, z\rangle - \rho\langle z, K^2 z\rangle = \langle z, g\rangle \geq 0,$$

and since K is self-adjoint, this yields

$$\langle Kz, Kz\rangle \leq \frac{1}{\rho}\langle z, z\rangle,$$

which is (8-9). It remains to check (8-11). As we noted before, all the results obtained earlier in this section apply to the kernel $k_2(x, s)$ and its integral operator K^2. In particular, we can replace K by K^2 in (ii) and in (8-1). Thus,

(ii) reads $K^2 = P + R$ and (8-1) becomes $(I - \lambda K^2)y = f$. Equations (8-2)–(8-8) hold as stated. In particular, applying these results when $f = 0$, so that also $F = 0$, if K^2 has no eigenvalues in $|\lambda| \leq \rho$, then (8-5) with $F = 0$ has only the trivial solution; consequently, $\det(I - C(\lambda)) \neq 0$ for $|\lambda| \leq \rho$. For g given by (8-10), Lemma 8-2 asserts that $h(\lambda) = \langle y, g \rangle$ is analytic in $|\lambda| \leq \rho$, where $y = y(x; \lambda)$ is the solution to $(I - \lambda K^2)y = g$. (We apply Lemma 8-2 with f in that lemma equal to g.) Now, for $|\lambda|$ small, the solution y to $(I - \lambda K^2)y = g$ is given by the Neumann series

$$y = \sum_0^\infty \lambda^n K^{2n} g$$

and hence

$$h(\lambda) = \langle y, g \rangle = \left\langle \sum_0^\infty \lambda^n K^{2n} g, g \right\rangle = \sum_0^\infty \lambda^n \langle K^{2n} g, g \rangle.$$

Since K is self-adjoint,

$$h(\lambda) = \sum_0^\infty \langle K^n g, K^n g \rangle \lambda^n.$$

Although this expansion was obtained for small $|\lambda|$, the power series must converge for $|\lambda| \leq \rho$ because $h(\lambda)$ is analytic in this disk. In particular, when $\lambda = \rho$ we have

$$\langle y, g \rangle = \sum_0^\infty \langle K^n g, K^n g \rangle \rho^n \geq 0, \tag{8-12}$$

where y solves

$$(I - \rho K^2)y = g. \tag{8-13}$$

Since ρ is not an eigenvalue of K^2, comparison of (8-10) and (8-13) reveals that $y = z$. Then (8-12) gives $(z, g) \geq 0$, which confirms (8-11) and proves the lemma.

Theorem 8-5. Assume that $k(x, s)$ is self-adjoint and that the corresponding integral operator $K \neq 0$. Then $k(x, s)$ has at least one eigenvalue.

Proof. It suffices (Lemma 8-4) to show that $k_2(x, s)$ has an eigenvalue. Assume that $k_2(x, s)$ has no eigenvalues. Then by Lemma 8-5 $\langle Kz, Kz \rangle \leq \langle z, z \rangle / \rho$ for all functions z in the domain of K and any $\rho > 0$. Let $\rho \to \infty$ to obtain

$$0 = \langle Kz, Kz \rangle = \int_a^b |Kz(x)|^2 \, dx.$$

Thus, $Kz = 0$ in $C[a, b]$ or $L_2(a, b)$, according as $k(x, s)$ is continuous, weakly singular, or square integrable. Since z is an arbitrary element in the domain of K, $K = 0$, the zero operator, which is a contradiction. Therefore, $k_2(x, s)$ and hence $k(x, s)$ has at least one eigenvalue.

Here is another fundamental consequence of Lemma 8-5.

Theorem 8-6. Assume that $k(x, s)$ is self-adjoint and $K \neq 0$. Then $k(x, s)$ has an eigenvalue of smallest modulus, say λ_1, and

$$\frac{1}{\lambda_1^2} = \max \frac{\langle Ky, Ky\rangle}{\langle y, y\rangle},$$

where the maximum is over all $y \neq 0$ in the domain of K.

Proof. We know that $k(x, s)$ has an eigenvalue by Th. 8-5. All its eigenvalues are real (Th. 8-2). Theorem 8-4 guarantees that there is an eigenvalue of smallest modulus because if there are infinitely many eigenvalues their moduli tend to $+\infty$. Let λ_1 be the eigenvalue of minimum modulus. Lemma 8-4 reveals that λ_1^2 is the eigenvalue of $k_2(x, s)$ with minimum modulus. Fix $\rho < \lambda_1^2$ and apply Lemma 8-5 to infer that $\langle Ky, Ky\rangle \leq \langle y, y\rangle/\rho$ for all y in the domain of K. Let $\rho \to \lambda_1^2$ to find $\langle Ky, Ky\rangle \leq \langle y, y\rangle/\lambda_1^2$. Thus,

$$\frac{\langle Ky, Ky\rangle}{\langle y, y\rangle} \leq \frac{1}{\lambda_1^2}$$

for $y \neq 0$. However, if $y = \phi_1$ an eigenfunction corresponding to λ_1, we find equality in the previous inequality. This proves Th. 8-6.

Corollary 8-1. Let $k(x, s)$ be self-adjoint, square integrable, and K: $L_2(a, b) \to L_2(a, b)$ be nonzero. Then

$$\|K\| = \frac{1}{|\lambda_1|}.$$

You are asked to check this in Prob. 8.

A Comment on Terminology

Terminology concerning eigenvalues is not uniform in the literature. Let $k(x, s)$ be a kernel and K be the corresponding integral operator. We have defined the eigenvalues λ and eigenfunctions y of the kernel $k(x, s)$ in the standard way: $\lambda Ky = y$, $y \neq 0$. On the other hand, it is also standard to define the *eigenvalues of the operator* K as those complex numbers μ such that $Ky = \mu y$ for some $y \neq 0$ (see Prob. 11 of Sec. 7-6). Thus, if λ is an eigenvalue of the kernel $k(x, s)$, then $\mu = 1/\lambda$ is an eigenvalue of the operator K. Conversely, if $\mu \neq 0$ is an eigenvalue of the operator K, then $\lambda = 1/\mu$ is an eigenvalue of the kernel $k(x, s)$. The eigenfunctions for λ and μ are clearly the same. This "confusion" in notation is standard.

PROBLEMS

1. Let $k(x, s)$ be continuous or square integrable and λ_0 be an eigenvalue of this kernel. Show that λ_0 has finite multiplicity as follows: Let $y_1, \ldots, y_p$ be orthonormal eigenfunctions. Apply Bessel's inequality to $f(x) = k(x, s)$ using the orthonormal set $y_1(s), \ldots, y_p(s)$. Simplify. Then integrate from a to b.

2. Confirm that the proofs and results of this section go through as is for a kernel $k(x, s)$ defined for x, s in D a domain in n-space. Assume that D is closed and bounded for a continuous or weakly singular kernel.
3. Let λ be an eigenvalue of $k(x, s)$. Then $\bar{\lambda}$ is an eigenvalue of $k^*(x, s)$ and both λ and $\bar{\lambda}$ have the same multiplicity.
4. Let λ [resp., μ] be an eigenvalue of $k(x, s)$ [resp., $k^*(x, s)$] with $\mu \neq \bar{\lambda}$. Show that eigenfunctions corresponding to these eigenvalues are orthogonal.
5. A kernel $k(x, s)$ is *skew Hermitian* if $k^*(x, s) = -k(x, s)$, equivalently, $K^* = -K$. Formulate and prove the analogues of Ths. 8-2, 8-3, and 8-5 for such kernels.
6. Check that $k_2(x, s)$ is continuous, weakly singular, or square integrable when $k(x, s)$ has the same property.
7. If $T = 1$ the Green's function in Prob. 3 of Sec. 7-2 is

$$g(x, s) = \begin{cases} x(1 - s), & 0 \leq x \leq s \leq 1, \\ s(1 - x), & 0 \leq s \leq x \leq 1. \end{cases}$$

   Observe that $g(x, s)$ is symmetric. Find its eigenvalues and eigenfunctions. Compare the results with the general theorems of this section. In particular, use the corollary of Th. 8-6 to compute $\|G\|$, where G is regarded as an integral operator on $L_2(0, 1)$. *Hint.* Use the differential equation $-y'' = \lambda y$ and $y(0) = y(1) = 0$ to find the eigenvalues and eigenfunctions.
8. Prove Cor. 8-1.
9. The Green's function for the Sturm–Liouville problem in the Example of Sec. 7-2 is $g(x, s) = \min(x, s)$ for $0 \leq x, s \leq 1$, when $k = 1$. Find the eigenvalues and eigenfunctions for this kernel.
10. Let $k(x, s)$ be continuous on $a \leq x, s \leq b$ with a and b finite. We can regard K: $C[a, b] \to C[a, b]$ or K: $L_2(a, b) \to L_2(a, b)$.
    (a) In the latter case, show that the range of K is actually contained in $C[a, b]$. *Hint.* Apply the Schwarz inequality to $Ky(x) - Ky(x')$.
    (b) Use part (a) to conclude that the eigenvalues and eigenfunctions are the same for K: $C[a, b] \to C[a, b]$ and K: $L_2(a, b) \to L_2(a, b)$.

7-9. Eigenfunction Expansions

One useful way to represent the solution to an integral equation with a self-adjoint kernel or the solution to a boundary value problem which is equivalent to such an integral equation is to use eigenfunction expansions. The fundamental theorem of Hilbert and Schmidt, which is presented below, justifies this procedure.

In this section $k(x, s)$ is nonzero, self-adjoint, and either continuous, weakly singular, or square integrable on $a \leq x, s \leq b$. As usual, a and b are finite except possibly in the square-integrable case. The eigenvalues of $k(x, s)$ will always be listed as

$$\lambda_1, \lambda_2, \ldots$$

with each eigenvalue listed to multiplicity. Corresponding orthonormal eigenfunctions are denoted by

$$\phi_1(x), \phi_2(x), \ldots.$$

To be definite, we order the eigenvalues by increasing moduli,

$$|\lambda_1| \le |\lambda_2| \le \cdots.$$

To say that the eigenvalues are *listed to multiplicity*, means: If a given complex number z is an eigenvalue of $k(x, s)$ with multiplicity m, then the number z appears m times in the list $\lambda_1, \lambda_2, \ldots$. Thus, the corresponding list of eigenfunctions $\phi_1(x), \phi_2(x), \ldots$ will contain a basis for the eigenspace of each eigenvalue.

First, we show that a (nonzero) self-adjoint kernel has an infinite number of eigenvalues, unless it is degenerate. This will complete the confirmation of all the general properties of eigenvalues mentioned in Sec. 7-8. To this end, we introduce a new kernel $k^{(n)}(x, s)$ defined by

$$k^{(n)}(x, s) = k(x, s) - \sum_{j=1}^{n} \frac{\phi_j(x)\overline{\phi_j(s)}}{\lambda_j}, \tag{9-1}$$

provided that $k(x, s)$ has the indicated eigenvalues an eigenfunctions. This new kernel is self-adjoint because $k(x, s)$ is and each λ_j is real.

Lemma 9-1. Any eigenfunction $\psi(x)$ of $k^{(n)}(x, s)$ is also an eigenfunction of $k(x, s)$. In fact, $\psi = \mu K^{(n)}\psi$ implies that

$$\langle \psi, \phi_r \rangle = 0, \qquad r = 1, 2, \ldots, n, \tag{9-2}$$

and $\psi = \mu K\psi$.

Proof. Let $K^{(n)}$ be the integral operator corresponding to the kernel $k^{(n)}(x, s)$. Then from (9-1),

$$K^{(n)}z = Kz - \sum_{j=1}^{n} \frac{\langle z, \phi_j \rangle \phi_j}{\lambda_j} \tag{9-3}$$

for any function z in the domain of K (which is also the domain of $K^{(n)}$). If

$$\psi = \mu K^{(n)}\psi \tag{9-4}$$

for some constant μ and $\psi \ne 0$, use of (9-3) gives

$$\langle \psi, \phi_r \rangle = \mu \langle K^{(n)}\psi, \phi_r \rangle = \mu \langle K\psi, \phi_r \rangle - \mu \sum_{j=1}^{n} \frac{\langle \psi, \phi_j \rangle \langle \phi_j, \phi_r \rangle}{\lambda_j}$$

$$= \mu \langle \psi, K\phi_r \rangle - \mu \frac{\langle \psi, \phi_r \rangle}{\lambda_r} = \frac{\mu}{\lambda_r} \langle \psi, \phi_r \rangle - \frac{\mu}{\lambda_r} \langle \psi, \phi_r \rangle = 0$$

for $r = 1, 2, \ldots, n$. Thus, (9-2) holds. This orthogonality result and (9-3) yield $K^{(n)}\psi = K\psi$; hence, $\psi = \mu K^{(n)}\psi = \mu K\psi$ and ψ is an eigenfunction of K with corresponding eigenvalue μ. This proves Lemma 9-1.

Theorem 9-1. Let $k(x, s)$ be a self-adjoint kernel which is continuous, weakly singular, or square integrable. Then $k(x, s)$ is degenerate if and only if it has only a finite number of eigenvalues.

Proof. The eigenvalue problem for a degenerate kernel (self-adjoint or not) is equivalent to a matrix problem. Thus, a degenerate kernel has only a finite number of eigenvalues.

Suppose, conversely, that the self-adjoint kernel $k(x, s)$ has only a finite number of eigenvalues, say $\lambda_1, \lambda_2, \dots, \lambda_n$ listed to multiplicity. Form the related kernel $k^{(n)}(x, s)$ of (9-1). We claim that $k^{(n)}(x, s) = 0$, in which case

$$k(x, s) = \sum_{j=1}^{n} \frac{\phi_j(x)\overline{\phi_j(s)}}{\lambda_j} \tag{9-5}$$

and $k(x, s)$ is degenerate. It remains to show that $k^{(n)}(x, s) = 0$. Assume the contrary. Then the self-adjoint kernel $k^{(n)}(x, s)$ has an eigenvalue μ and corresponding eigenfunction ψ. By Lemma 9-1, ψ is also an eigenfunction of $k(x, s)$. Consequently,

$$\psi = \sum_{j=1}^{n} a_j\phi_j$$

because the eigenvalues of $k(x, s)$ are listed to multiplicity. On the other hand, $\langle\psi, \phi_r\rangle = 0$ for $r = 1, 2, \dots, n$ (again by Lemma 9-1), so the preceding equation yields

$$0 = \langle\psi, \phi_r\rangle = \sum_{j=1}^{n} a_j\langle\phi_j, \phi_r\rangle = a_r.$$

Thus, $\psi = 0$, a contradiction. Consequently, $k^{(n)}(x, s) = 0$ and (9-5) holds.

Next, we describe the relationship between the eigenvalues of $k(x, s)$ and $k^{(n)}(x, s)$ when the original kernel is not degenerate.

Proposition 9-1. Let $k(x, s)$ be self-adjoint and not degenerate with eigenvalues $\lambda_1, \lambda_2, \dots, \lambda_n, \dots$, listed to multiplicity and corresponding orthonormal eigenfunctions $\phi_1(x), \phi_2(x), \dots, \phi_n(x), \dots$. Then the self-adjoint kernel $k^{(n)}(x, s)$ has eigenvalues $\lambda_{n+1}, \lambda_{n+2}, \dots$, listed to multiplicity and corresponding orthonormal eigenfunctions $\phi_{n+1}(x), \phi_{n+2}(x), \dots$. Moreover,

$$\lim_{n\to\infty} \langle K^{(n)}g, K^{(n)}g\rangle = 0$$

for every function g in the domain of K.

Proof. Let $r > n$ and apply (9-3) to get $K^{(n)}\phi_r = K\phi_r$ because $\langle\phi_r, \phi_j\rangle = 0$ for $j = 1, 2, \dots, n$. Then

$$\phi_r = \lambda_r K\phi_r = \lambda_r K^{(n)}\phi_r.$$

Thus, $\lambda_{n+1}, \lambda_{n+2}, \dots$ are eigenvalues of $k^{(n)}(x, s)$ and $\phi_{n+1}(x), \phi_{n+2}(x), \dots$ are corresponding orthonormal eigenfunctions. We maintain that this gives all the eigenvalue of $k^{(n)}(x, s)$ to multiplicity. Indeed, if this were not so, $k^{(n)}(x, s)$ would have an eigenfunction ψ *not* in the span of $\phi_{n+1}(x), \phi_{n+2}(x), \dots$. If $\psi = \mu K^{(n)}\psi$, Lemma 9-1 implies that $\psi = \mu K\psi$, so ψ is an eigenfunction of $k(x, s)$. Since the eigenvalues of $k(x, s)$ are listed to multiplicity and ψ is not a linear combination of $\phi_{n+1}(x), \phi_{n+2}(x), \ldots$, it must be a linear combination of $\phi_1(x), \ldots, \phi_n(x)$,

$$\psi = \sum_{j=1}^{n} a_j \phi_j(x).$$

Now, just as before, the orthogonality relations (9-2) yield $a_j = 0$ and $\psi = 0$, a contradiction. Thus, all but the final assertion in the proposition has been proved. The last result follows from Th. 8-6 applied to the self-adjoint kernel $k^{(n)}(x, s)$, which implies that

$$\langle K^{(n)}g, K^{(n)}g\rangle \leq \frac{1}{\lambda_{n+1}^2}\langle g, g\rangle \to 0 \tag{9-6}$$

as $n \to \infty$ because $\lambda_{n+1}^2 \to +\infty$.

The rest of this section is devoted to establishing various forms of the Hilbert–Schmidt theorem. We assume in the proofs that $k(x, s)$ is not degenerate because the desired conclusion is trivial for a degenerate kernel (see Prob. 10).

Theorem 9-2 (Hilbert–Schmidt Theorem). Let $k(x, s)$ be a nonzero, self-adjoint kernel which is continuous, weakly singular, or square integrable on $a \leq x, s \leq b$. (As usual, $a = -\infty$ or $b = +\infty$ is permissible in the square-integrable case.) Each function in the range of the corresponding integral operator K has an eigenfunction expansion which converges in the least-squares (L_2) sense. That is, if $f = Kg$, where g is continuous or square integrable depending on the kernel, then the Fourier series

$$\sum_{n=1}^{\infty} \langle f, \phi_n\rangle \phi_n$$

converges to f in the least-squares sense.

Proof. Let

$$s_n = \sum_{j=1}^{n} \langle f, \phi_j\rangle\phi_j = \sum_{j=1}^{n} \langle Kg, \phi_j\rangle\phi_j = \sum_{j=1}^{n} \langle g, K\phi_j\rangle\phi_j = \sum_{j=1}^{n} \frac{\langle g, \phi_j\rangle\phi_j}{\lambda_j}$$

be the nth partial sum of the Fourier series. Then [compare (9-3)]

$$f - s_n = Kg - \sum_{j=1}^{n} \frac{\langle g, \phi_j\rangle\phi_j}{\lambda_j} = K^{(n)}g,$$

and so

$$\langle f - s_n, f - s_n\rangle = \langle K^{(n)}g, K^{(n)}g\rangle \to 0$$

as $n \to \infty$ by (9-6). That is,

$$\lim_{n\to\infty} \int_a^b \left| f(x) - \sum_{j=1}^{n} \langle f, \phi_j\rangle\phi_j(x)\right|^2 dx = 0, \tag{9-7}$$

which is the asserted least-squares convergence of the Fourier series of f.

Under slightly stronger hypotheses, the Fourier series of $f = Kg$ will converge absolutely and uniformly. In fact, no additional hypotheses are needed when the kernel $k(x, s)$ is continuous. If the kernel is square integrable and also satisfies

(9-8) $$\int_a^b |k(x, s)|^2 \, ds \leq B,$$

a fixed bound, for all x in $[a, b]$, then the Fourier series is absolutely and uniformly convergent. Condition (9-8) holds for a weakly singular kernel $k(x, s) = m(x, s)/|x - s|^r$ when $0 \leq r < \frac{1}{2}$.

To confirm these statements, assume first that $k(x, s)$ is continuous; so $K: C[a, b] \to C[a, b]$. Let $f = Kg$ for g continuous and consider the Fourier series

(9-9) $$\sum_1^\infty \langle f, \phi_j \rangle \phi_j = \sum_1^\infty \frac{\langle g, \phi_j \rangle \phi_j}{\lambda_j}.$$

Fix x in $[a, b]$. The eigenvalue relation

$$\frac{\phi_j(x)}{\lambda_j} = \int_a^b k(x, s)\phi_j(s) \, ds$$

shows that $\phi_j(x)/\lambda_j$ can be interpreted as the jth Fourier coefficient of the function of s, $k(x, s)$, with respect to the orthonormal set $\overline{\phi_1(s)}, \overline{\phi_2(s)}, \ldots$. Consequently, Bessel's inequality gives

(9-10) $$\sum_1^\infty \left| \frac{\phi_j(x)}{\lambda_j} \right|^2 \leq \int_a^b |k(x, s)|^2 \, ds.$$

Also,

(9-11) $$\sum_1^\infty |(g, \phi_j)|^2 \leq \int_a^b |g(s)|^2 \, ds$$

by Bessel's inequality. Apply the Schwarz inequality to deduce that

$$\sum_n^\infty \left| \langle g, \phi_j \rangle \frac{\phi_j(x)}{\lambda_j} \right| \leq \left[\sum_n^\infty |\langle g, \phi_j \rangle|^2 \right]^{1/2} \left[\sum_n^\infty \left| \frac{\phi_j(x)}{\lambda_j} \right|^2 \right]^{1/2}.$$

Then use (9-10) to conclude that

(9-12) $$\sum_n^\infty \left| \langle g, \phi_j \rangle \frac{\phi_j(x)}{\lambda_j} \right| \leq \left[\sum_n^\infty |\langle g, \phi_j \rangle|^2 \right]^{1/2} \left[\int_a^b |k(x, s)|^2 \, ds \right]^{1/2}.$$

Since $k(x, s)$ is continuous, $|k(x, s)| \leq M$ for some M and all $a \leq x, s \leq b$. Then (9-12) gives

$$\sum_n^\infty |\langle f, \phi_j \rangle \phi_j(x)| \leq M(b - a)^{1/2} \left[\sum_n^\infty |\langle g, \phi_j \rangle|^2 \right]^{1/2} \to 0$$

as $n \to \infty$, independent of x in $[a, b]$ by (9-11). Thus, the Fourier series is absolutely and uniformly convergent on $[a, b]$. Since each $\phi_j(x)$ is continuous, the sum of the series is continuous. Because of the uniform convergence, we can take the limit under the integral in (9-7) to get

$$\int_a^b \left| f(x) - \sum_1^\infty \langle f, \phi_j \rangle \phi_j(x) \right|^2 dx = 0. \tag{9-13}$$

Since the integrand is continuous,

$$f(x) - \sum_1^\infty \langle f, \phi_j \rangle \phi_j(x) = 0$$

for all x in $[a, b]$. This proves

Theorem 9-3. Let $k(x, s)$ be nonzero, self-adjoint, and continuous. Let $f = Kg$, where g is continuous. Then the eigenfunction expansion of f converges absolutely and uniformly and

$$f(x) = \sum_{j=1}^\infty \langle f, \phi_j \rangle \phi_j(x).$$

Theorem 9-4. Let $k(x, s)$ be nonzero, self-adjoint, and square integrable. Let $f = Kg$, where g is square integrable. Assume that condition (9-8) holds. Then the eigenfunction expansion of f converges absolutely and uniformly and

$$f(x) = \sum_{j=1}^\infty \langle f, \phi_j \rangle \phi_j(x)$$

for almost all x in $[a, b]$.

Proof. The reasoning leading to (9-12) holds in the square-integrable case. Thus, using (9-8),

$$\sum_n^\infty |\langle f, \phi_j \rangle \phi_j(x)| \le B^{1/2} \left(\sum_n^\infty |\langle g, \phi_j \rangle|^2 \right)^{1/2} \to 0$$

as $n \to \infty$ independent of x in $[a, b]$ by (9-11). As before, $\sum \langle f, \phi_j \rangle \phi_j(x)$ converges absolutely and uniformly on $[a, b]$. On the other hand, a basic result of Lebesgue integration asserts that if a sequence converges in the least-squares sense, a subsequence will converge for almost all x. But (9-7) says that the sequence of partial sums of the Fourier series converges in the least-squares sense to $f(x)$. Thus, a subsequence of these partial sums converges to $f(x)$ for almost all x. But we just showed that the full sequence of partial sums converges; so the full sequence also converges to $f(x)$ for almost all x, which completes the proof of Th. 9-4.

Theorem 9-5. Let $k(x, s)$ be nonzero, self-adjoint, and weakly singular of the form $k(x, s) = m(x, s)/|x - s|^r$ with $m(x, s)$ continuous and $0 \le r < \frac{1}{2}$. Let $f = Kg$ for g continuous. Then the eigenfunction expansion of f converges absolutely and uniformly on $[a, b]$ and

$$f(x) = \sum_{j=1}^\infty \langle f, \phi_j \rangle \phi_j(x).$$

Proof. Let $M = \max |m(x, s)|$ for $0 \le x, s \le b$. Then

$$\int_a^b |k(x, s)|^2 \, ds \le M^2 \int_a^b \frac{ds}{|x - s|^{2r}} \le \frac{2M^2(b-a)^{1-2r}}{1-2r} \equiv B.$$

Consequently, by Th. 9-4, the Fourier series of f is absolutely and uniformly convergent. Since $K: C[a, b] \to C[a, b]$, the eigenfunctions are continuous and, in view of the uniform convergence, so is $\sum \langle f, \phi_j \rangle \phi_j(x)$. Now the reasoning leading to (9-13) holds and we infer that

$$f(x) = \sum_1^\infty \langle f, \phi_j \rangle \phi_j(x),$$

just as before.

Remark. It is worth noting that the proofs of Ths. 9-3, 9-4, and 9-5 show that the series of absolute values

$$\sum_{n=1}^\infty |\langle f, \phi_n \rangle \phi_n(x)|$$

is uniformly convergent on $a \le x \le b$.

Theorem 9-6. Let $k(x, s)$ satisfy the hypotheses of either Th. 9-3, 9-4, or 9-5. Then

$$\sum_{j=1}^\infty \frac{1}{\lambda_j^2} < \infty.$$

Proof. Under these assumptions (9-10) holds and the kernel $k(x, s)$ is square integrable. From (9-10),

$$\sum_{j=1}^n \left| \frac{\phi_j(x)}{\lambda_j} \right|^2 \le \int_a^b |k(x, s)|^2 \, ds.$$

Integration from a to b yields

$$\sum_{j=1}^n \frac{1}{\lambda_j^2} \le \int_a^b \int_a^b |k(x, s)|^2 \, ds \, dx < \infty$$

because ϕ_j has norm 1 and $k(x, s)$ is square integrable. Let $n \to \infty$ to reach the conclusion of the theorem.

We close this section with some remarks about eigenfunction expansions of the kernel $k(x, s)$. Equation (9-5) for a degenerate kernel suggests that we may have

$$k(x, s) = \sum_{j=1}^\infty \frac{\phi_j(x)\overline{\phi_j(s)}}{\lambda_j} \tag{9-14}$$

for a nondegenerate kernel. This always holds in the least-squares sense,

$$\lim_{n \to \infty} \int_a^b \int_a^b \left| k(x, s) - \sum_{j=1}^n \frac{\phi_j(x)\overline{\phi_j(s)}}{\lambda_j} \right|^2 ds \, dx = 0$$

for any continuous, weakly singular (with $r < \frac{1}{2}$), or square-integrable kernel.

In a number of practical applications, for example Sturm–Liouville problems, we know $k(x, s)$ is continuous and *positive semidefinite*, $\langle Kg, g\rangle \geq 0$ for all continuous $g \neq 0$ (see Prob. 7). Then (9-14) holds with absolute and uniform convergence by

Theorem 9-7 (Mercer's Theorem). Let $k(x, s)$ be continuous and positive semidefinite on $a \leq x, s \leq b$. Then the expansion (9-14) holds and the series converges absolutely and uniformly on $a \leq x, s \leq b$.

We omit the proofs of these kernel expansions.

PROBLEMS

1. Confirm that the results of this section hold with only minor changes in reasoning for a kernel $k(x, s)$ defined for x, s in D, a domain in n-space. As usual, D is closed and bounded except in the square-integrable case. In Th. 9-5, $0 \leq r < n/2$.

2. Let D be a bounded domain in 2- or 3-space with smooth bounding curve or surface B. The Green's function $g(x, s)$ for the Poisson equation $\Delta u = f$ in D with Dirichlet boundary condition $u = 0$ on B has the form $g(x, s) = r(x, s) + h(x, s)$, where $h(x, s) = h(s, x)$ is harmonic (i.e., $\Delta h = 0$) in D and continuous in $D + B$ and $r(x, s) = -\ln|x - s|$ or $|x - s|^{-1}$ according as $n = 2$ or 3. The solution to $\Delta u = f$ in D, $u = 0$ on B can be written as $u = Gf$. Show that Th. 9-5 applies so that u has an eigenfunction expansion.

In the following problems, $k(x, s)$ is self-adjoint with eigenvalues $\lambda_1, \lambda_2, \ldots$ listed to multiplicity, $\phi_1, \phi_2, \ldots$ are corresponding orthonormal eigenfunctions, and $|\lambda_1| \leq |\lambda_2| \leq \cdots$. To be definite, assume that $k(x, s)$ is weakly singular with $0 \leq r < n/2$. The results below can also be established for a square-integrable kernel, in which case convergence is in the L_2-sense.

3. Suppose that λ is not an eigenvalue of $k(x, s)$. Then $y = f + \lambda Ky$ has a unique solution. Apply the Hilbert–Schmidt theorem (Th. 9-5) to Ky to find $Ky = \sum \langle y, \phi_j\rangle \phi_j/\lambda_j$. Use this to establish the eigenfunction expansion

$$y(x) = f(x) + \lambda \sum_1^\infty \frac{\langle f, \phi_n\rangle}{\lambda_n - \lambda} \phi_n(x),$$

where the series is absolutely and uniformly convergent on $[a, b]$.

4. Suppose that λ is an eigenvalue of $k(x, s)$. If $\lambda = \lambda_{p+1} = \cdots = \lambda_{p+m}$ and if $y = f + \lambda Ky$ has a solution y, show that $\langle f, \phi_j\rangle = 0$ for $j = p + 1, \ldots, p + m$. (Do not apply the Fredholm alternative.) Conversely, given these orthogonality conditions, show that all solutions to $y = f + \lambda Ky$ are

$$y(x) = f(x) + \lambda \sum_j{}' \frac{\langle f, \phi_j\rangle}{\lambda - \lambda_j} \phi_j(x) + \sum_{p+1}^{p+m} a_j\phi_j(x),$$

where the a_j are arbitrary constants and $\sum'$ indicates a sum over all indices $j \neq p + 1, \ldots, p + m$.

5. Use the Hilbert–Schmidt theorem to show that

$$\langle Kg, g\rangle = \sum_{j=1}^\infty \frac{|\langle g, \phi_j\rangle|^2}{\lambda_j}$$

for any continuous function $g(x)$. Then deduce

(a) $|\langle Kg, g\rangle| \le (1/|\lambda_1|) \sum_j |\langle g, \phi_j\rangle|^2 \le \langle g, g\rangle/|\lambda_1|$;

(b) $\max |\langle Kg, g\rangle|/\langle g, g\rangle = 1/|\lambda_1|$, where the maximum is over all continuous functions g which are not identically zero.

6. Proceeding from Prob. 5, show that

(a) $\langle g, \phi_1\rangle = 0$ implies that $|\langle Kg, g\rangle| \le \langle g, g\rangle/|\lambda_2|$;

(b) $\max_{g \perp \phi_1} |\langle Kg, g\rangle|/\langle g, g\rangle = 1/|\lambda_2|$, where the maximum is over all nonzero continuous functions g orthogonal to ϕ_1.

(c) Characterize $1/|\lambda_n|$ as a maximum.

7. In analogy with the matrix case, the self-adjoint, integral operator $K: C[a, b] \to C[a, b]$ is called *positive definite* (resp., *positive semidefinite*) if $\langle Kg, g\rangle > 0$ (resp., $\langle Kg, g\rangle \ge 0$) for all nonzero continuous functions g. Show that K is positive definite (resp., positive semidefinite) if and only if all its eigenvalues are positive (resp., nonnegative). *Hint.* Expanded $\langle Kg, g\rangle$ as in Prob. 5.

8. Refer to Prob. 7 of Sec. 7-8. If the string has a given continuous, transverse load $f(x)$, its deflection y satisfies

$$y(x) = \int_0^1 g(x, s) f(s)\, ds.$$

Find explicitly the eigenfunction expansion provided by the Hilbert–Schmidt theorem.

9. The Green's function for the steady-state heat conduction problem described in the Example of Sec. 7-2 where $k = 1$ is $g(x, s) = \min(x, s)$ for $0 \le x, s \le 1$. If the initial temperature in the rod is given by the continous function $f(x)$, the steady-state temperature is

$$y(x) = \int_0^1 \min(x, s) f(s)\, ds.$$

Find explicitly the eigenfunction expansion provided by the Hilbert–Schmidt theorem.

10. Let $k(x, s)$ be a degenerate kernel as in (9-5) and let K be the corresponding integral operator. If $f = Kg$, show that

$$f(x) = \sum_{j=1}^{n} \langle f, \phi_j\rangle \phi_j(x).$$

7-10. Applications of Eigenfunction Expansions

In Secs. 7-1 and 7-2 we introduced Sturm–Liouville problems and explained their role in the separation-of-variables technique. Recall that the Sturm–Liouville eigenvalue problem is

$$\begin{cases} -(p(x)y')' + [q(x) - \lambda r(x)]y = 0, & 0 < x < 1, \\ y(0) - h_0 y'(0) = 0, & y(1) + h_1 y'(1) = 0 \end{cases} \tag{10-1}$$

We assume that p, p', q, r are continuous and make the further physically realistic assumptions

$$p(x) > 0, \qquad r(x) > 0, \qquad q(x) \ge 0, \qquad h_0, h_1 \ge 0. \tag{10-2}$$

With these assumptions, Th. 2-3 ensures that there is a symmetric, continuous Green's function $g(x, s)$ such that (10-1) is equivalent to the eigenvalue problem

$$y(x) = \lambda \int_0^1 g(x, s)r(s)y(s)\, ds. \tag{10-3}$$

for the kernel $g(x, s)r(s)$ [see (2-15)]. The change of variables $z(x) = \sqrt{r(x)}\,y(x)$ transforms (10-3) into the equivalent eigenvalue problem

$$z(x) = \lambda \int_0^1 k(x, s)z(s)\, ds \tag{10-4}$$

for the continuous, symmetric kernel

$$k(x, s) = g(x, s)\sqrt{r(x)r(s)}.$$

[Compare with (2-15)–(2-17).] The associated self-adjoint integral operator is

$$K: C[a, b] \to C[a, b].$$

The following results answer Basic Problems 1 and 2 formulated in Sec. 7-1.

Theorem 10-1. Let $p(x)$, $q(x)$, $r(x)$, h_0, and h_1 satisfy (10-2) with p, p', q, and r continuous on $[0, 1]$. Then the Sturm-Liouville problem (10-1) has an infinite number of eigenvalues $\lambda_1, \lambda_2, \ldots$ and corresponding eigenfunctions $y_1(x)$, $y_2(x), \ldots$. These eigenvalues are real and the eigenfunctions can be chosen real and orthonormal with weight function $r(x)$, that is,

$$\int_0^1 y_m(x)y_n(x)r(x)\, dx = \begin{cases} 1, & m = n, \\ 0, & m \neq n. \end{cases}$$

Furthermore, if $f(x)$ is any function with a continuous second derivative on $[0, 1]$ and which satisfies the boundary conditions in (10-1), then $f(x)$ has the absolutely and uniformly convergent eigenfunction expansion

$$f(x) = \sum_{n=1}^{\infty} c_n y_n(x),$$

where

$$c_n = \int_0^1 f(x)y_n(x)r(x)\, dx.$$

Proof. The Green's function $g(x, s) \not\equiv 0$. See Prob. 3 for a mathematical discussion of this physically obvious fact. Consequently, $k(x, s)$ is continuous, symmetric, and not identically zero. Thus, (10-4) has a nonempty but possibly finite set of eigenvalues $\lambda_1, \lambda_2, \ldots$ and corresponding eigenfunctions $z_1(x), z_2(x), \ldots$. These eigenvalues must be real (Th. 8-2) and the eigenfunctions can be chosen real and orthonormal (Th. 8-3). Thus,

$$\int_0^1 z_m(x)z_n(x)\, dx = \begin{cases} 1, & m = n, \\ 0, & m \neq n. \end{cases} \tag{10-5}$$

Therefore, the eigenvalue problem (10-3), equivalently the Sturm–Liouville problem (10-1), has only the real eigenvalues $\lambda_1, \lambda_2, \ldots$ and corresponding real eigenfunctions $y_1(x) = z_1(x)/\sqrt{r(x)}$, $y_2(x) = z_2(x)/\sqrt{r(x)}, \ldots$. In view of (10-5), these eigenfunctions satisfy the orthogonality relations

$$\int_0^1 y_m(x)y_n(x)r(x)\,dx = \begin{cases} 1, & m = n, \\ 0, & m \neq n. \end{cases} \tag{10-6}$$

Since the integral operator $K: C[0, 1] \to C[0, 1]$ has a continuous, symmetric kernel, the Hilbert–Schmidt theorem (Th. 9-2) implies that each function in the range of K has an absolutely and uniformly convergent expansion in terms of the eigenfunctions $z_1(x), z_2(x), \ldots$. Let $f(x)$ have a continuous second derivative on $[0, 1]$ and satisfy the boundary conditions in (10-1). Define

$$h(x) = -(p(x)f'(x))' + q(x)f(x).$$

Then $y = f(x)$ solves the boundary value problem

$$-(p(x)y')' + q(x)y = h(x), \qquad 0 < x < 1,$$
$$y(0) - h_0 y'(0) = 0, \qquad y(1) + h_0 y(1) = 0,$$

and consequently using the Green's function [see (2-1) and (2-2)],

$$y = f(x) = \int_0^1 g(x, s)h(s)\,ds$$

or

$$\tilde{f}(x) = \int_0^1 k(x, s)\left(\frac{h(s)}{\sqrt{r(s)}}\right) ds,$$

where $\tilde{f}(x) = \sqrt{r(x)}\,f(x)$. Apply the Hilbert–Schmidt theorem to $\tilde{f}(x)$ to get

$$\sqrt{r(x)}\,f(x) = \sum \langle \tilde{f}, z_n \rangle z_n(x)$$

with absolute and uniform convergence on $[0, 1]$. Since $r(x) > 0$, $1/\sqrt{r(x)}$ is continuous on $[0, 1]$ and

$$f(x) = \sum \langle \tilde{f}, z_n \rangle \frac{z_n(x)}{\sqrt{r(x)}} = \sum \langle \tilde{f}, z_n \rangle y_n(x)$$

with absolute and uniform convergence. Now

$$\langle \tilde{f}, z_n \rangle = \int_0^1 f(x)\sqrt{r(x)}\sqrt{r(x)}\,y_n(x)\,dx = \int_0^1 f(x)y_n(x)r(x)\,dx.$$

Thus,

$$f(x) = \sum c_n y_n(x) \quad \text{and} \quad c_n = \int_0^1 f(x)y_n(x)r(x)\,dx,$$

as asserted.

At this point, it is still possible that $\lambda_1, \lambda_2, \ldots$ and $y_1(x), y_2(x), \ldots$ are finite sets. If these sets are finite, then, by the expansion theorem just proven, there are only a finite number of linearly independent functions $f(x)$ with a continuous second derivative and which satisfy the boundary conditions in (10-1). (Indeed, all such functions are linear combinations of the finite number of eigenfunctions.) However, it is obvious that there are infinitely many linearly independent functions of the type just described. For example, the polynomials

$$P_n(x) = x^{2n}(1-x)^{2n}$$

for $n = 1, 2, \ldots$ have the required properties. Hence, there must be infinitely many eigenvalues and eigenfunctions. This completes the proof of Th. 10-1.

Here are two important additional features of the Sturm–Liouville eigenvalue problem which do not follow from general properties of integral equations.

Theorem 10-2. Under the assumptions in Th. 10-1, all the eigenvalues of the Sturm–Liouville problem (10-1) are positive and each eigenvalue is simple—has multiplicity 1.

Proof. Multiply the eigenvalue relation

$$\lambda_n r_n(x) y_n(x) = -(p(x)y_n')' + q(x)y_n$$

by $y_n(x)$, integrate from 0 to 1, and use integration by parts and the boundary conditions to get

$$\text{(10-7)} \quad \lambda_n = p(0)h_0 y_n'(0)^2 + p(1)h_1 y_n'(1)^2 + \int_0^1 [p(x)y_n'(x)^2 + q(x)y_n(x)^2]\,dx.$$

Evidently, $\lambda_n \geq 0$ and $\lambda_n > 0$, except possibly when $y_n'(x) \equiv 0$. If $y_n'(x) \equiv 0$, then $y_n(x) = y_n(0) = y_n(0) - h_0 y'(0) = 0$, which contradicts $y_n(x) \not\equiv 0$. Thus, $y_n'(x) \not\equiv 0$ and $\lambda_n > 0$ for all n.

If $y_n(x)$ and $\tilde{y}_n(x)$ are eigenfunctions belonging to λ_n, their Wronskian at $x = 0$ is

$$\begin{vmatrix} y_n(0) & \tilde{y}_n(0) \\ y_n'(0) & \tilde{y}_n'(0) \end{vmatrix} = \begin{vmatrix} h_0 y_n'(0) & h_0 \tilde{y}_n'(0) \\ y_n'(0) & \tilde{y}_n'(0) \end{vmatrix} = 0.$$

Thus, $y_n(x)$ and $\tilde{y}_n(x)$ are linearly dependent and the eigenvalue λ_n is simple.

Heat Conduction in a Rod

The developments above put us in the position to justify the formal, separation-of-variables solution to the heat flow problem (1-1)–(1-3). In the notation of Sec. 7-1, we have shown that the Sturm–Liouville problem (1-4) has an infinite number of *positive, simple* eigenvalues $\lambda_1, \lambda_2, \ldots$ and corresponding eigenfunctions $X_1(x), X_2(x), \ldots$, which are orthonormal with weight function

$$r(x) = \rho(x)c(x) > 0.$$

Each of the functions

$$u_n(x, t) = e^{-\lambda_n t} X_n(x)$$

satisfies the heat equation (1-1) and the homogeneous boundary conditions (1-2). At this point, notice that (1-2) and (1-3) suggest the natural compatibility condition that the initial temperature distribution $f(x)$ in the rod should satisfy the boundary conditions of the Sturm–Liouville problem (1-4). We assume this and also the added smoothness condition that $f(x)$ has a continuous second derivative. Then $f(x)$ will have the absolutely and uniformly convergent eigenfunction expansion

$$f(x) = \sum_{n=1}^{\infty} c_n X_n(x),$$

where

$$c_n = \int_0^1 f(x) X_n(x) r(x)\, dx. \tag{10-8}$$

With this choice of constants,

$$u(x, t) = \sum_{n=1}^{\infty} c_n e^{-\lambda_n t} X_n(x) \tag{10-9}$$

is absolutely and uniformly convergent in the strip $t \geq 0$ and $0 \leq x \leq 1$ (note that $e^{-\lambda_n t} < 1$) and satisfies the boundary conditions (1-2) and initial condition (1-3). It will also satisfy the heat equation (1-1) if the appropriate term-by-term differentiations are permissible for $t > 0$ and $0 < x < 1$.

We justify term-by-term differentiation once with respect to t and twice with respect to x by showing that resulting series,

$$\sum_n c_n(-\lambda_n) e^{-\lambda_n t} X_n(x), \quad \sum_1 c_n e^{-\lambda_n t} X_n'(x), \quad \sum_1 c_n e^{-\lambda_n t} X_n''(x), \tag{10-10}$$

are uniformly convergent for $t \geq t_0$ and $0 \leq x \leq 1$, where $t_0 > 0$ is arbitrary. To this end we show there are constants, A, A', and A'', such that

$$|X_n(x)| \leq A\lambda_n, \qquad |X_n'(x)| \leq A'\lambda_n, \quad \text{and} \quad |X_n''(x)| \leq A''\lambda_n^2. \tag{10-11}$$

Assume that these estimates hold. Then the terms in the series in (10-10) can be bounded by

$$|c_n \lambda_n e^{-\lambda_n t} X_n(x)| \leq A|c_n| \lambda_n^2 e^{-\lambda_n t_0},$$
$$|c_n e^{-\lambda_n t} X_n'(x)| \leq A'|c_n| \lambda_n e^{-\lambda_n t_0},$$
$$|c_n e^{-\lambda_n t} X_n''(x)| \leq A''|c_n| \lambda_n^2 e^{-\lambda_n t_0}$$

for $t \geq t_0$ and $0 \leq x \leq 1$. The Fourier coefficients c_n are bounded (in fact, $c_n \to 0$), $0 < \lambda_n \to +\infty$, and $\sum 1/\lambda_n^2 < \infty$ (Th. 9-6), so the numerical series

$$\sum |c_n| \lambda_n^2 e^{-\lambda_n t_0}, \quad \text{and hence} \quad \sum |c_n| \lambda_n e^{-\lambda_n t_0},$$

converges. See Prob. 6. Thus, the Weierstrass M-test implies the absolute and uniform convergence of the series in (10-10). This justifies differentiating (10-9) term by term once with respect to t and twice with respect to x for $t > 0$ and $0 \leq x \leq 1$ because $t_0 > 0$ was arbitrary. Therefore, (10-9) satisfies the heat equation (1-1).

It remains to establish the bounds in (10-11). In the present notation (10-3) reads

$$X_n(x) = \lambda_n \int_0^1 g(x, s) r(s) X_n(s)\, ds. \tag{10-12}$$

By the Schwarz inequality,

$$|X_n(x)| \leq \lambda_n \left[\int_0^1 g(x, s)^2 r(s)\, ds \right]^{1/2} \left[\int_0^1 X_n(s)^2 r(s)\, ds \right]^{1/2}.$$

Since $X_n(x)$ has norm 1 relative to the weight function $r(s)$, we have

$$|X_n(x)| \leq A\lambda_n \quad \text{with} \quad A = \max_{0 \leq x \leq 1} \left[\int_0^1 g(x, s)^2 r(s)\, ds \right]^{1/2},$$

which is the first estimate in (10-11). Next, from (10-12),

$$X_n'(x) = \lambda_n \int_0^1 \frac{\partial g}{\partial x}(x, s) r(s) X_n(s)\, ds$$

and another application of the Schwarz inequality gives

$$|X_n'(x)| \leq A'\lambda_n \quad \text{with} \quad A' = \max_{0 \leq x \leq 1} \left[\int_0^1 \frac{\partial g}{\partial x}(x, s)^2 r(s)\, ds \right]^{1/2},$$

which is the second estimate in (10-11). Finally, the differential equation in (1-4) yields

$$X_n''(x) = -\frac{k'(x)}{k(x)} X_n'(x) + \left[\frac{l(x)}{k(x)} - \frac{r(x)}{k(x)} \lambda_n \right] X_n(x)$$

where the coefficients are continuous and the thermal conductivity $k(x) > 0$. In view of the first two estimates in (10-11), we infer that $|X''(x)| \leq A''\lambda_n^2$ for some constant A''. Thus, (10-11) is confirmed. In summary, we have established

Theorem 10-3. Assume that the density $\rho(x)$ and the thermal coefficients $c(x)$, $k(x)$, $k'(x)$, and $l(x)$ are continuous on $0 \leq x \leq 1$ with $\rho(x), c(x), k(x) > 0$, and $l(x) \geq 0$. Let $h_0, h_1 \geq 0$ and $f(x)$, the initial temperature distribution, have a continuous second derivative and satisfy the boundary conditions (1-2) imposed on the rod. Then the initial, boundary value problem (1-1)–(1-3) for the heat equation has solution

$$u(x, t) = \sum_{n=1}^{\infty} c_n e^{-\lambda_n t} X_n(x)$$

expressed in terms of the eigenvalues and eigenfunctions of the Sturm–Liouville problem (1-4) and with

$$c_n = \int_0^1 f(x)X_n(x)r(x)\,dx,$$

where $r(x) = \rho(x)c(x)$.

Waves on a String

Consider the small transverse vibrations of a flexible string. If the string is inhomogeneous and has fixed ends, the motion is governed by the initial, boundary value problem

$$\begin{cases} u_{tt} = c^2(k(x)u_x)_x, & t > 0, \quad 0 < x < 1, \\ u(0, t) = 0, \quad u(1, t) = 0, & t \geq 0, \\ u(x, 0) = f(x), \quad u_t(x, 0) = g(x), & 0 \leq x \leq 1, \end{cases} \tag{10-13}$$

where $f(x)$ and $g(x)$ specify the initial position and velocity of the string and $k(x) > 0$. Applying separation of variables (see Prob. 5 of Sec. 7-1), we find the formal solution

$$u(x, t) = \sum_{n=1}^{\infty} \left(a_n \cos c\sqrt{\lambda_n}\,t + \frac{b_n}{c\sqrt{\lambda_n}} \sin c\sqrt{\lambda_n}\,t \right) X_n(x), \tag{10-14}$$

$$a_n = \int_0^1 f(x)X_n(x)\,dx, \qquad b_n = \int_0^1 g(x)X_n(x)\,dx, \tag{10-15}$$

where the eigenvalues λ_n and eigenfunctions $X_n(x)$ are determined by the Sturm–Liouville problem

$$\begin{cases} LX - \lambda X = 0, & 0 < x < 1, \\ X(0) = 0, & X(1) = 0. \end{cases} \tag{10-16}$$

Here L is the Sturm–Liouville operator given by

$$Ly = -(ky')'. \tag{10-17}$$

The coefficient of $\sin(c\sqrt{\lambda_n}t)$ in (10-14) is written as $b_n/c\sqrt{\lambda_n}$ instead of just b_n because this leads to the symmetrical formulas in (10-15).

The Sturm–Liouville problem (10-16) is (10-1) with the choices $p(x) = k(x)$, $q(x) \equiv 0$, $r(x) \equiv 1$, and $h_0 = h_1 = 0$. By Ths. 10-1 and 10-2 the Sturm–Liouville problem (10-16) has an infinite sequence of positive, simple eigenvalues $\lambda_1, \lambda_2, \ldots$ and corresponding real-valued orthonormal eigenfunctions $X_1(x), X_2(x), \ldots$. So the formal solution (10-14) can be constructed. It obviously satisfies the boundary condition in (10-13). We also have $u(x, 0) = f(x)$ provided that $f(x)$ has a continuous second derivative and satisfies $f(0) = f(1) = 0$, again by Th. 10-1. Notice that these boundary conditions on $f(x)$ are natural compatibility conditions required by the boundary and initial data in (10-13). Similarly, if (10-14) can be differentiated term by term with respect to t, then $u_t(x, 0) = g(x)$ provided that

$g(x)$ has a continuous second derivative and $g(0) = g(1) = 0$. Finally, (10-14) will satisfy the wave equation if we can justify term-by-term differentiation twice with respect to both x and t. We justify this in the standard way by showing that the series obtained from term-by-term differentiation are uniformly convergent. To show this, we need to impose further smoothness restrictions on the initial data $f(x)$ and $g(x)$. (Compare with Sec. 4-2.) Here we assume that both $f(x)$ and $g(x)$ have a continuous fourth derivative and satisfy the boundary conditions

$$f(0) = f(1) = 0, \qquad g(0) = g(1) = 0$$
$$(kf')'(0) = (kf')'(1) = 0, \qquad (kg')'(0) = (kg')'(1) = 0.$$

Notice that (10-14) can be regarded as the sum of two solutions to (10-13). The series of terms with coefficients a_n [resp., b_n] corresponds to the initial data $f(x)$ arbitrary, $g(x) \equiv 0$ [resp., $f(x) \equiv 0$ and $g(x)$ arbitrary]. We treat the case $f(x)$ arbitrary and $g(x) \equiv 0$. The entirely similar reasoning for the other case is left for Prob. 9. With $g(x) \equiv 0$, (10-14) becomes

$$u(x, t) = \sum_{n=1}^{\infty} a_n X_n(x) \cos(c\sqrt{\lambda_n}\, t). \tag{10-18}$$

Term-by-term differentiation twice with respect to x and t produces the four series

$$\sum_1^{\infty} a_n X_n'(x) \cos(c\sqrt{\lambda_n}\, t), \qquad \sum_1^{\infty} a_n X_n''(x) \cos(c\sqrt{\lambda_n}\, t), \tag{10-19}$$

$$-\sum_1^{\infty} c a_n \sqrt{\lambda_n}\, X_n(x) \sin(c\sqrt{\lambda_n}\, t), \qquad -\sum_1^{\infty} c^2 a_n \lambda_n X_n(x) \cos(c\sqrt{\lambda_n}\, t). \tag{10-20}$$

We develop estimates which show that these series are uniformly convergent. First, just as for the heat equation,

$$|X_n(x)| \le A\lambda_n, \qquad |X_n'(x)| \le A'\lambda_n$$

for constants A and A' independent of n. Second,

$$a_n = \int_0^1 f(x) X_n(x)\, dx = \int_0^1 f(x) \frac{1}{\lambda_n} L X_n(x)\, dx.$$

Integration by parts twice and use of $f(0) = f(1) = 0$ yield

$$a_n = \frac{1}{\lambda_n} \int_0^1 (Lf)(x) X_n(x)\, dx = \frac{1}{\lambda_n} \langle Lf, X_n \rangle \tag{10-21}$$

(see Prob. 7). Since Lf also satisfies $Lf(0) = Lf(1) = 0$, integration by parts exactly as above gives

$$a_n = \frac{1}{\lambda_n^2} \int_0^1 (L^2 f)(x) X_n(x)\, dx = \frac{1}{\lambda_n^2} \langle L^2 f, X_n \rangle. \tag{10-22}$$

We can bound the terms in series (10-18) by

$$|a_n X_n(x) \cos(c\sqrt{\lambda_n}\, t)| \le |a_n| A\lambda_n \le \frac{|A\langle L^2 f, X_n\rangle|}{\lambda_n}.$$

Similarly, the terms of the first series in (10-19) are bounded by

$$|a_n X_n'(x) \cos (c\sqrt{\lambda_n}\, t)| \le A' \frac{|\langle L^2 f, X_n \rangle|}{\lambda_n}$$

The Schwarz inequality gives

$$\sum_1^\infty \langle L^2 f, X_n \rangle \frac{1}{\lambda_n} \le \left(\sum_1^\infty \langle L^2 f, X_n \rangle^2 \right)^{1/2} \left(\sum_1^\infty \frac{1}{\lambda_n^2} \right)^{1/2} < \infty$$

by Bessel's inequality and Th. 9-6. The M-test implies that (10-18) and the first series in (10-19) are absolutely and uniformly convergent.

The same is true for the second series in (10-19), but the reasoning is more delicate. Since

$$X_n'' = -\frac{k'}{k} X_n' - \frac{\lambda_n}{k} X_n$$

the second series in (10-19) can be expressed as

$$\text{(10-23)} \qquad -\sum_1^\infty \frac{k'(x)}{k(x)} a_n X_n'(x) \cos (c\sqrt{\lambda_n}\, t) - \sum_1^\infty \frac{\lambda_n}{k(x)} a_n X_n(x) \cos (c\sqrt{\lambda_n}\, t).$$

Since $k'(x)/k(x)$ is continuous and $k(x) > 0$, the absolute and uniform convergence of the first series in (10-23) follows just as for the first series in (10-19). Next, use (10-21) to write the second series in (10-23) as

$$\text{(10-24)} \qquad -\sum_1^\infty \frac{1}{k(x)} \langle Lf, X_n \rangle X_n(x) \cos (c\sqrt{\lambda_n}\, t).$$

The terms in this series are bounded by a constant times the terms of the series

$$\text{(10-25)} \qquad \sum_1^\infty |\langle Lf, X_n \rangle X_n(x)|,$$

which converges uniformly by the Hilbert–Schmidt theorem (see the Remark following Th. 9-5) because Lf is in the range of G, the integral operator corresponding to the Green's function associated with the Sturm–Liouville problem (10-16). To see this, define $\tilde{f} = L^2 f = L(Lf)$. By assumption, Lf satisfies the boundary data $Lf(0) = Lf(1) = 0$. Thus, $y = Lf$ solves $Ly = \tilde{f}$, $y(0) = y(1) = 0$; hence, $Lf = y = G\tilde{f}$ and Lf is in the range of G, as asserted. The uniform convergence in (10-25) implies the absolute and uniform convergence of the series in (10-24). Therefore, (10-23), that is, the second series in (10-19), is absolutely and uniformly convergent.

Finally, the series in (10-20) can be expressed as

$$-\sum_1^\infty \frac{c}{\sqrt{\lambda_n}} \langle Lf, X_n \rangle X_n(x) \sin (c\sqrt{\lambda_n}\, t), \qquad -\sum_1^\infty c^2 \langle Lf, X_n \rangle X_n(x) \cos (c\sqrt{\lambda_n}\, t).$$

The uniform convergence of (10-25) immediately implies the absolute and uniform convergence of these two series.

In review, all the series in (10-18)–(10-20) are absolutely and uniformly convergent. The required term-by-term differentiations are justified and we have proven

Theorem 10-4. Assume that $k(x) > 0$ and $k'(x)$ are continuous on $0 \leq x \leq 1$, that the initial data $f(x)$ and $g(x)$ have continuous fourth derivatives, and that

$$f(0) = f(1) = 0, \qquad (kf')'(0) = (kf')'(1) = 0,$$
$$g(0) = g(1) = 0, \qquad (kg')'(0) = (kg')'(1) = 0.$$

Then the initial, boundary value problem (10-13) for the wave equation has solution

$$u(x, t) = \sum_{n=1}^{\infty} \left[a_n \cos (c\sqrt{\lambda_n}\, t) + \frac{b_n}{c\sqrt{\lambda_n}} \sin (c\sqrt{\lambda_n}\, t) \right] X_n(x)$$

expressed in terms of the eigenvalues and eigenfunctions of the Sturm–Liouville problem (10-16) and with

$$a_n = \int_0^1 f(x)X_n(x)\, dx \quad \text{and} \quad b_n = \int_0^1 g(x)X_n(x)\, dx.$$

PROBLEMS

1. Let $p(x) = r(x) = 1$, $q(x) = 0$, and $h_0 = h_1 = 0$ in (10-1). What are the eigenvalues and eigenfunctions? What familiar result does the eigenfunction expansion of Th. 10-1 yield?
2. Refer to Th. 2-1 and consider the Sturm–Liouville operator $Ly \equiv -(p(x)y')' + q(x)y$ with the boundary conditions given in (10-1). Show that this operator has a Green's function if and only if 0 is not an eigenvalue of (10-1). Then show that the Green's function will exist if $p(x) > 0$, $q(x) \geq 0$, $h_0, h_1 \geq 0$, or if all these inequalities are reversed. *Hint.* (10-7).
3. Assume that there is a Green's function $g(x, s)$ for (10-1). Show $g(x, s) \not\equiv 0$. *Hint.* Consider (2-1) and (2-2). Let $f(x) = 1$ in (2-1) and show that $y(x) \not\equiv 0$. What does (2-2) say if $g(x, s) \equiv 0$? Alternatively, apply (iii) of Th. 2-2.
4. Consider the Sturm–Liouville operator Ly of Prob. 2 but this time with the periodic boundary conditions $y(0) = y(1)$, $y'(0) = y'(1)$. (See Prob. 6 of Secs. 7-1 and 7-2.) The eigenvalue problem is $Ly - \lambda ry = 0$, $y(0) = y(1)$, $y'(0) = y'(1)$.
   (a) State and prove an analogue of Th. 10-1 for this case.
   (b) State and prove an analogue of Th. 10-2 for this case.
   (c) Your result in part (a) should include the case $p(x) = q(x) = r(x) = 1$. What familiar expansion do you get from part (a)?
5. Check the third estimate, $|X_n''(x)| \leq A''\lambda_n^2$, in (10-11).
6. If $\{c_n\}$ is a bounded sequence, $0 < \lambda_n \to +\infty$, $\sum 1/\lambda_n^2 < \infty$, and $t_0 > 0$, then $\sum |c_n|\lambda_n^2 e^{-\lambda_n t_0}$ converges. *Hint.*

$$\sum |c_n|\lambda_n^2 e^{-\lambda_n t_0} \leq B \sum e^{-\lambda_n t_0/2}$$

for some constant B. Now show that $\sum e^{-\lambda_n t_0/2} < \infty$ using $x^2 e^{-x} \leq 1$ for $x \geq 0$.

7. Verify (10-21) and (10-22).

8. Show that the boundary conditions imposed on $f(x)$ and $g(x)$ in Th. 10-4 are all natural compatibility requirements if we ask that the wave equation holds for $t \geq 0$ and $0 \leq x \leq 1$.

9. Complete the proof of Th. 10-4 by treating the case $f(x) \equiv 0$ and $g(x)$ arbitrary, in which case the formal solution (10-14) is

$$u(x, t) = \sum_{n=1}^{\infty} \frac{b_n}{c\sqrt{\lambda_n}} X_n(x) \sin (c\sqrt{\lambda_n}\, t).$$

10. Replace the wave equation in (10-13) by the more general form $r(x)u_{tt} = (k(x)u_x)_x$, where the constant c^2 is absorbed into $k(x)$ for simplicity and $r(x)$ is continuous with $r(x) > 0$. Establish the analogue of Th. 10-4 in this context.

8

Potential Theory

8-1. The Boundary Value Problems of Potential Theory

In equation (3-8) of Chap. 1 we found that the temperature u in a homogeneous, isotropic medium satisfies the equation

$$u_t = a\,\Delta u + F,$$

where $a > 0$, and F describes heat generation in the medium. If the heat flow is independent of time, u satisfies the *Poisson equation*

$$\Delta u = h, \tag{1-1}$$

where $h = -F/a$. If there are no internal heat sources or sinks, then $h = 0$ and u satisfies the *Laplace equation*

$$\Delta u = 0. \tag{1-2}$$

Equation (1-1) is also called the inhomogeneous Laplace equation. Solutions to (1-2) with continuous second derivatives are called *harmonic functions.*

Equations (1-1) and (1-2) govern the distribution of heat inside the medium in the time-independent case. Evidently, we must also specify the nature of heat flow at the boundary of the medium to obtain a complete description of the physical situation. To be definite, suppose that the heat-conducting medium is a bounded domain D in 3-space and let B be its boundary. (A domain is a nonempty connected, open set.) Let $x = (x_1, x_2, x_3)$ be a typical point in 3-space. One common way to control the heat flow at B is to maintain a constant temperature there. Then the time-independent heat flow is governed by either (1-1) or (1-2) and the Dirichlet boundary condition

$$u = f(x), \qquad x \in B, \qquad f \text{ given.} \tag{1-3}$$

The resulting problem is called the *Dirichlet problem* or *first boundary value problem* of potential theory for D.

A second way to control heat flow at B is to prescribe the heat flux there: Let $\nu = \nu(x)$ be the exterior unit normal at a point x of B and q be the heat flow vector (see Sec. 1-3). Then $q \cdot (-\nu)$ gives the rate of heat flow into D across B. By Fourier's law, $q \cdot (-\nu) = k \nabla u \cdot \nu = k\, \partial u/\partial \nu$. In this case we wish to solve (1-1) or (1-2) subject to the Neumann boundary condition

$$\text{(1-4)} \qquad \nabla u \cdot v = \frac{\partial u}{\partial \nu} = f(x), \qquad x \in B, \qquad f \text{ given.}$$

This problem is called the *Neumann* or *second boundary value problem.* Since $q \cdot \nu$ is the component of heat flux in the direction ν, no heat will cross B when $q \cdot \nu = 0$. We say that the body is insulated and, in this case, $f = 0$ in (1-4).

A third possibility is to describe heat radiation at the boundary B. Suppose that the portion of space surrounding D is maintained at a prescribed temperature U. Then $u - U$ measured on B is the temperature difference between the body and its surroundings. We assume that the rate of heat flow across B is proportional to this temperature difference,

$$\text{(1-5)} \qquad q \cdot \nu = \gamma(u - U),$$

where $\gamma > 0$ is the proportionality coefficient [which may vary over B (see Prob. 5)]. Since $q = -k \nabla u$, we are led to solving (1-1) or (1-2) subject to the boundary condition

$$\text{(1-6)} \qquad \frac{\partial u}{\partial \nu} + \alpha(x)u = \beta(x), \qquad x \in B,$$

where $\alpha = \gamma/k > 0$ and $\beta = \gamma U/k$. The resulting problem is called the *Robin* or *third boundary value problem.*

The law of radiation (1-5) is called *Newton's law of cooling.* A more accurate law is the nonlinear *Stefan's law,*

$$\text{(1-7)} \qquad q \cdot \nu = \mu(u^4 - U^4),$$

where temperatures are expressed in absolute units. Since

$$u^4 - U^4 = (u^3 + u^2U + uU^2 + U^3)(u - U),$$

if the temperature u does not vary too much from U, then (1-7) linearizes to (1-5) with $\gamma = 4U^3\mu$.

Boundary value problems for the Poisson and Laplace equations also arise for unbounded domains. Consider the case of a fluid in a very large basin or an ocean. Although technically these domains are bounded, disturbances taking place at large distances do not affect appreciably processes taking place near the points of observation so that mathematically the domain in question may be thought of as unbounded. Now let us assume that the fluid is irrotational. This means there are no eddies; that is, the average velocity around every closed path C is zero:

$$\frac{1}{l(C)} \int_C v \cdot \boldsymbol{s}\, ds = 0,$$

where σ is the tangent vector to C and v is the fluid's velocity. Here $l(C)$ is the length of the path and ds is the element of arc length. Stokes' theorem and a standard argument yield

$$\operatorname{curl} v = 0.$$

Consequently, there is a potential function u such that

$$v = \nabla u. \tag{1-8}$$

If we further assume that the fluid is incompressible, then div v vanishes, so that combining this fact with (1-8), we find that u must satisfy (1-2). Again the Dirichlet, Neumann, and Robin problems can be formulated for unbounded domains. However, to guarantee the unique solvability of the resulting problems, additional conditions which describe the behavior "at infinity" must be imposed. These will be discussed later.

PROBLEMS

1. Show that in polar coordinates, $x_1 = r\cos\theta$, $x_2 = r\sin\theta$, the two-dimensional Laplace operator has the form

$$\Delta u = \frac{1}{r}\frac{\partial}{\partial r}\left(r\frac{\partial u}{\partial r}\right) + \frac{1}{r^2}\frac{\partial^2 u}{\partial \theta^2}.$$

2. Show that in cylindrical coordinates $x_1 = r\cos\theta$, $x_2 = r\sin\theta$, $x_3 = z$, the three-dimensional Laplace operator has the form

$$\Delta u = \frac{1}{r}\frac{\partial}{\partial r}\left(r\frac{\partial u}{\partial r}\right) + \frac{1}{r^2}\frac{\partial^2 u}{\partial \theta^2} + \frac{\partial^2 u}{\partial z^2}.$$

3. Show that in spherical coordinates $x_1 = r\cos\theta\sin\phi$, $x_2 = r\sin\theta\sin\phi$, $x_3 = r\cos\phi$, the Laplace operator takes the form

$$\Delta u = \frac{1}{r^2}\left[\frac{\partial}{\partial r}\left(r^2\frac{\partial u}{\partial r}\right) + \frac{1}{\sin^2\phi}\frac{\partial^2 u}{\partial \theta^2} + \frac{1}{\sin\phi}\frac{\partial}{\partial \phi}\left(\sin\phi\frac{\partial u}{\partial \phi}\right)\right]$$

or

$$\Delta u = \frac{\partial^2 u}{\partial r^2} + \frac{2}{r}\frac{\partial u}{\partial r} + \frac{1}{r^2\sin^2\phi}\frac{\partial^2 u}{\partial \theta^2} + \frac{1}{r^2}\frac{\partial^2 u}{\partial \phi^2} + \frac{\cot\phi}{r^2}\frac{\partial u}{\partial \phi}.$$

4. In two dimensions the real and imaginary parts of an analytic function satisfy the Laplace equation. Set $x_1 = x$, $x_2 = y$, $x = r\cos\theta$, $y = r\sin\theta$, and show directly that the following functions satisfy the Laplace equation.
   (a) $r^n\cos(n\theta)$, $r^n\sin(n\theta)$.
   (b) $e^{nx}\cos(ny)$, $e^{nx}\sin(ny)$.
   (c) $x^2 - y^2$, $2xy$.
   (d) $\exp(x^2 - y^2)\cos(2xy)$, $\exp(x^2 - y^2)\sin(2xy)$.
5. Explain on physical grounds why $\gamma \geq 0$ in (1-5).

6. Suppose that ϕ is harmonic in a simply connected domain in two dimensions. Construct a function ψ by solving the Cauchy-Riemann equations

$$\frac{\partial \phi}{\partial x} = \frac{\partial \psi}{\partial y}, \qquad \frac{\partial \psi}{\partial x} = -\frac{\partial \phi}{\partial y}.$$

(a) Show that ψ is also harmonic.
(b) Show that the curves $\phi(x, y) = C$ and $\psi(x, y) = K$, where C and K are arbitrary constants, intersect at right angles.
(c) If $\nabla\phi(x, y)$ gives the velocity of a particle at the point (x, y), show that $\psi(x, y) = K$ gives the flow path. *Remark.* The curves $\phi(x, y) = C$ are called the equipotential lines and the curves $\psi(x, y) = K$ the streamlines.

7. In view of Prob. 6, the real and imaginary parts of an analytic function can be given interpretations as flows.
Show that
(a) $f(z) = z$, that is, $\phi(x, y) = x$ and $\psi(x, y) = y$, can be interpreted as plane parallel flow.
(b) Show that $f(z) = z^2$ can be interpreted as flow around a corner, $x > 0$, $y > 0$.
(c) Show that $f(z) = z + 1/z$, $|z| > 1$, can be interpreted as flow around a cylinder.
(d) Show that $f(z) = \alpha \log z$, α a constant, can be interpreted as flow out of a source or into a sink at $z = 0$.

8. Suppose that a mass M is located at y and a mass m is located at x in three dimensions. Regard y as fixed. The force on m exerted by M is

$$F = -G\frac{Mm}{r^3}(x - y), \qquad r = |x - y|.$$

The field generated by M is

$$E = \frac{1}{m}F = -GM\frac{x - y}{r^3}.$$

(a) Show that for $r \neq 0$, $F = \nabla\phi$ where $\phi = GMm/r$.
(b) Suppose that masses $M_1, \ldots, M_n$ are located at points $y^{(1)}, \ldots, y^{(n)}$. Apply the superposition principle to show that the field generated by these masses is

$$E = \frac{1}{m}F = -G\sum_{i=1}^{n} M_i \frac{x - y^{(i)}}{r_i^3}$$

where $r_i = |x - y^{(i)}|$. Find the potential ϕ for this force field F.
(c) Show that ϕ for parts (a) and (b) is harmonic for $x \neq y, y^{(i)}$.
(d) Let a continuous mass distribution with density $\rho = \rho(y)$ occupy a bounded domain D in space. Show that the field generated by this mass distribution is

$$E = \frac{1}{m}F = -G\int_D \frac{x - y}{r^3}\rho(y)\,dy,$$

and that F has potential

$$\phi(x) = Gm\int_D \frac{\rho(y)}{r}\,dy,$$

where $r = |x - y|$ and x lies outside D.

(e) Show that ϕ is harmonic. Assume that differentiation under the integral sign is permissible.

9. *Coulomb's law,*

$$F = \gamma \frac{qQ}{r^3}(x - y), \qquad r = |x - y|,$$

describes the force exerted by a charge Q located at y on a charge q located at x. Find a potential for this force. Establish the analogues of parts (d) and (e) of Prob. 8.

10. Consider the Laplace operator in n dimensions,

$$\Delta u = u_{x_1x_1} + u_{x_2x_2} + \cdots + u_{x_nx_n}.$$

Suppose that $u = u(r)$ depends only on the distance, $r = |x|$, of x from the origin. That is, u is a radially symmetric function.

(a) Show that

$$\Delta u = u'' + \frac{n-1}{r} u',$$

where the prime indicates derivatives with respect to r.

(b) Let u be radially symmetric and harmonic. Show that

$$u = L \log r + M \qquad (n = 2),$$

$$u = \frac{L}{r^{n-2}} + M \qquad (n > 2).$$

11. Suppose that a fluid is completely enclosed by a container of finite volume. Suppose further that the fluid is set in motion by slowly shaking the container so that the velocity is described by a potential, call it u. Suppose that the velocity of the surface at a point x is V. Argue that at the boundary $(u - V) \cdot \nu = 0$, where ν is the exterior unit normal. Then the velocity potential satisfies the Laplace equation in the interior of the container and at the boundary it satisfies $\partial u/\partial \nu = f$, where $f = V \cdot \nu$ is known.

12. This problem shows how to obtain hydrological parameters from certain pump tests. Let $(x_1, x_2, x_3) = (x, y, z)$ be the usual rectangular coordinates. Consider an aquifer bounded by two parallel, impermeable planes b units apart. Recall from Chap. 1 that the flow is governed by

$$q = -\frac{k}{\mu}(\nabla p + \rho \vec{g}) = -T\nabla h \quad \text{and} \quad \operatorname{div} q = 0,$$

where

$$T = \frac{k\rho g}{\mu}, \qquad h = \frac{p}{\rho g} + z.$$

Here ρ, k, g, and μ are assumed to be constants. In general, when a well is drilled into an aquifer, k is unknown and must first be determined in order to predict the future behavior of the system. Let us assume that a fully penetrating well is located at $x = y = 0$. Assume that water is pumped at the constant rate Q and that for large $x^2 + y^2$, $0 \leq z \leq b$, h is equal to the constant H (see Fig. 8-1). In view of these assumptions, we introduce cylindrical coordinates (r, θ, z) and assume that the flow is independent of the angle θ.

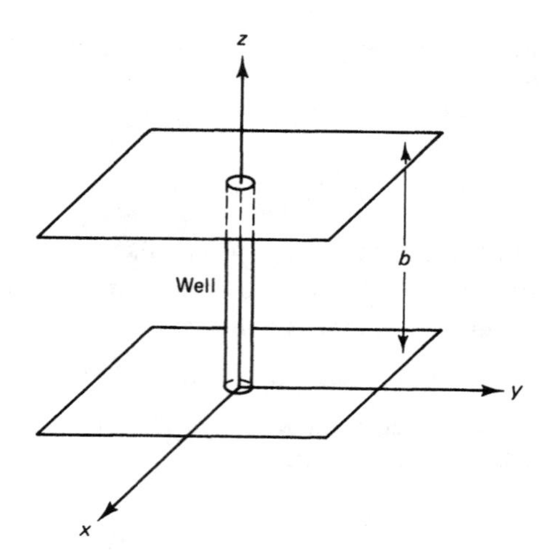

Figure 8-1

(a) Show that h satisfies

$$\Delta h = 0, \qquad r > 0,$$
$$h_z = 0 \quad \text{on} \quad z = 0, \quad \text{and} \quad z = b,$$
$$h(R, z) = H \quad \text{for some large value of} \quad R,$$
$$\lim_{r \to 0} 2\pi T r h_r(r, z) = f(z),$$

where f represents the amount of water which has flowed into the well at the height z. It is, in general, unknown in a field situation.

(b) Let

$$u(r) = \frac{1}{b} \int_0^b h(r, z)\, dz$$

and show that $u(r)$ satisfies

$$\frac{1}{r}(ru')' = 0, \qquad \lim_{r \to 0} ru'(r) = \frac{Q}{2\pi b T}, \qquad u(R) = H.$$

(c) Show that the solution is

$$u(r) = \frac{Q}{2\pi b T} \log \frac{R}{r} + H.$$

(d) Suppose that u can be measured at another "monitoring" well a distance a from the pumping well. Determine T given the thickness of the aquifer and the parameters ρ, μ, and g. *Remark.* This method of determining T and hence k is called *Thiem's test.*

8-2. Exact Solutions

The methods of separation of variables and integral transforms can be used in some cases to construct explicit solutions to the boundary value problems of potential theory. We now have at least two space dimensions; consequently, the boundaries of the domains involved can be very irregular and the methods of separation of variables and integral transforms can break down. Then new methods of solution must be developed. Nevertheless, when the old methods are applicable, they do work very well. Two examples and the following problems illustrate how these familiar techniques can be applied.

Example 1. Find the steady-state temperature in a circular plate of radius a. Assume that the top and bottom of the plate are insulated and that its circumference is kept at a prescribed temperature f.

Assume that the plate lies in the xy plane and that u is the temperature. Then u satisfies Laplace's equation and the Dirichlet boundary condition $u = f$ on the circumference. It is natural to use polar coordinates and express these conditions as

$$\frac{\partial^2 u}{\partial r^2} + \frac{1}{r}\frac{\partial u}{\partial r} + \frac{1}{r^2}\frac{\partial^2 u}{\partial \theta^2} = 0, \qquad 0 \le r < a, \tag{2-1}$$

$$u(a, \theta) = f(\theta). \tag{2-2}$$

Since (r, θ) and $(r, \theta + 2\pi)$ represent the same point in the plate, it is useful to regard f as 2π periodic, $f(\theta) = f(\theta + 2\pi)$. We seek separated solutions

$$u(r, \theta) = R(r)\Theta(\theta) \tag{2-3}$$

to (2-1). For the reason just given, we regard u as 2π periodic in θ; equivalently, Θ is 2π periodic. For (2-3) to satisfy (2-1) we must have

$$\left(R'' + \frac{1}{r}R'\right)\Theta + \frac{1}{r^2}R\Theta'' = 0.$$

Divide this equation by $R\Theta/r^2$ to obtain the pair of equations

$$\Theta'' + c\Theta = 0, \qquad \Theta(\theta) = \Theta(\theta + 2\pi), \tag{2-4}$$

$$R'' + \frac{1}{r}R' - \frac{c}{r^2}R = 0, \tag{2-5}$$

where c is the separation constant. If $c < 0$, the differential equation in (2-4) has only real exponential solutions which are not periodic. Thus, we must have $c \ge 0$ (see also Prob. 9). If $c = 0$, Θ is linear and the only periodic function of this sort is a constant:

$$\Theta_0 = \tilde{a}_0. \tag{2-6}$$

Finally, if $c > 0$,

$$\Theta = \tilde{a}\cos(\sqrt{c}\,\theta) + \tilde{b}\sin(\sqrt{c}\,\theta)$$

where $\tilde{a}$ and $\tilde{b}$ are constants, and this function is 2π periodic only if $c = n^2$ for $n = 1, 2, \ldots$. For such n,

$$\Theta_n = \tilde{a}_n\cos(n\theta) + \tilde{b}_n\sin(n\theta) \tag{2-7}$$

satisfies (2-4).

Next, consider (2-5). If $c = 0$,

$$R_0 = \alpha_0 \log r + \beta_0,$$

where α_0 and β_0 are constants. Since $u_0 = R_0\Theta_0$ is supposed to be a twice-differential function for $0 \le r < a$ we must choose $\alpha_0 = 0$ to eliminate the logarithmic singularity at $r = 0$. Thus,

$$R_0(r) = \beta_0. \tag{2-8}$$

If $c = n^2$, (2-5) is an Euler equation which has general solution

$$R_n = \alpha_n\left(\frac{r}{a}\right)^{-n} + \beta_n\left(\frac{r}{a}\right)^n.$$

As above, we must take $\alpha_n = 0$ so that

$$R_n = \beta_n\left(\frac{r}{a}\right)^n. \tag{2-9}$$

Combine (2-6) with (2-8) and (2-7) with (2-9) to obtain the separated solutions

$$u_0(r, \theta) = \frac{a_0}{2},$$

$$u_n(r, \theta) = \left(\frac{r}{a}\right)^n [a_n \cos(n\theta) + b_n \sin(n\theta)],$$

where a_0, a_n, and b_n are arbitrary constants and a_0 has been divided by two in anticipation of subsequent developments. Superpose the solutions just found to obtain

$$u(r, \theta) = \frac{a_0}{2} + \sum_{n=1}^{\infty}\left(\frac{r}{a}\right)^n [a_n \cos(n\theta) + b_n \sin(n\theta)], \tag{2-10}$$

which is our formal solution to (2-1). It will also satisfy (2-2) if

$$u(a, \theta) = \frac{a_0}{2} + \sum_{n=1}^{\infty} [a_n \cos(n\theta) + b_n \sin(n\theta)] = f(\theta). \tag{2-11}$$

Clearly, we want $f(\theta)$ to have a Fourier series expansion, and we choose

$$a_n = \frac{1}{\pi}\int_0^{2\pi} f(\phi)\cos(n\phi)\,d\phi, \qquad b_n = \frac{1}{\pi}\int_0^{2\pi} f(\phi)\sin(n\phi)\,d\phi. \tag{2-12}$$

Thus, (2-10) and (2-12) provide the formal solution to (2-1) and (2-2). We leave the verification that this formal solution is a solution for Prob. 10, where you are asked to prove:

Theorem 2-1. Suppose that f is continuous and 2π periodic. Then $u(r, \theta)$ defined by (2-10) satisfies (2-1) for $r < a$. In addition, if f has a continuous derivative, (2-10) also satisfies (2-2).

We show later that the added differentiability condition is not needed in Th. 2-1. For this, an alternative representation of the solution to (2-1) and (2-2) is needed. Insert (2-12) into (2-10) to obtain

$$(2\text{-}13)\quad u(r, \theta) = \frac{1}{\pi}\int_0^{2\pi} f(\phi)\left\{\frac{1}{2} + \sum_{n=1}^{\infty}\left(\frac{r}{a}\right)^n [\cos(n\phi)\cos(n\theta) + \sin(n\phi)\sin(n\theta)]\right\} d\phi$$

$$= \frac{1}{\pi}\int_0^{2\pi} f(\phi)\left\{\frac{1}{2} + \sum_{n=1}^{\infty}\left(\frac{r}{a}\right)^n \cos[n(\phi - \theta)]\right\} d\phi.$$

The series in braces can be summed in closed form. Indeed, let $\rho = r/a$ and $\alpha = \phi - \theta$. Then

$$(2\text{-}14)\qquad \frac{1}{2} + \sum_{n=1}^{\infty} \rho^n \cos n\alpha = \frac{1}{2}\frac{1 - \rho^2}{1 - 2\rho\cos\alpha + \rho^2}.$$

(see Prob. 11). Thus, (2-13) takes the form

$$(2\text{-}15)\qquad u(r, \theta) = \frac{1}{2\pi}\int_0^{2\pi} \frac{a^2 - r^2}{a^2 + r^2 - 2ar\cos(\phi - \theta)} f(\phi)\, d\phi$$

for $|r| < a$, which is known as *Poisson's formula*. Poisson's formula is very convenient for numerical calculations. Even the crudest quadrature formulas lead to excellent numerical results.

To illustrate the use of integral transforms, consider the steady-state temperature in the upper half of 3-space.

***Example* 2.** Solve

$$(2\text{-}16)\qquad \Delta u = 0, \qquad x_3 > 0, \qquad -\infty < x_1, x_2 < +\infty,$$

$$(2\text{-}17)\qquad u(x_1, x_2, 0) = f(x_1, x_2), \qquad -\infty < x_1, x_2 < +\infty,$$

where f is radially symmetric. That is,

$$f(x_1, x_2) = f(r) \quad \text{for} \quad r = (x_1^2 + x_2^2)^{1/2}.$$

Symmetry suggests that the solution u will be radially symmetric. Thus, we introduce cylindrical coordinates (r, θ, z) and seek a solution $u = u(r, z)$ independent of θ. In this case, (2-16) and (2-17) become

$$(2\text{-}18)\qquad \frac{1}{r}\frac{\partial}{\partial r}\left(r\frac{\partial u}{\partial r}\right) + \frac{\partial^2 u}{\partial z^2} = 0, \qquad z > 0, \qquad r \geq 0,$$

$$(2\text{-}19)\qquad u(r, 0) = f(r), \qquad r \geq 0.$$

We use the *Hankel transform* to solve this problem. From Prob. 8 of Sec. 3-4, the Hankel transform of $u(r, z)$ with respect to r is

$$(2\text{-}20)\qquad \hat{u}(\rho, z) = \int_0^{\infty} u(r, z) J_0(\rho r) r\, dr$$

and u is recovered from its transform by

$$u(r, z) = \int_0^\infty \hat{u}(\rho, z) J_0(r\rho)\rho \, d\rho. \tag{2-21}$$

To employ these results, multiply (2-18) by $J_0(\rho r)r$ and integrate from 0 to ∞ with respect to r. The first term in (2-18) transforms as follows, upon integration by parts:

$$\begin{aligned}
\int_0^\infty \frac{\partial}{\partial r}\left(r \frac{\partial u}{\partial r}\right) J_0(\rho r)\, dr &= -\rho \int_0^\infty r \frac{\partial u}{\partial r} J_0'(\rho r)\, dr \\
&= \rho \int_0^\infty u[\rho r J_0''(\rho r) + J_0'(\rho r)]\, dr \\
&= \int_0^\infty \frac{u}{r}[\rho^2 r^2 J_0''(\rho r) + \rho r J_0'(\rho r)]\, dr \\
&= \int_0^\infty \frac{u}{r}[-\rho^2 r^2 J_0(\rho r)]\, dr = -\rho^2 \hat{u}(\rho, z),
\end{aligned}$$

where we assume the integrated terms vanish at infinity and we have used $J_0'(0) = 0$ and Bessel's equation $\sigma^2 J_0''(\sigma) + \sigma J_0'(\sigma) + \sigma^2 J_0(\sigma) = 0$. Therefore, the transform of (2-18) is

$$-\rho^2 \hat{u}(\rho, z) + \hat{u}_{zz}(\rho, z) = 0.$$

The boundary condition (2-19) transforms into $\hat{u}(\rho, 0) = \hat{f}(\rho)$. Solve the ordinary differential equation for $\hat{u}$ to get

$$\hat{u}(\rho, z) = Ae^{\rho z} + Be^{-\rho z}.$$

Since we intend to recover u from $\hat{u}$ via (2-21), we want $\hat{u}$ integrable, which requires that $A = 0$. Then $\hat{u}(\rho, 0) = \hat{f}(\rho)$ gives $B = \hat{f}(\rho)$ and $\hat{u}(\rho, z) = \hat{f}(\rho)e^{-\rho z}$. Finally, the inversion formula (2-21) yields the formal solution

$$u(r, z) = \int_0^\infty \hat{f}(\rho) e^{-\rho z} J_0(r\rho)\rho \, d\rho. \tag{2-22}$$

PROBLEMS

1. In two dimensions, consider

$$\begin{aligned}
&\Delta u = 0, \qquad 0 < x < a, \qquad 0 < y < b, \\
&u(x, 0) = f(x), \qquad u(x, b) = 0, \qquad 0 \le x \le a, \\
&u(0, y) = u(a, y) = 0, \qquad 0 \le y \le b.
\end{aligned}$$

(a) Give a physical interpretation of this problem.
(b) Find a formal solution.
(c) Find reasonable restrictions on $f(x)$ so that the formal solution is a solution.

2. In Prob. 1 set $f(x) \equiv 0$ and replace the boundary condition $u(a, y) = 0$ by $u(a, y) = g(y)$ for $0 \le y \le b$. Now answer parts (a), (b), and (c) for this problem. *Hint.* Use the results from Prob. 1 and the essential symmetry of the geometry.
3. Find a formal solution to

$$\Delta u = 0, \qquad 0 < x < 1, \qquad 0 < y < 2,$$
$$u(x, 0) = x, \qquad u_y(x, 2) = 0, \qquad 0 \le x \le 1,$$
$$u(1, y) = u_x(0, y) = 0, \qquad 0 \le y \le 2.$$

Also, give a physical interpretation.

4. Consider

$$\frac{1}{r}\frac{\partial}{\partial r}\left(r\frac{\partial u}{\partial r}\right) + \frac{1}{r^2}\frac{\partial^2 u}{\partial \theta^2} = 0, \qquad 0 \le r < a,$$
$$\frac{\partial}{\partial r}u(a, \theta) + \alpha u(a, \theta) = f(\theta),$$

with f a 2π periodic continuous function and $\alpha > 0$ a constant.
(a) Solve formally by separation of variables.
(b) Find reasonable conditions on f so that the formal solution is a solution.

5. In two dimensions, use Fourier transforms to solve formally

$$\Delta u = 0, \qquad y > 0, \qquad -\infty < x < \infty,$$
$$u(x, 0) = f(x), \qquad -\infty < x < \infty,$$

and so find, assuming suitable behavior of u at infinity, that

$$u(x, y) = \frac{y}{\pi}\int_{-\infty}^{\infty} \frac{f(s)\, ds}{(x - s)^2 + y^2}$$

for $y > 0$.

6. Refer to Prob. 5.
(a) For $y > 0$ and $-\infty < x < \infty$ check directly that the solution $u(x, y)$ satisfies $\Delta u = 0$.
(b) Prove $u(x, y) \to f(x)$ as $y \to 0+$.
7. Solve problem (2-16) and (2-17) using Fourier transforms.
8. Use the Hankel transform to solve formally

$$\frac{1}{r}\frac{\partial}{\partial r}\left(r\frac{\partial u}{\partial r}\right) + \frac{\partial^2 u}{\partial z^2} = 0, \qquad 0 \le r, \qquad 0 < z < a,$$
$$u(r, 0) = f(r), \qquad u(r, a) = 0, \qquad 0 \le r.$$

9. Give a mathematical proof that $c \ge 0$ for any nontrivial, 2π periodic solution Θ to $\Theta'' + c\Theta = 0$. *Hint.* Multiply the differential equation by $\bar{\Theta}$, the complex conjugate of Θ, and integrate by parts from 0 to 2π.
10. Prove Th. 2-1. *Hint.* For $0 \le r \le b < a$ show that the series (2-10) and all its term-by-term derivatives are uniformly convergent.
11. Verify (2-14). *Hint.* $\cos n\alpha = (e^{in\alpha} + e^{-in\alpha})/2$.
12. Justify the interchange of summation and integration in passing from (2-10) to (2-13).
13. Establish reasonable sufficient conditions on f so that (2-22) solves (2-18) and (2-19). *Hint.* Use the M-test to justify differentiating under the integral in (2-22). Refer to Prob. 8 of Sec. 3-4 to determine when (2-22) satisfies the initial data.

8-3. Integral Identities and Green's Functions

Separation of variables and integral transform methods are excellent techniques to use when the geometries of the domains in question are simple. In order to treat more difficult problems and obtain more detailed information about solutions, deeper methods are needed. We begin with some convenient notation which will be used throughout this chapter. D always denotes a domain, B its boundary, and $\bar{D} = D \cup B$ its closure. A ball centered at c with radius ρ, $\{x:|x - c| < \rho\}$, is denoted by $K_\rho(c)$ or simply by K_ρ if c is understood. The boundary of $K_\rho(c)$ is $S_\rho(c) = \{x:|x - c| = \rho\}$ and its closure is $\bar{K}_\rho = K_\rho \cup S_\rho$. Throughout the rest of this chapter, we restrict ourselves to the case of three dimensions and leave the discussion of the two-dimensional case for the problems. Of course, in two dimensions K_ρ is also called a disk and S_ρ is its circumference.

In the integrals we are about to deal with, dx and dy denote volume elements and $d\sigma$ an element of surface area. The integrands often depend on two variables, say x and y. Then we write $d\sigma_x$ or $d\sigma_y$ to indicate whether x or y is the variable of integration. The *exterior unit normal* to the boundary B at a point y on B is $\nu = \nu_y$. Finally, $r = r_{xy} = |x - y|$ is the distance between x and y.

Boundary value problems for potential theory arise with both bounded and unbounded domains. *In Sec. 8-3 we consider only bounded domains.* Unbounded domains are discussed in Sec. 8-5.

Thus, assume that D is a bounded domain with boundary B. We always assume that the Gauss divergence theorem holds for D:

$$\int_D \operatorname{div} F \, dx = \int_B F \cdot \nu \, d\sigma. \tag{3-1}$$

Domains for which the divergence theorem holds are called *normal domains.* Examples of normal domains include balls, solid ellipsoids, cones, parallelepipedes, and regions of space bounded by a finite number of piecewise-smooth surfaces. In short, most domains that occur in practical problems are normal. [Consult O. D. Kellogg, *Foundations of Potential Theory* (Springer-Verlag, Berlin, 1929), for a detailed discussion of the validity of (3-1).] The vector field F in (3-1) is assumed to have continuous first-order partials in D, denoted $F \in C^1(D)$, and F is assumed continuous on $\bar{D}$, denoted $F \in C(\bar{D})$. In general, $C^k(R)$ denotes the set of (possibly vector-valued) functions on R all of whose partial derivatives of order k or less are continuous on R.

Suppose that $u, v \in C^2(D) \cap C^1(\bar{D})$ and set F in (3-1) equal, respectively, to $v \nabla u$, $v \nabla u - u \nabla v$, and ∇u to obtain (see Prob. 1)

$$\int_D [v \Delta u + \nabla v \cdot \nabla u] \, dx = \int_B v \frac{\partial u}{\partial \nu} \, d\sigma, \tag{3-2}$$

$$\int_D [v \Delta u - u \Delta v] \, dx = \int_B \left[v \frac{\partial u}{\partial \nu} - u \frac{\partial v}{\partial \nu} \right] d\sigma, \tag{3-3}$$

$$\int_D \Delta u \, dx = \int_B \frac{\partial u}{\partial \nu} \, d\sigma. \tag{3-4}$$

Relations (3-2)–(3-4) are known, respectively, as *Green's first*, *second*, and *third identities*. These identities play a fundamental role throughout potential theory. In particular, they imply uniqueness theorems for the three boundary value problems of potential theory and lead to integral representations for their solutions.

It is worthwhile to notice that the three identities of Green hold under somewhat milder assumptions on u and v. Since only normal derivatives of u and v occur in the boundary integrals, we can replace the requirement that $u, v \in C^1(\bar{D})$ by the weaker condition that $u, v \in C_\nu^1(\bar{D})$. A function $u \in C_\nu^1(\bar{D})$ if u is continuous in $\bar{D}$, has continuous first-order partials in D, and the following normal derivative limit exists uniformly for x in B,

$$\lim_{\substack{z \to x \in B \\ z \in L_x \cap D}} \frac{\partial u(z)}{\partial \nu_x},$$

where ν_x is the unit normal to B at x and L_x is the line in space through x determined by ν_x (see Fig. 8-2). A simple application of the mean value theorem shows that the function u has a normal derivative at $x \in B$ and that $\partial u(x)/\partial \nu_x$ is the value of the preceding limit. In applications to the Neumann problem

$$\frac{\partial u(x)}{\partial \nu_x} = f(x) \quad \text{for} \quad x \in B$$

and a solution in $C_\nu^1(\bar{D})$ will satisfy

$$\lim_{\substack{z \to x \in B \\ z \in L_x \cap D}} \frac{\partial u(z)}{\partial \nu_x} = f(x)$$

uniformly on B. Informally, the normal derivatives computed just inside B approach their prescribed values on B.

The identities (3-2)–(3-4) hold for functions $u, v \in C^2(D) \cap C_\nu^1(\bar{D})$ and for twice continuously differentiable surfaces B. Indeed, in this case, for each $h > 0$ and small there is a parallel surface B_h to B at a distance h from B (see Fig. 8-2). Let D_h be the domain bounded by B_h and apply (3-2)–(3-4) to D_h. Since the values of u and v and their normal derivatives on B_h converge uniformly to their values on B, a limit passage as $h \to 0$ leads to the conclusion that (3-2)–(3-4) hold for $u, v \in C^2(D) \cap C_\nu^1(\bar{D})$.

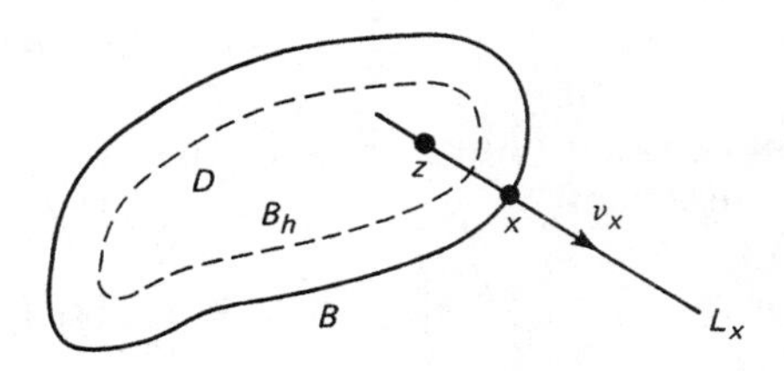

Figure 8-2

We turn now to uniqueness theorems for the three boundary value problems of potential theory. The uniqueness result for the Dirichlet problem will be strengthened in the next section.

Theorem 3-1. Let D be a bounded, normal domain with a twice continuously differentiable boundary B. Then there is at most one solution to the Dirichlet or Robin problem. That is, there is at most one function u in $C^2(D) \cap C^1_\nu(\bar{D})$ such that

$$\Delta u = h(x), \qquad x \in D,$$

and either

$$u = f(x), \qquad \text{or} \quad \frac{\partial u}{\partial \nu} + \alpha(x)u = f(x), \qquad x \in B,$$

where $h(x)$, $f(x)$ are given functions and $\alpha(x) \geq 0$ is bounded, continuous, and not identically zero.

Proof. Consider the Robin problem and suppose that u_1 and u_2 are both solutions. Then $w = u_1 - u_2$ satisfies

$$\Delta w = 0, \qquad x \in D$$

$$\frac{\partial w}{\partial \nu} + \alpha w = 0, \qquad x \in B.$$

Take $v = u = w$ in (3-2) to find

$$\int_D |\nabla w|^2 \, dx = \int_B w \frac{\partial w}{\partial \nu} \, d\sigma = -\int_B \alpha w^2 \, d\sigma.$$

Since the integrands in the first and third integrals are nonnegative, we must have $\nabla w = 0$ in D. Since $\nabla w = 0$, w is a constant in $\bar{D}$ that must be zero because w constant in D gives $\partial w/\partial v = 0$ on B since $w \in C^1_\nu(\bar{D})$; then the Robin condition yields $\alpha w = 0$ on B and so w must equal zero at some point on B. Thus, $w = 0$ on $\bar{D}$ or $u_1 = u_2$ in $\bar{D}$, which proves uniqueness. Similar reasoning, left for Prob. 3, proves uniqueness for the Dirichlet problem.

Theorem 3-2. Let D be a bounded, normal domain. Then any two solutions $u \in C^2(D) \cap C^1_\nu(\bar{D})$, to the Neumann problem

$$\Delta u = h(x), \qquad x \in D,$$

$$\frac{\partial u}{\partial \nu} = f(x), \qquad x \in B,$$

differ by a constant. Furthermore, a necessary condition for the existence of a solution is that

$$\int_D h(x) \, dx = \int_B f(x) \, d\sigma.$$

Proof. The uniqueness argument follows the lines used in the proof of Th. 3-1. The necessary condition follows immediately from (3-4).

We can relax the smoothness assumption on the boundary B in Ths. 3-1 and 3-2 by strengthening our assumption on the differentiability of a solution. Thus, if D is a bounded, normal domain and we assume that $u \in C^2(D) \cap C^1(\bar{D})$ in Ths. 3-1 and 3-2, the same proofs lead to uniqueness in this context.

Next, we use Green's first two identities to derive integral representations for solutions to the boundary value problems of potential theory. For this purpose we use the radially symmetric solutions to the Laplace equation,

$$w(r) = \frac{L}{r} + M$$

(see Prob. 10 of Sec. 8-1). With a view toward subsequent applications, we normalize w to be zero at infinity so that $M = 0$, and set $L = 1/4\pi$. Thus,

$$w(r) = \frac{1}{4\pi r} = \frac{1}{4\pi|x - y|}. \tag{3-5}$$

The function $w(r)$ is called the *fundamental solution* to the Laplace equation. It is harmonic in x or in y for $x \neq y$ and has a singularity at $x = y$. Of course, (3-5) is just the potential at x due to a unit, point mass, or charge at y, in suitable units.

Let $u \in C^2(D) \cap C^1(\bar{D})$. We wish to apply (3-3) to u and $v = 1/4\pi r$ where $x \in D$ is fixed and the integration is with respect to y. Green's second identity cannot be applied directly because of the singularity of v at $y = x$. Therefore, we exclude from D a small ball $\bar{K}_\varepsilon(x)$ about x of radius $\varepsilon > 0$ and then apply (3-3). Fix $\varepsilon > 0$ so that $\bar{K}_\varepsilon(x) \subset D$. Now apply (3-3) to u and $v = 1/4\pi r$ on the domain $D_\varepsilon = D - \bar{K}_\varepsilon(x)$ with boundary $B_\varepsilon = B \cup S_\varepsilon(x)$ to get

$$\begin{aligned} \int_{D_\varepsilon} \frac{1}{4\pi r} \Delta u \, dy &= \int_{D_\varepsilon} \left(\frac{1}{4\pi r} \Delta u - u \Delta \frac{1}{4\pi r} \right) dy \\ &= \int_{B_\varepsilon} \left(\frac{1}{4\pi r} \frac{\partial u}{\partial \nu} - u \frac{\partial}{\partial \nu} \frac{1}{4\pi r} \right) d\sigma_y. \end{aligned} \tag{3-6}$$

Let $\varepsilon \to 0$ in (3-6). It is elementary to check that

$$\int_{D_\varepsilon} \frac{1}{4\pi r} \Delta u \, dy \to \int_D \frac{1}{4\pi r} \Delta u \, dy$$

as $\varepsilon \to 0$ (Prob. 5). The integral over B_ε consists of the integral over the boundary B of D which is independent of ε and the integral over $S_\varepsilon(x)$. Consider the latter integral. Since u has continuous partials in D, $|\partial u/\partial \nu| = |\nabla u \cdot \nu| \leq M$ for some constant M and all points y in a fixed neighborhood of x. Then

$$\left| \int_{S_\varepsilon} \frac{1}{4\pi r} \frac{\partial u}{\partial \nu} d\sigma_y \right| \leq \frac{M}{4\pi\varepsilon} \int_{S_\varepsilon} d\sigma_y = \frac{M}{4\pi\varepsilon} (4\pi\varepsilon^2) \to 0$$

as $\varepsilon \to 0$. Finally, consider the integral

$$\int_{S_\varepsilon} u \frac{\partial}{\partial \nu} \frac{1}{4\pi r} d\sigma_y.$$

The outer unit normal to D_ε at points on S_ε is $\nu_y = (x - y)/r$ and

$$\frac{\partial}{\partial \nu_y}\frac{1}{r} = \nabla_y\left(\frac{1}{r}\right)\cdot \nu_y = \frac{x-y}{r^3}\cdot\frac{x-y}{r} = \frac{1}{r^2}.$$

Use spherical coordinates with origin at x to find

$$\int_{S_\varepsilon} u \frac{\partial}{\partial \nu}\frac{1}{4\pi r}\, d\sigma = \int_0^{2\pi}\int_0^{\pi} u(x+\varepsilon\omega)\frac{1}{4\pi\varepsilon^2}\varepsilon^2 \sin\phi\, d\phi\, d\theta,$$

where $\omega = (\cos\theta \sin\phi, \sin\theta\sin\phi, \cos\phi)$. The limit as $\varepsilon \to 0$ of this integral is

$$\frac{1}{4\pi}\int_0^{2\pi}\int_0^{\pi} u(x)\sin\phi\, d\phi\, d\theta = u(x).$$

Let $\varepsilon \to 0$ in (3-6) and use the observations above to conclude that

$$u(x) = \int_B \frac{1}{4\pi r}\frac{\partial u}{\partial \nu}\, d\sigma_y - \int_B u\frac{\partial}{\partial \nu}\frac{1}{4\pi r}\, d\sigma_y - \int_D \frac{1}{4\pi r}\Delta u\, dy. \tag{3-7}$$

Equation (3-7) represents any function u in $C^2(D) \cap C^1(\bar{D})$ in terms of integrals over B and D. The first integral over B has the form of a *single layer potential* with density $\partial u/\partial \nu$. The second integral over B is called a *double layer* (or *dipole*) *potential* (see Prob. 12). The integral over D is called a *volume potential*. Thus, the representation (3-7) is sometimes expressed as

Theorem 3-3. Every function $u \in C^2(D) \cap C^1(\bar{D})$ is the sum of a single-layer potential, a double-layer potential, and a volume potential.

Take $u \equiv 1$ in $\bar{D}$ to deduce from (3-7) the useful formula

$$\int_B \frac{\partial}{\partial \nu_y}\frac{1}{4\pi r_{xy}}\, d\sigma_y = \begin{cases} -1, & x \in D, \\ 0, & x \in c\bar{D}, \end{cases} \tag{3-8}$$

where $c\bar{D}$ is the complement of the set $\bar{D}$. The value zero is obtained directly from (3-3) applied to the harmonic functions $u \equiv 1$ and $v = 1/4\pi r$ in D.

Consider (3-7) as a representation for the solution to the Dirichlet problem

$$\Delta u = h, \qquad x \in D, \tag{3-9}$$

$$u = f(x), \qquad x \in B. \tag{3-10}$$

The second and third integrals are known because $u = f$ on B and $\Delta u = h$ in D. Unfortunately, the first integral is not known because $\partial u/\partial \nu$ is not given on B. What we need is a representation like (3-7) but without the first integral. We can get such a representation by first generalizing (3-7). Let g be any harmonic function in D with $g \in C^2(D) \cap C^1(\bar{D})$ and u any function in $C^2(D) \cap C^1(\bar{D})$. Then by (3-3),

$$0 = -\int_B \left(g\frac{\partial u}{\partial \nu} - u\frac{\partial g}{\partial \nu}\right) d\sigma + \int_D g\,\Delta u\, dy.$$

Add this to (3-7) to obtain

$$u(x) = \int_B \left(\frac{1}{4\pi r} - g\right)\frac{\partial u}{\partial \nu}\, d\sigma_y - \int_B u \frac{\partial}{\partial \nu}\left(\frac{1}{4\pi r} - g\right) d\sigma_y - \int_D \Delta u\left(\frac{1}{4\pi r} - g\right) dy, \tag{3-11}$$

for any harmonic function $g \in C^2(D) \cap C^1(\bar{D})$. Now it is clear how to get a representation for the solution u to the Dirichlet problem (3-9), (3-10). We choose a harmonic function g such that the first integral in (3-11), which contains the unknown values $\partial u/\partial \nu$, vanishes. That is, for each fixed $x \in D$, we choose a function $g = g_x(y) = g(x, y)$ such that

$$\begin{cases} g \in C^2(D) \cap C^1(\bar{D}), \\ \Delta_y g = 0, \qquad y \in D, \\ g(y) = \dfrac{1}{4\pi r_{xy}}, \qquad y \in B. \end{cases} \tag{3-12}$$

Then (3-11) yields the solution formula

$$u(x) = -\int_B \left[\frac{\partial}{\partial \nu_y} G(x, y)\right] f(y)\, d\sigma_y - \int_D G(x, y)h(y)\, dy \tag{3-13}$$

for the Dirichlet problem (3-9), (3-10), where we have set

$$G(x, y) = \frac{1}{4\pi|x - y|} - g(x, y). \tag{3-14}$$

The function defined by (3-12) and (3-14) is called the *Green's function for the Dirichlet problem* in D.

We derived (3-13) under the assumption that the Dirichlet problem has a solution and that the Green's function exists. For the Green's function to exist, we must be able to solve the special Dirichlet problem (3-12). We shall see later that this special problem is solvable for most domains of physical interest and that (3-13) does in fact solve (3-9) and (3-10).

Green provided a simple physical existence "proof" for the Green's function. Consider a unit point charge located at x inside a conductor which is grounded. In appropriate electrical units, the electrical potential $G(x, y)$ of such a conductor is harmonic inside the conductor except at x, vanishes on the conductor because it is grounded, and $(1/4\pi r_{xy}) - G(x, y)$ is harmonic for $y = x$. Thus,

$$g(x, y) = \frac{1}{4\pi r_{xy}} - G(x, y)$$

solves (3-12) and the Green's function is just the conductor potential induced by a point charge at x.

In general, finding the Green's function explicitly is very difficult. Explicit solutions require special geometries. Consider, for example, the Green's function for the ball $K_a(0)$. We seek a harmonic function $g_x(y) = g(x, y)$ such that $G(x, y) = (4\pi|x - y|)^{-1} - g(x, y)$ vanishes on the sphere $S_a(0)$. We use the *method of images*; that is, we try to locate a point charge $q/4\pi$ at x^* so that

$$g(x, y) = q(4\pi|x^* - y|)^{-1}$$

will be the required harmonic function of y. Since $|x^* - y|^{-1}$ is harmonic inside $K_a(0)$ for any x^* outside this ball, all we need to do is choose x^* outside $K_a(0)$ and the charge $q/4\pi$, so that

$$\frac{1}{4\pi|x - y|} = \frac{q}{4\pi|x^* - y|} \quad \text{for} \quad y \in S_a(0). \tag{3-15}$$

This requires that $|x^* - y|^2 = q^2|x - y|^2$, or

$$|x^*|^2 + a^2 - q^2(|x|^2 + a^2) = - 2[q^2 x - x^*] \cdot y \tag{3-16}$$

for all $y \in S_a(0)$. Since the left side is independent of y, we conclude that $q^2 x - x^* = 0$ or

$$x^* = q^2 x, \tag{3-17}$$

and the image charge is collinear with 0 and x. Use this in (3-16) to find

$$q^4|x|^2 + a^2 - q^2(|x|^2 + a^2) = 0, \qquad q = \frac{a}{|x|},$$

where we used $q^2 \neq 1$, which follows from (3-17) and the fact that x^* is outside $K_a(0)$. Thus,

$$g(x, y) = \frac{q}{4\pi|x^* - y|}, \qquad q = \frac{a}{|x|}, \qquad x^* = q^2 x,$$

and the Green's function for the ball $K_a(0)$ is

$$G(x, y) = \frac{1}{4\pi|x - y|} - \frac{q}{4\pi|x^* - y|}, \qquad q = \frac{a}{|x|}, \qquad x^* = q^2 x. \tag{3-18}$$

In the derivation above, we tacitly assumed that $x \neq 0$. When $x \to 0$, $g(x, y) \to 1/4\pi a$ and

$$G(0, y) = \frac{1}{4\pi|y|} - \frac{1}{4\pi a}. \tag{3-19}$$

As an application of these results, we observe that the solution to the Dirichlet problem

$$\Delta u = 0, \qquad |x| < a, \tag{3-20}$$

$$u = f(x), \qquad |x| = a, \tag{3-21}$$

is, by (3-13),

$$u(x) = -\int_{S_a} \left[\frac{\partial G}{\partial \nu_y}(x, y)\right] f(y)\, d\sigma_y.$$

This formula can be written out more explicitly: When $|y| = a$, $|x| < a$, we have $v_y = y/a$ and [using (3-15), (3-17) and $q = a/|x|$]

$$\frac{\partial}{\partial v_y} G(x, y) = \frac{|x|^2 - a^2}{4\pi a r^3}.$$

Therefore, (3-20) and (3-21) has solution

$$u(x) = \int_{S_a} \frac{a^2 - |x|^2}{4\pi a r^3} f(y)\, d\sigma_y, \tag{3-22}$$

where $r = |x - y|$, which is *Poisson's formula for the ball.*

We conclude this section by stating an important symmetry property of the Green's function for the Dirichlet problem.

Theorem 3-4. Let D be a bounded, normal domain, and suppose that the Green's function $G(x, y)$ for the Dirichlet problem exists. Then

$$G(x, y) = G(y, x)$$

for $x, y \in D$. Consequently, in addition to $\Delta_y G = 0$ for $y \neq x$, we also have $\Delta_x G = 0$ for $x \neq y$.

Proof. See Prob. 13.

PROBLEMS

1. Verify the three identities (3-2), (3-3), and (3-4).

2. Confirm (3-2)–(3-4) for two dimensions, where D is a two-dimensional domain and B is its bounding curve.

3. Complete the proof of Th. 3-1 by showing that the Dirichlet problem has a unique solution.

4. Prove Th. 3-2 in detail.

5. Verify the limit relation following (3-6). *Hint.* Use spherical coordinates with origin at x.

6. Prove the two dimensional analogues of Ths. 3-1 and 3-2.

7. In two dimensions, the radially symmetric solutions to Laplace's equation are $w(r) = L \log r + M$ by Prob. 10 of Sec. 8-1. Let $L = -1/2\pi$, $M = 0$ so $w(r) = (1/2\pi) \log(1/r)$ is the *fundamental solution.* Use this fundamental solution to obtain the two-dimensional analogue of (3-7).

8. Consider the Dirichlet problem (3-9) and (3-10) in two dimensions. Develop the analogue of (3-11)–(3-13) and so obtain a representation for the solution to the Dirichlet problem in this setting.

9. Suppose that $k > 0$ is a constant and that $Lu = \Delta u - k^2 u$. Formulate Dirichlet, Neumann, and Robin problems for this operator and prove that each problem has at most one solution.

10. Find Green's function for $K_a(0)$ in two dimensions and use it to solve the Dirichlet problem (3-20) and (3-21). *Hint.* Use the method of images with logarithmic potentials. You should end up with Poisson's formula (2-15).

11. Let D be a bounded, normal domain in 3-space. Show that one can define (a) a Neumann function, $N(x, y)$ and (b) a Robin function, $R(x, y)$ so that a representation similar to (3-13) is obtained.

12. Let v be a fixed unit vector that specifies a direction in space. Let q and $-q$ be point charges located at $y + hv$ and y respectively. A *dipole* consists of two charges in this configuration, where $h > 0$ is very small. The direction v is the *axis of the dipole*. Dipoles are very common in nature. In particular, water molecules are dipoles. Here is a useful mathematical idealization of this situation. In appropriate units, the potential of the dipole is

$$u(x) = \frac{q}{4\pi|x - (y + h)|} - \frac{q}{4\pi|x - y|} = \frac{qh}{4\pi}\left\{\frac{1}{h}\left[\frac{1}{|x - (y + vh)|} - \frac{1}{|x - y|}\right]\right\}.$$

Show that (by definition) the expression in braces has limit the directional derivative $\partial(|x - y|^{-1})/\partial v_y$ as $h \to 0$. Assume that $h \to 0$ and the charge q varies so that $qh \to p$ as $h \to 0$. Conclude that the limiting potential is

$$u(x) = p\frac{\partial}{\partial v_y}\frac{1}{4\pi|x - y|},$$

where p is called the *moment of the dipole*.

13. Prove Th. 3-4. *Hint*. Fix $x, x' \in D$. Fix $\varepsilon > 0$ so small that $K_\varepsilon(x)$, $K_\varepsilon(x') \subset D$ and $\varepsilon < \frac{1}{2}|x - x'|$. Apply (3-3) to

$$D_\varepsilon = D - (K_\varepsilon(x) \cup K_\varepsilon(x')), \quad u(y) = G(x, y), \quad \text{and} \quad v(y) = G(x', y).$$

Use the fact that u and v vanish on B and adjust the reasoning leading from (3-6) to (3-7) to find $G(x, x') = G(x', x)$, in the limit as $\varepsilon \to 0$.

8-4. The Maximum–Minimum Principle and Consequences

Laplace's equation governs the steady-state temperature distribution of a body. If the body is in thermal equilibrium, there can be no internal "hot" or "cold" spots because heat energy would flow away from a hot spot and toward a cold spot. In other words, the equilibrium temperature cannot have a maximum or minimum at an interior point of the body, unless of course the temperature is constant throughout the entire body. When formulated for any harmonic function, this assertion is known as the strong form of the maximum–minimum principle, or maximum principle for short. It has far-reaching consequences, several of which are presented in this section. The results to follow hold in n-space for $n = 1, 2, \ldots,$ and the proofs are essentially the same whatever the dimension. However, as is our custom, we continue to work in the three-dimensional setting. The domain in question is denoted by D and B is its boundary.

We begin by establishing a key fact:

Lemma 4-1. Let $u \in C^2(D) \cap C(\bar{D})$, where D is a bounded domain.
(i) If $\Delta u \leq 0$ in D and $u \geq 0$ on B, then $u \geq 0$ in D.
(ii) If $\Delta u \geq 0$ in D and $u \leq 0$ on B, then $u \leq 0$ in D.

Proof. Consider (i) and suppose that there were values of x in D with $u(x) < 0$. Then u achieves a *negative* minimum value at some point, say c, in D. Let $\varepsilon > 0$ and form

$$v(x) = u(x) - \varepsilon|x - c|^2.$$

Since $u \geq 0$ on B we can fix ε so that

$$v(x) > -\varepsilon|x - c|^2 \geq -\varepsilon d^2 > u(c) = v(c)$$

for all $x \in B$, where $d = \max\{|x - y| : x, y \in \bar{D}\}$ is the diameter of D. The preceding inequality shows that v assumes its minimum at some point $p \in D$. Since $v(p)$ is a minimum, the second derivative test gives $0 \leq \Delta v(p)$ and leads to the contradiction

$$0 \leq \Delta v(p) = \Delta u(p) - 6\varepsilon < 0.$$

Our original supposition that u could assume negative values must be false. Thus, $u \geq 0$ in D. Part (ii) follows from (i) applied to $-u$. (See Prob. 1 for some added insight into this proof.)

Theorem 4-1 (The Maximum–Minimum Principle). Let D be bounded and $u \in C^2(D) \cap C(\bar{D})$:
(i) If $\Delta u \geq 0$ in D, then $u(x) \leq \max_{y \in B} u(y)$ for $x \in D$.
(ii) If $\Delta u \leq 0$ in D, then $u(x) \geq \min_{y \in B} u(y)$ for $x \in D$.
(iii) If $\Delta u = 0$ in D, then $\min_{y \in B} u(y) \leq u(x) \leq \max_{y \in B} u(y)$ for x in D.

Proof. If $\Delta u \geq 0$ in D and $M = \max u(y)$ for $y \in B$, then $\Delta(M - u) \leq 0$ in D and $M - u \geq 0$ on B. By Lemma 4-1, $M - u \geq 0$ in D which confirms (i). Assertions (ii) and (iii) are left for the problems.

As a first application of the maximum–minimum principle, we strengthen our previous uniqueness result (Th. 3-1) for the Dirichlet problem.

Theorem 4-2. Let D be a bounded domain. There is at most one solution to the Dirichlet problem

$$\begin{aligned} &u \in C^2(D) \cap C(\bar{D}), \\ &\Delta u = h(x), \qquad x \in D, \\ &u = f(x), \qquad x \in B. \end{aligned}$$

Proof. Suppose that u_1 and u_2 are solutions to the Dirichlet problem above. Their difference $u = u_1 - u_2$ satisfies $\Delta u = 0$ in D and $u = 0$ on B. By the maximum–minimum principle $u \equiv 0$, equivalently $u_1 = u_2$ in $\bar{D}$.

Two other important consequences of the maximum principle follow.

Theorem 4-3. Suppose that $u \in C^2(D) \cap C(\bar{D})$ and satisfies

$$\Delta u = h \quad \text{in} \quad D, \qquad u = f \quad \text{on} \quad B.$$

Suppose that $h(x)$ is bounded by the constant H in D and $f(y)$ is bounded by F on B. Then there is a constant C depending only on the domain D such that $|u(x)| \leq F + CH$.

Proof. The idea is to find an auxiliary function w such that

(4-1) $$\Delta w \leq -H \quad \text{in} \quad D \quad \text{and} \quad w \geq \pm u \quad \text{on} \quad B.$$

Indeed, given such a function,

$$\Delta(w \pm u) \leq -H \pm h \leq 0 \quad \text{in} \quad D \quad \text{and} \quad w \pm u \geq 0 \quad \text{on} \quad B,$$

and Lemma 4-1 yields $w \pm u \geq 0$ in $\bar{D}$ or $w \geq |u|$ in $\bar{D}$. Thus, any function $w(x)$ satisfying (4-1) leads to the a priori estimate $|u(x)| \leq w(x)$. Trial and error leads to the following choice for $w(x)$. Recall that the origin lies in D and let $c = (c_1, c_2, c_3)$ be a point such that the plane $x_1 = c_1$ lies to the left of D. That is, $c_1 < x_1$ for all $x = (x_1, x_2, x_3)$ in $\bar{D}$. Fix a number d so that the plane $x_1 = c_1 + d$ lies to the right of D. Define

(4-2) $$w(x) = F + H[e^d - e^{(x_1 - c_1)}], \qquad x \in \bar{D}.$$

Obviously, $\Delta w \leq -H$ and $w \geq F \geq \pm u$ on B. Thus, (4-1) holds for the choice (4-2) and we conclude

$$|u(x)| \leq w(x) \leq F + CH,$$

where $C = e^d$.

Theorem 4-4. Suppose that the function $u_i \in C^2(D) \cap C(\bar{D})$, $i = 1, 2$, satisfies

$$\Delta u_i = h_i \quad \text{on} \quad D, \qquad u_i = f_i \quad \text{on} \quad B,$$

where h_i and f_i are given. If

$$|h_1(x) - h_2(x)| < \varepsilon \quad \text{for} \quad x \in D$$

and

$$|f_1(x) - f_2(x)| < \varepsilon \quad \text{for} \quad x \in B$$

then

$$|u_1(x) - u_2(x)| \leq (1 + C)\varepsilon,$$

where C is a constant determined by the domain D.

Proof. See Prob. 4.

In Secs. 8-2 and 8-3 we obtained Poisson's formula for the solution to

(4-3) $$\Delta u = 0 \quad \text{in} \quad K_a(0) \quad \text{and} \quad u = f \quad \text{on} \quad S_a(0).$$

In Sec. 8-2 we showed that (2-15) satisfied $\Delta u = 0$ for $|r| < a$, but we did not look closely at the behavior of the solution as $r \to a$. In Sec. 8-3 we were even more informal. We *assumed* that (4-3) had a solution and obtained the Poisson formula

(3-22). We pause now to tighten up our earlier discussions, and thus prepare the way for a sharpened form of the maximum principle.

Theorem 4-5 (Poisson). Suppose that f is continuous on $S_a(0)$. Then the function

$$u(x) = \begin{cases} \displaystyle\int_{S_a} \frac{a^2 - |x|^2}{4\pi a r^3} f(y)\, d\sigma_y, & x \in K_a(0) \\ f(x), \quad x \in S_a(0). \end{cases} \tag{4-4}$$

where $r = |x - y|$ satisfies Laplace's equation for $|x| < a$ and is continuous in $|x| \leq a$.

Proof. For $x \in K_a(0)$, it is routine to check that repeated differentiation under the integral is permissible and that $\Delta u(x) = 0$ (see Prob. 6). Clearly, u satisfies the boundary condition on S_a. It remains to check the continuity of the solution (4-4). Since we already know that $u \in C^2(K_a)$, we need only show that

$$\lim_{\substack{x \to x^\circ \\ |x| < a}} u(x) = f(x^\circ) \quad \text{for} \quad x^\circ \in S_a.$$

Since $v \equiv 1$ is harmonic with boundary values 1 on S_a, (3-22) does hold with $u = v$ and $f = 1$:

$$1 = \int_{S_a} \frac{a^2 - |x|^2}{4\pi a r^3}\, d\sigma_y, \qquad x \in K_a. \tag{4-5}$$

Therefore, from (4-4) and (4-5),

$$u(x) - f(x^\circ) = \int_{S_a} \frac{a^2 - |x|^2}{4\pi a r^3} [f(y) - f(x^\circ)]\, d\sigma_y \tag{4-6}$$

for $x \in K_a$, and we must show this expression has limit 0 when $x \to x^\circ$. Let $\varepsilon > 0$. Use the continuity of f to secure a $\delta > 0$ so that

$$|f(y) - f(x^\circ)| < \frac{\varepsilon}{2} \quad \text{for} \quad y \in S_a(0) \cap K_\delta(x^\circ) = B_1.$$

Let $B_2 = S_a - B_1$. The integral in (4-6) is the sum of the corresponding integrals over B_1 and B_2. Estimate each of these integrals as follows. First,

$$\left| \int_{B_1} \frac{a^2 - |x|^2}{4\pi a r^3} [f(y) - f(x^\circ)]\, d\sigma_y \right| \leq \frac{\varepsilon}{2} \int_{S_a} \frac{a^2 - |x|^2}{4\pi a r^3}\, d\sigma_y = \frac{\varepsilon}{2},$$

by (4-5). Second, since f is continuous on S_a, $|f(y)| \leq M$ for some constant M and all $y \in S_a$. Thus,

$$\left| \int_{B_2} \frac{a^2 - |x|^2}{4\pi a r^3} [f(y) - f(x^\circ)]\, d\sigma_y \right| \leq \frac{2M}{4\pi a} (a^2 - |x|^2) \int_{B_2} \frac{1}{r^3}\, d\sigma_y.$$

Now restrict x so that $x \in K_a(0) \cap K_{\delta/2}(x^\circ)$. Then for $y \in B_2$, $r = |x - y| \geq \delta/2$ and the right member above is at most

$$\frac{2M(a^2 - |x|^2)}{4\pi a}\left(\frac{2}{\delta}\right)^3 4\pi a^2 \leq \frac{16aM}{\delta^3}(a + |x|)(a - |x|) \leq \frac{32Ma^2}{\delta^3}(a - |x|).$$

Combining these estimates, we have

$$|u(x) - f(x^\circ)| \leq \frac{\varepsilon}{2} + \frac{32Ma^2}{\delta^3}(a - |x|)$$

for $x \in K_a(0) \cap K_{\delta/2}(x^\circ)$. Since $|x| \to |x^\circ| = a$ as $x \to x^\circ$, there is a $\delta' > 0$, which we may take less than $\delta/2$, such that $32Ma^2(a - |x|)\delta^{-3} < \varepsilon/2$ for $x \in K_{\delta'}(x^\circ)$. Therefore,

$$x \in K_a(0) \cap K_{\delta'}(x^\circ) \quad \text{implies that} \quad |u(x) - f(x^\circ)| < \varepsilon,$$

and $u(x) \to f(x^\circ)$, as claimed.

The Poisson formula has several interesting consequences. For example, suppose that u is harmonic in $K_a(0)$ and $\rho < a$. Then the Poisson formula applied to u on $K_\rho(0)$ yields

$$u(x) = \int_{S_\rho} \frac{\rho^2 - |x|^2}{4\pi\rho r^3} u(y)\, d\sigma_y$$

because the boundary values of u on S_ρ are $f(y) = u(y)$. In particular,

$$u(0) = \int_{S_\rho} \frac{\rho^2}{4\pi\rho^4} u(y)\, d\sigma_y = \frac{1}{4\pi\rho^2}\int_{S_\rho} u(y)\, d\sigma_y.$$

That is, the value of the harmonic function u at the origin equals its mean value computed over the sphere S_ρ. Since any point in space can be selected as the origin of coordinates, this observation leads to: Let u be harmonic in a domain D and $x^\circ \in D$. If $K_\rho(x^\circ) \subset D$, then (imagine the origin placed at x°) $u(x^\circ)$ is the mean value of u over S_ρ. That is,

$$u(x^\circ) = \frac{1}{4\pi\rho^2}\int_{S_\rho(x^\circ)} u(y)\, d\sigma_y, \tag{4-7}$$

which is called the *mean value property for harmonic functions*. There is an equivalent formulation of the mean value property in terms of volume integrals,

$$u(x^\circ) = \frac{3}{4\pi\rho^3}\int_{K_\rho(x^\circ)} u(y)\, dy. \tag{4-8}$$

(see Prob. 8). In summary, we have established

Theorem 4-6. Suppose that u is harmonic in a domain D. Then u satisfies the mean value property (4-7), equivalently (4-8), at every point in D.

Notice that the discussion leading to Th. 4-6 does not require D to be bounded.

Theorem 4-6 allows us to strengthen the maximum–minimum principle. Let us say that a continuous function u defined in a domain D satisfies the *mean value property in D* if (4-8) holds at each point $x^\circ \in D$, for all balls $K_\rho(x^\circ) \subset D$. We claim that u cannot assume its maximum at a point, say $x^\circ \in D$, unless $u \equiv u(x^\circ)$ in $K_\rho(x^\circ)$, for any ball $K_\rho(x^\circ) \subset D$. Indeed, assume $u(y) \le u(x^\circ)$ for $y \in D$ and let $K_\rho(x^\circ) \subset D$. If there is a point $y^\circ \in K_\rho(x^\circ)$ with $u(y^\circ) < u(x^\circ)$, then there is a ball K° centered at y° and contained in $K_\rho(x^\circ)$ on which $u(y) < [u(y^\circ) + u(x^\circ)]/2$. Then

$$\begin{aligned} u(x^\circ) &= \frac{3}{4\pi\rho^3}\int_{K_\rho(x^\circ)} u(y)\,dy = \frac{3}{4\pi\rho^3}\left[\int_{K_\rho - K^\circ} u(y)\,dy + \int_{K^\circ} u(y)\,dy\right] \\ &\le \frac{|K_\rho - K^\circ|}{|K_\rho|}\,u(x^\circ) + \frac{|K^\circ|}{|K_\rho|}\,\frac{u(y^\circ) + u(x^\circ)}{2} \\ &< \frac{|K_\rho - K^\circ|}{|K_\rho|}\,u(x^\circ) + \frac{|K^\circ|}{|K_\rho|}\,u(x^\circ) = u(x^\circ), \end{aligned}$$

a contradiction. (Here we have written $|S|$ for the volume of the set S.) Thus, $u \equiv u(x^\circ)$ in $K_\rho(x^\circ)$, as claimed.

Next, we sharpen this conclusion: If u achieves its maximum at $x^\circ \in D$, then $u \equiv u(x^\circ)$ in D. To see this let x be any point in D and let Γ be a continuous curve in D joining x° to x. We can express Γ as $y = \phi(t)$ for $0 \le t \le 1$ with $\phi(0) = x^\circ$, $\phi(1) = x$, and ϕ continuous. By what we just proved, $u \equiv u(x^\circ)$ near x°. In particular, $u(\phi(t)) = u(x^\circ)$ for all $t \ge 0$ and near 0. Consequently, there is a *largest* $t' \in [0, 1]$ such that $u(\phi(t)) = u(x^\circ)$ for $0 \le t \le t'$. If $t' < 1$, we can apply our previous result to see that $u \equiv u(\phi(t')) \equiv u(x^\circ)$ in a small ball centered at $x' = \phi(t')$. In particular, $u(\phi(t)) = u(x^\circ)$ for all $t \ge t'$ but near t', which contradicts the definition of t'. Thus, $t' = 1$ and $u(x) = u(\phi(1)) = u(x^\circ)$. Since x was any point in D, $u \equiv u(x^\circ)$, as asserted.

Entirely similar reasoning applies if $u(x^\circ)$ is a minimum. Therefore, we have

Theorem 4-7 (Strong Form of the Maximum–Minimum Principle). Suppose that u is continuous and satisfies the mean value property in a domain D. In particular, this will be so if u is harmonic in D.

(i) If u assumes either its maximum or minimum value at a point in D, then u is identically constant in D.

(ii) If D is bounded and u is continuous on $\bar{D}$, then u assumes its maximum and minimum values on B, the boundary of D.

We have seen that harmonic functions have the mean value property. Somewhat surprisingly, the converse is true. That is, we have

Theorem 4-8 (Koebe). Suppose that u is continuous and satisfies the mean value property in a domain D. Then u is harmonic in D.

Proof. Let $x^\circ \in D$ and $\bar{K}_\rho(x^\circ) \subset D$. Since u is continuous on $S_a(x^\circ)$, we can solve the Dirichlet problem

$$\Delta v = 0 \quad \text{in} \quad K_\rho(x^\circ), \qquad v = u(x) \quad \text{on} \qquad S_\rho(x^\circ).$$

Since v is harmonic in $K_\rho(x^\circ)$ it has the mean value property in this ball. Also, by construction, $v = u$ on $S_\rho(x^\circ)$. Evidently, the difference of two functions with the mean value property also has this property. Therefore, $u - v$ has the mean value property in $K_\rho(x^\circ)$ and $u - v = 0$ on $S_\rho(x^\circ)$. By Th. 4-7, $u - v$ assumes its maximum and minimum on $S_\rho(x^\circ)$, and we infer that $u - v = 0$ or $u = v$ on $K_\rho(x^\circ)$. Thus, u is harmonic in a neighborhood of x° for any x° in D.

We conclude our discussion of the maximum principle by stating without proof the following useful complement to Th. 4-7.

Theorem 4-9. Let u be a nonconstant harmonic function in a bounded domain D. Suppose that $u \in C^1(\bar{D})$ and assume that each point $y^\circ \in B$ lies on the surface of some ball that lies in $\bar{D}$. Let $x^\circ \in B$ be a point at which u achieves it maximum. Then $\partial u(x^\circ)/\partial v > 0$.

Theorems 4-1 through 4-9 give important general properties of harmonic functions. Some additional and equally important qualitative properties are explored in the problems.

PROBLEMS

1. Let $v \in C^2(D) \cap C(\bar{D})$ for some bounded domain D.

(i) If $\Delta v < 0$ in D and $v \geq 0$ on B, then $v \geq 0$ in D.
(ii) If $\Delta v > 0$ in D and $v \leq 0$ on B, then $v \leq 0$ in D.

Prove these elementary statements by a direct application of the second derivative test. [The idea in the proof of Lemma 4-1, where only $\Delta u \leq 0$ (or $\Delta u \geq 0$) is known, is to perturb u into v where the sharp inequalities hold.]

2. Prove (ii) and (iii) of Th. 4-1.

3. State and prove the analogue of Th. 4-2 for the Robin problem with $\alpha = \alpha(x) > 0$. *Hint*. The difference u of two solutions is harmonic and $u(y) = -\alpha(y)^{-1}\, \partial u(y)/\partial v$ for $y \in B$. Use the strong form of the maximum principle to deduce that u achieves its maximum, say at y°, and its minimum, say at y^1, in B. Use the boundary condition to deduce $u(y^\circ) \leq 0$ and $0 \leq u(y^1)$.

4. Prove Th. 4-4.

5. What single character must be changed in the proof of Lemma 4-1 to obtain the corresponding assertion in dimension $n \neq 3$? Now, convince yourself that the proofs of Ths. 4-1 through 4-4 apply for $n \neq 3$.

6. Refer to the Poisson formula in (4-4). Observe that the integrand is infinitely differentiable as a function of x and y for $x \neq y$, and check that

$$\Delta_x \frac{a^2 - |x|^2}{4\pi a r^3} = 0 \quad \text{for} \quad x \in K_a(0), \qquad y \in S_a(0).$$

[You can verify this last equation directly or more painlessly by using

$$\frac{a^2 - |x|^2}{4\pi a r^3} = -\frac{\partial G(x, y)}{\partial v_y}$$

and (3-18), which expresses G as the sum of harmonic functions.] Now state a theorem that permits differentiation under the integral in (4-4), and conclude that Poisson's integral can be differentiated indefinitely under the integral sign and that $\Delta u = 0$ for $x \in K_a(0)$ and u given by (4-4).

7. Let u be harmonic in a domain D. Show that u has partial derivatives of all orders in D. *Hint*. Fix $x^\circ \in D$ and $\rho > 0$ such that $K_\rho(x^\circ) \subset D$. Then for $x \in K_\rho(x^\circ)$,

$$u(x) = \int_{S_\rho(x^\circ)} \frac{\rho^2 - |x|^2}{4\pi \rho r^3} u(y)\, d\sigma_y.$$

Now use Prob. 6.

8. Refer to (4-7).
(a) Use spherical coordinates (r, θ, ϕ) with origin at x° and integrate (4-7) with respect to r from 0 to ρ to obtain (4-8).
(b) Assume that (4-8) holds, express the volume integral in spherical coordinates, multiply (4-8) by $4\pi\rho^3$, and differentiate with respect to ρ to recover (4-7).

9. Establish the mean value property (4-7) for harmonic functions by the following reasoning based on Green's identities instead of Poisson's formula. Assume that u is harmonic in $K_a(x^\circ)$ and let $\rho < a$. Use (3-4) to find $\int_{S_\rho} (\partial u/\partial v)\, d\sigma = 0$. In spherical coordinates with origin at x° this becomes

$$\int_0^\pi \int_0^{2\pi} \left[\frac{\partial}{\partial \rho} u(\rho, \theta, \phi)\right] \rho^2 \sin\phi \, d\theta\, d\phi = 0.$$

Use this to show

$$\int_0^\pi \int_0^{2\pi} u(\rho, \theta, \phi) \sin\phi \, d\theta\, d\phi = \text{constant}.$$

Find the constant by setting $\rho = 0$ so that $u(0, \theta, \phi) = u(x^\circ)$.

10. State and prove the analogues of (4-7) and (4-8) for two dimensions.

11. Let D be a bounded domain and assume that Green's function, $G(x, y) = (4\pi|x - y|)^{-1} - g(x, y)$, for the Dirichlet problem exists. Prove that $G(x, y) > 0$ for $x \neq y$ and $x, y \in D$. *Hint*. Fix $x \in D$. Use the continuity of g in y to determine $\varepsilon > 0$ so that $G(x, y) > 0$ on $S_\varepsilon(x)$. Apply the maximum principle to $G(x, y)$ on $D_\varepsilon = D - \overline{K_\varepsilon(x)}$ to find that $G > 0$ on D_ε.

12. Let D be a bounded domain. Show that the Green's function is unique (assuming that it exists). *Hint*. Assume that there were two Green's functions $G_i(x, y) = (4\pi|x - y|)^{-1} - g_i(x, y)$ with g_i satisfying (3-12). Show that, for each x,

$$u(y) = G_1(x, y) - G_2(x, y)$$

satisfies $\Delta u = 0$ in D, $u = 0$ on B.

13. Repeat Probs. 11 and 12 for two dimensions where $(4\pi|x - y|)^{-1}$ is replaced by $(1/2\pi) \log |x - y|^{-1}$.

14. Confirm (ii) of Th. 4-7.

15. Let u be harmonic in a bounded domain D and $|\nabla u| > 0$ in D.
(a) Show that $|\nabla u|$ takes its maximum on B the boundary of D.
(b) Interpret this result in terms of fluid flow using (1-8). Is this result clear on physical grounds?

16. Let $u \in C^2(K_a(0)) \cap C(\bar{K}_a(0))$ be harmonic and nonnegative, $u \geq 0$ in $\bar{K}_a$. Establish *Harnack's inequality*,

$$\frac{(a - |x|)a}{(a + |x|)^2} u(0) \leq u(x) \leq \frac{(a + |x|)a}{(a - |x|)^2} u(0).$$

Hint. For $y \in S_a(0)$, $a - |x| \leq |y - x| = r \leq a + |x|$. Use Poisson's formula followed by the mean value property.

17. Let u be harmonic in the entire three-dimensional space. Assume that u is bounded above (or below). Then u is a constant. This is called *Liouville's theorem*. *Hint*. Assume that $u \leq M$ and fix x. Apply Harnack's inequality to $M - u \geq 0$ in $K_a(0)$ where $a > |x|$. Let $a \to +\infty$.

18. Prove *Rellich's theorem*: Let u be harmonic in 3-space and satisfy

$$\int_{R_3} |u(y)|^2 \, dy = M < \infty.$$

Then $u \equiv 0$. *Hint*. Apply the mean value property at 0 and the Schwarz inequality on $K_a(0)$. Let $a \to \infty$. Observe that the origin can be any point in space.

19. Prove *Harnack's first theorem*: Let $\{u_n(x)\}$ be a sequence of harmonic functions in a domain D. Suppose that $u_n(x) \to u(x)$ uniformly on each closed, bounded subset of D. Then $u(x)$ is harmonic in D. *Hint*. Show that u is continuous and that u has the mean property in D because each u_n does.

20. Use Th. 4-9 to establish uniqueness for the Robin problem when $\alpha(x) \geq 0$, instead of the stronger assumption $\alpha(x) > 0$ used in Prob. 3. Of course, now you must assume that the geometric condition in Th. 4-9 holds for points on B.

8-5. Boundary Value Problems in Unbounded Domains

Although we have mentioned boundary value problems in unbounded domains, we have avoided some extra difficulties inherent in this setting. We have not faced squarely just what happens "at infinity." For example, in Ex. 2 of Sec. 8-2, we found a formal solution by assuming the functions involved decayed sufficiently rapidly at infinity so that all improper integrals converged and all boundary terms in certain integrations by parts vanished at infinity.

What happens "at infinity" is important. Indeed, the boundary value problems of potential theory will not have unique solutions unless some assumptions at infinity are made. For example, consider the Dirichlet problem

$$\Delta u = 0 \quad \text{in} \quad |x| > 1, \qquad u = 0 \quad \text{on} \quad |x| = 1.$$

In two dimensions, $u = C \log r$ solves this problem for any constant C. In three dimensions, corresponding solutions are $u = C(1 - 1/r)$. In two dimensions, the physically realistic assumption that u be bounded leads to the unique solution $u = 0$, while the reasonable requirement that the solution tend to zero at infinity singles out the unique solution $u = 0$ in three dimensions. As we shall see, assumptions such as these are precisely what is needed to guarantee uniqueness of solutions in unbounded domains.

As usual, we restrict our discussion to the three-dimensional case. Some corresponding results for dimension $n \neq 3$ appear in the problems. Also, we consider only so-called *exterior domains* D. That is, domains D such that $D = c\bar{D}_1$ where D_1 is a bounded domain and $c\bar{D}_1$ is the complement of $\bar{D}_1$. We always assume that $D = c\bar{D}_1$ throughout this section and that the origin of coordinates lies in D_1.

The three identities of Green play an important role in the study of the exterior boundary value problems. We call an exterior domain D *normal* if $D = c\bar{D}_1$ and D_1 is a normal domain. For such domains we have

Theorem 5-1. Let D be a normal, exterior domain. Assume that $u, v \in C^2(D) \cap C^1_\nu(\bar{D})$ and satisfy

$$u, v = O(|x|^{-1}) \quad \text{and} \quad \nabla u, \nabla v = O(|x|^{-2}) \quad \text{as} \quad |x| \to \infty.$$

Then the identities of Green, (3-2)–(3-4), hold.

Proof. Recall that $w = O(|x|^{-p})$ as $|x| \to \infty$ means there are constants M and a such that

$$|w(x)| \leq M|x|^{-p} \quad \text{for} \quad |x| \geq a.$$

Apply this to each of the functions, u, v, ∇u, and ∇v to obtain constants M and a so that

$$|u(x)|, |v(x)| \leq M|x|^{-1}, \qquad |\nabla u(x)|, |\nabla v(x)| \leq M|x|^{-2}$$

for $|x| \geq a$. Now, fix $R > a$ so that $D_1 \subset K_R(0)$ and apply (3-2) to the bounded, normal domain $D_R = D \cap K_R(0)$ with boundary $B_R = B \cup S_R(0)$, where B is the boundary of D, to find

$$\int_{D_R} [v\,\Delta u + \nabla v \cdot \nabla u]\,dx = \int_B v\frac{\partial u}{\partial \nu}\,d\sigma + \int_{S_R} v\frac{\partial u}{\partial \nu}\,d\sigma. \tag{5-1}$$

The integral over the sphere is easily estimated,

$$\left|\int_{S_R} v\frac{\partial u}{\partial \nu}\,d\sigma\right| \leq \frac{M^2}{R^3}(4\pi R^2) \to 0$$

as $R \to \infty$ because $|\partial u/\partial \nu| = |\nabla u \cdot \nu| \leq M/R^2$ on S_R. Thus, (5-1) gives

$$\int_D [v\,\Delta u + \nabla v \cdot \nabla u]\,dx = \lim_{R\to\infty} \int_{D_R} [v\,\Delta u + \nabla v \cdot \nabla u]\,dx = \int_B v\frac{\partial u}{\partial \nu}\,d\sigma.$$

The identities (3-3) and (3-4) follow similarly (see Prob. 1).

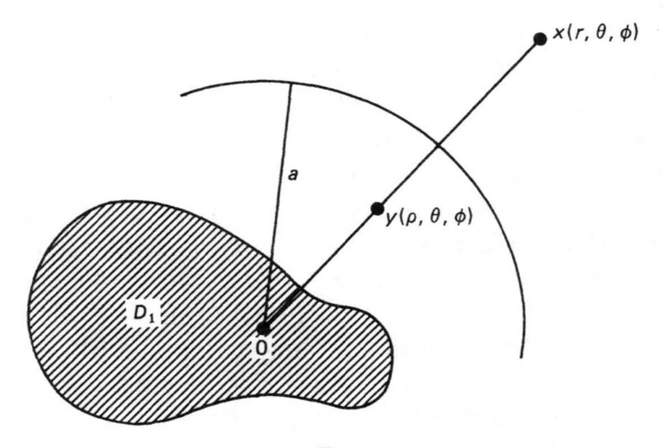

The exterior domain is $D = c\overline{D}_1$ and D_1 is shaded.

Figure 8-3

Let u be harmonic in an exterior domain D. To study the behavior of u at infinity, we use the *Kelvin transformation* which relates the values of the harmonic function $u(x)$ for large $|x|$ to the values of a related harmonic function $v(y)$ for $|y|$ near zero. The transform is based on *inversion in a sphere.* Let x be an arbitrary point outside the sphere $S_a(0)$. We map x to the point y inside the sphere which is collinear with O and x and also satisfies $|x|\,|y| = a^2$. Clearly, $y = a^2x/|x|^2$ and the inverse transformation is given by $x = a^2y/|y|^2$. Evidently, $y = a^2x/|x|^2$ maps the domain $|x| > a$ one-to-one, onto the domain $0 < |y| < a$. Also, as $|x| \to \infty$, $y \to 0$. Informally, points near infinity map into points near the origin. [Notice that this is exactly the transformation that determined the image point (charge) in the construction of the Green's function for the ball; see (3-18).]

Let u be harmonic in an exterior domain D. Fix $a > 0$ so that $D_1 \subset K_a(0)$. For $x \in D$ let $y = a^2x/|x|^2$ be its inversion in $S_a(0)$ (see Fig. 8-3). Represent x by spherical coordinates (r, θ, ϕ) and its image y by spherical coordinates (ρ, θ, ϕ). Then $r\rho = a^2$. The *Kelvin transform* of $u = u(x)$ for $|x| \geq a$ is

$$v = v(y) = \frac{a}{\rho}\, u\left(\frac{a^2}{\rho}, \theta, \phi\right), \qquad 0 < |y| \leq a. \tag{5-2}$$

A routine calculation (Prob. 2) shows that

$$\left(\frac{\rho}{a}\right)^5 \Delta v = \Delta u, \tag{5-3}$$

where the Laplacian on the left is computed in terms of (ρ, θ, ϕ) and on the right in terms of (r, θ, ϕ). Since u is harmonic in $|x| > a$, v is harmonic in $0 < |y| < a$. Similarly, since u is continuous on $|x| = a$, its transform v is continuous on $|y| = a$. Also, since $u(x) = \rho v(y)/a$ from (5-2), the behavior of u for large $|x|$ is given in terms of the behavior of v for $|y|$ near zero.

The example at the beginning of this section suggests that we should require u to have limit zero at infinity to obtain unique solutions to the boundary value problems for exterior domains. Therefore, we assume that u is harmonic in an exterior domain D and *tends to zero uniformly as* $|x| \to \infty$. This uniform convergence means that there is a function $\mu(r)$ such that $\mu(r) \to 0$ as $r \to +\infty$ and $|u(x)| \leq \mu(|x|)$. Under this assumption, we can define the Kelvin transform v at $y = 0$ so that v is harmonic in the entire ball $|y| < a$. To see this, we use the maximum principle and Poisson's formula. Let $w = w(y)$ be the harmonic function in $|y| < a$ which equals $v(y)$ on $|y| = a$. (The function w is given by Poisson's formula.) We claim that $w(y) = v(y)$ on $0 < |y| \leq a$. Given this we need only define $v(0) = w(0)$ to make v harmonic in $|y| < a$. To check the claim, note that v is bounded on $|y| = a$, say by A. Then, of course, w is bounded by A in $|y| \leq a$. From (5-2) and the decay restriction on u, $|v(y)| \leq a\mu(a^2/|y|)/|y|$, where $\mu(a^2/|y|) \to 0$ as $|y| \to 0$. Fix $\varepsilon > 0$. Since $\mu(1/\alpha) \to 0$ as $\alpha \to 0$,

$$\frac{a\mu(a^2/\alpha)}{\alpha} + A \leq \varepsilon\left(\frac{1}{\alpha} - \frac{1}{a}\right) \tag{5-4}$$

for all $\alpha > 0$ sufficiently small. For such α, consider the harmonic function

$$\phi(y) = \pm[v(y) - w(y)] - \varepsilon\left(\frac{1}{|y|} - \frac{1}{a}\right), \qquad \alpha \leq |y| \leq a.$$

On $|y| = a$ we have $\phi(y) = 0$, while on $|y| = \alpha$ the estimate (5-4) implies that $\phi(y) \leq 0$. Thus, Lemma 4-1 gives

$$\pm[v(y) - w(y)] \leq \varepsilon\left(\frac{1}{|y|} - \frac{1}{a}\right), \qquad |v(y) - w(y)| \leq \varepsilon\left(\frac{1}{|y|} - \frac{1}{a}\right) \tag{5-5}$$

for $\alpha \leq |y| \leq a$. Since this reasoning applies for all $\alpha > 0$ sufficiently small, we conclude that (5-5) holds for $0 < |y| \leq a$. Finally, since $\varepsilon > 0$ is arbitrary in (5-5), we infer that $|v(y) - w(y)| = 0$ or $v(y) = w(y)$ in $0 < |y| \leq a$. Thus, w is the harmonic extension of v to $|y| < a$.

We summarize these observations as follows: Let u be harmonic in an exterior domain D and tend to zero uniformly at infinity. Then the Kelvin transform v of u defined by (5-2) has a harmonic extension to the ball $|y| \leq a$. Consequently, if $|v(y)| \leq A$ on $|y| = a$, then $|v(y)| \leq A$ in $|y| \leq a$. This conclusion leads to important and somewhat surprising results on the behavior of u at infinity. Indeed, from the qualitative assumption that $|u(x)| \leq \mu(|x|)$, where μ simply has limit zero at infinity, we deduce

Theorem 5-2. Suppose that u is harmonic in an exterior domain D and u tends uniformly to zero at infinity. Then there are constants $M \geq 0$ and r_0 such that

$$|u(x)| \leq \frac{M}{|x|}, \qquad |\nabla u| \leq \frac{M}{|x|^2}, \qquad \left|\frac{\partial^2 u}{\partial x_i \partial x_j}\right| \leq \frac{M}{|x|^3}$$

for $|x| \geq r_0$.

Proof. Fix a so that $D_1 \subset K_a(0)$ and let A be the bound on v in $|y| \leq a$. Then from (5-2),

$$|u(x)| = \left|\frac{\rho}{a} v(y)\right| \leq \frac{a}{r} A = \frac{M_0}{|x|}$$

for $|x| \geq a$. We estimate the gradient of u in a similar way. We have $u(x) = |y|v(y)/a$, where $y = a^2x/|x|^2$, so

$$\frac{\partial u}{\partial x_i} = \frac{1}{a} \sum_j \left[\frac{\partial}{\partial y_j} |y|v(y)\right] \frac{\partial y_j}{\partial x_i}$$

$$= \sum_j \left[\frac{|y|}{a} \frac{\partial v}{\partial y_j} + \frac{y_j}{a|y|} v\right]\left[\frac{a^2}{|x|^2} \delta_{ji} - \frac{2a^2 x_j x_i}{|x|^4}\right],$$

where $\delta_{ij} = 1$ if $i = j$ and is zero otherwise. Now the first-order partials of the harmonic function v are bounded, say by L, in $|y| \leq a/2$. Since we also have $|y_j| \leq |y|$ and $|x_j| \leq |x|$, the formula for $\partial u/\partial x_i$ yields

$$\left|\frac{\partial u}{\partial x_i}\right| \leq 3\left(\frac{1}{2} L + \frac{A}{a}\right)\left[\frac{a^2}{|x|^2} + \frac{2a^2}{|x|^2}\right] = \frac{M_1}{|x|^2}$$

for $|x| \geq 2a$. Hence, $|\nabla u| \leq \sqrt{3}\, M_1/|x|^2$ for $|x| \geq 2a$. Finally, corresponding calculations (Prob. 3) produce the stated bound on $\partial^2 u/\partial x_i \partial x_j$. In the theorem, we may take $r_0 = 2a$ and M the largest of the computed bounds M_0, $\sqrt{3}\, M_1$, and M_2.

We are now in a position to formulate reasonable boundary value problems for exterior domains. Let D be an exterior domain. The *exterior Dirichlet problem* for D is to find a function $u \in C^2(D) \cap C(\bar{D})$, that satisfies

$\Delta u = h(x)$ in D, $\quad u = f(x)$ on B, $\quad u$ tends to zero uniformly at infinity.

The *exterior Neumann problem* for D is to find $u \in C^2(D) \cap C^1_\nu(\bar{D})$, such that

$\Delta u = h(x)$ in D, $\quad \dfrac{\partial u}{\partial \nu} = f(x)$ on B, $\quad u$ tends to zero uniformly at infinity.

The *exterior Robin problem* is to find u that satisfies Poisson's equation, vanishes at infinity and satisfies

$$\frac{\partial u}{\partial \nu} + \alpha(x)u = f(x) \quad \text{on} \quad B,$$

where $\alpha(x) \geq 0$ on B, $\alpha \not\equiv 0$.

Theorem 5-3. Let D be an exterior domain. Then the exterior Dirichlet problem has at most one solution.

Proof. Let $w = u_1 - u_2$, where u_1, u_2 solve the Dirichlet problem. Let $\varepsilon > 0$ and fix a point $x°$ in D. Choose $a > |x°|$ so large that $D_1 \subset K_a(0)$ and $|w| < \varepsilon$

on S_a, which is possible by the uniform convergence to zero at infinity. Since $w = 0$ on B, the maximum principle gives $|w(x)| < \varepsilon$ for $x \in D \cap K_a$. In particular, $|w(x^\circ)| < \varepsilon$. Since ε is arbitrary, we conclude that $w(x^\circ) = 0$, and then that $w \equiv 0$ because x° was any point in D. Thus, $u_1 = u_2$ on $\bar{D}$.

Theorem 5-4. Let D be an exterior domain with a twice-continuously differentiable boundary B. Then the exterior Neumann and Robin problems have at most one solution.

Proof. As usual, let $w = u_1 - u_2$ be the difference of two solutions to the exterior Neumann problem. Since w tends to zero uniformly at infinity, Ths. 5-1 and 5-2 imply that Green's first identity (3-2) applies with $v = u = w$. Thus,

$$\int_D |\nabla w|^2\, dx = \int_B w \frac{\partial w}{\partial \nu}\, d\sigma = 0,$$

and $\nabla w = 0$ in D. We conclude w is a constant which must be zero in view of the behavior at infinity. See Prob. 4 for the Robin problem.

PROBLEMS

1. Complete the proof of Th. 5-1 by establishing (3-3) and (3-4).
2. A point x in n-dimensional space can be represented either by rectangular coordinates $(x_1, x_2, \ldots, x_n)$ or by "polar" coordinates $(r, \theta_1, \ldots, \theta_{n-1})$, where $r = |x|$ and $\theta_1, \ldots, \theta_{n-1}$ are the angles between x and $n - 1$ of the coordinate axes. For example, when $n = 2$ [resp., $n = 3$] and the rectangular coordinates are (x, y) [resp., (x, y, z)], the customary choices are $\theta = \theta_1 = \arccos(x/r)$ [resp., $\theta = \theta_1 = \arccos(x/r)$ and $\phi = \theta_2 = \arccos(z/r)$]. In n dimensions, the Laplace operator has the polar form

$$\Delta = \frac{\partial^2}{\partial r^2} + \frac{n-1}{r}\frac{\partial}{\partial r} + \frac{1}{r^2}\Lambda(\theta_1, \ldots, \theta_{n-1}), \tag{5-6}$$

where $\Lambda = \Lambda(\theta_1, \ldots, \theta_{n-1})$ is a differential operator which only involves partial derivatives with respect to $\theta_1, \ldots, \theta_{n-1}$.

(a) Find Λ when $n = 2, 3$ (see Probs. 1 and 3 of Sec. 8-1).

(b) In n dimensions, inversion in the sphere $S_a(0)$ is defined by $y = a^2x/|x|^2$, where x and y are n-dimensional vector variables. The Kelvin transform of $u = u(x)$ for $|x| \geq a$ is given by

$$v = v(y) = \left(\frac{a}{\rho}\right)^{n-2} u\left(\frac{a^2}{\rho}, \theta_1, \ldots, \theta_{n-1}\right)$$

$0 < |y| \leq a$. If x has polar coordinates (r, θ) and y has polar coordinates (ρ, θ), where $\theta = (\theta_1, \ldots, \theta_{n-1})$ so that $\rho r = a^2$, show that

$$v_{\rho\rho}(\rho, \theta) + \frac{n-1}{\rho} v_\rho(\rho, \theta) = \left(\frac{a}{\rho}\right)^{n+2}\left[u_{rr}(r, \theta) + \frac{n-1}{r} u_r(r, \theta)\right],$$

$$\Lambda(\theta)v = \left(\frac{a}{\rho}\right)^{n-2} \Lambda(\theta)u, \quad \text{and hence} \quad \left(\frac{\rho}{a}\right)^{n+2} \Delta v = \Delta u.$$

(c) Write out the Kelvin transform when $n = 2$. Note that this is the only case when u is not scaled by a power of ρ^{-1}.
(d) Establish (5-6) for $n > 3$.

3. Establish the bound for $\partial^2 u/\partial x_i\, \partial x_j$ in Th. 5-2.

4. Prove uniqueness for the exterior Robin problem, and so complete the proof of Th. 5-4. Assume that $\alpha(x)$ is bounded and integrable.

5. Let D be an exterior domain in two dimensions and u a bounded harmonic function in D.
(a) Fix $a > 0$ so that $D_1 \subset K_a(0)$, and let $v(y)$, $0 < |y| \leq a$, be the Kelvin transform of $u(x)$ for $|x| \geq a$. Show that v has a harmonic extension to $|y| < a$. Hint. If M is a bound for u on D, replace (5-4) by $2M \leq \varepsilon[\alpha^{-1} - a^{-1}]$ to determine α.
(b) Conclude that $u(x)$ has a limit as $|x| \to \infty$.

6. Let D be an exterior domain in two dimensions and u be a bounded, harmonic function in D. Derive the estimates $|\nabla u| \leq M/|x|^2$ and $|\partial^2 u/\partial x_i\, \partial x_j| \leq M/|x|^3$ for all $|x|$ large.

7. Let D be the complement of the disk $\bar{K}_a(0)$ in two dimensions. Consider the exterior Dirichlet problem $\Delta u = 0$ in D, $u = f(x)$ on $S_a(0)$, and u bounded in D.
(a) Solve formally by separation of variables.
(b) Convert the series in part (a) into a Poisson integral representation.
(c) Get this new Poisson formula directly from the old one using the Kelvin transform.

8. Do part (c) of Prob. 7 for a ball in 3-space.

9. Let D be an exterior domain in two dimensions. Show that there exists at most one solution to the Dirichlet problem in the case that the solutions are bounded at infinity. *Hint.* Use Lemma 4-1 and the comparison function $v = M(\log r/R)/(\log a/R)$. Here, if u_1 and u_2 are two solutions, M is a constant greater than $|u_1| + |u_2|$, R is the radius of a circle lying in D_1 centered at the origin, and a is the radius of a concentric circle which is so large that D_1 is contained in its interior. Show that in $\bar{D} \cap \bar{K}_a$, $|u_1 - u_2| \leq v$ and then let $a \to \infty$.

10. Let D be an exterior domain in two dimensions. Show that any two bounded solutions to the Neumann problem differ by a constant.

11. Let D be an exterior domain in two dimensions. Show that there exists at most one bounded solution to the Robin problem.

12. For the proof of Th. 5-2, we needed to show that a function u harmonic in a domain D except at one point x° in D could, under certain conditions, be extended to be harmonic at x°. Prove that such a harmonic extension exists if

$$\lim_{x \to x^\circ} |x - x^\circ| u(x) = 0$$

uniformly as $x \to x^\circ$.

13. Let u be harmonic in a domain D in two dimensions, except at a point $x^\circ \in D$. Show that if $|u(x)| \leq \mu(r) \log 1/r$, where $\mu(r) \to 0$ as $r \to 0$ and $r = |x - x^\circ|$, then u may be defined at x° so that u is harmonic throughout D. Show that the same conclusion holds if u is bounded. [You have already confirmed this last assertion if you solved part (a) of Prob. 5.]

14. Let D be an exterior domain. A function u harmonic in D is called *harmonic at infinity* if its Kelvin transform v can be extended to be harmonic at the origin.
(a) Show that $u(x) = k$, a constant, is not harmonic at infinity for dimensions $n > 2$, but is harmonic at infinity when $n = 2$.
(b) Show that $u(x) = 1/|x|$ is harmonic at infinity when $n = 3$.

(c) Refer to the Kelvin transform in Prob. 2. If u is harmonic at infinity, show that it is natural to define $u(\infty) = 0$ for $n \geq 3$ and $u(\infty) = v(0)$ for $n = 2$.

(d) Prove this version of the maximum principle: Let u be harmonic in $D \cup \{\infty\}$, where D is an exterior domain. Then u takes its maximum and minimum values either at ∞ or on B, the boundary of D. *Hint.* Assume not and reduce to a finite domain in which the maximum principle is contradicted.

-6. Properties of Potentials

If D is a bounded, normal domain and u is a function with continuous second-order partials in D which also has continuous first-order partials in $\bar{D}$, then

$$u(x) = \frac{1}{4\pi}\int_B \frac{1}{r}\frac{\partial u}{\partial v}\,d\sigma_y - \frac{1}{4\pi}\int_B u\frac{\partial}{\partial v}\left(\frac{1}{r}\right)d\sigma_y - \frac{1}{4\pi}\int_D \frac{1}{r}\,\Delta u\,dy. \tag{6-1}$$

This representation, which is Eq. (3-7), also holds for an exterior, normal domain if we assume that

$$u(x) = O(|x|^{-1}), \qquad \nabla u = O(|x|^{-2}), \quad \text{and} \quad \Delta u = O(|x|^{-3}) \tag{6-2}$$

as $|x| \to \infty$ (see Prob. 1). Notice that if u is harmonic is an exterior domain and tends to zero uniformly at infinity, then conditions (6-2) hold by Th. 5-2 and we obtain the representation

$$u(x) = \frac{1}{4\pi}\int_B \left[\frac{1}{r}\frac{\partial u}{\partial v} - u\frac{\partial}{\partial v}\left(\frac{1}{r}\right)\right]d\sigma_y, \tag{6-3}$$

both for a bounded and an exterior domain.

We turn now to the study of the single layer, double layer, and volume potentials represented in (6-1). First, consider a typical volume potential,

$$w(x) = \frac{1}{4\pi}\int_D \frac{1}{r_{xy}} f(y)\,dy = \frac{1}{4\pi}\int_D \frac{f(y)}{|x-y|}\,dy, \tag{6-4}$$

for $x \in R_3$. We assume throughout that D is a bounded domain and that f is continuous in D. If D is an exterior domain, we assume in addition that

$$|f(y)| = O\left(\frac{1}{|y|^3}\right) \quad \text{as} \quad |y| \to \infty.$$

For points $x \notin D$ the integrand in (6-4) has derivatives of all orders with respect to both x and y, and by standard results, differentiation under the integral sign is permissible. Thus, $w(x)$ is infinitely differentiable on $c\bar{D}$ and $\Delta w = 0$ there. Our main concern is with the behavior of $w(x)$ for x in $\bar{D}$. For such an x the integral in (6-4) is improper, but convergent (Prob. 2).

Proposition 6-1. Let f be piecewise continuous in D. Then the volume potential $w(x)$ in (6-4) belongs to $C^1(R_3)$.

Proof. You may have noticed that w is the integral transform of f relative to the weakly singular kernel $|x-y|^{-1}$ (see Sec. 7-7). To study this transform, we "smooth" the kernel. Define

$$r_n(r) = \begin{cases} \dfrac{1}{r}, & r \geq \dfrac{1}{n}, \\ \dfrac{n}{2}(3 - n^2r^2), & 0 \leq r \leq \dfrac{1}{n}. \end{cases}$$

It is easy to check that $r_n(r)$ is continuously differentiable. Set

$$w_n(x) = \frac{1}{4\pi}\int_D r_n(r_{xy})f(y)\,dy. \tag{6-5}$$

The integrand in (6-5) is continuously differentiable in both x and y and derivatives may be taken under the integral sign. We use these properties to infer that the volume potential is continuously differentiable. First,

$$\begin{aligned} |w(x) - w_n(x)| &= \left|\frac{1}{4\pi}\int_D \left[\frac{1}{r} - r_n(r)\right] f(y)\,dy\right| \\ &\leq \frac{M}{4\pi}\int_{K_{1/n}(x)} \left[\frac{1}{r} - \frac{n}{2}(3 - n^2r^2)\right] dy \\ &\leq M\int_0^{1/n} \left[\frac{1}{r} - \frac{n}{2}(3 - n^2r^2)\right] r^2\,dr = \frac{M}{10n^2}, \end{aligned}$$

where M bounds $|f(y)|$ on D and spherical coordinates were used. Thus, $w_n(x) \to w(x)$ uniformly in R_3. Since each w_n is continuous, we conclude that $w(x)$ is continuous in R_3. Second, we show that w has continuous partial derivatives given by

$$\frac{\partial w}{\partial x_i}(x) = \frac{1}{4\pi}\int_D \frac{\partial}{\partial x_i}\left(\frac{1}{r_{xy}}\right) f(y)\,dy. \tag{6-6}$$

Problem 3 shows that this improper integral converges. We know that $\partial w_n/\partial x_i$ can be calculated by differentiation under the integral sign. Hence,

$$\begin{aligned} \left|\frac{\partial w_n}{\partial x_i}(x) - \int_D \frac{\partial}{\partial x_i}\left(\frac{1}{r_{xy}}\right) f(y)\,dy\right| &= \left|\frac{1}{4\pi}\int_{K_{1/n}(x)} f(y)\left[\frac{\partial r_n}{\partial x_i} - \frac{\partial}{\partial x_i}\left(\frac{1}{r}\right)\right] dy\right| \\ &= \left|\frac{1}{4\pi}\int_{K_{1/n}(x)} f(y)\frac{x_i - y_i}{r}\left[-n^3r + \frac{1}{r^2}\right] dy\right| \\ &\leq M\int_0^{1/n}\left[n^3r + \frac{1}{r^2}\right] r^2\,dr = \frac{5M}{4n}. \end{aligned}$$

Consequently,

$$\frac{\partial w_n}{\partial x_i}(x) \to \frac{1}{4\pi}\int_D \frac{\partial}{\partial x_i}\left(\frac{1}{r_{xy}}\right) f(y)\,dy \equiv v_i(x) \tag{6-7}$$

uniformly on R_3. Let x° be fixed in D. Then

$$w_n(x) - w_n(x^\circ) = \int_{x^\circ}^{x} \nabla w_n(z) \cdot dz,$$

where the line integral may be taken along any path from x° to x lying in D. In view of the uniform convergence in (6-7),

$$w(x) - w(x^\circ) = \int_{x^\circ}^{x} \sum_i v_i(z)\, dz_i,$$

and the fundamental theorem of calculus implies that $\partial w/\partial x_i$ exists and

$$\frac{\partial w}{\partial x_i}(x) = v_i(x) = \frac{1}{4\pi}\int_D \frac{\partial}{\partial x_i}\left(\frac{1}{r_{xy}}\right) f(y)\, dy,$$

as claimed.

We observed that $\Delta w = 0$ outside $\bar{D}$. Inside D we have

Theorem 6-1. Suppose that $f \in C^1(\bar{D})$. Then the volume potential w in (6-4) belongs to $C^2(D) \cap C^1(\bar{D})$ and satisfies the Poisson equation $\Delta w = -f(x)$ for $x \in D$.

Proof. $w \in C^1(\bar{D})$ by Prop. 6-1. The confirmation that $w \in C^2(D)$ is more delicate. Indeed, direct differentiation under the integral sign in (6-6) leads to possibly divergent integrals and does not give the correct expressions for the second-order partials of w in D. To overcome this problem, we first rewrite (6-6) using the Gauss divergence theorem and the useful observation,

$$\frac{\partial}{\partial x_i}\left(\frac{1}{r_{xy}}\right) = -\frac{\partial}{\partial y_i}\left(\frac{1}{r_{xy}}\right), \tag{6-8}$$

to obtain

$$\begin{aligned}
\frac{\partial w}{\partial x_i}(x) &= \frac{1}{4\pi}\int_D \frac{\partial}{\partial x_i}\left(\frac{1}{r_{xy}}\right) f(y)\, dy = -\frac{1}{4\pi}\int_D \frac{\partial}{\partial y_i}\left(\frac{1}{r_{xy}}\right) f(y)\, dy \\
&= -\frac{1}{4\pi}\int_D \frac{\partial}{\partial y_i}\left[\frac{1}{r_{xy}} f(y)\right] dy + \frac{1}{4\pi}\int_D \frac{1}{r_{xy}} f_{y_i}(y)\, dy \\
&= -\frac{1}{4\pi}\int_B \frac{1}{r_{xy}} f(y) v_i\, d\sigma_y + \frac{1}{4\pi}\int_D \frac{1}{r_{xy}} f_{y_i}(y)\, dy,
\end{aligned}$$

where v_i is the dot product of the outer unit normal v and the ith coordinate unit basis vector. This representation shows that $\partial w/\partial x_i$ is continuously differentiable because the integral over D is a volume potential and the surface integral has an infinitely differentiable integrand, since $x \in D$. Thus,

$$\frac{\partial^2 w}{\partial x_i^2}(x) = \frac{1}{4\pi}\int_B \left(\frac{\partial}{\partial y_i}\frac{1}{r_{xy}}\right) f(y) v_i\, d\sigma_y - \frac{1}{4\pi}\int_D \left(\frac{\partial}{\partial y_i}\frac{1}{r_{xy}}\right) f_{y_i}(y)\, dy.$$

Addition and manipulation of integrals leads to (see Prob. 5)

(6-9) $\Delta w(x)$

$$= \frac{1}{4\pi} \int_B \left(\frac{\partial}{\partial v_y} \frac{1}{r_{xy}} \right) f(y)\, d\sigma_y - \frac{1}{4\pi} \int_D \nabla_y \frac{1}{r_{xy}} \cdot \nabla_y [f(y) - f(x)]\, dy$$

$$= \frac{1}{4\pi} \int_B \left(\frac{\partial}{\partial v_y} \frac{1}{r_{xy}} \right) f(y)\, d\sigma_y - \frac{1}{4\pi} \int_D \operatorname{div}_y \left\{ [f(y) - f(x)] \nabla_y \frac{1}{r_{xy}} \right\} dy$$

$$= \frac{1}{4\pi} \int_B \left(\frac{\partial}{\partial v_y} \frac{1}{r_{xy}} \right) f(y)\, d\sigma_y - \frac{1}{4\pi} \int_B \left(\frac{\partial}{\partial v_y} \frac{1}{r_{xy}} \right) [f(y) - f(x)]\, d\sigma_y$$

$$= \frac{f(x)}{4\pi} \int_B \frac{\partial}{\partial v_y} \frac{1}{r_{xy}}\, d\sigma_y = -f(x)$$

by (3-8).

We turn next to the single- and double-layer potentials. The more difficult of the two is the double-layer potential, which we consider first. Suppose, then, that D is a bounded domain and set

$$u(x) = \int_B \left(\frac{\partial}{\partial v_y} \frac{1}{4\pi r_{xy}} \right) \mu(y)\, d\sigma_y, \tag{6-10}$$

where $\mu(y)$ is a continuous function defined on the surface B. Obviously, if $x \in cB$, the double-layer potential $u(x)$ is infinitely differentiable and satisfies $\Delta u = 0$. The critical questions relate to what happens to u when $x \in B$ or when x approaches B from inside or outside D. To answer these questions, we must be very precise about the nature of the boundary B of D.

We make the following hypotheses concerning the surface B.

Assumption A. Let $x^\circ \in B$. Then there is a ball $K_\delta(x^\circ)$ such that the portion of surface $B \cap K_\delta(x^\circ)$ is given by a twice-continuously differentiable function. That is, $B \cap K_\delta$ is given by an equation of the form $x_3 = \phi(x_1, x_2)$ [or $x_1 = \phi(x_2, x_3)$ or $x_2 = \phi(x_1, x_3)$], where ϕ has continuous second-order partials. Furthermore, there is a constant C such that

$$\int_B \left| \frac{\partial}{\partial v_y} \frac{1}{4\pi r_{xy}} \right| d\sigma_y \le C \quad \text{for all} \quad x. \tag{6-11}$$

A few comments are in order. The smoothness restrictions on ϕ (hence on B) guarantee that the integral in (6-10) is convergent for points $x \in B$ (see Prop. 6-2). Thus, the double-layer potential in (6-10) is defined for all $x \in R_3$. The final requirement in Assumption A places a geometric restriction on B. For example, if B bounds a convex domain, then (6-11) holds with $C = 4\pi$. More generally, if a ray from any point cuts B at most p times, then (6-11) holds with $C = 4\pi p$ (see Probs. 6–8).

We turn now to the existence of double-layer potentials.

Proposition 6-2. Let μ be continuous on a surface B that satisfies Assumption A with the exception of (6-11). Then the double-layer potential $u(x)$ in

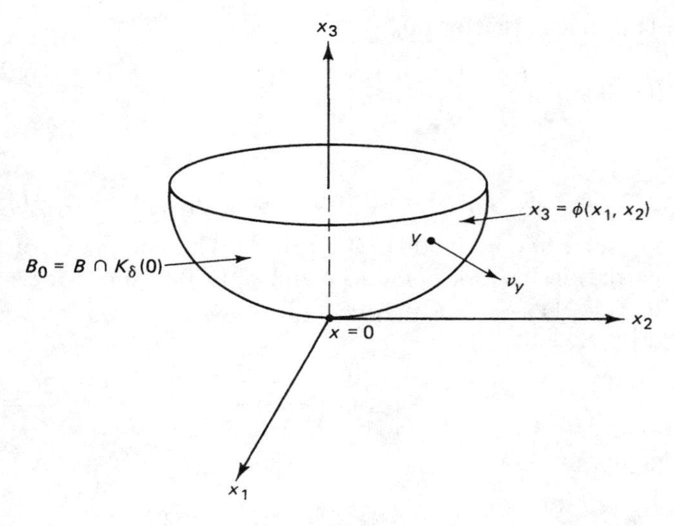

Figure 8-4

(6-10) exists for all $x \in R_3$ and is infinitely differentiable on cB, where it satisfies $\Delta u = 0$.

Proof. We have already confirmed all of these assertions except for the existence of the improper integral when $x \in B$. So, fix $x \in B$ and let $K_\delta(x)$ be a ball as mentioned in Assumption A. We set up coordinates so that the origin is at x, the x_1- and x_2-axes are in the tangent plane to B at x, the positive x_3-axis is taken along the *inner* normal to B at x, and $B_0 = B \cap K_\delta(0)$ has the equation

$$x_3 = \phi(x_1, x_2), \qquad x_1^2 + x_2^2 \leq \delta^2$$

(see Fig. 8-4). By our choice of coordinates

$$\phi(0, 0) = \phi_{x_1}(0, 0) = \phi_{x_2}(0, 0) = 0.$$

Let $B_1 = B - B_0$ and M be a bound for μ on B. Then

$$\int_B \left|\left(\frac{\partial}{\partial \nu_y} \frac{1}{r_{xy}}\right) \mu(y)\right| d\sigma_y \leq M \int_{B_0} \left|\frac{\partial}{\partial \nu_y} \frac{1}{r_{xy}}\right| d\sigma_y + M \int_{B_1} \left|\frac{\partial}{\partial \nu_y} \frac{1}{r_{xy}}\right| d\sigma_y,$$

where $x = 0$. The integral over B_1 exists; indeed, it has a continuous integrand because $|x - y| \geq \delta > 0$ on B_1. To show that the integral in (6-10) is absolutely convergent, we need only show that the integral over B_0 is finite. This requires a few estimates. Observe that

$$\nu_y = \frac{(\phi_{y_1}, \phi_{y_2}, -1)}{(1 + |\nabla\phi|^2)^{1/2}}, \qquad d\sigma_y = (1 + |\nabla\phi|^2)^{1/2}\, dy_1\, dy_2$$

and

$$\frac{\partial}{\partial \nu_y} \frac{1}{r_{0y}} = \nabla_y \frac{1}{r_{0y}} \cdot \nu_y = -\frac{(y_1, y_2, y_3)}{r_{0y}^3} \cdot \nu_y.$$

Thus (recall that $x = 0$),

$$\text{(6-12)} \quad \int_{B_0} \left| \frac{\partial}{\partial \nu_y} \frac{1}{r_{xy}} \right| d\sigma_y \le \int_{y_1^2 + y_2^2 \le \delta^2} \frac{|y_1 \phi_{y_1}| + |y_2 \phi_{y_2}| + |y_3|}{r_{xy}^3} \, dy_1 \, dy_2.$$

Now

$$\phi_{y_1}(y_1, y_2) = \phi_{y_1}(0, 0) + \phi_{y_1 y_1}(\eta_1, \eta_2) y_1 + \phi_{y_1 y_2}(\eta_1, \eta_2) y_2$$

for some (η_1, η_2) with $\eta_1^2 + \eta_2^2 \le \delta^2$. The continuity of the second-order partials in the disk of radius δ and $\phi_{y_1}(0, 0) = 0$ provide a constant C_1 such that $|\phi_{y_1}(y_1, y_2)| \le C_1(|y_1| + |y_2|)$. Since $r_{xy} = (y_1^2 + y_2^2 + y_3^2)^{1/2}$, we have $|y_i| \le r_{xy}$ and

$$\left| \frac{y_1 \phi_{y_1}}{r_{xy}^3} \right| \le C_1 \left(\frac{|y_1|^2}{r_{xy}^2} + \frac{|y_1||y_2|}{r_{xy}^2} \right) \frac{1}{r_{xy}} \le \frac{2C_1}{r_{xy}}.$$

Similarly,

$$\left| \frac{y_2 \phi_{y_2}}{r_{xy}^3} \right| \le \frac{2C_2}{r_{xy}}.$$

Finally, use a second-order Taylor expansion to find

$$\begin{aligned} \left| \frac{y_3}{r_{xy}^3} \right| &= \left| \frac{\phi(y_1, y_2)}{r_{xy}^3} \right| \\ &= \left| \frac{\phi_{y_1 y_1}(\eta_1, \eta_2) y_1^2 + 2\phi_{y_1 y_2}(\eta_1, \eta_2) y_1 y_2 + \phi_{y_2 y_2}(\eta_1, \eta_2) y_2^2}{r_{xy}^3} \right| \le \frac{4C_3}{r_{xy}}, \end{aligned}$$

where C_3 bounds the second partials of ϕ in the disk of radius δ. Use of these bounds in (6-12) leads to

$$\int_{B_0} \left| \frac{\partial}{\partial \nu_y} \frac{1}{r_{xy}} \right| d\sigma_y \le \int_0^{2\pi} \int_0^{\delta} \frac{C_4}{r} \, r \, dr \, d\theta = 2\pi \delta C_4 < \infty,$$

where polar coordinates were used. This completes the verification that the integral in (6-10) converges even when $x \in B$ and completes the proof of Prop. 6-2.

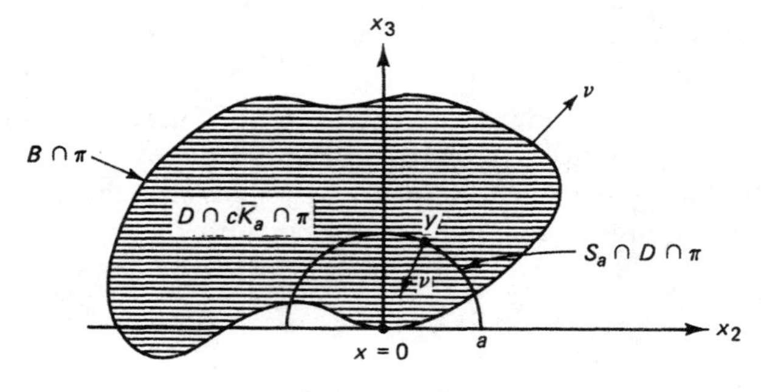

Figure 8-5

Although the double-layer potential exists for all x and is infinitely differentiable on cB, we cannot expect it to be continuous across B. Indeed, if $\mu(y) \equiv 1$, (3-8) yields $u(x) = -1$ for $x \in D$ and $u(x) = 0$ for $x \in c\bar{D}$. In the case $\mu \equiv 1$, we can even evaluate u on B. To this end, fix a point $x \in B$. We choose coordinates as in Fig. 8-4 so $x = 0$ and the figure shows the portion of B in $K_\delta(0)$. Figure 8-5 shows the intersection of all of B and the intersection of the ball $K_a(0)$, $0 < a \le \delta$, with the x_2x_3 plane. Apply the Gauss divergence theorem to the domain $D \cap c\bar{K}_a$ to obtain

$$0 = \int_{D \cap c\bar{K}_a} \operatorname{div}_y \left(\nabla_y \frac{1}{r_{xy}}\right) dy = \int_{B \cap cK_a} \left(\frac{\partial}{\partial v_y} \frac{1}{r_{xy}}\right) d\sigma_y + \int_{S_a \cap D} \left(\frac{\partial}{\partial v_y} \frac{1}{r_{xy}}\right) d\sigma_y,$$

$$\int_{B \cap c\bar{K}_a} \left(\frac{\partial}{\partial v_y} \frac{1}{r_{xy}}\right) d\sigma_y = -\int_{S_a \cap D} \left(\frac{\partial}{\partial v_y} \frac{1}{r_{xy}}\right) d\sigma_y.$$

On $S_a \cap D$ we have $v_y = (0 - y)/a$ and

$$\frac{\partial}{\partial v_y} \frac{1}{r_{xy}} = \frac{-y}{a^3} \cdot \frac{-y}{a} = \frac{a^2}{a^4} = \frac{1}{a^2}.$$

Consequently,

$$\int_{B \cap c\bar{K}_a} \left(\frac{\partial}{\partial v_y} \frac{1}{4\pi r_{xy}}\right) d\sigma_y = -\int_{S_a \cap D} \frac{1}{4\pi a^2}\, d\sigma_y = -\frac{|S_a \cap D|}{4\pi a^2},$$

where $|S_a \cap D|$ is the area of $S_a \cap D$. It is reasonably clear from Fig. 8-5 that

$$\frac{|S_a \cap D|}{4\pi a^2} \to \frac{2\pi a^2}{4\pi a^2} = \frac{1}{2}$$

as $a \to 0$. See Prob. 9 for the mathematical details. Therefore,

$$\int_B \left(\frac{\partial}{\partial v_y} \frac{1}{4\pi r_{xy}}\right) d\sigma_y = \lim_{a \to 0} \int_{B \cap c\bar{K}_a} \left(\frac{\partial}{\partial v_y} \frac{1}{4\pi r_{xy}}\right) d\sigma_y = -\frac{1}{2}.$$

Together with (3-8) this establishes

Theorem 6-2 (Gauss). Let D be a bounded domain with boundary B which satisfies Assumption A except for (6-11). Then

$$\int_B \left(\frac{\partial}{\partial v_y} \frac{1}{4\pi r_{xy}}\right) d\sigma_y = \begin{cases} -1, & x \in D, \\ -\frac{1}{2}, & x \in B, \\ 0, & x \in c\bar{D}. \end{cases}$$

Gauss' theorem enables us to prove the fundamental jump relations of potential theory. For double-layer potentials we have

Theorem 6-3. Let D be a bounded domain whose boundary B satisfies Assumption A. Suppose that μ is continuous on B. Then the double-layer potential

$$u(x) = \int_B \left(\frac{\partial}{\partial \nu_y} \frac{1}{4\pi r_{xy}}\right) \mu(y)\, d\sigma_y$$

satisfies the following jump relations at points $x^\circ \in B$.

(i) $\lim\limits_{\substack{x \to x^\circ \\ x \in D}} u(x) = u(x^\circ) - \frac{1}{2}\mu(x^\circ)$.

(ii) $\lim\limits_{\substack{x \to x^\circ \\ x \in cD}} u(x) = u(x^\circ) + \frac{1}{2}\mu(x^\circ)$.

Proof. We claim that the function

$$g(x) = \int_B \left(\frac{\partial}{\partial \nu_y} \frac{1}{4\pi r_{xy}}\right) [\mu(y) - \mu(x^\circ)]\, d\sigma_y \text{ is continuous.} \tag{6-13}$$

Assume this momentarily. Since

$$g(x) = u(x) - \frac{\mu(x^\circ)}{4\pi} \int_B \left(\frac{\partial}{\partial \nu_y} \frac{1}{r_{xy}}\right) d\sigma_y$$

is continuous, the Gauss theorem yields, as $x \to x^\circ$ with $x \in D$,

$$u(x) = g(x) - \mu(x^\circ) \to g(x^\circ) - \mu(x^\circ) = u(x^\circ) - \tfrac{1}{2}\mu(x^\circ),$$

which proves (i). Conclusion (ii) follows in the same way.

To establish (6-13), write

$$g(x) - g(x^\circ) = \int_B \left[\frac{\partial}{\partial \nu_y}\left(\frac{1}{4\pi r_{xy}} - \frac{1}{4\pi r_{x^\circ y}}\right)\right] [\mu(y) - \mu(x^\circ)]\, d\sigma_y. \tag{6-14}$$

Let $\varepsilon > 0$ and choose $\delta > 0$ so that $|\mu(y) - \mu(x^\circ)| < \varepsilon/4C$ for $y \in B \cap K_\delta(x^\circ) = B_0$, where C is the constant in Assumption A. Let $x \in K_{\delta/2}(x^\circ)$ and let $h(x)$ be the integral over $B_1 = B - B_0$ of the integrand in (6-14). Obviously, $h(x)$ is continuous on $K_{\delta/2}(x^\circ)$ and $h(x^\circ) = 0$. Now

$$|g(x) - g(x^\circ)| \le \frac{\varepsilon}{4C} \int_{B_0} \left(\left|\frac{\partial}{\partial \nu} \frac{1}{4\pi r_{xy}}\right| + \left|\frac{\partial}{\partial \nu} \frac{1}{4\pi r_{x^\circ y}}\right|\right) d\sigma_y + |h(x)| < \frac{\varepsilon}{2} + |h(x)| < \varepsilon$$

if $|x - x^\circ|$ is the small enough because $h(x) \to h(x^\circ) = 0$. Thus, (6-13) holds and the proof is complete.

Finally, we turn to the single-layer potential

$$v(x) = \int_B \frac{1}{4\pi r_{xy}} \rho(y)\, d\sigma_y. \tag{6-15}$$

Reasoning more elementary than that in the proof of Prop. 6-2 establishes

Proposition 6-3. Let ρ be continuous on a surface B which satisfies Assumption A except for (6-11). Then the single-layer potential $v(x)$ in (6-15) exists and is continuous for all x in R_3 and is infinitely differentiable on cB, where it satisfies $\Delta v = 0$.

The details of the proof are left for Prob. 12. Although v is continuous on R_3 and infinitely differentiable on cB, the normal derivative of $v(x)$ jumps as x passes across B.

Theorem 6-4. Let D be a bounded domain whose boundary B satisfies Assumption A. Suppose that ρ is continuous on B. Then the single-layer potential

$$v(x) = \int_B \frac{1}{4\pi r_{xy}} \rho(y)\, d\sigma_y$$

satisfies the following jump relations. Let $x^\circ \in B$ and L_0 be the line through x° parallel to $\nu^\circ = \nu(x^\circ)$, the unit outer normal to B at x°. Then

$$\text{(i)} \quad \lim_{\substack{x \to x^\circ \\ x \in L_0 \cap D}} \frac{\partial v}{\partial \nu^\circ}(x) = \int_B \left(\frac{\partial}{\partial \nu^\circ} \frac{1}{4\pi r_{x^\circ y}} \right) \rho(y)\, d\sigma_y + \frac{1}{2}\rho(x^\circ);$$

$$\text{(ii)} \quad \lim_{\substack{x \to x^\circ \\ x \in L_0 \cap c\bar{D}}} \frac{\partial v}{\partial \nu^\circ}(x) = \int_B \left(\frac{\partial}{\partial \nu^\circ} \frac{1}{4\pi r_{x^\circ y}} \right) \rho(y)\, d\sigma_y - \frac{1}{2}\rho(x^\circ).$$

Proof. Let $x^\circ \in B$. We proceed as in the proof of Prop. 6-2 and Fig. 8-4 to choose coordinates so that $x^\circ = (0, 0, 0)$ and $B_0 = B \cap K_\delta(x^\circ)$ is described by

$$x_3 = \phi(x_1, x_2) \quad \text{for} \quad x_1^2 + x_2^2 \le \delta^2 \quad \text{with} \quad \nu^\circ = (0, 0, -1)$$

and

$$\phi(0, 0) = \phi_{x_1}(0, 0) = \phi_{x_2}(0, 0) = 0.$$

With this choice of coordinates, L_0 is the x_3-axis. The normal ν_y and $d\sigma_y$ may be computed exactly as before,

$$\nu_y = \frac{(\phi_{y_1}, \phi_{y_2}, -1)}{N} \quad \text{and} \quad d\sigma_y = N\, dy_1\, dy_2,$$

where $N = (1 + \phi_{y_1}^2 + \phi_{y_2}^2)^{1/2}$. For $x \to x^\circ$ along L_0 we have $x = (0, 0, x_3)$ with $x_3 \to 0$. Now, for $x \ne x^\circ$,

$$\text{(6-16)} \quad \begin{aligned} \frac{\partial v}{\partial \nu^\circ}(x) &= \int_B \left(\frac{\partial}{\partial \nu^\circ} \frac{1}{4\pi r_{xy}} \right) \rho(y)\, d\sigma_y \\ &= \int_B \left[\left(\frac{\partial}{\partial \nu^\circ} + \frac{\partial}{\partial \nu_y} \right) \frac{1}{4\pi r_{xy}} \right] \rho(y)\, d\sigma_y \\ &\quad - \int_B \left(\frac{\partial}{\partial \nu_y} \frac{1}{4\pi r_{xy}} \right) \rho(y)\, d\sigma_y, \\ &= h(x) - u(x), \end{aligned}$$

where $h(x)$ is the first integral and $u(x)$ is the second, a double-layer potential with density ρ. We claim that $h(x)$ is continuous on $L_0 \cap K_\delta(x^\circ)$. Granted this, the jump relations for double-layer potentials and (6-16) immediately yield the jump relations for single-layer potentials (Prob. 13). The continuity of $h(x)$ for $x \in L_0 \cap K_\delta(x^\circ)$ and $x \neq x^\circ$ is clear. To check continuity at x°, we express the integral defining $h(x)$ as a sum of two integrals, $h_0(x)$ over B_0 and $h_1(x)$ over $B_1 = B - B_0$. Then

$$|h(x) - h(x^\circ)| \leq |h_0(x)| + |h_0(x^\circ)| + |h_1(x) - h_1(x^\circ)|.$$

We shall prove that given $\varepsilon > 0$, $|h_0(x)| < \varepsilon/3$ for all $x \in K_\delta(x^\circ) \cap L_0$ provided that δ is fixed suitably small. Since $h_1(x)$ involves integration over B_1, it is continuous for $x \in K_\delta(x^\circ) \cap L_0$ and consequently $|h_1(x) - h_1(x^\circ)| < \varepsilon/3$ for $|x - x^\circ|$ small enough. Combining these remarks gives $|h(x) - h(x^\circ)| < \varepsilon$ for $|x - x^\circ|$ sufficiently small, and $h(x)$ is continuous on $L_0 \cap K_\delta(x^\circ)$, as claimed.

It remains to prove that $|h_0(x)|$ can be made arbitrarily small on $L_0 \cap K_\delta(x^\circ)$ by taking δ small enough. For $x \in L_0 \cap K_a(x^\circ)$ [recall that $\partial/\partial v^\circ$ operates on $x = (0, 0, x_3)$]

$$\text{(6-17)} \quad \left(\frac{\partial}{\partial v^\circ} + \frac{\partial}{\partial v_y}\right)\frac{1}{4\pi r_{xy}} = \frac{x - y}{4\pi r_{xy}^3} \cdot v_0 - \frac{x - y}{4\pi r_{xy}^3} \cdot v_y$$

$$= \frac{x - y}{4\pi r_{xy}^3} \cdot \left[(0, 0, -1) - \frac{(\phi_{y_1}, \phi_{y_2}, -1)}{N}\right]$$

$$= \frac{y_1\phi_{y_1} + y_2\phi_{y_2} + (1 - N)(x_3 - y_3)}{4\pi r_{xy}^3 N}.$$

Just as in the proof of Prop. 6-2, we have

$$\left|\frac{y_1\phi_{y_1}}{r_{xy}^3}\right| \leq \frac{2C_1}{r_{xy}} \quad \text{and} \quad \left|\frac{y_2\phi_{y_2}}{r_{xy}^3}\right| \leq \frac{2C_2}{r_{xy}}.$$

Also,

$$\left|\frac{1 - N}{N}\right| = \frac{N^2 - 1}{N(N + 1)} = \frac{\phi_{y_1}^2 + \phi_{y_2}^2}{N(N + 1)} \leq \phi_{y_1}^2 + \phi_{y_2}^2,$$

and we established earlier that $|\phi_{y_i}(y_1, y_2)| \leq C_i(|y_1| + |y_2|)$. Thus,

$$\left|\frac{(1 - N)(x_3 - y_3)}{4\pi r_{xy}^3 N}\right| \leq (C_1^2 + C_2^2)\frac{(|y_1| + |y_2|)^2}{4\pi r_{xy}^2}\frac{x_3 - y_3}{r_{xy}} \leq C_5,$$

a constant. Use of these estimates in (6-17) leads to

$$\left|\left(\frac{\partial}{\partial v^\circ} + \frac{\partial}{\partial v_y}\right)\frac{1}{4\pi r_{xy}}\right| \leq \frac{C_6}{r_{xy}}$$

for some constant C_6. Finally, we use this estimate in the integral defining $h_0(x)$ to find

$$|h_0(x)| \leq \int_{B_0} \frac{C_6}{r_{xy}} |\rho(y)|\, d\sigma_y \leq \int_{y_1^2 + y_2^2 \leq \delta^2} \frac{C_6}{r_{xy}} |\rho(y)| N(y)\, dy_1\, dy_2$$

$$\leq C_7 \int_{y_1^2 + y_2^2 \leq \delta^2} \frac{1}{r_{xy}}\, dy_1\, dy_2 = C_7 \int_0^{2\pi}\int_0^\delta \frac{1}{r} r\, dr\, d\theta = 2\pi C_7 \delta.$$

Hence, $|h_0(x)|$ can be made uniformly small on $L_0 \cap K_\delta(x^\circ)$ by choosing δ small enough, and the proof is complete.

The jump relations established in Ths. 6-3 and 6-4 hold uniformly for $x^\circ \in B$. To confirm that the limits are uniform with respect to $x^\circ \in B$, you need only look closely at the proofs given earlier. One key point is the uniform bound on the total solid angle of Assumption A. We omit the details.

PROBLEMS

1. Show that (6-1) is valid for exterior domains when (6-2) holds. *Hint.* Apply the reasoning used to prove (3-7) to $D_R = D \cap K_R(0)$ and then let $R \to \infty$.
2. Show that (6-4) is a convergent improper integral for any $x \in \bar{D}$. *Hint.* $D = (K_\delta(x) \cap D) \cup D_1$, where $D_1 = D - (K_\delta(x) \cap D)$. The integral over D_1 is proper. Use spherical coordinates to estimate the other integral.
3. Show that the integral in (6-6) converges.
4. Check (6-8).
5. Refer to (6-9). The divergence theorem was applied to $F = [f(y) - f(x)]\, \nabla_y r_{xy}^{-1}$, which is singular at $y = x$.
   (a) Justify the use of the divergence theorem in this case by applying the usual theorem to F on $D - K_\delta(x)$ for small δ. Then let $\delta \to 0$.
   (b) Note that the divergence theorem does not apply to $F = f(y)\, \nabla_y r_{xy}^{-1}$.

Problems 6–8 deal with the geometric significance of (6-11) in Assumption A.

6. Let B be a bounded smooth surface in space and x a fixed point not on B. Suppose that each ray from x which intersects B does so exactly once. The union of the line segments from x to all points of B forms a cone C_x, with vertex x. Let $K_\delta(x)$, δ small, be a ball centered at x. The *solid angle* ω_x of the cone C_x is $|C_x \cap S_\delta(x)|/\delta^2$, where $|C_x \cap S_\delta|$ is the area of $C_x \cap S_\delta$. Show that the solid angle of C_x is given by

$$(*) \qquad \omega_x = -\int_B \frac{\partial}{\partial \nu_y} \frac{1}{r_{xy}}\, d\sigma_y.$$

   Hint. Apply the divergence theorem to $C_x \cap c\bar{K}_\delta(x)$ and $F = \nabla_y(1/r_{xy})$. Notice this shows that the definition of solid angle is independent of δ. We also refer to ω_x as the *solid angle subtended by B from x.* We take (*) as the definition of solid angle even when the rays from x cut B more than once.
7. Let ω_x denote the solid angle subtended by B from x.
   (a) Show that

$$\omega_x = \int_B \frac{\cos(\nu_y, y - x)}{r_{xy}^2}\, d\sigma_y,$$

   where $\cos(\nu_y, y - x)$ is the cosine of the angle between the normal ν_y and the vector $y - x$. Notice that the solid angle is a net quantity because $\cos(\nu_y, y - x)$ may change sign as y varies over B.
   (b) The bound in (6-11) is on the total solid angle. Use the geometric description of solid angle in Prob. 6 to explain why a surface B, which is intersected at most p times by any line, satisfies (6-11) with $C = 4\pi p$.

8. Repeat Prob. 6 in two dimensions, where B is replaced by a smooth curve C and the solid angle is just the ordinary angle subtended by C at x. You should find

$$\omega_x = \int_C \frac{\partial}{\partial v_y} \log \frac{1}{r_{xy}} \, d\sigma_y.$$

9. Refer to Fig. 8-5 and recall that $x_3 = \phi(x_1, x_2)$ is the equation for $B \cap K_\delta(0)$. In the figure $0 < a \le \delta$. Show that

$$|x_3| = |\phi(x_1, x_2)| \le C_3(x_1^2 + 2x_1x_2 + x_2^2) \le 2C_3(x_1^2 + x_2^2)$$

for $x_1^2 + x_2^2 \le a^2$ where C_3 bounds the second partials of ϕ in $x_1^2 + x_2^2 \le \delta^2$. Conclude that

$$S_a \cap D \cap \{x_3 \ge 2C_3a^2\} \subset S_a \cap D \subset S_a \cap D \cap \{x_3 \ge 0\}$$

and that

$$\frac{|S_a \cap D|}{4\pi a^2} \to \frac{2\pi a^2}{4\pi a^2} = \frac{1}{2}$$

as $a \to 0$.

10. Prove Gauss' theorem in two dimensions: If D is a bounded domain in the plane with twice-continuously differentiable boundary curve B, then

$$\frac{1}{2\pi} \int_B \frac{\partial}{\partial v_y} \log \frac{1}{r_{xy}} \, ds_y = \begin{cases} -1, & x \in D, \\ -\frac{1}{2}, & x \in B, \\ 0, & x \in c\bar{D}. \end{cases}$$

11. Let B be twice continuously differentiable in two dimensions. Prove that $(\partial/\partial v_y) \log 1/r_{xy}$ is continuous for $x, y \in B$. *Hint.* L'Hospital's rule. Conclude that the analogue of (6-11) in two dimensions,

$$\frac{1}{2\pi} \int_B \left| \frac{\partial}{\partial v_y} \log \frac{1}{r_{xy}} \right| d\sigma_y \le C,$$

is automatic.

12. Prove Prop. 6-3. *Hint.* For the continuity argue in a fashion similar to that used to verify (6-13).

13. Given the continuity of $h(x)$ in (6-16), confirm the jump relations in Th. 6-4.

14. Formulate and prove the jump relations (Ths. 6-3 and 6-4) for two dimensions. *Hint.* $(1/2\pi) \log (1/r_{xy})$ replaces $1/4\pi r_{xy}$.

15. Prove Poisson's theorem (Th. 6-1),

$$\Delta \int_D \left(\frac{1}{2\pi} \log \frac{1}{r_{xy}} \right) f(y) \, dy = -f(x),$$

for a domain D in the plane.

8-7. The Integral Equations of Potential Theory

In this section we establish that the three boundary value problems of potential theory have solutions, provided that the domains in question have reasonably smooth boundaries. We assume throughout that D is a normal, bounded domain whose boundary B satisfies Assumption A of Sec. 8-6. Note that $c\bar{D}$ is a typical exterior domain. It turns out to be natural to treat the exterior boundary value problems at the same time that we study the problems for bounded (or interior) domains. Thus, we write $E = c\bar{D}$ for a typical exterior domain. As usual, ν denotes the outer, unit normal to D. The outer unit normal to E is $\nu^* = -\nu$.

The existence of a solution to the interior and exterior Dirichlet and Neumann problems can be reduced to the solution of related Fredholm integral equations. We begin this reduction procedure by considering the interior Dirichlet problem. Given a continuous function f on B, we wish to find a function $u \in C^2(D) \cap C(\bar{D})$ such that

$$\Delta u = 0 \quad \text{in} \quad D \quad \text{and} \quad u = f \quad \text{on} \quad B. \tag{7-1}$$

We seek to solve this problem in terms of a double-layer potential. That is, for points $x \in D$, we set

$$u(x) = \int_B \left(\frac{\partial}{\partial \nu_y} \frac{1}{4\pi r_{xy}}\right) \mu(y)\, d\sigma_y, \tag{7-2}$$

where μ is a density function to be determined. To motivate the fact that (7-2) is a reasonable form for a solution to the Dirichlet problem, assume that u is a solution and use (3-7) to get

$$u(x) = -\int_B \left(\frac{\partial}{\partial \nu_y} \frac{1}{4\pi r_{xy}}\right) f(y)\, d\sigma_y + \int_B \frac{1}{4\pi r_{xy}} \frac{\partial u}{\partial \nu_y}\, d\sigma_y.$$

Since $\partial u/\partial \nu$ is unknown, this expression is useless as a solution formula. However, if we drop the unknown integral and attempt to compensate for this by modifying the given density $f(y)$ in the first integral, we are led to (7-2).

Evidently, u defined by (7-2) is twice continuously differentiable in D and satisfies $\Delta u = 0$ there, for any choice of μ. Since we seek a solution to the Dirichlet problem which is continuous on $\bar{D}$, u given by (7-2) must satisfy

$$\lim_{\substack{z \to x \in B \\ z \in D}} u(z) = f(x),$$

which, in view of the jump relations for a double-layer potential (Th. 6-3), means that

$$\int_B \left(\frac{\partial}{\partial \nu_y} \frac{1}{4\pi r_{xy}}\right) \mu(y)\, d\sigma_y - \frac{1}{2}\mu(x) = f(x), \qquad x \in B. \tag{7-3}$$

If the integral equation (7-3) has a continuous solution for μ, then the function $u(x)$ given by (7-2) for $x \in D$ and by $f(x)$ for $x \in B$ solves the interior Dirichlet problem.

Consider the exterior Neumann problem: Find a function $v \in C^2(E) \cap C^1_\nu(\bar{E})$ such that

$$\Delta v = 0 \quad \text{in} \quad E, \qquad \frac{\partial v}{\partial \nu^*} = f(x) \quad \text{on} \quad B, \tag{7-4}$$

and v tends to zero uniformly at infinity. This time we seek a solution in the form of a single-layer potential,

$$v(x) = \int_B \frac{1}{4\pi r_{xy}} \rho(y)\, d\sigma_y \tag{7-5}$$

for $x \in \bar{E}$. Since v tends to zero uniformly at infinity and $\Delta v = 0$ in E for any continuous density ρ, (7-5) will solve the Neumann problem if

$$\lim_{\substack{z \to x \\ z \in E}} \frac{\partial v}{\partial v_x^*}(z) = \int_B \left(\frac{\partial}{\partial v_x^*} \frac{1}{4\pi r_{xy}} \right) \rho(y)\, d\sigma_y + \frac{1}{2}\rho(x) = f(x)$$

for $x \in B$ [see (ii) of Th. 6-4]. Thus, since $v = -v^*$, (7-5) solves the exterior Neumann problem if $\rho(y)$ is a continuous solution to

$$\int_B \left(\frac{\partial}{\partial v_x} \frac{1}{4\pi r_{xy}} \right) \rho(y)\, d\sigma_y - \frac{1}{2}\rho(x) = -f(x), \qquad x \in B. \tag{7-6}$$

Notice that the kernels of the integral equations (7-3) and (7-6) are transposes of each other. Also, these kernels have weak singularities because the estimates following (6-12) show that

$$\left| \frac{\partial}{\partial v_y} \frac{1}{4\pi r_{xy}} \right| \le \frac{C}{r_{xy}}$$

for $x, y \in B$ with $x \neq y$ (see Prob. 2 of Sec. 7-7). Consequently, by the Fredholm alternative, (7-3) and (7-6) will each have a unique solution for all continuous $f(x)$ provided that the homogeneous equation

$$\int_B \left(\frac{\partial}{\partial v_x} \frac{1}{4\pi r_{xy}} \right) \rho(y)\, d\sigma_y - \frac{1}{2}\rho(x) = 0, \qquad x \in B, \tag{7-7}$$

corresponding to (7-6) has only the trivial solution $\rho \equiv 0$. Now, if ρ satisfies (7-7), the single-layer potential (7-5) solves the exterior Neumann problem

$$\Delta v = 0 \quad \text{in} \quad E, \qquad \frac{\partial v}{\partial v^*} = 0 \quad \text{on} \quad B,$$

and v tends to zero uniformly at infinity. Thus, $v(x) = 0$ in E. Since the single-layer potential is continuous on $\bar{E}$, $v(x) = 0$ on B. Then v solves the interior Dirichlet problem

$$\Delta v = 0 \quad \text{in} \quad D, \qquad v = 0 \quad \text{on} \quad B,$$

and we conclude that $v(x) = 0$ in D. Therefore, $\partial v(z)/\partial v_x = 0$ for $z \in D$ and the first jump relation in Th. 6-4 yields

$$0 = \int_B \left(\frac{\partial}{\partial v_x} \frac{1}{4\pi r_{xy}} \right) \rho(y)\, d\sigma_y + \frac{1}{2}\rho(x), \qquad x \in B. \tag{7-8}$$

Subtract (7-7) from (7-8) to get

$$\rho(x) = 0 \quad \text{for all} \quad x \in B.$$

In view of the Fredholm alternative, each of (7-3) and (7-6) has a unique solution and we have proved

Theorem 7-1. Let f be continuous on B. The interior Dirichlet problem (7-1) has a unique solution that can be expressed as a double-layer potential (7-2). The exterior Neumann problem (7-4) has a unique solution that can be expressed as a single-layer potential (7-5).

We turn our attention now to the interior Neumann problem: Find a function $v \in C^2(D) \cap C_\nu^1(\bar{D})$ such that

$$\Delta v = 0 \quad \text{in} \quad D, \qquad \frac{\partial v}{\partial \nu} = f(x) \quad \text{on} \quad B. \tag{7-9}$$

As for the exterior Neumann problem, we seek a solution in the form (7-5) of a single-layer potential. Just as before, we use the jump relations to conclude that (7-5) will solve (7-9) provided that $\rho(y)$ is continuous and solves the integral equation

$$\int_B \left(\frac{\partial}{\partial \nu_x} \frac{1}{4\pi r_{xy}}\right) \rho(y) + \frac{1}{2}\rho(x) = f(x), \qquad x \in B. \tag{7-10}$$

To determine whether a solution exists, we apply the Fredholm alternative and so consider the corresponding homogeneous equation

$$\int_B \left(\frac{\partial}{\partial \nu_x} \frac{1}{4\pi r_{xy}}\right) \rho(y)\, d\sigma_y + \frac{1}{2}\rho(x) = 0, \qquad x \in B. \tag{7-11}$$

The adjoint integral equation for (7-11) is

$$\int_B \left(\frac{\partial}{\partial \nu_y} \frac{1}{4\pi r_{xy}}\right) \mu(y)\, d\sigma_y + \frac{1}{2}\mu(x) = 0, \qquad x \in B, \tag{7-12}$$

and by Gauss' theorem (Th. 6-2), this equation has the nontrivial solution $\mu_0(x) \equiv 1$. Thus, (7-11) must also have at least one nontrivial solution, say $\rho_0(y)$. The next question to answer is how many linearly independent eigenfunctions there are. Suppose that $\rho_1(y)$ is also an eigenfunction of (7-11) and let

$$v_i(x) = \int_B \frac{1}{4\pi r_{xy}}\, \rho_i(y)\, d\sigma_y$$

for $i = 0, 1$. Both v_0 and v_1 satisfy Laplace's equation in D and their normal derivatives vanish on B by (7-11). Thus, both v_0 and v_1 are constants in D, say a_0 and a_1. Neither of these constants is zero. Indeed, assume that $v_0 = a_0 = 0$ in $\bar{D}$. Then $\Delta v_0 = 0$ in E, $v_0 = 0$ on B, and v_0 tends to zero uniformly at infinity. Consequently, $v_0(x) = 0$ in E. Calculating $\partial v_0(z)/\partial \nu_x$ as $z \to x \in B$ with $z \in E$ we find that

$$\int_B \left(\frac{\partial}{\partial \nu_x} \frac{1}{4\pi r_{xy}}\right) \rho_0(y) - \frac{1}{2}\rho_0(x) = 0, \qquad x \in B.$$

Subtract this from (7-11) with $\rho = \rho_0$ to find $\rho_0(x) \equiv 0$ on B, a contradiction. Thus, $a_0 \neq 0$ and by symmetry, $a_1 \neq 0$.

Now, consider the function

$$a_1 v_0(x) - a_0 v_1(x) = \int_B \frac{1}{4\pi r_{xy}} [a_1 \rho_0(y) - a_0 \rho_1(y)] \, d\sigma_y.$$

This function vanishes on B and hence solves both the homogeneous interior and exterior Dirichlet problems. Therefore, $a_1 v_0(x) - a_0 v_1(x) \equiv 0$ in R_3. Finally, we can use the jump relations for the normal derivatives of a single-layer potential to conclude as above that $a_1 \rho_0(y) - a_0 \rho_1(y) = 0$ for all $y \in B$. That is, ρ_0 and ρ_1 are linearly dependent. Thus, the adjoint equation also has, up to nonzero multiples, only one eigenfunction, namely $\mu_0(x) \equiv 1$. The Fredholm alternative asserts that (7-10) is solvable for ρ if and only if f is orthogonal to $\mu_0(x) \equiv 1$. That is,

$$\int_B f(x) \, d\sigma_x = 0.$$

Assume that this orthogonality condition holds and let ρ_p be a particular solution to (7-10). The general solution to (7-10) is then $\rho_p + C_0 \rho_0$, where C_0 is any constant. These densities yield the following solutions to the interior Neumann problem:

$$v(x) = \int_B \frac{1}{4\pi r_{xy}} [\rho_p(y) + C_0 \rho_0(y)] \, d\sigma_y = \int_B \frac{1}{4\pi r_{xy}} \rho_p(y) \, d\sigma_y + C, \tag{7-13}$$

where C is an arbitrary constant and we have used the fact that $v_0(x)$ is a constant. In summary, we have proved

> Theorem 7-2. Let $f(x)$ be continuous on B. Then the interior Neumann problem (7-9) has a solution if and only if $\int_B f(x) \, d\sigma_x = 0$, in which case there are infinitely many solutions of the form (7-13).

We solved the interior Dirichlet problem with a double-layer potential. Consider the exterior Dirichlet problem of finding a function $u \in C^2(E) \cap C(\bar{E})$ such that

$$\Delta u = 0 \quad \text{in} \quad E, \qquad u = f \quad \text{on} \quad B, \tag{7-14}$$

and u tends to zero uniformly at infinity. The double-layer potential (7-2) will solve this problem if

$$\int_B \left(\frac{\partial}{\partial \nu_y} \frac{1}{4\pi r_{xy}} \right) \mu(y) \, d\sigma_y + \frac{1}{2} \mu(x) = f(x), \qquad x \in B. \tag{7-15}$$

The kernel in (7-15) is the transpose of that in (7-10), which was used to solve the interior Neumann problem. The homogeneous equation (7-12) corresponding to (7-15) has the nontrivial solution $\mu_0(x) \equiv 1$. Thus, (7-15) is not always solvable, and when it is, there are infinitely many solutions. At first glance, one might suspect that the exterior Dirichlet problem is not always solvable. This is actually not the case. The problem is that the double-layer potential (7-2) decays too rapidly at infinity to represent solutions to all possible exterior Dirichlet problems. The double-layer kernel satisfies

$$\left|\frac{\partial}{\partial \nu_y} \frac{1}{r_{xy}}\right| = \left|\frac{x-y}{r_{xy}^3} \cdot \nu_y\right| \leq \frac{1}{r_{xy}^2},$$

which tends to zero like $|x|^{-2}$ for $y \in B$ as $|x| \to \infty$. However, the requirement that a solution u should tend to zero at infinity only implies that u tends to zero like $|x|^{-1}$ as $|x| \to \infty$ (see Th. 5-2). Thus, we modify (7-2) and assume a solution of the form

$$u(x) = \int_B \left[\left(\frac{\partial}{\partial \nu_y^*} \frac{1}{4\pi r_{xy}}\right) + \frac{1}{4\pi|x|}\right] \mu(y)\, d\sigma_y, \tag{7-16}$$

which has the proper decay rate at infinity. Clearly, $\Delta u = 0$ in E and u tends to zero uniformly at infinity. Thus, (7-16) will solve (7-14) if

$$\int_B \left[\left(\frac{\partial}{\partial \nu_y^*} \frac{1}{4\pi r_{xy}}\right) + \frac{1}{4\pi|x|}\right] \mu(y)\, d\sigma_y - \frac{1}{2}\mu(x) = f(x), \qquad x \in B. \tag{7-17}$$

This equation is uniquely solvable for each f if the only solution to

$$\int_B \left(\frac{\partial}{\partial \nu_y^*} \frac{1}{4\pi r_{xy}}\right) \mu(y)\, d\sigma_y + \frac{1}{4\pi|x|} \int_B \mu(y)\, d\sigma_y - \frac{1}{2}\mu(x) = 0 \tag{7-18}$$

is the trivial solution $\mu(y) = 0$ for $y \in B$. As usual, we assume that μ satisfies (7-18) and define

$$w(x) = \int_B \left[\left(\frac{\partial}{\partial \nu_y^*} \frac{1}{4\pi r_{xy}}\right) + \frac{1}{4\pi|x|}\right] \mu(y)\, d\sigma_y. \tag{7-19}$$

Then w satisfies the Laplace equation in E, tends to zero uniformly at infinity, and it vanishes on B by (7-18). By uniqueness, $w(x) \equiv 0$ in E. Now, multiply (7-19) by $|x|$ and recall that the integral of the first term in (7-19) decays like $|x|^{-2}$ as $|x| \to \infty$ to deduce that

$$\int_B \mu(y)\, d\sigma_y = 0 \tag{7-20}$$

Then (7-18) reduces to (7-12). As we have seen, (7-12) has solution $\mu(y) = c$, a constant. By (7-20) this constant is $c = 0$, and we have proved

> Theorem 7-3. Let f be continuous on B. Then there exists a unique solution to the exterior Dirichlet problem. This solution can be represented in the form (7-16).

We can show that the interior and exterior Robin problems have solutions expressible as single-layer potentials by reasoning similar to that used above. See Prob. 2.

Theorem 7-4. The interior and exterior Robin problems both have unique solutions that can be expressed as single-layer potentials.

These integral equations formulations of the classical problems of potential theory do not yield the most general existence theorems, but they do have some real advantages over other methods. First, from a practical standpoint, they reduce the dimension of the problem to be solved by one. Second, once the density has been obtained, the solution can be found at arbitrary points by integration, using the representations (7-2), (7-5), or (7-16) as appropriate. Here are two useful sources for the numerical implementation of these integral equations methods applied to engineering problems: P. S. Huyakorn and G. F. Pindar, *Computational Methods in Subsurface Flow* (Academic Press, Inc., New York, 1983) and J. A. Liggett and P. L.-F. Liu, *The Boundary Integral Equation Method for Porus Media Flow* [George Allen & Unwin (Publisher) Ltd., London, 1983]. Finally, we note that, in addition to the points already mentioned, the integral equations methods of this section treat exterior domains and interior domains with equal ease.

The existence theorems developed here have counterparts for domains in any n-dimensional space. The case $n = 2$ is left for the problems. We close with two interesting examples.

Consider the Dirichlet problem for the disk in two dimensions,

$$\Delta u = 0 \quad \text{in} \quad K_a(0), \qquad u = f(x) \quad \text{on} \quad S_a(0).$$

We assume a solution in the form of a double layer

$$u(x) = \frac{1}{2\pi} \int_{S_a} \left(\frac{\partial}{\partial \nu_y} \log \frac{1}{r_{xy}} \right) \mu(y)\, d\sigma_y.$$

This harmonic function will assume the desired boundary conditions if μ satisfies

$$f(x) = \frac{1}{2\pi} \int_{S_a} \left(\frac{\partial}{\partial \nu_y} \log \frac{1}{r_{xy}} \right) \mu(y)\, d\sigma_y - \frac{1}{2}\mu(x), \qquad x \in S_a.$$

For $x, y \in S_a$,

$$\frac{\partial}{\partial \nu_y} \log \frac{1}{r_{xy}} = \frac{x - y}{r_{xy}^2} \cdot \frac{y}{a} = \frac{x \cdot y - a^2}{a(2a^2 - 2x \cdot y)} = -\frac{1}{2a}$$

and the equation for μ becomes

$$f(x) = -\frac{1}{4\pi a} \int_{S_a} \mu(y)\, d\sigma_y - \frac{1}{2}\mu(x). \tag{7-21}$$

Integrate (7-21) with respect to x to find

$$\int_{S_a} f(x)\, d\sigma_x = \left(-\frac{2\pi a}{4\pi a} - \frac{1}{2} \right) \int_{S_a} \mu(y)\, d\sigma_y.$$

Thus,

$$\int_{S_a} \mu(y)\, d\sigma_y = -\int_{S_a} f(x)\, d\sigma_x$$

and from (7-21)

$$\mu(x) = \frac{1}{2\pi a}\int_{S_a} f(z)\,d\sigma_z - 2f(x).$$

The solution to the Dirichlet problem is, therefore,

$$\begin{aligned} u(x) &= \frac{1}{2\pi}\int_{S_a}\left(\frac{\partial}{\partial \nu_y}\log\frac{1}{r_{xy}}\right)\left[\frac{1}{2\pi a}\int_{S_a} f(z)\,d\sigma_z - 2f(y)\right]d\sigma_y \\ &= -\frac{1}{2\pi a}\int_{S_a} f(z)\,d\sigma_z - \frac{2}{2\pi}\int_{S_a}\left(\frac{\partial}{\partial \nu_y}\log\frac{1}{r_{xy}}\right)f(y)\,d\sigma_y \\ &= -\frac{1}{2\pi a}\int_{S_a}\left(1 + 2a\frac{\partial}{\partial \nu_y}\log\frac{1}{r_{xy}}\right)f(y)\,d\sigma_y, \end{aligned}$$

where Gauss' theorem was used in the second step. Finally, for $x \in K_a$ and $y \in S_a$,

$$\begin{aligned} 1 + 2a\frac{\partial}{\partial \nu_y}\log\frac{1}{r_{xy}} &= 1 + 2a\frac{y}{a}\cdot\frac{x-y}{r_{xy}^2} \\ &= \frac{(x-y)\cdot(x-y) + 2y\cdot(x-y)}{|x-y|^2} \\ &= \frac{(x-y)\cdot(x+y)}{|x-y|^2} = \frac{|x|^2 - a^2}{|x-y|^2}, \end{aligned}$$

which yields the familiar Poisson formula.

A final example shows that the methods of this section can be used to produce formal solutions, whose validity can be checked after they are found. For example, consider the Neumann problem for the unbounded domain,

$$\Delta u = 0 \quad \text{in} \quad x_3 > 0,$$

$$-\frac{\partial u}{\partial x_3}(x_1, x_2, 0) = f(x_1, x_2) \quad \text{on} \quad x_3 = 0,$$

u tends to zero uniformly at infinity.

This is not an exterior domain of the type we have considered. Nonetheless, we assume a solution in the form of a single-layer potential.

$$u(x) = \int_{-\infty}^{\infty}\int_{-\infty}^{\infty}\frac{1}{4\pi[(x_1 - y_1)^2 + (x_2 - y_2)^2 + x_3^2]^{1/2}}\rho(y_1, y_2)\,dy_1\,dy_2.$$

We calculate

$$-\frac{\partial u}{\partial x_3}(x_1, x_2, x_3) = \int_{-\infty}^{\infty}\int_{-\infty}^{\infty}\frac{x_3}{4\pi[(x_1 - y_1)^2 + (x_2 - y_2)^2 + x_3^2]^{3/2}}\rho(y_1, y_2)\,dy_1\,dy_2.$$

Now let $x_3 \to 0$ and apply the jump relation to obtain the equation

$$f(x_1, x_2) = 0 + \tfrac{1}{2}\rho(x_1, x_2).$$

Therefore, we obtain the (formal) solution

$$u(x_1, x_2, x_3) = \frac{1}{2\pi}\int_{-\infty}^{\infty}\int_{-\infty}^{\infty} \frac{1}{[(x_1 - y_1)^2 + (x_2 - y_2)^2 + x_3^2]^{1/2}} f(y_1, y_2)\, dy_1\, dy_2.$$

PROBLEMS

1. In the existence results of this section, we have used the Laplace equation $\Delta u = 0$ with inhomogeneous boundary data. Extend these results to the Poisson equation $\Delta u = h$ in D with inhomogeneous boundary data, where $h \in C^1(\bar{D})$. *Hint.* Write $u = w + v$, where v is the volume potential

$$v(x) = -\int_D \frac{1}{4\pi r_{xy}} h(y)\, dy$$

and apply Poisson's theorem (Th. 6-1).

2. Assume that α is constant and prove that there exist solutions to the interior and exterior Robin problems. *Hint.* Assume a solution in the form of a single-layer potential. Use the jump relations to find the integral equation for the density of the potential. If ρ_0 is a solution to the corresponding homogeneous integral equation, prove that

$$v(x) = \int_B \frac{1}{4\pi r_{xy}} \rho_0(y)\, d\sigma_y$$

is identically zero.

3. Express the solution to the (upper half space) Dirichlet problem

$$\Delta u = 0 \quad \text{in} \quad x_3 > 0,$$
$$u(x_1, x_2, 0) = f(x_1, x_2),$$
$$u \text{ tends to zero uniformly at infinity,}$$

as a double-layer potential.

4. Prove that the formula you find in Prob. 3 is a solution under reasonable assumptions on $f(x_1, x_2)$.

5. Suppose that we seek a solution to the exterior Dirichlet problem (7-14) in the form of a double layer and the appropriate orthogonality condition $\int_B f(x)\, d\sigma_x = 0$ is met. Then (7-15) has infinitely many solutions. Why doesn't this contradict the uniqueness theorem for the exterior Dirichlet problem?

In Probs. 6–15 we restrict ourselves to two dimensions. As usual, D denotes a bounded domain and $E = c\bar{D}$ an exterior domain. The boundary B of D is asssumed to be twice continuously differentiable.

6. Let E be an exterior domain. Show that a necessary condition for the solvability of

$$u \in C^2(E) \cap C_\nu^1(\bar{E}), \qquad u \quad \text{bounded in} \quad E,$$

$$\Delta u = 0 \quad \text{in} \quad E, \quad \text{and} \quad \frac{\partial u}{\partial \nu^*} = f \quad \text{on} \quad B$$

is that

$$\int_B f(y)\, d\sigma_y = 0.$$

Hint. Apply one of Green's identities to $E \cap K_R(0)$ for R large. Let $R \to \infty$ and show that the integral over $S_R(0)$ has limit zero.

7. Prove that the exterior Neumann problem (Prob. 6) has a solution. Use the following steps.

(a) Assume a solution in the form of a single layer

$$u(x) = \frac{1}{2\pi} \int_B \left(\log \frac{1}{r_{xy}} \right) \rho(y)\, d\sigma_y,$$

and note that $\Delta u = 0$ in E for any continuous ρ.

(b) In order for $u(x)$ in part (a) to assume the desired boundary values, show that ρ must satisfy

$$-f(x) = \frac{1}{2\pi} \int_B \left(\frac{\partial}{\partial \nu_x} \log \frac{1}{r_{xy}} \right) \rho(y)\, d\sigma_y - \frac{1}{2} \rho(x), \qquad x \in B.$$

(c) Show that any solution ρ to the integral equation in part (b) automatically satisfies

$$\int_B \rho(y)\, d\sigma_y = 0.$$

Hint. Integrate the integral equation in part (b) over B, note that the integral of f is zero, and use Gauss' theorem to show the double integral vanishes.

(d) Use part (c) to show that $u(x)$ in part (a) can be expressed as

$$u(x) = \frac{1}{2\pi} \int_B \left(\log \frac{|x|}{r_{xy}} \right) \rho(y)\, d\sigma_y$$

whenever ρ solves part (b). Conclude that if ρ is a continuous solution of the integral equation in part (b), then $u(x)$ is a bounded harmonic function in E and solves the exterior Neumann problem. Note that u tends to zero as x tends to infinity.

(e) Show that the integral equation in part (b) has a solution. *Hint.* Let ρ_0 be a solution to the corresponding homogeneous equation with $f = 0$. Then

$$v(x) = \frac{1}{2\pi} \int_B \left(\log \frac{1}{r_{xy}} \right) \rho_0(y)\, d\sigma_y$$

is a bounded [see parts (c) and (d)] harmonic function in E such that $\partial v/\partial \nu^* = 0$ and v has limit 0 at infinity. Thus, $v = 0$ in E. Proceed to find $v \equiv 0$ and then $\rho_0(y) = 0$ for $y \in B$.

8. Let D be a bounded domain and f continuous on B. Show that the Dirichlet problem $u \in C^2(D) \cap C(\bar{D})$,

$$\Delta u = 0 \quad \text{in} \quad D, \qquad u = f \quad \text{on} \quad B$$

has a solution, expressible as a double-layer potential in D. *Hint.* The kernel of the integral operator is the transpose of the one for the exterior Neumann problem.

9. Let D be a bounded domain, f be continuous on B, and

$$\int_B f(x)\, d\sigma_x = 0.$$

Prove that the interior Neumann problem, $u \in C^2(D) \cap C_\nu^1(\bar{D})$,

$$\Delta u = 0 \quad \text{in} \quad D \quad \text{and} \quad \frac{\partial u}{\partial \nu} = f \quad \text{on} \quad B,$$

has a solution.

10. Let E be an exterior domain and f be continuous on B. Prove that the exterior Dirichlet problem, $u \in C^2(E) \cap C(\bar{E})$,

$$\Delta u = 0 \quad \text{in} \quad E, \qquad u = f \quad \text{on} \quad B, \quad \text{and} \quad u \quad \text{bounded},$$

has a solution. *Hint.* Proceed as in the three-dimensional case and assume a solution in the form

$$u(x) = \frac{1}{2\pi} \int_B \left(\frac{\partial}{\partial \nu_y} \log \frac{1}{r_{xy}} + 1 \right) \sigma(y)\, d\sigma_y.$$

11. Formulate and solve the Robin problem for (a) a bounded domain D and (b) an exterior domain E.

12. Find an explicit formula for the solution to

$$\Delta u = 0 \quad \text{in} \quad K_a(0) \quad \text{and} \quad \frac{\partial u}{\partial \nu} = f \quad \text{on} \quad S_a(0)$$

in two dimensions starting with a single layer potential.

13. Let D be a bounded, simply connected domain in the plane with boundary curve B given parametrically by $(\phi(t), \psi(t))$ for $t \in [0, 1]$. Assume that ϕ, ψ are 1-periodic and continuously differentiable. Assume that the Dirichlet problem can be solved for D. Show that the Neumann problem can be reduced to the Dirichlet problem. *Hint.* Let u be such that

$$\Delta u = 0 \quad \text{in} \quad D \quad \text{and} \quad \frac{\partial u}{\partial \nu} = f \quad \text{on} \quad B,$$

where the compatibility condition $\int_B f\, d\sigma_y = 0$ holds. Here

$$\frac{\partial u}{\partial \nu} = \frac{u_x \psi' - u_y \phi'}{N} = f,$$

where $N = \sqrt{\phi'^2 + \psi'^2}$ and $\nu = (\psi'/N, -\phi'/N)$. Let v be the solution to the Dirichlet problem

$$\Delta v = 0 \quad \text{in} \quad D,$$

$$v(\phi(t), \psi(t)) = \int_0^t f(\tau) N\, d\tau, \qquad 0 \le t \le 1.$$

Now construct u as the solution to

$$u_x = v_y, \qquad u_y = -v_x$$

Observe that $(d/dt)v(\phi(t), \psi(t)) = v_x \phi' + v_y \psi'$ and obtain the desired result.

Many problems in two-dimensional potential theory can be solved by means of the *Riemann mapping theorem* with its extension by Carathéodory: Any two simply connected domains excluding the entire plane can be mapped one-to-one conformally onto each other. If these domains are bounded and their boundaries are simple closed curves, the conformal mapping can be extended by continuity so that the boundary points are mapped one-to-one onto each other. A conformal mapping is given by a differentiable function $\phi(z)$ of the complex variable $z = x + iy$. Let $\phi(z) = \xi + i\eta$, where $\xi = \xi(x, y)$ and $\eta = \eta(x, y)$. Then

$$\phi'(z) = \xi_x + i\eta_x = \eta_y - i\xi_y.$$

14. If $\phi'(z) \neq 0$, show that

$$\begin{vmatrix} \xi_x & \xi_y \\ \eta_x & \eta_y \end{vmatrix} \neq 0$$

so that x and y can be expressed as functions of ξ and η in a neighborhood of any point. Thus, a function $u(x, y)$ can be expressed as $u(\xi, \eta)$. Show that

$$\Delta_{x,y}u = |\phi'(z)|^2 \Delta_{\xi,\eta}u,$$

where $\Delta_{x,y}u = u_{xx} + u_{yy}$. Conclude that one Laplacian vanishes if and only if the other does.

15. Suppose that D is a bounded, simply connected domain enclosed by a simple closed curve. Let $\phi: D \to \bar{K}_1(0)$ be the conformal mapping provided by the Riemann mapping theorem. Show how to solve

$$\Delta u = 0 \quad \text{in} \quad D, \qquad u = f \quad \text{on} \quad B$$

using this mapping.

Conformal mapping techniques are efficient, widely used, and extensively tabulated. They have effective numerical implementations. From the theoretical point of view, the existence of the Riemann mapping function for general domains is equivalent to the existence theorem for the Dirichlet problem.

9

Parabolic Equations

9-1. The Initial Value Problem

We have seen that an equation of the form

$$u_t = a\,\Delta u + h(x, t), \qquad x \in D, \qquad t > 0, \tag{1-1}$$

describes diffusion and heat flow phenomena in a domain D, where $h(x, t)$ represents internal sources or sinks, $a > 0$ is a constant, and $u = u(x, t)$ is the concentration or temperature at position x and time t. We studied (1-1) in the case of one space variable x in Chap. 5. In this chapter we treat the parabolic problem (1-1) in three space dimensions. Thus, $x = (x_1, x_2, x_3)$ represents a point in three-dimensional Euclidean space R_3 and $\Delta u = u_{x_1x_1} + u_{x_2x_2} + u_{x_3x_3}$ in rectangular coordinates. Just as for Poisson's equation, the results derived here for three space variables have analogues for other space dimensions. The case of two space dimensions is left for the problems.

In order to determine the diffusion of a substance, governed by (1-1), as it evolves in time, we must know its initial state. This is specified by

$$u(x, 0) = f(x), \qquad x \in D. \tag{1-2}$$

In addition, if D is bounded by the surface B, we must describe how the substance diffuses across B. This leads to boundary conditions analogous to those in the Dirichlet, Neumann, and Robin problems of potential theory.

We begin our study of parabolic problems with the initial value problem for the homogeneous diffusion or heat conduction equation. That is, we seek a function $u = u(x, t)$ which is continuous in $t \geq 0$, $x \in R_3$ and satisfies

$$\begin{cases} u_t = a\,\Delta u, & x \in R_3, \qquad t > 0, \\ u(x, 0) = f(x), & x \in R_3. \end{cases} \tag{1-3}$$

This problem can be solved formally using Fourier transforms just as we did in Sec. 5-4, where there was only one space variable. The result (see Prob. 1) is

$$u(x, t) = \int_{R_3} k(x - y, at) f(y)\, dy, \tag{1-4}$$

where

$$k(x, t) = (4\pi t)^{-3/2} \exp\left(-\frac{|x|^2}{4t}\right), \qquad x \in R_3, \qquad t > 0. \tag{1-5}$$

The function $k(x, t)$ is called the *fundamental solution for the diffusion or heat equation,* and plays the same role in the solution theory for (1-1) as the fundamental solution to the Laplace equation does in potential theory.

We will confirm shortly that the formal solution (1-4) does solve the initial value problem (1-3) under reasonable restrictions on the initial distribution $f(x)$. The solution formula (1-4) leads to an important physical interpretation of $k(x, t)$: The fundamental solution $k(x, t)$ is the concentration of mass at position x and time t which results from the spread of a unit mass concentrated at the origin at time $t = 0$ under the diffusion process with $a = 1$. The reasoning (Prob. 3) needed to confirm this is virtually the same as for one space variable. Similarly, $k(x - x_0, at)$ is the concentration which arises from a unit mass concentrated at x_0 at time $t = 0$. By conservation of mass, we expect that the integral of $k(x - y, at) = k(y - x, at)$ with respect to y will be one.

Pursuing this physical reasoning further, note that $k(x - y, a(t - t_0))$ is the concentration at x at time $t > t_0$ due to a unit mass which diffused from y at time t_0. Another way to determine the concentration at x at time t is to select an intermediate time τ, $t_0 < \tau < t$, as a new initial time. The original unit of mass at y at time t_0 has diffused into the mass distribution $k(\xi - y, a(\tau - t_0))$ at point ξ and time τ. Use of this as initial data in (1-4) with the time t replaced by $t - \tau$, to translate the time origin to τ, yields

$$\int_{R_3} k(x - \xi, a(t - \tau)) k(\xi - y, a(\tau - t_0))\, d\xi$$

for the concentration at x and time t due to the original unit of mass located at y at time t_0. Thus, the integral above equals $k(x - y, a(t - t_0))$. These physical arguments make plausible

Lemma 1-1. The fundamental solution $k(x, t)$ of the heat equation satisfies: For $x, y \in R_3$ and $t > \tau > t_0$

(i) $\left(\dfrac{\partial}{\partial t} - a\Delta_x\right) k(x - y, a(t - \tau)) = 0, \qquad \left(\dfrac{\partial}{\partial \tau} + a\Delta_y\right) k(x - y, a(t - \tau)) = 0.$

(ii) $\displaystyle\int_{R_3} k(x - y, a(t - \tau))\, dy = 1.$

(iii) $\displaystyle k(x - y, a(t - t_0)) = \int_{R_3} k(x - \xi, a(t - \tau)) k(\xi - y, a(\tau - t_0))\, d\xi.$

The physical arguments motivating Lemma 1-1 are, of course, not proofs. Nevertheless, physical arguments, such as those used above, are extremely useful in formulating correct results and in suggesting how precise mathematical arguments might go. In the case of Lemma 1-1 the stated results are easily verified by direct calculation (Prob. 4) once they are suspected.

The justification that (1-4) solves (1-3) rests on a technical lemma.

Lemma 1-2. Let $R, T, \delta > 0$ be any numbers with $T > 2\delta$. Then for each set

$$\mathcal{M} = \{(x, t): |x| \le R, \delta \le t \le T - \delta\}$$

there is a constant M dependent on $\mathcal{M}$ and a such that

$$0 < k(x - y, at) \le Mk(y, aT),$$

$$\left|\frac{\partial}{\partial x_i} k(x - y, at)\right| \le Mk(y, aT),$$

$$\left|\frac{\partial^2}{\partial x_i\, \partial x_j} k(x - y, at)\right| \le Mk(y, aT),$$

$$\left|\frac{\partial}{\partial t} k(x - y, at)\right| \le Mk(y, aT),$$

for $(x, t) \in \mathcal{M}$, $y \in R_3$ and $i, j = 1, 2$, and 3.

The proof is a simple computation (Prob. 5) just as in Lemma 4-1 in Sec. 5-4. With the aid of the foregoing lemmas, we can prove

Theorem 1-1. Suppose that $f(x)$ is continuous and bounded on R_3, say $|f(x)| \le L$. Then (1-3) has a solution $u(x, t)$ which is continuous for $x \in R_3$ and $t \ge 0$. Moreover, $u(x, t)$ is given by (1-4) and (1-5) for $t > 0$ and $|u(x, t)| \le L$ for $x \in R_3$ and $t \ge 0$.

Proof. We claim that

$$u(x, t) = \begin{cases} \displaystyle\int_{R_3} k(x - y, at) f(y)\, dy, & x \in R_3, \quad t > 0, \\ f(x), & x \in R_3, \quad t = 0, \end{cases} \tag{1-6}$$

is such a continuous solution. Clearly, (1-6) has the correct initial data. Lemma 1-2, the Weierstrass M-test, and Prop. 5-4 of Chap. 3 permit us to differentiate under the integral sign in (1-6) and so obtain

$$\left(\frac{\partial}{\partial t} - \Delta_x\right) u(x, t) = 0$$

for $x \in R_3$ and $t > 0$, in view of (i) of Lemma 1-1. Again by Lemma 1-1,

$$\left|\int_{R_3} k(x - y, at) f(y)\, dy\right| \le L \int_{R_3} k(x - y, at)\, dy = L$$

so (1-6) yields $|u(x, t)| \leq L$. Only the continuity of (1-6) remains to be shown. We must verify

$$\lim_{(x,t)\to(x^\circ,0+)} \int_{R_3} k(x-y, at)f(y)\,dy = f(x^\circ) \tag{1-7}$$

for each $x^\circ \in R_3$. The proof of (1-7) is much as for one space dimension. Let $\varepsilon > 0$ and choose $\delta > 0$ so that

$$|f(y) - f(x^\circ)| < \frac{\varepsilon}{2} \quad \text{for} \quad y \in K_\delta(x^\circ).$$

From Lemma 1-1,

$$\begin{aligned}\int_{R_3} k(x-y, at)f(y)\,dy - f(x^\circ) &= \int_{R_3} k(x-y, at)[f(y) - f(x^\circ)]\,dy \\ &= \left(\int_{K_\delta} + \int_{cK_\delta}\right) k(x-y, at)[f(y) - f(x^\circ)]\,dy \\ &= I_1 + I_2,\end{aligned}$$

where I_1 is the integral over K_δ and I_2 the integral over cK_δ. Lemma 1-1 (ii) implies that $|I_1| < \varepsilon/2$. Next, take $x \in K_{\delta/2}(x^\circ)$. Then for $y \in cK_\delta(x^\circ)$,

$$\begin{aligned}k(x-y, at) &= (4\pi at)^{-3/2} \exp\left(-\frac{|x-y|^2}{8at}\right) \exp\left(-\frac{|x-y|^2}{8at}\right) \\ &\leq 2^{3/2} \exp\left(-\frac{\delta^2}{32at}\right) k(x-y, 2at)\end{aligned}$$

and hence

$$\begin{aligned}|I_2| &\leq \int_{cK_\delta} k(x-y, at)[|f(y)| + |f(x^\circ)|]\,dy \\ &\leq 2^{5/2} L \exp\left(-\frac{\delta^2}{32at}\right) \int_{cK_\delta} k(x-y, 2at)\,dy < 2^{5/2} L \exp\left(-\frac{\delta^2}{32at}\right).\end{aligned}$$

This expression has limit 0 as $t \to 0+$ and so can be made less than $\varepsilon/2$ for $t > 0$ and close to 0. Thus, for $x \in K_{\delta/2}(x^\circ)$ and $t > 0$ sufficiently close to 0, we have

$$\left|\int_{R_3} k(x-y, at)f(y)\,dy - f(x^\circ)\right| \leq |I_1| + |I_2| < \varepsilon,$$

which establishes (1-7) and completes the proof of Th. 1-1.

Remark. Theorem 1-1 covers most physical applications. However, as for one space dimension (see Prob. 11) there is a stronger existence theorem. If f satisfies a growth restriction of the form $|f(x)| \leq B \exp(b|x|^\beta)$ with $0 \leq \beta < 2$, then (1-6) still represents a solution to (1-3). The proof is essentially the same as for Th. 1-1, although the supporting computations are a little more tedious.

We show next that the bounded solution in Th. 1-1 is unique.

Theorem 1-2. There is at most one continuous function $u(x, t)$ defined for $x \in R_3$ and $t \geq 0$ and bounded on any bounded time interval which satisfies

$$\begin{cases} u_t = a\,\Delta u + h(x, t), & x \in R_3, \quad t > 0, \\ u(x, 0) = f(x), & x \in R_3. \end{cases}$$

Proof. Let $u = u_1 - u_2$ be the difference of two such solutions. Then u is continuous, bounded on any finite time interval, and satisfies

$$u_t - a\,\Delta u = 0, \qquad u(x, 0) = 0. \tag{1-8}$$

We must prove that u satisfying (1-8) is identically zero. The verification of this follows lines reminiscent of uniqueness arguments for the exterior problems in potential theory. Fix a point (x, t) with $t > 0$. We show that u is zero at this point, using an argument based on the divergence theorem. In order to work on bounded domains, we introduce a cutoff function $\zeta_R(y)$ for $R > 0$ which satisfies $0 \leq \zeta_R(y) \leq 1$ for all y,

$$\zeta_R(y) = 1 \quad \text{for} \quad |y| \leq R, \qquad \zeta_R(y) = 0 \quad \text{for} \quad |y| \geq R + 1,$$

and ζ_R twice continuously differentiable with its first- and second-order partials bounded independent of R. Consider the identity

$$\begin{aligned} &(u_\tau - a\,\Delta_y u)\zeta_R v + u(\zeta_R v_\tau + a\,\Delta_y(\zeta_R v)) \\ &\quad = (u\zeta_R v)_\tau - a \operatorname{div}\,[\zeta_R v\,\nabla u - u\,\nabla(\zeta_R v)]. \end{aligned} \tag{1-9}$$

We let $u = u(y, \tau)$ be the function satisfying (1-8) and $v(y, \tau) = k(x - y, a(t - \tau))$. Recall that x and t are fixed. Fix δ with $0 < \delta < t - \delta$ and $R > |x|$. Integrate (1-9) with respect to (y, τ) for $y \in K_{R+2}(0)$ and $\delta \leq \tau \leq t - \delta$. Since $v_\tau + a\,\Delta_y v = 0$ by Lemma 1-1 the left side of (1-9) yields

$$\begin{aligned} &\int_\delta^{t-\delta} \int_{K_{R+2}} (auv\,\Delta\zeta_R + 2au\,\nabla\zeta_R \cdot \nabla v)\,dy\,d\tau \\ &\quad = \int_\delta^{t-\delta} d\tau \int_{R \leq |y| \leq R+1} (auv\,\Delta\zeta_R + 2au\,\nabla\zeta_R \cdot \nabla v)\,dy. \end{aligned} \tag{1-10}$$

Upon integration and use of the Gauss divergence theorem, the right side of (1-9) yields

$$\int_\delta^{t-\delta} \int_{K_{R+2}} (u\zeta_R v)_\tau\,dy\,d\tau = \int_{K_{R+2}} u\zeta_R v\Big|_{\tau=\delta}^{\tau=t-\delta} dy. \tag{1-11}$$

Let $R \to \infty$ in (1-10) and (1-11). In view of the rapid decay of v and its derivatives as $R \to \infty$ and the boundedness of the other quantities in (1-10), the expression in (1-10) has limit zero and (1-11) has limit $\int_{R_3} uv|_{\tau=\delta}^{\tau=t-\delta}\,dy$. Equating these results gives

$$\int_{R_3} [u(y, t - \delta)v(y, t - \delta) - u(y, \delta)v(y, \delta)]\,dy = 0.$$

Recall that $v(y, \tau) = k(x - y, a(t - \tau))$, let $\delta \to 0$, and use (1-7) and $u(y, 0) = 0$ to find $u(x, t) = 0$ (see Prob. 12). Hence, Th. 1-2 is established.

The identity (1-9) leads to a solution formula to the inhomogeneous initial value problem

$$\text{(1-12)} \qquad \begin{cases} u_t = a\,\Delta u + h(x, t), & x \in R_3, \quad t > 0, \\ u(x, 0) = f(x), & x \in R_3. \end{cases}$$

Assume u solves (1-12) and let v be as above. This time the left side of (1-9) yields

$$\int_\delta^{t-\delta} \int_{K_{R+2}} h(y, \tau)\zeta_R(y)v(y, \tau)\, dy\, d\tau + \int_\delta^{t-\delta} \int_{R \le |y| \le R+1} [auv\, \Delta\zeta_R + 2au\, \nabla\zeta_R \cdot \nabla v]\, dy\, d\tau$$

after integration, and the right side is again

$$\int_{K_{R+2}} \zeta_R(y)u(y, \tau)v(y, \tau)\Big|_{\tau=\delta}^{\tau=t-\delta} dy.$$

Equate these expressions; let $R \to \infty$ and then $\delta \to 0$ to obtain (see also Prob. 13)

$$\text{(1-13)} \quad u(x, t) = \int_0^t \int_{R_3} k(x - y, a(t - \tau))h(y, \tau)\, dy\, d\tau + \int_{R_3} k(x - y, at)f(y)\, dy.$$

Since we derived (1-13) under the assumption that (1-12) has a solution, we must now confirm that this natural candidate for the solution is in fact a solution. This verification is similar to the proof of Poisson's theorem (Th. 6-1 of Chap. 8). We already know that the second term in (1-13) satisfies the homogeneous heat equation and assumes the initial data in (1-12) as $t \to 0$, while the first term in (1-13) has limit zero as $t \to 0$, assuming that h is bounded on finite time intervals. Thus, (1-13) will solve (1-12) if

$$\text{(1-14)} \qquad \left(\frac{\partial}{\partial t} - a\Delta\right) \int_0^t \int_{R_3} k(x - y, a(t - \tau))h(y, \tau)\, dy\, d\tau = h(x, t).$$

To verify (1-14), we need two lemmas.

Lemma 1-3. Suppose that α and σ are fixed with $0 < \alpha < 1$ and $\sigma \ge 0$. Then there is a constant M depending only upon α and σ such that

$$z^\sigma e^{-z} \le M e^{-\alpha z}, \qquad z \ge 0.$$

The proof amounts to the observation that the nonnegative continuous function $z^\sigma e^{-(1-\alpha)z}$ has limit zero as $z \to \infty$ and hence has a positive maximum M on $[0, \infty)$.

This lemma implies the useful estimates in

Lemma 1-4. Let $\lambda > 1$ be given. Then there is a constant M such that

$$\left|\frac{\partial}{\partial x_i} k(x-y, at)\right| \le \frac{M}{\sqrt{t}} k(x-y, a\lambda t),$$

$$\left|\frac{\partial^2}{\partial x_i \partial x_j} k(x-y, at)\right| \le \frac{M}{t} k(x-y, a\lambda t),$$

$$\left|\frac{\partial}{\partial t} k(x-y, at)\right| \le \frac{M}{t} k(x-y, a\lambda t)$$

for $i, j = 1, 2$, and 3.

Proof. Let $\lambda > 1$ and use Lemma 1-3 to obtain

$$\begin{aligned}\left|\frac{\partial}{\partial x_i} k(x-y, at)\right| &= \left| -\frac{2(x_i - y_i)}{4at}(4\pi at)^{-3/2} \exp\left(-\frac{|x-y|^2}{4at}\right)\right| \\ &\le \frac{2(4\pi at)^{-3/2}}{(4at)^{1/2}}\left(\frac{|x-y|^2}{4at}\right)^{1/2} \exp\left(-\frac{|x-y|^2}{4at}\right) \\ &\le \frac{M_1\lambda^{3/2}}{\sqrt{a}\sqrt{t}}(4\pi a\lambda t)^{-3/2} \exp\left(-\frac{|x-y|^2}{4a\lambda t}\right) \\ &= \frac{M_1\lambda^{3/2}a^{-1/2}}{\sqrt{t}} k(x-y, a\lambda t).\end{aligned}$$

Similar estimates lead to corresponding bounds on the other derivatives in Lemma 1-4 and we obtain the desired bounds with M the largest constant that occurs.

The estimates in Lemmas 1-3 and 1-4 guarantee that we can differentiate under the integral sign in the following calculations because the resulting integrals converge absolutely and uniformly in the neighborhood of any point (x, t) with $x \in R_3$ and $t > 0$. Thus,

$$\begin{aligned}&\frac{\partial}{\partial x_i}\int_0^t \int_{R_3} k(x-y, a(t-\tau))h(y, \tau)\, dy\, d\tau \\ &\quad = \int_0^t \int_{R_3} \frac{\partial}{\partial x_i} k(x-y, a(t-\tau))h(y, \tau)\, dy\, d\tau \\ &\quad = -\int_0^t \int_{R_3} \frac{\partial}{\partial y_i} k(x-y, a(t-\tau))h(y, \tau)\, dy\, d\tau\end{aligned}$$

provided that, say, h is bounded on bounded time intervals. Moreover, if h is differentiable with respect to y and has bounded derivatives, we may integrate by parts to obtain

$$\frac{\partial}{\partial x_i}\int_0^t \int_{R_3} k(x-y, a(t-\tau))h(y, \tau)\, dy\, d\tau = \int_0^t \int_{R_3} k(x-y, a(t-\tau))h_{y_i}(y, \tau)\, dy\, d\tau.$$

This expression shows that we can differentiate again to find

$$\frac{\partial^2}{\partial x_i^2}\int_0^t\int_{R_3} k(x-y, a(t-\tau))h(y,\tau)\,dy\,d\tau = \int_0^t\int_{R_3} k_{x_i}(x-y, a(t-\tau))h_{y_i}(y,\tau)\,dy\,d\tau.$$

We wish now to integrate by parts to transfer all derivatives to k. This is a delicate step because the second partials, $k_{x_ix_i}$, grow like $(t-\tau)^{-1}$ (see Lemma 1-4). We use the smoothness of $h(y, \tau)$ to reduce this growth to the integrable singularity $(t-\tau)^{-1/2}$: The integral on the right above can be written as

$$\begin{aligned}\int_0^t\int_{R_3} & k_{x_i}(x-y, a(t-\tau))[h(y,\tau)-h(x,\tau)]_{y_i}\,dy\,d\tau \\ &= -\int_0^t\int_{R_3} k_{x_iy_i}(x-y, a(t-\tau)][h(y,\tau)-h(x,\tau)]\,dy\,d\tau \\ &= \int_0^t\int_{R_3} k_{x_ix_i}(x-y, a(t-\tau))[h(y,\tau)-h(x,\tau)]\,dy\,d\tau\end{aligned}$$

provided that this last integral converges. To see that it does, apply the mean value theorem and use the boundedness of the partials of h to find a constant H such that

$$|h(y,\tau)-h(x,\tau)| \le H|y-x|.$$

Then an application of Lemma 1-4 followed by the estimates used in the proof itself gives

$$\begin{aligned}|k_{x_ix_i}&(x-y, a(t-\tau))[h(y,\tau)-h(x,\tau)]| \\ &\le \frac{HM|x-y|}{t-\tau}k(x-y, a\lambda(t-\tau)) \le \frac{\tilde{M}}{\sqrt{t-\tau}}k(x-y, a\tilde{\lambda}(t-\tau))\end{aligned}$$

for $\tilde{\lambda} > \lambda > 1$, which ensures convergence of the integral above. Combining these results, we infer that

$$\begin{aligned}\text{(1-15)}\qquad \Delta_x & \int_0^t\int_{R_3} k(x-y, a(t-\tau))h(y,\tau)\,dy\,d\tau \\ &= \int_0^t\int_{R_3} \Delta_x k(x-y, a(t-\tau))[h(y,t)-h(x,t)]\,dy\,d\tau.\end{aligned}$$

Reasoning in the same fashion and using (ii) of Lemma 1-1 we find that

$$\begin{aligned}\text{(1-16)}\quad \frac{\partial}{\partial t} & \int_0^t\int_{R_3} k(x-y, a(t-\tau))h(y,\tau)\,dy\,d\tau \\ &= \frac{\partial}{\partial t}\left\{\int_0^t\int_{R_3} k(x-y, a(t-\tau))[h(y,\tau)-h(x,\tau)]\,dy\,d\tau + \int_0^t h(x,\tau)\,d\tau\right\} \\ &= \int_0^t\int_{R_3} k_t(x-y, a(t-\tau))[h(y,\tau)-h(x,\tau)]\,dy\,d\tau + h(x,t).\end{aligned}$$

Combine (1-15) and (1-16) and use the fact that k satisfies the (homogeneous) heat equation to arrive at (1-14). We summarize this discussion as

Theorem 1-3. Suppose that $f(x)$ is bounded and continuous on R_3. Assume that $h(x, t)$ is continuous, and that h and its first-order partials with respect to its space variables are bounded on bounded time intervals. Then there is a unique continuous solution $u(x, t)$ to (1-12) which is given by (1-13) for $x \in R_3$ and $t > 0$. Moreover, (1-13) shows that

$$|u(x, t)| \leq M + LT, \quad \text{for} \quad x \in R_3 \quad \text{and} \quad 0 \leq t \leq T$$

if $|f(x)| \leq M$ and $|h(x, t)| \leq L$ for $x \in R_3$ and $0 \leq t \leq T$. Finally, (1-13) shows that (1-12) is well-posed over any finite time interval.

PROBLEMS

1. Use Fourier transforms to obtain the formal solution (1-4) and (1-5) to (1-3).
2. Find a formal solution to

$$u_t = a\,\Delta u, \qquad x \in R_2, \qquad t > 0, \qquad u(x, 0) = f(x), \qquad x \in R_2,$$

where $a > 0$. Your solution will lead you to the fundamental solution to the heat equation in two dimensions,

$$k(x, t) = (4\pi t)^{-1} \exp\left(-\frac{|x|^2}{4t}\right).$$

3. Review the argument in Sec. 5-4 which establishes that, in one space dimension, $k(x - x_0, at)$ is the concentration at position x at time t due to a unit mass concentrated at x_0 at time $t = 0$. Extend this reasoning to three space dimensions.
4. Prove Lemma 1-1.
5. Prove Lemma 1-2.
6. State and prove the analogue of Th. 1-1 for two dimensions.
7. Consider the Cauchy problem

$$u_t = \frac{1}{r}\frac{\partial}{\partial r}\left(r\frac{\partial u}{\partial r}\right) + \frac{\partial^2 u}{\partial z^2}, \qquad r \geq 0, \qquad |z| < \infty, \qquad t > 0,$$

$$u(r, z, 0) = f(r, z), \qquad r \geq 0, \qquad |z| < \infty.$$

(a) Find a formal solution by transforming the z variable with the one-dimensional Fourier transform and the r variable with the Hankel transform.

(b) Show that the solution can also be obtained from (1-4) and (1-5) by introducing cylindrical coordinates and observing that since f is independent of the angular variable, the solution u will depend only upon (r, z, t).

8. Solve

$$u_t = a\,\Delta u + h(x, t), \qquad u(x, 0) = 0,$$

in one, two, or three space dimensions using *Duhamel's principle*. That is, solve for any τ,

$$w_t = \Delta a w, \qquad w(x, 0) = h(x, \tau)$$

to obtain $w = w(x, t, \tau)$, where τ is a parameter. Now show that

$$u(x, t) = \int_0^t w(x, t - \tau, \tau)\, d\tau$$

solves the original problem.

9. Let $a(t) \geq a_0 > 0$, $v(t)$, and $c(t)$ be continuous functions of t, with $v(t)$ vector-valued. Solve formally

$$u_t = a(t)\, \Delta u + v(t) \cdot \nabla u + c(t)u, \qquad u(x, 0) = f(x),$$

in one, two, or three space dimensions.

10. Verify (1-9).

11. Establish the analogue of Th. 4-2 of Chap. 5 for two and three space dimensions. That is, if $|f(x)| \leq B \exp(b|x|^2)$, show that (1-4) and (1-5) solve

$$u_t = a\, \Delta u, \qquad x \in R_n, \qquad 0 < t \leq T, \qquad u(x, 0) = f(x), \qquad x \in R_n,$$

for T suitably small and $n = 2, 3$.

12. Check carefully the argument leading from

$$\int_{R_3} [u(y, t - \delta)v(y, t - \delta) - u(y, \delta)v(y, \delta)]\, dy = 0$$

to $u(x, t) = 0$ in the limit as $\delta \to 0$. See the discussion in the paragraph preceding (1-12).

13. Suppose that h in (1-12) can be regarded as a point source concentrated at the point $x = 0$, say, which introduces mass or heat at the rate $Q(t)$.

(a) Use physical arguments, in the spirit of Prob. 3, to deduce that

$$u(x, t) = \int_0^t k(x, a(t - \tau))Q(\tau)\, d\tau + \int_{R_3} k(x - y, at)f(y)\, dy.$$

(b) Argue, in the same spirit, that (1-13) should solve (1-12) when $h(x, t)$ is any continuous source distribution.

14. Suppose that $f(x)$ is bounded, continuous, and integrable over R_3. Let $u(x, t)$ be the solution to (1-3). Explain on physical grounds why

$$\int_{R_3} u(x, t)\, dx = \int_{R_3} f(x)\, dx$$

should hold for all t. Prove that this is so.

15. Suppose in (1-3) that f depends only on the distance r from the origin and that we seek a solution u of the same form. Then (1-3) reduces to

$$u_t = \frac{a}{r^2} \frac{\partial}{\partial r}\left(r^2 \frac{\partial u}{\partial r}\right), \qquad u(r, 0) = f(r).$$

(a) Let $u = w/r$ and show

$$w_t = aw_{rr}, \qquad w(r, 0) = rf(r).$$

(b) Find w by solving the one-dimensional heat equation and then determine u.

(c) Solve again by introducing spherical coordinates in (1-4) and (1-5).

16. In applied problems it can be extremely difficult to determine the coefficients in a differential equation. In this problem we describe the *Theis method*, which is used in groundwater hydrology to determine such coefficients. (Compare with Thiem's method in Prob. 13 of Sec. 8-1.) Assume that the density ρ of water is constant and that porosity $\phi = \phi(p) = \alpha p + \phi_0$, where p is the pressure and ϕ_0 is a constant. Let $h = x_3 + p/\rho g$. Then, as for Theim's method, Darcy's law gives

$$q = -\frac{k}{\mu}(\nabla p + \rho g \nabla x_3) = -K \nabla h,$$

where $K = k\rho g/\mu$. The continuity equation is $\alpha p_t + \operatorname{div} q = 0$.

(a) Show that h satisfies $\alpha \rho g h_t = K \Delta h$.

(b) Suppose that the aquifer is bounded by parallel impermeable planes given by $x_3 = 0$ and $x_3 = b$. Suppose at $t = 0$ that $h = H$ and as $r = \sqrt{x_1^2 + x_2^2} \to \infty$, $h \to H$. Let

$$u(x_1, x_2, t) = \frac{1}{b}\int_0^b h(x_1, x_2, x_3, t)\, dx_3 - H$$

and show that

$$(\alpha b g \rho) u_t = (Kb)\, \Delta u$$
$$u(x_1, x_2, 0) = 0, \qquad u \to 0 \quad \text{as} \quad r \to \infty,$$

where $\Delta = \partial^2/\partial x_1^2 + \partial^2/\partial x_2^2$.

(c) Suppose that a fully penetrating pump is located at $(x_1, x_2) = (0, 0)$, $0 \le x_3 \le b$ and pumps at the rate $Q(t)$. Assume radial symmetry and argue that

$$\lim_{r \to 0} r u_r(r, t) = \frac{Q}{2\pi K b}.$$

(d) Let $S = \alpha \rho g b$, $T = Kb$ and conclude that u must satisfy

$$S u_t = T \frac{1}{r}(r u_r)_r, \qquad r \ge 0, \qquad t > 0,$$
$$u(r, 0) = 0, \qquad r \ge 0,$$
$$u(r, t) \to 0 \quad \text{as} \quad r \to \infty,$$
$$\lim_{r \to 0} r u_r(r, t) = \frac{Q(t)}{2\pi T}.$$

S is called the storativity and T the transmissivity.

(e) Show that the problem in part (d) has solution

$$u(r, t) = -\frac{1}{S}\int_0^t k(r, a(t - \tau)) Q(\tau)\, d\tau$$

for $a = T/S$ and $r > 0$ and $k(r, t)$ the fundamental solution in two space dimensions. *Hint.* Reason that the condition $r u_r(r, t) \to Q(t)/2\pi T$ represents a point sink and apply Prob. 13.

(f) In the field, the pumping rate $Q(t)$ is known. Also, $u(r, t) + H$ and H can be measured at observation wells. Show how this information can be used to find S and a, and hence S and T.

2. Explicit Solutions to Initial, Boundary Value Problems

If the geometry of the underlying domain is simple enough, explicit solutions can be constructed using separation of variables, integral transforms, or a combination of these methods. The basic ideas are the same as for problems with only one space variable. The principal difference between the higher-dimensional case and the one-dimensional case is that the computations become much more complicated. Consequently, a great deal of patience is required. We present three examples that illustrate how such problems are treated.

Example 1. Find the temperature $u = u(x, y, z, t)$ in the region between two parallel planes, say $z = 0$ and $z = b$ with $b > 0$, if $u = 0$ on each plane and initially $u(x, y, z, 0) = f(x)$.

The initial boundary value problem in question is

$$u_t = a\,\Delta u,$$
$$u(x, y, 0, t) = u(x, y, b, t) = 0,$$
$$u(x, y, z, 0) = f(x, y, z).$$

To solve this problem, we apply the two-dimensional Fourier transform to the x and y variables to obtain

$$\hat{u}_t = -a(\omega_1^2 + \omega_2^2)\hat{u} + a\hat{u}_{zz},$$
$$\hat{u}(\omega_1, \omega_2, 0, t) = \hat{u}(\omega_1, \omega_2, b, t) = 0,$$
$$\hat{u}(\omega_1, \omega_2, z, 0) = \hat{f}(\omega_1, \omega_2, z),$$

where ω_1 and ω_2 are the Fourier transform variables. This transformed problem can be solved by separation of variables. We regard ω_1 and ω_2 as parameters, set $\alpha = \omega_1^2 + \omega_2^2$, and let $\hat{u} = Z(z)T(t)$ to obtain $ZT' = -\alpha aZT + aZ''T$. Thus,

$$\frac{T'}{aT} + \alpha = \frac{Z''}{Z} = c,$$

a separation constant. The boundary conditions lead to the eigenvalue problem

$$Z'' - cZ = 0, \qquad Z(0) = Z(b) = 0.$$

To obtain nontrivial solutions, we conclude as in Chaps. 4 and 5 that $-c = \lambda_n^2 = (n\pi/b)^2$ and that $Z = Z_n(z) = \sin(\lambda_n z)$, up to a nonzero multiple. A simple calculation gives

$$T_n = \exp[-(a\alpha + a\lambda_n^2)t],$$

and we use superposition to obtain the formal solution

$$\hat{u}(\omega_1, \omega_2, z, t) = \sum_{n=1}^{\infty} \hat{\phi}_n(\omega_1, \omega_2) \exp[-(a\alpha + a\lambda_n^2)t] \sin(\lambda_n z)$$

where

$$\hat{\phi}_n(\omega_1, \omega_2) = \frac{2}{b}\int_0^b \hat{f}(\omega_1, \omega_2, z)\sin(\lambda_n z)\,dz.$$

Finally, we take inverse transforms (term by term) with respect to ω_1 and ω_2 to obtain

$$u(x, y, z, t) = \sum_{n=1}^{\infty}\left[\iint_{R_2} k(x-\xi, y-\eta, at)\phi_n(\xi, \eta)\,d\xi\,d\eta\right]e^{-a\lambda_n^2 t}\sin(\lambda_n z),$$

where $\phi_n(\xi, \eta)$ is the inverse transform of $\hat{\phi}_n(\omega_1, \omega_2)$ and k is the fundamental solution in two space variables. Note that we used the convolution theorem for Fourier transforms and the basic fact that $\hat{k}(x, at) = e^{-at|\omega|^2}$ for $x \in R_n$, which follows from the case $n = 1$ given in Sec. 3-4.

Example 2. Find the temperature $u = u(x, y, t)$ in the upper half plane $y > 0$ if $u = 0$ on the x axis and initially $u = f(x, y)$ is prescribed.

We must solve

$$\begin{aligned} &u_t = a\,\Delta u, \qquad y > 0, \qquad -\infty < x < \infty, \qquad t > 0,\\ &u(x, 0, t) = 0, \qquad -\infty < x < \infty, \qquad t \geq 0,\\ &u(x, y, 0) = f(x, y), \qquad y \geq 0, \qquad -\infty < x < \infty. \end{aligned}$$

This time we take Fourier transforms with respect to x and find that

$$\text{(2-1)} \qquad \begin{cases} \hat{u}_t = -a\omega^2\hat{u} + a\hat{u}_{yy}, \qquad y > 0, \qquad t > 0\\ \hat{u}(\omega, 0, t) = 0, \qquad t \geq 0,\\ \hat{u}(\omega, y, 0) = f(\omega, y), \qquad y \geq 0, \end{cases}$$

where ω is the transform variable. We can solve this problem by the methods of Sec. 5-4 to obtain

$$\hat{u}(\omega, y, t) = e^{-a\omega^2 t}\int_0^{\infty}[k_1(y-\eta, at) - k_1(y+\eta, at)]\hat{f}(\omega, \eta)\,d\eta,$$

where k_1 is the one-dimensional fundamental solution (see Prob. 11). Finally, take inverse transforms to get the solution

$$\text{(2-2)} \qquad u(x, y, t) = \int_{-\infty}^{\infty}\int_0^{\infty}[k(x-\xi, y-\eta, at) - k(x-\xi, y+\eta, at)] \times f(\xi, \eta)\,d\xi\,d\eta,$$

where k is the two-dimensional fundamental solution.

Example 3. Find the temperature inside the solid cylinder $x^2 + y^2 \leq \alpha^2$, $0 \leq z \leq b$, if $u = 0$ on the three sides of the cylinder and initially $u = f(r, z)$, $r = \sqrt{x^2 + y^2}$, is a given radially symmetric function.

This time, the problem is

$$u_t = a\,\Delta u, \qquad x \in D, \qquad t > 0,$$
$$u(x, t) = 0, \qquad x \in B, \qquad t \geq 0,$$
$$u(x, 0) = f(x), \qquad x \in \bar{D},$$

where D is the cylinder and B its bounding surface. We apply separation of variables with $u(x, t) = v(x)T(t)$ and are led to the eigenvalue problem

$$\begin{cases} \Delta v - cv = 0, & x \in D, \\ v = 0, & x \in B, \end{cases} \tag{2-3}$$

and the time equation

$$T' - acT = 0, \qquad t > 0,$$

where c is again the separation constant.

As usual, we need nontrivial solutions of (2-3). Now, an easy integration by parts argument shows that the eigenvalues c of (2-3) must be negative (see Prob. 12). Thus, we set $c = -\lambda^2$ with $\lambda > 0$. In view of the radial symmetry of f, we seek a radially symmetric solution u. Then v will be radially symmetric, $v = v(r, z)$ in cylindrical coordinates and (2-3) can be written as

$$\begin{cases} \dfrac{1}{r}\dfrac{\partial}{\partial r}\left(r\dfrac{\partial v}{\partial r}\right) + \dfrac{\partial^2 v}{\partial z^2} + \lambda^2 v = 0, & (r, z) \in D, \\ v(r, z) = 0, & (r, z) \in B. \end{cases} \tag{2-3$'$}$$

We solve (2-3)$'$ also by separation of variables. Let $v(r, z) = R(r)Z(z)$ to find

$$\frac{1}{r}(rR')'Z + RZ'' + \lambda^2 RZ = 0,$$

and, upon separating variables,

$$\frac{1}{r}(rR')' - \tilde{c}R = 0, \qquad R(\alpha) = 0, \tag{2-4}$$

together with

$$Z'' + (\lambda^2 + \tilde{c})Z = 0, \qquad Z(0) = Z(b) = 0. \tag{2-5}$$

Our usual integration by parts argument shows that (2-4) can have nontrivial solutions only if $\tilde{c} < 0$, say $\tilde{c} = -\nu^2$ with $\nu > 0$ (see Prob. 13). Note that the differential equation in (2-4), which we now express as

$$R'' + \frac{1}{r}R' + \nu^2 R = 0, \tag{2-4$'$}$$

is singular when $r = 0$, but that physically meaningful solutions must be twice continuously differentiable in $0 \leq r \leq \alpha$. Equation (2-4)$'$ is Bessel's equation and has only one bounded solution (up to constant multiples)

$R(r) = J_0(\nu r)$, where $J_0(z)$ is Bessel's function of order zero. Recall that

$$J_n(z) = \sum_{k=0}^{\infty} \frac{(-1)^k}{k!(n+k)!} \left(\frac{z}{2}\right)^{n+2k},$$

and that J_n has an infinite number of simple zeros in $0 < z < \infty$. To satisfy the boundary condition in (2-4), we require that $R(\alpha) = J_0(\nu\alpha) = 0$. Thus, if $\mu_1 < \mu_2 < \cdots < \mu_n < \cdots$ are the positive roots of $J_0(z)$, we must choose $\nu = \nu_j = \mu_j/\alpha$. In summary, (2-4) has (twice differentiable) nontrivial solutions when $\tilde{c} = -\nu_j^2 = -\mu_j^2/\alpha^2$ and these solutions are

$$R_j(r) = J_0\left(\frac{\mu_j r}{\alpha}\right), \qquad j = 1, 2, \ldots. \tag{2-6}$$

With $\tilde{c} = -\mu_j^2/\alpha^2$ determined from (2-4), problem (2-5) becomes

$$Z'' + \left(\lambda^2 - \frac{\mu_j^2}{\alpha^2}\right)Z = 0, \qquad Z(0) = Z(b) = 0.$$

For each j we know that this problem has eigenvalues

$$\lambda^2 - \frac{\mu_j^2}{\alpha^2} = \left(\frac{n\pi}{b}\right)^2, \qquad n = 1, 2, \ldots,$$

and corresponding eigenfunctions

$$Z = Z_{nj}(z) = \sin\left(\frac{n\pi z}{b}\right).$$

In this way we have determined nontrivial solutions

$$v_{nj}(r, z) = R_j(r)Z_{nj}(z) = J_0\left(\frac{\mu_j r}{\alpha}\right)\sin\left(\frac{n\pi z}{b}\right)$$

to (2-3)′. The time equation $T' - acT = 0$ with $c = -\lambda^2 = -(\mu_j^2/\alpha) - (n\pi/b)^2$ has solution

$$T_j(t) = \exp\left(-\frac{a\mu_j^2 t}{\alpha}\right)\exp\left(-\frac{n^2\pi^2}{b^2}at\right),$$

and $u_{nj}(r, z, t) = v_{nj}(r, z)T_j(t)$ are nontrivial solutions of the heat equation and the homogeneous boundary condition. Thus, our formal solution is

$$u(r, z, t) = \sum_{j,n=1}^{\infty} c_{jn} \exp\left(-\frac{a\mu_j^2 t}{\alpha}\right)\exp\left(-\frac{n^2\pi^2}{b^2}at\right)J_0\left(\frac{\mu_j r}{\alpha}\right)\sin\left(\frac{n\pi z}{b}\right), \tag{2-7}$$

where the coefficients c_{jn} must be chosen so that

$$u(r, z, 0) = f(r, z) = \sum_{j,n=1}^{\infty} c_{jn}J_0\left(\frac{\mu_j r}{\alpha}\right)\sin\left(\frac{n\pi z}{b}\right). \tag{2-8}$$

To find the coefficients c_{jn}, we use orthogonality relations for the eigenfunctions $R_j(r) = J_0(\mu_j r/\alpha)$. If $p \neq q$, we multiply the differential equation satisfied by R_p by R_q, and vice versa, to obtain

$$R_q(rR'_p)' + v_p^2 rR_qR_p = 0 \quad \text{and} \quad R_p(rR'_q)' + v_q^2 rR_pR_q = 0.$$

Subtract and integrate to find

$$(v_p^2 - v_q^2) \int_0^\alpha rR_pR_q\,dr = \int_0^\alpha (rR'_qR_p - rR'_pR_q)'\,dr = 0,$$

$$\int_0^\alpha rR_pR_q\,dr = 0 \quad \text{for} \quad p \neq q.$$

If $p = q$, multiply (2-4) by $r^2R'_p$ and recall $\tilde{c} = -v_p^2$ to obtain

$$v_p^2 r^2 R_p R'_p = -(rR'_p)(rR'_p)' = -\frac{1}{2}\frac{d}{dr}(rR'_p)^2,$$

$$v_p^2 \int_0^\alpha r^2 R_p R'_p\,dr = -\frac{1}{2}\alpha^2 R'^2_p(\alpha).$$

Integrate by parts to find

$$v_p^2 \int_0^\alpha r^2 R_p R'_p\,dr = v_p^2 \int_0^\alpha \frac{r^2}{2}\left(\frac{d}{dr}R_p^2\right)dr = -v_p^2 \int_0^\alpha rR_p^2\,dr.$$

Combine these results to get

$$\int_0^\alpha rR_p^2\,dr = \frac{\alpha^2}{2v_p^2} R'^2_p(\alpha) = \frac{\alpha^2}{2} J_1^2(\mu_p)$$

because $J'_0(z) = -J_1(z)$. Therefore,

$$\text{(2-9)} \qquad \int_0^\alpha rR_p(r)R_q(r)\,dr = \int_0^\alpha rJ_0\left(\frac{\mu_p r}{\alpha}\right)J_0\left(\frac{\mu_q r}{\alpha}\right)dr = \begin{cases} 0, & p \neq q, \\ \dfrac{\alpha^2}{2} J_1^2(\mu_p), & p = q. \end{cases}$$

Now it is easy to determine the coefficients in (2-7). Multiply (2-8) by $rJ_0(\mu_p r/\alpha) \sin(q\pi z/b)$ and integrate with respect to r from 0 to α and with respect to z from 0 to b to obtain

$$\text{(2-10)} \qquad c_{pq} = \frac{4}{\alpha^2 b J_1^2(\mu_p)} \int_0^\alpha \int_0^b f(r, z) rJ_0\left(\frac{\mu_p r}{\alpha}\right) \sin\left(\frac{q\pi z}{b}\right) dr\,dz.$$

Thus, the solution to Example 3 is given by (2-7) with the coefficients given by (2-10). This formal solution can be justified rigorously just as we did for problems in the one space dimension in Chaps. 4 and 5. No really new ideas are involved but the derivation of the estimates required to justify term-by-term differentiations and eigenfunction expansions become tedious. We omit them. On the other hand, solutions such as we have found here are easily evaluated on a computer and can be evaluated with great precision.

Initial, boundary value problems with inhomogeneous differential equations can be treated similarly. The ideas introduced in one space dimension in Chaps.

4 and 5 extend immediately to higher dimensions, but practical implementation is complicated. To see what is involved, consider the problem

$$u_t = a\,\Delta u + h(x, t), \qquad x \in D, \qquad t > 0,$$
$$u(x, t) = 0, \qquad x \in B, \qquad t > 0,$$
$$u(x, 0) = f(x),$$

where D is a bounded domain in R_3 with boundary B. We assume that the corresponding problem with $h(x, t) \equiv 0$ can be solved by separation of variables. As in Ex. 3 we set $u(x, t) = v(x)T(t)$, obtain the eigenvalue problem (2-3), and find that $c = -\lambda^2$. Applying the theory of integral equations (Prob. 14), we find that (2-3) has a countable number of eigenvalues

$$0 < \lambda_1^2 \le \lambda_2^2 \le \cdots \le \lambda_n^2 \le \cdots$$

and corresponding orthonormal eigenfunctions $v_1(x), v_2(x), \ldots$:

$$\int_D v_i(x)v_j(x)\,dx = \begin{cases} 0, & i \neq j, \\ 1, & i = j, \end{cases}$$

Now, expand u, h, and f in eigenfunction expansions.

$$u(x, t) = \sum_{n=1}^{\infty} u_n(t)v_n(x),$$
(2-11)
$$f(x) = \sum_{n=1}^{\infty} f_n v_n(x), \qquad h(x, t) = \sum_{n=1}^{\infty} h_n(t)v_n(x),$$

where $u_n(t) = \int_D u(x, t)v_n(x)\,dx$ is unknown and

$$f_n = \int_D f(x)v_n(x)\,dx, \qquad h_n(t) = \int_D h(x, t)v_n(x)\,dx.$$

Substitute the series for u and h into the inhomogeneous heat equation and equate coefficients of $v_n(x)$ to obtain $u_n'(t) + a\lambda_n^2 u_n(t) = h_n(t)$. Thus,

$$u_n(t) = u_n(0)e^{-a\lambda_n^2 t} + \int_0^t h_n(\tau)e^{-a\lambda_n^2(t-\tau)}\,d\tau. \tag{2-12}$$

The initial data give

$$\sum_{n=1}^{\infty} u_n(0)v_n(x) = u(x, 0) = f(x) = \sum_{n=1}^{\infty} f_n v_n(x),$$

so we must have $u_n(0) = f_n$. Therefore, (2-12) determines $u_n(t)$ in terms of known data and (2-11) solves the given initial, boundary value problem.

We omit a detailed discussion of initial, boundary value problems along these lines. For general domains this approach is useful for theoretical purposes, but has limited practical applications because it is difficult to construct the eigenvalues and eigenfunctions. There are important applications where knowledge about the smallest eigenvalue is critical (see Prob. 15).

PROBLEMS

1. Interpret physically and solve

$$u_t = a\frac{1}{r}\frac{\partial}{\partial r}\left(r\frac{\partial u}{\partial r}\right), \qquad 0 \le r < \alpha, \qquad t > 0,$$
$$u(\alpha, t) = 0, \qquad t \ge 0, \qquad u(r, 0) = f(r), \qquad 0 \le r \le \alpha.$$

2. Interpret physically and solve

$$u_t = a\frac{1}{r^2}\frac{\partial}{\partial r}\left(r^2\frac{\partial u}{\partial r}\right), \qquad 0 \le r < \alpha, \qquad t > 0,$$
$$u(\alpha, t) = 0, \qquad t \ge 0, \qquad u(r, 0) = f(r), \qquad 0 \le r \le \alpha.$$

3. Let D be the box $0 < x < \alpha$, $0 < y < \beta$, $0 < z < \gamma$ and B its boundary. Interpret physically and solve

$$u_t = a(u_{xx} + u_{yy} + u_{zz}), \qquad (x, y, z) \in D,$$
$$u(x, y, z, t) = 0, \qquad (x, y, z) \in B, \qquad t \ge 0,$$
$$u(x, y, z, 0) = f(x, y, z), \qquad (x, y, z) \in \bar{D}.$$

4. (a) Set up the initial boundary value problem for the temperature u in a circular plate of radius α. Assume that the top and bottom of the plate are insulated and that heat only flows parallel to the plate, so a model with two space dimensions can be used.
   (b) Solve this problem. *Hint.* Use polar coordinates.
5. (a) Find the initial, boundary value problem for the temperature $u = u(x, y, t)$ in the region between the two parallel lines $y = 0$ and $y = b$, if $u = 0$ on each line and initially $u(x, y, 0) = f(x, y)$.
   (b) Solve this problem.
6. Interpret physically and solve

$$u_t = a\frac{1}{r}\frac{\partial}{\partial r}\left(r\frac{\partial u}{\partial r}\right), \qquad 0 \le r < \alpha, \qquad t > 0,$$
$$u_r(\alpha, t) = 0, \qquad t \ge 0, \qquad u(r, 0) = f(r), \qquad 0 \le r \le \alpha.$$

7. Interpret physically and solve

$$u_t = a(u_{xx} + u_{yy}), \qquad 0 < x < \alpha, \qquad 0 < y < \beta,$$
$$u(x, 0, t) = u_y(x, \beta, t) = 0, \qquad 0 \le x \le \alpha, \qquad t \ge 0,$$
$$u_x(0, y, t) = u_x(\alpha, y, t) = 0, \qquad 0 \le y \le \beta, \qquad t \ge 0,$$
$$u(x, y, 0) = f(x, y).$$

8. Interpret physically and solve

$$u_t = a\left[\frac{1}{r}\frac{\partial}{\partial r}\left(r\frac{\partial u}{\partial r}\right) + \frac{\partial^2 u}{\partial z^2}\right], \qquad r \ge 0, \qquad 0 < x < b, \qquad t > 0,$$
$$u(r, z, 0) = f(r, z), \qquad r \ge 0, \qquad 0 \le z \le b,$$
$$u_z(r, 0, t) = u_z(r, b, t) = 0, \qquad r \ge 0, \qquad t > 0.$$

9. Find the temperature $u = u(x, y, z, t)$ in the region between the two parallel planes $y = 0$ and $y = b$, if $u = 0$ on the plane $y = 0$, there is no heat flux across the plane $y = b$, and initially $u(x, y, z, 0) = f(x, y, z)$.

10. Suppose that

$$v_t = a\,\Delta v, \qquad x \in D, \qquad t > 0,$$
$$v(x, 0) = f(x), \qquad x \in \bar{D}, \qquad v(x, t) = 0, \qquad x \in B, \qquad t \geq 0,$$

can be solved for any initial values $f(x)$. Here D is a (bounded or unbounded) domain with boundary B in a space of any finite dimension. Consider the problem

$$u_t = \Delta u + h(x, t), \qquad x \in D, \qquad t > 0,$$
$$u(x, 0) = 0, \qquad x \in \bar{D}, \qquad u(x, t) = 0, \qquad x \in B, \qquad t \geq 0.$$

Show that this problem can be solved by *Duhamel's principle*: Let $w = w(x, t, \tau)$ be the solution to

$$w_t = a\,\Delta w, \qquad x \in D, \qquad t > 0,$$
$$w(x, 0, \tau) = h(x, \tau), \qquad x \in \bar{D}, \qquad w(x, t, \tau) = 0, \qquad x \in B, \qquad t \geq 0,$$

where τ is regarded as a parameter. Then

$$u(x, t) = \int_0^t w(x, t - \tau, \tau)\, d\tau$$

solves the problem for u.

11. Refer to (2-1).

(a) Define $v(\omega, y, t)$ by $\hat{u} = e^{bt}v$, where $\hat{u}$ solves (2-1). Find the problem that v satisfies, and show that the choice $b = -a\omega^2$ gives

$$v_t = av_{yy}, \qquad y > 0, \qquad t > 0,$$
$$v(\omega, 0, t) = 0, \qquad t \geq 0, \qquad v(\omega, y, 0) = \hat{f}(\omega, y), \qquad y \geq 0.$$

(b) Now use the results of Sec. 5-4 to find v, hence $\hat{u}$, and confirm (2-2).

12. Let D be a bounded, normal domain with boundary B. Use one of Green's identities to show that the eigenvalue problem (2-3) can only have negative eigenvalues.

13. Show that (2-4) can have nontrivial solutions only if $\tilde{c} < 0$. Recall solutions to (2-4) must be differentiable, hence bounded at $r = 0$.

14. Assume that the Green's function for D in (2-3) exists. Express (2-3) as an integral equation, eigenvalue problem. Use the results of Chap. 7 to discuss the existence of eigenvalues and eigenfunctions for (2-3).

15. In a chain reaction, the number of neutrons in any volume increases at a rate proportional to the number present. Let $u = u(x, t)$ be the neutron density per unit volume, so that $N = \int_V u\, dx$ is the number of neutrons in a given volume V.

(a) Argue that

$$\frac{d}{dt}\int_V u\, dx = -\int_B q \cdot \nu\, d\sigma + \beta \int_V u\, dx,$$

where q is the neutron flux vector and β is a proportionality factor, assumed to be constant.

(b) Assume that $q = -a\,\nabla u$, $a > 0$ a constant. Discuss the physical basis for this assumption and show that, if it holds,

$$u_t = a\,\Delta u + \beta u.$$

(c) Make the change of variable $u = e^{\beta t}v$ to find $v_t = a\,\Delta v$. Let λ_n^2, $n = 1, 2, \ldots$, with $0 < \lambda_1 \le \lambda_2 \le \cdots$ denote the eigenvalues and $v_n(x)$ the corresponding eigenfunctions of (2-3) where D is any bounded domain with boundary B. Show that the solution to

$$u_t = a\,\Delta u + \beta u, \qquad x \in D, \qquad t > 0,$$
$$u(x, t) = 0, \qquad x \in B, \qquad t > 0, \qquad u(x, 0) = f(x), \qquad x \in \bar{D},$$

is

$$u(x, t) = \sum_{n=1}^{\infty} c_n e^{(\beta - a\lambda_n^2)t} v_n(x)$$

and give the formula for c_n.

(d) Observe that as long as $c_1 \neq 0$,

$$\lim_{t\to\infty} u(x, t) = \begin{cases} \infty & \text{if } \beta > a\lambda_1^2, \\ c_1 v_1(x) & \text{if } \beta = a\lambda_1^2, \\ 0 & \text{if } \beta < a\lambda_1^2. \end{cases}$$

Obviously, it is important for given a and β to determine λ_1^2 for different reactor geometries. Determine λ_1^2 for a sphere, a cylinder, and a rectangular box.

16. Consider an ocean with a long, straight coastline. Ignore the tide, wind, current, and so on, so the water can be considered pretty much at rest. Suppose that a pipeline carries a pollutant far out into the ocean and empties at a distance b from the shore into an initially clean ocean. Let u be the concentration of pollutant.

(a) Suppose that pollutant is pumped into the ocean at the rate Q and that pollutant spreads out according to a diffusion process with constant a. Assume a two-dimensional process and set up the differential equation which describes the process.

(b) If the shoreline is given by $y = 0$, $-\infty < x < \infty$, and the ocean occupies the region $y > 0$, argue that the boundary conditions should be

$$u(x, 0, t) = 0, \qquad -\infty < x < \infty, \qquad t > 0,$$

and that initially $u(x, y, 0) = 0$ for $-\infty < x < \infty$, $y > 0$.

(c) The rate at which pollutant comes onto the beach is $q \cdot (0, -1)$ where $q = -a\,\nabla u$. Obtain an expression for the amount of pollutant which has washed onto the beach during the time interval $0 \le t \le T$.

17. Repeat Prob. 16, except replace the pipeline by an oil tanker which (instantaneously) discharges an amount Q of oil at a single point b units from shore.

3. Properties of Solutions of Parabolic Equations

Since parabolic and elliptic equations both describe diffusion and heat conduction processes in the time-dependent and time-independent cases, respectively, it should not be surprising that the properties of the respective solutions are often quite similar. In this section we discuss several key features of solutions to parabolic equations in three-dimensional space. Corresponding results hold in other dimensions and, just as for elliptic equations, the arguments are essentially the same for all dimensions.

For the rest of Chap. 9, unless explicit mention to the contrary is made, D will denote a bounded domain in R_3 and B will be its boundary. Again, the restriction to bounded domains is not essential as a rule, but it does avoid the need for special hypotheses about the behavior of solutions at infinity. The *space–time cylinder* built on D with height T is

$$Q_T = \{(x, t): x \in D, 0 < t \leq T\},$$

its lateral surface is

$$S_T = \{(x, t): x \in B, 0 < t \leq T\},$$

and the so-called *parabolic boundary* is the set $B_T = S_T \cup (\bar{D} \times \{0\})$. The functions we are considering depend on both space and time. Different smoothness requirements are appropriate for these variables. Consequently, we shall write $u \in C^{2,1}(Q_T)$ to indicate that u has continuous second-order partials with respect to its spatial variables and has a continuous derivative with respect to time. We write $C(Q_T) = C^{0,0}(Q_T)$ for the set of functions continuous on Q_T.

The next few results extend the maximum principle of Chap. 5 to three space variables.

Lemma 3-1. Suppose that $u \in C^{2,1}(Q_T)$, $a > 0$,

$$u_t - a\,\Delta u > 0 \quad \text{on} \quad Q_T.$$

Then u cannot assume a local minimum at any point in Q_T.

Proof. See Prob. 1.

Proposition 3-1. Let $u \in C(\bar{Q}_T) \cap C^{2,1}(Q_T)$. Assume that

$$u_t - a\,\Delta u \geq 0 \quad \text{in} \quad Q_T \quad \text{and} \quad u \geq 0 \quad \text{on} \quad B_T.$$

Then

$$u \geq 0 \quad \text{throughout} \quad \bar{Q}_T.$$

Similarly, the pair of inequalities

$$u_t - a\,\Delta u \leq 0 \quad \text{on} \quad Q_T \quad \text{and} \quad u \leq 0 \quad \text{on} \quad B_T,$$

imply that $u \leq 0$ throughout $\bar{Q}_T$.

Proof. Consider the first assertion and form the auxiliary function $w = u + \varepsilon t$, where $\varepsilon > 0$. We have

$$w_t - a\,\Delta w = u_t - a\,\Delta u + \varepsilon > 0 \quad \text{on} \quad Q_T.$$

By Lemma 3-1, w assumes its minimum on B_T. Thus,

$$w = u + \varepsilon t \geq 0 \quad \text{on} \quad \bar{Q}_T.$$

Since $0 \leq t \leq T$, we let $\varepsilon \to 0$ and conclude that $u \geq 0$ on $\bar{Q}_T$, as asserted. The second assertion follows from the first applied to $-u$.

As a simple consequence of Prop. 3-1 we obtain the *maximum–minimum principle* for parabolic equations:

Theorem 3-1. Suppose that $u \in C(\bar{Q}_T) \cap C^{2,1}(Q_T)$ and that

$$u_t - a\,\Delta u = 0 \quad \text{in} \quad Q_T.$$

Then

$$\min_{B_T} u(x, t) \le u(x, t) \le \max_{B_T} u(x, t).$$

That is, u always takes its maximum and minimum values somewhere on the parabolic boundary B_T.

Proof. Let

$$M = \max_{(x,t) \in B_T} u(x, t) \quad \text{and} \quad m = \min_{(x,t) \in B_T} u(x, t).$$

Then the function $M - u(x, t)$ satisfies the first pair of inequalities in Prop. 3-1; hence, $M \ge u(x, t)$ throughout $\bar{Q}_T$. Similarly, $m \le u(x, t)$ on $\bar{Q}_T$.

The maximum–minimum principle implies uniqueness of solutions to the Dirichlet problem

$$\text{(3-1)} \qquad \begin{cases} u_t - a\,\Delta u = h(x, t), & (x, t) \in Q_T, \\ u(x, t) = \phi(x, t), & (x, t) \in S_T, \\ u(x, 0) = f(x), & x \in \bar{D}. \end{cases}$$

Theorem 3-2. The Dirichlet problem (3-1) has at most one solution $u \in C(\bar{Q}_T) \cap C^{2,1}(Q_T)$.

Proof. See Prob. 2.

The maximum-minimum principle also implies an a priori bound on solutions to (3-1).

Theorem 3-3. Suppose that $u \in C(\bar{Q}_T) \cap C^{2,1}(Q_T)$ satisfies (3-1), that $|h(x, t)| \le N$ on Q_T, that $|\phi(x, t)| \le M$ on S_T, and that $|f(x)| \le M$ on $\bar{D}$. Then

$$|u(x, t)| \le M + NT \quad \text{on} \quad \bar{Q}_T.$$

Proof. The function $M + Nt \mp u(x, t)$ satisfies the first pair of inequalities in Prop. 3-1. Consequently, $M + Nt \ge \pm u(x, t)$ and the conclusion of the theorem follows.

Corollary 3-1. The solution (assumed to exist) to the Dirichlet problem (3-1) depends continuously on the data.

The maximum principle can also be used to show uniqueness for the exterior Dirichlet problem. To see this, let $D = cD_1$ be a typical exterior domain and define

Q_T and S_T just as for bounded domains. Suppose that there were two *bounded* solutions, u_1 and u_2, to

$$\text{(3-2)} \qquad \begin{cases} u \in C(\bar{Q}_T) \cap C^{2,1}(Q_T) \\ u_t - a\,\Delta u = h(x, t), & (x, t) \in Q_T, \\ u(x, t) = \phi(x, t), & (x, t) \in S_T, \\ u(x, 0) = f(x), & x \in \bar{D}. \end{cases}$$

Then $w = u_1 - u_2$ satisfies the same type of problem except that all the data are zero. The maximum principle cannot be used directly to conclude that $w \equiv 0$ because D is unbounded. However, we can reduce to a bounded domain as follows. There is a constant $M \geq |u_1| + |u_2| \geq |w|$. Choose R so large that $D_1 \subset K_R(0)$ and define

$$W(x, t) = \frac{6M}{R^2}\left(\frac{|x|^2}{6} + at\right).$$

Notice that $W_t - a\,\Delta W = 0$ and define

$$Q_{R,T} = \{(x, t) : x \in K_R \cap D, 0 < t \leq T\}.$$

Consider $v = W \mp w$. Clearly, $v_t - a\,\Delta v = 0$ in $Q_{R,T}$ and $v = W \mp 0 \geq 0$ on $\bar{D} \cup \{0 \leq t \leq T\}$ and on S_T. For $|x| = R$ and $t \geq 0$, $W \geq M \geq |w|$, so $v \geq 0$ for such x and t. In summary, $v = W \mp w \geq 0$ on the parabolic boundary of $Q_{R,T}$ and hence, by the maximum principle, $W \mp w \geq 0$ in $\bar{Q}_{R,T}$. Hence,

$$|w(x, t)| \leq \frac{6M}{R^2}\left(\frac{|x|^2}{6} + at\right)$$

in $\bar{Q}_{R,T}$. Fix $x \in D$ and $t \geq 0$. Let $R \to \infty$ to deduce that $w(x, t) = 0$, which proves uniqueness.

Theorem 3-4. There exists at most one bounded solution to the exterior Dirichlet problem (3-2).

PROBLEMS

1. Prove Lemma 3-1. *Hint.* Review the proof of Prop. 2-1 of Chap. 5.
2. Prove Th. 3-2.
3. Formulate Cor. 3-1 more precisely and prove it.
4. Prove that Lemma 3-1, the maximum–minimum principle, and Ths. 3-2 through Th. 3-4 hold in n space dimensions.
5. Let D be a bounded, normal domain. Show there is at most one solution $u \in C^{2,1}(Q_T) \cap C^{1,0}(\bar{Q}_T)$ to the Neumann problem

$$u_t - a\,\Delta u = h(x, t), \qquad (x, t) \in Q_T,$$

$$\frac{\partial u}{\partial \nu} = \psi(x, t), \qquad (x, t) \in S_T, \qquad u(x, 0) = f(x), \qquad x \in \bar{D}.$$

Hint. Let w be the difference of two solutions, form

$$I(t) = \frac{1}{2}\int_D [w(x, t)]^2\, dx,$$

and show that $I'(t) \leq 0$.

6. In the context of Prob. 5 show that there is at most one solution to the Robin problem, where the boundary condition becomes

$$\frac{\partial u}{\partial \nu} + b(x, t)u = \psi(x, t), \qquad (x, t) \in S_T,$$

with $b(x, t) \geq 0$.

7. Let $a = a(x, t)$, $b = b(x, t)$, and $c = c(x, t)$ be given functions on Q_T with $a > 0$ and $c \geq 0$

(a) If $u \in C^{2,1}(Q_T)$ and

$$u_t - a\,\Delta u + b \cdot \nabla u + cu > 0 \quad \text{on} \quad Q_T$$

prove that u cannot assume a negative minimum in Q_T.

(b) Now draw the same conclusion assuming that

$$u_t - a\,\Delta u + b \cdot \nabla u + cu \geq 0 \quad \text{on} \quad Q_T.$$

8. Let a, b, and c be as in Prob. 7. Prove: If $u \in C(\bar{Q}_T) \cap C^{2,1}(Q_T)$ satisfies

$$u_t - a\,\Delta u + b \cdot \nabla u + cu \geq 0 \quad \text{on} \quad Q_T, \qquad u \geq 0 \quad \text{on} \quad B_T.$$

then $u(x, t) \geq 0$ throughout $\bar{Q}_T$.

9. Let a, b, and c be as in Prob. 7. If $u \in C^{2,1}(Q_T)$ satisfies

$$u_t = a\,\Delta u - b \cdot \nabla u - cu \quad \text{on} \quad Q_T,$$

show that u cannot achieve a positive maximum or a negative minimum in Q_T. Conclude that if $u \in C(\bar{Q}_T)$ and u has either a positive maximum or a negative minimum, then these values must be assumed on B_T.

10. Let a, b, and c be as in Prob. 7. Show that there is at most one solution $u \in C^{2,1}(Q_T) \cap C(\bar{Q}_T)$ to the Dirichlet problem

$$u_t - a\,\Delta u + b \cdot \nabla u + cu = h(x, t), \qquad (x, t) \in Q_T,$$
$$u(x, t) = \phi(x, t), \qquad (x, t) \in S_T, \qquad u(x, 0) = f(x), \qquad x \in \bar{D}.$$

11. Consider the Dirichlet problem in Prob. 10 with the assumption that $c(x, t)$ is bounded on Q_T instead of $c(x, t) \geq 0$. Prove there is at most one solution to the Dirichlet problem in this context. *Hint.* Let w be the difference of two solutions and $M \geq |c(x, t)|$ on Q_T. Let $w = e^{Mt}v$ and consider the problem that v solves.

12. Let $(x, t) \in Q_T$ be fixed, $u \in C^{2,1}(Q_T) \cap C^{1,0}(\bar{Q}_T)$, and $v(y, \tau) = k(x - y, a(t - \tau))$ be the fundamental solution to the heat equation. Integrate the identity (1-10) with $\zeta_R \equiv 1$ over the domain $y \in D$, $0 < \varepsilon < \tau < t - \varepsilon$, let $\varepsilon \to 0$ and use basic properties of the fundamental solution to find that

$$\begin{aligned} u(x, t) = &\int_D k(x - y, at)u(y, 0)\, dy \\ &+ a\int_{S_t}\left[k(x - y, a(t - \tau))\frac{\partial u}{\partial \nu_y} - u\frac{\partial}{\partial \nu_y}k(x - y, a(t - \tau))\right] d\sigma_y \\ &+ \int_{Q_t} k(x - y, a(t - \tau))[u_\tau - a\,\Delta_y u]\, dy\, d\tau. \end{aligned}$$

13. Show that if $g(x, y, a(t-\tau))$ is any smooth solution to

$$\left(\frac{\partial}{\partial t} + a\,\Delta_y\right) g(x, y, a(t-\tau)) = 0,$$

then the representation of Prob. 12 holds with k replaced by $k(x-y, a(t-\tau)) - g(x, y, a(t-\tau))$.

14. Consider the Dirichlet problem (3-1) for $u \in C^{2,1}(Q_T) \cap C(\bar{Q}_T)$. Find conditions on the function g in Prob. 13 so that

$$\begin{aligned} u(x, t) = &\int_D G(x, y, at) f(y)\, dy \\ &- a \int_{S_t} \phi(y, \tau) \frac{\partial}{\partial \nu_y} G(x, y, a(t-\tau))\, d\sigma_y\, d\tau \\ &+ \int_{Q_t} G(x, y, a(t-\tau)) h(y, \tau)\, dy\, d\tau, \end{aligned}$$

where $G = k - g$ is called the *Green's function* for the Dirichlet problem (3-1).

15. Show that the Green's function of Prob. 14 satisfies:
(a) $G(x, y, a(t-\tau)) > 0$, $(x, y) \in D$, $t > \tau$.
(b) $G(x, y, a(t-\tau)) = G(y, x, a(t-\tau))$.
(c) $(\partial/\partial t - a\,\Delta_x) G(x, y, a(t-\tau)) = 0$ for $x, y \in D$ and $t > \tau$.

16. Show that the Green's function of Prob. 14 is unique, given that it exists.

17. Show that the Green's function for the two-dimensional domain $D = \{(x_1, x_2): -\infty < x_1 < \infty, x_2 > 0\}$ is given by (assume that $a = 1$)

$$G(x, y, t) = k(x_1 - y_1, x_2 - y_2, t) - k(x_1 - y_1, x_2 + y_2, t)$$

where $k(x, t) = (4\pi t)^{-1} \exp(-|x|^2/4t)$ is the fundamental solution. Now, solve the corresponding Dirichlet problem.

18. Suppose that $u(x, t)$ solves

$$\begin{aligned} &u_t - \Delta u = 1, \qquad (x, t) \in Q_T, \\ &u(x, t) = 0, \qquad (x, t) \in S_T, \qquad u(x, 0) = 0,\ x \in \bar{D}, \end{aligned}$$

where D is the ball of radius 1 and center at the origin.
(a) Argue on physical grounds that this problem should have a steady-state, radially symmetric solution, say $\mathscr{A}(r)$, where (r, θ, ϕ) are spherical coordinates, and that $u(r, \theta, \phi, t) \le \mathscr{A}(r)$.
(b) Show that $\mathscr{A}(r) = (1 - r^2)/6$ and prove that

$$u(r, \theta, \phi, t) \le \frac{1 - r^2}{6}.$$

(c) Determine the best choice of $\alpha, \beta, > 0$ so that

$$\beta(1 - e^{-\alpha t}) \frac{(1 - r^2)}{6} \le u(r, \theta, \phi, t).$$

They are $\alpha = 6$ and $\beta = 1$.
(d) Use parts (b) and (c) to estimate how many units of time it takes for u to differ from its steady state by less than 1%.

-4. Thermal Potentials

In this section we investigate the so-called *thermal potentials* which are analogous in many respects to the single- and double-layer potentials used to study the Laplace equation. As in Chap. 8, we shall be led to jump relations and integral equations which yield solutions to the boundary value problems for the heat equation. Motivated by Prob. 12 of Sec. 9-3, it is natural to introduce the *single-layer (thermal) potential*

$$v(x, t) = a \int_{S_t} k(x - y, a(t - \tau))\rho(y, \tau)\, d\sigma_y\, d\tau, \tag{4-1}$$

and the *double-layer (thermal) potential*

$$w(x, t) = a \int_{S_t} \left[\frac{\partial}{\partial v_y} k(x - y, a(t - \tau))\right] \mu(y, \tau)\, d\sigma_y\, d\tau. \tag{4-2}$$

We assume that ρ and μ are continuous for $x \in B$ and $t \geq 0$.

To establish the jump relations, we must investigate the behavior of these potentials as x approaches a point $x^\circ \in B$, the boundary of a bounded domain D. As in Chap. 8, we assume that B is twice continuously differentiable and satisfies Assumption A of Sec. 8-6. The arguments to follow are rather similar to those used for the potentials in Chap. 8. Thus, we will be brief and merely highlight the key steps.

Consider first the double-layer potential, $w(x, t)$, and take $x \in D$ and $x^\circ \in B$. We express $w(x, t)$ as the sum of two integrals

$$I(x, t) = a \int_{S_t} \left[\frac{\partial}{\partial v_y} k(x - y, a(t - \tau))\right] \mu(y, t)\, d\sigma_y\, d\tau$$

and

$$J(x, t) = a \int_{S_t} \frac{\partial}{\partial v_y} k(x - y, a(t - \tau))[\mu(y, \tau) - \mu(y, t)]\, d\sigma_y\, d\tau.$$

As usual, let $r = |x - y|$. Then $I(x, y)$ can be expressed as

$$\begin{aligned}
I(x, y) &= a \int_B \mu(y, t)\, d\sigma_y \frac{\partial}{\partial v_y} \int_0^t [4\pi(t - \tau)]^{-3/2} \exp\left[-\frac{r^2}{4a(t - \tau)}\right] d\tau \\
&= a \int_B \mu(y, t)\, d\sigma_y v_y \cdot (x - y) \int_0^t \frac{2 \exp[-r^2/4a(t - \tau)]}{4a(t - \tau)[4\pi(t - \tau)]^{3/2}}\, d\tau \\
&= a^{3/2} \int_B \mu(y, t)\, d\sigma_y v_y \cdot \frac{x - y}{4\pi r^3}\left[\frac{2}{\sqrt{\pi}} \int_{r^2/4at}^{\infty} e^{-\lambda}\lambda^{1/2}\, d\lambda\right] \\
&= a^{3/2} \int_B \mu(y, t)\phi(r, t)\left(\frac{\partial}{\partial v_y} \frac{1}{4\pi r}\right) d\sigma_y
\end{aligned}$$

where

$$\phi(r, t) = \frac{2}{\sqrt{\pi}} \int_{r^2/4at}^{\infty} e^{-\lambda}\lambda^{1/2}\, d\lambda.$$

Notice that $\phi(r, t) \to 2\pi^{-1/2}\Gamma(3/2) = 1$ as $r \to 0$. Apply the jump relation for the double-layer potential $\partial((4\pi r)^{-1})/\partial\nu$ to obtain

$$\lim_{\substack{x \to x^\circ \in B \\ x \in D}} I(x, t) = -\frac{1}{2}\mu(x^\circ, t) + a^{3/2}\int_B \mu(y, t)\phi(r, t)\left(\frac{\partial}{\partial\nu_y}\frac{1}{4\pi r}\right) d\sigma_y,$$

where $r = |x^\circ - y|$. Reverse the change of variables made above to find

$$\text{(4-3)} \quad \lim_{\substack{x \to x^\circ \in B \\ x \in D}} I(x, t) = -\frac{1}{2}\mu(x^\circ, t) + a\int_{S_t}\left[\frac{\partial}{\partial\nu_y}k(x^\circ - y, a(t-\tau))\right]\mu(y, t)\, d\sigma_y\, d\tau.$$

Similarly,

$$\text{(4-4)} \quad \lim_{\substack{x \to x^\circ \in B \\ x \in cD}} I(x, t) = \frac{1}{2}\mu(x^\circ, t) + a\int_{S_t}\left[\frac{\partial}{\partial\nu_y}k(x^\circ - y, a(t-\tau))\right]\mu(y, t)\, d\sigma_y\, d\tau.$$

We have tacitly assumed that the integrals over S_t exist when $x^\circ \in B$. The existence of these integrals follows much as for corresponding integrals in Chap. 8, where in this situation Lemma 1-4 is used to estimate the fundamental solution. This will be clear from the calculations which follow and establish that

$$\text{(4-5)} \quad J(x, t) \to J(x^\circ, t) \quad \text{as} \quad x \to x^\circ \in B, \qquad x \notin B.$$

To establish (4-5) we need to make several careful estimates of the integrand as τ approaches t and the fundamental solution becomes singular. Let $\varepsilon > 0$ and fix t_1 with $0 < t_1 < t$ sufficiently close to t so that

$$|\mu(y, t) - \mu(y, \tau)| < \varepsilon \quad \text{for} \quad t_1 \le \tau \le t \quad \text{and} \quad y \in B.$$

We express $J(x, t)$ as the sum of $J_1(x, t)$ and $J_2(x, t)$, where

$$J_1(x, t) = a\int_{S_{t_1}}\left[\frac{\partial}{\partial\nu_y}k(x - y, a(t-\tau))\right][\mu(y, \tau) - \mu(y, t)]\, d\sigma_y\, d\tau$$

and

$$J_2(x, t) = a\int_{t_1}^{t}\int_B\left[\frac{\partial}{\partial\nu_y}k(x - y, a(t-\tau))\right][\mu(y, \tau) - \mu(y, t)]\, d\sigma_y\, d\tau.$$

Now the integrand of $J_1(x, y)$ is nonsingular and we infer that

$$\text{(4-6)} \quad \lim_{\substack{x \to x^\circ \\ x \notin B}} J_1(x, t) = J_1(x^\circ, t).$$

Thus, to establish (4-5) we need the corresponding limit for $J_2(x, t)$.

Consider the difference $J_2(x, t) - J_2(x^\circ, t)$ for $x^\circ \in B$ fixed. Without loss in generality we can choose the coordinate system so that $x^\circ = 0$. As in Chap. 8, let $K_\delta(x^\circ)$ be a ball chosen so that for $y_1^2 + y_2^2 \le \delta^2$ the surface B is described by $y_3 = \phi(y_1, y_2)$ with ϕ twice continuously differentiable and

$$\phi(0, 0) = \phi_{y_1}(0, 0) = \phi_{y_2}(0, 0) = 0.$$

Now

$$
\begin{aligned}
&|J_2(x, t) - J_2(x^\circ, t)| \\
&= \left| a \int_{t_1}^{t} \int_B \left[\frac{\partial}{\partial v_y} k(x - y, a(t - \tau)) - \frac{\partial}{\partial v_y} k(y, a(t - \tau)) \right] [\mu(y, \tau) - \mu(y, t)] \, d\sigma_y \, d\tau \right| \\
&\le a\varepsilon \int_{t_1}^{t} \int_{K_\delta \cap B} \left[\left| \frac{\partial}{\partial v_y} k(x - y, a(t - \tau)) \right| + \left| \frac{\partial}{\partial v_y} k(y, a(t - \tau)) \right| \right] d\sigma_y \, d\tau \\
&\quad + a\varepsilon \int_{t_1}^{t} \int_{cK_\delta \cap B} \left[\left| \frac{\partial}{\partial v_y} k(x - y, a(t - \tau)) \right| + \left| \frac{\partial}{\partial v_y} k(y, a(t - \tau)) \right| \right] d\sigma_y \, d\tau \\
&= a\varepsilon(J' + J'' + J'''),
\end{aligned}
$$

where J' and J'' are the integrals over $[t_1, t] \times (K_\delta \cap B)$ of the two normal derivatives in the first integral and J''' is the entire remaining integral.

With the coordinate system we have chosen,

$$
y = (y_1, y_2, \phi(y_1, y_2)), \qquad d\sigma_y = \sqrt{1 + \phi_{y_1}^2 + \phi_{y_2}^2} \, dy_1 \, dy_2,
$$

and

$$
v_y = (\phi_{y_1}, \phi_{y_2}, -1)/\sqrt{1 + \phi_{y_1}^2 + \phi_{y_2}^2}
$$

on $K_\delta \cap B$. Thus, the estimates in Prop. 6-2 of Chap. 8 yield

$$
|y \cdot v_y| \, d\sigma_y \le M(y_1^2 + y_2^2) \, dy_1 \, dy_2
$$

for some constant M. Since

$$
\frac{\partial k}{\partial v_y}(x - y, a(t - \tau)) = \frac{(x - y) \cdot v_y}{2a(t - \tau)} k(x - y, a(t - \tau)), \tag{4-7}
$$

we can introduce polar coordinates (ρ, θ) with center at $x^\circ = 0$ in J'' and obtain

$$
\begin{aligned}
J'' &\le \int_{t_1}^{t} \int_0^{2\pi} \int_0^{\delta} \frac{M\rho^2}{2a(t - \tau)} [4\pi a(t - \tau)]^{-3/2} \exp\left[-\frac{\rho^2 + \phi^2}{4a(t - \tau)} \right] \rho \, d\rho \, d\theta \, d\tau \\
&\le \frac{M}{2a(4\pi a)^{3/2}} \int_{t_1}^{t} \int_0^{2\pi} \int_0^{\infty} \frac{\rho^3}{(t - \tau)^{5/2}} \exp\left[-\frac{\rho^2}{4a(t - \tau)} \right] d\rho \, d\theta \, d\tau \\
&= \frac{2M(t - t_1)^{1/2}}{\sqrt{\pi a}},
\end{aligned}
$$

where the integral was evaluated using the substitution $\lambda = \rho^2/4a(t - \tau)$.

Since in (4-6), $x \notin B$ we have $y - x \ne 0$ in the integrand for J'. Thus, from (4-7),

$$
\begin{aligned}
J' &\le \int_{t_1}^{t} \int_{K_\delta \cap B} \left| \frac{(x - y) \cdot v_y}{2a(t - \tau)} \right| k(x - y, a(t - \tau)) \, d\sigma_y \, d\tau \\
&\le \int_{K_\delta \cap B} \left| \frac{(x - y) \cdot v_y}{r^3} \right| \frac{\bar{a}}{2\pi^{3/2}} \int_{r^2/4a(t - t_1)}^{\infty} e^{-\lambda} \lambda^{1/2} \, d\lambda \, d\sigma_y,
\end{aligned}
$$

where the change of variables $\lambda = r^2/4a(t-\tau)$ with $r = |x-y|$ was used. Thus,

$$J' \le \frac{\sqrt{a}}{2\pi^{3/2}} \Gamma\left(\frac{3}{2}\right) \int_B \left|\frac{\partial}{\partial v_y} \frac{1}{r}\right| d\sigma_y = \sqrt{a} \int_B \left|\frac{\partial}{\partial v_y} \frac{1}{4\pi r}\right| d\sigma_y \le \sqrt{a}\, C,$$

in view of Assumption A (see Eq. (6-11) of Sec. 8-6).

Finally, consider J'''. Since we are interested in the limit as $x \to x^\circ = 0$, we can assume that $x \in K_{\delta/2}(0)$. Then $x - y \neq 0$ for $y \in cK_\delta \cap B$ and of course $y \neq 0$ for such y. Thus, the change of variables used to estimate J' applies, once with $r = |x-y|$ and once with $r = |x^\circ - y| = |y|$, and we find

$$J''' \le 2\sqrt{a}\, C.$$

Combining these estimate gives

$$|J_2(x,t) - J_2(x^\circ, t)| \le a\varepsilon\left[\frac{2M(t-t_1)^{1/2}}{\sqrt{\pi a}} + 2\sqrt{a}\, C\right]$$

for $x \in K_{\delta/2}(x^\circ)$ and $x \notin B$. Thus,

$$J_2(x,t) \to J_2(x^\circ, t) \quad \text{as} \quad x \to x^\circ \in B, \qquad x \notin B.$$

This result and (4-6) establish (4-5). Then (4-3)–(4-5) yield the jump relations for double-layer thermal potentials.

Theorem 4-1. Suppose that the surface B bounding D is twice continuously differentiable, satisfies Assumption A of Sec. 8-6, and that $\mu(y,\tau)$ is continuous on $B \times [0, \infty)$. Then the double-layer potential

$$w(x,t) = a \int_{S_t} \left[\frac{\partial}{\partial v_y} k(x-y, a(t-\tau))\right] \mu(y,\tau)\, d\sigma_y\, d\tau$$

is defined for all $x \in R_3$ and $t \ge 0$ and satisfies

$$w_t = a\,\Delta w, \qquad x \notin B, \qquad t > 0,$$

$$w(x,0) = 0, \qquad x \notin B,$$

$$\lim_{\substack{x \to x^\circ \in B \\ x \in D}} w(x,t) = -\tfrac{1}{2}\mu(x^\circ, t) + w(x^\circ, t)$$

$$\lim_{\substack{x \to x^\circ \in B \\ x \in cD}} w(x,t) = \tfrac{1}{2}\mu(x^\circ, t) + w(x^\circ, t).$$

Proof. The only assertion left to prove is that $w(x,t)$ satisfies the heat equation. This follows easily (Prob. 7) from the fact that the fundamental solution satisfies the heat equation and that when differentiating the integral over S_t with respect to the upper limit t, $k(x-y, a(t-\tau))$ and all its derivatives tend to zero as $\tau \to t$. (Recall that $x - y \neq 0$ because $x \notin B$.)

The corresponding results for single-layer thermal potentials are established in a manner similar to that used for single-layer potentials for Laplace's equation. We simply record the result.

Theorem 4-2. Suppose that the surface B bounding D is twice continuously differentiable, satisfies Assumption A of Sec. 8-6, and that $\rho(y, \tau)$ is continuous on $B \times [0, \infty)$. Then the single-layer potential

$$v(x, t) = a \int_{S_t} k(x - y, a(t - \tau))\rho(y, \tau)\, d\sigma_y\, d\tau$$

is continuous for all $x \in R_3$ and $t \geq 0$ and satisfies

$$v_t = a\,\Delta v, \qquad x \notin B, \qquad t > 0,$$

$$v(x, 0) = 0, \qquad x \notin B,$$

$$\lim_{\substack{x \to x^\circ \in B \\ x \in D}} \frac{\partial v}{\partial \nu^\circ}(x, t) = \frac{1}{2}\rho(x^\circ, t) + \frac{\partial v}{\partial \nu^\circ}(x^\circ, t),$$

$$\lim_{\substack{x \to x^\circ \in B \\ x \in cD}} \frac{\partial v}{\partial \nu^\circ}(x, t) = -\frac{1}{2}\rho(x^\circ, t) + \frac{\partial v}{\partial \nu^\circ}(x^\circ, t),$$

where in both limits x approaches x° along the line determined by the unit outer normal ν° to B at x°.

PROBLEMS

1. Consider the heat equation in one spatial dimension

$$u_t = a u_{xx}, \qquad x > 0, \qquad t > 0,$$

$$u(x, 0) = 0, \qquad x > 0, \qquad u(0, t) = f(t), \qquad t > 0,$$

with $f(0) = 0$. Solve this problem by assuming a solution in the form of a double-layer potential

$$u(x, t) = a \int_0^t \left(-\frac{\partial}{\partial y}\right) k(x - y, a(t - \tau))\bigg|_{y=0} \mu(\tau)\, d\tau$$

and determine $\mu(t)$. *Hint.* Show that the heat equation and initial condition are satisfied for any $\mu(\tau)$. Then show that

$$\lim_{x \to 0+} u(x, t) = \lim_{x \to 0+} a \int_0^t k_x(x, a(t - \tau))\mu(\tau)\, d\tau = -\frac{1}{2}\mu(t).$$

Compare with the approach to this problem given at the end of Sec. 5-4.

2. In one spatial dimension solve

$$u_t = a u_{xx}, \qquad x > 0, \qquad t > 0,$$

$$u_x(0, t) = f(t), \qquad t > 0, \qquad u(x, 0) = 0, \qquad x > 0,$$

where $f(0) = 0$. *Hint.* Assume a solution in the form of a single-layer potential,

$$u(x, t) = a \int_0^t k(x, a(t - \tau))\rho(\tau)\, d\tau.$$

3. Solve in one spatial dimension

$$u_t = a u_{xx}, \qquad x > 0, \qquad t > 0,$$

$$-u_x(0, t) + \alpha u(0, t) = f(t), \qquad t > 0, \qquad u(x, 0) = 0, \qquad x > 0,$$

where $\alpha > 0$ is constant. *Hint.* Assume a solution in the form of a single-layer potential as in Prob. 2. Show that ρ must satisfy

$$\frac{1}{2}\rho(t) + \alpha a \int_0^t [4\pi a(t-\tau)]^{-1/2}\rho(\tau)\,d\tau = f(t).$$

This equation can be solved by Laplace transforms. It is also a Volterra integral equation and can be solved by iteration as in Chap. 7.

4. Consider

$$\begin{aligned} &u_t = \Delta u, \qquad x_1 > 0, \qquad -\infty < x_2, x_3 < \infty, \qquad t > 0,\\ &u(0, x_2, x_3, t) = f(x_2, x_3, t), \qquad -\infty < x_2, x_3 < \infty, \qquad t > 0,\\ &u(0, x_2, x_3, 0) = 0, \qquad -\infty < x_2, x_3 < \infty. \end{aligned}$$

Solve formally by assuming a solution as a double-layer potential. Observe that the integral over the plane $x_1 = 0$ vanishes when evaluated at a point in that plane.

5. Prove Th. 4-2.

6. Consider the single-layer potential

$$v(x, t) = a \int_{S_t} k(x - y, a(t-\tau))\rho(y, \tau)\,d\sigma_y\,d\tau$$

in two spatial dimensions.

(a) Show that

$$a\int_0^t \frac{1}{4\pi a(t-\tau)} \exp\left[-\frac{r^2}{4a(t-\tau)}\right] d\tau$$
$$= \frac{e^{-r^2/4at}}{2\pi}\log\frac{1}{r} + \frac{e^{-r^2/4at}}{4\pi}\log(4at) + \frac{1}{4\pi}\int_{r^2/4at}^{\infty} e^{-\lambda}\log\lambda\,d\lambda.$$

Hint. Set $\lambda = r^2/4a(t-\tau)$ and integrate by parts.

(b) Prove Th. 4-1 for two spatial dimensions. *Hint.* Observe that

$$\int_{S_t}\left[\frac{\partial}{\partial \nu_y} k(x - y, a(t-\tau))\right]\mu(y, t)\,d\sigma_y\,d\tau$$
$$= \int_B \mu(y, t)\left\{\frac{\partial}{\partial \nu_y}\int_0^t k(x - y, a(t-\tau))\,d\tau\right\} d\sigma_y.$$

Apply part (a) and jump relations for potentials for the Laplace equation.

(c) Prove Th. 4-2 for two spatial dimensions.

7. Confirm that the double-layer thermal potential in Th. 4-1 satisfies the heat equation.

9-5. Solution of the Initial, Boundary Value Problem

We work with three space variables and consider the Dirichlet problem

$$\text{(5-1)} \qquad \begin{cases} u \in C^{2,1}(Q_T) \cap C(\bar{Q}_T),\\ u_t = a\,\Delta u, \qquad (x, t) \in Q_T,\\ u(x, 0) = 0, \qquad x \in \bar{D},\\ u(x, t) = f(x, t), \qquad (x, t) \in \bar{S}_T, \end{cases}$$

where $T > 0$ is fixed, $f \in C(\bar{S}_T)$, and the compatibility condition $f(x, 0) = 0$ holds. We know (Sec. 9-3) that this problem has at most one solution. Our work with the Laplace equation suggests that we seek a solution to the interior Dirichlet problem in the form of a double-layer thermal potential,

$$\text{(5-2)} \qquad u(x, t) = a \int_{S_t} \left[\frac{\partial}{\partial \nu_y} k(x - y, a(t - \tau)) \right] \mu(y, \tau) \, d\sigma_y \, d\tau$$

for $x \in D$, $t \geq 0$, where μ is a continuous density that must be determined. When $t = 0$ we interpret in (5-2) to be 0, which is its limit as $t \to 0+$ (see Lemma 5-1). Thus, u given by (5-2) is continuous for $x \in D$ and $t \geq 0$. For any choice of μ, (5-2) satisfies the heat equation and homogeneous initial condition. As $x \in D$ approaches a point on the boundary of D, we require, by continuity, that $u(x, t)$ in (5-2) approaches $f(x, t)$. Thus, the interior jump relation for the double-layer potential leads to the requirement that

$$\text{(5-3)} \qquad f(x, t) = -\frac{1}{2} \mu(x, t) + a \int_{S_t} \left[\frac{\partial}{\partial \nu_y} k(x - y, a(t - \tau)) \right] \mu(y, \tau) \, d\sigma_y \, d\tau$$

for $x \in B$ and $t \geq 0$.

Notice that (5-3) is a Volterra-like integral equation for $\mu(x, t)$ because the time limit of integration is variable. From our experience with similar equations in Secs. 4-6 and 4-7, Sec. 7-3 (Prob. 4), and Sec. 7-4 (Prob. 9), we should expect that (5-3) is solvable by iteration. Therefore, let $\mu_0(x, t)$ be an initial guess at a solution of (5-3) and define the sequence $\{\mu_n(x, t)\}$ recursively by

$$\text{(5-4)} \qquad \mu_{n+1}(x, t) = -2f(x, t) + 2a \int_{S_t} \left[\frac{\partial}{\partial \nu_y} k(x - y, a(t - \tau)) \right] \mu_n(y, \tau) \, d\sigma_y \, d\tau$$

for $x \in B$, $t \geq 0$, and $n = 0, 1, 2, \ldots.$ We will show that (5-4) defines functions $\mu_n(x, t)$ which are continuous on $\bar{S}_T$ and that the series

$$\text{(5-5)} \qquad \mu_0(x, t) + \sum_{n=1}^{\infty} [\mu_n(x, t) - \mu_{n-1}(x, t)]$$

converges uniformly on $\bar{S}_T$. Thus,

$$\mu(x, t) = \lim_{n \to \infty} \mu_n(x, t) = \lim_{n \to \infty} \left\{ \mu_0(x, t) + \sum_{k=1}^{n} [\mu_k(x, t) - \mu_{k-1}(x, t)] \right\}$$

is a continuous solution to (5-3) and (5-2) solves the Dirichlet problem. Since we have used a jump relation from Sec. 9-3, our results apply only to domains D bounded by twice-continuously differentiable surfaces B which satisfy Assumption A of Sec. 8-6. *We assume throughout this section that Assumption A holds.*

The key to the convergence assertions above is the following technical estimate.

Lemma 5-1. Let $h \in C(\bar{S}_T)$ and suppose that $|h(y, \tau)| \leq Mt^\alpha$ for some constants M and $\alpha > -1$. Then, for $x \in B$ and $0 \leq \tau \leq t \leq T$,

$$\left| 2a \int_{S_t} \left[\frac{\partial}{\partial \nu_y} k(x - y, a(t - \tau)) \right] h(y, \tau) \, d\sigma_y \, d\tau \right| \leq LM \frac{\Gamma(\frac{1}{2})\Gamma(\alpha + 1)}{\Gamma(\alpha + \frac{3}{2})} t^{\alpha + (1/2)}$$

for some constant L independent of h.

Proof. We have

$$\left|2a\int_{S_t}\left[\frac{\partial}{\partial \nu}k(x-y, a(t-\tau))\right]h(y,\tau)\,d\sigma_y\,d\tau\right|$$

$$\leq 2aM\int_0^t \tau^{\alpha}\int_B\left|\frac{\partial}{\partial \nu_y}k(x-y, a(t-\tau))\right|d\sigma_y\,d\tau$$

We estimate the inner integral over B by expressing it as a sum of two integrals, I_1 and I_2, with I_1 the integral over $K_\delta(x)\cap B$ and I_2 the integral over the rest of B. We introduce local coordinates near x with origin at x exactly as we did near x° in the argument following (4-6). Thus, for $y\in K_\delta(x)\cap B$, we have $y_3=\phi(y_1, y_2)$, $\phi(0,0)=\phi_{y_1}(0,0)=\phi_{y_2}(0,0)=0$, B has normal $(\phi_{y_1}, \phi_{y_2}, -1)/N$ at y where

$$N=\sqrt{1+\phi_{y_1}^2+\phi_{y_2}^2},\qquad x-y=(-y_1, -y_2, -\phi(y_1, y_2)),$$

and we obtain the estimate

$$|y\cdot \nu_y|\,d\sigma_y\leq L_1(y_1^2+y_2^2)\,dy_1\,dy_2,$$

where L_1 is a bound determined by the second derivatives of ϕ on $\bar{K}_\delta(x)$. Consequently, using (4-7) and introducing polar coordinates with center x, just as in the estimate following (4-7), we find that

$$|I_1|\leq\int_0^{2\pi}\int_0^{\delta}\frac{L_1\rho^2}{2a(t-\tau)}\frac{1}{[4\pi a(t-\tau)]^{3/2}}\exp\left[-\frac{\rho^2}{4a(t-\tau)}\right]\rho\,d\rho\,d\theta$$

$$\leq\frac{4L_1}{[4\pi a(t-\tau)]^{3/2}}\int_0^{\delta}\rho\exp\left[-\frac{\rho^2}{8a(t-\tau)}\right]d\rho$$

because

$$\frac{\rho^2}{8a(t-\tau)}\exp\left[-\frac{\rho^2}{4a(t-\tau)}\right]\leq\exp\left[-\frac{\rho^2}{8a(t-\tau)}\right].$$

Replace the upper limit in the last integral by ∞ to get

$$|I_1|\leq\frac{2L_1}{\pi^{3/2}a^{1/2}}(t-\tau)^{-1/2}. \tag{5-6}$$

A similar estimate holds for I_2. Indeed,

$$|I_2|\leq\int_{cK_\delta\cap B}\left|\frac{\partial}{\partial\nu_y}k(x-y, a(t-\tau))\right|d\sigma_y$$

and the integrand is bounded, say by $L_2=L_2(B,\delta)$, because $|x-y|=|y|\geq\delta$. Then

$$|I_2|\leq L_2|B|\leq L_2|B|T^{1/2}(t-\tau)^{-1/2} \tag{5-7}$$

for $0\leq t-\tau\leq T$, where $|B|$ is the surface area of B. The estimates (5-6) and

(5-7) yield

$$\left|2a \int_{S_t} \left[\frac{\partial}{\partial \nu_y} k(x - y, a(t - \tau))\right] h(y, \tau)\, d\sigma_y\, d\tau\right|$$

$$\le LM \int_0^t \tau^\alpha (t - \tau)^{-1/2}\, d\tau$$

$$= LMt^{\alpha + (1/2)} \int_0^1 \lambda^\alpha (1 - \lambda)^{-1/2}\, d\lambda = LM \frac{\Gamma(\frac{1}{2})\Gamma(\alpha + 1)}{\Gamma(\alpha + \frac{3}{2})} t^{\alpha + (1/2)},$$

where

$$L = 4a \max \left\{\frac{2L_1}{\pi^{3/2} a^{1/2}}, L_2 |B| T^{1/2}\right\},$$

and $\Gamma(z)$ is the Euler gamma function. This establishes Lemma 5-1.

We return to the iteration scheme (5-4). Let $\mu_0 \in C(\bar{S}_T)$. Then (5-4) and Lemma 5-1 with $\alpha = 0$ show that $\mu_1 \in C(\bar{S}_T)$. Similarly, $\mu_2, \mu_3, \ldots$ are all continuous on $\bar{S}_T$. Now, let

$$M = \max_{\bar{S}_T} |\mu_1(x, t) - \mu_0(x, t)|.$$

Then

$$\mu_2(x, t) - \mu_1(x, t) = 2a \int_{S_t} \frac{\partial}{\partial \nu_y} k(x - y, a(t - \tau))[\mu_1(y, \tau) - \mu_0(y, \tau)]\, d\sigma_y\, d\tau.$$

From Lemma 5-1 with $\alpha = 0$,

$$|\mu_2(x, t) - \mu_1(x, t)| \le 2LMt^{1/2}.$$

Using this estimate and Lemma 5-1 with $\alpha = \frac{1}{2}$ leads to

$$|\mu_3(x, t) - \mu_2(x, t)| \le L(2LM) \frac{\Gamma(\frac{1}{2})\Gamma(\frac{3}{2})}{\Gamma(2)} t = \frac{(L\sqrt{\pi})^2 Mt}{\Gamma(2)}.$$

A simple induction argument establishes that

$$|\mu_n(x, t) - \mu_{n-1}(x, t)| \le \frac{M(L\sqrt{\pi})^{n-1}}{\Gamma\left(\frac{n+1}{2}\right)} t^{(n-1)/2}$$

for $n = 1, 2, \ldots$. Consequently, the terms in the series in (5-5) are bounded by

$$|\mu_n(x, t) - \mu_{n-1}(x, t)| \le \frac{M(L\sqrt{\pi T})^{n-1}}{\Gamma\left(\frac{n+1}{2}\right)}$$

on $\bar{S}_T$, and the Weierstrass M-test implies the absolute and uniform convergence of series (5-5) on $\bar{S}_T$. As mentioned earlier, this proves that the Dirichlet problem is solvable.

Theorem 5-1. The Dirichlet problem (5-1) with $f \in C(\bar{S}_T)$ and $f(x, 0) = 0$ has a unique solution which can be represented by the double-layer potential (5-2) with density determined from the integral equation (5-3).

This existence result assumes homogeneous initial values and no source term. Consider the general Dirichlet problem

$$\text{(5-8)} \qquad \begin{cases} u \in C^{2,1}(Q_T) \cap C(\bar{Q}_T), \\ u_t = a\,\Delta u + F(x, t), & (x, t) \in Q_T, \\ u(x, 0) = g(x), & x \in \bar{D}, \\ u(x, t) = f(x, t), & (x, t) \in \bar{S}_T, \end{cases}$$

where the data are continuous and for compatibility $f(x, 0) = g(x)$. The function

$$\text{(5-9)} \qquad w(x, t) = \int_{Q_T} k(x - y, a(t - \tau))F(y, \tau)\,dy\,d\tau + \int_D k(x - y, at)g(y)\,dy$$

satisfies

$$\text{(5-10)} \qquad \begin{aligned} &w_t = a\,\Delta w + F(x, t), && (x, t) \in Q_T, \\ &w(x, 0) = g(x), && x \in \bar{D}. \end{aligned}$$

Consequently, u satisfies (5-8) if and only if $v = u - w$ satisfies a problem of the form (5-1). Theorem 5-1 guarantees that v exists and $u = v + w$ solves (5-8) (see Prob. 2).

The other boundary value problems corresponding to Neumann and Robin boundary conditions can be reduced to integral equations and solved in a similar way. There are no surprises. The iteration schemes lead to series like (5-5) which converge very rapidly, as the estimate preceding Th. 5-1 suggests. Although the computations needed to carry out the iteration procedure can be very laborious, the form of the final solution as a single- or double-layer potential is extremely useful and lends itself to accurate numerical computations. Also, the representations in terms of single and double layers often reveal significant qualitative properties of solutions.

PROBLEMS

1. Confirm the estimate

$$|\mu_n(x, t) - \mu_{n-1}(x, t)| \le \frac{M(L\sqrt{\pi})^{n-1}}{\Gamma\left(\dfrac{n+1}{2}\right)} t^{(n-1)/2}$$

for $n = 1, 2, \ldots$.

2. Check carefully that (5-9) solves (5-10). Then formulate the problem that $v = u - w$ must satisfy if u is to solve (5-8). Check that Th. 5-1 does apply to the problem for v.
3. Show that (5-1) has a solution in the case of two space variables.

4. Let D be a bounded domain in R_3 and $Q_T = \{(x, t): x \in c\bar{D},\ 0 < t \leq T\}$. Prove that there is a unique bounded solution to

$$u \in C^{2,1}(Q_T) \cap C(\bar{Q}_T), \qquad u_t = a\,\Delta u, \qquad (x, t) \in Q_T,$$
$$u(x, 0) = 0, \qquad x \in cD, \qquad u(x, t) = f(x, t), \qquad (x, t) \in \bar{S}_T,$$

where $f(x, t)$ is bounded, continuous, and $f(x, 0) = 0$.

5. Check that the uniqueness results in Prob. 5 and 6 in Sec. 9-3 hold when B satisfies Assumption A and a solution $u \in C^{2,1}(Q_T) \cap C_\nu^{1,0}(\bar{Q}_T)$.

6. Let Q_T be as in Prob. 4. Show there is at most one solution $u \in C^{2,1}(Q_T) \cap C_\nu^{1,0}(\bar{Q}_T)$ to

$$u_t = a\,\Delta u + F(x, t), \qquad (x, t) \in Q_T,$$
$$u(x, 0) = g(x), \qquad x \in cD, \qquad \frac{\partial u}{\partial \nu}(x, t) = f(x, t), \qquad (x, t) \in \bar{S}_T,$$
$$|u(x, t)| \leq \frac{M}{|x|}, \qquad |\nabla u| \leq \frac{M}{|x|^2},$$

where M is a constant.

7. Establish the analogue of Prob. 6 in two space dimensions. This time assume that

$$|u(x, t)| \leq M \quad \text{and} \quad |\nabla u| \leq \frac{M}{|x|^2}.$$

8. Prove that the exterior Neumann problem above with $g(x) \equiv 0$, $F(x, t) \equiv 0$, and $(\partial g/\partial \nu)(x) = f(x, 0)$ for $x \in B$ has a solution u in the form of a single-layer potential. Show that u satisfies the conditions of the uniqueness theorem in Prob. 6.

9. Consider the exterior Dirichlet problem (Prob. 4). Suppose that $|f(x, t)| \leq C$ a constant. Let $x^\circ \in cD$ be a point very far from B so that

$$\operatorname{dist}(x^\circ, B) = \min_{y \in B} |x^\circ - y| \approx \max_{y \in B} |x^\circ - y| \equiv \delta.$$

Use the representation of u as a double-layer potential to obtain an estimate on $u(x^\circ, t)$.

10

Hyperbolic Equations

10-1. Waves

Wave motion in one space variable was studied in Chap. 4. Now we turn to analogous problems in two and three spatial dimensions. The idea of a wave is quite general and refers to disturbances that propagate at finite speed, such as sound waves in a gas, the vibrations of a string or membrane, radio waves, and general electromagnetic phenomena. We restrict our attention to disturbances that can be described by the inhomogeneous wave equation

$$u_{tt} = c^2 \Delta u + F, \tag{1-1}$$

where $F(x, t)$ is a given function and $c > 0$ is a constant. If $F \equiv 0$, we obtain the (homogeneous) wave equation

$$u_{tt} = c^2 \Delta u. \tag{1-2}$$

We saw in Sec. 1-7 that this equation describes the variation in density and pressure of a gas; see (7-9) and Prob. 1 in that section. The wave equation also describes the small transverse vibrations in a membrane (see Prob. 4 of Sec. 1-2). We will refer to solutions of (1-1) or (1-2) as *waves*. More precisely, in this chapter a wave will be a twice-continuously differentiable solution of (1-1) or (1-2). In Chapter 11 we extend the notion of a solution to the wave equation to include waves without smooth profiles. However, we should point out that other wave-like motions are described by different equations. For example, the vibrations of a plate are governed by

$$u_{tt} + c^2 \Delta^2 u = 0, \tag{1-3}$$

which is called the biharmonic wave equation. Needless to say, the solutions to (1-2) and (1-3) have quite different properties.

As usual, we treat the three-dimensional case in the text and leave analogous results in two dimensions for the problems. However, there are some significant differences between wave propagation in even and in odd dimensions. In such instances we develop results for both two and three dimensions in the text itself. Occasionally, we specialize to one spatial variable and extend and complement some results from Chap. 4. We begin by looking for some special solutions to the wave equation (1-2). These simple solutions give considerable physical insight.

In Sec. 4-2 we superposed standing waves to find the motion of a vibrating string with pinned ends. In three spatial dimensions, *standing waves* are separated solutions to (1-2) of the form

$$u(x, t) = v(x)T(t). \tag{1-4}$$

Routine reasoning shows that (1-4) solves (1-2) precisely when

$$T(t) = \cos(c\lambda t + \delta), \tag{1-5}$$

up to a constant multiple, and $v(x)$ satisfies

$$\Delta v + \lambda^2 v = 0, \tag{1-6}$$

where $-\lambda^2$ is the separation constant. The physical interpretation of standing waves is easy: The disturbance $u(x, t)$ at a fixed point x executes simple harmonic motion with period $2\pi/c\lambda$ and amplitude $|v(x)|$. The surfaces on which $v(x) = 0$ are of special interest. Points on these so-called *nodal surfaces* remain at rest throughout the motion.

Recall from Sec. 4-1 that the general solution of the one-dimensional wave equation $u_{tt} = c^2 u_{xx}$ is the superposition of a wave $A(x + ct)$ traveling to the left at speed c and a wave $B(x - ct)$ traveling to the right at speed c. We seek analogues for these traveling waves in three dimensions. That is, we seek solutions to (1-2) in the form

$$u(x, t) = h(x \cdot \omega - ct), \tag{1-7}$$

where h and ω are to be determined. For (1-7) to solve (1-2), we must have

$$u_{tt} - c^2 \Delta u = c^2 h''(x \cdot \omega - ct) - c^2|\omega|^2 h''(x \cdot \omega - ct) = 0,$$

and this is the case for any unit vector ω and any twice-differentiable function $h(z)$. For such h and ω, (1-7) provides solutions to (1-2) called *plane waves*. These plane waves have an important physical interpretation which allows us to identify c as the *speed of wave propagation*: Let $u(x, t)$ be given by (1-7) with $\omega \cdot \omega = 1$. Then $u(x, t)$ represents the wave disturbance at position x and time t. Set $\xi = x - ct\omega$. Then

$$u(x, t) = h(x \cdot \omega - ct) = h((x - ct\omega) \cdot \omega) = h(\xi \cdot \omega) = u(\xi, 0).$$

This means that the wave disturbance which occurs at position ξ at time 0 has moved to location x, which is ct units from ξ in the direction of the unit vector ω. Thus, the wave propagates through space so that the disturbance at any given point moves along the direction of the vector ω and at speed c. In particular, the disturbances that occur at points p of the plane $\omega \cdot x = 0$, which is the plane through the origin with normal ω, will occur at the corresponding points $q = p + ct\omega$ of the plane $\omega \cdot x = ct$ at time t. This explains why solutions of the form (1-7) are called plane waves.

If $h(z) = A \cos(z + \delta)$ in (1-7), where A and δ are given constants, we obtain solutions

$$u(x, t) = A \cos(\omega \cdot x - ct + \delta),$$

called monochromatic plane waves or harmonic plane waves. We also obtain cylindrical or spherical waves by seeking solutions u to (1-2) which have the form $u = u(r, t)$, where $r = (x_1^2 + x_2^2)^{1/2}$ for a cylindrical wave and $r = (x_1^2 + x_2^2 + x_3^2)^{1/2}$ for a spherical wave. Such waves are discussed in the problems. Occasionally, the moving boundary of a disturbance passing through a medium is called a wave, or a wave front, or just a front.

PROBLEMS

1. Confirm that (1-2) has standing-wave solutions of the form (1-4) if and only if (1-5) and (1-6) hold, where λ is a (possibly complex) constant.
2. Show that $u(x, t) = h(x \cdot \omega + ct)$, where $|\omega|^2 = 1$, represents a plane wave which travels at speed c in the $-\omega$ direction.
3. Let (r, θ, ϕ) be spherical coordinates and $u = u(r, t)$ be a radially symmetric solution to (1-2).
   (a) Show that
   $$u_{tt} = c^2 \frac{1}{r^2} \frac{\partial}{\partial r}\left(r^2 \frac{\partial u}{\partial r}\right).$$
   (b) Express the result in part (a) as
   $$(ru)_{tt} = c^2(ru_{rr} + 2u_r) = c^2(ru)_{rr},$$
   and conclude that part (a) has general solution
   $$u(r, t) = \frac{F(r + ct)}{r} + \frac{G(r - ct)}{r},$$
   where F and G are arbitrary, twice-differentiable functions.
4. Continue in the context of Prob. 3.
   (a) Let $G(r) = 1$ for $r \le 1$ and 0 for $r > 1$. For several values of $t > 0$, sketch the regions where $u(r, t) = G(r - ct)/r$ is nonzero and conclude that $u(r, t)$ represents an outgoing wave.
   (b) Give a physical description of the waves $G(r - ct)/r$ for a general G.
   (c) Let $F(r) = 1$ for $r > 100$ and 0 for $0 \le r \le 100$. Sketch the regions where $u(r, t) = F(r + ct)/r$ is nonzero for different values of t and conclude that $u(r, t)$ is an incoming wave.
   (d) Give a physical description of the waves $F(r + ct)/r$ for a general F.
5. Show directly that the functions $(1/r)e^{ik(r \pm ct)}$, where $r = |x|$, satisfy (1-2) for any constant k.

6. Verify that $u(x, t) = (c^2t^2 - x_1^2 - x_2^2)^{-1/2}$ satisfies (1-2) in two spatial dimensions inside the cone $x_1^2 + x_2^2 < c^2t^2$.
7. Seek solutions to the wave equation in three spatial dimensions of the form $u(x, t) = f(r/ct)$, $r = |x|$.
8. Seek solutions to the wave equation (1-2) of the form $u(x, t) = f(c^2t^2 - |x|^2)$ in the case of (a) two spatial dimensions and (b) three spatial dimensions.
9. Solve formally

$$u_{tt} = c^2\,\Delta u, \qquad x \in R_3, \qquad t > 0,$$
$$u(x, 0) = f(x), \qquad u_t(x, 0) = g(x),$$

using Fourier transforms in (a) two spatial dimensions and (b) three spatial dimensions.
10. Consider Maxwell's equations in free space

$$\frac{1}{c}\frac{\partial B}{\partial t} + \nabla \times E = 0 \quad \text{and} \quad \frac{1}{c}\frac{\partial E}{\partial t} - \nabla \times B = 0,$$
$$\nabla \cdot B = 0 \quad \text{and} \quad \nabla \cdot E = 0,$$

where E is the electric field strength and B the magnetic field strength. Eliminate B from the first-order partial differential equations to get $\partial^2 E/\partial t^2 = c^2\nabla \times (\nabla \times E)$. Use a vector identity to conclude that $\partial^2 E/\partial t^2 = c^2\,\Delta E$, and derive the corresponding result for B. Thus, each component of the E and B fields satisfies the wave equation (1-2).
11. Consider the wave equation in cylindrical coordinates

$$u_{tt} = c^2\left(u_{rr} + \frac{1}{r}u_r + \frac{1}{r^2}u_{\theta\theta} + u_{zz}\right).$$

(a) Seek a solution of the form $u = T(t)R(r)$ and so obtain the cylindrical waves $e^{ic\lambda t}J_0(\lambda r)$, where $J_0(z)$ is Bessel's function of order 0. *Hint.* Use separation constant $-\lambda^2$.

(b) Seek solutions of the form $u = T(t)R(r)\Theta(\theta)$ and so obtain the solutions $e^{ic\lambda t}e^{in\theta}J_n(\lambda r)$ where $J_n(z)$ is Bessel's function of order n.
12. Write out the (homogeneous) wave equation in spherical coordinates (r, θ, ϕ). Seek separated solutions of the form $u = T(t)R(r)$ and so obtain the spherical waves

$$e^{i\lambda ct}\frac{\sin(\lambda r)}{r} \quad \text{and} \quad e^{i\lambda ct}\frac{\cos(\lambda r)}{r}.$$

10-2. Separation of Variables

We saw in Sec. 10-1 that separation of variables leads to standing-wave solutions to the wave equation. We can superpose these standing waves to construct solutions to initial, boundary value problems for the wave equation when the domains in question are simple enough. No new ideas come up when separation of variables is used in this context, so we restrict ourselves to two typical examples.

Example 1. The small transverse vibrations $u(x, y, t)$ of a rectangular membrane which is pinned along its edges and whose initial position and velocity are prescribed is described by

$$\begin{aligned}
&u_{tt} = c^2 \Delta u, \qquad 0 < x < a, \qquad 0 < y < b, \qquad t > 0,\\
&u(x, y, 0) = f(x, y), \qquad 0 \le x \le a, \qquad 0 \le y \le b,\\
&u_t(x, y, 0) = g(x, y), \qquad 0 \le x \le a, \qquad 0 \le y \le b,\\
&u(x, 0, t) = u(x, b, t) = 0, \qquad 0 \le x \le a, \qquad t \ge 0,\\
&u(0, y, t) = u(a, y, t) = 0, \qquad 0 \le y \le b, \qquad t \ge 0.
\end{aligned}$$

We seek separated solutions

$$u(x, y, t) = v(x, y)T(t), \tag{2-1}$$

and observe that v and T must satisfy

$$T'' + c^2\lambda^2 T = 0, \qquad t > 0, \tag{2-2}$$

and

$$\begin{cases} \Delta v + \lambda^2 v = 0, \qquad 0 < x < a, \qquad 0 < y < b,\\ v = 0 \quad \text{when} \quad x = 0 \quad \text{or} \quad a \quad \text{and} \quad y = 0 \quad \text{or} \quad b, \end{cases} \tag{2-3}$$

where the separation constant is $-\lambda^2$. We have written $-\lambda^2$ with $\lambda > 0$ because the only separation constants that lead to nontrivial solutions are negative. Compare with (2-3) of Sec. 9-2.

To find v, we separate variables once more. Thus, let $v(x, y) = X(x)Y(y)$ and use (2-3) to find that

$$\frac{X''}{X} = -\left(\frac{Y''}{Y} + \lambda^2\right) = -\mu^2$$

and then

$$\begin{aligned}
&X''(x) + \mu^2 X(x) = 0, \qquad X(0) = X(a) = 0,\\
&Y''(y) + (\lambda^2 - \mu^2)Y(y) = 0, \qquad Y(0) = Y(b) = 0.
\end{aligned}$$

The boundary value problem for X implies that $\mu > 0$, as we already anticipated. In fact,

$$\mu = \mu_n = \frac{n\pi}{a} \quad \text{and} \quad X_n(x) = \sin(\mu_n x), \qquad n = 1, 2, \ldots.$$

Now we infer from the problem for Y that $\lambda^2 - \mu_n^2 \equiv \nu^2$ must be $(m\pi/b)^2$ for $m = 1, 2, \ldots,$ so that

$$\nu_m = \frac{m\pi}{b} \quad \text{and} \quad Y_m(y) = \sin(\nu_m y).$$

Therefore, the eigenvalues of (2-3) are

$$\lambda^2 = \lambda_{nm}^2 = \mu_n^2 + \nu_m^2 = \left(\frac{n\pi}{a}\right)^2 + \left(\frac{m\pi}{b}\right)^2$$

and the corresponding eigenfunctions are

$$v = v_{nm}(x) = \sin(\mu_n x)\sin(\nu_m y).$$

For each m and n we obtain from (2-2)

$$T = T_{nm}(t) = A_{nm}\cos(\lambda_{nm}ct) + B_{nm}\sin(\lambda_{nm}ct).$$

Using these formulas for v and T in (2-1) and superposing, we expect the solution u to the initial, boundary value problem to be

$$u(x, t) = \sum_{n,m=1}^{\infty} [A_{nm}\cos(\lambda_{nm}ct) + B_{nm}\sin(\lambda_{nm}ct)]\sin(\mu_n x)\sin(\nu_m y), \tag{2-4}$$

where the constants A_{nm} and B_{nm} are chosen so that

$$u(x, y, 0) = f(x, y) = \sum_{n,m=1}^{\infty} A_{nm}\sin(\nu_n x)\sin(\nu_m y)$$

and

$$u_t(x, y, 0) = g(x, y) = \sum_{n,m=1}^{\infty} \lambda_{nm}cB_{nm}\sin(\mu_n x)\sin(\nu_m y).$$

The coefficients in these two-dimensional, Fourier series can be obtained using orthogonality relations analogous to the corresponding one-dimensional relations (see Prob. 1). The results are

$$A_{nm} = \frac{4}{ab}\int_0^a\int_0^b f(x, y)\sin(\mu_n x)\sin(\nu_m y)\,dx\,dy, \tag{2-5}$$

and

$$B_{nm} = \frac{4}{abc\lambda_{nm}}\int_0^a\int_0^b g(x, y)\sin(\mu_n x)\sin(\nu_m y)\,dx\,dy. \tag{2-6}$$

Thus, (2-4)–(2-6) constitute the formal solution to the initial boundary value problem in Ex. 1. As usual, this formal solution is a solution if $f(x, y)$ and $g(x, y)$ are suitably smooth and satisfy certain compatibility conditions (see Prob. 2).

For a second example, consider the vibrations of a circular drum of radius a.

Example 2. Solve (in two spatial dimensions)

$$\begin{aligned} &u_{tt} = c^2\,\Delta u = c^2\left[\frac{1}{r}\frac{\partial}{\partial r}\left(r\frac{\partial u}{\partial r}\right) + \frac{1}{r^2}\frac{\partial^2 u}{\partial\theta^2}\right], \qquad 0 \le r < a, \qquad t > 0, \\ &u(r, \theta, 0) = 0, \qquad u_t(r, \theta, 0) = f(r, \theta), \qquad u(a, \theta, t) = 0. \end{aligned} \tag{2-7}$$

Since (r, θ) and $(r, \theta + 2\pi)$ mark the same point, all functions are 2π periodic in θ, $-\infty < \theta < \infty$. We seek separated solutions

$$u(r, \theta, t) = v(r, \theta)T(t)$$

and obtain the eigenvalue problem

$$\begin{cases} \Delta v + \lambda^2 v = 0 & \text{in} \quad 0 \le r < a, \\ v = 0 & \text{on} \quad r = a, \end{cases} \tag{2-8}$$

for v and relations

$$T'' + c^2\lambda^2 T = 0, \qquad T(0) = 0, \tag{2-9}$$

for T. Just as for (2-3) we deduce that $\lambda > 0$, let $v(r, \theta) = R(r)\Theta(\theta)$, and find that

$$\frac{(1/r)(rR')'}{(1/r^2)R} + \lambda^2 r^2 = -\frac{\Theta''}{\Theta} = k,$$

a constant. The function Θ must satisfy $\Theta'' + k\Theta = 0$ and be 2π periodic in θ. Nontrivial solutions only exist for $k = n^2$, $n = 0, 1, 2, \ldots$, in which case corresponding eigenfunctions are

$$\tfrac{1}{2};\ \cos\theta,\ \sin\theta;\ \cos 2\theta,\ \sin 2\theta;\ \ldots;\ \cos n\theta,\ \sin n\theta;\ \ldots. \tag{2-10}$$

This notation indicates that when $k = 0$ the eigenfunction is $\frac{1}{2}$ (up to a constant multiple) and when $k = n$ there are two linearly independent eigenfunctions $\cos n\theta$ and $\sin n\theta$.

With $k = n^2$ we see that R must satisfy

$$R'' + \frac{1}{r}R' + \left(\lambda^2 - \frac{n^2}{r^2}\right)R = 0, \qquad R(a) = 0. \tag{2-11}$$

Since this differential equation becomes singular at $r = 0$, we emphasize that we seek a twice-differentiable solution $R(r)$ on $0 \le r < a$, because $v = R\Theta$ must satisfy (2-8). The change of variables $r = x/\lambda$, $R(r) = y(x)$ leads immediately to

$$y'' + \frac{1}{x}y' + \left(1 - \frac{n^2}{x^2}\right)y = 0,$$

which is Bessel's equation of order n. The general solution is

$$y = AJ_n(x) + BN_n(x),$$

where $J_n(x)$ and $N_n(x)$ are the nth-order Bessel functions. $N_n(x)$ becomes infinite as $x \to 0$ and $J_n(x)$ is given by a rapidly convergent series (see Sec. 9-2). Therefore, the differential equation in (2-11) has general solution $R = AJ_n(\lambda r) + BN_n(\lambda r)$ and the requirement that R be twice differentiable on $0 \le r < a$ forces $B = 0$. Then for R to vanish when $r = a$, we require that $J_n(\lambda a) = 0$. The Bessel function $J_n(x)$ has an infinite number of positive simple zeros, say

$$\mu_{n1} < \mu_{n2} < \cdots < \mu_{nj} < \cdots.$$

Consequently, the eigenvalues of (2-8) are given by

$$\lambda = \lambda_{nj} = \frac{\mu_{nj}}{a},$$

for $n = 0, 1, 2, \ldots$ and $j = 1, 2, 3, \ldots$, and the corresponding eigenfunctions are

$$\text{(2-12)} \qquad \tfrac{1}{2}J_0(\lambda_{0j}r), \qquad J_n(\lambda_{nj}r)\cos(n\theta), \qquad J_n(\lambda_{nj}r)\sin(n\theta).$$

Finally, (2-9) yields

$$T = T_{nj}(t) = \sin(c\lambda_{nj}t)$$

up to a constant multiple. Thus, we arrive at the following formal solution to (2-7):

$$\text{(2-13)} \qquad \begin{aligned} u(r, \theta, t) &= \sum_{j=1}^{\infty} \frac{1}{2} a_{0j} J_0(\lambda_{0j}r)\sin(c\lambda_{0j}t) \\ &+ \sum_{n,j=1}^{\infty} [a_{nj}J_n(\lambda_{nj}r)\cos(n\theta) + b_{nj}J_n(\lambda_{nj}r)\sin(n\theta)]\sin(c\lambda_{nj}t). \end{aligned}$$

This formal solution satisfies the wave equation, the initial condition $u(r, \theta, 0) = 0$, and the boundary condition $u(a, \theta, t) = 0$. The coefficients a_{nj} and b_{nj} still must be chosen so that $u_t(r, \theta, 0) = f(r, \theta)$. This determination is left for Probs. 3–6. The tedious verification that (2-13) is a solution under reasonable assumptions on the initial velocity profile $f(r, \theta)$ is omitted.

PROBLEMS

1. Use the orthonormality relations

$$\frac{2}{L}\int_0^L \sin\left(\frac{n\pi z}{L}\right)\sin\left(\frac{m\pi z}{L}\right) dz = \begin{cases} 0, & n \neq m, \\ 1, & n = m \end{cases}$$

and integrate term by term to deduce (2-5) and (2-6).

2. The vibrating membrane in Ex. 1 was treated in strict analogy with the vibrating string of Sec. 4-2. Review that section. Then formulate and prove the analogue of Th. 2-1 of Sec. 4-2.

3. In both Exs. 1 and 2, we encountered the eigenvalue problem

$$\Delta v + \lambda^2 v = 0, \qquad x \in D, \qquad v = 0, \qquad x \in B,$$

where D is a domain with boundary B. In the examples $D \subset R_2$. However, consider both two and three space dimensions here. If λ_1^2 and λ_2^2 are distinct eigenvalues with corresponding eigenfunctions $v_1(x)$ and $v_2(x)$, derive the orthogonality relation

$$\int_D v_1(x)v_2(x)\,dx = 0.$$

Hint. $(\lambda_1^2 - \lambda_2^2)v_1v_2 = -\text{div}\,[v_2\,\nabla v_1 - v_1\,\nabla v_2]$.

4. Each eigenvalue λ_{nj} of (2-8) with $n \neq 0$ has two eigenfunctions

$$v^{(1)} = J_n(\lambda_{nj}r)\cos(n\theta) \quad \text{and} \quad v^{(2)} = J_n(\lambda_{nj}r)\sin(n\theta).$$

Show that these eigenfunctions are orthogonal on D, the disk of radius a centered at the origin. Use this observation and Prob. 3 to conclude that any pair of distinct eigenfunctions v_1 and v_2 selected from (2-13) are orthogonal.

5. Check that $R_{nj}(r) = J_n(\lambda_{nj} r)$ satisfies

$$(rR'_{nj})' + \left(\lambda_{nj}^2 r - \frac{n^2}{r}\right)R_{nj} = 0, \qquad R_{nj}(0) = 0, \qquad R_{nj}(a) = 0.$$

Multiply the differential equation by rR'_{nj}, note that $R_{nj}R'_{nj} = (R_{nj}^2/2)'$, and integrate from 0 to a. Simplify and integrate by parts to get an expression for $\int_0^a rR_{nj}^2(r)\,dr$ in terms of $R'_{nj}(a)$. Now, if v is any eigenfunction in (2-13), find a formula for

$$\int_D v^2\,dx = \int_0^{2\pi}\int_0^a v^2(r, \theta)r\,dr\,d\theta.$$

6. Use Probs. 3–5 to find formulas for the coefficients in the expansion (2-13).
7. Solve (in two spatial dimensions)

$$\begin{aligned}
&u_{tt} = c^2\,\Delta u, \quad 0 < x < a, \quad 0 < y < b,\\
&u_y(x, 0, t) = u_y(x, b, t) = 0, \quad 0 \le x \le a, \quad t \ge 0,\\
&u(0, y, t) = u(a, y, t) = 0, \quad 0 \le y \le b, \quad t \ge 0,\\
&u(x, y, 0) = f(x, y), \quad 0 \le x \le a, \quad 0 \le y \le b,\\
&u_t(x, y, 0) = g(x, y), \quad 0 \le x \le a, \quad 0 \le y \le b.
\end{aligned}$$

8. Solve (in two spatial dimensions)

$$\begin{aligned}
&u_{tt} = c^2\,\Delta u, \quad 0 < x < a, \quad 0 < y < b,\\
&u_y(x, 0, t) = u_y(x, b, t) = 0, \quad 0 \le x \le a, \quad t \ge 0,\\
&u_x(0, y, t) = u_x(a, y, t) = 0, \quad 0 \le y \le b, \quad t \ge 0,\\
&u(x, y, 0) = f(x, y) \quad 0 \le x \le a, \quad 0 \le y \le b,\\
&u_t(x, y, 0) = g(x, y), \quad 0 \le x \le a, \quad 0 \le y \le b.
\end{aligned}$$

9. Consider

$$\begin{aligned}
&u_{tt} = c^2\frac{1}{r}\frac{\partial}{\partial r}\left(r\frac{\partial u}{\partial r}\right), \quad 0 \le r < a, \quad t > 0,\\
&u(r, 0) = f(r), \quad u_t(r, 0) = g(r), \quad 0 \le r \le a,\\
&u(a, t) = 0, \quad t \ge 0.
\end{aligned}$$

This is a radially symmetric drum problem. Solve it by specializing Ex. 2, and also solve it from scratch.

10. Solve

$$\begin{aligned}
&u_{tt} = c^2\frac{1}{r}\frac{\partial}{\partial r}\left(r\frac{\partial u}{\partial r}\right) + f(r, t) \quad 0 \le r < a, \quad t > 0,\\
&u(r, 0) = u_t(r, 0) = 0, \quad 0 \le r \le a,\\
&u(a, t) = 0, \quad t \ge 0.
\end{aligned}$$

Hint. Let $R_n(r)$ be the eigenfunctions of Prob. 9, assume a solution of the form

$$u(r, t) = \sum_{n=1}^{\infty} \phi_n(t)R_n(r),$$

expand $f(r, t) = \sum_{n=1}^{\infty} f_n(t)R_n(r)$, and determine $\phi_n(t)$ as in Sec. 4-3.

11. Solve

$$u_{tt} = c^2 \frac{1}{r}\frac{\partial}{\partial r}\left(r\frac{\partial u}{\partial r}\right) + f(r, t), \qquad 0 \le r < a, \qquad t > 0,$$

$$u(r, 0) = u_t(r, 0) = 0, \qquad 0 \le r \le a,$$

$$u_r(a, t) = 0, \qquad t \ge 0.$$

12. Solve

$$u_{tt} = c^2 \frac{1}{r^2}\frac{\partial}{\partial r}\left(r^2\frac{\partial u}{\partial r}\right), \qquad 0 \le r < a, \qquad t > 0,$$

$$u(r, 0) = 0, \qquad u_t(r, 0) = f(r), \qquad 0 \le r \le a,$$

$$u(a, t) = 0, \qquad t \ge 0.$$

13. Solve

$$u_{tt} + ku_t = c^2\,\Delta u, \qquad 0 < x < a, \qquad 0 < y < b, \qquad t > 0,$$

together with the same initial and boundary data as in Ex. 1.

10-3. Uniqueness Theory

In Sec. 10-2 we used separation of variables to construct solutions to certain initial, boundary value problems for the wave equation. In subsequent sections we develop solution formulas for initial value problems for the wave equation. The results of this section show that the solutions to these problems are unique.

We start with the initial, boundary value problem,

$$\text{(3-1)} \qquad \begin{cases} u_{tt} = c^2\,\Delta u + F(x, t), & (x, t) \in Q_T, \\ u(x, 0) = f(x), \quad u_t(x, 0) = g(x), & x \in D, \\ u(x, t) = \phi(x, t), & (x, t) \in S_T, \end{cases}$$

where D is a bounded, normal domain with boundary B in two or three-dimensional space and we have used the notation from Sec. 9-3. As usual, let w be the difference of two solutions to (3-1). Then w satisfies the homogeneous problem corresponding to (3-1). Problem (3-1) will have at most one solution if we can prove that $w \equiv 0$. To confirm this and arrive at an important energy relation, multiply the wave equation $w_{tt} = c^2\,\Delta w$ by w_t, integrate with respect to x over the domain D, apply the Gauss divergence theorem, and use the fact that $w(x, t) = 0$ for $x \in B, 0 \le t \le T$ to obtain

$$\int_D w_t w_{tt}\,dx = -c^2 \int_D \nabla w_t \cdot \nabla w\,dx$$

or

$$\frac{d}{dt}\int_D \left(\frac{1}{2}w_t^2 + \frac{c^2}{2}|\nabla w|^2\right)dx = 0.$$

We integrate from 0 to t to find that

$$\int_D \left[\frac{1}{2} w_t(x, t)^2 + \frac{c^2}{2} |\nabla w(x, t)|^2\right] dx = 0 \tag{3-2}$$

for all t in $[0, T]$, because $w_t(x, 0) = 0$ and $w(x, 0) = 0$, so $\nabla w(x, 0) = 0$. Thus, w_t, $\nabla w \equiv 0$, and consequently w is a constant which must be zero by the initial data. We have proved Th. 3-1.

Theorem 3-1. The initial, boundary value problem (3-1) has at most one solution u in $C^{2,2}(Q_T) \cap C^{1,1}(\bar{Q}_T)$.

Of course, this uniqueness result is physically obvious. Indeed, we can regard (3-1) as the model for the vibrations of a membrane driven by the external force F and subject to the indicated initial and boundary conditions. Clearly, only one motion is possible for the membrane.

The reasoning leading to (3-2) deserves a closer look. Consider (3-1) as a model for the small transverse vibrations of a two-dimensional membrane. Then $c^2 = \tau/\rho$, where τ is the tension and ρ the density. We write the wave equation in (3-1) as

$$\rho u_{tt} = \tau \, \Delta u + \rho F(x, t), \qquad (x, t) \in Q_T, \tag{3-3}$$

and assume that $\phi \equiv 0$ so that the boundary of the membrane remains fixed for all time. Multiply (3-3) by u_t and proceed as above to find

$$\frac{d}{dt} \int_D \left(\frac{\rho}{2} u_t^2 + \frac{\tau}{2} |\nabla u|^2\right) dx = \int_D \rho F u_t \, dx. \tag{3-4}$$

The expressions

$$\int_D \frac{\rho}{2} u_t^2 \, dx \quad \text{and} \quad \int_D \frac{\tau}{2} |\nabla u|^2 \, dx$$

are, respectively, the kinetic and potential energies. Thus, (3-4) states that the time rate of change of kinetic plus potential energy is equal to the rate at which work is being done on the membrane by the external forces, or put another way, is equal to the power that is being supplied to the membrane. Let

$$E(t) = \int_D \left(\frac{1}{2} \rho u_t^2 + \frac{1}{2} \tau |\nabla u|^2\right) dx$$

and integrate (3-4) from 0 to t to obtain

$$E(t) = E(0) + \int_{Q_t} \rho F u_t \, dx \, dt, \tag{3-5}$$

where

$$E(0) = \int_D \left[\frac{1}{2} \rho g(x)^2 + \frac{1}{2} \tau |\nabla f(x)|^2\right] dx.$$

Equation (3-5) states that the energy at time t equals the initial energy plus the net work done on the membrane by the external forces. If no external forces act, $F \equiv 0$ and (3-5) asserts that the total energy remains constant. Similar results are formulated in three spatial dimensions in Prob. 3.

There is also an energy inequality for the initial value problem for the wave equation

$$u_{tt} = c^2 \Delta u. \tag{3-6}$$

To motivate the discussion, let (x°, t_0) be an arbitrary point in space-time. The subsequent analysis applies in any number of spatial dimensions. For definiteness, we assume three dimensions. Suppose that at time $t = 0$ a disturbance governed by (3-6) commences. Since the disturbance propagates with speed c, the resultant disturbance at position x° at time t_0, $u(x^\circ, t_0)$, can be caused only by the effects of the initial disturbance due to points in space in the closed ball centered at x° with radius ct_0. See Fig. 10-1, where the ball in question is shaded in the x_1–x_n plane. As Fig. 10-1 shows, the point (x°, t_0) is the vertex of a (space-time) cone extending backward in time and cutting from the plane $t = 0$ the ball of radius ct_0 centered at x°. This cone, called the *retrograde cone*, has the equation

$$(x - x^\circ) \cdot (x - x^\circ) - c^2(t - t_0)^2 = 0. \tag{3-7}$$

We denote the intersection of the cone (3-7) with the plane perpendicular to the time axis at t by D_t. So D_t is the ball centered at x° with radius $c(t_0 - t)$ in

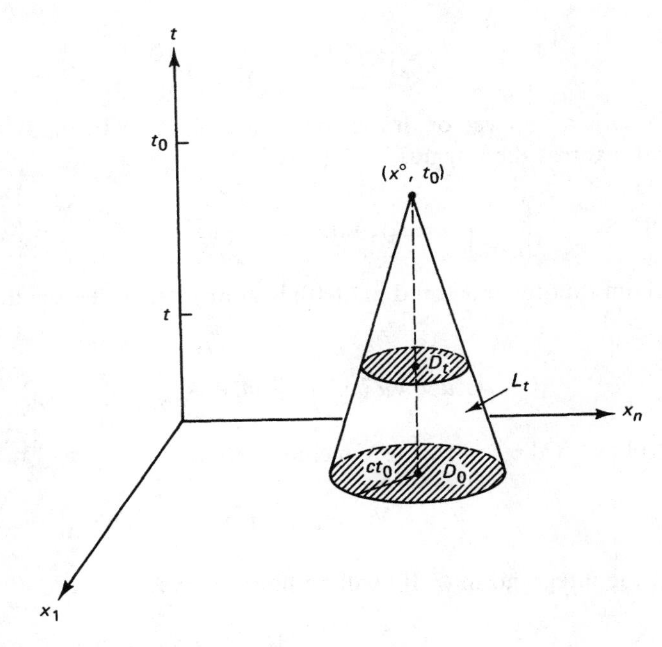

Figure 10-1

the plane at level t. We set

$$E(t) = \int_{D_t} \left(\frac{1}{2} u_t^2 + \frac{c^2}{2} |\nabla u|^2 \right) dx,$$

which is the energy integral at level t in the cone. Let C_t be the portion of space–time cut from the retrograde cone by the planes $t = 0$ and $t = t$. The region C_t is bounded by D_0, D_t, and L_t the lateral surface of the cone between D_0 and D_t.

In order to relate $E(t)$ and $E(0)$, we integrate the identity

$$\begin{aligned} u_\tau[u_{\tau\tau} - c^2 \Delta_y u] &= \tfrac{1}{2}(u_\tau^2)_\tau - c^2 \operatorname{div}_y(u_\tau \nabla u) + c^2 \nabla u_\tau \cdot \nabla u \\ &= \left(\frac{1}{2} u_\tau^2 + \frac{c^2}{2} |\nabla u|^2 \right)_\tau + \operatorname{div}_y(-c^2 u_\tau \nabla u) \\ &= \operatorname{div}_{y,\tau}\left(-c^2 u_\tau \nabla u, \frac{1}{2} u_\tau^2 + \frac{c^2}{2} |\nabla u|^2 \right) \end{aligned}$$

and use the (four-dimensional) Gauss divergence theorem. Note that the outer unit normal on D_t is $(0, 0, 0, 1)$ and on D_0 is $(0, 0, 0, -1)$, so that we obtain

$$0 = E(t) - E(0) + \int_{L_t} \left[-c^2 u_\tau \nabla u \cdot v_y + \left(\frac{1}{2} u_\tau^2 + \frac{c^2}{2} |\nabla u|^2 \right) v_\tau \right] d\sigma_{y\tau}, \tag{3-8}$$

where $v = (v_y, v_\tau)$ is the outer unit normal to L_t. A short calculation using (3-7) reveals that

$$v = (v_y, v_\tau) = \left(\frac{\omega}{\sqrt{1 + c^2}}, \frac{c}{\sqrt{1 + c^2}} \right), \tag{3-9}$$

where ω is a unit vector in the space variables (see Prob. 5). Since $v_\tau > 0$ on L_t we can express the integral in (3-8) as

$$\int_{L_t} \frac{1}{v_\tau} \left[-c^2 u_\tau v_\tau \nabla u \cdot v_y + \left(\frac{1}{2} u_\tau^2 + \frac{c^2}{2} |\nabla u|^2 \right) v_\tau^2 \right] d\sigma_{y\tau}. \tag{3-10}$$

We claim that the integrand in (3-10) is nonnegative. To see this, observe that for any $\varepsilon > 0$

$$|c^2 u_\tau v_\tau \nabla u \cdot v_y| \le \frac{\varepsilon c^2}{2} u_\tau^2 v_\tau^2 + \frac{1}{2\varepsilon} c^2 |\nabla u \cdot v_y|^2$$

by Prob. 6. Take $\varepsilon = 1/c^2$ to conclude that

$$|c^2 u_\tau v_\tau \nabla u \cdot v_y| \le \frac{1}{2} u_\tau^2 v_\tau^2 + \frac{c^4}{2} |\nabla u \cdot v_y|^2.$$

Thus, the integrand in (3-10) will be nonnegative if

$$\frac{c^4}{2} |\nabla u \cdot v_y|^2 \le \frac{c^2}{2} |\nabla u|^2 v_\tau^2.$$

This inequality follows from (3-9) because

$$c^4|\nabla u \cdot v_y|^2 \le c^4|\nabla u|^2|v_y|^2 \le c^4|\nabla u|^2 \frac{1}{1+c^2} = c^2|\nabla u|^2 v_\tau^2.$$

Finally, since the integrand in (3-10) is nonnegative, (3-8) yields

$$E(t) \le E(0) \tag{3-11}$$

for $t > 0$.

The energy inequality (3-11) yields the following uniqueness result:

Theorem 3-2. There exists at most one solution to the initial value problem

$$u_{tt} = c^2\,\Delta u + F(x, t), \qquad x \in R_3, \qquad t > 0,$$
$$u(x, 0) = f(x), \qquad u_t(x, 0) = g(x), \qquad x \in R_3,$$

Proof. See Prob. 7.

Here is another useful consequence of (3-11):

Proposition 3-1. Suppose that u satisfies

$$u_{tt} = c^2\,\Delta u, \qquad x \in R_3, \qquad t > 0,$$
$$u(x, 0) = f(x), \qquad u_t(x, 0) = g(x), \qquad x \in R_3,$$

Assume that $\nabla f(x)$ and $g(x)$ are continuous and square integrable over R_3. Then the solution $u(x, t)$ has first-order partials which are square integrable over R_3 at each time t.

Proof. Fix a time t. Choose $t_0 > t$, $x^\circ \in R_3$, and form the retrograde cone as in Fig. 10-1. Then from (3-11):

$$\int_{D_t}\left[\frac{1}{2}\,u_t(x, t)^2 + \frac{c^2}{2}\,|\nabla u(x, t)|^2\right]dx \le \int_{D_0}\left[\frac{1}{2}\,g(x)^2 + \frac{c^2}{2}\,|\nabla f(x)|^2\right]dx.$$

The right member is less than the integral taken over all of R_3 and by assumption this integral is finite. The integral on the left is over the ball with center x° and radius $c(t_0 - t)$. Let $t_0 \to \infty$ to conclude that

$$\int_{R_3}\left[\frac{1}{2}\,u_t(x, t)^2 + \frac{c^2}{2}\,|\nabla u(x, t)|^2\right]dx \le \int_{R_3}\left[\frac{1}{2}\,g(x)^2 + \frac{c^2}{2}\,|\nabla f(x)|^2\right]dx < \infty.$$

PROBLEMS

1. The energy considerations applied to the membrane apply to other initial, boundary value problems for the wave equation: Suppose that u satisfies

$$u_{tt} = c^2\,\Delta u + F(x, t), \qquad (x, t) \in Q_T$$
$$u(x, 0) = f(x), \qquad u_t(x, 0) = g(x), \qquad x \in D,$$

where D is a bounded domain with boundary B.

(a) Show that

$$\frac{d}{dt}\int_D \left(\frac{1}{2}u_t^2 + \frac{c^2}{2}|\nabla u|^2\right)dx = c^2 \int_B u_t \frac{\partial u}{\partial \nu}\, d\sigma + \int_D u_t F\, dx.$$

(b) Show that there is at most one solution u which also satisfies the Neumann boundary condition

$$\frac{\partial u}{\partial \nu} = \phi(x, t), \qquad (x, t) \in S_T.$$

(c) Show that there is at most one solution u which also satisfies the Robin boundary condition

$$\frac{\partial u}{\partial \nu} + a(x)u = \phi(x, t), \qquad (x, t) \in S_T,$$

where $a(x) \geq 0$.

2. Consider the initial, boundary value problem

$$\begin{aligned} &u_{tt} + ku_t = c^2\, \Delta u + F(x, t), \qquad (x, t) \in Q_T,\\ &u(x, 0) = f(x), \qquad u_t(x, 0) = g(x), \qquad x \in D,\\ &u(x, t) = 0, \qquad (x, t) \in S_T. \end{aligned}$$

(a) Show that

$$\frac{d}{dt}\int_D \left[\frac{1}{2}u_t^2 + \frac{c^2}{2}|\nabla u|^2\right]dx = -k\int_D u_t^2\, dx + \int_D F u_t\, dx.$$

(b) In the case of a membrane, $c^2 = \tau/\rho$ and

$$\frac{d}{dt}\int_D \left[\frac{1}{2}\rho u_t^2 + \frac{\tau}{2}|\nabla u|^2\right]dx = -2k\int_D \frac{1}{2}\rho u_t^2\, dx + \int_D \rho F u_t\, dx.$$

Argue that this result means that the rate of change of kinetic plus potential energy is equal to the power supplied to the membrane minus the rate at which energy is dissipated.

(c) Show that there is at most one solution to this initial, boundary value problem.

3. Consider (3-1) in any number of spatial variables with $\phi \equiv 0$. Let

$$E(t) = \int_D \left[\frac{1}{2}u_t^2 + \frac{c^2}{2}|\nabla u|^2\right]dx$$

Show that $E(t) = E(0) + \int_{Q_T} F u_t\, dx\, dt$.

4. Consider (3-1) with $\phi \equiv 0$. Use the results in Prob. 3 to formulate and prove that, given existence, the solution to (3-1) depends continuously on the data.

5. Verify (3-9).

6. Show that $ab \leq (\varepsilon/2)a^2 + (1/2\varepsilon)b^2$ for any $\varepsilon > 0$ and $a, b \geq 0$.

7. Prove Th. 3-2.

8. In Sec. 1-7 we saw that the propagation of sound waves in a gas is described by

$$u_t + \operatorname{div} v = 0, \qquad v_t = -c^2\, \nabla u, \qquad c^2 = \frac{\lambda p_0}{\rho_0},$$

where $\rho/\rho_0 = 1 + u$ and $p/p_0 = 1 + \lambda u$. Here v is the velocity and u describes the variation in density and pressure from the reference levels ρ_0 and p_0. Suppose that the sound waves propagate in a closed room D from which no gas can escape. Then

$$v \cdot \nu = 0, \qquad x \in B, \qquad t \geq 0,$$

where B is the boundary of D. Suppose that the sound is initiated by a sudden change in density, for example by snapping one's fingers, from the reference density ρ_0 of the quiet room. Argue that appropriate initial conditions are

$$u(x, 0) = 0, \qquad u_t(x, 0) = f(x), \qquad x \in D.$$

(a) Show that (v, u) satisfies the energy equality

$$\frac{d}{dt} \int_D \left(\frac{1}{2} \rho_0 |v|^2 + \frac{\lambda p_0}{2} u^2 \right) dx = 0.$$

(b) Prove there is at most one solution to the stated problem.

9. Consider wave propagation in an inhomogeneous medium,

$$\rho u_{tt} = \text{div}\,(c\,\nabla u) - qu + \rho F(x, t), \qquad x \in D, \qquad t > 0,$$
$$u(x, 0) = f(x), \qquad u_t(x, 0) = g(x), \qquad x \in D,$$
$$u(x, t) = \phi(x), \qquad x \in B, \qquad t \geq 0,$$

where the boundary condition is time independent and ρ, c, q are nonnegative functions of x with ρ, $c > 0$.

(a) Establish the energy relation

$$E(t) = E(0) + \int_0^t \int_D \rho F u_t \, dx \, d\tau$$

where

$$E(t) = \int_D \left(\frac{1}{2} \rho u_t^2 + \frac{1}{2} c |\nabla u|^2 + \frac{1}{2} q u^2 \right) dx.$$

(b) Show that the given problem has at most one solution.

(c) If $F(x, t) \equiv 0$, show that (given existence) the solution depends continuously on the initial and boundary data.

(d) Extend part (c) to the case $F \not\equiv 0$. Now the problem will be well-posed over any finite time interval $0 \leq t \leq T$. *Hint*. Use Prob. 6 to estimate the integral involving F.

10. Let D in Prob. 9 be an exterior domain. Formulate natural conditions on u, ρ, c, q, F, f, g, ϕ so that the energy relation in part (a) of Prob. 9 holds. Then prove a uniqueness theorem for the exterior problem. *Hint*. As usual integrate over $D_R = D \cap K_0(R)$ and then let $R \to \infty$.

11. Find the solution to

$$u_{tt} = c^2 \Delta u, \qquad x \in R_3, \qquad t > 0,$$
$$u(x, 0) = 1, \qquad u_t(x, 0) = \frac{1}{1 + |x|^2}, \qquad x \in R_3.$$

Hint. The symmetry of the initial data suggests that the solution should only depend on $|x| = r$. Remember Prob. 3 of Sec. 10-1.

10-4. The Initial Value Problem

To motivate the formulas that follow for the solution of the initial value problem for the wave equation, we begin with some physical reasoning. Imagine that sound waves are initiated by a sudden change in density in the air near some point, say $x = 0$, and that to begin with the air has a constant reference density. Then [see (7-9) of Sec. 1-7] the variations $u(x, t)$ in density from the reference level are governed by

$$\begin{cases} u_{tt} = c^2 \Delta u, & x \in R_3, \quad t > 0, \\ u(x, 0) = 0, & u_t(x, 0) = \delta_\varepsilon(x), \quad x \in R_3, \end{cases} \tag{4-1}$$

where $\delta_\varepsilon(x)$ describes the sudden change in density near $x = 0$. Let's suppose that the intensity of the disturbance is 1. Then $\delta_\varepsilon(x)$ should have the properties

$$\int_{R_3} \delta_\varepsilon(x)\, dx = 1 \quad \text{and} \quad \lim_{\varepsilon \to 0} \delta_\varepsilon(x) = 0 \quad \text{for all} \quad x \neq 0,$$

because the initial disturbance is to be localized at $x = 0$. This, of course, reminds us of similar considerations in Secs. 5-4 and 9-1 which led to the interpretation of the fundamental solution to the heat equation as the temperature at position x and time t due to a unit heat source concentrated at the origin at time $t = 0$. Letting $\varepsilon = at$ in the fundamental solution to the heat equation, we see that

$$\delta_\varepsilon(x) = (4\pi\varepsilon)^{-3/2} \exp\left(-\frac{|x|^2}{4\varepsilon}\right) \tag{4-2}$$

will satisfy the two requirements on $\delta_\varepsilon(x)$ given above. Make this choice for $\delta_\varepsilon(x)$ in the initial value problem (4-1).

Since the initial data in (4-1) are radially symmetric, we expect that the resultant variations in density $u = u(x, t)$ will share this property. From Prob. 3 of Sec. 10-1 we know that such radially symmetric solutions have the form

$$u = \frac{F(r + ct)}{r} + \frac{G(r - ct)}{r},$$

where $r = |x|$. The initial data in (4-1) require that

$$F(r) + G(r) = 0, \quad \text{and} \quad F'(r) - G'(r) = \frac{r}{c}\, \delta_\varepsilon.$$

Thus $F = -G$, $G' = -r\delta_\varepsilon/2c$,

$$G(r) = \int \frac{-r}{2c} (4\pi\varepsilon)^{-3/2} \exp\left(-\frac{r^2}{4\varepsilon}\right) dr = \frac{\varepsilon(4\pi\varepsilon)^{-3/2}}{c} \exp\left(-\frac{r^2}{4\varepsilon}\right) + \text{const},$$

and we find that

$$u(x, t) = \frac{1}{4\pi c r} \frac{1}{\sqrt{4\pi\varepsilon}} \exp\left[-\frac{(r - ct)^2}{4\varepsilon}\right] - \frac{1}{4\pi c r} \frac{1}{\sqrt{4\pi\varepsilon}} \exp\left[-\frac{(r + ct)^2}{4\varepsilon}\right].$$

For convenience, we write this result as

$$u(x, t) = I_\varepsilon(r, t) - J_\varepsilon(r, t) \tag{4-3}$$

where $I_\varepsilon(r, t)$ is an outgoing spherical wave and $J_\varepsilon(r, t)$ is an incoming spherical wave.

Now, consider the variations in density due to a general initial disturbance with intensity $g(x)$,

$$\begin{aligned} &u_{tt} = c^2 \Delta u, \qquad x \in R_3, \qquad t > 0, \\ &u(x, 0) = 0, \qquad u_t(x, 0) = g(x), \qquad x \in R_3, \end{aligned} \tag{4-4}$$

where $g(x)$ is continuous. Consider the effect of the part of the initial disturbance $g(y)$ as y varies over a small cube of volume Δy. Since Δy is small, $g(y)$ is nearly constant in the cube and by (4-3)

$$g(y)\, \Delta y[I_\varepsilon(r, t) - J_\varepsilon(r, t)],$$

where $r = |x - y|$, gives the subsequent variations in density due to the source localized in Δy with intensity $g(y)\,\Delta y$. Sum these effects over all of space to get

$$\int_{R_3} g(y)[I_\varepsilon(r, t) - J_\varepsilon(r, t)]\, dy$$

for the total response to the localized sources. Thus, the solution to (4-4) should be

$$u(x, t) = \lim_{\varepsilon \to 0} \int_{R_3} g(y)[I_\varepsilon(r, t) - J_\varepsilon(r, t)]\, dy.$$

Since $r + ct \geq ct > 0$ for all $t > 0$, the formula for J_ε reveals that J_ε tends to zero at an exponential rate as ε tends to zero, and we conclude that

$$u(x, t) = \lim_{\varepsilon \to 0} \int_{R_3} g(y) I_\varepsilon(r, t)\, dy.$$

We introduce spherical coordinates $y = x + \rho\omega$, where $\rho \geq 0$ and ω is a unit vector to evaluate the remaining limit:

$$\begin{aligned} u(x, t) &= \lim_{\varepsilon \to 0} \frac{1}{4\pi c} \int_0^\infty \frac{e^{-(\rho - ct)^2/4\varepsilon}}{\sqrt{4\pi\varepsilon}} \left[\frac{1}{\rho} \int_{|\omega|=1} g(x + \rho\omega)\rho^2\, d\sigma_\omega\right] d\rho \\ &= \frac{1}{4\pi c} \cdot ct \int_{|\omega|=1} g(x + ct\omega)\, d\sigma_\omega, \end{aligned}$$

where we have used the basic reproducing property [(4-11) of Sec. 5-4] of the fundamental solution to the heat equation. Thus,

$$u(x, t) = \frac{t}{4\pi} \int_{|\omega|=1} g(x + ct\omega)\, d\sigma_\omega. \tag{4-5}$$

Since the area element on a sphere of radius ct is $d\sigma_y = (ct)^2\, d\sigma_\omega$, we can also express this result as

$$u(x, t) = \frac{1}{4\pi c^2 t} \int_{S_{ct}(x)} g(y)\, d\sigma_y, \tag{4-6}$$

where as usual $S_{ct}(x)$ is the sphere with center x and radius ct.

These plausible arguments suggest Th. 4-1.

Theorem 4-1. Suppose that $g \in C^2(R_3)$. Then (4-5) is the unique solution to the initial value problem (4-4).

Proof. Uniqueness was settled in Sec. 10-3. We need only check that (4-5) is a solution to (4-4). Obviously, $u(x, 0) = 0$ and

$$u_t(x, t) = \frac{1}{4\pi} \int_{|\omega|=1} g(x + ct\omega)\, d\sigma_\omega + \frac{t}{4\pi} \int_{|\omega|=1} \nabla g(x + ct\omega) \cdot c\omega\, d\sigma_\omega \tag{4-7}$$

so that

$$u_t(x, 0) = \frac{1}{4\pi}\, g(x) 4\pi = g(x).$$

Thus, (4-5) satisfies the initial data in (4-4). Use the Gauss divergence theorem to find

$$\frac{ct}{4\pi} \int_{|\omega|=1} \nabla g(x + ct\omega) \cdot \omega\, d\sigma_\omega = \frac{1}{4\pi ct} \int_{S_{ct}(x)} \frac{\partial g}{\partial v}\, d\sigma_y = \frac{1}{4\pi ct} \int_{K_{ct}(x)} \Delta g(y)\, dy$$

and consequently,

$$u_t(x, t) = \frac{1}{4\pi} \int_{|\omega|=1} g(x + ct\omega)\, d\sigma_\omega + \frac{1}{4\pi ct} \int_0^{ct} \int_{S_\rho(x)} \Delta g(y)\, d\sigma_y\, d\rho,$$

$$\begin{aligned}
u_{tt}(x, t) &= \frac{c}{4\pi} \int_{|\omega|=1} \nabla g(x + ct\omega) \cdot \omega\, d\sigma_\omega - \frac{1}{4\pi ct^2} \int_{K_{ct}(x)} \Delta g(y)\, dy \\
&\quad + \frac{c}{4\pi ct} \int_{S_{ct}(x)} \Delta g(y)\, d\sigma_y \\
&= \frac{1}{4\pi ct^2} \int_{K_{ct}(x)} \Delta g(y)\, dy - \frac{1}{4\pi ct^2} \int_{K_{ct}(x)} \Delta g(y)\, dy \\
&\quad + \frac{1}{4\pi t} \int_{S_{ct}(x)} \Delta g(y)\, d\sigma_y \\
&= \frac{c^2 t}{4\pi} \int_{|\omega|=1} \Delta g(x + ct\omega)\, d\sigma_\omega.
\end{aligned}$$

Thus,

$$u_{tt}(x, t) = \frac{c^2 t}{4\pi} \int_{|\omega|=1} \Delta g(x + ct\omega)\, d\sigma_\omega. \tag{4-8}$$

The right member of (4-8) is $c^2 \Delta u(x, t)$ by (4-5), so (4-5) satisfies the wave equation and Th. 4-1 is established.

If $g \in C^3(R_3)$, then u given by (4-5) has continuous third-order partials and u_t satisfies

$$\left(\frac{\partial^2}{\partial t^2} - c^2\Delta\right)u_t = \frac{\partial}{\partial t}(u_{tt} - c^2 \Delta u) = 0$$

and

$$u_t(x, 0) = g(x), \qquad u_{tt}(x, 0) = 0$$

by (4-7) and (4-8). This observation allows us to formulate *Kirchhoff's solution* to the initial value problem.

Theorem 4-2. Let $f \in C^3(R_3)$ and $g \in C^2(R_3)$. Then the unique solution to the initial value problem

$$\begin{aligned} &u_{tt} = c^2 \Delta u, \qquad x \in R_3, \qquad t > 0, \\ &u(x, 0) = f(x), \qquad u_t(x, 0) = g(x), \qquad x \in R_3, \end{aligned}$$

is

$$u(x, t) = \frac{\partial}{\partial t}\left[\frac{t}{4\pi}\int_{|\omega|=1} f(x + ct\omega)\, d\sigma_\omega\right] + \frac{t}{4\pi}\int_{|\omega|=1} g(x + ct\omega)\, d\sigma_\omega,$$

or equivalently,

$$u(x, t) = \frac{\partial}{\partial t}\left[\frac{1}{4\pi c^2 t}\int_{S_{ct}(x)} f(y)\, d\sigma_y\right] + \frac{1}{4\pi c^2 t}\int_{S_{ct}(x)} g(y)\, d\sigma_y.$$

The solution to the inhomogeneous wave equation

$$\text{(4-9)} \qquad \begin{cases} u_{tt} = c^2 \Delta u + F(x, t), & x \in R_3, \quad t > 0, \\ u(x, 0) = 0, \quad u_t(x, 0) = 0, & x \in R_3, \end{cases}$$

can be obtained from Duhamel's principle. Let τ be a parameter and solve

$$w_{tt} = c^2 \Delta w, \qquad w(x, 0) = 0, \qquad w_t(x, 0) = F(x, \tau)$$

for $w = w(x, t, \tau)$. By Th. 4-1,

$$w(x, t, \tau) = \frac{t}{4\pi}\int_{|\omega|=1} F(x + ct\omega, \tau)\, d\sigma_\omega.$$

Now set

$$\text{(4-10)} \quad u(x, t) = \int_0^t w(x, t - \tau, \tau)\, d\tau = \int_0^t \frac{t - \tau}{4\pi}\int_{|\omega|=1} F(x + c(t - \tau)\omega, \tau)\, d\sigma_\omega\, d\tau.$$

Obviously, (4-10) satisfies the initial data in (4-9). It is a simple calculation (Prob. 3) to confirm that (4-10) solves the inhomogeneous wave equation under suitable restrictions on F.

Theorem 4-3. Suppose that $F \in C^{2,0}(R_3 \times [0, \infty))$. Then (4-10) solves (4-9).

The solution to the initial value problem for the wave equation in two spatial dimensions can be obtained from the three-dimensional case by the *method of descent*. To see how this is done, consider

$$\text{(4-11)} \qquad \begin{cases} u_{tt} = c^2 \Delta u, & x = (x_1, x_2) \in R_2, \quad t > 0, \\ u(x, 0) = 0, & u_t(x, 0) = g(x), \quad x \in R_2. \end{cases}$$

Mark points in R_3 by $(x, x_3) = (x_1, x_2, x_3)$ and define $G(x, x_3) = g(x)$. Then

$$\text{(4-12)} \qquad U(x, x_3) = \frac{1}{4\pi c^2 t} \int_{S_{ct}(x,x_3)} G(y, y_3)\, d\sigma_{(y,y_3)} = \frac{1}{4\pi c^2 t} \int_{S_{ct}(x,x_3)} g(y)\, d\sigma_{(y,y_3)}$$

solves

$$\text{(4-13)} \qquad \begin{cases} U_{tt} = c^2 \Delta U, & (x, x_3) \in R_3, \quad t > 0, \\ U(x, x_3, 0) = 0, & U_t(x, x_3, 0) = G(x, x_3) = g(x), \quad (x, x_3) \in R_3. \end{cases}$$

If we confirm that (4-12) does not depend on x_3, as the initial data suggest, then $u(x) = U(x, x_3)$ solves (4-11). Since the sphere $S_{ct}(x, x_3)$ is given by

$$y_3 = x_3 \pm [(ct)^2 - (y_1 - x_1)^2 - (y_2 - x_2)^2]^{1/2},$$

we have

$$d\sigma_{(y,y_3)} = \left[1 + \left(\frac{\partial y_3}{\partial y_1}\right)^2 + \left(\frac{\partial y_3}{\partial y_2}\right)^2\right]^{1/2} dy_1\, dy_2 = \frac{ct\, dy_1\, dy_2}{[(ct)^2 - (y_1 - x_1)^2 - (y_2 - x_2)^2]^{1/2}}$$

on both hemispheres, where $y = (y_1, y_2)$ ranges over the disk $K_{ct}(x)$ in R_2 given by $(y_1 - x_1)^2 + (y_2 - x_2)^2 \le (ct)^2$. Thus, (4-12) can be expressed as

$$U(x, x_3) = \frac{1}{2\pi c} \int_{K_{ct}(x)} \frac{g(y)}{\sqrt{(ct)^2 - |y - x|^2}}\, dy.$$

Evidently, U is independent of x_3 and we conclude:

Theorem 4-4. Suppose that $f \in C^3(R_2)$ and $g \in C^2(R_2)$. Then the solution to

$$u_{tt} = c^2 \Delta u, \qquad x \in R_2, \qquad t > 0,$$
$$u(x, 0) = f(x), \qquad u_t(x, 0) = g(x), \qquad x \in R_2,$$

is

$$u(x, t) = \frac{\partial}{\partial t} \frac{1}{2\pi c} \int_{K_{ct}(x)} \frac{1}{\sqrt{(ct)^2 - |x - y|^2}} f(y)\, dy + \frac{1}{2\pi c} \int_{K_{ct}(x)} \frac{1}{\sqrt{(ct)^2 - |x - y|^2}} g(y)\, dy.$$

The explicit solution formulas we have obtained for the initial value problem have several important consequences. For example, they show that the solution to the initial value problem depends continuously on the data (see Prob. 4). The solution formulas also reveal the domain of dependence for each solution, and this in turn discloses a significant difference between wave propagation in two and three spatial dimensions. The formulas in Ths. 4-2 and 4-4 both show that the solution (wave disturbance) at (x, t) depends only on the values of the initial data in the set $\{y : |y - x| \leq ct\}$. This conclusion is expected on physical grounds because waves propagate with speed c. It is also consistent with and sharpens the uniqueness discussion in Sec. 10-3, which shows that the solution at (x, t) is uniquely determined (Fig. 10-1) by its initial values at points on the plane $t = 0$ which are inside the retrograde cone.

We turn now to the significant difference between wave propagation in two and three spatial dimensions. Figure 10-2 portrays an initial value problem in three spatial dimensions where the initial data are nonzero only in a bounded, closed domain D. Consider the disturbance at the location x outside of D. at various times t. Let d_{min} and d_{max} be the minimum and maximum distances from x to points in D. If $ct < d_{\text{min}}$, the spherical surface $S_{ct}(x)$ does not intersect D and Th. 4-2 shows that $u(x, t) = 0$. For times t with $d_{\text{min}} \leq ct \leq d_{\text{max}}$, the initial values may produce a nonzero disturbance at x, and during this time interval the disturbance at x at time t is due solely to the initial data *on* the sphere with radius ct; it does not depend on any other portion of the initial data. For times t with $ct > d_{\text{max}}$, the disturbance at x returns to zero because the spherical surface $S_{ct}(x)$ does not meet D. If the initial disturbance is localized near a point so that $d_{\text{min}} \approx d_{\text{max}} \approx d$, the disturbance at x will be zero except for a wavefront that will arrive at time $t \approx d/c$. In other words, in three dimensions sharp signals are

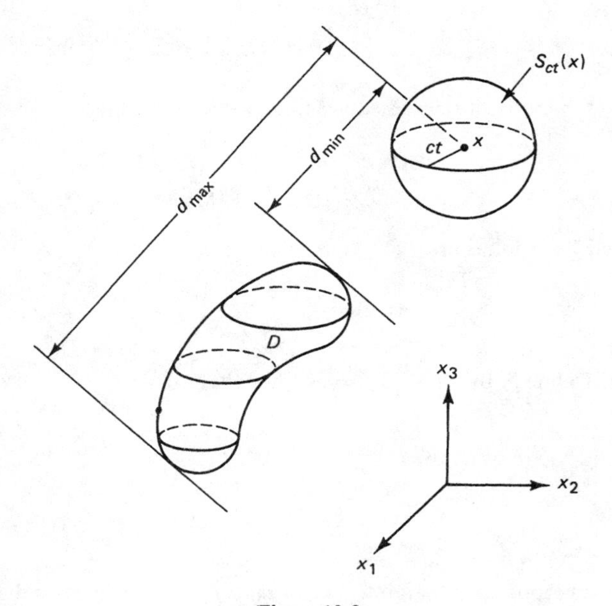

Figure 10-2

propagated from a point source. This phenomenon is called *Huygens' principle.* Stars in the night sky appear as points, rather than strips, because of this principle.

Now, consider the corresponding situation in two spatial dimensions. Imagine the two-dimensional analogue of Fig. 10-2 where D is a plane domain and the sphere about x is replaced by a solid disk with radius ct. As for three spatial dimensions, there is no disturbance at x for times t with $ct < d_{\min}$ by Th. 4-4. Now the situation changes. For times t with $d_{\min} \le ct \le d_{\max}$, the disturbance at x may be nonzero, and during this time interval the disturbance at x at time t is due to the entire portion of the initial data in D which is *on and inside* the circle $S_{ct}(x)$. Finally, for times t with $ct > d_{\max}$, the disturbance does not return to zero, as in three dimensions; rather the entire initial data in D continue to effect the solution because the integrals in Th. 4-4 are over the solid disk $\bar{K}_{ct}(x)$. Thus, even if D is localized near a point, the effects of the data in D are felt at x for *all* times $t > d_{\max}/c$. In other words, sharp signals are not propagated in two spatial dimensions and Huygens' principle does not hold.

This diffusion of waves in two spatial dimensions can be observed by dropping a stone into a large, undisturbed pond with a cork placed not too far from the point where the stone hits the water. The cork remains at rest until the initial disturbance engulfs it. From then on, the cork continues to bob up and down as the aftereffects of the initial wavefront persist.

Although we shall not prove it, Huygens' principle holds in all spaces of odd dimension $n \ge 3$ and fails in even dimensions. The one-dimensional case is special (see Prob. 8).

PROBLEMS

1. Provide plausibility arguments similar to those leading to (4-5) and (4-6) to infer that

$$u_{tt} = c^2\,\Delta u, \qquad x \in R_3, \qquad t > 0,$$
$$u(x, 0) = f(x), \qquad u_t(x, 0) = 0, \qquad x \in R_3,$$

should have solution

$$u(x, t) = \frac{\partial}{\partial t}\left[\frac{t}{4\pi}\int_{|\omega|=1} f(x + ct\omega)\, d\sigma_\omega\right].$$

Hint. Define δ_ε by (4-2). If $f = \delta_\varepsilon$ show that

$$u(x, t) = \frac{(r + ct)\delta_\varepsilon(r + ct) + (r - ct)\delta_\varepsilon(r - ct)}{2r}$$
$$= \frac{1}{cr}\frac{\partial}{\partial t}\,[\varepsilon\delta_\varepsilon(r - ct) - \varepsilon\delta_\varepsilon(r + ct)].$$

Now superpose to get an integral for $u(x, t)$ for a general f and small $\varepsilon > 0$. Introduce spherical coordinates and let $\varepsilon \to 0$.

2. Solve the initial value problem for a very large, damped membrane. That is, solve the two-dimensional damped wave equation

$$u_{tt} + ku_t = c^2 \Delta u, \qquad x \in R_2, \qquad t > 0,$$
$$u(x, 0) = f(x), \qquad u_t(x, 0) = g(x), \qquad x \in R_2,$$

as follows.

(a) Let $u(x, t) = e^{\alpha t}w(x, t)$ and show that w satisfies

$$w_{tt} = c^2 \Delta w + \frac{k^2}{4} w,$$

$$w(x, 0) = f(x), \qquad w_t(x, 0) = \frac{k}{2} f(x) + g(x)$$

when $\alpha = -k/2$.

(b) Now define v by

$$v(x_1, x_2, x_3, t) = w(x_1, x_2, t) \exp\left(\frac{k}{2c} x_3\right).$$

If w satisfies the problem in part (a), show that v satisfies

$$v_{tt} = c^2 \Delta v, \qquad x \in R_3, \qquad t > 0,$$

$$v(x, 0) = f(x_1, x_2) \exp\left(\frac{k}{2c} x_3\right),$$

$$v_t(x, 0) = \left[\frac{k}{2} + f(x_1, x_2) + g(x_1, x_2)\right] \exp\left(\frac{k}{2c} x_3\right).$$

(c) Find v and work back through the substitutions to get u.

3. Prove Th. 4-3 by checking that (4-10) does solve (4-9).

4. Show that the solution to

$$u_{tt} = c^2 \Delta u + F(x, t), \qquad x \in R_3, \qquad t > 0,$$
$$u(x, 0) = f(x), \qquad u_t(x, 0) = g(x), \qquad x \in R_3,$$

depends continuously on the data over any finite time interval. *Hint.* Find an explicit formula for the solution and make the reasonable continuity and differentiability assumptions needed to read off the desired result.

5. Consider the initial value problem in Prob. 4. Let $u(x, t) = v(x, t) + f(x)$ and show that v solves

$$v_{tt} = c^2 \Delta v + F(x, t) - c^2 \Delta f, \qquad x \in R_3, \qquad t > 0,$$
$$v(x, 0) = 0, \qquad v_t(x, 0) = g(x), \qquad x \in R_3.$$

Find the solution v and use it to get the expected formula for u.

6. Show that the solution (4-10) of (4-9) can be expressed as

$$u(x, t) = \int_{K_{ct}(x)} \frac{1}{4\pi r} F\left(y, t - \frac{r}{c}\right) dy,$$

where $r = |y - x|$. This expression for u is referred to as a *retarded potential.*

7. Use the method of descent to recover the d'Alembert solution to the initial value problem in one spatial dimension from the two-dimensional result in Th. 4-4.

8. Consider the wave equation in one spatial variable,

$$u_{tt} = c^2 u_{xx}, \qquad -\infty < x < \infty, \qquad t > 0.$$

(a) Show that Huygens' principle holds for the initial data

$$u(x, 0) = f(x), \qquad u_t(x, 0) = 0.$$

(b) Show that Huygens' principle does not hold for the data

$$u(x, 0) = 0, \qquad u_t(x, 0) = g(x).$$

9. The Kirchhoff solution in Th. 4-2 can be expressed in terms of mean values or averages of the initial data. Show that

$$u(x, t) = \frac{\partial}{\partial t}[t(M_{ct}f)(x)] + t(M_{ct}g)(x),$$

where $(M_r h)(x)$ is the mean value of the function h over the sphere with center x and radius r,

$$(M_r h)(x) = \frac{1}{4\pi r^2} \int_{S_r(x)} h(y)\, d\sigma_y.$$

The next two problems show how to use averaging and the simple solution to the wave equation in the spherically symmetric case to deduce Kirchhoff's formula.

10. Let $u(x, t)$ be a solution of the wave equation in three spatial dimensions. For each fixed x, define $v(r, t) = (M_r u)(x, t)$. Show that $v(r, t)$ satisfies the radially symmetric wave equation $v_{tt} = c^2(v_{rr} + (2/r)v_r)$. *Hint.* Express v explicitly in terms of spherical coordinates and then compute.

11. Conclude from Prob. 11 that

$$v(r, t) = \frac{F(r + ct)}{r} + \frac{G(r - ct)}{r}.$$

Assume that u satisfies the initial value problem in Th. 4-2.

(a) Argue that

$$u(x, t) = \lim_{r \to 0} v(r, t) = \lim_{r \to 0} \frac{F(r + ct) + G(r - ct)}{r}$$

and conclude that $F(ct) + G(-ct) = 0$ for this limit to exist.

(b) Use the result in part (a) to deduce that

$$u(x, t) = \lim_{r \to 0} \frac{F(r + ct) - F(ct - r)}{r} = 2F'(ct).$$

(c) Use the initial data to find

$$F'(r) = \frac{1}{2}\left[(rM_r f)_r + \frac{r}{c} M_r g\right].$$

(d) Finally, use parts (b) and (c) to obtain Kirchhoff's solution.

12. Although Huygens' principle fails in two dimensions, show that for x fixed and initial data which is nonzero only in a bounded set the disturbance $u(x, t)$ at x at time t diminishes over time. More precisely,

$$\lim_{t \to \infty} u(x, t) = 0.$$

10-5. The Delta Function and Fundamental Solutions

We obtained the solution to the general initial value problem for the wave equation by first restricting our attention to the special problem

$$u_{tt} = c^2\,\Delta u, \qquad x \in R_3, \qquad t > 0,$$
$$u(x, 0) = 0, \qquad u_t(x, 0) = \delta_\varepsilon(x), \qquad x \in R_3,$$

where $\delta_\varepsilon(x)$ described a sudden change in state localized near $x = 0$. Once the solution to this special problem was known, we were able to superpose the effects of such localized disturbances to solve the general initial value problem. In this section we look somewhat more carefully into this procedure.

Consider first just what we should mean by a localized disturbance, or point source or sink. This is a mathematical idealization and in some respects is a matter of scale. When an acoustic disturbance is initiated by a snap of fingers, the place from which the sound emanates is only a centimeter or two in extent. The resultant sound waves propagate throughout a room, say, of vastly greater size. With respect to the dimensions of the room, the initial disturbance has essentially zero extent and it is reasonable to model the disturbance as emanating from a single point. A similar situation occurs when a violin string is plucked. The resultant vibrations are due to a virtual point source. For another example, consider the flow of water in the neighborhood of a pump in a reservoir. The diameter of the pipe connected to the pump may be only a couple of inches, a negligible size in comparison to the overall dimensions of the reservoir. When water is drawn from the reservoir, we have a point sink. If water is pumped in, the pump represents a point source. Disturbances can also be localized in time, as when a soccer ball is kicked or a baseball is hit. The actual impact lasts a mere fraction of a second and is inconsequential in comparison with the time of flight of the ball.

The common thread running through these situations is that there is a highly localized disturbance which has sufficient intensity to have global effects. It is natural to model the localized disturbance as occurring at a representative point in the region where the physical action takes place. For the moment, we choose coordinates so the disturbance occurs at the origin, and try to describe the disturbance in terms of a function, call it $\delta(x)$, which gives its intensity per unit volume (or length or area in one or two spatial dimensions). Since the disturbance is localized at $x = 0$, we require that $\delta(x) = 0$ for $x \neq 0$. On the other hand, the total strength of the local disturbance $\int_{R_3} \delta(x)\,dx$ must be nonzero because something happens. For convenience, we normalize our idealized point source to have strength one. Thus, we seek a function $\delta(x)$ such that

$$\delta(x) = \begin{cases} 0, & x \neq 0 \\ +\infty, & x = 0 \end{cases} \quad \text{with} \quad \int_{R_n} \delta(x)\,dx = 1, \tag{5-1}$$

where $n = 1$, 2, or 3, depending on the physical situation. The choice $\delta(0) = +\infty$ is motivated by the feeling that the idealized intensity must be very large at $x = 0$ if it is to have a nonzero overall effect and still be zero when $x \neq 0$. (Consider, for example, the intensity of the forces acting during the instant a ball is hit.)

As you probably know, $\delta(x)$ is called the *Dirac delta function* and there is an immediate and obvious problem with the definition (5-1), within the traditional framework of calculus. Indeed, the definition (5-1) and the usual evaluation (definition) of improper integrals are at odds,

$$1 = \int_{R_n} \delta(x)\, dx = \lim_{\varepsilon \to 0} \int_{|x| > \varepsilon} \delta(x)\, dx = \lim_{\varepsilon \to 0} 0 = 0.$$

This contradiction makes it clear that $\delta(x)$ cannot be a function in the ordinary sense. Nevertheless, we have encountered functions which nearly satisfy (5-1), namely the functions

$$\delta_\varepsilon(x) = (4\pi\varepsilon)^{-n/2} \exp\left(-\frac{|x|^2}{4\varepsilon}\right) \tag{5-2}$$

gotten from the fundamental solution to the heat equation by setting $\varepsilon = at$. Recall that

$$\lim_{\varepsilon \to 0} \delta_\varepsilon(x) = \begin{Bmatrix} 0, & x \neq 0 \\ +\infty, & x = 0 \end{Bmatrix} \quad \text{and} \quad \int_{R_n} \delta_\varepsilon(x)\, dx = 1 \tag{5-3}$$

for each $\varepsilon > 0$, and that

$$\lim_{\varepsilon \to 0} \int_{R_n} \delta_\varepsilon(x) f(x)\, dx = f(0)$$

for any bounded function $f(x)$ which is continuous at $x = 0$. [See (4-11) of Sec. 5-4 and Lemma 1-1 of Sec. 9-1.] Since it is natural to think of $\delta(x)$ as the limit of the bona fide functions $\delta_\varepsilon(x)$, we define the Dirac delta function by

$$\int_{R_n} \delta(x) f(x)\, dx \equiv \lim_{\varepsilon \to 0} \int_{R_n} \delta_\varepsilon(x) f(x)\, dx = f(0). \tag{5-4}$$

We regard the integral on the left as a symbolic reminder of the limiting process on the right of (5-4). We sometimes write

$$\delta(x) = \lim_{\varepsilon \to 0} \delta_\varepsilon(x),$$

which is again just shorthand for (5-4).

If the source is at y, we define the Dirac delta function with source at y as

$$\delta(y - x) = \lim_{\varepsilon \to 0} \delta_\varepsilon(y - x),$$

which means

$$\int_{R_n} \delta(y - x) f(x)\, dx \equiv \lim_{\varepsilon \to 0} \int_{R_n} \delta_\varepsilon(y - x) f(x)\, dx = f(y) \tag{5-4$'$}$$

for $f(x)$ bounded and continuous at y.

A close look at (5-4) and (5-4)′ shows that we have not defined δ as a function in the usual sense. That is, given $x \in R_n$ we have not given a numerical value to $\delta(x)$. Rather, we have described the effect of δ on the class of functions f which are bounded and continuous at the origin [or at y in (5-4)′]. The Dirac delta function associates to each function f a number gotten through the limiting process in (5-4) or (5-4)′. Such mathematical operations are called *functionals.* Thus, the Dirac delta function is really a functional; however, we retain the familiar notations $\delta(x)$ and $\int_{R_n} \delta(x)f(x)\,dx$ associated with functions because many properties of the delta functional are most easily remembered in this notation and because we can define useful operations on the δ functional which are consistent with the operations of calculus. For these reasons it is customary to speak of δ as a function rather than a functional. In the final analysis, however, properties and operations on δ are deduced and defined by determining the corresponding effect on the functions $\delta_\varepsilon(x)$ and taking the limit as $\varepsilon \to 0$ in (5-4)′.

To illustrate these ideas, we first show that the delta function is even,

$$\delta(y - x) = \delta(x - y). \tag{5-5}$$

This follows from the fact that all the functions δ_ε are even, so that

$$\begin{aligned}\int_{R_n} \delta(x - y)f(x)\,dx &= \lim_{\varepsilon\to 0} \int_{R_n} \delta_\varepsilon(x - y)f(x)\,dx \\ &= \lim_{\varepsilon\to 0} \int_{R_n} \delta_\varepsilon(y - x)f(x)\,dx = \int_{R_n} \delta(y - x)f(x)\,dx.\end{aligned}$$

Thus,

$$\int_{R_n} \delta(x - y)f(x)\,dx = \int_{R_n} \delta(y - x)f(x)\,dx = f(y)$$

for all bounded functions f which are continuous at y. The equality (5-5) is shorthand for this result.

Similarly, we can define the Fourier transform of $\delta(x)$ by

$$\hat{\delta}(\omega) = \int_{R_n} e^{i\omega\cdot x}\delta(x)\,dx = \lim_{\varepsilon\to 0} \int_{R_n} e^{i\omega\cdot x}\delta_\varepsilon(x)\,dx = e^{i\omega\cdot 0} = 1.$$

We then expect that

$$\left(\frac{1}{2\pi}\right)^n \int_{R_n} e^{-ix\cdot\omega}\hat{\delta}(\omega)\,d\omega = \left(\frac{1}{2\pi}\right)^n \int_{R_n} e^{-ix\cdot\omega}\,d\omega = \delta(x).$$

The function $e^{-ix\cdot\omega}$ is not integrable over R_n; however, we interpret this inversion formula in the limiting sense that

$$\begin{aligned}\left(\frac{1}{2\pi}\right)^n \int_{R_n} e^{-ix\cdot\omega}\hat{\delta}(\omega)\,d\omega &= \lim_{\varepsilon\to 0} \left(\frac{1}{2\pi}\right)^n \int_{R_n} e^{-ix\cdot\omega}\hat{\delta}_\varepsilon(\omega)\,d\omega \\ &= \lim_{\varepsilon\to 0} \delta_\varepsilon(x) = \delta(x).\end{aligned}$$

Here is another useful property of the delta function in one dimension,

$$\int_{-\infty}^{x} \delta(s)\,ds = H(x), \tag{5-6}$$

where H is the *Heaviside step function*

$$H(x) = \begin{cases} 0, & x < 0, \\ 1, & x > 0. \end{cases}$$

In applications, only the jump at zero is important. Thus, there is no need to assign a value to H when $x = 0$. We confirm (5-6) in the standard way:

$$\text{(5-7)} \qquad \int_{-\infty}^{x} \delta(s)\, ds = \lim_{\varepsilon \to 0} \int_{-\infty}^{x} \delta_\varepsilon(s)\, ds = \lim_{\varepsilon \to 0} \int_{-\infty}^{\infty} \delta_\varepsilon(x) f(s, x)\, ds$$

where

$$f(s, x) = \begin{cases} 1, & -\infty < s < x, \\ 0, & x < s < \infty. \end{cases}$$

By (4-11) in Sec. 5-4, the right member of (5-7) equals $f(0, x) = 0$ for $x < 0$ and equals $f(0, x) = 1$ for $x > 0$, so (5-6) holds. Formula (5-6) suggests that $H'(x) = \delta(x)$. This result is indeed true when H and δ are regarded as functionals acting on appropriate "test" functions. The general procedure which puts all of this discussion on a firm foundation is called the theory of distributions, and in this setting H, δ, and other important functionals are called *distributions*. We do not have time to develop this theory in detail and so must be content with formal calculations. However, the solution formulas we shall obtain below using delta function arguments and other plausible reasoning can be checked directly once they are obtained. In Prob. 3 we ask you to confirm that

$$\text{(5-8)} \qquad \int_{a}^{b} \delta(s - y)\, ds = H(b - y) - H(a - y).$$

Let's return to the wave equation in one spatial dimension and consider the wave $k = k(x, y, t)$ generated by an initial disturbance $\delta(x - y)$ localized at y. Thus,

$$\text{(5-9)} \qquad \begin{cases} k_{tt} = c^2 k_{xx}, & -\infty < x < \infty, \quad t > 0, \\ k(x, 0) = 0, & k_t(x, 0) = \delta(x - y). \end{cases}$$

An application of d'Alembert's solution in Sec. 4-1 yields

$$k(x, y, t) = \frac{1}{2c} \int_{x-ct}^{x+ct} \delta(s - y)\, ds,$$

so by (5-8),

$$\text{(5-10)} \qquad k(x, y, t) = \frac{1}{2c} [H(x - y + ct) - H(x - y - ct)].$$

The function $k(x, y, t)$ given by (5-10) is called the *fundamental solution to the one-dimensional wave equation.* With the fundamental solution in hand, we turn to the more general problem

$$
\text{(5-11)} \qquad \begin{aligned} &u_{tt} = c^2 u_{xx}, \qquad -\infty < x < \infty, \qquad t > 0, \\ &u(x, 0) = 0, \qquad u_t(x, 0) = g(x), \end{aligned}
$$

of wave motion generated by a continuous distribution $g(x)$ of initial disturbances. We can think of g as the superposition of a host of impulses because

$$
g(x) = \int_{-\infty}^{\infty} \delta(x - y) g(y)\, dy.
$$

This suggests that the solution u to (5-11) is the corresponding superposition of the solutions $k(x, y, t)g(y)$ to (5-9) with $\delta(x - y)$ replaced by $\delta(x - y)g(y)$. That is,

$$
\text{(5-12)} \qquad u(x, t) = \int_{-\infty}^{\infty} k(x, y, t) g(y)\, dy
$$

with k given by (5-10). Since $k(x, y, t)$ as a function of y is $1/2c$ on the interval $x - ct < y < x + ct$ and is 0 otherwise, (5-12) reduces to

$$
u(x, t) = \frac{1}{2c} \int_{x-ct}^{x+ct} g(y)\, dy,
$$

which is just d'Alembert's solution found in Sec. 4-1. Thus, (5-12) does indeed solve (5-11).

Once we have a fundamental solution for the wave equation, we can reason much as we did for the heat equation in Sec. 5-4 to obtain a solution formula for

$$
\text{(5-13)} \qquad \begin{cases} u_{tt} = c^2 u_{xx} + F(x, t), & -\infty < x < \infty, \qquad t > 0, \\ u(x, 0) = f(x), & u_t(x, 0) = g(x). \end{cases}
$$

Fix (x, t) and consider the identity

$$
\begin{aligned} F(y, \tau) k(x, y, t - \tau) &= (u_{\tau\tau} - c^2 u_{yy}) k - u(k_{\tau\tau} - c^2 k_{yy}) \\ &= \frac{\partial}{\partial \tau}(u_\tau k - u k_\tau) - c^2 \frac{\partial}{\partial y}(u_y k - u k_y), \end{aligned}
$$

where we have used the fact that k satisfies the wave equation (except on the lines $x - y \pm c\tau = 0$). Integrate with respect to τ from 0 to t and with respect to y from $-R$ to R. Then let $R \to \infty$ and note that for R sufficiently large $k(x, \pm R, t - \tau) = 0$ for $0 \le \tau \le t$ to obtain

$$
\text{(5-14)} \qquad \int_0^t \int_{-\infty}^{\infty} k(x, y, t - \tau) F(y, \tau)\, dy\, d\tau = \int_{-\infty}^{\infty} (u_\tau k - u k_\tau)\, dy \Bigg|_{\tau=0}^{\tau=t}.
$$

Consider each integral separately. From (5-10),

$$
\int_0^t \int_{-\infty}^{\infty} k(x, y, t - \tau) F(y, \tau)\, dy\, d\tau = \frac{1}{2c} \int_0^t \int_{x-c(t-\tau)}^{x+c(t-\tau)} F(y, \tau)\, dy\, d\tau.
$$

Next,

$$\int_{-\infty}^{\infty} u_\tau k \, dy\bigg|_{\tau=t} = \int_{-\infty}^{\infty} u_\tau(y, t)k(x, y, 0)\, dy = 0$$

from (5-10), and

$$\int_{-\infty}^{\infty} uk_\tau \, dy\bigg|_{\tau=t} = \int_{-\infty}^{\infty} u(y, t)\frac{-c}{2c}[\delta(x-y) + \delta(x-y)]\, dy = -u(x, t)$$

since $H' = \delta$. Finally, at the lower limit

$$\int_{-\infty}^{\infty} u_\tau k \, dy\bigg|_{\tau=0} = \int_{-\infty}^{\infty} g(y)k(x, y, t)\, dy = \frac{1}{2c}\int_{x-ct}^{x+ct} g(y)\, dy,$$

as we have already seen, and

$$\begin{aligned}\int_{-\infty}^{\infty} uk_\tau \, dy\bigg|_{\tau=0} &= -\frac{1}{2}\int_{-\infty}^{\infty} f(y)[\delta(x-y+ct) + \delta(x-y+ct)]\, dy \\ &= -\frac{1}{2}[f(x+ct) + f(x-ct)].\end{aligned}$$

Use these evaluations in (5-14) to obtain

$$u(x, t) = \frac{1}{2}[f(x+ct) + f(x-ct)] + \frac{1}{2c}\int_{x-ct}^{x+ct} g(y)\, dy + \frac{1}{2c}\int_0^t \int_{x-c(t-\tau)}^{x+c(t-\tau)} F(y, \tau)\, dy\, d\tau, \tag{5-15}$$

which is the formula we developed in Probs. 6–8 of Sec. 4-1.

The same line of reasoning using fundamental solutions leads to solution formulas in higher dimensions. For example, consider three spatial dimensions. We determine the fundamental solution $k = k(x, y, t)$ as the solution to

$$\begin{cases} k_{tt} = c^2 \Delta k, & x \in R_3, \quad t > 0, \\ k(x,0) = 0, & k_t(x, 0) = \delta(x - y), \end{cases} \tag{5-16}$$

where y is a parameter. Refer to (4-1) and its solution to see that (5-16) with δ replaced by δ_ε has solution

$$k_\varepsilon = \frac{1}{4\pi cr}[\delta_\varepsilon(ct - r) - \delta_\varepsilon(ct + r)],$$

where $r = |x - y|$. Let $\varepsilon \to 0$ to find that

$$k(x, y, t) = \frac{1}{4\pi cr}[\delta(ct - r) + \delta(ct + r)].$$

Since $ct + r > 0$, $\delta(ct + r) = 0$ and the *fundamental solution for the wave equation in three spatial variables* is

$$k(x, y, t) = \frac{1}{4\pi cr}\delta(ct - r), \qquad r = |x - y|. \tag{5-17}$$

Notice that the fundamental solution represents an outgoing wave initiated at x and concentrated on the sphere $r = ct$. The physically uninteresting incoming wave is automatically eliminated because $\delta(r + ct) = 0$.

Proceeding exactly as in one spatial variable, the more general problem

$$\begin{cases} u_{tt} = c^2 \Delta u, & x \in R_3, \quad t > 0, \\ u(x, 0) = 0, & u_t(x, 0) = g(x), \end{cases} \tag{5-18}$$

has the solution

$$u(x, t) = \int_{R_3} k(x, y, t) g(y)\, dy \tag{5-19}$$

obtained by superposition. This solution also can be expressed as

$$u(x, t) = \int_0^\infty \frac{\delta(ct - r)}{4\pi c r} \int_{S_r(x)} g(y)\, d\sigma_y\, dr = \frac{1}{4\pi c^2 t} \int_{S_{ct}(x)} g(y)\, d\sigma_y,$$

which is just our previous solution formula (4-6). This confirms that (5-19) does solve the initial value problem (5-18).

We can continue in the same manner to recover the explicit solution to

$$\begin{cases} u_{tt} = c^2 \Delta u + F(x, t), & x \in R_3, \quad t > 0, \\ u(x, 0) = f(x), & u_t(x, 0) = g(x), \end{cases} \tag{5-20}$$

found in Sec. 10-4. The details are left for Prob. 7.

PROBLEMS

1. Explain intuitively why $\int_K \delta(x)\, dx = 1$ for any ball K with the origin inside K. Then establish this result with a limiting argument.

2. In this problem let δ_n be the delta function in n variables and δ_1 be the delta function in 1 variable. Show that

$$\delta_n(x) = \delta_1(x_1)\delta_1(x_2) \cdots \delta_1(x_n),$$

where $x = (x_1, \ldots, x_n)$. *Hint.* Factor $(\delta_n)_\varepsilon$.

3. Show that (5-8) holds.

4. Use the standard limiting procedure to show that $x\delta(x) = 0$. Do the same to determine $\phi(x)\delta(x)$, where ϕ is a given continuous function.

5. Here is another way to look at fundamental solutions, which is analogous to our earlier study of Green's functions. We define the fundamental solution to be the functional $k(x, y, t)$ such that the integral in (5-19) solves the initial value problem (5-18) for all bounded continuous data $g(x)$.
(a) Explain why the functional k is unique (if it exists).
(b) If k is defined as above, use (5-18) and (5-19) with $g(x) = \delta_\varepsilon(x)$ to show that k solves (5-16). So the two approaches to k are equivalent.

6. Use Prob. 5 to show that the wave equation in two spatial variables has fundamental solution

$$k(x, y, t) = \frac{H(c^2t^2 - r^2)}{2\pi c\sqrt{c^2t^2 - r^2}}, \qquad \text{where} \quad r = |x - y|.$$

7. Use the fundamental solution (5-17) and the following steps to solve (5-20).

(a) For fixed (x, t) verify that

$$\begin{aligned} k(x, y, t-\tau)F(y, \tau) &= k(u_{\tau\tau} - c^2\Delta_y u) - u(y, \tau)(k_{\tau\tau} - c^2\Delta_y k) \\ &= (ku_\tau - uk_\tau)_\tau - c^2 \operatorname{div}(k\nabla_y u - u\nabla_y k). \end{aligned}$$

(b) Integrate over $0 \le \tau \le t$ and over $K_R(x)$ and then let $R \to \infty$ to find

$$\int_0^t \int_{R_3} k(x, y, t-\tau)F(y, \tau)\, dy\, d\tau = \int_{R_3} (ku_\tau - uk_\tau)\, dy \Big|_{\tau=0}^{\tau=t}.$$

(c) Now evaluate each integral in part (b) much as we did in the text for one spatial variable. *Hint.* You may find it helpful to express integrals over $S_{c(t-\tau)}(x)$ in terms of integrals over $|\omega| = 1$. Also, note that formally $k_\tau = -k_t$.

The following problems lead you to the fundamental solution to the damped wave equation.

8. Let $k > 0$ be a constant. Show that u solves

$$\begin{aligned} &u_{tt} + ku_t = c^2\,\Delta u, \qquad x \in R_3, \qquad t > 0, \\ &u(x, 0) = 0, \qquad u_t(x, 0) = g(x), \end{aligned}$$

if and only if $v = e^{kt/2}u$ solves

$$\begin{aligned} &v_{tt} = c^2\,\Delta v + \frac{k^2}{4} v, \qquad x \in R_3, \qquad t > 0, \\ &v(x, 0) = 0, \qquad v_t(x, 0) = g(x). \end{aligned}$$

9. If $r = |x|$ and $v = v(r, t)$ is a radially symmetric solution to $v_{tt} = c^2\,\Delta v + k^2 v/4$, show that $w(r, t) = rv(r, t)$ satisfies

$$w_{tt} = c^2 w_{rr} + \frac{k^2}{4} w.$$

Then show that this equation has solutions of the form $w = f(\lambda)$, where $\lambda = c^2t^2 - r^2$ and $\lambda \ge 0$ if f satisfies

$$\lambda f'' + f' - \alpha^2 f = 0, \qquad \alpha = \frac{k}{4c}.$$

Finally, show that the substitution $\mu = \sqrt{\lambda}$ leads to Bessel's equation of order zero, and deduce that $f(\lambda) = AI_0(2\alpha\sqrt{\lambda}) + BK_0(2\alpha\sqrt{\lambda})$. Here I_0 and K_0 are modified Bessel functions of order zero. So $I_0(z) = J_0(iz)$ and $K_0(z)$ grows like $\log(2/z)$ as $z \to 0$.

10. Consider the function

$$V(x, t) = \int_{R_3} L(x, y, t)g(y)\, dy,$$

with

$$L(x, y, t) = \frac{AI_0(2\alpha\sqrt{\lambda}) + BK_0(2\alpha\sqrt{\lambda})}{r} H(\lambda),$$

where H is the Heaviside function, $\lambda = c^2t^2 - r^2$, and $r = |x - y|$.

(a) Show that

$$V(x, t) = \int_0^{ct} \int_{|\omega|=1} [AI_0(2\alpha\sqrt{c^2t^2 - r^2}) + BK_0(2\alpha\sqrt{c^2t^2 - r^2})]\rho g(x + r\omega)\, d\sigma_\omega\, dr.$$

(b) Show that for V_t to exist requires $B = 0$, and with this choice V satisfies

$$V_{tt} = c^2\, \Delta V + \frac{k^2 V}{4}.$$

(c) Assume that $B = 0$ from now on. Show that

$$V_t = \frac{\partial}{\partial t} \int_0^{ct} \int_{|\omega|=1} \frac{AI_0(2\alpha\sqrt{c^2t^2 - r^2})}{r} g(x + r\omega) r^2\, d\sigma_\omega\, dr$$

satisfies $V_t(x, 0) = 0$ and also $V_{tt}(x, 0) = g(x)$ if $A = 1/4\pi c^2$, which is assumed now. Conclude that $v = V_t$ solves the problem for v in Prob. 8.

(d) Use part (c) to deduce that the fundamental solution for $v_{tt} = c^2\, \Delta v + k^2 v/4$ is

$$k(x, y, t) = \frac{1}{4\pi c} \frac{\delta(ct - r)}{r} + \frac{\alpha t}{2\pi r} \frac{I_1(2\alpha\sqrt{c^2t^2 - r^2})}{\sqrt{c^2t^2 - r^2}} H(c^2t^2 - r^2).$$

Hint. Use Prob. 5. After differentiating the integral in part (c) introduce a delta function to express the result as in integral over R_3.

11. Use the fundamental solution of part (d) of Prob. 10 to solve

$$v_{tt} = c^2\, \Delta v + \frac{k^2}{4} v + F(x, t), \qquad x \in R_3, \qquad t > 0,$$

$$v(x, 0) = f(x), \qquad v_t(x, 0) = g(x).$$

12. Use the solution in Prob. 11 to solve the corresponding initial value problem for $u_{tt} + ku_t = c^2\, \Delta u + F(x, t)$.

13. Use *Duhamel's principle* to solve the inhomogeneous initial value problem (5-20) when $f(x) = g(x) = 0$.

10-6. Hyperbolic Potential Theory in One Spatial Variable

The fundamental solution to the wave equation can be used to define hyperbolic potentials which are used to construct solutions to various boundary value problems. This means of representing solutions is the natural extension of techniques used earlier for the Laplace equation and the heat equation. However, the fundamental solution for the wave equation contains Heaviside step functions in one dimension and delta functions in three dimensions, so care must be taken in interpreting the potentials.

To show more clearly what is involved, we consider the motion of a very long string which is at rest initially and subject to prescribed displacements at $x = 0$,

$$\text{(6-1)} \qquad \begin{cases} u_{tt} = c^2 u_{xx}, & x > 0, \quad t > 0, \\ u(x, 0) = u_t(x, 0) = 0, & x \geq 0, \\ u(0, t) = \phi(t), & t \geq 0. \end{cases}$$

At time $t = 0$ we assume that the compatibility conditions $\phi(0) = \phi'(0) = 0$ and extend the function $\phi(t)$ to be zero for $t < 0$. Since $u(0, t) = \phi(t)$ is a Dirichlet type boundary condition, we try to solve (6-1) with a double layer potential

$$u(x, t) = \int_0^t \left(-\frac{\partial}{\partial y}\right) k(x, y, t - \tau)\bigg|_{y=0} \sigma(\tau)\, d\tau, \tag{6-2}$$

where σ is to be determined. Since we have set $\phi(t) = 0$ for $t \leq 0$, we can let $t \leq 0$ in (6-1) and by uniqueness we have $u(x, t) = 0$ for $t \leq 0$. In view of (6-2) we define $\sigma(\tau) = 0$ for $\tau \leq 0$ and obtain

$$u(x, t) = \int_{-\infty}^t \left(-\frac{\partial}{\partial y}\right) k(x, y, t - \tau)\bigg|_{y=0} \sigma(\tau)\, d\tau.$$

Now

$$k(x, y, t - \tau) = \frac{1}{2c}[H(x - y + ct - c\tau) - H(x - y - ct + c\tau)]$$

and a formal differentiation gives

$$-\frac{\partial}{\partial y} k(x, y, t - \tau)\bigg|_{y=0} = \frac{1}{2c}[\delta(x + ct - c\tau) - \delta(x - ct + c\tau)].$$

Thus, we expect that

$$\begin{aligned} u(x, t) &= \int_{-\infty}^t \frac{1}{2c}[\delta(x + ct - c\tau) - \delta(x - ct + c\tau)]\sigma(\tau)\, d\tau \\ &= \frac{1}{2c^2}\int_x^\infty \delta(z)\sigma\left(\frac{x}{c} + t - \frac{z}{c}\right) dz - \frac{1}{2c^2}\int_{-\infty}^x \delta(w)\sigma\left(t - \frac{x}{c} + \frac{w}{c}\right) dw. \end{aligned}$$

For $x > 0$ the first integral on the right vanishes because $\delta(z)$ vanishes there, the second has value $\sigma(t - x/c)$, and we have found that

$$u(x, t) = -\frac{1}{2c^2}\sigma\left(t - \frac{x}{c}\right). \tag{6-3}$$

Finally, from (6-1) and (6-3),

$$\phi(t) = u(0, t) = -\frac{1}{2c^2}\sigma(t)$$

and (6-3) becomes

$$u(x, t) = \phi\left(t - \frac{x}{c}\right). \tag{6-4}$$

It is a simple matter to check that (6-4) solves (6-1) assuming that ϕ is twice differentiable.

We consider next a shorter string driven by prescribed displacements at each end,

$$\begin{cases} u_{tt} = c^2 u_{xx}, & 0 < x < L, \quad t > 0, \\ u(x, 0) = u_t(x, 0) = 0, & 0 \leq x \leq L, \\ u(0, t) = \phi(t), \quad u(L, t) = \psi(t), & t \geq 0. \end{cases} \tag{6-5}$$

Again we assume a solution in the form of a double layer with a contribution from each boundary point of the string,

(6-6)
$$u(x, t) = \int_{-\infty}^{t} \left(-\frac{\partial}{\partial y}\right) k(x, y, t-\tau)\bigg|_{y=0} \sigma(\tau)\, d\tau + \int_{-\infty}^{t} \left[\frac{\partial}{\partial y} k(x, y, t-\tau)\right]\bigg|_{y=L} \rho(\tau)\, d\tau.$$

As in the previous example, we assume the compatibility conditions $\phi(0) = \phi'(0) = \psi(L) = \psi'(L) = 0$ and extend ϕ and ψ to be zero for $t < 0$. The string is at rest for $t < 0$, so we take σ and ρ to be zero for $t \le 0$ in (6-6). We need to determine these densities for $t > 0$.

Just as before, formal calculations (Prob. 1) lead to

(6-7)
$$u(x, t) = -\frac{1}{2c^2}\sigma\left(t - \frac{x}{c}\right) - \frac{1}{2c^2}\rho\left(t - \frac{L-x}{c}\right).$$

In order for (6-7) to satisfy the boundary conditions in (6-5), we require that

(6-8)
$$u(0, t) = -\frac{1}{2c^2}\sigma(t) - \frac{1}{2c^2}\rho\left(t - \frac{L}{c}\right) = \phi(t)$$

and

(6-9)
$$u(L, t) = -\frac{1}{2c^2}\sigma\left(t - \frac{L}{c}\right) - \frac{1}{2c^2}\rho(t) = \psi(t).$$

Since σ and ρ are zero for negative arguments, we can determine σ and ρ by alternate substitutions from (6-8) and (6-9) with arguments successively decreased by L/c. For example,

$$\begin{aligned}\sigma(t) &= -2c^2\phi(t) - \rho\left(t - \frac{L}{c}\right) = -2c^2\phi(t) + 2c^2\psi\left(t - \frac{L}{c}\right) + \sigma\left(t - \frac{2L}{c}\right)\\ &= -2c^2\phi(t) + 2c^2\psi\left(t - \frac{L}{c}\right) - 2c^2\phi\left(t - \frac{2L}{c}\right) - \rho\left(t - \frac{3L}{c}\right)\\ &= -2c^2\phi(t) + 2c^2\psi\left(t - \frac{L}{c}\right) - 2c^2\phi\left(t - \frac{2L}{c}\right) + 2c^2\psi\left(t - \frac{3L}{c}\right) + \sigma\left(t - \frac{4L}{c}\right),\end{aligned}$$

and continuing in this way leads to

$$\sigma(t) = -2c^2 \sum_{n=0}^{\infty} \phi\left(t - \frac{2nL}{c}\right) + 2c^2 \sum_{n=0}^{\infty} \psi\left(t - \frac{(2n+1)L}{c}\right),$$

where the sums actually involve only a finite number of terms because ϕ and ψ vanish for negative arguments. Similarly,

$$\rho(t) = -2c^2 \sum_{n=0}^{\infty} \psi\left(t - \frac{2nL}{c}\right) + 2c^2 \sum_{n=0}^{\infty} \phi\left(t - \frac{(2n+1)L}{c}\right),$$

with only a finite number of nonzero terms. Insert the series expansions for σ and ρ into (6-7) to get

$$(6\text{-}10)\quad u(x, t) = \sum_{n=0}^{\infty} \left[\phi\left(t - \frac{x}{c} - \frac{2nL}{c}\right) - \psi\left(t - \frac{x}{c} - \frac{(2n+1)L}{c}\right) + \psi\left(t - \frac{L-x}{c} - \frac{2nL}{c}\right) - \phi\left(t - \frac{L-x}{c} - \frac{(2n+1)L}{c}\right) \right],$$

which is a finite summation for (x, t) in any bounded set. It is now routine to check that (6-10) solves (6-5) when ϕ and ψ are twice continuously differentiable.

Theorem 6-1. Let ϕ and ψ in (6-5) have continuous second derivatives and satisfy the compatibility conditions $\phi(0) = \phi'(0) = \psi(0) = \psi'(0) = 0$. Then the unique solution to the initial, boundary value problem (6-5) is given by (6-10).

Finally, we turn to the general initial, boundary value problem for the wave equation,

$$(6\text{-}11)\quad \begin{cases} u_{tt} = c^2 u_{xx} + F(x, t), & 0 < x < L, \quad t > 0, \\ u(x, 0) = f(x), \quad u_t(x, 0) = g(x), & 0 \le x \le L, \\ u(0, t) = \phi(t), \quad u(L, t) = \psi(t), & t \ge 0. \end{cases}$$

Extend the domains of F, f, and g to $-\infty < x < \infty$ and consider the initial value problem

$$(6\text{-}12)\quad \begin{cases} v_{tt} = c^2 v_{xx} + F(x, t), & -\infty < x < \infty, \quad t > 0, \\ v(x, 0) = f(x), \quad v_t(x, 0) = g(x), & -\infty < x < \infty. \end{cases}$$

This problem has twice continuously differentiable solution (5-15) provided that $f(x)$ has a continuous second derivative, g has a continuous first derivative, and F is continuously differentiable with respect to x and continuous in t. Given these assumptions, the function $w = u - v$ satisfies (6-5) with ϕ replaced by $\phi(t) - v(0, t)$ and ψ replaced by $\psi(t) - v(L, t)$. If these boundary conditions satisfy the compatibility conditions in Th. 6-1, w can be found and $u = v + w$ solves (6-11). This reasoning (see Prob. 7) leads to the following sharpened forms of results obtained in Secs. 4-2 and 4-3 by separation of variables.

Theorem 6-2. Suppose that $F(x, t)$ is continuously differentiable with respect to x on $-\infty < x < \infty$ and is continuous in t, that $f(x)$ is twice continuously differentiable on $-\infty < x < \infty$, and that $g(x)$ is continuously differentiable there. Assume also that $\phi(t)$ and $\psi(t)$ have continuous second derivatives for $t \ge 0$, and that the compatibility conditions

$$\phi(0) = f(0), \qquad \phi'(0) = g(0), \qquad \psi(L) = f(L), \quad \text{and} \quad \psi'(L) = g(L)$$

hold. Then (6-11) has a unique solution.

Note that the discussion culminating in Th. 6-2 provides an explicit representation for the solution. Also, the hypotheses in Th. 6-2 are significantly weaker than those required in our earlier discussion of the initial, boundary value problem for the wave equation in Secs. 4-2 and 4-3.

PROBLEMS

1. Derive (6-7) from (6-6).
2. Prove Th. 6-1 by checking that (6-10) does solve (6-5).
3. Solve

$$\begin{aligned}
&u_{tt} = c^2 u_{xx}, \qquad x > 0, \qquad t > 0\\
&u(x, 0) = u_t(x, 0) = 0, \qquad x \geq 0,\\
&u_x(0, t) = \phi(t), \qquad t \geq 0,
\end{aligned}$$

by assuming the solution in the form of a single-layer potential. State carefully what you assume about ϕ.

4. Use the following steps to solve

$$\begin{aligned}
&u_{tt} = c^2 u_{xx}, \qquad x > 0, \qquad t > 0,\\
&u(x, 0) = 0, \qquad u_t(x, 0) = 0, \qquad x \geq 0,\\
&-u_x(0, t) + au(0, t) = \phi(t), \qquad t \geq 0.
\end{aligned}$$

(a) Assume that

$$u(x, t) = \int_0^t k(x, 0, t - \tau)\sigma(\tau)\, d\tau,$$

where $\sigma = 0$ for $\tau \leq 0$, and show that σ must satisfy

$$\sigma(t) + ac \int_0^t \sigma(\tau)\, d\tau = 2c^2\phi(t).$$

(b) Let $\sum(t) = \int_0^t \sigma(\tau)\, d\tau$ in the integral equation in part (a). $\sum$ satisfies an ordinary differential equation. Solve for $\sum(t)$ and then $\sigma(t)$.

(c) Find $u(x, t)$ and state the restrictions on $\phi(t)$.

5. Solve

$$\begin{aligned}
&u_{tt} = c^2 u_{xx}, \qquad 0 < x < L, \qquad t > 0,\\
&u(x, 0) = u_t(x, 0) = 0, \qquad 0 \leq x \leq L,\\
&u_x(0, t) = \phi(t), \qquad u_x(L, t) = \psi(t), \qquad t \geq 0,
\end{aligned}$$

by assuming a solution in the form of a single-layer potential

$$u(x, t) = \int_0^t k(x, 0, t - \tau)\sigma(\tau)\, d\tau + \int_0^t k(x, L, t - \tau)\rho(\tau)\, d\tau$$

with σ and $\rho = 0$ for $\tau \leq 0$. State carefully what must be assumed about ϕ and ψ.

6. Solve

$$\begin{aligned}
&u_{tt} = c^2 u_{xx}, \qquad 0 < x < L, \qquad t > 0,\\
&u(x, 0) = u_t(x, 0) = 0, \qquad 0 \leq x \leq L,\\
&u(0, t) = \phi(t), \qquad u_x(L, t) = \psi(t), \qquad t \geq 0.
\end{aligned}$$

Either use previous problems or assume a solution as the sum of a double-layer potential at $x = 0$ and a single-layer potential at $x = L$.

7. Prove Th. 6-2 using the line of reasoning outlined after (6-11).

10-7. Hyperbolic Potential Theory in Three Spatial Dimensions

Hyperbolic potentials also can be used to represent solutions to the wave equation in more than one spatial variable. In this section we consider three spatial variables, so the fundamental solution is

$$k(x, y, t) = \frac{\delta(ct - r)}{4\pi cr}, \qquad r = |x - y|.$$

In analogy with the Laplace equation and heat equation, we define single- and double-layer potentials by

$$v(x, t) = \int_0^t \int_B k(x, y, c(t - \tau))\rho(y, \tau)\, d\sigma_y\, d\tau \tag{7-1}$$

and

$$w(x, t) = \int_0^t \int_B \frac{\partial}{\partial v_y} k(x, y, c(t - \tau))\mu(y, \tau)\, d\sigma_y\, d\tau, \tag{7-2}$$

respectively. Here B is the boundary of the domain D in which a solution is sought. Since the potentials in (7-1) and (7-2) involve delta functions we pause to clarify their meaning. First, as in Sec. 10-6, it is convenient to assume that ρ and μ are zero when $t < 0$. Proceeding formally, we find that

$$\begin{aligned} v(x, t) &= \int_0^t \int_B \frac{\delta(ct - c\tau - r)}{4\pi cr} \rho(y, \tau)\, d\sigma_y\, d\tau \\ &= \int_B \int_{-\infty}^{\infty} \frac{\delta(\eta)}{4\pi c^2 r} \rho\left(y, \frac{ct - r - \eta}{c}\right) d\eta\, d\sigma_y \end{aligned}$$

or

$$v(x, t) = \int_B \frac{1}{4\pi c^2 r} \rho\left(y, t - \frac{r}{c}\right) d\sigma_y, \tag{7-3}$$

which we take as the meaning of (7-1). Similarly,

$$\begin{aligned} w(x, t) &= \int_0^t \int_B \left[\frac{\partial}{\partial v_y} \frac{\delta(ct - c\tau - r)}{4\pi cr}\right] \mu(y, \tau)\, d\sigma_y\, d\tau \\ &= -\text{div}_x \int_B \int_{-\infty}^{t} \frac{\delta(ct - c\tau - r)}{4\pi cr} v_y \mu(y, \tau)\, d\tau\, d\sigma_y, \end{aligned}$$

and we interpret (7-2) to mean that

$$w(x, t) = -\text{div}_x \int_B \frac{1}{4\pi c^2 r} \mu\left(y, t - \frac{r}{c}\right) v_y\, d\sigma_y. \tag{7-4}$$

To illustrate the use of hyperbolic potentials, consider

$$\text{(7-5)} \qquad \begin{cases} u_{tt} = c^2 \Delta u, & x \in D, \quad t > 0, \\ u(x, 0) = u_t(x, 0) = 0, & x \in \bar{D}, \\ \dfrac{\partial u}{\partial v} = \phi(x, t), & x \in B, \quad t \geq 0, \end{cases}$$

where D is the half-space $\{x \in R_3 : x_1 > 0\}$ and B is its boundary $\{x \in R_3 : x_1 = 0\}$. As in our one dimensional examples, it is convenient to set $\phi = 0$ for $t < 0$. In view of the Neumann boundary condition, we assume a solution in the form of a single-layer potential

$$\text{(7-6)} \qquad u(x, t) = \int_{-\infty}^{\infty} \int_{-\infty}^{\infty} \frac{1}{4\pi c^2 r} \rho\left(y, t - \frac{r}{c}\right) dy_2\, dy_3, \qquad r = |x - y|.$$

Since $\partial/\partial v_x = -\partial/\partial x_1$, we find that

$$\frac{\partial u}{\partial v}(x, t) = \int_{-\infty}^{\infty} \int_{-\infty}^{\infty} \frac{1}{4\pi c^2} \frac{x_1}{r^3} \left[\rho\left(y, t - \frac{r}{c}\right) + r\rho_t\left(y, t - \frac{r}{c}\right)\right] dy_1\, dy_2.$$

This normal derivative vanishes for $x_1 = 0$. So when we apply the interior jump relation to the single-layer potential (7-6), we obtain

$$\phi(x, t) = \lim_{x_1 \to 0+} \frac{\partial u}{\partial v}(x, t) = \frac{1}{2c^2} \rho(x, t), \qquad x \in B.$$

Substitute this result in (7-6) to find the representation

$$\text{(7-7)} \qquad u(x, t) = \int_{-\infty}^{\infty} \int_{-\infty}^{\infty} \frac{1}{2\pi r} \phi\left(y, t - \frac{r}{c}\right) dy_2\, dy_3, \qquad r = |x - y|$$

for the solution to (7-5). Since $\phi = 0$ when its second argument is negative, the integration in (7-7) is actually over the bounded region

$$\{(y_2, y_3) : (y_2 - x_2)^2 + (y_3 - x_3)^2 \leq c^2 t^2 - x_1^2\}.$$

Given this observation it is easy to confirm that (7-7) solves (7-5) under the assumptions in the following theorem. Observe that $y = (0, y_2, y_3)$ in (7-7).

Theorem 7-1. If $\phi(x, t)$ is continuous on $-\infty < x_2, x_3, t < \infty$ and has a continuous second derivative with respect to t, then (7-7) is the unique solution to (7-5).

Now let D be a domain with boundary B, and consider

$$\text{(7-8)} \qquad \begin{cases} u_{tt} = c^2 \Delta u, & x \in D, \quad t > 0, \\ u(x, 0) = u_t(x, 0) = 0, & x \in \bar{D}, \\ u(x, t) = \phi(x, t), & x \in B, \quad t \geq 0. \end{cases}$$

This time we have a Dirichlet boundary condition and so assume a solution in the form of a double-layer potential

$$u(x, t) = -\operatorname{div}_x \int_B \frac{1}{4\pi c^2 r} \mu\left(y, t - \frac{r}{c}\right) \nu_y \, d\sigma_y$$

$$= \int_B \frac{1}{4\pi c^2} \left(\frac{\partial}{\partial \nu_y} \frac{1}{r}\right) \left[\mu\left(y, t - \frac{r}{c}\right) + \frac{r}{c} \mu_t\left(y, t - \frac{r}{c}\right)\right] d\sigma_y.$$

As usual, it is convenient to set $\phi(x, t) = 0$ for $t < 0$. Then $u(x, t) = 0$ for $t < 0$ and consequently we take $\mu = 0$ for $t < 0$. The problem is to determine the density μ for $t \geq 0$. Let x approach the boundary B of D and use Th. 6-3 of Chap. 8 to obtain the integral equation

$$\text{(7-9)} \qquad \phi(x, t) = -\frac{1}{2c^2} \mu(x, t) + \int_B \frac{1}{4\pi c^2} \left(\frac{\partial}{\partial \nu_y} \frac{1}{r}\right) \left[\mu\left(y, t - \frac{r}{c}\right) + \frac{r}{c} \mu_t\left(y, t - \frac{r}{c}\right)\right] d\sigma_y,$$

where $x \in B$ and $t \geq 0$, for the determination of μ. Note that we can use Th. 6-3 even if B is not bounded because the integrand in the double-layer potential is nonzero only for $r = |x - y| < ct$.

The general Fredholm theory does not apply to (7-9). However, we can solve (7-9) by iteration. That is, we set $\mu_0 \equiv 0$ and define

$$\text{(7-10)} \qquad \mu_{n+1}(x, t) = -2c^2 \phi(x, t) + \frac{1}{2\pi} \int_B \left(1 + \frac{r}{c} \frac{\partial}{\partial t}\right) \mu_n\left(y, t - \frac{r}{c}\right) \left(\frac{\partial}{\partial \nu_y} \frac{1}{r}\right) d\sigma_y$$

for $n \geq 0$. To investigate this iteration scheme, we introduce the following notation. For points x, $y^{(k)}$ in B, we set

$$r_1 = |x - y^{(1)}|, \qquad r_k = |y^{(k)} - y^{(k-1)}| \quad \text{for} \quad k \geq 2,$$

denote the unit normal at $y^{(k)}$ by ν_k, and let

$$d\Omega_k = \left(\frac{\partial}{\partial \nu_k} \frac{1}{r_k}\right) d\sigma_{y^{(k)}}.$$

Then from (7-10),

$$\mu_1(x, t) = -2c^2 \phi(x, t),$$

$$\mu_2(x, t) = -2c^2 \left[\phi(x, t) + \frac{1}{2\pi} \int_B \left(1 + \frac{r_1}{c} \frac{\partial}{\partial t}\right) \phi\left(y^{(1)}, t - \frac{r_1}{c}\right) d\Omega_1\right],$$

$$\mu_3(x, t) = -2c^2 \left[\phi(x, t) + \frac{1}{2\pi} \int_B \left(1 + \frac{r_1}{c} \frac{\partial}{\partial t}\right) \phi\left(y^{(1)}, t - \frac{r_1}{c}\right) d\Omega_1\right.$$

$$\left. + \frac{1}{2\pi} \int_B d\Omega_1 \frac{1}{2\pi} \int_B \left(1 + \frac{r_1}{c} \frac{\partial}{\partial t}\right) \left(1 + \frac{r_2}{c} \frac{\partial}{\partial t}\right) \phi\left(y^{(2)}, t - \frac{r_1}{c} - \frac{r_2}{c}\right) d\Omega_2\right].$$

Proceeding in this way we find that

$$\mu_n(x, t) = -2c^2\left[\phi(x, t) + \sum_{k=1}^{n-1} T_k(x, t)\right], \qquad n \geq 2, \tag{7-11}$$

where

$$T_k(x, t) = \left(\frac{1}{2\pi}\right)^k \int_B d\Omega_1 \cdots \int_B d\Omega_k \left[\prod_{j=1}^{k}\left(1 + \frac{r_j}{c}\frac{\partial}{\partial t}\right)\phi\left(y^{(k)}, t - \sum_{i=1}^{k}\frac{r_i}{c}\right)\right]. \tag{7-12}$$

Notice that (7-12) requires that $\phi(x, t)$ be infinitely often differentiable as a function of t, which we now assume. In view of (7-11), we must investigate the convergence of

$$\sum_{k=1}^{\infty} T_k(x, t), \tag{7-13}$$

and for this we must be more precise about out assumptions on the surface B and the boundary data $\phi(x, t)$.

Just as in Chaps. 8 and 9, we assume that B is twice continuously differentiable, and introduce at each $x^\circ \in B$ local coordinates with origin at x° and with the normal ν to B at x° given by $(0, 0, 1)$. We assume the following:

(A1) There is a constant $a > 0$ such that $\{x \in B: |x - x^\circ| \leq a\}$ can be expressed as $x_3 = h(x_1, x_2)$, where $x = (x_1, x_2, x_3)$ in local coordinates, h is twice continuously differentiable,

$$h(0, 0) = h_{x_1}(0, 0) = h_{x_2}(0, 0) = 0,$$

and the choice of a is independent of $x^\circ \in B$.

(A2) All the second-order partials of h are bounded in $\bar{K}_a(x^\circ)$ by a bound K_1 which depends only on B and is uniform over $x^\circ \in B$.

(A3) If $r' > r'' \geq a$ are fixed, then for $x \in B$,

$$\int_{r'' < r < r'} d\sigma_y \leq K_2(r' - r''), \qquad r = |x - y|,$$

where K_2 depends only on B.

Boundaries B that satisfy (A1), (A2), and (A3) will be called of class (A). Notice that the requirement $h(0, 0) = h_{x_1}(0, 0) = h_{x_2}(0, 0) = 0$ just means that we select the coordinates x_1 and x_2 in the tangent plane to B at x°.

We have already noted that $\phi(x, t)$ must be infinitely often differentiable if (7-11) is to provide a solution to (7-9) in the limit as $n \to \infty$. Our assumptions on the boundary data ϕ are as follows:

(B1) $\phi(x, t)$ is defined for $x \in B$, $-\infty < t \leq T$ for some $T > 0$ and $\phi = 0$ for $t \leq 0$.

(B2) $\phi(x, t)$ is infinitely often differentiable with respect to t and each derivative is continuous in (x, t).

(B3) For $x \in B$ and $-\infty < t \leq T$, there exists a constant $C > 0$ and $0 < \delta < 1$ such that

$$|\phi(x, t)| \leq C,$$
$$\left|\frac{\partial^n}{\partial t^n} \phi(x, t)\right| \leq C^n n^{(1+\delta)n}, \qquad n \geq 1.$$

Functions satisfying (B1), (B2), and (B3) are called of class (B). The constants C and δ in (B3) may depend on ϕ.

Our discussion of the convergence of the series (7-13) rests on

Lemma 7-1. Suppose that B is of class (A) and $\psi(r)$ is continuous for $r \geq 0$. Then for $x \in B$

$$\int_{r \leq R} |\psi(r)|\,|d\Omega_y| = \int_{|x-y| \leq R} \psi(|x-y|)\left|\frac{\partial}{\partial \nu_y}\frac{1}{r}\right| d\sigma_y \leq K \int_0^R |\psi(r)|\,dr,$$

where K is a constant that depends only upon B.

Proof. Determine $a > 0$ as in (A1). There are two cases, $R > a$ and $R \leq a$. Suppose first that $R > a$. Then

$$\int_{r \leq R} |\psi(r)|\,|d\Omega_y| = \int_{r \leq a} |\psi(r)|\,|d\Omega_y| + \int_{a < r \leq R} |\psi(r)|\,|d\Omega_y| = I_1 + I_2.$$

Consider I_2 and observe that

$$\left|\frac{\partial}{\partial \nu}\frac{1}{r}\right| \leq \frac{1}{r^2} \leq \frac{1}{a^2} \quad \text{for} \quad a < r,$$

so

$$I_2 \leq \frac{1}{a^2} \int_{a < r \leq R} |\psi(r)|\,d\sigma_y.$$

Now partition the interval $a < r \leq R$ as $a = \rho_0 < \rho_1 < \cdots < \rho_n = R$, where $\rho_j = a + j(R-a)/n$ and use (A3) to find

$$I_2 \leq \frac{1}{a^2} \sum_{j=1}^{n} \int_{\rho_{j-1} < r < \rho_j} |\psi(r)|\,d\sigma_y \leq \frac{K_2}{a^2} \sum_{j=1}^{n} m_j(\rho_j - \rho_{j-1}),$$

where $m_j = \max |\psi(r)|$ for $\rho_{j-1} \leq r \leq \rho_j$. Since ψ is continuous, we can let $n \to \infty$ to deduce that

$$I_2 \leq \frac{K_2}{a^2} \int_a^R |\psi(r)|\,dr \leq \frac{K_2}{a^2} \int_0^R |\psi(r)|\,dr.$$

To estimate I_1, recall from Sec. 8-6 that there is a constant C depending only on B such that

$$\left|\frac{\partial}{\partial \nu}\frac{1}{r}\right| \leq \frac{C}{r},$$

for $r \leq a$; see the discussion following (6-12) in Sec. 8-6. Consequently,

$$I_1 \leq C \int_{r \leq a} \frac{|\psi(r)|}{r} \, d\sigma_y = C \int_0^{2\pi} \int_0^a \frac{|\psi(r)|}{r} r \, dr \, d\theta \leq 2\pi C \int_0^R |\psi(r)| \, dr.$$

Thus,

$$\int_{r \leq R} |\psi(r)| \, |d\Omega_y| \leq I_1 + I_2 \leq K \int_0^R |\psi(r)| \, dr,$$

where $K = (K_2/a^2) + 2\pi C$ depends only on the surface B. In case $R \leq a$, we only encounter an integral of type I_1 and find $K = 2\pi C$, and the lemma is established.

Now, we are prepared to establish the absolute and uniform convergence of the series (7-13) for $0 \leq t \leq T$ and $x \in B$. First observe that $T_k(x, t)$ in (7-12) is zero unless $r_1 + r_2 + \cdots + r_j \leq ct$ for $j \leq k$ because otherwise the integrand is zero. Expand out the integrand in (7-12), invoke (B3), and use the arithmetic–geometric mean inequality to find

$$\begin{aligned}
\left| \prod_{j=1}^{k} \left(1 + \frac{r_j}{c} \frac{\partial}{\partial t}\right) \phi \right| &= \left| \phi + \sum_{s=1}^{k} \sum_{1 \leq j_1 < \cdots < j_s \leq k} \frac{1}{c^s} r_{j_1} \cdots r_{j_s} \frac{\partial^s \phi}{\partial t^s} \right| \\
&\leq |\phi| + \sum_{s=1}^{k} \sum_{1 \leq j_1 < \cdots < j_s \leq k} \frac{1}{c^s} \left(\frac{r_{j_1} + \cdots + r_{j_s}}{s} \right)^s \left| \frac{\partial^s \phi}{\partial t^s} \right| \\
&\leq C + \sum_{s=1}^{k} \frac{\binom{k}{s} t^s}{s^s} C^s s^{(1+\delta)s} \\
&\leq C + k^{\delta k} \sum_{s=0}^{k} \binom{k}{s} (Ct)^s = C + k^{\delta k}(1 + Ct)^k.
\end{aligned}$$

Hence, $|T_k(x, t)|$ is bounded by

$$\frac{C + k^{\delta k}(1 + Ct)^k}{(2\pi)^k} \int_{r_1 < ct} |d\Omega_1| \cdots \int_{r_1 + \cdots + r_i < ct} |d\Omega_i| \cdots \int_{r_1 + \cdots + r_k < ct} |d\Omega_k|.$$

By Lemma 7-1 the k-fold integral is bounded above by

$$K^k \int_0^{ct} dr_1 \cdots \int_0^{ct - (r_1 + \cdots + r_{i-1})} dr_i \cdots \int_0^{ct - (r_1 + \cdots + r_{k-1})} dr_k = \frac{(Kct)^k}{k!},$$

and we obtain the estimate

$$|T_k(x, t)| \leq \frac{C}{k!} \left(\frac{Kct}{2\pi} \right)^k + k^{\delta k} \left(\frac{1 + Ct}{2\pi} \right)^k \frac{(Kct)^k}{k!}.$$

Finally, the Taylor series for e^x gives $e^k \geq k^k/k!$ and we find that

$$|T_k(x, t)| \leq \frac{C}{k!} \left(\frac{KcT}{2\pi} \right)^k + \left[\frac{(1 + CT)(KcT)e}{2\pi} \right]^k \frac{1}{k^{(1-\delta)k}} = M_k, \tag{7-14}$$

for $x \in B$ and $0 \leq t \leq T$. Since $1 - \delta > 0$ it is easy (Prob. 3) to show that $\sum M_k < \infty$ and the Weierstrass M-test implies the absolute and uniform convergence of $\mu_n(x, t)$ in (7-11). Since, for fixed (x, t), the integration in (7-10) is only over the bounded set $r = |x - y| < ct$, we can pass to the limit under the integral sign in (7-10) to conclude that

$$\mu(x, t) = \lim_{n \to \infty} \mu_n(x, t)$$

solves the integral equation (7-9). This proves Th. 7-2.

Theorem 7-2. Suppose that the boundary B of D is of class (A) and that the boundary data $\phi(x, t)$ defined for $x \in B$ and $-\infty < t \leq T$ belongs to the class (B). Then there is a unique solution to

$$\begin{cases} u_{tt} = c^2 \Delta u, & (x, t) \in Q_T, \\ u(x, 0) = u_t(x, 0) = 0, & x \in \bar{D}, \\ u(x, t) = \phi(x, t), & x \in B, \quad 0 \leq t \leq T, \end{cases}$$

which can be represented in the form of a double-layer potential (7-4) where the density μ is the solution to the integral equation (7-9) and can be found by iteration.

The extension of Th. 7-2 to inhomogeneous initial data is done just as in one spatial variable (see Prob. 3).

A few concluding remarks are in order. Our discussion places heavy differentiability assumptions on the function ϕ and on other inhomogeneous data in Prob. 3. These restrictions will be relaxed in Chap. 11. Since ϕ is infinitely often differentiable in t, the same is true of the double-layer potential. This follows by differentiating (7-11) repeatedly with respect to t and applying the reasoning above to see that the resulting series are uniformly convergent for $x \in B$ and $0 \leq t \leq T$. Finally, our reasoning above does not require B or D to be bounded. In particular, Th. 7-2 applies to exterior and interior domains.

PROBLEMS

1. Check directly that (7-7) solves (7-5) under the assumptions in Th. 7-1.
2. Show that $\sum M_k < \infty$, where M_k is defined in (7-14).
3. Review the discussion in one spatial dimension preceding Th. 6-2, which reduces the general initial, boundary value problem to the case with homogeneous initial data. Use these ideas and Th. 7-2 to state and prove the analogue of Th. 6-2 for three spatial dimensions.
4. Solve explicitly

$$\begin{aligned} &u_{tt} = c^2 \Delta u, && x \in D = \{x \in R_3 : x_1 > 0\}, \quad t > 0, \\ &u(x, 0) = u_t(x, 0) = 0, && x \in \bar{D}, \\ &u(0, x_2, x_3, t) = \phi(x_2, x_3, t), && x \in B, \quad t \geq 0. \end{aligned}$$

5. Solve, by assuming a solution in the form of a single-layer potential,

$$u_{tt} = c^2 \Delta u, \qquad (x, t) \in Q_T,$$
$$u(x, 0) = u_t(x, 0) = 0, \qquad x \in \bar{D},$$
$$\frac{\partial u}{\partial \nu} = \psi(x, t), \qquad x \in B, \qquad t \geq 0.$$

Extend $\psi = 0$ for $t \leq 0$ and assume that B is of class (A).

6. Show that the analogues of Th. 7-2 and Prob. 5 for two spatial variables can be handled by the method of descent. *Hint.* Review the discussion following (4-11). The idea is to adjoin a third variable x_3 and extend the boundary data so that they are independent of x_3. Then the density μ will also be independent of x_3.

7. Show that (7-3) [resp., (7-4)] satisfies the wave equation for $x \notin B$, $t > 0$ and the homogeneous initial conditions $u(x, 0) = u_t(x, 0) = 0$ for any choice of $\rho(y, t)$ [resp., $\mu(y, t)$] which has a continuous second (resp., third)-order partial with respect to t.

8. Solve explicitly the problem

$$u_{tt} = c^2 \Delta u, \qquad x \in D = \{x \in R_2 : x_2 > 0\}, \qquad t > 0,$$
$$u(x, 0) = u_t(x, 0) = 0, \qquad x \in \bar{D},$$
$$\frac{\partial u}{\partial x_2} = \psi(x_1, 0, t), \qquad -\infty < x_1 < \infty, \qquad t \geq 0,$$

by defining an appropriate single-layer potential and using it.

11

Variational Methods

11-1. Introduction to the Calculus of Variations

The calculus of variations is a vast subject with a rich and important classical development. Owing to a lack of space, we only can hint at some of the topics usually covered in an introduction to this subject. An extensive treatment of such material may be found in either I. M. Gelfand and S. V. Fomin, *Calculus of Variations* (Prentice-Hall, Inc. Englewood Cliffs, N.J., 1963) or R. Weinstock, *Calculus of Variations with Applications to Physics and Engineering* (McGraw-Hill Book Company, New York, 1952). In this chapter we choose to focus on some modern aspects of the calculus of variations related to so-called weak formulations of certain physical problems. On the theoretical side, this approach leads to a natural relaxation of the requirement that solutions to partial differential equations must be smooth. The new solutions are called generalized solutions, but they are just as physically relevant as classical solutions. For example, we can say precisely what it means for a wave form with corners to be a solution to the wave equation. On the practical side, the numerical implementation of the solution procedures we shall develop includes the methods of Ritz–Galerkin and of finite elements.

The calculus of variations was developed, concurrently with the more familiar differential and integral calculus, in response to "minimum principles" that described the behavior of certain physical systems. For example, in about 1650, Fermat observed that the laws of reflection and refraction of light follow from the *principle of least time*: Light travels between two points along the path that minimizes transit time.

Another principle, the *principle of minimum potential energy*, states: A physical system in stable equilibrium assumes the configuration that minimizes its potential energy.

Let's take a closer look at Fermat's principle in the case where light rays travel in the xy plane through a medium with variable index of refraction. The speed of light at point (x, y) is $v(x, y)$ (see Fig. 11-1). We seek the path of the light ray which passes through two given points P and Q. Fermat's principle says the actual path of the light ray from P to Q can be found by minimizing the time of transit

$$T(u) = \int dt = \int \frac{ds}{v} = \int_a^b \frac{\sqrt{1 + u'(x)^2}}{v(x, u(x))}\, dx$$

over all curves $y = u(x)$ joining P to Q. The formula

$$T(u) = \int_a^b \frac{\sqrt{1 + u'(x)^2}}{v(x, u(x))}\, dx \tag{1-1}$$

associates to each function u, whose graph joins P to Q, the real number $T(u)$. Thus, $T(u)$ is a real-valued function whose domain consists of certain other functions. We call $T(u)$ a *functional.* (Compare this usage with our discussion of the delta functional in Sec. 10-5.) Fermat's principle asserts that the function u whose graph joins P to Q and which renders $T(u)$ a minimum gives the path of the light ray from P to Q.

The time functional in (1-1) is of the form

$$I(u) = \int_a^b F(x, u(x), u'(x))\, dx. \tag{1-2}$$

It turns out that the solutions of a host of physical and geometric problems can be characterized as follows: For a specific choice of F, determine the function u that minimizes or maximizes the functional $I(u)$ among all functions u which satisfy some natural constraints intrinsic to the given physical or geometric problem. If F in (1-2) is the integrand from (1-1), then a natural class of admissible functions is

$$\mathfrak{A} = \{u \in C^1[a, b] : u(a) = A \text{ and } u(b) = B\} \tag{1-3}$$

(see Fig. 11-1).

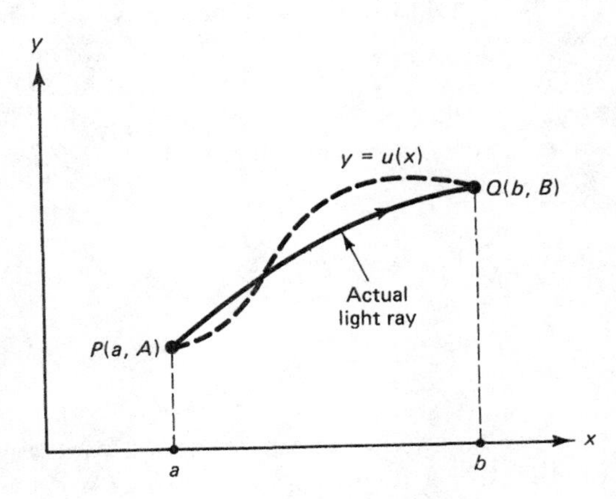

Figure 11-1

Suppose that the functional I in (1-2) is minimized by the function $u \in \mathfrak{A}$, a natural class of admissible functions. How can we use the fact that $I(u)$ is a minimum to determine u? Obviously, we would like to use differential calculus and to set some derivative to zero. We can do this as follows. Since u minimizes I,

$$I(u) \leq I(v) \quad \text{for all} \quad v \in \mathfrak{A}. \tag{1-4}$$

To get back to the familiar calculus setting where the functions have real variable arguments, we choose special comparison functions

$$v = u + \varepsilon\zeta \in \mathfrak{A} \tag{1-5}$$

where ε is a real parameter which varies in some interval containing the origin and $\zeta = \zeta(x)$ is any fixed function such that $u + \varepsilon\zeta \in \mathfrak{A}$. This places certain general restrictions on ζ because both u and $u + \varepsilon\zeta$ must be in the admissible function space $\mathfrak{A}$. For instance, in the light-ray problem with $\mathfrak{A}$ given by (1-3), $u \in \mathfrak{A}$ and $u + \varepsilon\zeta \in \mathfrak{A}$ plainly require that $\zeta \in C^1[a, b]$ and that $\zeta(a) = \zeta(b) = 0$.

From (1-4) and (1-5) we have $I(u) \leq I(u + \varepsilon\zeta)$ for each ζ such that $u + \varepsilon\zeta \in \mathfrak{A}$. Thus, the function of ε, $I(u + \varepsilon\zeta)$ has a minimum at $\varepsilon = 0$ and we deduce that

$$\frac{d}{d\varepsilon} I(u + \varepsilon\zeta)\bigg|_{\varepsilon=0} = 0.$$

The expression on the left is called the *first variation of I*, and we write

$$\delta I = \delta I(u) = \delta I(u)\zeta = \frac{d}{d\varepsilon} I(u + \varepsilon\zeta)\bigg|_{\varepsilon=0}. \tag{1-6}$$

Notice that δI is itself a functional which operates on the function ζ. Furthermore, we can regard δI as the "directional derivative" of I at u in the "direction" ζ because

$$\frac{d}{d\varepsilon} I(u + \varepsilon\zeta)\bigg|_{\varepsilon=0} = \lim_{\varepsilon \to 0} \frac{I(u + \varepsilon\zeta) - I(u)}{\varepsilon},$$

which is the usual formula for a directional derivative when the variables u and ζ are vectors.

In summary, if I has a minimum (or for that matter a maximum) at $u \in \mathfrak{A}$, then its first variation $\delta I(u)\zeta = 0$ for all functions ζ satisfying $u + \varepsilon\zeta \in \mathfrak{A}$. To better understand what this means, we shall express the condition $\delta I = 0$ in a more convenient form. We have

$$\begin{aligned}
\delta I(u)\zeta &= \frac{d}{d\varepsilon} \int_a^b F(x, u(x) + \varepsilon\zeta(x), u'(x) + \varepsilon\zeta'(x))\, dx\bigg|_{\varepsilon=0} \\
&= \int_a^b [F_u(x, u(x), u'(x))\zeta(x) + F_{u'}(x, u(x), u'(x))\zeta'(x)]\, dx,
\end{aligned}$$

where the subscripts denote partial differentiation with respect to the second and third variables of F. Consequently, if u minimizes I when $\mathfrak{A}$ is given by (1-3), then

u satisfies

$$\int_a^b [F_u(x, u, u')\zeta + F_{u'}(x, u, u')\zeta'] \, dx = 0 \tag{1-7}$$

for all functions $\zeta \in C^1[a, b]$ with $\zeta(a) = \zeta(b) = 0$. The requirement (1-7) still looks imposing, but we can put it in a more convenient form if we assume the minimizing function $u \in C^2[a, b]$. Given this, we integrate the second term in (1-7) by parts and obtain

$$F_{u'}(x, u(x), u'(x))\zeta(x)\Big|_{x=a}^{x=b} + \int_a^b \left[F_u(x, u, u') - \frac{d}{dx} F_{u'}(x, u, u')\right]\zeta \, dx = 0. \tag{1-8}$$

Since $\zeta(a) = \zeta(b) = 0$, we have

$$\int_a^b \left[F_u(x, u, u') - \frac{d}{dx} F_{u'}(x, u, u')\right]\zeta \, dx = 0 \tag{1-9}$$

for all functions $\zeta \in C^1[a, b]$ with $\zeta(a) = \zeta(b) = 0$. In typical applications, F is smooth enough so that the term in brackets in the integrand is continuous. Then by the fundamental lemma of the calculus of variations (Prob. 2), (1-9) implies that the term in brackets vanishes,

$$\frac{d}{dx} F_{u'}(x, u, u') = F_u(x, u, u'). \tag{1-10}$$

This second-order differential equation for u is called the *Euler(–Lagrange) equation* for the functional (1-2). Since $\mathfrak{A}$ is given by (1-3), we also have

$$u(a) = A \quad \text{and} \quad u(b) = B. \tag{1-11}$$

Thus, u may be determined as the solution to the boundary value problem (1-10) and (1-11). Typically, (1-10) is a nonlinear second-order differential equation and explicit solutions are hard to come by, except in some special cases. Numerical methods are usually needed to solve (1-10) and (1-11). We will not discuss explicit solution methods or numerical procedures because our primary interest in the rest of Chap. 11 will be with the basic variational equation $\delta I(u)\zeta = 0$ expressed in a form such as (1-7).

Now, suppose that u minimizes (1-2) when $\mathfrak{A} = C^1[a, b]$. Reasoning exactly as above, we arrive at (1-8) for all functions $\zeta \in C^1[a, b]$. Among these functions ζ are all those for which also $\zeta(a) = \zeta(b) = 0$. For such ζ, (1-9) holds and we deduce the Euler equation (1-10) as before. Then the integrand in (1-8) vanishes and we conclude that

$$F_{u'}(x, u(x), u'(x))\zeta(x)\Big|_{x=a}^{x=b} = 0$$

for all $\zeta \in C^1[a, b]$. In particular, the choices $\zeta(x) = x - b$ and $\zeta = x - a$ yield

$$F_{u'}(a, u(a), u'(a)) = 0 \quad \text{and} \quad F_{u'}(b, u(b), u'(b)) = 0, \tag{1-12}$$

which are called *natural boundary conditions* because they are generated directly from the variational equation $\delta I = 0$. In this case, u is determined by the boundary value problem (1-10) and (1-12).

We turn now to the principle of minimum potential energy. Consider a tightly stretched membrane that spans a simple closed curve Γ in the $x_1 x_2$ plane. We shall assume that the membrane has constant tension τ and occupies the region D bounded by Γ to which it is rigidly attached. Now suppose that a small external transverse load with force density $F = F(x_1, x_2)$ is applied to the membrane so that it assumes a new equilibrium position. Then the potential energy arising from the stretching will be proportional to the change in surface area of an element of the membrane; that is, it will be, ignoring higher-order terms,

$$\tau[\sqrt{1 + u_{x_1}^2 + u_{x_2}^2}\, \Delta x_1\, \Delta x_2 - \Delta x_1\, \Delta x_2] = \frac{\tau}{2} |\nabla u|^2\, \Delta x_1\, \Delta x_2,$$

where the constant τ is the tension. The stored elastic potential energy is given by the integral of this quantity over D. The work done by the membrane against the external forces, namely, $\int_D -Fu\, dx$, is also stored as potential energy. Therefore, the total potential energy is

$$V(u) = \int_D \left[\frac{\tau}{2} |\nabla u|^2 - uF \right] dx, \tag{1-13}$$

if the membrane has shape $u = u(x)$. The principle of minimum potential energy asserts that the membrane will assume the shape $u = u(x)$ which minimizes (1-13). Here one natural class of admissible functions for the domain of V is

$$\mathfrak{A} = \{u \in C^1(D) \cap C(\bar{D}) : u(x) = 0 \quad \text{for} \quad x \in \Gamma\}. \tag{1-14}$$

The functional (1-13) is typical of a variety of problems which can be formulated in the following way. Minimize (or maximize) a functional of the form

$$I(u) = \int_D F(x, u(x), \nabla u(x))\, dx, \tag{1-15}$$

where u varies over a natural set of admissible functions $\mathfrak{A}$, $x = (x_1, x_2, \ldots, x_n)$, $\nabla u(x) = (u_{x_1}(x), \ldots, u_{x_n}(x))$, and where $F = F(x, u, p)$ is a suitably smooth function of all its $2n + 1$ variables. Reasoning just as for (1-2), we find that if u minimizes I, then

$$\delta I = \delta I(u)\zeta = \frac{d}{d\varepsilon} I(u + \varepsilon\zeta)\bigg|_{\varepsilon = 0} = 0$$

must hold for all ζ with $u + \varepsilon\zeta \in \mathfrak{A}$. That is, from (1-15),

$$\int_D [F_u \zeta + F_{p_1} \zeta_{x_1} + \cdots + F_{p_n} \zeta_{x_n}]\, dx = 0$$

for all such ζ, where F_u and F_{p_i} are evaluated at $(x, u(x), \nabla u(x))$. If we assume the minimizing function $u(x)$ has continuous second-order partials, analogues of (1-8) and (1-10) can be obtained. See Prob. 17.

Return to the potential energy functional (1-13) for the membrane. In this case, we form $V(u + \varepsilon\zeta)$ and obtain

$$\delta V = \int_D [\tau \nabla u \cdot \nabla\zeta - F\zeta]\, dx$$

so that the variational equation is to solve

$$\int_D [\tau \nabla u \cdot \nabla\zeta - F\zeta]\, dx = 0 \tag{1-16}$$

for all $\zeta \in C^1(D) \cap C(\bar{D})$ with $\zeta = 0$ on Γ, the boundary of D. If we assume that u is twice continuously differentiable, an application of the Gauss divergence theorem in (1-16) yields

$$\int_\Gamma \zeta \nabla u \cdot n\, d\sigma - \int_D [\tau \Delta u + F]\zeta\, dx = 0.$$

Since $\zeta = 0$ on Γ, we find that $\int_D [\tau \Delta u + F]\zeta\, dx = 0$ for all $\zeta \in C^1(D) \cap C(\bar{D})$ with $\zeta = 0$ on Γ. Finally, the fundamental lemma of the calculus of variations (Prob. 5) gives

$$\tau \Delta u + F = 0.$$

Thus, the shape of the membrane is determined by the boundary value problem

$$\Delta u = \frac{-F}{\tau}, \qquad x \in D,$$

$$u(x) = 0, \qquad x \in \Gamma.$$

We close this section by updating Fermat's principle of least time and the principle of minimum potential energy. The correct formulation of Fermat's principle is not that light will always follow a path such that $T(u)$ is a minimum. Rather, light will follow a path u such that the first variation of the time functional T vanishes, $\delta T(u) = 0$. To see why this reformulation is needed, consider a medium with variable index of refraction such that light rays travel on circular arcs (see Fig. 11-2). Then light can travel from P to Q along either of the arcs indicated in Fig. 11-2. Both arcs will render $\delta T = 0$, but clearly only the shorter arc minimizes travel time if P is very close to Q.

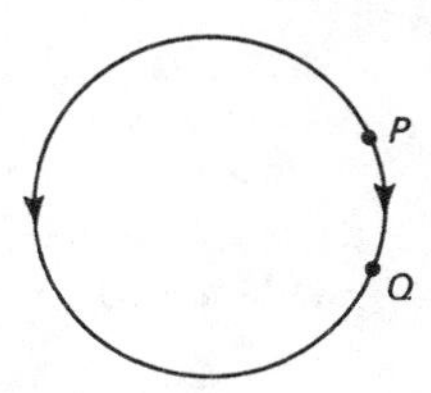

Figure 11-2

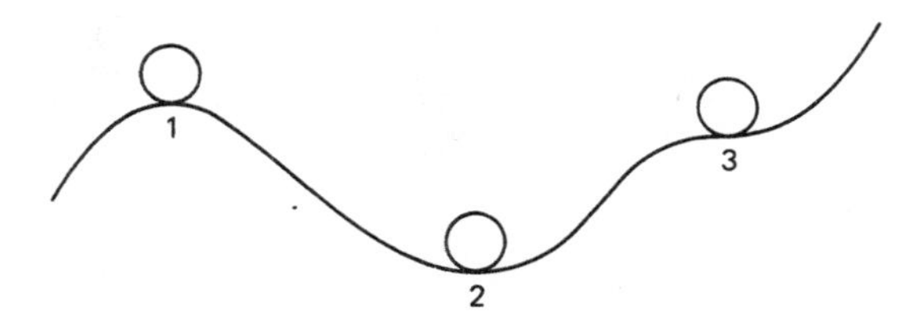

Figure 11-3

In general, if $I(u)$ is a functional defined on a class of admissible functions $\mathfrak{A}$, we say that the functional is *stationary* at u if $\delta I = \delta I(u) = 0$. Thus, Fermat's principle now asserts that light will follow a path which renders the time of transit functional stationary.

Similarly, the principle of minimum potential energy should be extended to say that the possible equilibrium configurations u of a physical system are those which make the potential energy functional stationary. Figure 11-3 shows why the principle must be extended. The figure shows equilibrium locations for a small ball in a gravitational field. At position 1 the potential energy is a maximum and the ball is in unstable equilibrium. Also, position 3 is an unstable equilibrium where the potential energy has an inflection point. Finally, position 2 minimizes potential energy and is a point of stable equilibrium. In all three cases, the potential energy is stationary.

PROBLEMS

1. Use Fermat's principle of least time to derive (a) the law of reflection and (b) Snell's law of refraction.

2. Suppose that $f(x)$ is continuous for $x \in [a, b]$ and satisfies

$$(*) \qquad \int_a^b f(x)\zeta(x)\,dx = 0$$

for all functions $\zeta \in C^1[a, b]$ with $\zeta(a) = \zeta(b) = 0$. Prove $f(x) = 0$ for all $x \in [a, b]$. *Hint.* Assume that $f(x_0) \neq 0$, say $f(x_0) > 0$, for some $x_0 \in [a, b]$. Use continuity to determine $x_1 < x_0 < x_2$ such that $f(x) \geq \frac{1}{2}f(x_0)$ on $[x_1, x_2]$. Now choose $\zeta_0(x) = (x - x_1)^2(x_2 - x)^2$ for $x \in [x_1, x_2]$ and $\zeta_0(x) = 0$ otherwise. Apply $(*)$ with $\zeta = \zeta_0$ to reach a contradiction.

3. In Prob. 2 assume $(*)$ holds only for $\zeta \in C^k[a, b]$ such that $\zeta = 0$ in a neighborhood of the endpoints of the interval $[a, b]$. Conclude that $f \equiv 0$, as before. Here k is fixed with $k \geq 1$.

4. Check that the reasoning used in Prob. 3 applies with $[a, b]$ replaced by any interval I, open, closed, half-open, bounded, or unbounded.

5. Prove this n-dimensional version of the fundamental lemma of the calculus of variations. Let D be a domain (possibly unbounded) and $f(x)$ be continuous for $x \in D$ and satisfy

$$\int_D f(x)\zeta(x)\,dx = 0$$

for all $\zeta \in C^k(D)$, for some $k \geq 1$, such that $\zeta = 0$ outside some closed, bounded set which does not intersect the boundary of D. (This closed, bounded set may depend on ζ.) Then $f(x) = 0$ for $x \in D$.

6. A homogeneous chain of constant density ρ and length L is suspended between two points P and Q. Let $y = u(x)$ be a possible shape of the hanging chain. Assume that gravity is the only force acting on the chain. Find the functional which must be minimized to find the actual equilibrium position of the chain. Describe the domain $\mathfrak{A}$ of this functional.

Problems 7–11 deal with the equilibrium position of an elastic beam.

7. Consider a beam of length L which extends from 0 to L along the x-axis. The beam deflects under a load density (force per unit length) $f(x)$. Assume that the deflections $u = u(x)$ are small. Show that the potential energy functional is

$$V(u) = \int_0^L \left[\frac{1}{2} EI(u'')^2 - f(x)u\right] dx.$$

Hint. See Sec. 6-2.

8. The class $\mathfrak{A}$ of admissible functions for the beam functional in Prob. 7 depends on how the ends of the beam are supported. If $u \in \mathfrak{A}$ and $u + \varepsilon\zeta \in \mathfrak{A}$, show that

$$\delta V(u)\zeta = \int_0^L [EIu''\zeta'' - f(x)\zeta]\, dx.$$

The variational equation for $u = u(x)$, the equilibrium shape of the beam, is

$$\int_0^L [EIu''\zeta'' - f(x)\zeta]\, dx = 0$$

for all ζ such that $u + \varepsilon\zeta \in \mathfrak{A}$.

9. In Probs. 7 and 8 it is natural to assume that $u \in C^2[0, L]$. Assume further that $u \in C^4[0, L]$ and deduce from Prob. 8 that

$$\Big[EIu''\zeta' - (EIu'')'\zeta\Big]_0^L + \int_0^L [(EIu'')'' - f(x)]\zeta\, dx = 0$$

for all ζ such that $u + \varepsilon\zeta \in \mathfrak{A}$. We allow the possibility of an inhomogeneous beam, so EI may depend on x.

10. Observe that no matter how the beam is supported at its ends, $u + \varepsilon\zeta \in \mathfrak{A}$ if $\zeta \in C^4[0, L]$ and vanishes near the endpoints 0 and L. Deduce from Prob. 9 that $(EIu'')'' = f(x)$, a fourth-order differential equation for the equilibrium shape of the beam.

11. In view of Prob. 10, you should expect four boundary conditions to go with the equilibrium beam equation. These side conditions come from the admissible class $\mathfrak{A}$ and the variational equation, which, in view of Probs. 9 and 10, is

$$(*) \qquad \Big[EIu''\zeta' - (EIu'')'\zeta\Big]_0^L = 0$$

for all ζ such that $u + \varepsilon\zeta \in \mathfrak{A}$. First, check this. Then show the following.

(a) If the beam is clamped at both ends, then

$$\mathfrak{A} = \{u \in C^4[0, L] : u(0) = u'(0) = u(L) = u'(L) = 0\}.$$

In this case $(*)$ reduces to $0 = 0$ and the boundary conditions are given by $\mathfrak{A}$.

(b) If the left end is clamped and the right end is free, then

$$\mathfrak{A} = \{u \in C^4[0, L]: u(0) = u'(0) = 0\}.$$

Show that (*) yields the natural boundary conditions

$$EIu'' = 0 \quad \text{and} \quad (EIu'')' = 0 \quad \text{at} \quad x = L.$$

(c) If the left end is clamped and the right simply supported,

$$\mathfrak{A} = \{u \in C^4[0, L]: u(0) = u'(0) = u(L) = 0\}.$$

Show that (*) yields the natural boundary condition

$$EIu'' = 0 \quad \text{at} \quad x = L.$$

Problems 12–16 shed some light on a famous conjecture called *Dirichlet's principle:* The Dirichlet problem of finding $u \in C^2(D) \cap C(\bar{D})$ such that

$$\Delta u = 0 \quad \text{in} \quad D \quad \text{and} \quad u = f \quad \text{on} \quad B$$

where D is a bounded domain in n-space and B its boundary is equivalent to the problem of minimizing

$$I(u) = \int_D |\nabla u|^2 \, dx$$

over $\mathfrak{A} = \{u \in C^1(D) \cap C(\bar{D}): u = f \text{ on } B\}$. Here f is a given continuous function on B.

12. If u minimizes I over $\mathfrak{A}$, show that

$$\int_D \nabla u \cdot \nabla \zeta \, dx = 0$$

for all $\zeta \in C^1(D) \cap C(\bar{D})$ with $\zeta = 0$ on B.

13. If $u \in \mathfrak{A}$ satisfies the variational equation in Prob. 12 and $v \in \mathfrak{A}$ show that $I(u) \le I(v)$. Conclude that u minimizes I over $\mathfrak{A}$. *Hint.* Write $v = u + (v - u) = u + \zeta$ and calculate $I(v)$. Check that ζ has the properties required to apply the variational equation.

14. Let $u \in \mathfrak{A}$ minimize I over $\mathfrak{A}$. Assume additionally that $u \in C^2(D)$. Prove

$$\Delta u = 0 \quad \text{in} \quad D, \qquad u = f \quad \text{on} \quad B.$$

15. If the Dirichlet problem in Prob. 14 has a solution $u \in C^2(D) \cap C(\bar{D})$ show that u minimizes I over $\mathfrak{A}$. *Hint.* Reverse the reasoning used in Prob. 14.

16. Do Probs. 12–15 prove the Dirichlet principle? If not, why not?

17. Establish the analogues of (1-8) and (1-10) for the functional (1-15). *Hint.* Apply the Gauss divergence theorem to the terms $F_{p_1}\zeta_{x_1} + \cdots + F_{p_n}\zeta_{p_n}$ in the variational equation following (1-15). Then use the fundamental lemma.

11-2. Hamilton's Principle

The partial differential equations we treated prior to Chap. 11 were derived by applying basic mechanical principles, such as Newton's second law, to the systems under study. Under certain circumstances, these equations of motion can be deduced from energy considerations and a rather general variational principle known as Hamilton's principle. This principle sheds new light on the equations of motion themselves, and in addition it leads naturally to equations of motion which apply when the classical partial differential equations are inappropriate. Furthermore, the variational principle suggests effective numerical solution procedures. The rest of this chapter is devoted to an introductory development of these ideas and methods.

Assume that we are studying the evolution of a mechanical system over time. We assume that the state of the system can be described by a function $u = u(x, t)$, where x denotes position and t time. For instance, u may be the deflection of a string or membrane from some reference position. The mechanical system may be subject to certain geometrical or kinematical constraints which impose certain general restrictions on $u(x, t)$. For example, for a vibrating string of length L with pinned ends we have $u(0, t) = 0$ and $u(L, t) = 0$ for all $t \geq 0$. For all states $u(x, t)$ consistent with these general restrictions, we assume that the mechanical system has a well-defined potential energy $V = V(t, u, \nabla_x u)$ and kinetic energy $T = T(t, u, u_t)$, where $\nabla_x u = (u_{x_1}, u_{x_2}, u_{x_3})$. The difference $L = T - V$ is called the *Lagrangian* (function) for the mechanical system. *Hamilton's principle* states that the mechanical system evolves from its state at any time t_1 to its state at any subsequent time t_2 in such a way that the integral

$$\int_{t_1}^{t_2} L\, dt \tag{2-1}$$

is stationary.

It is useful to call a state $u = u(x, t)$ a *virtual motion* of the mechanical system if this state is consistent with the general geometrical and kinematical constraints imposed by the physical situation. Then we can express Hamilton's principle as follows:

> The actual motion of a (nondissipative) physical system makes the integral (2-1) stationary with respect to all virtual motions which lead from the actual initial to the actual final state of the system in the same time.

To illustrate the use of Hamilton's principle, consider a one-dimensional, elastic string which is fastened at both ends and executes small transverse oscillations. As usual, we model the string in equilibrium as the segment, $0 \leq x \leq L$, of the x axis, and let $u(x, t)$ be the transverse deflection from equilibrium. The total kinetic energy of the string is

$$T = \frac{1}{2}\int_0^L \rho u_t^2(x, t)\, dx,$$

where ρ is the linear density, which we assume is constant. As in the case of an elastic membrane, the total potential energy is

$$V = \int_0^L \left[\frac{1}{2}\tau u_x^2(x, t) - u(x, t)F(x, t)\right] dx.$$

According to Hamilton's principle, the string will assume the shape $u = u(x, t)$, which makes the integral

(2-2) $$I(v) = \int_{t_1}^{t_2} \int_0^L \left(\frac{1}{2} \rho v_t^2 - \frac{1}{2} \tau v_x^2 + vF \right) dx\, dt$$

stationary among all (virtual motions) functions $v(x, t)$ such that

(2-3) $$v(0, t) = v(L, t) = 0 \quad \text{for} \quad t_1 \le t \le t_2.$$

As in Sec. 11-1, we write $v = u + \varepsilon\zeta$, where u is the actual motion of the string. For v to satisfy (2-3) we must require that

(2-4) $$\zeta(0, t) = \zeta(L, t) = 0 \quad \text{for} \quad t_1 \le t \le t_2.$$

Also, in order that v determine the actual state of the string at times t_1 and t_2, we must have $v = u$ and $v_t = u_t$ when $t = t_1$ and $t = t_2$. This requires that

(2-5) $$\zeta(x, t_i) = \zeta_t(x, t_i) = 0 \quad \text{for} \quad 0 \le x \le L \quad \text{and} \quad i = 1, 2.$$

Since u makes the functional I stationary, we must, of course, assume that $I(u)$ is finite. In view of the integrand in (2-2) and the Schwarz inequality, this is certainly the case if we assume that F, u, u_t, and u_x are square integrable over the rectangle in question. To be sure that $v = u + \varepsilon\zeta$ is square integrable, we could impose the same integrability restrictions on ζ. However, it turns out to be just as useful and simpler to assume that $\zeta(x, t)$ is smooth on the rectangle $0 \le x \le L$, $t_1 \le t \le t_2$. In fact, any order of smoothness can be assumed, but for our purposes it is enough to suppose that $\zeta(x, t)$ is continuously differentiable.

For such $v = u + \varepsilon\zeta$, the first variation of I at u is easily calculated to be

$$\delta I = \delta I(u)\zeta = \int_{t_1}^{t_2} \int_0^L (\rho u_t \zeta_t - \tau u_x \zeta_x + \zeta F)\, dx\, dt$$

(see Prob. 1). According to Hamilton's principle the string moves so that $\delta I = 0$; that is,

(2-6) $$\int_{t_1}^{t_2} \int_0^L (\rho u_t \zeta_t - \tau u_x \zeta_x + \zeta F)\, dx\, dt = 0$$

for all smooth functions ζ satisfying (2-4) and (2-5). Thus, Hamilton's principle leads to the following formulation of the vibrating string problem:

> Suppose that F, f, and g are square integrable. Find a function u such that u, u_t, u_x are square integrable over $0 \le x \le L$ and $t_1 \le t \le t_2$ (any finite time interval), $u(0, t) = u(L, t) = 0$, $u(x, 0) = f(x)$, $u_t(x, 0) = g(x)$ and such that (2-6) holds for all smooth functions ζ that satisfy (2-4) and (2-5).

Although this new formulation of the vibrating string problem may strike you as more complex or less appealing than the earlier one (Chap. 4) based on the classical wave equation, the present point of view has an important advantage. There are solutions to (2-6) which correspond to bone fide physical waves (e.g., wave profiles with corners or shocks) which do not satisfy the wave equation.

We give such an example shortly. First, however, we relate the new formulation of the vibrating string problem to our earlier one. To do so, we simply add to (2-6) the purely mathematical assumption that the solution u be twice continuously differentiable. Then we integrate by parts in (2-6) and use (2-4) and (2-5) to find

$$\int_{t_1}^{t_2} \int_0^L (\rho u_{tt} - \tau u_{xx} - F)\zeta \, dx \, dt = 0$$

for all smooth ζ satisfying (2-4) and (2-5). Now, the fundamental lemma of the calculus of variations gives

$$\rho u_{tt} - \tau u_{xx} - F = 0, \tag{2-7}$$

the familiar one-dimensional wave equation.

Here is a simple example of a wave motion that satisfies the variational equation derived from Hamilton's principle, but is not a solution of the wave equation. We take an infinite string which initially coincides with the x axis and hit it with a hammer. This gives initial data of the type

$$u(x, 0) = 0, \qquad u_t(x, 0) = g(x), \qquad -\infty < x < \infty,$$

where

$$g(x) = \left\{ \begin{matrix} 1, & -1 \le x \le 1, \\ 0, & |x| > 1, \end{matrix} \right\} = H(x + 1) - H(x - 1),$$

where H is the Heaviside step function. We assume there is no external load ($F = 0$), so (2-2) is replaced by

$$I(v) = \int_{t_1}^{t_2} \int_{-\infty}^{\infty} \left(\frac{1}{2} \rho v_t^2 - \frac{1}{2} \tau v_x^2 \right) dx \, dt \tag{2-8}$$

and the variational equation (2-6) becomes

$$\int_{t_1}^{t_2} \int_{-\infty}^{\infty} (\rho u_t \zeta_t - \tau u_x \zeta_x) \, dx \, dt = 0, \tag{2-9}$$

for all smooth functions ζ satisfying (2-5).

On the basis of d'Alembert's solution, we expect that the string will vibrate according to

$$u(x, t) = \frac{1}{2c} \int_{x-ct}^{x+ct} g(y) \, dy, \tag{2-10}$$

where $c^2 = \tau/\rho$. In fact, if there were a twice-continuously differentiable solution, it would be given by (2-10). However, $u(x, t)$ given by (2-10) is not twice continuously differentiable and there is no solution in the "classical" sense. Nonetheless, the string does vibrate, and (2-10) is the solution according to Hamilton's principle. That is, (2-10) solves the variational equation (2-9). To see this, recall that the value of a double integral is not affected by the values of its integrand on a finite number of smooth curves in the domain of integration. For (x, t) not on the lines

$x + ct = \pm 1$ and $x - ct = \pm 1$, we deduce from (2-10) that

$$u_t = \tfrac{1}{2}[H(x + ct + 1) - H(x + ct - 1) + H(x - ct + 1) - H(x - ct - 1)]$$

and

$$u_x = \frac{1}{2c}[H(x + ct + 1) - H(x + ct - 1) - H(x - ct + 1) + H(x - ct - 1)].$$

Clearly, u_t and u_x are square integrable over $\{(x, t)\colon -\infty < x < \infty, t_1 \le t \le t_2\}$. Also,

$$\begin{aligned}\int_{t_1}^{t_2}\int_{-\infty}^{\infty} u_t\zeta_t\,dx\,dt &= \frac{1}{2}\int_{t_1}^{t_2}\left(\int_{-ct-1}^{-ct+1}\zeta_t\,dx + \int_{ct-1}^{ct+1}\zeta_t\,dx\right)dt\\ &= \frac{c}{2}\int_{t_1}^{t_2}[\zeta(-ct+1, t) - \zeta(-ct-1, t)\\ &\qquad - \zeta(ct+1, t) + \zeta(ct-1, t)]\,dt,\end{aligned}$$

because $\zeta(x, t_1) = \zeta(x, t_2) = 0$. Next,

$$\begin{aligned}\int_{t_1}^{t_2}\int_{-\infty}^{\infty} u_x\zeta_x\,dx\,dt &= \frac{1}{2c}\int_{t_1}^{t_2}[\zeta(-ct+1, t) - \zeta(-ct-1, t)\\ &\qquad - \zeta(ct+1, t) + \zeta(ct-1, t)]\,dt.\end{aligned}$$

Since $c^2 = \tau/\rho$ combining these expressions yields

$$\int_{t_1}^{t_2}\int_{-\infty}^{\infty}(\rho u_t\zeta_t - \tau u_x\zeta_x)\,dx\,dt = \rho\int_{t_1}^{t_2}\int_{-\infty}^{\infty}(u_t\zeta_t - c^2 u_x\zeta_x)\,dx\,dt = 0.$$

Therefore, (2-10) does indeed solve the variational equation (2-9) which arises from Hamilton's principle.

We close this section by observing that Hamilton's principle implies the principle of minimum potential energy. Indeed, if we seek equilibrium states there is no kinetic energy so $T = 0$, the admissible states $u = u(x)$ are time independent, and the potential energy can only depend on u. Then

$$\int_{t_1}^{t_2} L\,dt = -(t_1 - t_2)V(u),$$

and Hamilton's principle amounts to making the potential energy functional stationary.

PROBLEMS

1. Calculate $\delta I = \delta I(u)\zeta$ for the functional (2-2) and so confirm (2-6).

2. Refer to Sec. 11-1.

(a) Find the Lagrangian functional for the small transverse vibrations $u = u(x, t) = u(x_1, x_2, t)$ of an elastic membrane. Assume that the membrane is fixed at its boundary and that a load $F = F(x, t)$ is applied.

(b) Use Hamilton's principle to determine the variational equation for the vibrations. State the natural assumptions on u and the requirements on $\zeta = \zeta(x, t)$.

(c) In addition to the assumptions in part (b), suppose that u is twice continuously differentiable. Derive the classical equation

$$\rho u_{tt} = \tau \, \Delta u + F$$

for the small transverse vibrations of a membrane.

3. Consider the transverse vibrations of a beam of length L which extends from $0 \le x \le L$ when in equilibrium. Let $u = u(x, t)$ be its transverse deflection.

(a) Show that the Lagrangian functional is

$$I(v) = \int_{t_1}^{t_2} \int_0^L \left(\frac{1}{2} \rho v_t^2 - \frac{1}{2} EIv_{xx}^2 + fv \right) dx \, dt$$

where $\rho = \rho(x)$, $E = E(x)$, $I = I(x)$, and $f = f(x, t)$ are, respectively, the density, Young's modulus, the moment of inertia of the cross section at x, and the external load.

(b) Find the variational equation for the vibrating beam by applying Hamilton's principle.

4. Let $\mathfrak{A}$ specify the virtual motions for the beam in Prob. 3. The exact specification of $\mathfrak{A}$ depends on how the beam is supported. Assume that the actual motion u of the beam is four times continuously differentiable. If $u + \varepsilon\zeta \in \mathfrak{A}$, derive the relation

$$\delta I = \delta I(u)\zeta = \int_{t_1}^{t_2} [-EIu_{xx}\zeta_x + (EIu_{xx})_x \zeta] \, dt$$
$$+ \int_{t_1}^{t_2} \int_0^L [-\rho u_{tt} - (EIu_{xx})_{xx} + f]\zeta \, dx \, dt.$$

5. Use the result of Prob. 4 to derive the classical beam equation

$$\rho u_{tt} + (EIu_{xx})_{xx} = f$$

and proceed as in Prob. 11 of Sec. 11-1 to develop appropriate boundary conditions for the beam.

6. Consider the small transverse vibrations of a string as in the text.

(a) If the ends of the string are not pinned but subject to elastic restoring forces, argue that the terms $h_0 u(0, t)^2/2$ and $h_L u(L, t)^2/2$, where h_0, h_L are positive constants, must be added to the potential energy.

(b) Assume that motion is twice continuously differentiable and apply Hamilton's principle to find that the actual motion satisfies the wave equation (2-7) and the natural boundary conditions

$$-\tau u_x(0, t) + h_0 u(0, t) = 0, \qquad \tau u_x(L, t) + h_L u(L, t) = 0.$$

(c) If the ends of the string are free, find the natural boundary conditions.

7. Answer Prob. 6 for a membrane instead of a string. In part (a) show that the new potential energy term is

$$\int_\Gamma \frac{1}{2} h(x) u^2 \, d\sigma_x,$$

where $h(x) > 0$, Γ is the boundary of the membrane, which can move only vertically.

Probems 8 and 9 show that dissipative systems can sometimes be given a variational description.

8. Let

$$I(u) = \int_{t_1}^{t_2} \int_0^L e^{\alpha t} \left(\frac{1}{2} \rho u_t^2 - \frac{1}{2} \tau u_x^2 \right) dx\, dt$$

with $u(0, t) = u(L, t) = 0$. Calculate $\delta I(u)\zeta$ and show that the variational equation $\delta I = 0$ leads to the damped wave equation

$$\rho u_{tt} + \alpha \rho u_t = \tau u_{xx},$$

assuming that u is suitably smooth.

9. The Lagrangian functional for a damped vibrating plate D with fixed boundary is

$$I(u) = \int_{t_1}^{t_2} \int_D e^{\alpha t} \left[\frac{1}{2} \rho u_t^2 - \frac{1}{2} E(\Delta u)^2 \right] dx\, dt$$

where E is an elastic contant. Find the first variation of I and apply Hamilton's principle to find the differential equation.

10. Let

$$I(u) = \frac{1}{2} \int_D e^{-v \cdot x} |\nabla u|^2\, dx,$$

where v is a constant vector and $u = f$ on B, the boundary of D. Calculate $\delta I(u)$. If I is stationary at u and u is twice continuously differentiable, show that $\Delta u - v \cdot \nabla u = 0$.

11. Let

$$V(v) = \int_D \left(\frac{1}{2} |\nabla v|^2 - vF \right) dx$$

be the potential energy function for an n-dimensional membrane. Assume that D is bounded and $v = f$ on B, the boundary of D.

(a) If u minimizes V (or just makes V stationary), show that

$$\int_D (\nabla u \cdot \nabla \zeta - \zeta F)\, dx = 0$$

for all smooth ζ such that $\zeta = 0$ on B.

(b) If u satisfies the variational equation in part (a), show that $V(u) \leq V(v)$ for all v with $v = f$ on B. Thus, minimizing V is equivalent to solving the variational equation. *Hint.* Write $v = u + \zeta$, where $\zeta = v - u$, and expand $V(v) = V(u + \zeta)$.

(c) Show that there is at most one u that minimizes V.

11-3. Hilbert Space and Strong and Weak Convergence

This section and the next set the stage for showing that the variational equations developed in Secs. 11-1 and 11-2 have solutions. The ideas to be introduced also play a decisive role in analyzing generalizations of many of the problems we encountered in earlier chapters. A higher level of abstraction is required. Nevertheless, the numerical implementation of the solution precedures we shall discuss is actually quite straightforward and such numerical methods have proven to be accurate and flexible. Throughout this section we use basic properties of vector (or linear) spaces, norms, and inner products.

Recall that a vector space V is just a set of elements, called vectors, which obey the usual algebraic laws of vector arithmetic in ordinary Euclidean space. Since the vector spaces of primary interest to us are function spaces, we denote vectors in V by letters such as $f, g, \zeta, \ldots$ usually reserved for functions. The vector space V is called a normed vector space or normed linear space if each vector f in V is assigned a length or norm $\|f\|$ which satisfies properties (i)–(iii) of Sec. 7-4. In many cases of interest, the vector space V is an inner product space. That is, a real or complex number $\langle f, g \rangle$ is assigned to each pair of vectors f and g in V. The inner product $\langle , \rangle$ satisfies properties (i)–(iv) of Sec. 7-6, which mimic familiar properties of the dot product in 3-space. The inner product induces a norm on V given by

$$\|f\| = \sqrt{\langle f, f \rangle}. \tag{3-1}$$

Since the Schwarz inequality

$$|\langle f, g \rangle| \leq \|f\| \, \|g\| \tag{3-2}$$

holds in any inner product space, we can define the angle θ between two nonzero vectors f and g by

$$\theta = \arccos \frac{\langle f, g \rangle}{\|f\| \, \|g\|}.$$

Thus, we say that f and g are orthogonal if $\langle f, g \rangle = 0$. We also have the familiar result $\langle f, g \rangle = \|f\| \, \|g\| \cos \theta$.

Convergence in any normed space with norm $\|\cdot\|$ is defined as follows:

$$\lim_{n \to \infty} f_n = f \quad \text{or} \quad f_n \to f \quad \text{as} \quad n \to \infty$$

means

$$\lim_{n \to \infty} \|f_n - f\| = 0.$$

In this section we restrict ourselves to inner product spaces and use the norm (3-1). To distinguish this notion of convergence from a concept to be introduced shortly, we sometimes refer to this type of convergence as *strong convergence*. A sequence $\{f_n\}$ in an inner product space is a Cauchy sequence if for each $\varepsilon > 0$ there is an integer $N = N(\varepsilon)$ such that $\|f_n - f_m\| < \varepsilon$ whenever $n, m > N$. We noted in Lemma 4-1 of Sec. 7-4 that every convergent sequence is a Cauchy sequence. Unfortunately, Cauchy sequences are not necessarily convergent (see Prob. 6). An inner product space in which every Cauchy sequence converges is called complete. A complete inner product space is called a *Hilbert space*. In the rest of this section, H denotes a Hilbert space.

A set S is *dense* in H if for every f in H and $\varepsilon > 0$ there is a g in S with $\|g - f\| < \varepsilon$; equivalently, there is a sequence $\{g_n\}$ in S with $g_n \to f$ as $n \to \infty$. In other words, S is dense in H if the set $\bar{S}$ of all limit points of S actually equals

H. Let $\mathscr{B}$ be a set of linearly independent vectors in H and let $S = \text{sp}(\mathscr{B})$ be the set of all finite linear combinations of vectors in $\mathscr{B}$. If S is dense in H, that is if $\bar{S} = H$, we call $\mathscr{B}$ a basis for H. It follows from elementary linear algebra that if H has a basis $\mathscr{B}$ which is a finite set, with say d elements, then all bases for H have d elements and d is called the dimension of H. If H has a basis $\mathscr{B}$ which consists of a sequence of vectors, then all bases of H consist of sequences of vectors and we say that H is a *separable* Hilbert space. In this case, the dimension of H is said to be countably infinite. If $\mathscr{B}$ is a linearly independent set of vectors in H it is always possible to extend $\mathscr{B}$ to a basis for H. This familiar result from finite dimensions holds in any Hilbert space; however, we omit the proof. There are Hilbert spaces whose bases are uncountable. Such spaces are of no interest to us. From now on, we assume that all Hilbert spaces considered are separable. Here is a useful property about bases.

Lemma 3-1. Suppose that $\mathscr{B}$ is a basis for H and that $f \in H$. If $\langle f, \phi \rangle = 0$ for all ϕ in $\mathscr{B}$, then $f = 0$.

Proof. Given $\varepsilon > 0$ there is a finite linear combination of basis vectors $f_n = \sum_{k=1}^{n} a_k \phi_k$ such that $\|f - f_n\| < \varepsilon$. Since $\langle f, f_n \rangle = 0$,

$$\|f\|^2 = \langle f, f - f_n \rangle \leq \|f\| \, \|f - f_n\| \leq \varepsilon \|f\|,$$
$$\|f\| \leq \varepsilon,$$

and since $\varepsilon > 0$ is arbitrary, $f = 0$.

Let H be a Hilbert space with basis $\mathscr{B} = \{\phi_k\}_{k=1}^{\infty}$. Since the ϕ_k are linearly independent, we can apply the Gram–Schmidt orthogonalization procedure (see Sec. 7-6) to obtain an orthonormal set of vectors $\{e_j\}_{j=1}^{\infty}$ such that

$$\text{sp}\{\phi_1, \ldots, \phi_r\} = \text{sp}\{e_1, \ldots, e_r\}$$

for $r = 1, 2, \ldots$. Thus, $\mathscr{C} = \{e_i\}_{j=1}^{\infty}$ is an orthonormal basis for H.

Proposition 3-1. Every Hilbert space has an orthonormal basis.

Given an orthonormal set $\{e_j\}$, the number

$$c_j = \langle f, e_j \rangle$$

is called the jth *Fourier coefficient* of f relative to the orthonormal set $\{e_j\}$. If $a_1, \ldots, a_n$ are any scalars, a direct calculation (Prob. 9) yields

$$\left\| f - \sum_{j=1}^{n} a_j e_j \right\|^2 = \|f\|^2 - \sum_{j=1}^{n} |c_j|^2 + \sum_{j=1}^{n} |c_j - a_j|^2, \tag{3-3}$$

where c_j is the jth Fourier coefficient of f. Evidently, the expression on the left of (3-3) is uniquely minimized by the choice $a_j = c_j$. Thus,

$$\left\| f - \sum_{j=1}^{n} \langle f, e_j \rangle e_j \right\| \leq \left\| f - \sum_{j=1}^{n} a_j e_j \right\| \tag{3-4}$$

for any choice of coefficients a_j. In words, the best approximation to f from the subspace sp $\{e_1, \ldots, e_n\}$ is the nth partial sum of the Fourier series $\sum_{j=1}^{\infty} \langle f, e_j \rangle e_j$ of f relative to the orthornormal set $\{e_j\}$. These simple calculations lead to the following important result.

Theorem 3-1. Let $\mathscr{B} = \{e_j\}_{j=1}^{\infty}$ be an orthonormal set in a Hilbert space H. Then for any f in H.

$$\sum_{j=1}^{\infty} |\langle f, e_j \rangle|^2 \le \|f\|^2 \qquad \text{(Bessel's inequality)}.$$

Moreover, $\mathscr{B}$ is an orthonormal basis if and only if

$$\sum_{j=1}^{\infty} |\langle f, e_j \rangle|^2 = \|f\|^2 \qquad \text{(Parseval's equality)}$$

holds for every f in H.

Proof. To get Bessel's inequality apply (3-3) with $a_j = c_j = \langle f, e_j \rangle$ and then let $n \to \infty$. Next, (3-4) reveals that $\mathscr{B}$ is a basis precisely when

$$f = \sum_{j=1}^{\infty} \langle f, e_j \rangle e_j \tag{3-5}$$

for every f in H. Another application of (3-3) with $a_j = \langle f, e_j \rangle$ shows that (3-5) holds exactly when

$$\|f\|^2 = \sum_{j=1}^{\infty} |\langle f, e_j \rangle|^2,$$

which completes the proof.

Let M be a subset of H which is closed ($M = \bar{M}$) and is also a vector space. In this case, we all M a *subspace* of H. M is a Hilbert space in its own right (Prob. 11) and so has an orthonormal basis, say $\{e_j\}$. The set

$$N = \{f \in H : \langle f, e \rangle = 0 \quad \text{for all} \quad e \in M\},$$

which is called the *orthogonal complement of M*, is also a subspace (Prob. 12). Thus, N has an orthonormal basis, say $\{f_i\}$. We assert that

$$\mathscr{B} = \{e_1, e_2, \ldots, f_1, f_2, \ldots\}$$

is a basis for H. Indeed, given $g \in H$ it is elementary to check that $h = g - \sum \langle g, e_j \rangle e_j$ is in N and hence

$$g - \sum \langle g, e_j \rangle e_j = \sum \langle h, f_i \rangle f_i, \qquad g = \sum \langle g, e_j \rangle e_j + \sum \langle h, f_i \rangle f_i,$$

which shows that the closed span of $\mathscr{B}$ is H. This equation also shows that each g in H can be expressed as

$$g = e + f \quad \text{with} \quad e \in M \quad \text{and} \quad f \in N. \tag{3-6}$$

Furthermore, this representation is unique because

$$e' + f' = g = e + f \quad \text{with} \quad e, e' \in M \quad \text{and} \quad f, f' \in N$$

implies that $e' - e = f - f'$, which belongs to N. Then $\|e' - e\|^2 = \langle e' - e, f - f' \rangle = 0$, so $e' = e$, and consequently $f' = f$. We write $H = M \oplus N$ to express the unique decomposition in (3-6). The last calculation also shows that 0 is the only vector in both M and N.

Theorem 3-2. Let H be a Hilbert space, M a subspace, and N be the orthogonal complement of M. Then $H = M \oplus N$ and if $g = e + f$ according to this decomposition, we have $\|g\|^2 = \|e\|^2 + \|f\|^2$.

We have proved everything except the last equality, which is routine:

$$\begin{aligned} \|g\|^2 &= \langle e + f, e + f \rangle = \|e\|^2 + \langle e, f \rangle + \langle f, e \rangle + \|f\|^2 \\ &= \|e\|^2 + \|f\|^2. \end{aligned}$$

If $H = M \oplus N$ and $g = e + f$, we call e the *projection of g on M* and write

$$e = Pg = P_M g.$$

This defines a projection operator $P: H \to H$ with range M. It is easy to see that $e = Pg$ is the best approximation to g from M. Indeed, if m is any vector in M,

$$g - m = (Pg - m) + (g - Pg)$$

expresses $g - m$ as the sum of a vector in M and a vector in N so

$$\|g - m\|^2 = \|Pg - m\|^2 + \|g - Pg\|^2$$

by Th. 3-2. Thus,

$$\|g - Pg\| \leq \|g - m\| \quad \text{for all} \quad m \in M$$

with equality only if $m = Pg$.

In subsequent sections we use these ideas in the following context. We choose an orthonormal basis $\{e_1, e_2, \ldots\}$ for H so that every f in H has the Fourier series expansion (3-5). We take $M_n = \text{sp}\,\{e_1, \ldots, e_n\}$ and note that

$$f = \sum_{j=1}^{n} \langle f, e_j \rangle e_j + \sum_{j=n+1}^{\infty} \langle f, e_j \rangle e_j$$

is the sum of a vector in M_n and in N_n, its orthogonal complement. Therefore,

$$P_n f = \sum_{j=1}^{n} \langle f, e_j \rangle e_j$$

defines a projection operator with range M_n, and $P_n f$ is the best approximation to f from M_n.

In addition to the projection operators just introduced, certain other operators and functionals will be important to us. The simplest such mappings are the *continuous linear functionals* on H. These are the mappings, L, on H which associate with each $f \in H$ a scalar $L(f)$ and are linear $[L(\alpha f + \beta g) = \alpha L(f) + \beta L(g)]$ and continuous $[L(f_n) \to L(f)$ whenever $f_n \to f]$. The key facts we need about linear functionals are summarized in the next two theorems.

Theorem 3-3. Let L be a linear functional on H. The following are equivalent.
(i) L is (uniformly) continuous on H.
(ii) L is continuous at $f = 0$.
(iii) L is bounded; that is, there is a constant C such that

$$|L(f)| \leq C\|f\| \quad \text{for all} \quad f \quad \text{in} \quad H.$$

Proof. See Prob. 13.

Theorem 3-4. (*Riesz Representation Theorem*). Let L be a continuous linear functional on H. Then there is a unique element g in H such that

$$L(f) = \langle f, g \rangle \quad \text{for all} \quad f \quad \text{in} \quad H.$$

Proof. The uniqueness is easy. If

$$\langle f, g_1 \rangle = L(f) = \langle f, g_2 \rangle \quad \text{for all} \quad f \quad \text{in} \quad H,$$

then $\langle f, g_1 - g_2 \rangle = 0$ for all f in H. Choose $f = g_1 - g_2$ to conclude that $\|g_1 - g_2\| = 0$ and $g_1 = g_2$. To produce the required g we split H up into two subspaces

$$M = \{f : L(f) = 0\}$$

and its orthogonal complement. M is a subspace because L is both linear and continuous. If $M = H$, we take $g = 0$ and are done. If $M \neq H$, then its orthogonal complement $N \neq \{0\}$. Fix any h in N with $h \neq 0$. Then $L(h)f - L(f)h$ is in M and thus

$$0 = \langle L(h)f - L(f)h, h \rangle.$$

Solve for $L(f)$ to find

$$L(f) = \langle f, g \rangle$$

for $g = \overline{L(h)}h/\|h\|^2$, which completes the proof.

See Prob. 14 for a constructive proof of Th. 3-4.

We turn now to another mode of convergence in H which arises naturally in variational problems. First, take a finite-dimensional space H with basis $\{e_1, \ldots, e_r\}$. Then each f in H can be expressed as

$$f = \sum_{j=1}^{r} \langle f, e_j \rangle e_j,$$

and there is a corresponding expansion for each element f_n in a sequence $\{f_n\}$ in H. Now

$$\|f_n - f\|^2 = \sum_{j=1}^{r} |\langle f_n - f, e_j \rangle|^2.$$

So $f_n \to f$ in H if and only if $\langle f_n - f, e_j \rangle \to 0$ as $n \to \infty$ for each $j = 1, \ldots, r$. Since any vector ζ in H is a *finite* linear combination of $e_1, \ldots, e_r$, it follows that the previous inner products have limit zero exactly when $\langle f_n - f, \zeta \rangle \to 0$ as $n \to \infty$ for each ζ in H. In finite dimensions, we have

$$f_n \to f \quad \text{in} \quad H \leftrightarrow \langle f_n, \zeta \rangle \to \langle f, \zeta \rangle \quad \text{for each} \quad \zeta \quad \text{in} \quad H.$$

In infinite-dimensional Hilbert spaces this equivalence breaks down. The implication from left to right is always true (Prob. 15); however, the converse is not true. To see this, apply Bessel's inequality to each ζ in H to get

$$\sum_{n=1}^{\infty} |\langle e_n, \zeta \rangle|^2 \leq \|\zeta\|^2 < \infty,$$

$$\langle e_n, \zeta \rangle \to 0 = \langle 0, \zeta \rangle \quad \text{as} \quad n \to \infty.$$

On the other hand, e_n does not converge to 0 because $\|e_n - 0\| = \|e_n\| = 1$ for all n.

In view of the foregoing discussion, we say that $\{f_n\}$ *converges weakly* to f in H and write

$$f_n \rightharpoonup f \quad \text{as} \quad n \to \infty$$

if

$$\lim_{n \to \infty} \langle f_n, \zeta \rangle = \langle f, \zeta \rangle \quad \text{for each} \quad \zeta \quad \text{in} \quad H.$$

In finite dimensions (strong) convergence and weak convergence are equivalent; however, in infinite dimensions, convergence implies weak convergence, but weakly convergent sequences need not converge (strongly).

For the variational analysis in later sections, we need the following properties of weak convergence.

Theorem 3-5. Let $f_n, f \in H$ for $n = 1, 2, \ldots$.

(i) If $\{f_n\}$ converges weakly to f, then there is a constant C such that $\|f_n\| \leq C$ for all n.

(ii) If for each ζ in H the sequence of scalars $\{\langle f_n, \zeta \rangle\}$ converges, then $\{f_n\}$ converges weakly to some f in H.

(iii) If a sequence $\{f_n\}$ is uniformly bounded, that is there is a constant C such that $\|f_n\| \leq C$ for all n, then there is a subsequence $\{f_{n_k}\}$ which converges weakly.

Proof. To prove (i) it suffices to show that there is a constant C such that

$$\text{(3-7)} \qquad |\langle f_n, \zeta \rangle| \leq C\|\zeta\| \quad \text{for all} \quad \zeta \in H \quad \text{and} \quad n = 1, 2, \ldots.$$

For then we could just set $\zeta = f_n$ in (3-7) to get (i). The estimate (3-7) holds if there is a point ζ_0 in H and positive constants C_1 and δ such that

$$\text{(3-8)} \quad |\langle f_n, \zeta \rangle| \leq C_1\|\zeta\| \quad \text{for all} \quad \zeta \quad \text{with} \quad \|\zeta - \zeta_0\| \leq \delta \quad \text{and} \quad n = 1, 2, \ldots.$$

Indeed, suppose that (3-8) holds and let ξ be any nonzero element in H and apply (3-8) with $\zeta = a\xi + \zeta_0$, where $a = \delta/2\|\xi\|$, to get

$$|\langle f_n, a\xi + \zeta_0\rangle| \le C_1\|a\xi + \zeta_0\| \le \frac{C_1\delta}{2} + C_1\|\zeta_0\|.$$

Since $\langle f_n, a\xi\rangle = \langle f_n, a\xi + \zeta_0\rangle - \langle f_n, \zeta_0\rangle$, the triangle inequality gives

$$|\langle f_n, a\xi\rangle| \le \frac{C_1\delta}{2} + 2C_1\|\zeta_0\|,$$

$$|\langle f_n, \xi\rangle| \le \left[C_1 + \frac{4C_1}{\delta}\|\zeta_0\|\right]\|\xi\| = C\|\xi\|.$$

which establishes (3-7).

We prove (3-8) by contradiction. Suppose that no point ζ_0 in H and positive constants C_1 and δ exist satisfying (3-8). Pick any point ζ_0 in H and $\delta_0 > 0$. Then (3-8) cannot hold with $C_1 = 1$ and $\delta = \delta_0/2$. So there exist ζ_1 and n_1 such that $\|\zeta_1 - \zeta_0\| \le \delta_0/2$ and $|\langle f_n, \zeta_1\rangle| > \|\zeta_1\|$. By continuity of the inner product there is $\delta_1 > 0$ such that

$$\delta_1 \le \frac{\delta_0}{2} \quad \text{and} \quad |\langle f_{n_1}, \zeta\rangle| > \|\zeta\| \quad \text{for} \quad \|\zeta - \zeta_1\| \le \delta_1.$$

Similarly, since (3-8) fails with $\zeta_0 = \zeta_1$, $C_1 = 2$, $\delta = \delta_1/2$ there exist ζ_2 and n_2 with $\|\zeta_2 - \zeta_1\| \le \delta_1/2$ and $|\langle f_{n_2}, \zeta_2\rangle| > 2\|\zeta_2\|$. By continuity there is a $\delta_2 > 0$ such that

$$\delta_2 \le \frac{\delta_1}{2} \quad \text{and} \quad |\langle f_{n_2}, \zeta\rangle| > 2\|\zeta\| \quad \text{for} \quad \|\zeta - \zeta_2\| \le \delta_2.$$

Continuing in this way we construct sequences $\{\zeta_k\}$, $\{\delta_k\}$, and $\{n_k\}$ such that

$$\|\zeta_k - \zeta_{k-1}\| \le \frac{\delta_{k-1}}{2}, \qquad \delta_k \le \frac{\delta_{k-1}}{2},$$

and

$$|\langle f_{n_k}, \zeta\rangle| > k\|\zeta\| \quad \text{for} \quad \|\zeta - \zeta_k\| \le \delta_k.$$

The points $\{\zeta_k\}$ form a Cauchy sequence and hence converge to say ζ^* in H. Furthermore, for each k, $\|\zeta^* - \zeta_k\| \le \delta_k$ (see Prob. 16) so that

$$|\langle f_{n_k}, \zeta^*\rangle| > k\|\zeta^*\|. \tag{3-9}$$

The left-hand side is bounded because the numerical sequence $\langle f_{n_k}, \zeta^*\rangle$ is convergent in view of the weak convergence of $\{f_n\}$. The right-hand side is unbounded as $k \to \infty$ because (3-9) shows that $\zeta^* \neq 0$. This contradiction proves (3-8) and (i) follows.

In the proof of (i), the only property of weak convergence used was that $\{\langle f_n, \zeta\rangle\}$ converges for every ζ. Thus, under the hypothesis in (ii), the bound (3-7) still holds. Set

$$L(\zeta) = \lim_{n\to\infty} \langle \zeta, f_n\rangle \quad \text{for} \quad \zeta \text{ in } H. \tag{3-10}$$

Clearly, L is linear and from (3-7) $|L(\zeta)| \leq C\|\zeta\|$, so L is continuous by Th. 3-3. Then by Th. 3-4 there is a vector f in H such that $L(\zeta) = \langle \zeta, f \rangle$ and (3-10) yields $\langle \zeta, f_n \rangle \to \langle \zeta, f \rangle$ as $n \to \infty$ for each ζ in H. That is, $f_n \rightharpoonup f$ as $n \to \infty$.

To prove (iii) suppose that $\|f_n\| \leq C$ for some constant C and $n = 1, 2, \ldots$. Let M be the closure of sp $\{f_1, f_2, \ldots\}$, so M is a subspace of H. Let N be the orthogonal complement of M. The numerical sequence $\{\langle f_n, f_1 \rangle\}$ is bounded, because $|\langle f_n, f_1 \rangle| \leq C\|f_1\|$ by the Schwarz inequality, and hence contains a convergent subsequence, say $\{\langle f_{n,1}, f_1 \rangle\}$. Now the sequence $\{\langle f_{n,1}, f_2 \rangle\}$ is bounded and therefore contains a convergent subsequence, say $\{\langle f_{n,2}, f_2 \rangle\}$. Continuing in this way, at the kth step we find a subsequence $\{f_{n,k}\}$ with

$$\{f_{n,k}\} \subset \{f_{n,k-1}\} \subset \cdots \subset \{f_{n,1}\} \subset \{f_n\}$$

and such that

$$\{\langle f_{n,k}, f_j \rangle\}_{n=1}^{\infty} \quad \text{converges for} \quad j = 1, \ldots, k.$$

Therefore, the "diagonal" sequence $\{f_{n,n}\} \subset \{f_n\}$ converges for each h in sp $\{f_1, f_2, \ldots\}$. That is, $\{\langle f_{n,n}, h \rangle\}$ converges. Since $\|f_{n,n}\| \leq C$, it follows (Prob. 17) that $\{\langle f_{n,n}, g \rangle\}$ converges for every g in M. Finally, each ζ in H can be expressed as $\zeta = g + g'$ with g in M and g' in N. Since $g' \in N$, $\langle f_{n,n}, g' \rangle = 0$ for all n and $\langle f_{n,n}, \zeta \rangle = \langle f_{n,n}, g \rangle$. Therefore, $\{\langle f_{n,n}, \zeta \rangle\}$ converges for all ζ in H and (ii) guarantees the weak convergence of the subsequence $\{f_{n,n}\}$ of $\{f_n\}$.

PROBLEMS

1. Prove that the inner product is continuous in each variable separately and, in fact, is jointly continuous. That is, show: If $f_n \to f$, $g_n \to g$, and $h \in H$, then

$$\langle f_n, h \rangle \to \langle f, h \rangle, \qquad \langle h, g_n \rangle \to \langle h, g \rangle,$$
$$\langle f_n, g_n \rangle \to \langle f, g \rangle,$$

as $n \to \infty$.

2. Let $f_n \in H$. Recall that $\sum_{n=1}^{\infty} f_n$ converges if its sequence of partial sums $\{s_n\}$ converges, where $s_n = f_1 + \cdots + f_n$. If $s_n \to s$, we call s the sum of the series and write $\sum_{n=1}^{\infty} f_n = s$. Let $\sum_{n=1}^{\infty} f_n$ and $\sum_{n=1}^{\infty} g_n$ be convergent and let $h \in H$. Prove

$$\langle \textstyle\sum f_n, h \rangle = \sum \langle f_n, h \rangle, \qquad \langle h, \sum g_n \rangle = \sum \langle h, g_n \rangle,$$

$$\left\langle \sum_n f_n, \sum_m g_m \right\rangle = \sum_{n,m} \langle f_n, g_m \rangle.$$

3. Suppose that $f_n \rightharpoonup f$ and $\zeta_n \to \zeta$. Show that $\langle f_n, \zeta_n \rangle \to \langle f, \zeta \rangle$.
4. Suppose that $f_n \rightharpoonup f$ and $\|f_n\| \to \|f\|$. Show $f_n \to f$.
5. Show that weak limits are unique. That is, $f_n \rightharpoonup f$ and $f_n \rightharpoonup g$ implies that $f = g$.

6. $C[0, 2]$ with $\langle f, g\rangle = \int_0^2 f(x)\overline{g(x)}\,dx$ is an inner product space (see Prob. 4 of Sec. 7-6). Show that $f_n(t) = t^n, 0 \leq t \leq 1$ and $f_n(t) = 1, 1 \leq t \leq 2$ defines a sequence $\{f_n\}$ in $C[0, 2]$. Now, show that $\{f_n\}$ is a Cauchy sequence in this inner product space and that there is no $f \in C[0, 2]$ such that $f_n \to f$.
7. Let $\{e_j\}$ be an orthonormal sequence in a Hilbert space H. Show that $\sum_{j=1}^{\infty} a_j e_j$ converges if and only if $\sum_{j=1}^{\infty} |a_j|^2 < \infty$.
8. Let l_2 be the spaces of sequences of complex numbers $a = \{a_j\}$, $b = \{b_j\}$, and so on, such that $\sum |a_j|^2 < \infty$. Show that $\langle a, b\rangle = \sum a_j \overline{b_j}$ is well-defined and makes l_2 a Hilbert space.
9. Expand out the left side of (3-3) and hence confirm this equation.
10. Let $\{e_j\}$ be an orthonormal basis for H. Show that

$$\langle f, g\rangle = \sum_{j=1}^{\infty} \langle f, e_j\rangle \overline{\langle g, e_j\rangle}$$

for all f, g in H.
11. Verify that a subspace M of a Hilbert space H is itself a Hilbert space. (The point at issue is the completeness of M.)
12. Let M be a subspace of H and N be the orthogonal complement of M. Prove that N is a subspace.
13. Prove Th. 3-3. *Hint.* Refer to Prop. 4-3 of Sec. 7-4.
14. The proof in the text of Th. 3-4 works in any Hilbert space, separable or not. Confirm the following proof for a separable Hilbert space. Let $a_j = L(e_j)$, where $\{e_j\}$ is an orthonormal basis. Show that $\sum a_j e_j$ converges, say with sum f. Then show that $L(\zeta) = \langle \zeta, f\rangle$. (To see why this proof is natural, write $\zeta = \sum \zeta_j e_j$, where $\zeta_j = \langle \zeta, e_j\rangle$ and expand $L(\zeta)$.)
15. If $f_n \to f$ in H, then $\langle f_n, \zeta\rangle \to \langle f, \zeta\rangle$ for each ζ in H. That is, strong convergence implies weak convergence.
16. In the proof of (i) of Th. 3-5, show that $\zeta_k \to \zeta^*$, where

$$\zeta^* = \zeta_0 + \sum_{k=1}^{\infty} (\zeta_k - \zeta_{k-1}) = \zeta_k + \sum_{j=k+1}^{\infty} (\zeta_j - \zeta_{j-1}).$$

Conclude that $\|\zeta^* - \zeta_k\| \leq \delta_k$.
17. In the proof of (iii) of Th. 3-5 show that $\{\langle f_{n,n}, h\rangle\}$ convergent for h in sp $\{f_1, f_2, \ldots\}$ implies that $\{\langle f_{n,n}, g\rangle\}$ converges for g in M. *Hint.* If $g \in M$, there is $h \in$ sp $\{f_1, f_2, \ldots\}$ such that $\|g - h\| < \varepsilon/3$. Observe that

$$\langle f_{n,n}, g\rangle - \langle f_{m,m}, g\rangle = \langle f_{n,n}, g - h\rangle + [\langle f_{n,n}, h\rangle - \langle f_{m,m}, h\rangle] - \langle f_{m,m}, g - h\rangle.$$

18. Show that a set of vectors $f_1, \ldots, f_n$ in an inner product space is linearly independent if and only if the Gram determinant $\det([\langle f_i, f_j\rangle]_{n\times n}) \neq 0$.
19. Let L be a continuous linear functional on H. Then $|L(f)| \leq C\|f\|$ for some constant C and all f in H. The smallest C with this property is called the norm of L and is denoted $\|L\|$. (Compare with the discussion of linear operators in Sec. 7-4.) Show that

$$\|L\| = \sup_{\|\zeta\|=1} |L(\zeta)| = \sup_{\|\zeta\|\leq 1} |L(\zeta)| = \sup_{\zeta\neq 0} \frac{|L(\zeta)|}{\|\zeta\|}.$$

20. Let L be a continuous linear functional on H so that $L(\zeta) = \langle \zeta, f\rangle$ for a unique f in H. Prove that $\|L\| = \|f\|$.

11-4. The Function Spaces H_0, H_1, and $\mathring{H}_1$

The general properties of Hilbert spaces and weak and strong convergence covered in Sec. 11-3 will be applied in the context of three concrete function spaces which have come to play a central role in the modern study of partial differential equations and variational problems.

Consider, for example, the problem of minimizing the integral

$$I(u) = \frac{1}{2}\int_D |\nabla u|^2 \, dx, \tag{4-1}$$

where D is a closed bounded domain and $u = f$, a given function on B, the boundary of D. We shall minimize (4-1), and treat more general problems, by first constructing a sequence $\{u_n\}$ of approximate solutions which turns out to be a Cauchy sequence in the norm derived from the inner product

$$\langle u, v\rangle = \int_D [u\bar{v} + \nabla u \cdot \overline{\nabla v}] \, dx. \tag{4-2}$$

A glance at (4-1) and (4-2) suggests that a reasonable function space to work in would be $C^1(\bar{D})$, where, instead of using the usual maximum norm, we equip $C^1(\bar{D})$ with the inner product (4-2). We denote this inner product space by $\mathscr{H}_1$. Now a fundamental problem arises. Even if we construct the sequence of approximate solution $\{u_n\}$ so that they belong to $\mathscr{H}_1$ and are a Cauchy sequence in $\mathscr{H}_1$, we have no guarantee that the sequence $\{u_n\}$ converges because $\mathscr{H}_1$ is not complete. On the other hand, the approximate solution procedure strongly suggests that the Cauchy sequence $\{u_n\}$ should have a limit that minimizes (4-1). Therefore, we need to enlarge, or complete, the space $\mathscr{H}_1$ so that it contains all the limits of Cauchy sequences in $\mathscr{H}_1$. In effect, we enlarge the function space $\mathscr{H}_1$ so that it will contain the minimizing function we seek.

This enlargement or completion process is a standard mathematical procedure. For example, in the early history of mathematics rational numbers were introduced to solve a variety of problems. Then Pythagoras discovered that no rational number gave the diagonal of a unit square; that is, the equation $x^2 = 2$ has no rational number solution. We can come arbitrarily close to solving this equation with rational numbers. One familiar sequence of approximations is

$$\{1, 1.4, 1.41, 1.414, 1.4142, \ldots\}, \tag{4-3}$$

which is a Cauchy sequence of rational numbers. This Cauchy sequence is identified with a new nonrational real number which we denote by $\sqrt{2}$. Of course, there are other rational sequences which approximate the solution to $x^2 = 2$, for example,

$$\{2, 1.5, 1.42, 1.415, 1.4143, \ldots\}. \tag{4-4}$$

This Cauchy sequence is also identified with the new real number $\sqrt{2}$. In fact, all the Cauchy sequences of rational numbers that approximate the solution to $x^2 = 2$ arbitrarily closely are identified with the real number $\sqrt{2}$.

All real numbers can be obtained as limits of Cauchy sequences of rational numbers by the process just described. That is, we regard two Cauchy sequences of rationals, say $\{x_n\}$ and $\{y_n\}$, as equivalent if $|x_n - y_n| \to 0$ as $n \to \infty$, as in (4-3) and (4-4). We identify all these Cauchy sequences with a real number , say x, which is the common limit of these Cauchy sequences. The distance between two real numbers x and x' is defined by

$$|x - x'| = \lim_{n\to\infty} |x_n - x'_n|,$$

where $\{x_n\}$ and $\{x'_n\}$ are Cauchy sequences of rationals associated with x and x', respectively. If Q denotes the set of rational numbers and R the set of reals, then $Q \subset R$ because each rational number q can be identified with the Cauchy sequence $(q, q, q, \ldots\}$. Also, Q is dense in R by construction.

The completion process we have just described can be carried out in any normed space, in particular, in any inner product space. We choose not to enter into the details of the proof, which simply amount to carrying out the construction indicated above with a little more care. Here is the result.

Theorem 4-1. Let $\mathscr{H}$ be a separable inner product space. Then there is a complete inner product space (Hilbert space) H, called the completion of $\mathscr{H}$, which satisfies:

(i) $\mathscr{H} \subset H$.
(ii) $\mathscr{H}$ is dense in H.
(iii) The completion is unique in the sense that if H' and H'' satisfy (i) and (ii) there is a one-to-one correspondence $u \leftrightarrow v$ between elements $u \in H'$ and $v \in H''$ such that $\langle u_1, u_2\rangle_{H'} = \langle v_1, v_2\rangle_{H''}$.

Let $u, v \in H$ the completion of $\mathscr{H}$. Then by (ii) there are sequences $\{u_n\}$ and $\{v_n\}$ in $\mathscr{H}$ with $u_n \to u$ and $v_n \to v$. Thus,

$$\langle u, v\rangle = \lim_{n\to\infty} \langle u_n, v_n\rangle \tag{4-5}$$

and

$$\|u\| = \lim_{n\to\infty} \|u_n\|, \tag{4-6}$$

by the continuity of the inner product. This means that computations in H can be reduced to limits of computations in $\mathscr{H}$. This fact, in turn, implies that properties of H may often be established by first proving them for $\mathscr{H}$ and then passing to the limit. (For a typical example, see the proof of Th. 4-2.)

As a first application of Th. 4-1 consider the function space $\mathscr{H}_0$, which consists of continuous functions $u, v, \ldots$ defined on a closed, bounded domain $\bar{D}$ and equipped with the inner product

$$\langle u, v\rangle = \int_D u(x)\overline{v(x)}\, dx. \tag{4-7}$$

It is elementary to check that $\mathscr{H}_0$ is an inner product space. This space is not complete (see Prob. 6 of Sec. 11-3). Its completion, provided by Th. 4-1, is denoted by either $L_2 = L_2(D)$ or $H_0 = H_0(D)$ and is the space of Lebesgue square-integrable functions in D.

Next, suppose that the boundary B of D is at least once continuously differentiable and consider the function space $\mathscr{H}_1$ with inner product (4-2). This space is also incomplete (see Prob. 1). Its completion is denoted by H_1 and is called a *Sobolev space.*

The final space we need is denoted by $\mathring{H}_1$. With D and B as above, let $C_0^1(\bar{D})$ be the space of continuously differentiable functions on D which vanish outside some closed, bounded subset of D which does not meet the boundary B. [If $u \in C_0^1(\bar{D})$, the set outside of which u vanishes may depend on u.] Once again, we equip $C_0^1(\bar{D})$ with the inner product (4-2). The completion of this space is $\mathring{H}_1$. Think of $\mathring{H}_1$ as those functions in H_1 which (in a certain sense) vanish on the boundary of D.

Notice that $\mathring{H}_1 \subset H_1 \subset H_0$ as sets; however, the norms used in H_1 and H_0 are different. Although we shall not go into an extensive account of the properties of the spaces $\mathring{H}_1$, H_1 and H_0, it is useful to point out that there is considerable flexibility in the choice of the spaces which are completed in the constructions leading to $\mathring{H}_1$, H_1, and H_0. For example, to construct $\mathring{H}_1$ we started with $C_0^1(\bar{D})$ and completed it relative to the inner product (4-2). We would obtain exactly the same space $\mathring{H}_1$ starting with $C_0^2(\bar{D})$ or even $C_0^\infty(\bar{D})$ and completing these spaces with respect to (4-2). For our purposes, the main points to keep in mind are that $H_0 = L_2$, H_1, and $\mathring{H}_1$ are all complete (i.e., are Hilbert spaces) and that we can operate with functions in these spaces much as for smooth functions. Moreover, differentiable functions are dense in all these spaces, and in view of (4-5) and (4-6), computations involving nonsmooth functions in one of these spaces can be expressed as limits of corresponding computations on smooth functions. (The situation here is strictly analogous to concrete computations with irrational numbers, such as $\sqrt{2}$, which are reduced to suitably accurate rational approximations.)

The space H_1 carries with it a generalization of the classical notion of derivative. To see this, fix a function f in H_1. Since $C^1(\bar{D})$ is dense in H_1 there is a sequence of functions f_n in $C^1(\bar{D})$ such that $f_n \to f$ in H_1. The sequence of continuously differentiable functions $\{f_n\}$ is a Cauchy sequence in H_1. A glance at the norms corresponding to (4-2) and (4-7) reveals that both $\{f_n\}$ and $\{\nabla f_n\}$ are Cauchy sequences in $L_2(D)$. By completeness of L_2 there are functions $\tilde{f}$ and $g_1, \ldots, g_r$ in $L_2(D)$, where r is the dimension of D, such that $f_n \to \tilde{f}$ and

$$\text{(4-8)} \qquad \frac{\partial f_n}{\partial x_j} \to g_j \quad \text{in} \quad L_2(D) \quad \text{for} \quad j = 1, \ldots, r.$$

By uniqueness of limits in L_2, $\tilde{f} = f$, but what is interesting is that (4-8) allows us to define the jth (*generalized*) *partial derivative* of f in H_1 by

$$\text{(4-9)} \qquad \frac{\partial f}{\partial x_j} = g_j \quad \text{for} \quad j = 1, \ldots, r.$$

Note that these partials are functions in $L_2(D)$. As the notation (4-9) suggests, the generalized derivatives g_j depend only on f and not on the particular choice of the sequence $\{f_n\} \subset C^1(\bar{D})$ which approximates f (see Prob. 3). In Prob. 4 you are asked to show that if a function f in H_1 is actually in $C^1(\bar{D})$, then the generalized derivatives and classical derivatives agree. Also, see Probs. 8–15 for another approach to generalized derivatives.

We close this section with an important estimate which is called the *Poincaré lemma.*

Theorem 4-2. Let D be a bounded domain. Then there is a constant C depending only on D such that

$$\int_D |u(x)|^2\,dx \le C\int_D |\nabla u(x)|^2\,dx \tag{4-10}$$

for all u in $\mathring{H}_1$.

Proof. To begin with, let $u \in C_0^1(\bar{D})$ be real-valued and $x^\circ \in D$. Apply the Gauss divergence theorem to obtain

$$\begin{aligned} 0 &= \int_D \operatorname{div}\,[u(x)^2(x - x^0)]\,dx \\ &= r\int_D u(x)^2\,dx + 2\int_D [(x - x^0)\cdot\nabla u(x)]u(x)\,dx, \end{aligned}$$

where r is the dimension of the underlying Euclidean space. Let $d = \max|x - y|$ as x and y vary over $\bar{D}$. Then by the Schwarz inequality

$$\begin{aligned} r\int_D u(x)^2\,dx &= -2\int_D [(x - x^\circ)\cdot\nabla u(x)]u(x)\,dx \\ &\le 2\int_D |x - x^\circ|\,|\nabla u|\,|u|\,dx \\ &\le 2d\left(\int_D |\nabla u|^2\,dx\right)^{1/2}\left(\int_D |u|^2\,dx\right)^{1/2}. \end{aligned}$$

For any $\varepsilon > 0$ and $a, b \ge 0$, we have

$$2ab < \varepsilon a^2 + \frac{1}{\varepsilon}b^2$$

because $(\sqrt{\varepsilon}a - b/\sqrt{\varepsilon})^2 \ge 0$. Use of this estimate yields

$$r\int_D u(x)^2\,dx \le \varepsilon d\int_D |\nabla u|^2\,dx + \frac{d}{\varepsilon}\int_D |u|^2\,dx.$$

Choose $\varepsilon = 2d/r$ to get

$$\frac{r}{2}\int_D u(x)^2\,dx \le \frac{2d^2}{r}\int_D |\nabla u|^2\,dx,$$

which establishes the theorem with $C = 4d^2/r^2$ when $u \in C_0^1(\bar{D})$ is real-valued. The complex-valued case follows by applying the real-valued result to the real and imaginary parts of u. Finally, (4-10) follows for any $u \in \mathring{H}^1(D)$ by a simple limit argument (Prob. 7) because there is a sequence $\{u_n\}$ in $C_0^1(\bar{D})$ which converges to u in $\mathring{H}_1$.

PROBLEMS

1. Show that $\mathscr{H}_1$ is not complete. *Hint.* Consider $g_n(x) = \int_0^x f_n(t)\,dt$, where f_n is given in Prob. 6 of Sec. 11-3.
2. Show that $H_1 \subset H_0 = L_2$. That is, each function in H_1 is square integrable.
3. Let $f \in H_1$. Suppose that $\{\tilde{f}_n\} \subset C^1(\bar{D})$ and $\tilde{f}_n \to f$ in H_1. Following the discussion of generalized derivatives in the text,

$$\frac{\partial \tilde{f}_n}{\partial x_j} \to \tilde{g}_j \quad \text{in} \quad L_2(D) \quad \text{for} \quad j = 1, 2, \ldots, r.$$

Show that $\tilde{g}_j = g_j$ in $L_2(D)$, and hence confirm that the definition of generalized derivatives makes sense.
4. Let $f \in C^1(\bar{D})$. Then $f \in H_1(D)$, so there are two notions of derivative for f. Show that the classical derivative and generalized derivative are the same. *Hint.* Use Prob. 3 and a very simple sequence $\{f_n\} \subset C^1(\bar{D})$ which approximates f.
5. Let D be a bounded domain. Show that there is a constant α such that

$$\alpha \|u\|_{\mathring{H}_1} \leq \|\nabla u\|_{L_2}$$

for all u in $\mathring{H}_1$. Conclude that a sequence $\{u_n\}$ in $\mathring{H}_1$ converges if the sequence $\{\nabla u_n\}$ converges in the L_2 norm.
6. Show that $\mathring{H}_1$ is a subspace of H_1. (Recall that subspaces must be closed.)
7. Use continuity of the L_2 norm to complete the proof of Th. 4-2.

Problems 8–15 give another natural way to introduce generalized derivatives.

8. Let $f \in C^1(\bar{D})$. Show that

$$\left\langle \frac{\partial f}{\partial x_j}, \zeta \right\rangle = -\left\langle f, \frac{\partial \zeta}{\partial x_j} \right\rangle$$

for all $\zeta \in C_0^1(\bar{D})$.
9. The result of Prob. 8 suggest the following definition of a *weak derivative.* Let $f \in L_2$. We call $h_j \in L_2$ the weak derivative of f with respect to x_j if

$$\left\langle f, \frac{\partial \zeta}{\partial x_j} \right\rangle = -\langle h_j, \zeta \rangle$$

for all $\zeta \in C_0^1(\bar{D})$. In this problem and those that follow, we refer to the generalized partial derivatives $\partial f/\partial x_j$ of functions in H_1 as strong derivatives. Show that $f \in H_1$ has weak derivatives h_j, and that $h_j = \partial f/\partial x_j$. *Hint.* Choose a sequence $\{f_n\} \in C^1(\bar{D})$ which converges to f in H_1 and use Prob. 8.
10. If $f \in L_2$ has a weak derivative, it is unique.
11. Let $\phi(x) \geq 0$ be a function on R_r which vanishes for $|x| \geq 1$, has classical derivatives of all orders, and satisfies $\int_{R_r} \phi(x)\,dx = 1$.
(a) Check that

$$\phi(x) = \begin{cases} C \exp\left(\dfrac{-1}{1 - |x|^2}\right), & |x| < 1, \\ 0, & |x| \geq 1, \end{cases}$$

is such a function where C is chosen so that $\int_{R_r} \phi(x)\,dx = 1$.

(b) Let $u = u(x)$ be integrable and vanish outside a closed bounded set $K \subset D$ with $K \cap B \neq \phi$. Define

$$u_\varepsilon(x) = \varepsilon^{-n} \int_D u(y)\phi\left(\frac{x-y}{\varepsilon}\right) dy$$

for $0 < \varepsilon < \text{dist}(K, B)$. Show that $u_\varepsilon \in C_0^\infty(\bar{D})$ and

$$u_\varepsilon(x) = \int_D \phi(y)u(x - \varepsilon y)\, dy.$$

Also show that $u_\varepsilon(x) \to u(x)$ uniformly on D as $\varepsilon \to 0$, provided that u is continuous.

12. Suppose that u vanish off K as in Prob. 11 and that $u \in H_0 = L_2$. Show that

$$\|u_\varepsilon\|_{L_2} \leq \|u\|_{L_2} \quad \text{and} \quad \lim_{\varepsilon \to 0} \|u_\varepsilon - u\|_{L_2} = 0.$$

Hint. Apply the Schwarz inequality to get

$$|u_\varepsilon(x)|^2 \leq \left(\varepsilon^{-n} \int_D \phi\left(\frac{x-y}{\varepsilon}\right) dy\right) \varepsilon^{-n} \int_D \phi\left(\frac{x-y}{\varepsilon}\right) |u(y)|^2\, dy.$$

Note that the term in parentheses is 1. Finally, integrate with respect to x. For the second part, given $\delta > 0$, find $v \in C_0(D)$ such that $\|u - v\|_{L_2} < \delta$. Form v_ε and estimate

$$\|u_\varepsilon - u\|_{L_2} = \|u_\varepsilon - v_\varepsilon + v_\varepsilon - v + v - u\|_{L_2}.$$

13. Let $h = h(x)$ be the weak derivative of u with respect to some x_j. Show that $(\partial/\partial x_j)u_\varepsilon(x) = h_\varepsilon(x)$.

14. Suppose that u has a weak derivative h with respect to x_j in D. Show that in every closed, bounded subset $K \subset D$ there is a sequence of continuously differentiable functions $\{u_n\}$ such that $u_n \to u$ and $\{\partial u_n/\partial x_j\}$ converges in H_0. *Hint.* Consider u_{ε_n}.

15. Prove: If u is weakly differentiable in D, then u is strongly differentiable in D.

11-5. The Ritz–Galerkin Method

We return now to the problem of minimizing the potential energy functional

$$I(u) = \frac{1}{2} \int_D (|\nabla u|^2 + 2F(x)u)\, dx, \tag{5-1}$$

where u is subject to the boundary condition

$$u = f \quad \text{on} \quad B. \tag{5-2}$$

Here F and f are given functions and B is the boundary of a bounded domain D.

In Prob. 11 of Sec. 11-2, we found that u minimizes $I(u)$ if and only if the variational equation

$$\int_D [\nabla u \cdot \nabla \zeta + F\zeta]\, dx = 0 \tag{5-3}$$

for all $\zeta \in C_0^1(D)$, and hence for all $\zeta \in \mathring{H}_1$. Thus, the problem of finding $u \in H_1$ which minimizes $I(u)$ subject to (5-2) is equivalent to solving the variational equation (5-3) for $u \in H_1$ which satisfies (5-2). In this section we show that (5-3) has a solution u which satisfies (5-2). One advantage of working with (5-2) and (5-3) rather than (5-1) and (5-2) is that other problems for which variational principles are not immediately apparent often can be converted to a form similar to (5-3) (see Prob. 1).

We refer to (5-2) and (5-3) as the *generalized Dirichlet problem* because if $u \in C^2(D) \cap C(\bar{D})$ and solves (5-2) and (5-3), then $\Delta u = F$ in D and $u = f$ on B (see Sec. 11-1).

In the boundary condition (5-2), f need only be defined on B. We assume that f is defined throughout $\bar{D}$ and that $f \in H_1(D)$. In any practical situation, there are many ways to extend the domain of f from B to $\bar{D}$ such that the extended function, still denoted by f, belongs to H_1. Problem 2 shows that we get the same solution to (5-2) and (5-3) no matter which extension we use. With f defined throughout $\bar{D}$ and $f \in H_1$, we can express the boundary condition (5-2) as $u - f \in \mathring{H}_1$. Therefore, we seek a function $u \in H_1$ such that

(5-4) $$\int_D \nabla u \cdot \nabla \zeta \, dx = -\int_D F\zeta \, dx \quad \text{for all} \quad \zeta \in \mathring{H}_1$$

and

(5-5) $$u - f \in \mathring{H}_1.$$

We summarize the foregoing remarks in Th. 5-1.

Theorem 5-1. The variational problem of minimizing (5-1) over functions $u \in H_1$ which satisfy $u - f \in \mathring{H}_1$ is equivalent to the solution of the generalized Dirichlet problem,

$$\int_D [\nabla u \cdot \nabla \zeta + F\zeta] \, dx = 0, \qquad u \in H_1, \qquad \zeta \in \mathring{H}_1, \qquad u - f \in \mathring{H}_1,$$

in the sense that $u \in H_1$ minimizes (5-1) subject to $u - f \in \mathring{H}_1$ if and only if u solves the generalized Dirichlet problem. Moreover, there exists at most one solution to the generalized Dirichlet problem.

Proof. Only the uniqueness assertion remains to be proved. Suppose that u_1 and u_2 both solve the generalized Dirichlet problem and let $\zeta_0 = u_1 - u_2$. Then $\zeta_0 \in \mathring{H}_1$ and

$$\int_D [\nabla u_i \cdot \nabla \zeta_0 + F\zeta_0] \, dx = 0 \quad \text{for} \quad i = 1, 2.$$

Subtract the equation with $i = 2$ from the equation with $i = 1$ to conclude that

$$\int_D |\nabla \zeta_0|^2 \, dx = 0.$$

By Th. 4-2 we infer that $\int_D |\zeta_0|^2 \, dx = 0$ and hence that $\zeta_0 = 0$ in $\mathring{H}_1$ or $u_1 = u_2$.

We solve (5-4) and (5-5) by an approximation scheme. Let $\phi_1, \phi_2, \ldots$ be a basis for the Hilbert space $\mathring{H}_1$. That is, the set $\{\phi_i\}$ is linearly independent and the closure of sp $\{\phi_1, \phi_2, \ldots\}$ is $\mathring{H}_1$. The set $\{\phi_i\}$ need not be orthonormal, and in many applications it is not. Then

$$H^{(k)} = \text{sp}\,\{\phi_1, \ldots, \phi_k\}$$

is a k-dimensional subspace of $\mathring{H}_1$. Clearly,

$$\text{(5-6)} \qquad H^{(1)} \subset H^{(2)} \subset \cdots \subset H^{(k)} \subset \cdots \quad \text{and} \quad \overline{\bigcup_{k=1}^{\infty} H^{(k)}} = \mathring{H}_1.$$

If (5-4) and (5-5) has a solution $u \in H_1$, then $u = f + w$, where $w = u - f \in \mathring{H}_1$. Our plan is to construct approximate solutions $u_k = f + w_k$ to $u = f + w$. We choose $w_k \in \mathring{H}_1$ so that $u_k - f \in \mathring{H}_1$ and so that u_k satisfies (5-4) for all $\zeta \in H^{(k)}$. In view of (5-6), it is plausible that the sequence $\{w_k\}$ so constructed will converge to w in $\mathring{H}_1$ and that

$$u_k = f + w_k \to f + w = u,$$

which will solve (5-4) and (5-5).

The sequences $\{u_k\}$ and $\{w_k\}$ are constructed as follows. Since $w_k \in H^{(k)}$ it has the form

$$\text{(5-7)} \qquad w_k = \sum_{j=1}^{k} a_j \phi_j$$

and of course $u_k = f + w_k$. For any choice of $a_1, \ldots, a_k$ we have $u_k - f \in \mathring{H}_1$. In order that (5-4) hold for u_k and all $\zeta \in H^{(k)}$, we must have

$$\int_D \nabla w_k \cdot \nabla \zeta \, dx = -\int_D (\nabla f \cdot \nabla \zeta + F\zeta)\, dx$$

for $\zeta \in H^{(k)}$. Since $\{\phi_1, \ldots, \phi_k\}$ is a basis for $H^{(k)}$, the previous condition is equivalent to

$$\text{(5-8)} \qquad \int_D \nabla w_k \cdot \nabla \phi_i \, dx = -\int_D (\nabla f \cdot \nabla \phi_i + F\phi_i)\, dx$$

for $i = 1, 2, \ldots, k$, where w_k is given by (5-7). Substitute (5-7) into (5-8) to obtain the system of equations

$$\text{(5-9)} \qquad \sum_{j=1}^{k} B(\phi_i, \phi_j) a_j = L(\phi_i), \qquad i = 1, 2, \ldots, k,$$

where

$$\text{(5-10)} \qquad B(\phi, \psi) = \int_D \nabla \phi \cdot \nabla \psi \, dx$$

and

$$\text{(5-11)} \qquad L(\phi) = -\int_D (\nabla f \cdot \nabla \phi + F\phi)\, dx.$$

In review, $u_k = f + w_k$ with w_k given by (5-7) will satisfy $u_k - f \in \mathring{H}_1$ and u_k will satisfy (5-4) for all ζ in $H^{(k)}$ provided that the system of linear equations (5-9) can be solved. The next theorem guarantees that we can construct w_k and hence u_k.

Theorem 5-2. Let $f \in H_1(D)$ and $F \in H_0(D)$.

(i) The linear system (5-9) has a unique solution for each $k = 1, 2, \ldots$.

(ii) There is a constant M depending on f, F, and D such that

$$\|w_k\|_{\mathring{H}_1} = \left[\int_D (|w_k|^2 + |\nabla w_k|^2)\, dx\right]^{1/2} \le M.$$

(iii) For each $k = 1, 2, \ldots$ the approximations $u_k = f + w_k$ satisfy $I(u_{k+1}) \le I(u_k)$.

(iv) There is a constant $K \ge 0$ which depends on $\|f\|_{H_1}$ and $\|F\|_{H_0}$ such that $I(v) \ge -K$ for all $v \in H_1$ with $v - f \in \mathring{H}_1$.

Proof. (i) We must show that

$$\sum_{j=1}^{k} B(\phi_i, \phi_j)a_j = 0, \qquad i = 1, \ldots, k,$$

has only the trivial solution $a_1 = \cdots = a_k = 0$. By the linearity of B in its second variable

$$B\left(\phi_i, \sum_{j=1}^{k} a_j\phi_j\right) = 0, \qquad i = 1, \ldots, k.$$

Multiply this equation by a_i and add the results to get

$$B\left(\sum_{i=1}^{k} a_i\phi_i, \sum_{j=1}^{k} a_j\phi_j\right) = 0.$$

In view of (5-10) and the Poincaré lemma $\sum_{i=1}^{k} a_i\phi_i = 0$. Finally, $a_1 = \cdots = a_k = 0$ because $\{\phi_1, \ldots, \phi_k\}$ are linearly independent.

(ii) Multiply both sides of (5-8) by a_i and sum to obtain

$$\int_D |\nabla w_k|^2\, dx = -\int_D (w_k F + \nabla w_k \cdot \nabla f)\, dx.$$

Apply the Schwarz inequality to deduce that

$$\int_D |\nabla w_k|^2\, dx \le \left(\int_D |w_k|^2\, dx\right)^{1/2}\left(\int_D |F|^2\, dx\right)^{1/2} + \left(\int_D |\nabla w_k|^2\, dx\right)^{1/2}\left(\int_D |\nabla f|^2\, dx\right)^{1/2}.$$

Since $ab \le (\varepsilon a^2/2) + (b^2/2\varepsilon)$ for $a, b \ge 0$ and any $\varepsilon > 0$, we find

$$\int_D |\nabla w_k|^2\, dx \le \frac{\varepsilon}{2}\int_D |w_k|^2\, dx + \frac{1}{2\varepsilon}\int_D |F|^2\, dx + \frac{\delta}{2}\int_D |\nabla w_k|^2\, dx + \frac{1}{2\delta}\int_D |\nabla f|^2\, dx,$$

for any $\varepsilon, \delta > 0$. Choose $\delta = 1$ to get

$$\int_D |\nabla w_k|^2\, dx \le \varepsilon\int_D |w_k|^2\, dx + \frac{1}{\varepsilon}\int_D |F|^2\, dx + \int_D |\nabla f|^2\, dx.$$

The Poincaré lemma yields

$$\int_D |\nabla w_k|^2\,dx \geq \frac{1}{2}\int_D |\nabla w_k|^2\,dx + \frac{1}{2C}\int_D |w_k|^2\,dx.$$

Combine these inequalities to find

$$\left(\frac{1}{2C} - \varepsilon\right)\int_D |w_k|^2\,dx + \frac{1}{2}\int_D |\nabla w_k|^2\,dx \leq \frac{1}{\varepsilon}\int_D |F|^2\,dx + \int_D |\nabla f|^2\,dx.$$

Finally, let $\varepsilon = 1/4C$ to obtain

$$\int_D \left(|w_k|^2 + |\nabla w_k|^2\right) dx \leq \mu \int_D \left(|F|^2 + |\nabla f|^2\right) dx$$

after a short calculation, where

$$\mu = \frac{\max\{4C, 1\}}{\min\{1/2, 1/4C\}}.$$

This establishes the bound in (ii). Statements (iii) and (iv) are not used later. We leave their proofs for Probs. 4 and 5.

The construction of the sequence of approximations $\{u_k\}$ given above is called the *Ritz–Galerkin method.* There are many names associated with variants of this method, including Rayleigh, Ritz, Galerkin, Hilbert, Courant, Tonelli, and Trefftz. These methods are among the so-called "direct methods" of the calculus of variations which employ constructive procedures to solve variational problems. If the basis $\phi_1, \phi_2, \ldots$ is chosen in a special way, the approximation scheme used above is known as the finite element method. For a detailed discussion of the finite element method, see G. J. Fix and G. Strang, *An Analysis of the Finite Element Method* (Prentice-Hall, Inc., Englewood Cliffs N.J., 1973) or G. F. Pinder and P. S. Huyakorn, *Computational Methods in Subsurface Flow* (Academic Press, Inc., New York, 1983).

We turn now to the proof that the generalized Dirichlet problem (5-4) and (5-5) has a solution. The argument we follow is due to Hildebrandt and Wienholtz. First, note that the generalized Dirichlet problem can be expressed as

(5-12) $$B(u - f, \zeta) = L(\zeta) \quad \text{for all} \quad \zeta \in \mathring{H}_1 \quad \text{where} \quad u - f \in \mathring{H}_1$$

and that the Ritz–Galerkin procedure leads to approximations $u_k = f + w_k$ such that

(5-13) $$B(w_k, \zeta) = L(\zeta) \quad \text{for all} \quad \zeta \in H^{(k)} \quad \text{and} \quad w_k \in \mathring{H}_1 \quad \text{for } k = 1, 2, \ldots.$$

By (ii) of Th. 5-2 the sequence $\{w_k\}$ is uniformly bounded in $\mathring{H}_1$. Consequently, there is a subsequence $\{w_{k_m}\}$ of $\{w_k\}$ such that w_{k_m} converges weakly to some $w \in \mathring{H}_1$. We shall prove that the full sequence $\{w_k\}$ converges to w and that $u = f + w$ solves the generalized Dirichlet problem (5-12). To this end, observe that if $\zeta \in \mathring{H}_1$ is arbitrary, we have

$$\zeta = \zeta_k + \zeta_k',$$

where $\zeta_k = P_k\zeta$ is the projection of ζ on $H^{(k)}$ and ζ'_k is its projection on the orthogonal complement of $H^{(k)}$. Since $H^{(k)} = \text{sp}\,\{\phi_1, \ldots, \phi_k\}$ and $\{\phi_j\}$ is a basis for $\mathring{H}_1$, we infer that $\zeta_k \to \zeta$ in $\mathring{H}_1$. Now, from (5-13) we have

$$B(w_{k_m}, \zeta_{k_n}) = L(\zeta_{k_n}) \quad \text{for} \quad m > n.$$

Let $m \to \infty$ and use the weak convergence $w_{k_m} \rightharpoonup w$ in $\mathring{H}_1$ to find

$$B(w, \zeta_{k_n}) = L(\zeta_{k_n}),$$

and then let $n \to \infty$ to arrive at

$$B(w, \zeta) = L(\zeta) \quad \text{for} \quad \zeta \in \mathring{H}_1. \tag{5-14}$$

We conclude from (5-12) that $u = f + w$ solves the generalized Dirichlet problem.

In summary, the generalized Dirichlet problem (5-4) and (5-5) has a solution $u = f + w$, where w is the weak limit of the subsequence $\{w_{k_m}\}$. Now, the solution u is unique by Th. 5-1 and hence $w = u - f$ is unique. The arguments above applied to $\{w_k\}$ can be applied starting with any subsequence of $\{w_k\}$. Thus, we conclude that any subsequence of $\{w_k\}$ itself contains a weakly convergent subsequence whose limit must be $u - f$, where u is the unique solution to (5-12). It follows that the full sequence $\{w_k\}$ converges weakly to $w = u - f$. For otherwise, there would exist $\zeta_1 \in \mathring{H}_1$ and $\varepsilon_1 > 0$ and a subsequence $\{w_{k_m}\}$ such that

$$|\langle w_{k_m}, \zeta_1\rangle - \langle w, \zeta_1\rangle| \geq \varepsilon.$$

Since w_{k_m} contains a subsequence which converges weakly to w, we let $k_m \to \infty$ through this subsequence to reach the contradiction

$$|\langle w, \zeta_1\rangle - \langle w, \zeta_1\rangle| \geq \varepsilon.$$

At this point, we know that $\{w_k\}$ and hence $\{u_k\}$ converge weakly to $w = u - f$ and to u, respectively, where u is the unique solution to the generalized Dirichlet problem. For practical applications it is important to realize that these sequences are strongly convergent. To see this, use (5-13) and (5-14) to get

$$B(w_k, \zeta) = B(w, \zeta) \quad \text{for all} \quad \zeta \in H^{(k)}.$$

The choice $\zeta = w_k - P_k w$ leads to

$$\begin{aligned} B(w_k, w_k - P_k w) &= B(w, w_k - P_k w), \\ B(w_k - w, w_k - P_k w) &= 0. \end{aligned}$$

Next by the Poincaré lemma (see Prob. 5 of Sec. 11-4) there is a constant $\alpha > 0$ such that

$$\begin{aligned} \alpha\|w_k - P_k w\|^2 &\leq B(w_k - P_k w, w_k - P_k w) \\ &= B(w_k - w, w_k - P_k w) + B(w - P_k w, w_k - P_k w) \\ &= B(w - P_k w, w_k - P_k w) \\ &\leq \|w - P_k w\|\,\|w_k - P_k w\|, \end{aligned}$$

where the norms are all taken in the $\mathring{H}_1$ sense and the Schwarz inequality was used in the last step. Thus,

$$\|w_k - P_k w\| \le \frac{1}{\alpha} \|w - P_k w\| \to 0$$

as $k \to \infty$, and therefore

$$\|w - w_k\| \le \|w - P_k w\| + \|P_k w - w_k\| \to 0$$

as $k \to \infty$. That is, $w_k \to w$ strongly in $\mathring{H}_1$; therefore, $u_k \to u$ in H_1.

Theorem 5-3. Let $f \in H_1$ and $F \in H_0$. Then there is a unique solution $u \in H_1$ to the generalized Dirichlet problem (5-4) and (5-5). This solution can be constructed by the Ritz–Galerkin procedure and the sequence of approximations u_k converges to u in the H_1 norm.

PROBLEMS

1. The partial differential equation

$$(*) \qquad \Delta u + \frac{\partial}{\partial x_1}(au) + \frac{\partial}{\partial x_2}(bu) = F \quad \text{in} \quad D$$

with $u = f$ on B arises in diffusion–dispersion problems and an associated variational formulation is not immediately apparent. Multiply the differential equation by $\zeta \in C_0^1$ and integrate by parts to find that

$$(**) \qquad \int_D [\nabla u \cdot \nabla \zeta + uv \cdot \nabla \zeta + F\zeta]\, dx = 0$$

for all $\zeta \in C_0^1$ and hence all $\zeta \in \mathring{H}_1$, where $v = (a, b)$. This equation has a form similar to (5-3) and the techniques of this section could be employed to seek a solution $u \in H_1$ to (**) such that $u = f$ on B. Assume that $u \in C^2(D) \cap C(\bar{D})$ satisfies (**) and $u = f$ on B. Show that u solves (*).

2. Assume that $f_i \in H_1(D)$ for $i = 1, 2$ and that $f_i = f$ on B. Let u_i solve

$$\int_D [\nabla u_i \cdot \nabla \zeta + F\zeta]\, dx = 0, \qquad u_i - f_i \in \mathring{H}_1$$

for all $\zeta \in \mathring{H}_1$. Show that $u_1 = u_2$. *Hint.* Observe that $f_1 - f_2 \in \mathring{H}_1$ so that $u_1 - u_2 = (u_1 - f_1) - (u_2 - f_2) + (f_1 - f_2) \in \mathring{H}_1$.

3. Let $u_k = f + w_k$ with w_k given by (5-7). Then $I(u_k) = g(a_1, a_2, \ldots, a_k)$ is a real-valued function of $a_1, \ldots, a_k$. To minimize $I(u)$ for $u \in H^{(k)}$ with $u = f$ on B calculate $\partial g/\partial a_j$, set it equal to zero and arrive at (5-9).

4. Prove (iii) of Th. 5-2. *Hint.* Show that, as in Prob. 11 of Sec. 11-2, v minimizes $I(v)$ for $v \in \mathring{H}^{(k)}$ with $v = f$ on B if and only if (5-3) holds for v with $\zeta \in H^{(k)}$.

5. Prove (iv) of Th. 5-2. *Hint.* Let $v \in H_1$ with $v - f \in \mathring{H}_1$ be arbitrary. Then

$$I(v) = I(v - f + f)$$
$$= \frac{1}{2}\int_D [|\nabla(v-f)|^2 + 2\nabla(v-f)\cdot\nabla f + |\nabla f|^2 + 2F(v-f) + 2Ff]\,dx.$$

Now make estimates as in the proof of (ii) of Th. 5-2.

6. Approximate the solution to

$$\Delta u = 0 \quad \text{in} \quad D, \qquad u = f \quad \text{on} \quad B,$$

where D is the triangle cut from the first quadrant by the line $x + y = 1$. Assume that $f(0, y) = 0$, $f(x, 1 - x) = 0$, and $f(x, 0) = x(1 - x)$. Take $k = 1$, $f(x, y) = x(1 - x - y)$ for the extension of f to D, and $\phi_1 = xy(1 - x - y)$.

7. Consider the one-dimensional problem

$$u'' - \alpha(x)u = F(x), \qquad 0 < x < 1,$$
$$u(0) = u(1) = 0,$$

where $\alpha(x) \geq 0$. As in Prob. 1, obtain the related problem for u:

$$\int_0^1 [u'\zeta' + \alpha u\zeta + F\zeta]\,dx = 0, \qquad u(0) = u(1) = 0.$$

Derive equations analogous to (5-9) for this problem.

8. Suppose in Prob. 7 that $\alpha(x) = \alpha > 0$ a constant. Choose $\phi_j(x) = \sin(j\pi x)$ and determine the solution u using the Ritz–Galerkin method.

9. In (5-9) let $k = 9$, say, and consider Prob. 7. Partition the interval $[0, 1]$ into 10 subintervals $[x_j, x_{j+1}]$ with $x_0 = 0$ and $x_{10} = 1$. Let $\phi_j(x) = l_j(x)$, $j = 1, \ldots, 9$, be the hat functions shown in Fig. 7-3 and write down the system of equations (5-9). Observe that $\phi_j \in \mathring{H}_1$ but is not differentiable everywhere.

A general principle known as the *Lax–Milgram theorem* underlies the proof of Th. 5-3. The next problem outlines a proof of this result.

10. Let H be a Hilbert space and $B(u, v)$ be a real-valued, symmetric bilinear functional on H. That is,

$$B(u + v, w) = B(u, w) + B(v, w), \qquad B(\alpha u, v) = \alpha B(u, v), \quad \text{and} \quad B(u, v) = B(v, u).$$

Assume that there are constants α and β with $\alpha > 0$ such that

(1) $$|B(u, v)| \leq \beta\|u\|\,\|v\|$$

(2) $$\alpha\|u\|^2 \leq B(u, u).$$

Then given any bounded linear functional $L(\zeta)$ on H there is a $u \in H$ such that

$$B(u, \zeta) = L(\zeta) \quad \text{for all} \quad \zeta \in H.$$

Prove this using the following steps.

(a) Use Th. 3-4 to reduce to proving that for each $s \in H$ there exists a σ in H such that $B(\sigma, \zeta) = \langle \zeta, s \rangle$ for all $\zeta \in H$.

(b) Let $S = \{s\}$ of elements in H such that there is a σ in H for which $B(\sigma, \zeta) = \langle \zeta, s \rangle$ holds for all $\zeta \in H$. (For example, $0 \in S$ and the corresponding σ is $\sigma = 0$.) Show that S is a subspace of H. Remember that subspaces are closed. *Hint*. Let $s_n \in S$ converge to $h \in H$ and let σ_n correspond to s_n. Use $B(\sigma_n - \sigma_m, \zeta) = \langle \zeta, s_n - s_m \rangle$ and (2) to deduce that $\{\sigma_n\}$ is a Cauchy sequence in H. Then $\sigma_n \to \sigma$ in H. Show that $B(\sigma, \zeta) = \langle \zeta, h \rangle$, so $h \in S$.

(c) Show that $S = H$. Assume the contrary. Then there would exist $h \in H$ with $h \notin S$. Then $L_1(s) = \langle s, h \rangle$ defines a continuous linear functional on the Hilbert space S. Thus, $(s, h) = \langle s, t \rangle$ for some $t \in S$ and all $s \in S$ or

$$\langle s, h - t \rangle = 0 \quad \text{for all} \quad s \in S.$$

Show that

$$B(h - t, \zeta) = \langle \zeta, r \rangle$$

for all $\zeta \in H$ and some $r \in H$. Conclude that $r \in S$ and $\langle r, h - t \rangle = 0$. Use (2) to deduce $\|h - t\| = 0$. Then $h = t \in S$, a contradiction.

11-6. The Dirichlet Problem for the Laplace Equation

In Sec. 11-5 we saw that the problem of minimizing the potential energy functional,

$$I(u) = \frac{1}{2} \int_D |\nabla u|^2 \, dx, \qquad u = f \quad \text{on} \quad B$$

has a unique solution $u \in H_1$ with $u - f \in \mathring{H}_1$. If our aim is simply to minimize the potential energy, the problem is solved. However, we have already seen in Secs. 11-1 and 11-2 that there is a close connection between minimizing this potential energy function and the Dirichlet problem for the Laplace equation (see Probs. 12–16 of Sec. 11-1). We shall show that the solution to the minimum problem is a solution to the Dirichlet problem.

Consider the problem of finding $u \in C^2(D) \cap C(\bar{D})$ such that

$$\Delta u = 0 \quad \text{in} \quad D, \tag{6-1}$$

$$u = f \quad \text{on} \quad B, \tag{6-2}$$

the boundary of D. As usual, we assume that D is a bounded domain in 3-space whose boundary B is once continuously differentiable. The case $n = 2$ is treated in the problems. Finally, assume that $f \in H_1(D)$ and that its restriction to B is a continuous function. We have restricted our attention to the Laplace equation (6-1) rather than the more general Poisson equation $\Delta v = F$ because by Th. 6-1 of Sec. 8-6 $w = -\int_D (4\pi r)^{-1} F(y)\, dy$ solves $\Delta w = F$; hence, the change of variables $v = u + w$ reduces the Poisson problem to (6-1) and (6-2).

To solve (6-1) and (6-2) let $\zeta \in C_0^2(D)$ be arbitrary, multiply (6-1) by ζ, and integrate by parts to find

$$\int_D \nabla u \cdot \nabla \zeta \, dx = 0, \qquad u = f \quad \text{on} \quad B. \tag{6-3}$$

We know from Th. 5-3 that (6-3) has a solution $u \in H_1(D)$ with $u - f \in \mathring{H}_1$ and that this solution minimizes the potential energy functional $I(u)$. If in addition,

$u \in C^2(D)$, it will satisfy (6-1); see (1-16) and the subsequent discussion. The solution $u \in H_1$ of (6-3) has this extra smoothness by virtue of *Weyl's lemma*, which we record as

Theorem 6-1. Suppose that $u \in H_0(D)$ and that

$$\int_D u \, \Delta\zeta \, dx = 0 \quad \text{for all} \quad \zeta \in C_0^2(D).$$

Then $u \in C^2(D)$ and $\Delta u = 0$ in D.

Remark. To be very precise in Th. 6-1, note that a function $u \in H_0$ is only determined up to a set of measure zero. The assertion $u \in C^2(D)$ means that u can be redefined (if need be) on a set of measure zero so that it becomes twice continuously differentiable.

Assume for the moment that Weyl's lemma holds. Let $u \in H_1$ be the solution to (6-3) and integrate by parts to obtain

$$0 = \int_D \nabla u \cdot \nabla\zeta \, dx = -\int_D u \, \Delta\zeta \, dx.$$

Thus, $u \in C^2(D)$ and $\Delta u = 0$ in D.

We turn to the proof of Weyl's lemma. The key idea is to show that $\int_D u \, \Delta\zeta \, dx = 0$ for all $\zeta \in C_0^2(D)$ implies that u is continuous and has the mean value property in D. Then u is harmonic in D, by Th. 4-8 of Sec. 8-4. To establish the mean value property, fix a point $x^\circ \in D$, and suppose that the ball $\bar{K}_R(x^\circ) \subset D$. Let $\zeta \in C_0^2(D)$ be defined by

$$\zeta(r) = \begin{cases} (R^2 - r^2)^3, & 0 \le r \le R, \\ 0, & \text{otherwise}, \end{cases}$$

where $r = |x - x^\circ|$ and introduce spherical coordinates with origin at x°. A simple calculation gives

$$\Delta\zeta = \begin{cases} 6(3R^4 - 10R^2r^2 + 7r^4), & 0 \le r \le R, \\ 0, & \text{otherwise}. \end{cases}$$

Apply the condition $\int_D u \, \Delta\zeta \, dx = 0$ with this choice of ζ to conclude that

$$\int_0^R \int_{|\omega|=1} u(x^\circ + r\omega)[3R^4 - 10R^2r^2 + 7r^4]r^2 \, d\sigma_\omega \, dr = 0$$

for all $R > 0$ sufficiently small. Differentiate with respect to R and note that the integrand vanishes when $r = R$ to find that

$$\int_0^R \int_{|\omega|=1} u(x^\circ + r\omega)[12R^3 - 20Rr^2]r^2 \, d\sigma_\omega \, dr = 0,$$

$$R^2 \int_{K_R} u \, dx = \frac{5}{3} \int_{K_R} r^2 u \, dx.$$

Differentiate again with respect to R, using

$$\frac{d}{dR}\int_{K_R} u\,dx = \int_{|x|=R} u\,d\sigma,$$

to deduce that

$$2R\int_{K_R} u\,dx + R^2\int_{|x|=R} u\,d\sigma = \frac{5}{3}\int_{|x|=R} R^2 u\,d\sigma,$$

$$\int_{K_R} u\,dx = \frac{1}{3}R\int_{|\omega|=R} u\,d\sigma,$$

or

$$U = \frac{1}{3}R\frac{dU}{dR},$$

where $U = \int_{K_R(x^\circ)} u\,dx$. The differential equation for U is easily solved as $U = c_1R^3$ for some constant c_1. Thus

$$\text{(6-4)} \qquad \frac{3}{4\pi R^3}\int_{K_R(x^\circ)} u\,dx = c \quad \text{for} \quad R > 0 \quad \text{and small,}$$

where c is some constant. Now it is a basic property of integrable functions that the expression on the left of (6-4) has limit $u(x^\circ)$ as $R \to 0$ for almost all $x^\circ \in D$. So $c = u(x^\circ)$ for such x°. We redefine $u(x^\circ) = c$ for the possible set of measure zero where this equality does not already hold and obtain

$$\text{(6-5)} \qquad \frac{3}{4\pi R^3}\int_{K_R(x^\circ)} u\,dx = u(x^\circ).$$

Finally, the left member of (6-5) is continuous in x°, so u is continuous in D and has the mean value property there. Thus, u is harmonic and Weyl's lemma is established.

At this point we know that the solution $u \in H_1(D)$ of (6-3) is in $C^2(D)$, $u - f \in \mathring{H}_1$, and (6-1) holds. Next, we show that u assumes the given boundary values; that is, for each $x^\circ \in B$, $u(x) \to f(x^\circ)$ as $x \to x^\circ$ through points in D. To this end, we assume that $f \in C^1(\bar{D})$ and that each boundary point $x^\circ \in B$ has the following property: There is a $\rho_0 > 0$ such that the ball $\bar{K}_0 = \bar{K}_{\rho_0}(x^\circ - \rho_0 v)$ lies in D except for the point $x^\circ \in B$ and the ball $\bar{K}_{\rho_0}(x^\circ + \rho_0 v)$ lies in cD except for the point $x^\circ \in B$. Here v is the outer unit normal at x° (see Fig. 11-4 on page 474).

For the moment regard $x \in D$ as fixed. Choose $\varepsilon > 0$ so that $\bar{K}_\varepsilon(x) \subset D$ and set $D_\varepsilon = D - \bar{K}_\varepsilon(x)$. Apply Green's first identity (3-2) of Sec. 8-3 with $v \in C_0^2(D)$ converging to a general function ϕ in $\mathring{H}_1(D)$ to find that

$$\int_{D_\varepsilon} [\phi\,\Delta\zeta + \nabla\phi \cdot \nabla\zeta]\,dy = \int_{S_\varepsilon(x)} \phi\frac{\partial\zeta}{\partial v}\,d\sigma_y$$

for all $\phi \in \mathring{H}_1$ and $\zeta \in C^2(D_\varepsilon)$. The choice $\zeta = 1/4\pi r$ with $r = |x - y|$ yields

$$\int_{D_\varepsilon} \nabla\phi \cdot \nabla\left(\frac{1}{4\pi r}\right)dy = \int_{S_\varepsilon(x)} \phi\frac{1}{4\pi r^2}\,d\sigma_y.$$

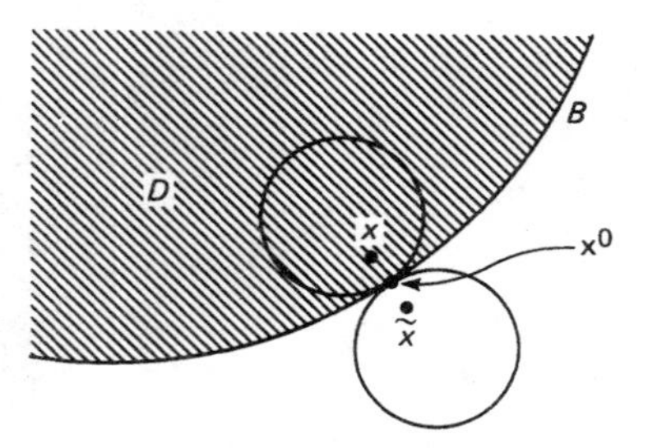

Figure 11-4

Assuming that ϕ is continuous at x, we can pass to the limit and find that

$$\phi(x) = \int_D \nabla\phi \cdot \nabla\left(\frac{1}{4\pi r}\right) dy.$$

Since $\phi = u - f$ is in $\mathring{H}_1(D)$ and is continuous at x,

$$u(x) - f(x) = \int_D \nabla(u - f) \cdot \nabla\left(\frac{1}{4\pi r}\right) dy. \tag{6-6}$$

Return to (3-2) of Sec. 8-3. This time fix $\tilde{x} \in cD$, let v converge to $\phi = u - f$ as above, and set $u = 1/4\pi\tilde{r}$, where $\tilde{r} = |\tilde{x} - y|$, to deduce that

$$0 = \int_D \nabla(u - f) \cdot \nabla\left(\frac{1}{4\pi\tilde{r}}\right) dy. \tag{6-7}$$

From (6-6) and (6-7) we have

$$u(x) - f(x) = \frac{1}{4\pi}\int_D \nabla(u - f) \cdot \nabla\left(\frac{1}{r} - \frac{1}{\tilde{r}}\right) dy. \tag{6-8}$$

Now, let $x \in K_0$ be near x° and let $\tilde{x}$, which belongs to the exterior ball, be the inverse point to x via the Kelvin transform (Sec. 8-5); see Fig. 11-4. Also, as $x \to x^\circ$ we have $\tilde{x} \to x^\circ$ and it follows from (6-8) that $u(x) - f(x) \to 0$ as $x \to x^\circ$. Since $f \in C^1(\bar{D})$ we conclude that

$$u(x) = f(x^\circ) + (f(x) - f(x^\circ)) + (u(x) - f(x)) \to f(x^\circ)$$

as x approaches x°.

Theorem 6-2. Suppose that the boundary B of D satisfies the sphere condition above and that $f \in C^1(\bar{D})$. Then the solution $u \in H_1(D)$, $u - f \in \mathring{H}_1(D)$ to

$$\int_D \nabla u \cdot \nabla\zeta \, dx = 0 \quad \text{for all} \quad \zeta \in C_0^2(D)$$

satisfies $u \in C^2(D) \cap C(\bar{D})$, $\Delta u = 0$ in D, and $u(x) \to f(x^\circ)$ as $x \to x^\circ$ in D.

PROBLEMS

1. Prove the analogue of Th. 6-1 for two dimensions.
2. Prove the analogue of Th. 6-2 for two dimensions.
3. This problem is really an invitation to further study. Let D be a bounded domain in two or three dimensions and consider

$$\Delta u - au = 0 \quad \text{in} \quad D$$
$$u = f \quad \text{on} \quad B$$

where $a > 0$ is a constant. Consider finding a weak solution using the Ritz–Galerkin method. Show that this leads to finding a $u \in H_1$ with $u - f \in \mathring{H}_1$ such that

$$\int_D [\nabla u \cdot \nabla \zeta + au\zeta]\, dx = 0$$

for all $\zeta \in C_0^2(D)$. Use Ritz–Galerkin arguments to show there is a weak solution. Then the fact that $u \in C^2(D)$ requires an extension of Weyl's lemma. For such results, consult Hellwig—see the Suggestions for Further Reading.

11-7. Generalized Solutions for Time-Dependent Problems

Hamilton's principle is the key for generalizing the concept of a solution for time-dependent problems as well as for static ones. Recall from Sec. 11-2 that Hamilton's principle applied to an unloaded, stretched, infinite string leads to the functional

$$I(u) = \int_{t_1}^{t_2} \int_{-\infty}^{\infty} \left(\frac{1}{2}\rho u_t^2 - \tau u_x^2\right) dx\, dt, \tag{7-1}$$

which is to be made stationary subject to the initial conditions

$$u(x, 0) = f(x), \qquad u_t(x, 0) = g(x), \qquad -\infty < x < \infty. \tag{7-2}$$

Natural restrictions on u are that it be continuous on $Q_T = \{(x, t): -\infty < x < \infty, 0 < t \le T\}$ and that u_t and u_x be square integrable. These conditions mean that the string is unbroken and has finite energy. Here $T > 0$ specifies the time interval over which the string will be observed. We summarize the integrability conditions on u by writing $u \in H_{1,1}(Q_T)$, where the first subscript refers to a generalized derivative in space and the second to a derivative in time.

As in Sec. 11-2, we find (for $0 \le t_1 < t_2 \le T$) that

$$\delta I(u) = \int_{t_1}^{t_2} \int_{-\infty}^{\infty} (\rho u_t \zeta_t - \tau u_x \zeta_x)\, dx\, dt,$$

where $\zeta = \zeta(x, t)$ is smooth and vanishes outside a closed bounded subset of $\{(x, t): -\infty < x < \infty, 0 < t < T\}$. In view of these restrictions on ζ, we can write

$$\delta I = \int_{Q_T} (\rho u_t \zeta_t - \tau u_x \zeta_x)\, dx\, dt. \tag{7-3}$$

In the classical treatment of the vibrating string, we integrate by parts to get rid of derivatives of ζ and obtain

$$\delta I = -\rho \int_{Q_T} (u_{tt} - c^2 u_{xx})\zeta \, dx \, dt, \tag{7-4}$$

where $c = (\tau/\rho)^{1/2}$. Then the condition $\delta I = 0$ and the fundamental lemma of the calculus of variations yields the classical wave equation. The drawback to this approach is that waves with corners are excluded because we must assume u is twice differentiable. On the other hand, no additional physical assumptions are needed to integrate by parts to get rid of derivatives of u and obtain

$$\delta I = -\rho \int_{Q_T} u(\zeta_{tt} - c^2 \zeta_{xx}) \, dx \, dt. \tag{7-5}$$

Hamilton's principle asserts that the shape of the string will be given by the function u such that

$$\int_{Q_T} u(\zeta_{tt} - c^2 \zeta_{xx}) \, dx \, dt = 0 \tag{7-6}$$

for all $\zeta \in C_0^2(Q_T)$.

To better understand (7-6) and the functions u which may satisfy it, consider the following approximation scheme. Let $\{f_n\}$ and $\{g_n\}$ be smooth functions approximating the given initial data f and g, which may not be smooth. Determine the classical solution $u_n = u_n(x, t)$ to

$$\frac{\partial^2 u_n}{\partial t^2} = c^2 \frac{\partial^2 u_n}{\partial x^2}, \qquad u_n(x, 0) = f_n(x), \quad \text{and} \quad \frac{\partial u_n}{\partial t}(x, 0) = g_n(x).$$

Under appropriate assumptions, which will be made precise below, the convergence of $\{f_n\}$ and $\{g_n\}$ will imply the uniform convergence of $\{u_n\}$ to a continuous function u. In view of (7-4), $\delta I(u_n) = 0$. Thus, (7-6) holds with u replaced by u_n and the uniform convergence yields (7-6) for u itself because the term in parentheses vanishes off a closed, bounded set. The solution u of (7-6) determined in this manner as the limit of classical solutions is called a *generalized solution* of the initial value problem

$$\begin{aligned} &u_{tt} = c^2 u_{xx}, && -\infty < x < \infty, && 0 < t \le T, \\ &u(x, 0) = f(x), && u_t(x, 0) = g(x), && -\infty < x < \infty. \end{aligned}$$

Theorem 7-1. Let $f \in H_1(-\infty, \infty) \cap C(-\infty, \infty)$ and $g \in L_2(-\infty, \infty)$. Suppose that there are sequences $\{f_n\}$ and $\{g_n\}$ of smooth functions with $f_n \to f$ uniformly on each bounded subset of $(-\infty, \infty)$ and $g_n \to g$ in L_2. Then the classical solutions $u_n = u_n(x, t)$ to the initial value problem with f and g replaced by f_n and g_n converges uniformly to a continuous function u which satisfies (7-6), and so is a generalized solution to the initial value problem for the wave equation.

Proof. The D'Alembert solution gives

$$u_n(x, t) = \frac{1}{2}[f_n(x + ct) + f_n(x - ct)] + \frac{1}{2c}\int_{x-ct}^{x+ct} g_n(s)\, ds.$$

The hypotheses on f_n and g_n guarantee that $u_n(x, t)$ converges to

$$u(x, t) = \frac{1}{2}[f(x + ct) + f(x - ct)] + \frac{1}{2c}\int_{x-ct}^{x+ct} g(s)\, ds$$

and that the convergence is uniform on bounded sets. This uniform convergence and the discussion preceding the theorem show that u satisfies (7-6).

The preceding discussion clarifies a point that comes up frequently. Namely, suppose that the initial conditions are given by $g(x) \equiv 0$ and

$$f(x) = \begin{cases} 0, & |x| > 1, \\ x + 1, & -1 \le x \le 0, \\ -x + 1, & 0 \le x \le 1, \end{cases}$$

so that f is a triangular wave. The "solution" to the wave equation must be

$$u(x, t) = \tfrac{1}{2}[f(x + ct) + f(x - ct)] \tag{7-7}$$

but $u(x, t)$ does not satisfy the classical wave equation at the points (x, t) with $x + ct = 0$ or $x - ct = 0$. On the other hand, the function u in (7-7) obviously belongs to $C \cap H_{1,1}$ and it is easy to construct a sequence of smooth functions $\{f_n\}$ which converge uniformly to f. The corresponding classical solutions $u_n(x, t)$ converge uniformly to $u(x, t)$ on Q_T, and therefore u is a generalized solution to the wave equation (see Prob. 1).

It should be clear from the previous example that the terminology "generalized solution" is not meant to imply a mathematical solution that is somehow physically unrealistic. Generalized solutions are the solutions of physical interest. When the data permit a classical solution, the two notions of a solution coincide. Of course, the discussion above can be applied to higher-dimensional problems and other equations. Just integrate by parts in the variational functional to transfer all derivatives from u to ζ. The result is the analogue of (7-6) for the problem at hand.

Consider next an initial, boundary value problem.

Theorem 7-2. Let D be a bounded domain and f, g, $\nabla f \in L_2(D)$, and $F \in L_2(Q_T)$. If $u \in C^{2,2}(Q_T) \cap C^{0,1}(\bar{Q}_T)$ is a solution to

$$u_{tt} = c^2\, \Delta u + F(x, t), \qquad (x, t) \in Q_T,$$
$$u(x, 0) = f(x), \qquad u_t(x, 0) = g(x), \qquad x \in D,$$
$$u(x, t) = 0, \qquad x \in B, \qquad 0 < t \le T,$$

then

$$\int_D (c^2|\nabla u|^2 + u_t^2)\, dx \le e^T \left[\int_D (c^2|\nabla f|^2 + |g|^2)\, dx + \frac{1}{2}\int_0^t \int_D |F(x,\tau)|^2\, dx\, d\tau \right]$$

and

$$\int_0^T \int_D (c^2|\nabla u|^2 + |u_t|^2)\, dx\, dt \le 2T \int_D [c^2|\nabla f|^2 + |g|^2]\, dx + \frac{T^2}{2}\int_0^T \int_D |F|^2\, dx\, dt.$$

Proof. Let

$$E(t) = \frac{1}{2}\int_D (c^2|\nabla u|^2 + u_t^2)\, dx.$$

Differentiate with respect to t and integrate by parts to get

$$\frac{d}{dt} E(t) = \int_D (c^2\, \nabla u \cdot \nabla u_t + u_t u_{tt})\, dx$$
$$= \int_D u_t(u_{tt} - c^2\, \Delta u)\, dx = \int_D u_t F\, dx.$$

Apply the Schwarz inequality to find

$$\text{(7-8)} \qquad \frac{dE}{dt} \le \left(\int_D |u_t|^2\, dx\right)^{1/2} \left(\int_D |F|^2\, dx\right)^{1/2}$$
$$\le \sqrt{2}\left[\frac{1}{2}\int_D (|u_t|^2 + c^2|\nabla u|^2)\, dx\right]^{1/2} \left(\int_D |F|^2\, dx\right)^{1/2}$$
$$\le \sqrt{2}E^{1/2}\left(\int_D |F|^2\, dx\right)^{1/2} \le \sqrt{2}\,\frac{\varepsilon}{2}\, E + \frac{\sqrt{2}}{2\varepsilon}\int_D |F|^2\, dx,$$

for any $\varepsilon > 0$. Take $\varepsilon = \sqrt{2}$ to conclude that

$$\frac{dE}{dt} - E \le \frac{1}{2}\int_D |F|^2\, dx, \qquad \frac{d}{dt}(e^{-t}E) \le \frac{1}{2}\int_D |F|^2\, dx.$$

Integrate with respect to t from 0 to t to obtain

$$E(t) \le e^t\left[E(0) + \frac{1}{2}\int_0^t \int_D |F|^2\, dx\, d\tau \right].$$

Since $e^t \le e^T$ the first inequality in the theorem follows.

To obtain the second estimate, note from (7-8) that

$$E' \le \sqrt{2}E^{1/2}\left(\int_D |F|^2\, dx\right)^{1/2}.$$

Integrate from 0 to t to get

$$E(t)^{1/2} \le E(0)^{1/2} + \frac{1}{\sqrt{2}} \int_0^t \left(\int_D |F|^2 \, dx \right)^{1/2} d\tau.$$

Square and make additional estimates to find that

$$\begin{aligned} E(t) &\le E(0) + \sqrt{2}\, E(0)^{1/2} \int_0^t \left(\int_D |F|^2 \, dx \right)^{1/2} d\tau \\ &\quad + \frac{1}{2} \left[\int_0^t \left(\int_D |F|^2 \, dx \right)^{1/2} d\tau \right]^2 \\ &\le E(0) + \sqrt{2}\frac{\varepsilon}{2}\, E(0) + \left(\sqrt{2}\frac{1}{2\varepsilon} + \frac{1}{2} \right) \left[\int_0^t \left(\int_D |F|^2 \, dx \right)^{1/2} d\tau \right]^2 \\ &\le \left(1 + \frac{\varepsilon}{\sqrt{2}} \right) E(0) + \left(\frac{1}{\sqrt{2}\varepsilon} + \frac{1}{2} \right) \int_0^t d\tau \int_0^t \int_D |F|^2 \, dx \, d\tau, \end{aligned}$$

by the Schwarz inequality. Finally, set $\varepsilon = \sqrt{2}$ and integrate again from 0 to $t \le T$ to obtain

$$\int_0^T E(t) \, dt \le 2TE(0) + \frac{T^2}{2} \int_0^T \int_D |F(x, t)|^2 \, dx \, dt,$$

which is the second inequality in the theorem.

We extend the notion of a solution to the initial, boundary value problem in Th. 7-2 as follows. We say that a function $u \in \mathring{H}_{1,1}$ is a *generalized solution* to the problem

$$\text{(7-9)} \qquad \begin{cases} u_{tt} = c^2 \, \Delta u + F(x, t), & (x, t) \in Q_T, \\ u(x, 0) = f(x), \quad u_t(x, 0) = g(x), & x \in \bar{D}, \\ u(x, t) = 0, \quad x \in B, & 0 < t \le T, \end{cases}$$

where $F \in L_2(Q_T)$, $g \in L_2(D)$, $f \in \mathring{H}_1(D)$ if the following holds:

1. There are sequences $\{f_n\}$, $\{g_n\}$, and $\{F_n\}$ of smooth functions which converge to f, g, and F in $\mathring{H}_1(D)$, $L_2(D)$, and $L_2(Q_T)$, respectively, and are such that the problem (7-9) with f, g, F replaced by f_n, g_n, F_n has a classical solution u_n. [That is, $u_n \in C^{2,2}(Q_T) \cap C^{0,1}(\bar{Q}_T)$ and satisfies the appropriate initial, boundary value problem.]
2. The sequence $\{u_n\}$ converges to u in $\mathring{H}_{1,1}$.

The notation used above has the expected meaning. Thus, $H_{1,1}$ is the space of functions which have generalized first-order partials with respect to the spatial and time variables. Also, $u \in \mathring{H}_{1,1}$ means that for almost all t in $[0, T]$ the function $u(x, t)$ is an element of $\mathring{H}_1(D)$.

An immediate consequence of this definition and Th. 7-2 is the uniqueness of generalized solutions.

Theorem 7-3. There exists at most one generalized solution to the initial, boundary value problem (7-9).

Proof. See Prob. 2.

We turn now to the existence of a generalized solution for

$$\text{(7-10)}\quad \begin{cases} u_{tt} = c^2\,\Delta u + F(x, t), & (x, t) \in Q_T, \\ u(x, 0) = f(x), \quad u_t(x, 0) = g(x), & x \in D, \\ u(x, t) = \phi(x, t), \quad x \in B, & 0 \le t \le T. \end{cases}$$

We assume that ϕ can be extended to $\bar{D} \times (-\infty, T]$ so that it is twice continuously differentiable and vanishes when $t \le 0$. Assume that $F \in L_2(D)$, $f \in \mathring{H}_1(D)$, and $g \in L_2(D)$. Set $u = \phi + v$ in (7-10) to reduce to solving

$$\text{(7-11)}\quad \begin{cases} v_{tt} = c^2\,\Delta v + F - (\phi_{tt} - c^2\,\Delta\phi), & (x, t) \in Q_T, \\ v(x, 0) = f(x), \quad v_t(x, 0) = g(x), & x \in D, \\ v(x, t) = 0, \quad x \in B, & 0 \le t \le T. \end{cases}$$

Thus, we must show that (7-11) has a generalized solution. Now there are sequences $\{f_n\}$, $\{g_n\}$, and $\{F_n\}$ of smooth functions which converge to f, g, and $F - (\phi_{tt} - c^2\,\Delta\phi)$ in $\mathring{H}_1(D)$, $L_2(D)$, and $L_2(Q_T)$. These functions can be chosen to be infinitely often differentiable with f_n and g_n vanishing near the boundary of D and such that $F_n(x, t)$ is a function of the class (B) introduced in Sec. 10-7. Consider the problem (7-11) with f, g, with $F - (\phi_{tt} - c^2\,\Delta\phi)$ replaced by f_n, g_n, and F_n. This problem has a classical solution v_n. Indeed, write $v_n = w_n + z_n$, where z_n is the solution to the initial value problem

$$\begin{aligned} (z_n)_{tt} &= c^2\,\Delta z_n + F_n, \qquad x \in R_3, \qquad t > 0, \\ z_n(x, 0) &= f(x), \qquad (z_n)_t(x, 0) = g(x), \qquad x \in R_3, \end{aligned}$$

which can be obtained from Kirchhoff's formula and Duhamel's principle as in Sec. 10-4. Thus, w_n must satisfy

$$\text{(7-12)}\quad \begin{cases} (w_n)_{tt} = c^2\,\Delta w_n, & (x, t) \in Q_T, \\ w_n(x, 0) = (w_n)_t(x, 0) = 0, & x \in \bar{D}, \\ w_n(x, t) = -z_n(x, t), \quad x \in B, & t \ge 0. \end{cases}$$

In view of the explicit solution formula for $z_n(x, t)$ and our assumptions on f_n, g_n, and F_n, the solution $z_n(x, t)$ is a function in the class (B). Thus, (7-12) has a classical solution by Th. 7-2 of Sec. 10-7, and $v_n = w_n + z_n$ is a classical solution to (7-11) with f, g, and $F - (\phi_{tt} - c^2\,\Delta\phi)$ replaced by f_n, g_n, and F_n.

Finally, Th. 7-2 yields the estimate

$$\|v_n - v_m\|_{\mathring{H}_{1,1}(Q_T)} \le 2T \int_D [c^2|\nabla(f_n - f_m)|^2 + |(g_n - g_m)|^2]\,dx + \frac{T^2}{2}\int_0^T \int_D |F_n - F_m|^2\,dx\,dt,$$

where the right member has limit zero as $m, n \to \infty$. Since $\mathring{H}_{1,1}(Q_T)$ is a Hilbert space, the Cauchy sequence $\{v_n\}$ converges to some v in $\mathring{H}_{1,1}(Q_T)$. Thus, v is a generalized solution to (7-11) and $u = \phi + v$ is the generalized solution to (7-10). We summarize this discussion as follows.

Theorem 7-4. Let the boundary of D be of the class (A) of Sec. 10-7 and assume that f, g, F, and ϕ in (7-10) satisfy the assumptions following (7-10). Then (7-10) has a unique generalized solution.

Generalized solutions also can be constructed for hyperbolic and parabolic problems using the Ritz–Galerkin method. We shall show informally how this is done for the heat equation. Consider the problem

$$\text{(7-13)} \qquad \begin{cases} u_t = a\,\Delta u + F(x, t), & (x, t) \in Q_T, \\ u(x, 0) = f(x), & x \in D, \\ u(x, t) = 0, & x \in B, \quad 0 < t \le T, \end{cases}$$

where D is a bounded domain, $f \in \mathring{H}_1(D)$, and $F \in L_2(Q_T)$. Suppose that $\zeta \in C_0^1(D)$ and multiply the heat equation by ζ and integrate by parts to obtain

$$\text{(7-14)} \qquad \int_D u_t\zeta\, dx = -\int_D a\,\nabla u \cdot \nabla\zeta\, dx + \int_D \zeta F(x, t)\, dx.$$

By a generalized solution u of (7-13) we mean a function $u \in \mathring{H}_{1,1}(Q_T)$ such that $u(x, 0) = f(x)$ and (7-14) holds for all $\zeta \in C_0^\infty(D)$ and almost all t in $[0, T]$.

Theorem 7-5. There is at most one generalized solution to (7-13).

Proof. Suppose that u_1 and u_2 were both generalized solutions and set $u = u_1 - u_2$. Then $u \in \mathring{H}_{1,1}$ and

$$\int_D u_t\zeta\, dx = -a\int_D \nabla u \cdot \nabla\zeta\, dx$$

for all $\zeta \in C_0^\infty(D)$, and hence for all $\zeta \in \mathring{H}_1$. The choice $\zeta = u$ yields

$$\frac{d}{dt}\int_D \frac{1}{2}u^2\, dx = -a\int_D |\nabla u|^2\, dx,$$

$$\frac{1}{2}\int_D u(x, t)^2\, dx + a\int_0^t\int_D |\nabla u|^2\, dx = 0,$$

and uniqueness follows.

Now, we use (7-14) as the basis for generating a sequence of functions which will converge to a generalized solution of (7-13). To this end, let $e_1, e_2, \ldots$ be a dense set of functions in $\mathring{H}_1(D)$ and set

$$\text{(7-15)} \qquad u_n(x, t) = \sum_{j=1}^{n} a_j(t)e_j(x),$$

where, for each n, the functions $a_j(t)$ are to be determined. Evidently, $a_j(t)$ also depends on n, but we suppress explicit mention of this fact in what follows. Let f_n and F_n to be the projections of f and F onto sp $\{e_1, \ldots, e_n\}$, where for f the projection is in $\mathring{H}_1(D)$ and for F in $L_2(D)$ for each t. Thus,

$$f_n(x) = \sum_{j=1}^{n} \alpha_j e_j(x)$$

and

$$F_n(x) = \sum_{j=1}^{n} \phi_j(t) e_j(x).$$

We choose the functions $a_j(t)$ in (7-15) so that (7-14) with F replaced by F_n and u replaced by u_n holds for all ζ in sp $\{e_1, \ldots, e_n\}$. That is (see Prob. 5), we require

$$\text{(7-16)} \qquad \sum_{j=1}^{n} a_j'(t) \int_D e_j(x)e_k(x)\,dx = -a \sum_{j=1}^{n} a_j(t) \int_D \nabla e_j(x) \cdot \nabla e_k(x)\,dx + \sum_{j=1}^{n} \phi_j(t) \int_D e_j(x)e_k(x)\,dx$$

for $k = 1, 2, \ldots, n$, and

$$\text{(7-17)} \qquad a_j(0) = \alpha_j, \qquad j = 1, 2, \ldots, n,$$

where the last stipulation guarantees that $u_n(x, 0) = f_n(x)$. The determinant of the matrix with entries

$$A_{jk} = \int_D e_j(x)e_k(x)\,dx$$

is nonzero because $e_1, \ldots, e_n$ are linearly independent. Hence the system (7-16) can be solved for $a_1'(t), \ldots, a_n'(t)$ and then integrated using (7-17) to obtain $a_1(t), \ldots, a_n(t)$. This determines the sequence $\{u_n\}$ in (7-15).

We shall prove that $\{u_n\}$ converges weakly to a function u which is a generalized solution to (7-13). The first step is to establish an a priori estimate on the u_n.

Theorem 7-6. Let $f \in \mathring{H}_1(D)$, $F \in L_2(Q_T)$, and u_n be constructed as above in (7-15). Then there is a constant C such that

$$\begin{aligned} \|u_n\|^2_{\mathring{H}_{1,1}(Q_T)} &= \int_0^T \int_D \left(|u_n|^2 + |\nabla u_n|^2 + \left|\frac{\partial u_n}{\partial t}\right|^2 \right) dx\,dt \\ &\le C\left[\int_D [|f|^2 + |\nabla f|^2]\,dx + \int_0^T \int_D |F|^2\,dx\,dt \right] \\ &= C\left[\|f\|^2_{\mathring{H}_1(D)} + \|F\|^2_{L_2(Q_T)} \right]. \end{aligned}$$

Proof. Observe that (7-16) can be written as

$$\int_D \frac{\partial u_n}{\partial t} e_k \, dx = -a \int_D \nabla u_n \cdot \nabla e_k \, dx + \int_D F_n e_k \, dx. \tag{7-18}$$

Multiply by $a_k(t)$ and sum over k to obtain

$$\frac{1}{2}\int_D \frac{\partial u_n^2}{\partial t} dx = -a \int_D |\nabla u_n|^2 \, dx + \int_D F_n u_n \, dx.$$

Integrate with respect to t and apply the Schwarz inequality to find

$$\frac{1}{2}\int_D u_n^2 \, dx + a \int_0^T \int_D |\nabla u_n|^2 \, dx \, dt$$
$$\leq \frac{1}{2}\int_D f_n^2 \, dx + \left(\int_0^T \int_D |F_n|^2 \, dx \, dt\right)^{1/2} \left(\int_0^T \int_D u_n^2 \, dx \, dt\right)^{1/2}.$$

Now the integral of f_n^2 over D is less than $\|f_n\|_{\mathring{H}_1}^2$, which in turn is less than $\|f\|_{\mathring{H}_1}^2$ because f_n is the projection of f onto sp $\{e_1, \ldots, e_n\}$. Similarly, the integral of F_n^2 is bounded by the integral of F^2. Thus, dropping the integral of u_n^2 over D on the left side of the previous inequality, we get

$$a \int_0^T \int_D |\nabla u_n|^2 \, dx \, dt \leq \frac{1}{2}\int_D [f^2 + |\nabla f|^2] \, dx$$
$$+ \frac{1}{2\varepsilon}\int_0^T \int_D |F|^2 \, dx \, dt + \frac{\varepsilon}{2}\int_0^T \int_D |u_n|^2 \, dx \, dt,$$

for any $\varepsilon > 0$. Finally, an application of the Poincaré inequality gives

$$\int_0^T \int_D \left|\frac{\partial u_n}{\partial t}\right|^2 dx \, dt \leq C_2(\|f\|_{\mathring{H}_1}^2 + \|F\|_{L_2(Q_D)}^2), \tag{7-19}$$

for some constant C_1 independent of n (see Prob. 6).

To bound $\partial u_n/\partial t$, multiply (7-18) by $a_k'(t)$ and sum over k to deduce that

$$\int_D \left|\frac{\partial u_n}{\partial t}\right|^2 dx = -\frac{a}{2}\frac{\partial}{\partial t}\int_D |\nabla u_n|^2 \, dx + \int_D F_n \frac{\partial u_n}{\partial t} dx.$$

Integrate with respect to t from 0 to T and proceed (Prob. 7) as in the derivation of (7-19) to conclude that

$$\int_0^T \int_D \left|\frac{\partial u_n}{\partial t}\right|^2 dx \, dt \leq C_1(\|f\|_{\mathring{H}_1}^2 + \|F\|_{L_2(Q_T)}^2), \tag{7-20}$$

for some constant C_2. Combine (7-19) and (7-20) to complete the proof.

It is now easy to show that (7-13) has a generalized solution.

Theorem 7-7. Let $f \in \mathring{H}_1(D)$ and $F \in L_2(Q_T)$. Then (7-13) has a generalized solution u. Furthermore, u can be obtained as the weak limit of the sequence of Ritz–Galerkin approximations $\{u_n\}$ defined in (7-15).

Proof. Theorem 7-6 shows that $\{u_n\}$ is a bounded sequence in $\mathring{H}_{1,1}$. Therefore, there is a weakly convergent subsequence, say $\{u_{n_k}\}$, which converges weakly to some $u \in \mathring{H}_{1,1}$. Actually, the full sequence $\{u_n\}$ converges to u and u is a generalized solution, as you are asked to check in Prob. 8.

PROBLEMS

1. Exhibit a sequence of smooth functions $\{f_n\}$ which approximate the triangular function f defined before (7-7), as required by Th. 7-1. This completes the verification that (7-7) is a generalized solution.
2. Prove Th. 7-3. *Hint*. Use Th. 7-2.
3. Consider the initial, boundary value problem from Chap. 4

$$\begin{aligned} &u_{tt} = u_{xx}, \qquad 0 < x < L, \qquad t > 0 \\ &u(x, 0) = f(x), \qquad u_t(x, 0) = g(x), \qquad 0 \le x \le L, \\ &u(0, t) = u(L, t) = 0, \qquad t \ge 0. \end{aligned}$$

   We obtained a formal solution by separation of variables. What natural conditions must you impose on f and g so that this formal solution is a generalized solution? Compare these conditions with those required to get a classical solution.
4. Show that a generalized solution u to (7-9) satisfies the equations

$$\int_0^T \int_D (\zeta_t u_t - c^2\, \nabla\zeta \cdot \nabla u + F\zeta)\, dx\, dt = 0$$

   and

$$\int_0^T \int_D u(\zeta_{tt} - c^2\, \Delta\zeta)\, dx\, dt = \int_0^T \int_D F\zeta\, dx\, dt$$

   for all $\zeta \in C_0^2(Q_T)$.
5. Show that (7-16) determines the coefficients in (7-15) so that (7-14) holds for all ζ in sp $\{e_1, \ldots, e_n\}$ with u replaced by u_n and F replaced by F_n.
6. Complete the verification of (7-19). *Hint*. Use the Poincaré lemma. If need be, consult the proof of (ii) of Th. 5-2.
7. Establish (7-20), reasoning as for (7-19).
8. Complete the proof of Th. 7-7 using the following steps.
   (a) Show that $\{u_n\}$ converges weakly to u. *Hint*. See Sec. 11-5 where the corresponding result is established for the Ritz–Galerkin approximations for the generalized Dirichlet problem.
   (b) Integrate from 0 to $t \le T$ in (7-18) and let $n \to \infty$ to deduce that

$$\int_0^t \left[\int_D (u_t\zeta + c\, \nabla u \cdot \nabla\zeta - F\zeta)\, dx \right] d\tau = 0$$

   for all $\zeta \in C_0^\infty(D)$.
   (c) Differentiate with respect to t to get (7-14) for almost all t.

9. Consider

$$u_t = u_{xx}, \qquad 0 < x < 1, \qquad t > 0,$$
$$u(x, 0) = x(1 - x), \qquad 0 \le x \le 1,$$
$$u(0, t) = u(1, t) = 0, \qquad t \ge 0.$$

Solve by separation of variables to obtain

$$u(x, t) = \sum_{n=1}^{\infty} a_n e^{-\lambda_n t} \sin(\lambda_n x), \qquad \lambda_n = n\pi,$$

as in Chap. 5. Now assume that $n = 1$ in the Ritz–Galerkin scheme, let $e(x) = e_1(x) = x(1 - x)$, and determine the Ritz–Galerkin approximation $u_1(x, t)$. Evaluate $u(x, t)$ and $u_1(x, t)$ for various x and t. Compare plots of $u(x, t)$ and $u_1(x, t)$ with respect to x for various time values.

10. Show that the Ritz–Galerkin procedure outlined for the heat equation can be applied to the wave equation. Assume that

$$u_n(x, t) = \sum_{j=1}^{n} a_j(t) e_j(x)$$

and formulate the second-order, ordinary differential equation and initial data used to find the coefficients $a_j(t)$.

11. In the context of Prob. 10, state and prove the analogues of (a) Th. 7-5, (b) Th. 7-6, and (c) Th. 7-7.

12

Further Applications

12-1. Initial Value Problems in Linear Continuum Mechanics

The study of macroscopic phenomena often leads to systems of partial differential equations. These systems are linear or can be linearized under certain simplifying assumptions. We shall consider the initial value problem for several systems of this type. The state of the system will be described by an unknown vector field. We will be able to determine the curl and divergence of this field. Thus, we begin this section with the problem of recovering a vector field from its curl and divergence. That is, if $\omega(x)$ and $\theta(x)$ are given, we wish to find a vector field $w = w(x)$ such that

$$\text{curl } w(x) = \omega(x), \qquad x \in R_3, \tag{1-1}$$

$$\text{div } w(x) = \theta(x), \qquad x \in R_3. \tag{1-2}$$

Since div curl $w = 0$, ω must satisfy the compatibility condition div $\omega = 0$ in order for a solution to exist. All vector fields in this section are assumed to have continuous second-order partial derivatives.

It is clear that (1-1) and (1-2) do not determine the vector field $w(x)$ uniquely because if w is a solution, so is w plus any constant vector. In typical applications, we expect the vector fields in question to decay at infinity. The results below pertain to such fields.

Theorem 1-1. There is at most one vector field $w(x)$ which satisfies (1-1) and (1-2), and tends to zero at infinity.

Proof. Assume that $w^{(1)}$ and $w^{(2)}$ both satisfy (1-1) and (1-2), and tend to zero at infinity. Then $w = w^{(1)} - w^{(2)}$ satisfies curl $w = 0$, div $w = 0$, and w tends to zero at infinity. Since curl $w = 0$ in R_3, $w = \nabla\phi$ for some potential.

Also, since div $w = 0$, we find $\Delta\phi = 0$. Then each component w_i of w is harmonic and tends to zero at infinity. By Prob. 17 of Sec. 8-4, w_i is a constant that must be zero in view of its behavior at infinity.

Next, consider the existence of a solution w to (1-1) and (1-2). To determine w from (1-1) and (1-2), we write $w = u + v$ and seek a u which satisfies

$$\text{curl } u = \omega, \qquad \text{div } u = 0,$$

and a v such that

$$\text{div } v = \theta, \qquad \text{curl } v = 0.$$

Since curl $v = 0$, there exists a function ϕ such that $v = \nabla\phi$. Then div $v = \theta$ becomes

$$\Delta\phi = \theta(x), \qquad x \in R_3.$$

From the results in Chap. 8, we expect to obtain ϕ as a volume potential

$$\phi(x) = -\int_{R_3} \frac{1}{4\pi r_{xy}}\,\theta(y)\,dy, \tag{1-3}$$

where we must impose sufficient restrictions on θ so that the integral exists and the analogue of Th. 6-1 of Sec. 8-6 holds (see Prob. 1).

To find u, recall the vector identity

$$\text{curl curl } u = \nabla \text{ div } u - \Delta u. \tag{1-4}$$

Since we require curl $u = \omega$ and div $u = 0$, this identity yields

$$\text{curl } \omega = \text{curl curl } u = \nabla \text{ div } u - \Delta u = -\Delta u,$$

and again we can express u as a volume potential,

$$u(x) = \int_{R_3} \frac{1}{4\pi r_{xy}}\,\text{curl } \omega(y)\,dy. \tag{1-5}$$

Combine (1-3) and (1-5) to obtain

$$w(x) = -\nabla_x \int_{R_3} \frac{1}{4\pi r_{xy}}\,\theta(y)\,dy + \int_{R_3} \frac{1}{4\pi r_{xy}}\,\text{curl } \omega(y)\,dy \tag{1-6}$$

as a candidate for a solution to (1-1) and (1-2). Once (1-6) has been obtained, it is a simple matter to check that the required differentiations under the integral sign can be carried out and that we can integrate by parts because $\partial r_{xy}/\partial x_i = -\partial r_{xy}/\partial y_i$ to confirm that (1-6) solves (1-1) and (1-2). The details are much the same as in Sec. 8-6, where now additional assumptions at infinity are needed to assure convergence of the integrals involved. We summarize this discussion in a theorem.

Theorem 1-2. Let $\theta \in C^1(R_3)$ and $\omega \in C^2(R_3)$ with div $\omega = 0$. Assume that $\theta(x)$ and curl $\omega(x)$ are $O(|x|^{-2-\varepsilon})$ for some $\varepsilon > 0$ as $|x| \to \infty$. Then (1-1) and (1-2) has a solution w given by (1-6). Moreover, this is the unique solution that tends to zero at infinity.

The representation (1-6) expresses w as a sum of two terms one for which the curl vanishes and one for which the divergence vanishes. This splitting of w is a general result known as *Helmholtz's theorem.*

Theorem 1-3. Let w be a vector field with $\theta = \text{div } w$ and $\omega = \text{curl } w$ satisfying the hypotheses of Th. 1-2. Assume that w has limit zero at infinity. Then there are functions ϕ and A such that

$$w = -\nabla\phi + \text{curl } A,$$

where

$$\phi = \int_{R_3} \frac{1}{4\pi r_{xy}} \text{div } w(y)\, dy, \quad \text{and} \quad A = \int_{R_3} \frac{1}{4\pi r_{xy}} \text{curl } w(y)\, dy.$$

The function ϕ is called the *scalar potential* of w, and A its *vector potential.*

Proof. By Th. 1-2,

$$w(x) = -\nabla_x \int_{R_3} \frac{1}{4\pi r_{xy}} \theta(y)\, dy + \int_{R_3} \frac{1}{4\pi r_{xy}} \text{curl } \omega(y)\, dy.$$

The components of curl $\omega(y)$ are linear expressions involving $\partial\omega_i/\partial y_j$ and

$$\int_{R_3} \frac{1}{4\pi r_{xy}} \frac{\partial\omega_i}{\partial y_j}\, dy = -\int_{R_3} \left(\frac{\partial}{\partial y_j} \frac{1}{4\pi r_{xy}}\right) \omega_i(y)\, dy = \frac{\partial}{\partial x_j} \int_{R_3} \frac{1}{4\pi r_{xy}} \omega_i(y)\, dy.$$

Thus,

$$\int_{R_3} \frac{1}{4\pi r_{xy}} \text{curl } \omega(y)\, dy = \text{curl}_x \int_{R_3} \frac{1}{4\pi r_{xy}} \omega(y)\, dy,$$

and the expression for $w(x)$ above is the one stated in the theorem.

The Equations of Linear Elasticity

The equations governing the motion of an infinite, isotropic, homogeneous elastic continuum are

$$\rho u_{tt} = \mu\, \Delta u + (\lambda + \mu)\nabla\, \text{div } u, \qquad x \in R_3, \qquad t > 0, \tag{1-7}$$

where ρ, μ, and λ are positive constants. The vector field $u = u(x, t) = (u_1(x, t), u_2(x, t), u_3(x, t))$ describes the small vibrations of the continuum, relative to some reference state. The internal stresses are given by

$$W = \mu\left(\frac{\partial u_i}{\partial x_j} + \frac{\partial u_j}{\partial x_i}\right) + \lambda(\text{div } u)I, \tag{1-8}$$

where I is the 3×3 identity matrix (tensor). Relation (1-8) postulates a linear relationship between the stress and strains $\partial u_i/\partial x_j$, and is called *Hooke's law.* The constants λ and μ characterize the elastic properties of the medium and are called the *Lamé constants.*

The system of equations (1-7) must be solved subject to given initial data which we express as

$$\begin{cases} u(x, 0) = f(x), \\ u_t(x, 0) = g(x), \end{cases} \tag{1-9}$$

for $x \in R_3$. We can solve the initial value problem (1-7) and (1-9) with the help of Th. 1-2 or 1-3. We shall apply Th. 1-3 and leave the approach using Th. 1-2 for Prob. 3. On the basis of Helmholtz's theorem, we assume a solution in the form of a scalar plus a vector potential,

$$u(x, t) = -\nabla p(x, t) + \text{curl } A(x, t). \tag{1-10}$$

We also represent the initial data in terms of scalar and vector potentials.

$$\begin{cases} f(x) = -\nabla\phi(x) + \text{curl } B(x), \\ g(x) = -\nabla\psi(x) + \text{curl } C(x). \end{cases} \tag{1-11}$$

Insert (1-10) into (1-7) and use (1-11) to conclude that (1-10) solves the initial value problem if

$$\begin{cases} \rho p_{tt} = (\lambda + 2\mu)\,\Delta p, \\ p(x, 0) = \phi(x), \\ p_t(x, 0) = \psi(x), \end{cases} \tag{1-12}$$

and

$$\begin{cases} \rho A_{tt} = \mu\,\Delta A, \\ A(x, 0) = B(x), \\ A_t(x, 0) = C(x). \end{cases} \tag{1-13}$$

We can use the Kirchhoff formula from Sec. 10-4 to construct solutions to (1-12) and (1-13). The function $p(x, t)$ represents a wave propagating with speed $[(\lambda + 2\mu)/\rho]^{1/2}$ and A is a wave with speed $(\mu/\rho)^{1/2}$. Observe also that $\theta \equiv \text{div } u = -\Delta p$. The mean normal stress is given by one-third the trace of W, namely,

$$\tfrac{1}{3}\,\text{tr}\,(W) = \tfrac{1}{3}(3\lambda + 2\mu)\,\text{div } u = \tfrac{1}{3}(3\lambda + 2\mu)\theta.$$

The quantity $\theta = -\Delta p$ represents a compression and is called a compression wave.

To summarize, if the initial conditions f and g are sufficiently smooth, we can construct the decompositions in (1-11) and obtain smooth enough data in (1-12) and (1-13) so that these wave equations can be solved by Kirchhoff's formula. Then u in (1-10) solves the initial value problem (1-7) and (1-9).

Finally, consider the uniqueness question. Let $w = u' - u''$, where u' and u'' are solutions to (1-7) and (1-9). Then w satisfies (1-7) and $w(x, 0) = w_t(x, 0) = 0$. Take the divergence of the differential equation for w to see that div w satisfies (1-12) with zero initial data. Consequently, div $w \equiv 0$. Similarly, curl w satisfies (1-13) with zero initial data and hence vanishes identically. If we require in addition that the vibrations decay to zero at infinity, then we can conclude that $w \equiv 0$. These considerations lead to the following result.

Theorem 1-4. Assume that $f \in C^4(R_3)$, $g \in C^3(R_3)$, and that the curl and divergence of f and g are $O(|x|^{-2-\varepsilon})$ as $|x| \to \infty$. Then the initial value problem (1-7) and (1-9) has a unique solution with limit zero at infinity.

The Linearized Equations for Viscous Incompressible Fluids

The flow of viscous, incompressible fluids is governed by the *Navier–Stokes equations*

$$\text{div } v = 0, \tag{1-14}$$

$$\rho[v_t + (v \cdot \nabla)v] = -\nabla p + \mu\,\Delta v + F. \tag{1-15}$$

Here ρ is a positive constant, $v = (v_1, v_2, v_3)$ is the velocity field, p is the hydrostatic pressure, and $(v \cdot \nabla)v$ is the vector field defined in Sec. 1-7. Also, F incorporates the external forces, and we assume these forces are given by a potential $F = -\nabla\Omega$. This would be the case, for example, if gravity were the only external force. Now, $-\nabla p + F = -\nabla(p + \Omega)$, and we write (1-15) as

$$\rho[v_t + (v \cdot \nabla)v] = -\nabla p + \mu\,\Delta v, \tag{1-16}$$

with the understanding that now p represents the hydrostatic pressure plus the potential from the external forces. For simplicity we still refer to p in (1-16) as the pressure.

The system (1-14), (1-16) is extremely difficult to solve in this generality. Certain special problems can be solved under suitable restrictions, and considerable information can be obtained. For one instance of this see Prob. 5. Here we assume that the flow is so slow that $(v \cdot \nabla)v$ can be neglected and then (1-16) reduces to the linear system

$$\rho v_t = -\nabla p + \mu\,\Delta v. \tag{1-17}$$

We seek a solution to (1-14), (1-17) subject to the initial condition

$$v(x, 0) = f(x), \tag{1-18}$$

where f satisfies the compatibility condition

$$\text{div } f(x) = \text{div } v(x, 0) = 0.$$

To obtain the solution set $\omega = \text{curl } v$ and take the curl of (1-17) to obtain the heat equation

$$\begin{cases} \rho\omega_t = \mu\,\Delta\omega, & x \in R_3, \quad t > 0, \\ \omega(x, 0) = \text{curl } f(x), & x \in R_3. \end{cases} \tag{1-19}$$

We obtain an expression for ω from Sec. 9-1. Next, we use Th. 1-2 to determine v from

$$\text{curl } v = \omega \quad \text{and} \quad \text{div } v = 0,$$

assuming, of course, that f is such that ω obtained from (1-19) satisfies the restrictions in Th. 1-2.

With v so determined, (1-17) requires that

$$\nabla p = -(\rho v_t - \mu \, \Delta v) \tag{1-20}$$

if a solution v, p does exist. Now, the heat equation in (1-19) assures that the right member of (1-20) has zero curl. Thus, (1-20) can be integrated and determines p up to an additive constant. These observations result in the following theorem.

Theorem 1-5. Assume that $f \in C^1(R_3)$, div $f(x) = 0$, and $\lim_{|x| \to \infty} \operatorname{curl} f(x) = 0$. Then there exists precisely one v with $\lim_{|x| \to \infty} \operatorname{curl} v = 0$ and p unique up to a constant such that

$$\begin{cases} \rho v_t = -\nabla p + \mu \, \Delta v, & x \in R_3, \quad t > 0, \\ \operatorname{div} v = 0, & x \in R_3, \quad t \geq 0, \\ v(x, 0) = f(x). \end{cases}$$

Maxwell's Equations

An electromagnetic field is described by vector fields giving the electric field strength E, the magnetic field strength H, and the electric and magnetic induction D and B. These quantities satisfy Maxwell's equations

$$\operatorname{curl} H = \frac{1}{c}\frac{\partial D}{\partial t} + \frac{4\pi}{c} i, \quad \operatorname{curl} B = -\frac{1}{c}\frac{\partial B}{\partial t}, \quad \operatorname{div} B = 0, \quad \operatorname{div} D = 4\pi\rho,$$

where i is the current and ρ is the charge density. The first and last of these equations imply the compatibility condition

$$\rho_t + \operatorname{div} i = 0,$$

which is just conservation of charge.

In addition to Maxwell's equations, constitutive relations among these quantities are needed. These relations depend on the medium in question. We take them to be

$$D = \varepsilon E, \qquad B = \mu H, \quad \text{and} \quad i = \sigma E$$

where ε, μ, and σ are constants known, respectively, as the dielectric constant, magnetic permeability, and the conductivity of the medium. Use of these relations to eliminate D, B, and i from Maxwell's equations yields

$$\operatorname{curl} H = \frac{\varepsilon}{c}\frac{\partial E}{\partial t} + \frac{4\pi\sigma}{c} E, \tag{1-21}$$

$$\operatorname{curl} E = -\frac{\mu}{c}\frac{\partial H}{\partial t}, \tag{1-22}$$

$$\operatorname{div} H = 0. \tag{1-23}$$

$$\operatorname{div} E = \frac{4\pi}{\varepsilon} \rho. \tag{1-24}$$

It is easy to find $\rho = \rho(x, t)$ when initial data are given for $E(x, 0)$. Indeed, let $\theta = \text{div } E$ and take divergence in (1-21) to find that

$$\frac{\varepsilon}{c}\theta_t + \frac{4\pi\sigma}{c}\theta = 0, \qquad \theta(x, 0) \equiv \theta_0(x) = \text{div } E(x, 0).$$

Thus,

$$\theta = \theta_0(x)\exp\left(-\frac{4\pi\sigma}{\varepsilon}t\right)$$

and from (1-24),

$$\rho(x, t) = \frac{\varepsilon}{4\pi} E(x, 0)\exp\left(-\frac{4\pi\sigma}{\varepsilon}t\right)$$

is known. Thus, we need to determine E and H from (1-21)–(1-24). Take the curl of (1-21) to obtain

$$\text{curl curl } H = \nabla \text{ div } H - \Delta H = \frac{\varepsilon}{c}\frac{\partial}{\partial t}\text{curl } E + \frac{4\pi\sigma}{c}\text{curl } E,$$

where (1-4) was used. Now, (1-22) and (1-23) yield the damped wave equation

$$\Delta H = \frac{\mu\varepsilon}{c^2}\frac{\partial^2 H}{\partial t^2} + \frac{4\pi\sigma\mu}{c^2}\frac{\partial H}{\partial t} \tag{1-25}$$

for H. In a vacuum $\mu = \varepsilon = 1$ and the speed of wave propagation is c.

The electric field strength also satisfies a wave equation. To get it take the curl of (1-22) and use (1-4) to obtain

$$\nabla \text{ div } E - \Delta E = -\frac{\mu}{c}\frac{\partial}{\partial t}\text{curl } H.$$

Then (1-21) and (1-24) yield

$$\Delta E = \frac{\mu\varepsilon}{c^2}\frac{\partial^2 E}{\partial t^2} + \frac{4\pi\sigma\mu}{c^2}\frac{\partial E}{\partial t} + \frac{4\pi}{\varepsilon}\nabla\rho, \tag{1-26}$$

an inhomogeneous, damped wave equation for E. In the event that E is zero initially, $\rho(x, t) \equiv 0$ and (1-26) reduces to

$$\Delta E = \frac{\mu\varepsilon}{c^2}\frac{\partial^2 E}{\partial t^2} + \frac{4\pi\sigma\mu}{c^2}\frac{\partial E}{\partial t}, \tag{1-27}$$

the same equation that H satisfies. The methods of Chap. 10, applied component by component, can be used to solve (1-25) and (1-26) or (1-27) once initial data are given for H, E, H_t, and E_t.

Concluding Remarks

We do not intend to give the impression that it is routinely possible to reduce general problems for systems of partial differential equations to the special solution formulas developed in earlier chapters. Many problems that arise are not initial value problems. Rather, they are boundary value problems for bounded domains. In the case of the linearized Navier–Stokes equations, for example, the natural boundary condition for a container at rest is $v = 0$ at the walls of the container. However, we derived an equation for the vorticity, curl v, and it is not clear what type of boundary condition should be imposed on it. Similarly, Maxwell's equations usually must be solved subject to a combination of conditions that do not readily separate in the manner above. General problems for the systems introduced here turn out to be very complex and entire monographs have been devoted to their solution, both in theoretical and approximate terms. We must be content with this brief introduction to the initial value problem for such systems.

PROBLEMS

1. Review the arguments in Sec. 8-6 through the proof of Poisson's theorem (Th. 6-1). Show that they apply with $D = R_3$ provided that $f(y) = O(|y|^{-2-\varepsilon})$ for some $\varepsilon > 0$.
2. Verify the identity (1-4).
3. Solve (1-7) and (1-9) by obtaining an equation for $\theta = \text{div } u$ and $\omega = \text{curl } u$. Confirm that these equations can be solved for θ and ω. Now use Th. 1-2 to get a representation for u.
4. Solve the initial value problem as indicated below.

$$\rho u_{tt} = \mu\,\Delta u + (\lambda + \mu)\nabla \text{ div } u + F(x, t), \qquad x \in R_3, \qquad t > 0,$$
$$u(x, 0) = f(x), \qquad u_t(x, 0) = g(x), \qquad x \in R_3.$$

To do this derive an inhomogeneous equation for $\theta = \text{div } u$ and show that θ can be found. Then get u from

$$\rho u_{tt} = \mu\,\Delta u + G(x, t),$$

where $G = (\lambda + \mu)\,\nabla\theta + F$ is known.

5. (*Hagen–Poiseuille flow*) Consider fluid flow in a right circular cylinder of length L and radius a. Seek a solution to the Navier–Stokes equations (1-14) and (1-15) assuming the flow is parallel to the axis (z-axis) of the cylinder so $v_1 = v_2 = 0$ and assume that $w = v_3(r, z)$ is axially symmetric. Assume also that v is time independent.

(a) Show that

$$\frac{\partial p}{\partial x_1} = \frac{\partial p}{\partial x_2} = 0 \quad \text{and} \quad \frac{\partial v_3}{\partial z} = 0; \quad \text{hence,} \quad w = w(r).$$

(b) Show that w satisfies

$$0 = -\frac{\partial p}{\partial z} + \mu\frac{1}{r}\frac{\partial}{\partial r}\left(r\frac{\partial w}{\partial r}\right), \qquad w(a) = 0,$$

where $w(a) = 0$ because of the so-called no-slip condition on the walls of the cylinder.

(c) Let $p(0) = p_0$, $p(L) = p_L$ and assume that $p_0 > p_L$. Then

$$p(z) = \alpha z + \beta, \qquad \alpha = \frac{p_L - p_0}{L},$$

and

$$w(r) = -\frac{\alpha}{4\mu}(a^2 - r^2) = \frac{p_0 - p_L}{4\mu L}(a^2 - r^2),$$

which is the characteristic parabolic profile for such flows.

(d) Use part (c) to show that the mass flux $\mathscr{M}$ of the fluid at $z = L$ is given by

$$\mathscr{M} = \frac{1}{\pi a^2}\int_0^{2\pi}\int_0^a \rho w(r) r \, dr \, d\theta = \frac{p_0 - p_L}{L}\frac{\rho a^2}{8\mu}.$$

This formula is known as the law of Hagen and Poiseuille.

6. State and prove an existence and uniqueness theorem for the initial value problem for Maxwell's equations (1-21)–(1-24). Don't forget to notice natural compatibility conditions.

12-2. One-Dimensional Gas Dynamics

Most problems in nonlinear continuum mechanics are extremely difficult to analyze. One problem which can be treated explicitly is the initial value problem for the flow of a one-dimensional isentropic (constant entropy) gas, which is described by

$$\rho_t + (\rho v)_x = 0, \tag{2-1}$$

$$\rho[v_t + v v_x] = -p_x, \tag{2-2}$$

and the equation of state

$$\frac{p}{p_0} = \left(\frac{\rho}{\rho_0}\right)^\lambda, \qquad \lambda > 1. \tag{2-3}$$

See Sec. 1-7 for a derivation of these equations. We have assumed that no external forces are acting on the gas. In view of (2-3) the pressure p can be eliminated from (2-2), and we can write (2-1), (2-2) as the system

$$\begin{bmatrix} \rho \\ v \end{bmatrix}_t + \begin{bmatrix} v & \rho \\ \dfrac{c^2}{\rho} & v \end{bmatrix} \begin{bmatrix} \rho \\ v \end{bmatrix}_x = \begin{bmatrix} 0 \\ 0 \end{bmatrix}, \tag{2-4}$$

where

$$c^2 = c^2(\rho) = \frac{dp}{d\rho} = \left(\frac{p_0 \lambda}{\rho_0^\lambda}\right)\rho^{\lambda - 1}. \tag{2-5}$$

Notice that $c(\rho)$ depends on the solution ρ, v to (2-4). The quantity c has dimensions of velocity and can be identified as the local speed of sound.

The system (2-4) is a special quasilinear, hyperbolic system. In Sec. 4-9 we used an iterative scheme to find local solutions of such systems. Here we present an elegant analysis of (2-4) due to Riemann. (See B. Riemann, "Ueber die

Fortpflanzung ebener Wellen von endlicher Schwingungsweite," in *Collected Works*, Dover Publications, Inc., New York, 1953.)

We begin by writing the system (2-4) in a more symmetric form. Multiply the first equation in (2-4) by $\pm c/\rho$ and add the result to the second equation to obtain

$$\text{(2-6)} \qquad \frac{c}{\rho}\left[\rho_t + (v + c)\rho_x\right] + \left[v_t + (v + c)v_x\right] = 0,$$

$$\text{(2-7)} \qquad \frac{c}{\rho}\left[\rho_t + (v - c)\rho_x\right] - \left[v_t + (v - c)v_x\right] = 0.$$

Suppose that $\rho = \rho(x, t)$ and $v = v(x, t)$ is a solution of (2-6) and (2-7). The coefficients of ρ_x and v_x are the eigenvalues of the matrix of coefficients in (2-4). Consequently, based on our previous discussion in Sec. 4-7, it appears reasonable to introduce characteristic coordinates. That is, set $\alpha = \phi(x, t)$ and $\beta = \psi(x, t)$, where

$$\text{(2-8)} \qquad \frac{dx}{dt} = v - c \quad \text{has solution} \quad \phi(x, t) = \text{const},$$

and

$$\text{(2-9)} \qquad \frac{dx}{dt} = v + c \quad \text{has solution} \quad \psi(x, t) = \text{const}.$$

Use the chain rule to find

$$0 = \psi_x x_\alpha + \psi_t t_\alpha, \qquad \frac{x_\alpha}{t_\alpha} = -\frac{\psi_t}{\psi_x} = \frac{dx}{dt} = v + c,$$

or

$$\text{(2-10)} \qquad \frac{\partial x}{\partial \alpha} = (v + c)\frac{\partial t}{\partial \alpha}.$$

Similarly,

$$\text{(2-11)} \qquad \frac{\partial x}{\partial \beta} = (v - c)\frac{\partial t}{\partial \beta}.$$

Now,

$$\frac{\partial \rho}{\partial \alpha} = \frac{\partial \rho}{\partial t}\frac{\partial t}{\partial \alpha} + \frac{\partial \rho}{\partial x}\frac{\partial x}{\partial \alpha} = \left[\frac{\partial \rho}{\partial t} + (v + c)\frac{\partial \rho}{\partial x}\right]\frac{\partial t}{\partial \alpha}$$

and there are similar expressions for $\partial v/\partial \alpha$, $\partial \rho/\partial \beta$, $\partial v/\partial \beta$. Use of these expressions in (2-6) and (2-7) yields

$$\text{(2-12)} \qquad \frac{c}{\rho}\frac{\partial \rho}{\partial \alpha} + \frac{\partial v}{\partial \alpha} = 0$$

and

$$\frac{c}{\rho}\frac{\partial \rho}{\partial \beta} - \frac{\partial v}{\partial \beta} = 0. \tag{2-13}$$

These equations can be simplified even further by introducing the function

$$f(\rho) = \int \frac{c}{\rho}\, d\rho = \frac{2}{\lambda - 1}\left(\frac{p_0^\lambda}{\rho_0^\lambda}\right)^{1/2} \rho^{(\lambda-1)/2} = \frac{2c}{\lambda - 1},$$

which allows us to write (2-12) and (2-13) as

$$\frac{\partial}{\partial \alpha}(f(\rho) + v) = 0, \tag{2-14}$$

$$\frac{\partial}{\partial \beta}(f(\rho) - v) = 0. \tag{2-15}$$

Thus,

$$f(\rho) + v = 2r = 2r(\beta), \tag{2-16}$$

$$f(\rho) - v = 2s = 2s(\alpha), \tag{2-17}$$

where r depends only on β and s only on α, as the notation indicates.

The functions r and s are called the *Riemann invariants* of the flow. These invariants can be used to reduce the one-dimensional, nonlinear gas dynamics equations to a single linear, second-order equation. To see this, regard x and t as functions of r and s. Then $x_\alpha = x_r r_\alpha + x_s s_\alpha = x_s s_\alpha$ because $r = r(\beta)$. Similarly,

$$x_\beta = x_r r_\beta, \qquad t_\alpha = t_s s_\alpha, \qquad t_\beta = t_r r_\beta,$$

and (2-10) and (2-11) can be written as

$$\frac{\partial x}{\partial s} = (v + c)\frac{\partial t}{\partial s}, \tag{2-18}$$

$$\frac{\partial x}{\partial r} = (v - c)\frac{\partial t}{\partial t}. \tag{2-19}$$

From (2-16), (2-17) we see that

$$v = r - s \tag{2-20}$$

and $f(\rho) = r + s$. Since $f(\rho) = 2c/(\lambda - 1)$, we find

$$c = \frac{\lambda - 1}{2}(r + s). \tag{2-21}$$

Now, (2-18)–(2-21) give

$$\frac{\partial x}{\partial s} = \left[(r - s) + \frac{\lambda - 1}{2}(r + s)\right]\frac{\partial t}{\partial s} \tag{2-22}$$

and

$$\frac{\partial x}{\partial r} = \left[(r - s) - \frac{\lambda - 1}{2}(r + s)\right]\frac{\partial t}{\partial r}, \tag{2-23}$$

which is a linear system for x and t. Differentiate the first equation with respect to r, the second with respect to s, and subtract the results to get

$$\frac{\partial^2 t}{\partial r\, \partial s} + \frac{\lambda + 1}{2(\lambda - 1)(r + s)}\left(\frac{\partial t}{\partial r} + \frac{\partial t}{\partial s}\right) = 0. \tag{2-24}$$

The initial value problem for the hyperbolic equation (2-24) can be solved by the methods of Sec. 4-6. Once $t = t(r, s)$ is determined, we obtain $x = x(r, s)$ from (2-22) and (2-23). Next, we determine r and s as functions of x and t. Then (2-20) yields v as a function of x and t. Finally, (2-21) and (2-5) give ρ as a function of x and t.

Evidently, the explicit execution of this program can be very tedious. Moreover, (2-24) must be solved subject to appropriate initial data on a curve Γ in the rs plane. We close this section with a description of how to find Γ and the data on it. Suppose that the given initial data is

$$v(x, 0) = v_0(x), \qquad \rho(x, 0) = \rho_0(x). \tag{2-25}$$

These data allows us to calculate

$$p(x, 0) = p_0(x) \quad \text{and} \quad c(\rho_0) = c_0(x)$$

using (2-3) and (2-5). Now, when $t = 0$, r and s satisfy

$$r + s = \frac{2c_0(x)}{\lambda - 1}, \qquad r - s = v_0(x),$$

by (2-20) and (2-21). Thus, the initial curve Γ in the rs plane is given parametrically by

$$\Gamma: \begin{cases} r = \dfrac{c_0(x)}{\lambda - 1} + \dfrac{1}{2} v_0(x), \\[2ex] s = \dfrac{c_0(x)}{\lambda - 1} - \dfrac{1}{2} v_0(x). \end{cases} \tag{2-26}$$

Now, calculate x_r and x_s on Γ to get

$$1 = \left[\frac{c_0'(x)}{\lambda - 1} + \frac{1}{2} v_0'(x)\right] x_r,$$

$$1 = \left[\frac{c_0'(x)}{\lambda - 1} - \frac{1}{2} v_0'(x)\right] x_s.$$

Insert these expressions for x_r and x_s on Γ into (2-18) and (2-19) with v and c replaced by $v_0(x)$ and $c_0(x)$ to obtain known initial data for t_r and t_s on Γ.

Not all nonlinear equations can be transformed into linear equations as we have done for the one-dimensional, gas dynamics equations. However, some other examples of this approach are given in the problems.

PROBLEMS

1. Confirm that (2-1)–(2-3) yield the system (2-4).
2. Suppose that the equation of state is $p/p_0 = \rho/\rho_0$. Develop Riemann's solution in this case.
3. Read Riemann's paper, cited earlier. He obtains information about shock waves; there is a wealth of other ideas; and the paper is readable. Notice his motivation for proceeding to the Riemann function.
4. Consider *Burgers' equation*

$$u_t + uu_x = \alpha u_{xx}, \qquad -\infty < x < \infty, \qquad t > 0,$$
$$u(x, 0) = f(x), \qquad -\infty < x < \infty,$$

where $\alpha > 0$ is a constant. Find u by following the steps below due to Hopf and Cole.

(a) Show that

$$\frac{\partial}{\partial t} u = \frac{\partial}{\partial x}\left(\alpha u_x - \frac{u^2}{2}\right),$$

and argue that there is a function $Q(x, t)$ such that

$$Q_x = u \quad \text{and} \quad Q_t = \alpha u_x - \frac{u^2}{2}.$$

(b) Show that

$$Q_t = \alpha Q_{xx} - \tfrac{1}{2} Q_x^2.$$

(c) For any real λ, show that $w = e^{\lambda Q}$ satisfies

$$\alpha w_{xx} - w_t = \lambda e^{\lambda Q}(\alpha Q_{xx} + \alpha \lambda Q_x^2 - Q_t).$$

(d) Now use parts (b) and (c) to determine an equation for w and solve it; hence get Q and then u.

5. Sometimes special solutions can be obtained that shed light on how more general solutions may behave. Consider $u_t = au_{xx} + bu_x^2$ with $a > 0$ and b given constants. Let $\xi = x^\alpha t^\beta$ and seek solutions of the form $u = f(\xi)$. Substitute this expression for u in the differential equation and determine α and β so that an ordinary differential equation whose coefficients depend on ξ results for $f(\xi)$. Now try to solve the equation for f.
6. Show that (2-24) has solutions of the form $t = \phi(\xi)$, where $\xi = r + s$, and find them.
7. Find the Riemann function for (2-24). The result is

$$R(r, s; \xi, \eta) = \frac{(r + s)^{2a}}{(s + \xi)^a (r + \eta)^a} F\left[a, a; 1; \frac{(\xi - r)(\eta - s)}{(s + \xi)(r + \eta)}\right],$$

where F is the hypergeometric function and $a = (\lambda + 1)/2(\lambda - 1)$.

3. Discontinuous Solutions and Shocks

In Sec. 12-2 we reduced the nonlinear gas dynamics equations to a linear problem by means of the Riemann invariants. In past chapters we usually treated linear problems, or in the case of nonlinear problems we restricted our attention to finding local solutions, defined in the neighborhood of the given initial data. However, in practice, we are interested in determining global solutions which are defined over the entire time interval of physical interest. The physical phenomenon does not stop just because the mathematics breaks down. On the other hand, we expect challenging mathematical problems in the face of unusual or unpleasant physical phenomena. In this section we give a brief introduction to the theory of shocks, which arise in nonlinear models.

To see more clearly the issues that must be faced, consider the nonlinear problem

$$\begin{cases} u_t + a(u)u_x = 0, & -\infty < x < \infty, \quad t > 0, \\ u(x, 0) = \phi(x). \end{cases} \tag{3-1}$$

It is natural to use the method of characteristics to solve (3-1). The initial curve can be parameterized as $t_0(s) = 0$, $x_0(s) = s$, $u_0(s) = \phi(s)$, and the characteristic curves are given by

$$\frac{dt}{d\tau} = 1, \qquad \frac{dx}{d\tau} = a(u), \qquad \frac{du}{d\tau} = 0.$$

Thus, $\tau = t$ and the characteristics are given by

$$\frac{dx}{dt} = a(u), \qquad x(s, 0) = s, \tag{3-2}$$

and u is determined along each characteristic by

$$\frac{du}{dt} = 0, \qquad u(s, 0) = \phi(s). \tag{3-3}$$

Equation (3-3) reveals that u is constant in time along the characteristics defined by (3-2), which propagates with speed $a(u)$. In analogy with acoustics we call $a(u)$ the sound speed. Integrating the system (3-2), (3-3) yields

$$u = u(s, t) = \phi(s), \tag{3-4}$$

$$x = x(s, t) = a(\phi(s))t + s. \tag{3-5}$$

Now, in order to evaluate u at a specific point (x, t), we must solve (3-5) for $s = s(x, t)$ in terms of x and t to find the characteristic which starts at $t = 0$ and $x = s$ and travels to (x, t). Since u has the constant value $\phi(s)$ at each point of the characteristic (3-5), $u(x, t) = \phi(s(x, t))$. The question is can we solve (3-5) for $s = s(x, t)$. The implicit function theorem allows us to solve (3-5) when

$$a'(\phi(s))\phi'(s)t + 1 \neq 0. \tag{3-6}$$

This condition surely holds if a' and ϕ' agree in sign or if a' and ϕ' are bounded and t is small enough. However, if a condition such as (3-6) is not satisfied, serious difficulties may arise. Consider an example.

Example 1. Solve

$$u_t + uu_x = 0, \qquad u(x, 0) = e^{-x}.$$

From (3-4) and (3-5), $u = e^{-s}$ along the characteristic $x = e^{-s}t + s$. In particular, if $s = 0$, then $u = 1$ along the characteristic $x = t$, while if $s = 1$, then $u = e^{-1}$ along the characteristic $x = e^{-1}t + 1$. But these straight lines intersect at $(e/(e-1), e/(e-1))$. Now, we have a dilemma. What is the value of u at this point?

We encounter the dilemma of Ex. 1 whenever two or more characteristics pass through the same point. In the context of (3-1), we can expect this difficulty to arise frequently. Indeed, the characteristics (3-5) are the straight lines $t = (x - s)/a(\phi(s))$ and unless special restrictions are placed on $a(u)$ and $\phi(s)$, some of these lines will intersect in the region $t > 0$. Obviously, we cannot extend a classical solution into the region where the characteristics intersect. To obtain global solutions, we must generalize our notion of a solution in a physically appropriate way. Several examples will help point the way.

To address the difficulty just raised and to obtain physically meaningful solutions, we return to first principles and review the origin of problems like (3-1). Consider a physical system without dissipation, whose evolution we seek to describe by a conservation law. We introduce a (scalar) density function u for the physical quantity under study and ask how the quantity described by u changes over time in a given control volume V. If there are no sources or sinks and f designates the flux of material across a given surface per unit area per unit time, we deduce that

$$\frac{d}{dt}\int_V u\, dx = -\int_S f \cdot n\, d\sigma, \tag{3-7}$$

where S is the surface bounding V. If f is smooth enough so that the Gauss divergence theorem applies and u has continuous partials, then (3-7) yields

$$u_t + \operatorname{div} f = 0. \tag{3-8}$$

Equations (3-7) and (3-8) are called conservation laws. For simplicity, we restrict our attention to the case of a single conservation law for a scalar density u which depends on a one-dimensional spatial variable. Then (3-7) and (3-8) can be expressed as

$$\frac{d}{dt}\int_\alpha^\beta u\, dx = -f\Big|_\alpha^\beta \tag{3-9}$$

and

$$u_t + f_x = 0, \tag{3-10}$$

respectively. Moreover, we assume that $f = f(u)$ and set $a(u) = f'(u)$. Then (3-10) can be written as $u_t + a(u)u_x = 0$, which is the quasilinear equation in (3-1).

Consider the possibility of a discontinuous solution u to (3-1). In this case, we must interpret the differential equation in (3-10) in the integral form (3-9). To be more specific, we seek a solution u with a single curve $x = \sigma(t)$ across which u may have jump discontinuities. We assume that u is a classical solution to the left and right of σ, but permit u to jump across $x = \sigma(t)$. Set

$$\lim_{x\to\sigma(t)-} u(x, t) = u_l \quad \text{and} \quad \lim_{x\to\sigma(t)+} u(x, t) = u_r \tag{3-11}$$

and denote the jump of u across the curve $x = \sigma(t)$ by

$$[u] = u_r - u_l. \tag{3-12}$$

Choose an interval $[\alpha, \beta]$ which contains $\sigma(t)$ at a fixed time t. Then by (3-9) we have

$$\frac{d}{dt}\int_\alpha^\beta u(x, t)\, dx = -\{f(u(\beta, t)) - f(u(\alpha, t))\}. \tag{3-13}$$

However, on $[\alpha, \sigma(t))$ and on $(\sigma(t), \beta]$ the solution u satisfies (3-10) and we can express the left member of (3-13) as

$$\begin{aligned}
&\frac{d}{dt}\left[\int_\alpha^{\sigma(t)-} u(x, t)\, dx + \int_{\sigma(t)+}^\beta u(x, t)\, dx\right] \\
&= \int_\alpha^{\sigma(t)-} u_t\, dx + \int_{\sigma(t)+}^\beta u_t\, dx + u_l\dot\sigma - u_r\dot\sigma \\
&= -[u]\dot\sigma - f(u)\Big|_\alpha^{\sigma-} - f(u)\Big|_{\sigma+}^\beta,
\end{aligned}$$

where $\dot\sigma = d\sigma/dt$. Substitute this result into (3-13) to conclude that

$$[u]\dot\sigma = [f], \tag{3-14}$$

where $[f] = f(u_r) - f(u_l)$. Equation (3-14) is called the *Rankine–Hugoniot jump condition.* It is tempting to define a discontinuous solution by the requirements that it satisfy (3-10) where it is smooth and (3-14) across a discontinuity. The following elaboration on Ex. 1 lends support to this idea.

***Example* 2.** Consider again the equation

$$u_t + uu_x = 0, \qquad -\infty < x < \infty, \qquad t > 0,$$

so that $f(u) = u^2/2$ and assume initially that

$$u(x, 0) = \phi(x) = \begin{cases} 1, & x \le 0, \\ 1 - x, & 0 < x < 1, \\ 0, & x \ge 1. \end{cases}$$

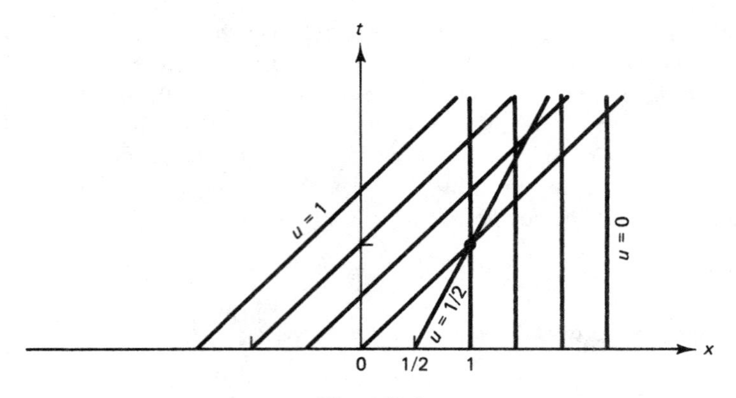

Figure 12-1

The classical solution is easily constructed for t near 0. The value of the solution u along characteristics given by $x = \phi(s)t + s$ is indicated in Fig. 12-1. All the characteristics emanating from points on the interval $0 < x < 1$ pass through (1, 1), and the value of u on such a characteristic is $1 - x$. The characteristic starting at $x = \frac{1}{2}$ is shown.

We seek to extend the classical solution beyond time $t = 1$. To this end, we wish to find a curve $x = \sigma(t)$ emanating from the point (1, 1) so that the solution is $u = 1$ in the region $t > 1$, which lies to the left of the curve, and is $u = 0$ to the right. Since $f(u) = u^2/2$, the jump relation (3-14) becomes $\dot{\sigma} = 1/2$ with $\sigma(1) = 1$. Thus, $x = \sigma(t) = (t + 1)/2$ and we obtain a solution in $t > 0$ given by the classical solution for $0 \leq t \leq 1$ and by $u = 1$ for $x < (t + 1)/2$ and $u = 0$ for $x > (t + 1)/2$ when $t > 1$.

Notice that other extensions of the classical solution into the region $t > 1$ are possible: for example, the extension indicated in Fig. 12-2, where the values of u are determined in the shaded triangular regions by the characteristics that emanate from $0 < x < 1$. The solution in Fig. 12-2 does not satisfy the Rankine–Hugoniot condition and can be rejected as physically uninteresting.

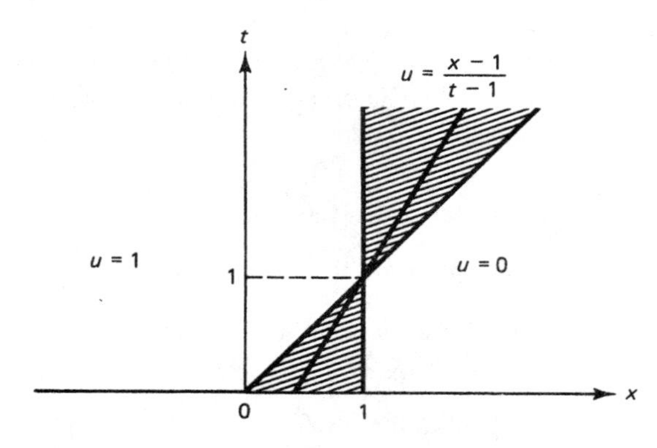

Figure 12-2

A variation of Ex. 2 leads to the realization that more is needed than the jump condition.

Example 3. Solve

$$u_t + uu_x = 0, \qquad -\infty < x < \infty, \qquad t > 0,$$

with initial data

$$u(x, 0) = \phi(x) = \begin{cases} 0, & x < 0, \\ 1, & x > 0. \end{cases}$$

This time there are no characteristics in the region $\{0 < x < t, t > 0\}$. The function $u = x/t$ interpolates the values $u = 0$ and $u = 1$ across this triangular region. Also, the extended solution satisfies the differential equation at all points where it is differentiable. It also satisfies the conservation law (3-13). On the other hand, we can use the jump relation (3-14) to join the values $u = 0$ and $u = 1$ across the curve $x = \sigma(t) = t/2$ and so obtain another solution to the problem in Ex. 3. The question is which solution is the physically relevant one.

To distinguish these solutions, consider the hypothetical situations shown in Fig. 12-3. A physically meaningful extension should be determined by its initial data and not by arbitrary extensions. Thus, we wish to exclude the situation on the far right in Fig. 12-3. We obviously wish to eliminate multiple intersections of the characteristic with a curve σ of discontinuity. The sketch on the left in Fig. 12-3 allows us to set up the Rankine–Hugoniot relation uniquely and each characteristic intersects $x = \sigma(t)$ precisely once. To guarantee this situation, $x = \sigma(t)$ must move to the right slower than the signals coming from the left with speed a, and must move left slower than signals coming from the right. Thus, we require that

$$a(u_r) < \dot{\sigma} < a(u_l). \tag{3-15}$$

This condition allows us to rule out the second solution in Ex. 3 because $u_r = 1$, $u_l = 0$ and $a(u) = u$ violate (3-15). Of course, it is possible that there may be still other solutions of Ex. 3 which also satisfy (3-15). The uniqueness we seek follows from Th. 3-1 below.

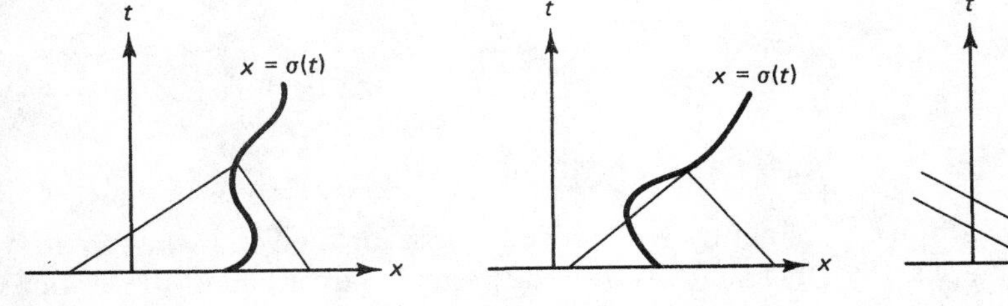

Figure 12-3

In analogy with the theory of gas dynamics, (3-15) is called the *entropy condition.* Although we shall not prove it, (3-15) guarantees that entropy increases as material crosses a discontinuity. A curve $x = \sigma(t)$ satisfying both the Rankine–Hugoniot jump relation and the entropy condition is called a *shock.* We say that an integrable function u is a *generalized solution* to

$$\begin{cases} u_t + f(u)_x = 0, & -\infty < x < \infty, \quad t > 0, \\ u(x, 0) = \phi(x), & -\infty < x < \infty \end{cases} \tag{3-16}$$

if (3-13) holds over any finite interval $[\alpha, \beta]$, $u(x, 0) = \phi(x)$, and the only discontinuities of the piecewise continuous function u are shocks.

Theorem 3-1. Suppose that $f''(u) > 0$, so that $a(u) = f'(u)$ is increasing and $\phi(x)$ is integrable on $-\infty < x < \infty$. Then there is at most one generalized solution to (3-16).

Proof. Assume that u and v are generalized solutions of (3-16) and set $w = u - v$. Form

$$I(t) = \int_{-\infty}^{\infty} |w(x, t)|\, dx.$$

We plan to show that $I(t)$ vanishes for all $t \geq 0$. To do this, we need only show that $I(t)$ is nonincreasing because then $0 \leq I(t) \leq I(0) = 0$. For each t, partition the x axis into subintervals $[\alpha_n, \alpha_{n+1}]$ such that w has a constant sign on $[\alpha_n, \alpha_{n+1}]$ and w changes sign on successive intervals. Choose the indexing so that the sign of w on $[\alpha_n, \alpha_{n+1}]$ is $(-1)^n$. Notice that $\alpha_n = \alpha_n(t)$. Then $I(t)$ can be written as

$$I(t) = \sum_{n=-\infty}^{\infty} (-1)^n \int_{\alpha_n}^{\alpha_{n+1}} w(x, t)\, dx,$$

and

$$\begin{aligned} I'(t) &= \sum_{n=-\infty}^{\infty} (-1)^n \left[\int_{\alpha_n}^{\alpha_{n+1}} w_t\, dx + w(\alpha_{n+1}, t)\dot{\alpha}_{n+1} - w(\alpha_n, t)\dot{\alpha}_n \right] \\ &= \sum_{n=-\infty}^{\infty} (-1)^n \left\{ -\int_{\alpha_n}^{\alpha_{n+1}} [f(u)_x - f(v)_x]\, dx \right. \\ &\quad \left. + w(\alpha_{n+1}, t)\dot{\alpha}_{n+1}, - w(\alpha_n, t)\dot{\alpha}_n \right\} \\ &= \sum_{n=-\infty}^{\infty} (-1)^n [-f(u(\alpha_{n+1}, t)) + f(u(\alpha_n, t)) + f(v(\alpha_{n+1}, t)) \\ &\quad - f(v(\alpha_n, t)) + w(\alpha_{n+1}, t)\dot{\alpha}_{n+1} + w(\alpha_n, t)\dot{\alpha}_n]. \end{aligned} \tag{3-17}$$

There are two possibilities to consider. First, for a given n, u and v may be continuous at (α_n, t) and (α_{n+1}, t). Then u and v are equal at these points, w

vanishes there, and the contribution to the sum in (3-17) due to such an index n is zero.

The other possibility is that u or v has a discontinuity (shock) at (α_n, t) or (α_{n+1}, t). Assume for definiteness that u has a shock at α_{n+1} while v remains continuous there, $w \geq 0$ on $[\alpha_n, \alpha_{n+1}]$, and $v \leq u_l$ at (α_{n+1}, t). Since $a(u)$ is increasing, the entropy condition gives $u_r < u_l$ at (α_{n+1}, t). Also, $v \geq u_r$ at (α_{n+1}, t) because of the sign change of w. Assume that u and v are continuous at (α_n, t). Since $w \geq 0$ on $[\alpha_n, \alpha_{n+1}]$, we have n even and the contribution to (3-17) in this case is

$$Q = f(v(\alpha_{n+1}, t)) - f(u(\alpha_{n+1}, t)) + (u_l - v(\alpha_{n+1}, t))\dot{\alpha}_{n+1}.$$

Now

$$\dot{\alpha}_{n+1} = \frac{f(u_r) - f(u_l)}{u_r - u_l}$$

by (3-14) so that, with $v_{n+1} = v(\alpha_{n+1}, t)$,

$$\begin{aligned} Q &= f(v_{n+1}) - f(u_l) + \frac{u_l - v_{n+1}}{u_r - u_l}[f(u_r) - f(u_l)] \\ &= f(v_{n+1}) - \left[\frac{u_l - v_{n+1}}{u_l - u_r} f(u_r) + \frac{v_{n+1} - u_r}{u_l - u_r} f(u_l)\right]. \end{aligned}$$

Since f is a convex function and $u_r \leq v_{n+1} \leq u_l$, its value at v_{n+1} is less than the corresponding value of the straight line joining $(u_r, f(u_r))$ and $(u_l, f(u_l))$, which is given in brackets above. Thus,

$$Q \leq f(v_{n+1}) - f(v_{n+1}) = 0.$$

The other cases concerning the behavior of u and v at (α_n, t) and (α_{n+1}, t) can be handled similarly. Thus, we always find $I'(t) \leq 0$ and uniqueness follows.

We now describe a solution procedure for obtaining classical solutions to (3-16) when $f'' > 0$ and ϕ is integrable. This procedure also yields generalized solutions. Let $u = u(x, t)$ be a solution to (3-16). The differential equation says that the vector field $(u, -f(u))$ has zero curl. Hence, there is a potential Q such that $Q_x = u$ and $Q_t = -f(u)$. In fact,

$$Q(x, t) = \int_{-\infty}^{x} u(y, t)\, dy$$

by the standard construction. So Q satisfies

$$Q_t + f(Q_x) = 0, \tag{3-18}$$

$$Q(x, 0) = \int_{-\infty}^{x} \phi(y)\, dy. \tag{3-19}$$

We need the following facts concerning convex functions.

Lemma 3-1. Assume that $f''(v) > 0$ for all v. Then:
(i) $f(v) \geq f(v_0) + f'(v_0)(v - v_0)$ for any v_0 and all v.
(ii) $f(v) \leq \dfrac{v_2 - v}{v_2 - v_1} f(v_1) + \dfrac{v - v_1}{v_2 - v_1} f(v_2)$ for any pair v_1, v_2 with $v_1 \leq v \leq v_2$.

The first condition asserts that the graph of f lies above its tangent lines and (ii) says that the part of the graph of f joining two of its points lies below the chord connecting these points. A function f which satisfies (ii) is called convex. See Prob. 1 for a proof of Lemma 3-1.

From the lemma $f(Q_x) \geq f(v) + f'(v)(Q_x - v)$ for any v. Hence, from (3-18) we find that

(3-20) $$Q_t + f'(v)Q_x \leq vf'(v) - f(v).$$

For fixed (x, t) and each v construct the straight line $\xi - x = f'(v)(\tau - t)$. Denote by $(y, 0)$ the point where this line crosses the ξ axis so that

(3-21) $$f'(v) = \frac{x - y}{t}.$$

Integrate the inequality (3-20) along the line from $(y, 0)$ to (x, t) to get

(3-22) $$Q(x, t) \leq Q(y, 0) + t[vf'(v) - v].$$

Now, $f'(v)$ is a strictly increasing function of v and so (3-21) can be solved for v in terms of $(x - y)/t$ using the inverse function, say b; thus,

(3-23) $$v = b\left(\frac{x - y}{t}\right).$$

Define

$$g(z) = vf'(v) - f(v) \qquad \text{where} \qquad v = b(z).$$

Then

$$g(z) = b(z)f'(b(z)) - f(b(z)) = zb(z) = f(b(z))$$

because f' and b are inverse functions. Moreover,

$$g'(z) = b(z) + zb'(z) - f'(b(z))b'(z) = b(z).$$

Note further that if $c = f'(0)$ and we normalize f by setting $f(0) = 0$, which has no effect on (3-16), then $b(c) = 0$ and

$$g(c) = cb(c) - f(b(c)) = 0 - 0 = 0.$$

Thus, g is determined by the initial value problem

$$g'(z) = b(z), \qquad g(c) = 0 \quad \text{where} \quad c = f'(0).$$

By the definition of g, we can use (3-22) to conclude that

$$Q(x, t) \le Q(y, 0) + tg\left(\frac{x - y}{t}\right). \tag{3-24}$$

This inequality holds for all y because the slope of the line $\xi - x = f'(v)(\tau - t)$ is arbitrary. In particular, if $u(x, t)$ is the solution of (3-16), we can choose $y = y(x, t)$ so that

$$u(x, t) = b\left(\frac{x - y}{t}\right).$$

That is, $v = u(x, t)$ in (3-23). Then equality holds in (3-20) and hence also in (3-24) for this choice of y. Therefore, we can characterize $y = y(x, t)$ as minimizing

$$Q(y, 0) + tg\left(\frac{x - y}{t}\right).$$

We summarize these arguments as a theorem.

Theorem 3-2. Suppose that $f''(v) > 0$, $f(0) = 0$, and ϕ is integrable. Set $c = f'(0)$ and let $u = u(x, y)$ be a classical solution to (3-16). Then

$$u(x, t) = b\left(\frac{x - y}{t}\right)$$

where $y = y(x, t)$ is chosen to minimize

$$h(y) = \int_{-\infty}^{y} \phi(\eta)\, d\eta + tg\left(\frac{x - y}{t}\right)$$

for each (x, t). Here g is determined by

$$g'(z) = b(z), \qquad g(c) = 0,$$

where b is the inverse of f'.

Example 4. Consider the problem

$$u_t + (e^u)_x = 0, \qquad u(x, 0) = \phi(x).$$

Here we take $f(v) = e^v - 1$ so that $f''(v) = e^v > 0$ and $f(0) = 0$. Also, $z = f'(v) = e^v$ so $v = b(z) = \log z$. Then g is found from $g'(z) = \log z$, $g(1) = 0$. Thus, $g(z) = z \log z - z + 1$. Now, $y = y(x, t)$ must be chosen so that

$$h(y) = \int_{-\infty}^{y} \phi(\eta)\, d\eta + tg\left(\frac{x - y}{t}\right)$$

is minimized. To minimize we set $h'(y) = 0$, which yields

$$\phi(y) - \log\left(\frac{x - y}{t}\right) = 0.$$

For bounded data, $\phi(y)$, this equation obviously has at least one solution because $\log z$ varies from $-\infty$ to $+\infty$. The number of roots depends on the behavior of ϕ. If there is a unique root, $y(x, t)$ will be smooth and so will $u(x, t) = b([x - y]/t)$. On the other hand, if there are multiple roots $y(x, t)$ is not unique and the possibility of discontinuous solutions arises.

On the basis of this example, it should seem plausible that the function $u(x, t)$ defined in Th. 3-2 is a possibly discontinuous solution to (3-16).

Theorem 3-3. Under the assumptions of Th. 3-2

$$u(x, t) = b\left(\frac{x - y}{t}\right)$$

is a generalized solution to (3-16).

We shall omit the proof of Th. 3-3. However, a few remarks are in order. The approach we have followed here is due to P. D. Lax and is a development of ideas going back to E. Hopf, who considered the initial value problem for $u_t + uu_x = 0$. The idea is to add a so-called viscosity term εu_{xx} to the right side and solve

$$u_t + uu_x = \varepsilon u_{xx}, \qquad u(x, 0) = \phi(x).$$

The solution is, by Prob. 4 of Sec. 12-2,

$$u_\varepsilon(x, t) = \frac{\displaystyle\int_{-\infty}^{\infty} \frac{x - y}{t} \exp\left\{-\frac{1}{2\varepsilon}\left[\frac{(x - y)^2}{2t} + \int_{-\infty}^{y} \phi(\eta)\, d\eta\right]\right\} dy}{\displaystyle\int_{-\infty}^{\infty} \exp\left\{-\frac{1}{2\varepsilon}\left[\frac{(x - y)^2}{2t} + \int_{-\infty}^{y} \phi(\eta)\, d\eta\right]\right\} dy}.$$

The solution is classical; there are no shocks. The next step is to let $\varepsilon \to 0$ and to prove that the limit is $(x - y(x, t))/t$, where $y(x, t)$ minimizes the function h in Th. 3-2. In the general case of (3-16) the solution $u_\varepsilon(x, t)$ is

$$u_\varepsilon(x, t) = \frac{\displaystyle\int_{-\infty}^{\infty} b\left(\frac{x - y}{t}\right) \exp\left[-\frac{1}{2\varepsilon} G(x, y, t)\right] dy}{\displaystyle\int_{-\infty}^{\infty} \exp\left[-\frac{1}{2\varepsilon} G(x, y, t)\right] dy}$$

where $G(x, y, t)$ for (x, t) fixed is the function $h(y)$ in Th. 3-2. Again the difficult point is to show that

$$u_\varepsilon(x, t) \to b\left(\frac{x - y(x, t)}{t}\right),$$

and that this limit solves (3-16).

PROBLEMS

1. Prove Lemma 3-1. *Hint.* For (i) use Taylor's formula with remainder. For (ii) let $v_1 < v_2$ and form

$$g(v) = \frac{v_2 - v}{v_2 - v_1} f(v_1) + \frac{v - v_1}{v_2 - v_1} f(v_2) - f(v).$$

Then $g(v_1) = g(v_2) = 0$. Since $g''(v) < 0$, g cannot have a negative minimum in $[v_1, v_2]$.

2. Find the functions f, b, g, h of Th. 3-2 for the equation $u_t + uu_x = 0$.

3. Consider $u_t + f(u)_x = 0$. Show that $w = f'(u)$ satisfies

$$w_t + ww_x = 0.$$

Hint. Multiply the equation for u by $f''(u)$.

4. Let ϕ, b, f, g, h be as in Th. 3-2. Fix (x, t). Show that $h(y)$ does have a minimum which is achieved at one or more points.

5. In this problem fix t and supress it in the notation to follow. Thus, write $y(x)$ for $y(x, t)$ a minimizer for $h(y)$. Show that

$$y_1 = y(x_1) \leq y(x_2) = y_2 \quad \text{for} \quad x_1 < x_2.$$

That is, $y(x)$ increases with x. [Note that $y(x)$ need not be unique.] *Hint.* Suppose that $x_1 < x_2$ and $y_2 < y_1$. For any $y < y_1$ show that $g'(x_1 - y) < g'(x_2 - y)$ by integrating $g''(z) > 0$ from $x_1 - y$ to $x_2 - y$. Replace y by η and integrate from y to y_1 to find

$$g\left(\frac{x_2 - y_1}{t}\right) + g\left(\frac{x_1 - y}{t}\right) < g\left(\frac{x_1 - y_1}{t}\right) + g\left(\frac{x_2 - y}{t}\right),$$

where the supressed t has been written explicitly for clarity. Let $G(x, y, t)$ be the right side of the formula for $h(y)$. Show that

$$G(x_2, y_1, t) + G(x_1, y, t) < G(x_1, y_1, t) + G(x_2, y, t).$$

Set $y = y_2$ to reach a contradiction.

6. Conclude from Prob. 5 that $y(x, t)$ is a nondecreasing function of x for each t and so has at most a countable set of discontinuities. If, in fact, $y(x, t)$ is continuous, show that the formula in Th. 3-2 solves (3-16).

7. Consider a long highway and suppose that cars enter it at $x = 0$. Let u be the number of cars per unit length, so that

$$N(t) = \int_a^b u(x, t)\,dx$$

gives the number of cars in $[a, b]$ at time t.

(a) If q is the flux of cars, show that $u_t + q_x = 0$.

(b) Suppose initially that $u(x, 0) = 0$. Suppose that $q = q(u)$. If the density is not too great, we can assume that $q(u) = au$ with $a > 0$ and constant. Note that a has units of velocity. If $\phi(t)$ is the rate at which vehicles enter the road at $x = 0$, argue that $au(0, t) = \phi(t)$. Solve the resulting initial, boundary value problem, and discuss the solution in the context of traffic flow.

(c) A more sophisticated model results by assuming that $q(u) = au(1 - b^{-1}u)$ where a is is the speed on an uncrowded highway and b is the density at which a traffic jam occurs. Let $g(u) = 0$ for $u \geq b$ and $u \leq 0$. Suppose initially that there are no cars on the road. If at $x = 0$ cars enter at the rate $\phi(t)$, obtain an initial, boundary value problem for u, try to solve it, and discuss the qualitative properties of the solution.

8. Consider the wave equation $u_{tt} = (a(x, t)u_x)_x$, where $a(x, t) > 0$ is a given continuously differentiable function. Suppose that u is continuous but that the first derivatives may jump along a curve $x = \sigma(t)$. Except on $x = \sigma(t)$, u is twice continously differentiable. The conservation law is

$$\frac{d}{dt}\int_{\alpha}^{\beta} u_t(x, t)\, dx = a(\beta, t)u_x(\beta, t) - a(\alpha, t)u_x(\alpha, t).$$

(a) If at time t, $\alpha < \sigma(t) < \beta$ use the conservation law to show that σ satisfies that Rankine-Hugoniot condition

$$[u_t]\dot{\sigma} + a(\sigma(t), t)[u_x] = 0,$$

where the brackets indicate the jump of the indicated function across $x = \sigma(t)$.

(b) Differentiate $u(\sigma(t), t)$ to find

$$\frac{d}{dt}u(\sigma(t), t) = u_x(\sigma(t)-, t)\dot{\sigma} + u_t(\sigma(t)-, t)$$

and a corresponding result with right limits. Combine these results to conclude that $[u_t] + [u_x]\dot{\sigma} = 0$.

(c) Suppose that at least one of the jumps $[u_t]$ or $[u_x]$ is nonzero. Use the results in parts (a) and (b) to deduce that $\dot{\sigma}^2 - a(\sigma(t), t) = 0$ and that singularities are propagated along characteristics.

12-4. Free Boundary Value Problems

Up until now we have considered problems where a fixed body was given and we have asked about conditions in that body at a certain location and perhaps at a given time. There are problems where the shape of the body is not known in advance but must be determined as part of the solution. The last three sections of Chap. 12 are devoted to such problems.

We begin with the so-called *Stefan problems*, which describes the melting of ice, solidification of metal alloys, and other change-of-phase problems. Consider an insulated pipe filled with ice. Heat is applied at the left end of the pipe and the other end is insulated. Our goal is to find the temperature distribution in the water–ice system. We take the x axis along the center of the pipe. Let $x = s(t)$ be the location of the ice–water interface at time t. Suppose initially that this interface is located at $x = 1$ and that the pipe has length 2. Evidently, the location of the interface changes with time and is determined by the temperature of the water and of the ice. Thus, the interface $x = s(t)$ is not known a priori and must be determined as part of the solution.

Denote the specific heat, density, thermoconductivity, and temperature of the water by c_w, ρ_w, k_w, and u_w. The corresponding quantities for the ice are c_i, ρ_i,

k_i, and u_i. Clearly, u_i and w_i must satisfy

$$\text{(4-1)} \qquad \begin{cases} \dfrac{\partial u_w}{\partial t} = a_w \dfrac{\partial^2 u_w}{\partial x^2}, & 0 < x < s(t), \quad t > 0, \\[2ex] \dfrac{\partial u_i}{\partial t} = a_i \dfrac{\partial^2 u_i}{\partial x^2}, & s(t) < x < 2, \quad t > 0, \end{cases}$$

where $a_w = k_w/c_w\rho_w$ and $a_i = k_i/c_i\rho_i$ are assumed to be constants. Of course, initial and boundary conditions must be given for u_w and u_i at the ends $x = 0$ and $x = 2$ of the rod. Furthermore, the assumption that temperature varies continuously across the interface yields

$$\text{(4-2)} \qquad u_w(s(t), t) = u_i(s(t), t) = 0,$$

where the freezing point of water has been taken as zero. But (4-2) is not enough to characterize $s(t)$. What is missing is an equation that describes the melting process. This is achieved by applying an energy balance. At time t the interface is located at $s(t)$, while at time $t + \Delta t$ it has moved to $s(t + \Delta t)$. The mass of water which has been converted from ice is $\rho_w A[s(t + \Delta t) - s(t)]$, where A is the constant cross-sectional area of the pipe. The energy absorbed by the element over the time interval Δt is $\lambda\rho_w A[s(t + \Delta t) - s(t)]$, where $\lambda > 0$ is the latent heat, and this must equal the net amount of heat which flowed into the element over the time interval Δt. In view of Fourier's law, this is

$$-A\left[q_w(-1)\Big|_{x=s(t)-} + q_i(+1)\Big|_{x=s(t)+} \right]\Delta t$$

$$= -A\left[k_w \frac{\partial u_w}{\partial x}\Big|_{x=s(t)-} - k_i \frac{\partial u_i}{\partial x}\Big|_{x=s(t)+} \right]\Delta t.$$

Equate these two expressions for the heat, divide by Δt, and let $\Delta t \to 0$, to obtain

$$\text{(4-3)} \qquad \lambda\rho_w s'(t) = -k_w \frac{\partial u_w}{\partial x}(s(t), t) + k_i \frac{\partial u_i}{\partial x}(s(t), t).$$

The melting problem as we have formulated it takes the following form. Find functions $u_w(x, t)$, $u_i(x, t)$, and $s(t)$ such that

$$\text{(4-4)} \qquad \frac{\partial u_w}{\partial t} = a_w \frac{\partial^2 u_w}{\partial x^2}, \qquad 0 < x < s(t), \qquad t > 0,$$

$$\text{(4-5)} \qquad \frac{\partial u_i}{\partial t} = a_i \frac{\partial^2 u_i}{\partial x^2}, \qquad s(t) < x < 2, \qquad t > 0,$$

$$\text{(4-6)} \qquad u_w(x, 0) = f_w(x), \qquad 0 \le x \le 1,$$

$$\text{(4-7)} \qquad u_i(x, 0) = f_i(x), \qquad 1 \le x \le 2,$$

$$\text{(4-8)} \qquad \frac{\partial u_w}{\partial x}(0, t) = -\phi(t), \qquad \phi(t) \ge 0, \qquad t > 0,$$

(4-9) $$\frac{\partial u_i}{\partial x}(2, t) = 0, \qquad t > 0,$$

(4-10) $$u_w(s(t), t) = u_i(s(t), t) = 0, \qquad t > 0,$$

(4-11) $$s'(t) = -\alpha_w \frac{\partial u_w}{\partial x}(s(t), t) + \alpha_i \frac{\partial u_i}{\partial x}(s(t), t), \qquad t > 0,$$

(4-12) $$s(0) = 1,$$

where a_w, a_i, $\alpha_w = k_w/\lambda\rho_w$, $\alpha_i = k_i/\lambda\rho_w$ are positive constants and f_i, f_w, and ϕ are given functions.

Problem (4-4)–(4-12) is called the *two-phase Stefan problem* because both phases, the water and ice, must be determined. Of course, other boundary conditions could replace (4-8) and (4-9). For example, the temperatures could be specified at $x = 0$ and $x = 2$.

It should be noted that this model, despite its nonlinear nature and complexity, does not handle the case of freezing in a confined medium. The reason is simple. The density of water is greater than that of ice, so the specific volumes satisfy $\rho_i^{-1} > \rho_w^{-1}$. This means that in freezing the ice will take up more volume than the water. Thus, as the freezing process begins, internal pressures build up. The freezing point depends on the pressure and decreases as pressure increases. Eventually, either the freezing process will stop or the container will burst, a common experience for one who has stored liquids in a freezer or seen roads buckle in the winter. On the other hand, when the densities in the two phases are the same, as happens for many metals, then the equations above describe both the melting and freezing processes.

In certain cases it is reasonable to assume that u_i is known. For example, the entire ice phase may be at or very near the freezing point. Then we may take $u_i = 0$ for $x > s(t)$ and the problem reduces to find $u = u_w$ and $s(t)$. This simplified problem is called the *one-phase Stefan problem.*

Many physical situations lead to one-dimensional, free boundary value problems. These situations include determining the level of a lake in the presence of evaporation (and hence the determination of the amount of water that has evaporated), the immiscible displacement of one fluid by another, and so on. The derivation of models for some of these situations will be explored in the problems.

Obviously, treating (4-4)–(4-12) in any degree of generality is difficult. However, there is one situation, the problem for the semi-infinite rod $x > 0$, which can be handled immediately and which gives some insights into the physics of the situation. We seek functions $u_w(x, t)$, $u_i(x, t)$, and $s(t)$ which satisfy the problem illustrated in Fig. 12-4. In the figure ϕ and T are positive constants. Because of the simplicity of the boundary and initial conditions, it is possible to construct a solution starting from elementary solutions of the heat equation.

Let

(4-13) $$e(x) = \frac{2}{\sqrt{\pi}} \int_0^x e^{-\lambda^2}\, d\lambda.$$

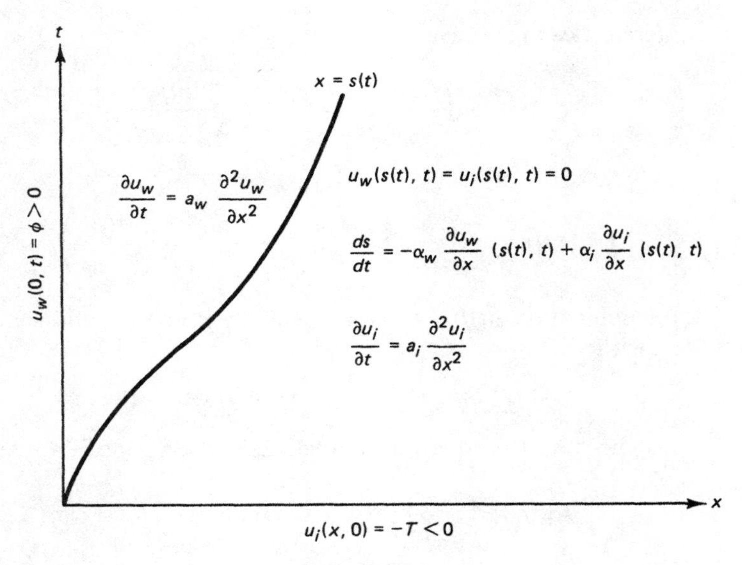

Figure 12-4

Then (Prob. 1) it is easy to confirm that

$$e(x, t) = e\left(\frac{x}{2\sqrt{at}}\right)$$

satisfies

$$e_t = ae_{xx} \tag{4-14}$$

and

$$e(0, t) = 0, \qquad \lim_{x\to\infty} e(x, t) = 1 \tag{4-15}$$

for each $t > 0$.

We seek a solution to the problem in Fig. 12-4 in the form

$$u_w(x, t) = A + Be\left(\frac{x}{2\sqrt{a_w t}}\right) \tag{4-16}$$

$$u_i(x, t) = C + De\left(\frac{x}{2\sqrt{a_i t}}\right), \tag{4-17}$$

where A, B, C, and D are constants to be determined. Obviously, u_w and u_i satisfy the requisite heat equations. We must specify the constants so that the other conditions in Fig. 12-4 are met. First, $u_w(0, t) = \phi$ yields $A = \phi$ from (4-16). Similarly, (4-17) gives

$$u_i(x, 0+) = C + D = -T.$$

At the interface we must have

$$\phi + Be\left(\frac{s(t)}{2\sqrt{a_w t}}\right) = C + De\left(\frac{s(t)}{2\sqrt{a_i t}}\right) = 0,$$

and

$$s'(t) = \frac{-\alpha_w B}{2\sqrt{a_w t}}\, e'\left(\frac{s(t)}{2\sqrt{a_w t}}\right) + \frac{\alpha_i D}{2\sqrt{a_i t}}\, e'\left(\frac{s(t)}{2\sqrt{a_i t}}\right).$$

Since the combination $s(t)/\sqrt{t}$ occurs repeatedly in the equations above, we seek a solution with

$$s(t) = \rho t^{1/2}, \qquad \rho > 0 \text{ and constant.}$$

Use of this choice in the equations above yields the algebraic system

$$\phi + Be\left(\frac{\rho}{2\sqrt{a_w}}\right) = 0, \qquad -D + De\left(\frac{\rho}{2\sqrt{a_i}}\right) = T,$$

$$-\frac{\alpha_w}{\sqrt{a_w}} B \exp\left(-\frac{\rho^2}{4a_w}\right) + \frac{\alpha_i}{\sqrt{a_i}} D \exp\left(-\frac{\rho^2}{4a_i}\right) = \rho.$$

The first two equations yield

$$B = -\frac{\phi}{e(\rho/2\sqrt{a_w})}, \qquad D = -\frac{T}{1 - e(\rho/2\sqrt{a_i})},$$

and the third can be expressed as

$$\text{(4-18)} \quad \frac{\alpha_w}{\sqrt{a_w}} \frac{\phi}{e(\rho/2\sqrt{a_w})} \exp\left(-\frac{\rho^2}{4a_w}\right) - \frac{\alpha_i T}{\sqrt{a_i}[1 - e(\rho/2\sqrt{a_i})]} \exp\left(-\frac{\rho^2}{4a_i}\right) = \rho.$$

If (4-18) can be solved for ρ, then the other constants are determined by the previous equations and u_w, u_i, and s are given explicitly by (4-16), (4-17), and $s = \rho t^{1/2}$. To see that (4-18) is solvable, notice that the left side of (4-18) varies from $+\infty$ to $-\infty$ as ρ varies from 0 to $+\infty$. Thus the graphs of the left and right sides of (4-18) must cross at least once. See Prob. 5, which also shows that there is a unique choice for ρ.

PROBLEMS

1. Verify that $e(x, t) = e(x/2\sqrt{at})$ with $e(x)$ given by (4-13) satisfies (4-14) and (4-15).
2. Consider the problem described in Fig. 12-4. Show that if $T = 0$, then $C = D = 0$ so that $u_i(x, t) \equiv 0$. Suppose that α_w and a_w are both 1 and that $\phi = 10$. Determine $s(t)$ and $u = u_w(x, t)$ explicitly. Sketch u as a function of x for several values of t.
3. A large, deep lake is lying in the sun. There is no wind and the temperature is sufficiently uniform that large temperature effects may be neglected. Ignore gravitational effects as

well. The air is fairly dry, so that water from the lake will evaporate. At time $t = 0$, suppose that the lake is located at $x = 0$. Assume that evaporation takes place by diffusion. Let g be the saturation vapor density, $u(x, t)$ the vapor density, and f the initial vapor density. As $x \to \infty$, assume that $u \to 0$.

(a) Derive the model

$$\begin{aligned} &u_t = Du_{xx}, \quad x > s(t), \quad t > 0, \\ &u(x, 0) = f(x), \quad x \geq 0, \\ &u(s(t), t) = g, \quad t \geq 0, \\ &(\rho - g)s' = Du_x(s(t), t), \\ &\lim_{x \to \infty} u(x, t) = 0. \end{aligned}$$

Hint. ρ is the density of water. To obtain the ordinary differential equation for the interface, make a mass balance.

(b) Assuming that f, g, ρ, and D are constants, solve this problem explicitly.

4. Consider the situation of one fluid displacing another in a tube filled with a porous medium. Assume that the medium is saturated, the fluids are flowing in contact, the first fluid completely displaces the second, the fluids are incompressible, and the medium is rigid. Distinguish between the two fluids with subscripts. Let u_1 and u_2 be the pressure in the two fluids. Assume that the pressure is controlled on the ends of the tube. Use Darcy's law and the continuity equation to show that
(a) $(\partial^2 u_1/\partial x^2) = 0, \quad 0 < x < s(t), \quad (\partial^2 u_2/\partial x^2) = 0, \quad s(t) < x < 2.$
(b) $u_1(0, t) = \phi(t), \quad u_2(2, t) = \psi(t).$
(c) $s'(t) = -\alpha_1(\partial u_1/\partial x)(s(t), t) = -\alpha_2(\partial u_2/\partial x)(s(t), t).$
(d) $u_1(s(t)) = u_2(s(t)).$
(e) $s(0) = 1$ (initial point of separation).
Solve this problem explicitly. It is known as the one dimensional Muskat problem for immiscible displacement.

5. Check the assertion that the left side of (4-18) varies from $+\infty$ to $-\infty$ as ρ varies from 0 to ∞. Check also that the left side is decreasing. Conclude that (4-18) has a unique solution.

5. The One-Phase Stefan Problem

If the ice phase is known to be at temperature zero (or very nearly so), then we can set $u_i \equiv 0$ and $u(x, t) = u_w(x, t)$ in the two-phase Stefan problem to obtain the one-phase Stefan problem

$$u_t = u_{xx}, \quad 0 < x < s(t), \quad t > 0, \tag{5-1}$$

$$u(x, 0) = f(x), \quad 0 \leq x \leq 1, \tag{5-2}$$

$$u(0, t) = \phi(t), \quad t > 0, \tag{5-3}$$

$$s'(t) = -u_x(s(t), t), \quad t > 0, \tag{5-4}$$

$$u(s(t), t) = 0, \quad t > 0, \tag{5-5}$$

$$s(0) = 1, \tag{5-6}$$

where we assume tnat $f(x) \geq 0$ and $\phi(t) \geq 0$, which is physically realistic for the problem of melting ice. Finally, all physical constants have been set equal to 1 for ease of exposition.

There are several techniques which can be used to treat (5-1)–(5-6). They are all rather difficult. We shall use the potential theoretic methods of Chap. 9. We always assume that either $f(x) \not\equiv 0$ or if $f(x) \equiv 0$, then $\phi(t) \not\equiv 0$ in any neighborhood $(0, \delta]$. Indeed, if $f(x) \equiv 0$ and $\phi(t) = 0$ in $(0, \delta]$, then $u \equiv 0$ in $0 \leq x \leq 1, 0 \leq t \leq \delta$ and $s(t) \equiv 1$ obviously solves (5-1)–(5-6). This trivial solution is the only one by a uniqueness theorem which is given below. So the assumptions just mentioned simply mean that we have a bona fide free boundary problem. Some ice will melt and the boundary will move.

Since melting occurs, we expect the water–ice interface to move to the right over time. Thus, we expect that $x = s(t)$ will not decrease over time, which follows immediately from the next result.

Lemma 5-1. If (5-1)–(5-6) has a solution $(u(x, t), s(t))$, then $s'(t) \geq 0$.

Proof. Since $u \geq 0$ on $x = 0$ and on $t = 0$ and $u = 0$ on $x = s(t)$, we conclude by the maximum principle (Prob. 5) that $u(x, t) \geq 0$ throughout the region $0 \leq x \leq s(t)$, $t \geq 0$. Consequently,

$$\frac{u(x, t) - u(s(t), t)}{x - s(t)} = \frac{u(x, t)}{x - s(t)} \leq 0$$

for all $x < s(t)$. Then by (5-4),

$$s'(t) = -\lim_{x \to s(t)-} \frac{u(x, t) - u(s(t), t)}{x - s(t)} \geq 0.$$

We establish the existence of a solution to the one-phase Stefan problem by recasting it as an integral equation. For this purpose, we introduce the Green's and Neumann's functions for the region $x > 0$, $t > 0$. Let

$$G(x, y, t - \tau) = k(x - y, t - \tau) - k(x + y, t - \tau)$$
$$N(x, y, t - \tau) = k(x - y, t - \tau) + k(x + y, t - \tau),$$

where

$$k(x, t) = (4\pi t)^{-1/2} \exp\left(-\frac{x^2}{4t}\right)$$

is the fundamental solution for the heat equation. Both G and N satisfy

$$\text{(5-7)} \qquad v_{xx} - v_t = 0, \qquad v_{yy} + v_\tau = 0, \qquad t \neq \tau$$

and the identities

$$\text{(5-8)} \qquad G_x = -N_y, \qquad G_y = -N_x,$$

which we shall need shortly.

Assume that (u, s) is a solution to (5-1)–(5-6), fix (x, t) with $0 < x < s(t)$, $t > 0$, and fix ε such that $0 < \varepsilon < t/4$. We follow a familiar path to a representation for the solution pair $(u(x, t), s(t))$. Integrate the identity

$$0 = [G_{yy} + G_\tau]u + [u_{yy} - u_\tau]G = [G_y u - u_y G]_y + (uG)_\tau$$

over $D_\varepsilon = \{(y, \tau): 0 < y < s(\tau), \varepsilon < \tau < t - \varepsilon\}$ and use Green's theorem to find

$$\int_{C_\varepsilon} [(G_y u - u_y G)\, d\tau - uG\, dy] = 0, \tag{5-9}$$

where C_ε is the boundary of D_ε. Since u vanishes on $y = s(\tau)$,

$$\int_{C_\varepsilon} G(x, y, t - \tau)u(y, \tau)\, dy = \int_0^{s(\varepsilon)} G(x, y, t - \varepsilon)u(y, \varepsilon)\, dy$$
$$- \int_0^{s(t-\varepsilon)} G(x, y, \varepsilon)u(y, t - \varepsilon)\, dy.$$

Now, using basic properties of the Green's function we find that

$$\lim_{\varepsilon \to 0} \int_{C_\varepsilon} G(x, y, t - \tau)u(y, \tau)\, dy = \int_0^1 G(x, y, t)f(y)\, dy - u(x, t).$$

Next, observe that, since G vanishes on $y = 0$,

$$\lim_{\varepsilon \to 0} \int_{C_\varepsilon} (G_y u - u_y G)\, d\tau = -\int_0^t u_y(s(\tau), \tau)G(x, s(\tau), t - \tau)\, d\tau$$
$$- \int_0^t G_y(x, 0, t - \tau)\phi(\tau)\, d\tau.$$

Use of these limit results in (5-9) leads to

$$u(x, t) = \int_0^1 G(x, y, t)f(y)\, dy + \int_0^t G_y(x, 0, t - \tau)\phi(\tau)\, d\tau$$
$$+ \int_0^t G(x, s(\tau), t - \tau)u_y(s(\tau), \tau)\, d\tau. \tag{5-10}$$

Notice that all terms except $u_y(s(\tau), \tau)$ and $s(\tau)$ are known on the right side of (5-10). Moreover, if

$$v(t) = u_x(s(t), t) \tag{5-11}$$

were known, then (5-4) and (5-6) would determine $s(t)$, and (5-10) would express $u(x, t)$ in terms of known functions.

With this observation in mind, we seek an equation for $v(t)$. Such an equation results by differentiating (5-10) with respect to x and allowing x to approach $s(t)$. To carry this out, we need to impose some additional assumptions on the data ϕ and f:

(5-12) Assume that ϕ and f are continuously differentiable on their domains and that $f(0) = \phi(0)$, $f(1) = 0$.

In view of (5-8), the derivatives with respect to x of the first two integrals on the right of (5-10) are

$$\int_0^1 G_x(x, y, t)f(y)\,dy = -\int_0^1 N_y(x, y, t)f(y)\,dy$$

$$= \int_0^1 N(x, y, t)f'(y)\,dy + N(x, 0, t)f(0)$$

and [using both (5-7) and (5-8)]

$$\int_0^t G_{xy}(x, 0, t-\tau)\phi(\tau)\,d\tau = \int_0^t N_\tau(x, 0, t-\tau)\phi(\tau)\,d\tau$$

$$= -\int_0^t N(x, 0, t-\tau)\phi'(\tau)\,d\tau - N(x, 0, t)\phi(0).$$

Thus,

$$u_x(x, t) = \int_0^1 N(x, y, t)f'(y)\,dy - \int_0^t N(x, 0, t-\tau)\phi'(\tau)\,d\tau + \int_0^t G_x(x, s(\tau), t-\tau)v(\tau)\,d\tau, \tag{5-13}$$

and we wish to let $x \to s(t)-$. The limits in the first two integrals are easily found by replacing x by $s(t)$ under the integral. The limit of the third integral as $x \to s(t)-$ is more delicate. Note that the integrand

$$G_x(x, s(\tau), t-\tau) = -\frac{x - s(\tau)}{2(t-\tau)} k(x - s(\tau), t-\tau) + \frac{x + s(\tau)}{2(t-\tau)} k(x + s(\tau), t-\tau).$$

In the second term, $x + s(\tau) \geq 1$ and we can pass to the limit under the integral sign without any difficulty. The problem is with the first term, where $s(t) - s(\tau) \to 0$ as $\tau \to t-$. Of course, the first term will still be integrable when $x = s(t)$ because $|s(t) - s(\tau)| \leq M(t - \tau)$ in view of the differentiability of s. The problem is not with the existence of the integral of $G_x(s(t), s(\tau), t-\tau)$, but rather with limiting behavior of the integral as $x \to s(t)$. This limiting behavior follows from a lemma of Holmgren.

Lemma 5-2. Let

$$w(x, t) = \int_0^t k_x(x - s(\tau), t-\tau)\rho(\tau)\,d\tau,$$

where, on any finite interval $0 \leq \tau \leq T$, $s(\tau)$ and $\rho(\tau)$ are continuous and there is a constant M (which may depend on T) such that $|\rho(\tau)| \leq M$ and $|s(t) - s(\tau)| \leq M|t - \tau|$. Then

$$\lim_{x \to s(t)-} w(x, t) = \frac{1}{2}\rho(t) + \int_0^t k_x(s(t) - s(\tau), t-\tau)\rho(\tau)\,d\tau.$$

Proof. We write $w(x, t)$ as a sum of integrals, each of which is dealt with separately. Thus,

$$\begin{aligned} w(x,t) &= \int_0^t k_x(x - s(t), t - \tau)\rho(t)\,dt \\ &+ \int_0^t k_x(x - s(t), t - \tau)[\rho(\tau) - \rho(t)]\,d\tau \\ &+ \int_0^t [k_x(x - s(\tau), t - \tau) - k_x(x - s(t), t - \tau)]\rho(\tau)\,d\tau \\ &= I_1 + I_2 + I_3, \end{aligned}$$

respectively. Now, we proceed with arguments analogous to those used in Sec. 9-4 on thermal potentials. The change of variables $\lambda = [x - s(t)]^2/4(t - \tau)$ in I_1 yields

$$I_1 = \frac{\rho(t)}{\sqrt{4\pi}} \int_{[x-s(t)]^2/4t}^{\infty} e^{-\lambda}\lambda^{-1/2}\,d\lambda \to \frac{\rho(t)}{\sqrt{4\pi}} \int_0^{\infty} e^{-\lambda}\lambda^{-1/2}\,d\lambda = \frac{1}{2}\rho(t),$$

as $x \to s(t)-$.

Next,

$$I_2 = -\int_0^t \frac{x - s(t)}{2(t - \tau)} k(x - s(t), t - \tau)[\rho(\tau) - \rho(t)]\,d\tau.$$

Let $\varepsilon > 0$ be given and choose δ so small that $|\rho(\tau) - \rho(t)| < \varepsilon$ for $|t - \tau| < \delta$. Express the integral above as a sum of two integrals, one from 0 to $t - \delta$ and the other from $t - \delta$ to t. The first integral obviously has limit zero as $x \to s(t)-$. A repeat of the argument used on I_1 shows that the second integral is bounded by $\varepsilon/2$. Therefore,

$$\lim_{x \to s(t)-} (I_1 + I_2) = \frac{1}{2}\rho(t). \tag{5-14}$$

The analysis of I_3 is more complicated. First,

$$\begin{aligned} I_3 &= -\int_0^t \left[\frac{x - s(\tau)}{2(t - \tau)} k(x - s(\tau), t - \tau) - \frac{x - s(\tau)}{2(t - \tau)} k(x - s(t), t - \tau)\right]\rho(\tau)\,d\tau \\ &= -\int_0^t \frac{s(t) - s(\tau)}{2(t - \tau)} k(x - s(\tau), t - \tau)\rho(\tau)\,d\tau \\ &\quad - \int_0^t \frac{x - s(t)}{2(t - \tau)} [k(x - s(\tau), t - \tau) - k(x - s(t), t - \tau)]\rho(\tau)\,d\tau \\ &= I_3' + I_3'', \end{aligned}$$

respectively. Since $|s(t) - s(\tau)| \le M|t - \tau|$ it is easy to confirm that the limit as $x \to s(t)-$ can be taken under the integral sign in I_3'. Thus,

$$\text{(5-15)} \qquad \lim_{x \to s(t)-} I_3' = \int_0^t k_x(s(t) - s(\tau), t - \tau)\rho(\tau)\, d\tau.$$

To complete the proof of the lemma, we must show that I_3'' has limit zero as $x \to s(t)-$. Note first that

$$I_3'' = -\int_0^t \frac{x - s(t)}{2(t-\tau)} k(x - s(t), t - \tau) \times \left\{\exp\left[-\frac{(x - s(\tau))^2}{4(t-\tau)} + \frac{(x - s(t))^2}{4(t-\tau)}\right] - 1\right\}\rho(\tau)\, d\tau.$$

Now

$$\frac{-(x - s(\tau))^2 + (x - s(t))^2}{4(t-\tau)} = \frac{[s(t) - s(\tau)][(s(t) - x) + (s(\tau) - x)]}{4(t-\tau)}$$
$$\le \frac{M}{4}[|s(t) - x| + |s(\tau) - x|]$$
$$\le \frac{M}{4}[2|s(t) - x| - |s(t) - s(\tau)|].$$

Once again write $I_3'' = J_1 + J_2$, where J_1 is the integral over 0 to $t - \delta$ and J_2 is the integral over $t - \delta$ to t. For any choice of $0 < \delta < t$, we clearly have $J_1 \to 0$ as $x \to s(t)-$. To estimate J_2 observe that $e^\alpha - 1 \le \alpha e^\alpha$ for any $\alpha > 0$. Consequently, assuming that $|x - s(t)| < 1$,

$$|J_2| \le \frac{M^2}{4} \int_{t-\delta}^t \frac{|x - s(t)|}{2(t-\tau)} [2|s(t) - x| + |s(t) - s(\tau)|] k(x - s(t), t - \tau)\, d\tau$$
$$\le M^2 \int_{t-\delta}^t \left[\frac{(x - s(t))^2}{4(t-\tau)} + \frac{M}{8}\right] \frac{\exp\{-[x - s(t)]^2/4(t-\tau)\}}{\sqrt{4\pi(t-\tau)}}\, d\tau$$
$$\le M^2\left(A + \frac{M}{8}\right) \int_{t-\delta}^t [4\pi(t-\tau)]^{-1/2} \exp\left[-h\frac{(x - s(t))^2}{4\pi(t-\tau)}\right] d\tau$$

for fixed h with $0 < h < 1$ and some constant A (determined by h). Since the exponential is less than 1, we deduce that

$$|J_2| \le M^2\left(A + \frac{M}{8}\right)\frac{\delta^{1/2}}{\sqrt{4\pi}}.$$

Since δ can be chosen arbitrarily small, we infer that

$$\lim_{x \to s(t)-} I_3'' = 0.$$

Combining this result with (5-14) and (5-15) establishes Lemma 5-2.

With the aid of Lemma 5-2 and the remarks following (5-13), we have established the following theorem.

Theorem 5-1. Suppose that $(u(x, t), s(t))$ is a solution to (5-1)–(5-6). Assume that f and ϕ have continuous derivatives and satisfy the compatibility conditions $\phi(0) = f(0)$, $f(1) = 0$. Then $u(x, t)$ is given by

$$u(x, t) = \int_0^1 G(x, y, t)f(y)\,dy + \int_0^t G_y(x, 0, t - \tau)\phi(\tau)\,d\tau$$
$$+ \int_0^t G(x, s(\tau), t - \tau)v(\tau)\,d\tau,$$

where $v(t) = u_x(s(t), t)$ satisfies

$$v(t) = 2\int_0^1 N(s(t), y, t)f'(y)\,dy - 2\int_0^t N(s(t), 0, t - \tau)\phi'(\tau)\,d\tau \tag{5-16}$$
$$+ 2\int_0^t G_x(s(t), s(\tau), t - \tau)v(\tau)\,d\tau$$

and

$$s(t) = 1 - \int_0^t v(\theta)\,d\theta, \tag{5-17}$$

so that (5-16) is a nonlinear Volterra integral equation for $v(t)$.

The natural question now is whether a solution to (5-16) will deliver a solution to (5-1)–(5-6). In other words, suppose that we can solve (5-16) for a function $v(t)$ when s in (5-16) is replaced by $1 - \int_0^t v(\theta)\,d\theta$. Then a simple integration yields $s(t)$ from (5-17) and $u(x, t)$ is defined by the formula in Th. 5-1. By the definition of the Green's function G, $u(x, t)$ so defined satisfies (5-1)–(5-3), and $u_x(s(t), t) = v(t)$. Thus, (5-4) and (5-6) hold. It remains to be seen that (5-5) also holds. We use a trick to verify this condition. The function $u(x, t)$ defined in Th. 5-1 satisfies (5-1)–(5-6) except possibly for (5-5). Apply the identity $[G_y u - u_y G]_y + (uG)_\tau = 0$ to this function to get (5-9). New contributions occur in the line integrals following (5-9) because we no longer know that u vanishes on $y = s(\tau)$. The reasoning following (5-9) leads to the expression

$$u(x, t) = \int_0^1 G(x, y, t)f(y)\,dy + \int_0^t G_y(x, 0, t - \tau)\phi(\tau)\,d\tau$$
$$+ \int_0^t G(x, s(\tau), t - \tau)v(\tau)\,d\tau$$
$$+ \int_0^t G(x, s(\tau), t - \tau)u(s(\tau), \tau)s'(\tau)\,d\tau$$
$$- \int_0^t G_y(x, s(\tau), t - \tau)u(s(\tau), \tau)\,d\tau.$$

Comparison with the formula for u in Th. 5-1 yields

$$\int_0^t G(x, s(\tau), t-\tau)u(s(\tau), \tau)s'(\tau)\, d\tau - \int_0^t G_y(x, s(\tau), t-\tau)u(s(\tau), \tau)\, d\tau = 0.$$

Since $G_y = -N_x$ we can use Lemma 5-2 to conclude that

$$u(s(\tau), \tau) = -2\int_0^t G(s(t), s(\tau), t-\tau)u(s(\tau), \tau)s'(\tau)\, d\tau + 2\int_0^t G_y(s(t), s(\tau), t-\tau)u(s(\tau), \tau)\, d\tau. \tag{5-18}$$

Now, by the triangle inequality,

$$|G_y(s(t), s(\tau), t-\tau)| \le \left|\frac{s(t)-s(\tau)}{2(t-\tau)} \frac{1}{\sqrt{4\pi(t-\tau)}}\right| + \left|\frac{s(t)+s(\tau)}{2(t-\tau)} \frac{e^{-4/4(t-\tau)}}{\sqrt{4\pi(t-\tau)}}\right| \le \frac{C}{\sqrt{t-\tau}}$$

for some constant C. Similarly, $G(s(t), s(\tau), t-\tau)s'(\tau)$ is bounded by a constant over $\sqrt{t-\tau}$ over any finite time interval $0 \le t \le T$. Let

$$\mu(t) = |u(s(t), t)|$$

for $0 \le t \le T$. Use the estimates above in (5-18) to infer that

$$\mu(t) \le C\int_0^t (t-\tau)^{-1/2}\mu(\tau)\, d\tau$$

for some suitably large constant C and $0 \le t \le T$. Replace t by θ in this inequality, multiply by $(t-\theta)^{-1/2}$ and integrate to find

$$\begin{aligned}\int_0^t (t-\theta)^{-1/2}\mu(\theta)\, d\theta &\le C\int_0^t d\theta \int_0^\theta (t-\theta)^{-1/2}(\theta-\tau)^{-1/2}\mu(\tau)\, d\tau \\ &= C\int_0^t \mu(\tau)\, d\tau \int_\tau^t (t-\theta)^{-1/2}(\theta-\tau)^{-1/2}\, d\theta \\ &= C\pi\int_0^t \mu(\tau)\, d\tau.\end{aligned}$$

Thus,

$$\mu(t) \le C\pi\int_0^t \mu(\tau)\, d\tau$$

or

$$0 \ge \frac{d}{dt}\left[e^{-C\pi t}\int_0^t \mu(\tau)\, d\tau\right].$$

Integrate to find that $\int_0^t \mu(\tau)\, d\tau = 0$. Then $0 \le \mu(t) \le 0$, so $u(s(t), t) = 0$ and (5-5) holds.

In summary we have proven the following theorem.

Theorem 5-2. The integral equation (5-16) is equivalent to (5-1)–(5-6) in the following sense. If $(u(x, t), s(t))$ is a solution of (5-1)–(5-6), then the function

$$u_x(s(t), t) \equiv v(t) \quad \text{with} \quad s(t) = 1 - \int_0^t v(\theta)\, d\theta$$

satisfies (5-16). Conversely, if $v(t)$ is a solution to (5-16), where $s(t)$ is replaced by $1 - \int_0^t v(\theta)\, d\theta$ in (5-16), then $u(x, t)$ defined in Th. 5-1 and $s(t) = 1 - \int_0^t v(\theta)\, d\theta$ solve (5-1)–(5-6).

Now, we use an iterative procedure to show that the integral equation (5-16) can be solved. Denote the right member of (5-16) by Tv,

$$Tv = Tv(t) = 2\int_0^1 N(s(t), y, t) f'(y)\, dy - 2\int_0^t N(s(t), 0, t - \tau)\phi'(\tau)\, d\tau + 2\int_0^t G_x(s(t), s(\tau), t - \tau) v(\tau)\, d\tau. \tag{5-19}$$

The integral equation (5-16) is then

$$v = Tv. \tag{5-20}$$

To find such a v, we choose v_0 as an initial guess and set

$$v_{n+1} = Tv_n, \qquad n = 0, 1, 2, \ldots. \tag{5-21}$$

As usual, we must show that (5-21) is well-defined and that $\{v_n\}$ converges to a solution of (5-20). The convergence of the sequence is most easily shown by studying the series

$$v_0 + \sum_{n=0}^{\infty} (v_{n+1} - v_n),$$

whose partial sums are $\{v_n\}$.

Our first step is to show that (5-21) is well-defined over a small enough time interval. Let $0 < t^* \le 1$ be fixed arbitrarily and set

$$M = \max\left\{ \max_{0 \le x \le 1} |f'(x)|,\ \max_{0 \le t \le t^*} |\phi'(t)| \right\}. \tag{5-22}$$

Fix $L > 2M$ and consider a continuous function $w(t)$ with $|w(t)| \le L$ on $0 \le t \le t^*$. We assert that $Tw(t)$ is also bounded by L on a suitably small time interval. It will be useful later to have bounds on the corresponding function

$$s(t) = 1 - \int_0^t w(\tau)\, d\tau.$$

Thus, we fix σ_1 with $0 < \sigma_1 < t^*$ so that

$$s(t) = 1 - \int_0^t w(\tau)\, d\tau \geq 1 - Lt \geq \frac{1}{2}$$

and

$$s(t) \leq 1 + Lt \leq \tfrac{3}{2}$$

for $0 \leq t \leq \sigma_1$. Notice that σ_1 depends only on L.

For $|w(t)| \leq L$ we have

$$(5\text{-}23) \qquad |Tw| \leq 2M \int_0^1 N(s(t), y, t)\, dy + 2M \int_0^t N(s(t), 0, t - \tau)\, d\tau + 2L \int_0^t |G_x(s(t), s(\tau), t - \tau)|\, d\tau.$$

Now

$$\int_0^t N(s(t), y, t)\, dy \leq \int_0^\infty N(s(t), y, t)\, dt = 1$$

and

$$\int_0^t N(s(t), 0, t - \tau)\, d\tau \leq 2 \int_0^t \frac{d\tau}{\sqrt{4\pi(t - \tau)}} = \frac{2}{\sqrt{\pi}}\, t^{1/2}.$$

Also,

$$\begin{aligned} |G_x(s(t), s(\tau), t - \tau)| &\leq \frac{|s(t) - s(\tau)|}{2(t - \tau)} \frac{e^{-(s(t) - s(\tau))^2/4(t - \tau)}}{\sqrt{4\pi(t - \tau)}} + \frac{|s(t) + s(t)|}{2(t - \tau)} \frac{e^{-(s(t) + s(\tau))^2/4(t - \tau)}}{\sqrt{4\pi(t - \tau)}} \\ &\leq \frac{L}{2} \cdot \frac{1}{\sqrt{4\pi(t - \tau)}} + \frac{3}{2\sqrt{4\pi(t - \tau)}} \frac{1}{t - \tau} \exp\left[-\frac{1}{4(t - \tau)}\right], \end{aligned}$$

where we have used the bounds $1 \leq 2s(t)$, $2s(\tau) \leq 3$ on $0 \leq \tau \leq t \leq \sigma_1$. Since $\alpha e^{-\alpha} \leq (2/e)e^{-\alpha/2}$, we see that

$$\begin{aligned} |G_x(s(t), s(\tau), t - \tau)| &\leq \frac{L}{2\sqrt{4\pi(t - \tau)}} + \frac{12}{e\sqrt{4\pi(t - \tau)}} \exp\left[-\frac{1}{8(t - \tau)}\right] \\ &\leq \frac{Le + 24}{2e\sqrt{4\pi(t - \tau)}} \end{aligned}$$

and

$$\int_0^t |G_x(s(t), s(\tau), t - \tau)|\, d\tau \leq \frac{Le + 24}{4e\sqrt{\pi}}\, t^{1/2}.$$

Use the foregoing estimates in (5-23) to find that

$$|Tw(t)| \leq 2M + \frac{4Mt^{1/2}}{\sqrt{\pi}} + \frac{L(Le + 24)}{2e\sqrt{\pi}} t^{1/2}$$

for $0 \leq t \leq \sigma_1$. The right member of this inequality is at most L if $0 \leq t \leq \sigma_2 \leq \sigma_1$, where σ_2 depends only on σ_1, L, and M. Thus, we conclude:

(5-24) If $|w(t)| \leq L$, then $|Tw(t)| \leq L$ for $0 \leq t \leq \min(\sigma_2, t^*)$.

With t so restricted and $v_0(t)$ bounded by L, (5-24) shows that (5-21) defines a sequence $\{v_n\}$ with $|v_n(t)| \leq L$ and with $\frac{1}{2} \leq s_n(t) \leq \frac{3}{2}$ where

$$s_n(t) = 1 - \int_0^t v_n(\tau)\, d\tau.$$

Let $\sigma = \min(\sigma_2, t^*)$ and assume that $0 \leq t \leq \sigma$ in what follows. We turn to the convergence of the series following (5-21). Let

$$\Delta = \max_{0 \leq t \leq \sigma} |v_1(t) - v_0(t)|.$$

Our plan is to estimate the differences $v_2 - v_1, v_3 - v_2, \ldots$ in terms of Δ. From (5-19) we have

$$\begin{aligned} v_2 - v_1 = Tv_1 - Tv_0 &= 2\int_0^1 [N(s_1(t), y, t) - N(s_0(t), y, t)]f'(y)\, dy \\ &\quad - 2\int_0^t [N(s_1(t), 0, t-\tau) - N(s_0(t), 0, t-\tau)]\phi'(\tau)\, d\tau \\ &\quad + 2\int_0^t [G_x(s_1(t), s_1(\tau), t-\tau)v_1(\tau) - G_x(s_0(t), s_0(\tau), t-\tau)v_0(\tau)]\, d\tau \\ &= I_1 + I_2 + I_3, \end{aligned}$$

respectively.

Apply the mean value theorem to obtain the estimate

$$|I_1| \leq 2\int_0^1 N_x(\xi, y, t)(s_1(t) - s_0(t))f'(y)\, dy.$$

Then observe that

$$\text{(5-25)} \qquad \begin{aligned} |N_x(\xi, y, t)| &\leq \frac{|\xi - y|}{2t} k(\xi - y, t) + \frac{|\xi + y|}{2t} k(\xi + y, t) \\ &\leq \frac{2}{\sqrt{6et}} k\left(\frac{1}{2}(\xi - y), t\right) + \frac{2}{\sqrt{6et}} k\left(\frac{1}{2}(\xi + y), t\right) \end{aligned}$$

because $\alpha e^{-3\alpha^2} \leq 1/\sqrt{6e}$ for $\alpha \geq 0$. We also have

$$|s_1(t) - s_0(t)| = \left| -\int_0^t [v_1(\tau) - v_0(\tau)]\, d\tau \right| \leq t\Delta.$$

Put these estimates together to obtain

$$|I_1| \le 2Mt\Delta \cdot \frac{2}{\sqrt{6et}} \cdot 2 = \frac{8M\Delta\sqrt{t}}{\sqrt{6e}}. \tag{5-26}$$

To estimate I_2, begin by using the mean value theorem to get

$$|I_2| \le 2Mt\Delta \int_0^t |N_x(\xi, 0, t-\tau)|\, d\tau. \tag{5-27}$$

Use (5-25) with $y = 0$ and t replaced by $t - \tau$ to deduce that

$$\begin{aligned} |N_x(\xi, 0, t-\tau)| &\le \frac{|\xi|}{t-\tau} k(\xi, t-\tau) \\ &\le \frac{\sqrt{2}}{t-\tau} \exp\left[-\frac{\xi^2}{8(t-\tau)}\right] \le \frac{\sqrt{2}}{t-\tau} \exp\left[-\frac{1}{32(t-\tau)}\right], \end{aligned}$$

where we have used $\alpha e^{-\alpha^2} \le 1$ and the fact that the intermediate value $\xi > 1/2$ because $s_0(t), s_1(t) \ge 1/2$. Put this last estimate in (5-27) to get

$$|I_2| \le 2Mt\Delta\sqrt{2} \int_0^t \frac{1}{t-\tau} \exp\left[-\frac{1}{32(t-\tau)}\right] d\tau = 2\sqrt{2}Mt\Delta \int_{1/32t}^{\infty} \frac{1}{\lambda} e^{-\lambda}\, d\lambda.$$

The integral on the right has limit zero when t tends to zero. Hence, we can find σ_3 dependent only upon M such that for $0 \le t \le \sigma_3$ the integral is less than $1/2\sqrt{2}M$. Thus,

$$|I_2| \le t\Delta. \tag{5-28}$$

Finally, to estimate I_3 we express it as

$$\begin{aligned} I_3 &= 2\int_0^t G_x(s_1(t), s_1(\tau), t-\tau)[v_1(\tau) - v_0(\tau)]\, d\tau \\ &\quad + 2\int_0^t [G_x(s_1(t), s_1(\tau), t-\tau) - G_x(s_0(t), s_0(\tau), t-\tau)]v_0(\tau)\, d\tau \\ &\equiv I_3' + I_3'', \end{aligned}$$

respectively. Now

$$|I_3'| \le \Delta \int_0^t |G_x(s_1(t), s_1(\tau), t-\tau)|\, d\tau \le \frac{Le + 24}{4e\sqrt{\pi}} t^{1/2}\Delta,$$

by the estimates used on the third integral in (5-23). As for I_3'', apply the mean value theorem to obtain

$$I_3'' = 2\int_0^t [G_{xx}(\xi, \eta, t-\tau)(s_1(t) - s_0(t)) + G_{xy}(\xi, \eta, t-\tau)(s_1(\tau) - s_0(\tau)]v_0(\tau)\, d\tau,$$

where (ξ, η) is on the line joining $(s_0(t), s_0(\tau))$ to $(s_1(t), s_1(\tau))$; that is,

$$(\xi, \eta) = \theta(s_0(t), s_0(\tau)) + (1-\theta)(s_1(t), s_1(\tau))$$

for some $\theta \in (0, 1)$. Hence, $\xi, \eta \geq \frac{1}{2}$ and

$$\xi - \eta = \theta(s_0(t) - s_0(\tau)) + (1 - \theta)(s_1(t) - s_1(\tau))$$

so that

$$|\xi - \eta| \leq L(t - \tau).$$

Since $k_{xx}(x, t) = [(x^2 - 2t)/(4t^2)]k(x, t)$ and by the estimate above $(\xi - \eta)^2/(t - \tau)^2 \leq L$, we can bound the terms G_{xx} and G_{xy} using the reasoning applied previously. Also,

$$|s_1(t) - s_0(t)|, |s_1(\tau) - s_0(\tau)| \leq t\Delta,$$

as we have seen earlier. Thus, we find a constant C_1 such that

$$|I_3''| \leq C_1 t\Delta \quad \text{for} \quad 0 \leq t \leq \sigma_4$$

where C_1 and σ_4 depend only upon M and L, and so

$$|I_3| \leq C_2 t^{1/2}\Delta \tag{5-29}$$

for some constant C_2. Combine (5-26), (5-28), and (5-29) to get

$$|v_2(t) - v_1(t)| \leq Ct^{1/2}\Delta$$

for $0 \leq t \leq \min(\sigma_2, \sigma_3, \sigma_4, t^*)$ and some constant C dependent only upon L and M.

Finally, fix

$$\sigma^* \leq \min(\sigma_2, \sigma_3, \sigma_4, t^*, 1/(2C)^2).$$

Then

$$|v_2(t) - v_1(t)| \leq \tfrac{1}{2}\Delta \quad \text{for} \quad 0 \leq t \leq \sigma^*. \tag{5-30}$$

The reasoning applied to v_0, v_1, and v_2 applies to any three consecutive v's so

$$|v_{n+1}(t) - v_n(t)| \leq \tfrac{1}{2} \max_{0 \leq t \leq \sigma^*} |v_n(t) - v_{n-1}(t)|.$$

It follows that

$$|v_{n+1}(t) - v_n(t)| \leq (\tfrac{1}{2})^n\Delta \quad \text{for} \quad 0 \leq t \leq \sigma^*,$$

and hence the series

$$v_0 + \sum_{n=0}^{\infty} (v_{n+1} - v_n)$$

is absolutely and uniformly convergent. That is, $\{v_n\}$ converges uniformly to v, which is continuous on $0 \leq t \leq \sigma^*$. Passing to the limit in (5-21) shows that (5-20) holds and v solves the integral equation (5-16). Moreover, if v and w both solve (5-20), the estimates above lead to

$$|v(t) - w(t)| \leq (\tfrac{1}{2})^n \max_{0 \leqslant t \leqslant \sigma^*} |v(t) - w(t)|$$

for all n. Hence, $v = w$ and we have shown:

Theorem 5-3. There is a number $\sigma^* > 0$ such that in the time interval $0 \leq t \leq \sigma^*$ there is a unique solution to the integral equation (5-20); consequently, if f and ϕ are continuously differentiable functions satisfying $f(0) = \phi(0)$ and $f(1) = 0$, then the free boundary value problem (5-1)–(5-6) has a unique solution.

This local existence theorem can be extended to a global result. This is done by repeated application of Th. 5-3 along parallel strips together with estimates rather like those used in the proof of the local theorem. We state the final result.

Theorem 5-4. Suppose that f and ϕ are continuously differentiable functions with $f(x) \geq 0$, $\phi(t) \geq 0$, $f(0) = \phi(0)$, and $f(1) = 0$. Then on any time interval $0 \leq t \leq T$, there exists a unique solution to (5-1)–(5-6).

This approach to free boundary value problems is due to Friedman. Additional problems are treated in his book listed in the Suggestions for Further Reading.

PROBLEMS

1. Let $x = s(t)$ be continuously differentiable with $s(t) > 0$ and $s(0) = 1$. Consider the Dirichlet problem

$$\begin{aligned} &u_t = u_{xx}, \qquad 0 < x < s(t), \qquad t > 0,\\ &u(0, t) = \phi(t), \qquad u(s(t), t) = \psi(t), \qquad t > 0,\\ &u(x, 0) = f(x), \qquad 0 \leq x \leq 1, \end{aligned}$$

where $\phi(t)$, $\psi(t)$, $f(x)$, and $s(t)$ are given functions. Show there is a unique solution to this problem.

2. Consider the free boundary value problem

$$\begin{aligned} &u_t = u_{xx}, \qquad 0 < x < s(t), \qquad t > 0,\\ &u_x(0, t) = -\phi(t), \qquad \phi(t) \geq 0, \qquad t > 0,\\ &u(x, 0) = f(x), \qquad 0 \leq x \leq 1,\\ &s'(t) = -u_x(s(t), t), \qquad u(s(t), t) = 0, \qquad t > 0,\\ &s(0) = 1. \end{aligned}$$

Show that there is a unique solution to this problem in a sufficiently small time interval. *Hint.* Replace $G(x, y, t)$ by $N(x, y, t)$ in the reasoning following (5-8).

3. Consider the free boundary value problem

$$\begin{aligned} &u_t = u_{xx}, \qquad 0 < x < s(t),\\ &u(0, t) = \phi(t), \qquad \phi(t) \geq 0, \qquad t > 0,\\ &u(x, 0) = f(x), \qquad 0 \leq x \leq 1,\\ &u(s(t), t) = \psi(t), \qquad s'(t) = -u_x(s(t), t), \qquad t > 0, \qquad \psi(t) \geq 0,\\ &s(0) = 1. \end{aligned}$$

Show that, in a sufficiently small neighborhood of $t = 0$, there exists a unique solution to this problem. Here ϕ, ψ, and f are given functions. You will need to determine compatibility conditions for these functions. Can the solution be extended for all time?

4. Suppose in (5-1)–(5-6) that $\phi(0) \neq f(0)$. Can one still obtain a solution?
5. Show that the maximum principle applies to the situation in Lemma 5-1.

-6. Free Boundary Value Problems for Hyperbolic Equations

We restrict ourselves to a problem with one spatial dimension. Thus, all variables are scalar-valued. Figure 12-5 shows a cylinder filled with gas. The left end of the cylinder $x = 0$ is fixed and there is a movable piston whose position at time t is denoted by $x = s(t)$. We suppose that the lateral walls of the cylinder are insulated and that gas can enter or leave the cylinder only at $x = 0$. In particular, there is no mass flux at the piston. In the model we are about to describe, the gas dynamics equations will be linearized, thus avoiding the more complicated aspects of nonlinear gas dynamics problems. Nevertheless, if we imagine a rod attached to the piston head, this model can be viewed as a first attempt to understand the behavior of an internal combustion engine, although we shall not study the case of an initial point source disturbance. This problem also arises in ballistics.

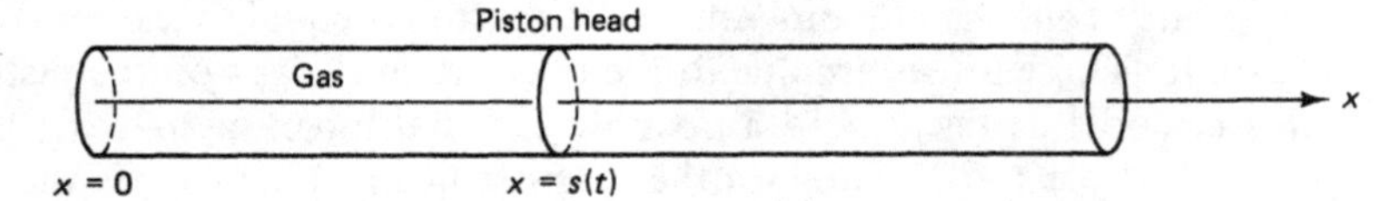

Figure 12-5

The gas in the cylinder must satisfy the continuity equation (see Sec. 1-7)

$$\rho_t + (u\rho)_x = 0 \tag{6-1}$$

and the equation of motion

$$\rho(u_t + uu_x) + p_x = 0, \tag{6-2}$$

where ρ is the density of the gas, u is the velocity, and p is the pressure. We also need an equation of state involving ρ, p, and the entropy. We assume that the entropy is constant and that the equation of state has the form

$$p = A\rho^\gamma, \tag{6-3}$$

where $A > 0$ is a constant and $\gamma = c_p/c_v$ is the ratio of specific heats at constant pressure and constant volume. For all gases $\gamma > 1$ and in most cases $\gamma < 2$. If we denote the speed of sound by

$$\frac{dp}{d\rho} = c^2(\rho) = \gamma A \rho^{\gamma-1}, \tag{6-4}$$

then p can be eliminated from (6-2) and we obtain

$$u_t + uu_x + \frac{c^2(\rho)}{\rho}\rho_x = 0. \tag{6-5}$$

Equations (6-1) and (6-5) describe the state of the gas. We turn now to the boundary conditions at $x = 0$ and at the free boundary $x = s(t)$. First, the velocity of the gas at $x = s(t)$ must equal the velocity of the piston, so

$$s'(t) = u(s(t), t). \tag{6-6}$$

Next, the total mass of gas in $0 < x < s(t)$ is

$$m(t) = \int_0^{s(t)} \rho(y, t)\, dy.$$

From (6-1) and (6-6) we find that

$$\begin{aligned}\frac{dm}{dt} &= \int_0^{s(t)} \rho_t(y, t)\, dt + \rho(s(t), t)s'(t)\\ &= -\int_0^{s(t)} (\rho u)_y\, dy + \rho(s(t), t)s'(t) = \rho(0, t)u(0, t).\end{aligned}$$

This rate, call it $r(t)$, at which mass enters or leaves the system at $x = 0$ must be given. Thus, we obtain the condition

$$\rho(0, t)u(0, t) = r(t). \tag{6-7}$$

The rate of change of momentum of the piston is equal to the sum of the forces acting on it. These forces are due to the pressure of the gas on the piston, friction which we model as linear, and a possible external force, say $-F_1(s(t), t)$, caused, for example, by a rod pushing on the cylinder head. Thus, if α is the area of the piston head and μ is its mass, we find

$$\mu s''(t) = \alpha p(s(t), t) - F_1(s(t), t) - ks'(t), \tag{6-8}$$

where $k > 0$ is the damping constant.

We linearize the gas dynamics equations just as we did in Sec. 1-7. We introduce the condensation σ by setting $\rho = \rho_0(1 + \sigma)$, where ρ_0 is the mean density about which the small variations occur. This leads to the system

$$\begin{cases}\sigma_t + u_x = 0,\\ u_t + A\gamma\rho_0^{\gamma-1}\sigma_x = 0.\end{cases}$$

Equation (6-6) remains in force, (6-7) becomes

$$u(0, t) = \frac{r(t)}{\rho_0},$$

and the differential equation for the free boundary becomes

$$s'' + \frac{k}{\mu} s' = \frac{\alpha A\rho_0^{\gamma}}{\mu}[1 + \gamma\sigma(s(t), t)] - \frac{1}{\mu} F_1(s(t), t). \tag{6-9}$$

It is convenient to choose units so that the mean density $\rho_0 = 1$ and $A\gamma = 1$. Then the speed of sound is $c^2 = A\gamma\rho_0^{\gamma-1} = 1$. We also set $\nu = k/\mu$ and $\lambda = \alpha/\mu$ so that (6-9) becomes

$$s'' + vs' = \lambda\sigma - \lambda\left(-A + \frac{F_1}{\mu\alpha}\right).$$

We define F to be the term in parentheses and obtain the following free boundary value problem for (u, σ, s), where the $s(0) = 1$ is the initial location of the cylinder:

(6-10) $\sigma_t + u_x = 0, \quad 0 < x < s(t), \quad t > 0,$

(6-11) $u_t + \sigma_x = 0, \quad 0 < x < s(t), \quad t > 0,$

(6-12) $u(x, 0) = \phi(x), \quad 0 \le x \le 1,$

(6-13) $\sigma(x, 0) = \psi(x), \quad 0 \le x \le 1,$

(6-14) $u(0, t) = r(t), \quad t > 0,$

(6-15) $s'(t) = u(s(t), t), \quad t > 0,$

(6-16) $s'' + vs' = \lambda\sigma(s(t), t) - \lambda F(s(t), t), \quad t > 0$

(6-17) $s(0) = 1, \quad s'(0) = u(s(0), 0) = u(1, 0) = \phi(1) \equiv \beta.$

Here $\lambda > 0$, $v \ge 0$ are given constants and ϕ, ψ, r, and F are given functions.

Recall that along the characteristics of (6-10) and (6-11), that is, along the lines with slopes ± 1, the functions $\sigma \pm u$ are constants. In fact, $\sigma \pm u$ are the Riemann invariants (see also Chap. 4). Thus, the functions

$$f(t - x) = \sigma(x, t) + u(x, t),$$
$$g(t + x) = \sigma(x, t) - u(x, t)$$

are well defined. The differential equation (6-16) can be expressed as

$$s'' + vs' = \lambda\sigma - \lambda F = \lambda(\sigma + u) - \lambda s' - \lambda F$$

by (6-15) or

(6-18) $$s'' + (v + \lambda)s' = \lambda f(t - s(t)) - \lambda F(s(t), t),$$

where s has to satisfy the initial conditions (6-17).

As long as $f(x)$ satisfies a Lipschitz condition and $F(x, t)$ satisfies a Lipschitz condition in x, there is a unique solution $s(t)$ to (6-17) and (6-18). Now, from an intuitive standpoint, we see that to set up (6-18) we must be able to construct σ and u. If different possibilities are drawn where the lines with slope ± 1 can intersect $x = s(t)$, we see that if $s'(t) > 1$, then only $u(s(t), t) = s'(t)$ is known on the free boundary; that is, $x = s(t)$ is moving too rapidly to the right and the solution to the partial differential equation is underdetermined. If $s'(t) < -1$, we have two equations for σ and u where a pair of characteristics meet $x = s(t)$, but we also have the third condition $u(s(t), t) = s'(t)$ and the system is overdetermined. Thus, we seek solutions to (6-17) and (6-18) with $|s'(t)| < 1$. This condition is natural also because 1 is the sound speed and our linear model assumes that the velocity of the gas and hence the velocity of the free boundary are small compared with the sound speed.

We assume from now on that $F(x, t)$ is continuous on $0 \le x < \infty$, $t \ge 0$ and is continuously differentiable with respect to x. We also assume that $\phi(x)$, $\psi(x)$

are continuously differentiable on $0 \le x \le 1$ and that $r(t)$ is continuously differentiable on $t \ge 0$. Finally, we assume that $|F(x, t)| \le L$ for some constant L and that $\lambda L \le \lambda + \nu$. Then

$$M \equiv \frac{\lambda + \nu - \lambda L}{\lambda}$$

is nonnegative.

Theorem 6-1. Let $|\beta| \le 1$. Assume that $f(y)$ is defined on the interval $-1 \le y \le Y$ for some $Y \le \infty$, satisfies $|f(y)| \le M$, and is locally Lipschitz there. Then there is a unique solution to (6-17) and (6-18) which is defined on a maximal time interval $[0, T]$ and satisfies $|s'(t)| < 1$ there.

Proof. Let

$$\Omega = \{(x, t): x \ge 0, t \ge 0, -1 \le t - x \le Y\}$$

and $H(s, t) = \lambda[f(t - s) - F(s, t)]$. Then (6-18) has the form

$$s'' + (\mu + \lambda)s' = H(s, t), \tag{6-19}$$

where $H(s, t)$ satisfies a Lipschitz condition. Consequently, by the basic existence and uniqueness theorem for the initial value problem for ordinary differential equations, there is a unique solution $s(t)$ for t sufficiently small. This local solution can be extended to its maximal domain of definition by repeated application of the local theorem as long as $(s(t), t)$ remains in Ω.

To obtain the critical bound on $s'(t)$, write (6-19) as

$$\frac{d}{dt} e^{(\nu + \lambda)t} s'(t) = e^{(\nu + \lambda)t} H(s, t),$$

which upon integration gives

$$s'(t) = \beta \exp[-(\nu + \lambda)t] + \int_0^t H(s(\tau), \tau) \exp[-(\lambda + \nu)(t - \tau)]\, d\tau. \tag{6-20}$$

By our assumptions on F and f and the choice of M,

$$\begin{aligned}|H(s, t)| &\le \lambda[|f(t - s)| + |F(s, t)|] \le \lambda[M + L] \\ &= \lambda + \nu - \lambda L + \lambda L = \lambda + \nu.\end{aligned}$$

Hence, from (6-20),

$$\begin{aligned}|s'(t)| &\le |\beta| \exp[-(\lambda + \nu)t] + (\lambda + \nu) \int_0^t \exp[-(\lambda + \nu)(t - \tau)]\, d\tau \\ &= |\beta| \exp[-(\lambda + \nu)t] + 1 - \exp[-(\lambda + \nu)t] < 1\end{aligned}$$

because $|\beta| < 1$, and the theorem is proven.

In view of the definition of Ω, the fact that $|s'(t)| < 1$, and the fact that $(t, s(t))$ hits the boundary of Ω at time $t = T$, we obtain the following corollary of Th. 6-1.

Corollary 6-1. Let $I = [0, T]$ be the maximal interval of existence of the solution in Th. 6-1. If $Y = +\infty$, then either
(i) $T = +\infty$ and $0 < s(t) < 1 + t$ for $t > 0$; or
(ii) $T < +\infty$ then $T > 1$ and $s(T) = 0$.
If $Y < \infty$ then one of the following three statements holds:
(i) $T = +\infty$ and $t - Y < s(t) < t + 1$;
(ii) $T < \infty$, $T > 1$, and $s(T) = 0$;
(iii) $T < \infty$ and $s(T) = T - Y$.

In the following, we give conditions on ϕ, ψ, and r, which guarantee that $|f(y)| \leq M$ on a hopefully large interval.

Lemma 6-1. Suppose that

$$|\phi(x)| \leq \tfrac{1}{2}M \quad \text{and} \quad |\psi(x)| \leq \tfrac{1}{2}M$$

for $0 \leq x \leq 1$. Then the assumptions about f in Th. 6-1 are satisfied for some $Y \geq 0$.

For the proof observe that $f(-x) = \phi(x) + \psi(x)$ is continuously differentiable for $0 \leq x \leq 1$ and $|f| \leq M$.

Lemma 6-2. Suppose the bounds in Lemma 6-1 hold and that

$$|r(t) - \tfrac{1}{2}(\psi(t) - \phi(t))| \leq \tfrac{1}{2}M \quad \text{for} \quad 0 \leq t \leq 1.$$

Then the assertions about f in Th. 6-1 hold for some $Y \geq 1$.

For the proof, use the initial and boundary data together with the expressions for f and g to find that

$$f(y) = \begin{cases} \phi(-y) + \psi(-y), & -1 \leq y \leq 0, \\ 2r(y) + \psi(y) - \phi(y), & 0 \leq y \leq 1. \end{cases}$$

The lemma follows at once.

The derivation of further conditions on $r(t)$ for $t > 1$ involves the behavior of the free boundary since the backward characteristic starting from $(0, t)$ meets the free boundary at a point earlier in time (see Fig. 12-6 on page 534). Let us suppose that $s(t)$ is given on $[0, T]$ with $T \geq 1$ and satisfies $|s'(t)| < 1$. Set $T_r = s(T) + T$. For each t in the interval $1 \leq t \leq T_r$ there is a unique point t^* such that the point $(s(t^*), t^*)$ lies on the backward characteristic emanating from $(0, t)$ (see Fig. 12-6). More precisely, the following holds.

Lemma 6-3. The equation

$$s(t^*) + t^* = t$$

has a unique solution $t^* = t^*(t)$ for each $t = [1, T_r]$. The function $t^*(t)$ is continuously differentiable and $0 \leq t^*(t) \leq T$.

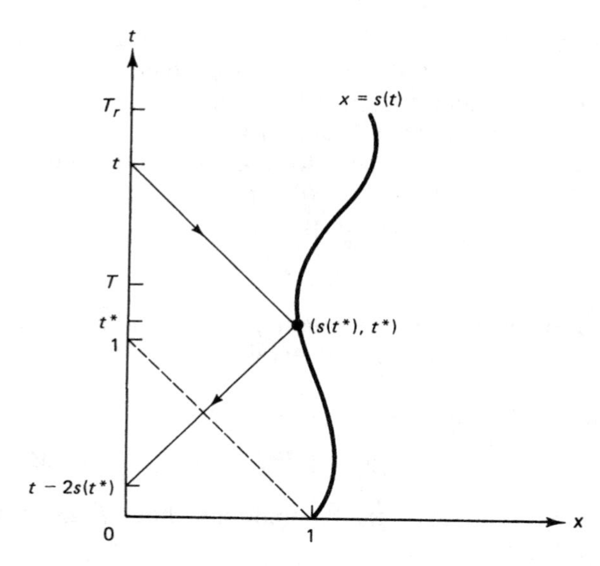

Figure 12-6

Proof. Consider the function

$$h(\tau, t) = s(\tau) + \tau - t, \qquad 0 \le \tau \le T, \qquad 1 \le t \le T_r.$$

We have $h(0, t) = s(0) - t = 1 - t \le 0$, $h(T, t) = s(T) + T - t = T_r - t \ge 0$, and $h_\tau(\tau, t) = s'(\tau) + 1 > 0$ because $|s'(\tau)| < 1$. Hence, $h(\tau, t)$ is strictly increasing in τ for each t and there is a unique point t^* such that $h(t^*, t) = 0$. Clearly, $t^* = t^*(t)$ is a function of t. We can now apply the implicit function theorem to deduce that $t^*(t)$ is a continuously differentiable function of t.

We use the function $t^*(t)$ to express $f(t)$ on $1 \le t \le T_r$ as follows (see Fig. 12-6):

$$\begin{aligned} f(t) &= \sigma(0, t) + u(0, t) + 2r(t) \\ &= \sigma(s(t^*), t^*) - u(s(t^*), t^*) + 2r(t) \\ &= \sigma(s(t^*), t^*) + u(s(t^*), t^*) + 2r(t) - 2s'(t^*) \\ &= f(t - 2s(t^*)) + 2r(t) - 2s'(t^*), \end{aligned} \tag{6-21}$$

where (6-15) has been used.

Lemma 6-4. Let the assumptions about f in Th. 6-1 hold for some finite $Y \ge 1$. Let $x = s(t)$ be the corresponding solution for the free boundary on the maximal interval $[0, T]$. Assume that $s(T) \ne 0$ and $T < \infty$. If $r(t)$ satisfies

$$|r(t) - [s'(t^*) - \tfrac{1}{2} f(t - 2s(t^*))]| \le \tfrac{1}{2} M \tag{6-22}$$

for

$$Y = s(T) - T \le t \le s(T) + T \equiv Y^*$$

and $t^* = t^*(t)$, then $f(y)$ satisfies a local Lipschitz condition on $-1 \le y \le Y^*$ and $|f(y)| \le M$ there.

Proof. The bound $|f(y)| \le M$ follows at once from (6-21) and (6-22). The local Lipschitz condition follows from the expression for $f(t)$ and the fact that ϕ, ψ, r, s', and t^* are continuously differentiable.

Combining Th. 6-1 and Lemmas 6-1–6-4 leads to the following result.

Theorem 6-2. Assume that $|s'(0)| = |\phi(1)| < 1$, $\phi(x)$ and $\psi(x)$ are continuously differentiable for $0 \le x \le 1$, and $r(t)$ is continuously differentiable for $t \ge 0$ with $r(0) = \phi(0)$. In addition, suppose that

$$|\phi(x)|, |\psi(x)| \le \tfrac{1}{2}M,$$
$$|r(t) + \tfrac{1}{2}(\psi(t) - \phi(t))| \le \tfrac{1}{2}M \quad \text{on} \quad 0 \le t \le 1,$$

and that

$$|r(t) - [s'(t^*) - \tfrac{1}{2}f(t - 2s(t^*)]| \le \tfrac{1}{2}M$$

for $t > 1$ as long as t^* and $s(t^*)$ are defined. Then there is a unique solution $x = s(t)$ for the free boundary and $|s'(t)| < 1$. The function $s(t)$ has a maximal interval of existence $[0, T]$ and if $T < \infty$ we have $s(T) = 0$. Moreover, u and σ are uniquely determined, so the free boundary value problem (6-10)–(6-17) has a unique solution.

Proof. Lemmas 6-1 and 6-2 ensure that the assumptions in Th. 6-1 hold at least for $Y = 1$. Thus, $s(t)$ exists on some interval of positive length. The second condition on r together with Lemma 6-4 implies that the solution can be extended as long as $s(t) \ne 0$.

The discussion above of the free boundary value problem for a piston is due to Jürgen Gerlach; see "Two linearized models for a hyperbolic free boundary value problem" in the *Journal of Applied Mathematics and Physics*, Vol. 35, pp. 180–192, for additional details.

PROBLEMS

1. Suppose that $x = s(t)$ is given with $|s'(t)| < 1$. Formulate an initial boundary value problem for the system

$$\begin{aligned} \sigma_t + u_x &= 0 \\ u_t + \sigma_x &= 0 \end{aligned} \quad \text{in} \quad 0 < x < s(t), \qquad t > 0.$$

2. Set up the problem analogous to (6-10)–(6-17) for the nonlinear system (6-1) and (6-5). What difficulties do you foresee in actually trying to extend the linear theory to this nonlinear situation?
3. Consider Ths. 6-1 and 6-2 for the case $v = 0$, $F \equiv 0$. Determine M. If the gas is brought in at $x = 0$ and not extracted, how do you expect the free boundary to behave if, say, initially the velocity is zero and the density is constant? Can you prove your conjecture?

4. Suppose that $\phi(x) = \psi(x) = \frac{1}{2}$ so that $s(0) = 1$, $s'(0) = \frac{1}{2}$. Show that $f(y) \equiv 1$ for $-1 \leq y \leq 0$. For small times (how small?) the effects of $r(t)$ on $s(t)$ will be zero. Take $\nu = 0$, $\lambda = 1$, $F = 0$. Then $M = 1$. Show that in the region $t - s(t) < 0$ the function $s(t)$ is determined by

$$s''(t) + s'(t) = 0, \qquad s(0) = 1, \qquad s'(0) = \tfrac{1}{2}.$$

Find $s(t)$ and show that $|s'(t)| < 1$, $t - s(t) \leq -\frac{1}{2}$, so in fact $s(t)$ never enters the domain of influence of $r(t)$.

5. The assumption that mass only enters or leaves the cylinder at $x = 0$ is equivalent to the velocity relation (6-6). Confirm this by calculating dm/dt without assuming (6-6). Show that the result depends only upon mass transfer at $x = 0$ when (6-6) holds.

6. Verify the formula for $f(y)$ in the proof of Lemma 6-2.

Solutions and Hints to Selected Problems

CHAPTER 1

Section 1-2:

1. Let $u = u(x, t)$ be the displacement, g the acceleration due to gravity and L the length of the string.

$$u_{tt} = g((L - x)u_x)_x.$$

4. Using a notation similar to that of the string, one obtains

$$u_{tt} + ku_t = c^2(u_{xx} + u_{yy}) + F, \qquad c^2 = \tau/\rho_0.$$

Section 1-3:

3. $\rho c u_t = \operatorname{div}(k\nabla u) + f.$

5. Let n denote the unit outer normal to B. Then $q \cdot n = -k\nabla u \cdot n = -k\partial u/\partial n$. The condition on u at the boundary is

$$-k\frac{\partial u}{\partial n} = \alpha(u - T_\infty).$$

Section 1-4:

3. $C_t = \operatorname{div}(D(C)\nabla C) - \operatorname{div}(Cv) + f.$

Section 1-6:

1. $v_{xx} - LCv_{tt} - (LG + RC)v_t - RGv = 0.$

2. If $R_xL - L_xR = 0$, $Lv_{xx} - L^2Cv_{tt} - (L^2G + RLC)v_t - L_xv_x - RLGv = 0.$
If $R_xL - L_xR \neq 0$, $Lv_{xxt} - L^2Cv_{ttt} + Rv_{xx} - (2RLC + L^2G)v_{tt} - L_xv_{xt}$
$-(2RLG + R^2C)v_t - R_xv_x - R^2Gv = 0.$

Section 1-7:

2. $u_{tt} = c^2\Delta u + gu_z.$

CHAPTER 2

Section 2-1:

4. $u(x, y) = y + \exp[x \exp(-y)]$.

5. $u(x, y) = \exp[x \exp(-\arctan y) + \arctan y]$.

6. $u(x, t) = f(t - x/v)$ for $t \geq x/v$, $u(x, t) = 0$ for $t \leq x/v$, where $f(z) = 0$ for $z \leq 0$.

Section 2-3:

1. The solution is given implicitly by $S = f(x - 2tS)$.

Section 2-4:

1. $u(x, y) = -1 + \sqrt{1 + 4xy} + \log(2x/(1 + \sqrt{1 + 4xy}))$.

2. If one takes $q_0(s) = \sqrt{1 - 3s^2}$, the solution in parametric form is

$$x = (2e^\tau - 1)s, \qquad y = (e^\tau - 1)\sqrt{1 - 3s^2},$$
$$u = (s^2 + 1)e^\tau, \qquad p = 2se^\tau, \qquad q = \sqrt{1 - 3s^2}e^\tau.$$

6. $u(x, y) = \sqrt{xy}(y/x)^{xy} = \sqrt{xy} \exp[xy \log(y/x)]$, $x > 0$, $y > 0$.

Section 2-6:

4. (a) hyperbolic if $x^4 + 4y > 0$, elliptic if $x^4 + 4y < 0$ and parabolic if $x^4 + 4y = 0$.
(b) hyperbolic if $4 + (x^2 + y^2)xy > 0$, elliptic if $4 + (x^2 + y^2)xy < 0$ and parabolic if $4 + (x^2 + y^2)xy = 0$.

CHAPTER 3

Section 3-1:

1. (a) The Fourier coefficients are
$a_0 = 10/3$, $a_n = (8/n^2\pi^2)(-1)^n$, $n \geq 1$,
$b_n = (-1/n\pi)(1 + 3(-1)^n) + (2/n\pi)^3((-1)^n - 1)$, $n \geq 1$.

3. The Fourier coefficients are
$a_0 = 2$, $a_1 = -1/2$, $a_n = 2(-1)^{n+1}/(n^2 - 1)$, $n \geq 2$,
$b_n = 0$, $n \geq 1$.

12. $b_n = \dfrac{4L^2}{(n\pi)^3}[1 - (-1)^n]$.

CHAPTER 4

ection 4-1:

8. $$u(x, t) = \frac{f(x + ct) + f(x - ct)}{2} + \frac{1}{2c}\int_{x-ct}^{x+ct} g(y)\,dy + \frac{1}{2c}\int_0^t \int_{x-c(t-\tau)}^{x+c(t-\tau)} F(y, \tau)\,dy\,d\tau.$$

ection 4-2:

1. $$u(x, t) = \sum_{n=0}^{\infty} [a_n \cos(\lambda_n ct) + b_n \sin(\lambda_n ct)] \sin(\lambda_n x),$$

where $\lambda_n = (n + 1/2)(\pi/L)$ and

$$a_n = \frac{2}{L}\int_0^L f(y) \sin(\lambda_n y)\,dy, \qquad b_n = \frac{2}{Lc\lambda_n}\int_0^L g(y) \sin(\lambda_n y)\,dy.$$

3. Let $\lambda_n = n\pi/L$ and suppose $k^2 - 4c^2\pi^2/L < 0$. Set $\mu_n = 4c^2\lambda_n^2 - k^2 > 0$.

$$u(x, t) = \exp(-kt/2) \sum_{n=1}^{\infty} [a_n \cos(\mu_n t) + b_n \sin(\mu_n t)] \sin(\lambda_n x),$$

$$a_n = \frac{2}{L}\int_0^L f(y) \sin(\lambda_n y)\,dy, \qquad b_n = \frac{2}{L}\int_0^L g(y) \sin(\lambda_n y)\,dy + \frac{k}{2}a_n.$$

4. $I'(x)S(t) = -V(x)[CT'(t) + GT(t)],$

$V'(x)T(t) = -I(x)[LS'(t) + RS(t)].$

$$\frac{I'(x)}{V(x)} = -\frac{CT'(t) + GT(t)}{S(t)} \equiv \alpha,$$

$$\frac{V'(x)}{I(x)} = -\frac{LS'(t) + RS(t)}{T(t)} \equiv \beta.$$

$I'(x) = \alpha V(x), \qquad V'(x) = \beta I(x),$

$I'' + \lambda^2 I = 0, \qquad \lambda^2 = -\alpha\beta, \qquad I(0) = I(L) = 0.$

$I_n(x) = a_n \sin(\lambda_n x), \qquad \lambda_n = n\pi/L.$

$V_n(x) = b_n\lambda_n \cos(\lambda_n x).$

$-(CT_n' + GT_n) = (a_n/b_n)S_n(t),$

$-(LS_n' + RS_n) = -(\lambda_n^2 b_n/a_n)T_n.$

One obtains the differential equation

$LCT_n'' + (LG + RC)T_n' + (RG + \lambda_n^2)T_n = 0.$

Now proceed as before.

Section 4-3:

1. Differentiate $E(t) = (1/2)\int_0^L[\rho_0 u_t^2 + \tau u_x^2]\,dx$ with respect to t, integrate the term containing $u_x u_{xt}$ by parts and conclude that $E'(t) = 0$ so $E(t) = E(0)$.
3. Consider first the problem

$$v_{tt} = c^2 v_{xx}, \qquad 0 < x < L, \qquad t > 0, \qquad v(0, t) = v_x(L, t) = 0, \qquad t > 0.$$

If one sets $v(x, t) = X(x)T(t)$, one finds $X_n(x) = \sin(\lambda_n x)$, $\lambda_n = (n + 1/2)(\pi/L)$, $n = 0, 1, 2, \ldots$. The solution is

$$u(x, t) = \sum_{n=0}^{\infty} a_n(t) \sin(\lambda_n x)$$

where $a_n = (1/c\lambda_n)\int_0^t \sin[c\lambda_n(t - \tau)]F_n(\tau)\,d\tau$ and $F_n(t) = (2/L)\int_0^L F(y, t)\sin(\lambda_n y)\,dy$.
5. The solution is of the form

$$u(x, t) = \sum_{n=0}^{\infty} a_n(t) \sin(\lambda_n x)$$

where $\lambda_n = (n + 1/2)(\pi/L)$ and $a_n(t)$ satisfies $a_n''(t) + ka_n'(t) + c^2\lambda_n^2 a_n(t) = F_n(t)$, $a_n(0) = a_n'(0) = 0$. $F_n(t)$ is as in Prob. 3.

Section 4-4:

1.
$$\begin{bmatrix} u_1 \\ u_2 \end{bmatrix}_t = \begin{bmatrix} 0 & 1 \\ c^2 & 0 \end{bmatrix} \begin{bmatrix} u_1 \\ u_2 \end{bmatrix}_x,$$

$$u_1(x, 0) = f'(x), \qquad u_2(x, 0) = g(x).$$

4. Let P be the matrix constructed from the $P_1, \ldots, P_n$ such that $P^{-1}AP = \Lambda$, where Λ is the diagonal matrix with $\lambda_1, \ldots, \lambda_n$ on the main diagonal and zeros elsewhere. Let $U = PV$. Then V satisfies $PV_t + APV_x = 0$ or $V_t + P^{-1}APV_x = 0$, i.e. $V_t + \Lambda V_x = 0$. The initial conditions transform to $P^{-1}U^0(x) = V^0(x)$. Let $V^0(x) = (g_1(x), \ldots, g_n(x))$. The system uncouples and we find for the jth component

$$\frac{\partial V_j}{\partial t} + \lambda_j \frac{\partial V_j}{\partial x} = 0, \qquad V_j(x, 0) = g_j(x) \quad \text{so} \quad V_j(x, t) = g_j(x - \lambda_j t).$$

Now transform back.

Section 4-5:

1. The eigenvalues of $\mathcal{A} = \begin{bmatrix} 4 & 1 \\ 3 & 2 \end{bmatrix}$ are 1, 5. The matrix that diagonalizes this matrix is obtained from the eigenvectors and we find with $P = \begin{bmatrix} 1 & 1 \\ -3 & 1 \end{bmatrix}$ that $P^{-1}\mathcal{A}P = \begin{bmatrix} 1 & 0 \\ 0 & 5 \end{bmatrix}$. Let $V = PU$. Then V satisfies $V_t + \Lambda V_x = 0$ and its general solution is $V(x, t) = \begin{bmatrix} \varphi(x - t) \\ \psi(x - 5t) \end{bmatrix}$. The values of $\varphi(x)$, $\psi(x)$ for $x \geq 0$ are obtained from the initial condi-

tions and for negative values of the argument from the boundary conditions at $x = 0$. The compatibility condition is obtained by checking $V(0, 0)$. Note that $x - 5t \leq x - t$, and so there are three cases for the solution: $x - 5t \geq 0$, $x - 5t \leq 0 \leq x - t$ and $x - t \leq 0$.

ction 4-6:

3. Let $u = e^{-kt/2}v$. The v satisfies $v_{tt} - (k^2/4)v = c^2 v_{xx}$, $v(x, 0) = 0$, $v_t(x, 0) = f(x)$. Introduce the new variables $x = Ly$, $t = T\tau$, $w(y, \tau) = v(x, t)$. Choose $T = 2/k$ and $L = cT = 2c/k$. Then w satisfies the problem

$$w_{\tau\tau} - w = w_{yy}, \qquad w(y, 0) = 0, \qquad w_\tau(y, 0) = F(y)$$

where $F(y) = Tf(Ly)$. The solution for w is

$$w(y, \tau) = \frac{1}{2}\int_{y-\tau}^{y+\tau} F(\eta)e^{-\eta/2} I_0\left(\frac{\sqrt{5}\tau}{2}\sqrt{1 - \left(\frac{x-\eta}{\tau}\right)^2}\right) d\eta,$$

where $I_0(z) = \sum_{j=0}^{\infty}(1/(j!)^2)(z/2)^{2j}$ is the modified Bessel function.

ction 4-8:

2. It is often simplest to make a substitution of the form $u(x, y) = h(x, y)v(x, y)$ and then choose $h(x, y)$ to eliminate the terms involving v_x and v_y. In this case one finds that a natural choice for $h(x, y)$ is e^{-xy}. If one sets $u(x, y) = e^{-xy}v(x, y)$, one finds that $v(x, y)$ must satisfy $v_{xy} - v = 0$.

CHAPTER 5

ction 5-1:

1. (a)

$$u(x, t) = \frac{\alpha_0}{2} + \sum_{n=1}^{\infty} \alpha_n e^{-\alpha\lambda_n^2 t} \cos(\lambda_n x),$$

where $\lambda_n = n\pi/L$ and $\alpha_n = (2/L)\int_0^L f(y)\cos(\lambda_n y)\, dy$.

3.

$$u(x, t) = \frac{\alpha_0}{2} + \sum_{n=1}^{\infty} \alpha_n e^{-\alpha\lambda_n^2 t}\cos(\lambda_n x), + \frac{\varphi_0(t)}{2} + \sum_{n=1}^{\infty}\varphi_n(t)\cos(\lambda_n x),$$

where $\alpha_0, \alpha_1, \ldots$ are defined as in Prob. 1(a). The φ's are defined as follows. Let

$$F_n(t) = \frac{2}{L}\int_0^L F(y, t)\cos(\lambda_n y)\, dy.$$

Then

$$\varphi_0(t) = \int_0^t F_0(\tau)\, d\tau, \qquad \varphi_n(t) = \int_0^t e^{-\alpha\lambda_n^2(t-\tau)}F_n(\tau)\, d\tau.$$

5. Let $\lambda_n > 0$, $n = 1, 2, \ldots$ be the roots of $\tan(\lambda L) = \mu/L$. The $X_n(x) = \cos(\lambda_n x)$ satisfy the orthogonality relations

$$\int_0^L X_n(x)X_p(x)\,dx = \begin{cases} 0, & n \neq p, \\ \dfrac{L}{2} + \dfrac{\sin(2\lambda_n L)}{4\lambda_n} \equiv \gamma_n, & n = p. \end{cases}$$

$$w(x, t) = \sum_{n=1}^{\infty} \alpha_n e^{-\alpha\lambda_n^2 t} \cos(\lambda_n x)$$

where $\alpha_n = (1/\gamma_n)\int_0^L [f(y) - U] \cos(\lambda_n y)\,dy$ and $u = w + U$.

10. Let $\lambda_n = (n + 1/2)(\pi/L)$ for $n = 0, 1, 2, \ldots$. Then

$$u(x, t) = \sum_{n=0}^{\infty} \alpha_n e^{-\alpha\lambda_n^2 t} \sin(\lambda_n x), \qquad \alpha_n = \frac{2}{L}\int_0^L f(y) \sin(\lambda_n y)\,dy.$$

Section 5-3:

4. Let $u = e^{\alpha x + \beta t} v$ and choose first $\alpha = -b/2a$ and then $\beta = a\alpha^2 + b\alpha + c$. Then v satisfies $v_t = av_{xx}$ and the same boundary conditions as u.

Section 5-4:

11. Let $A(t) = \int_0^t a(\tau)\,d\tau$, $B(t) = \int_0^t b(\tau)\,d\tau$, $C(t) = \int_0^t c(\tau)\,d\tau$. Then

$$\hat{u}(\omega, t) = \hat{f}(\omega) \exp[-A(t)\omega^2 - B(t)i\omega + C(t)].$$

$$u(x, t) = \exp(C(t)) \int_{-\infty}^{\infty} k(x - y + B(t), A(t)) f(y)\,dy.$$

12.

$$\hat{u}(\omega, t) = \hat{f}(\omega - it) \exp[-\omega^2 t + i\omega t^2 + t^3/3],$$

$$u(x, t) = \int_{-\infty}^{\infty} e^{(yt + t^3/3)} k(t^2 - x + y, t) f(y)\,dy.$$

13. The transformed equation is

$$\hat{u}_t + \omega \frac{\partial \hat{u}}{\partial \omega} = -(\omega^2 + 1)\hat{u}, \qquad \hat{u}(\omega, 0) = \hat{f}(\omega).$$

$$\hat{u}(\omega, t) = \hat{f}(\omega e^{-t}) \exp\left[-\frac{\omega^2}{2} - t + \frac{\omega^2}{2} e^{-2t}\right].$$

Let $B(t) = e^{-t}$ and $A(t) = (1 - e^{-2t})/2$. Then

$$u(x, t) = B(t) \int_{-\infty}^{\infty} k(x - B(t)y, A(t)) f(y)\,dy.$$

ction 5-6:

8. The variance of u from Prob. 4 is $V(u) = Dt$.

CHAPTER 6

ction 6-2:

5. $z_1 = 3.9266$, $z_2 = 7.0686$, $z_3 = 10.2102$, $z_4 = 13.3518$.

10. Assume there were two solutions, u_1 and u_2, and let $u = u_1 - u_2$. Integrate the expression $u_t(u_{tt} + c^2u_{xxxx}) = 0$ by parts to derive the energy equation

$$\frac{d}{dt}\int_0^L \left[\frac{1}{2}u_t^2(x, t) + \frac{c^2}{2}u_{xx}^2(x, t)\right] dx = 0.$$

13. Let $f(z) = \tan z - \tanh z$. Then $f'(z) = \sec^2 z - \operatorname{sech}^2 z$. $f'(0) = 0$, $\sec^2 z > 1$ for $0 < z < \pi/2$, and $\operatorname{sech}^2 z < 1$ there so that $f'(z) > 0$ for $0 < z < \pi/2$.

ction 6-3:

1.
$$u(x, t) = \frac{A}{\sqrt{\pi\alpha}}\int_0^\infty e^{-\omega^2/4\alpha}\cos(\alpha\omega^2 t)\cos(\omega x)\, d\omega.$$

ction 6-4:

5. Let $c^2 = \tau/\rho_0$ and $k = \epsilon/\rho_0$. Separate variables to find

$$X_n'' + \lambda_n^2 X_n = 0, \qquad T_n'' + \lambda_n^2 k T_n' + \lambda_n^2 c^2 T_n = 0, \qquad X_n(0) = X_n(L) = 0.$$

9. Separation of variables leads to the eigenvalue problem

$$X^{(iv)} - cX = 0, \qquad X(0) = X'(0) = 0, \qquad X(L) = X'(L) = 0.$$

The corresponding problem for T is

$$T'' + \epsilon c T' + a^2 c T = 0.$$

It is readily shown that $c > 0$ so set $c = \lambda^4$. Then

$$X = A\cosh(\lambda x) + B\cos(\lambda x) + C\sinh(\lambda x) + D\sin(\lambda x).$$

$X(0) = 0$ implies that $B = -A$ and $X'(0) = 0$ implies that $D = -C$. The conditions at $x = L$ lead to the eigenvalues and corresponding eigenfunctions for X.

ection 6-5:

12. The solution to Prob. 9 is, formally,

$$u(x, t) = \frac{1}{2\pi}\int_{-\infty}^\infty e^{-i\omega x - \alpha\omega^2 t/(1+b\omega^2)}\left\{\int_{-\infty}^\infty e^{i\omega y} f(y)\, dy\right\} d\omega.$$

To verify that this is a solution, one must place severe growth and differentiability restrictions on f, which will in turn force $\hat{f}(\omega)$ to be sufficiently well behaved that the integrals converge. Assume for example that $f(y)$ is four times continuously differentiable and that $f(y), \ldots, f^{(iv)}(y)$ are absolutely integrable.

CHAPTER 7

Section 7-1:

5. (a) $T(t) = A \cos \sqrt{\lambda} t + B \sin \sqrt{\lambda} t$

(b) $c^2(k(x)X')' + \lambda X = 0, \qquad X(0) = X(1) = 0$

(c) $u(x, t) = \Sigma_{n=1}^{\infty} (A_n \cos \sqrt{\lambda_n} t + B_n \sin \sqrt{\lambda_n} t) X_n(x)$,

$A_n = \int_0^1 f(x) X_n(x)\, dx / \int_0^1 X_n(x)^2\, dx$,

$B_n = \lambda^{-1/2} \int_0^1 g(x) X_n(x)\, dx / \int_0^1 X_n(x)^2\, dx$.

6. (a) Temperature and heat flux are continuous across the surface where the ends of the rod are joined together.

Section 7-2:

3. $g(x, s)$ given by (2-11) with $v_0 = Ax$, $v_1 = B(1 - x)$, and $AB = 1/T$.

4. Assume $k, l > 0$. Also assume a rod whose left end obeys Newton's law of cooling and is in contact with surroundings at temperature zero. The right end is held at temperature zero. Lateral radiation takes place according to Newton's law of cooling with radiation coefficient l. Let $\alpha = \sqrt{l/k}$ and $\beta = (1 - \alpha)/(1 + \alpha)$. The Green's function $g(x, s)$ is given by (2-11) with $v_0 = A(e^{\alpha x} - \beta e^{-\alpha x})$, $v_1 = B(e^{\alpha x} - e^{2\alpha - \alpha x})$ and $AB = -1/2k\alpha(e^{\alpha} - \beta)$.

5. $g(x, s) = (1/2)e^{-|x-s|}$.

7. $g(x, s) = (v_0(s)v_1(x) - v_0(x)v_1(s))$ for $a \le s \le x \le b$ where v_0 and v_1 satisfy $Lv_0 = 0$, $v_0(0) = 0$; $Lv_1 = 0$, $v_1'(0) = 0$; $pW = -1$.

12. Theorems 2-1 and 2-2 together guarantee that the Sturm–Liouville problem has a unique solution if and only if the corresponding homogeneous equation has only the trivial solution. What happens when the corresponding homogeneous equation has nontrivial solutions is not discussed.

Section 7-3:

2. $k_n(x, s) = e^{x-s}(b - a)^{n-1}$, $\gamma(x, s; \lambda) = e^{x-s}(1 - \lambda(b - a))^{-1}$, $y(x) = f(x) + \lambda(1 - \lambda(b - a))^{-1} \int_a^b e^{x-s} f(s)\, ds$.

3. $k_n(x, s) = (1/3)^{n-1} xs$, $\gamma(x, s; \lambda) = xs/(1 - \lambda/3)^{-1}$, $y(x) = f(x) + \lambda(1 - \lambda/3)^{-1} \int_0^1 xs f(s)\, ds$.

5. Hint: Use the reasoning in the 1-dimensional case in the text as a guide.

8. The integral equation for $\theta(t)$ is

$$\theta(t) = \int_0^1 g(t, \tau)[a^2 \sin \theta(\tau) - h(\tau)]\, d\tau$$

where $g(t, \tau) = (1 - t)\tau$ for $0 \le \tau \le t \le 1$ and $g(t, \tau) = (1 - \tau)t$ for $0 \le t \le \tau \le 1$.

ction 7-4:

1. Hint: To check the triangle inequality for the least squares norm write $\|u + v\|^2 = (u + v) \cdot (u + v)$ and expand.

5. Hints: (a) Show that $\|K\| \le \max_{a \le x \le b} \int_a^b \|k(x, s)\|\, dx \equiv B$ by easy estimates. (b) For each x, assume $k(x, s)$ has only a finite number of zeros in $[a, b]$. Use the functions $v_x(s) = \operatorname{sgn} k(x, s)$ to show that $\|K\| \ge B$. So $\|K\| = B$ for such kernels. (c) Use the Weierstrass approximation theorem to approximate any continuous kernel $k(x, s)$ by a sequence of polynomials $k_n(x, s)$ that converge uniformly to $k(x, s)$ on the square $a \le x, s \le b$.

ction 7-5:

3. Hints: (a) Use the addition formulas for the sine and cosine to see that the truncated Fourier series for g gives a degenerate kernel $g(x - s)$. (b) Use Th. 1-3 in Sec. 3-1 to get uniform convergence of the Fourier series on $[a, b]$. (c) Show this gives operator norm convergence of the degenerate kernels to the given kernel.

5. (b) $y(x) = -2$, (c) $y(x) = 1$.

ction 7-6:

5. If v_n is a linear combination of $v_1, \ldots, v_{n-1}$ then $w_n = 0$ and u_n cannot be defined.

9. $c_n = (2/(n^2 + 1))^{1/2}$, $\langle f, v_n \rangle = c_n \int_0^1 (f(x) \sin n\pi x + nf'(x) \cos n\pi x)\, dx$.

11. Hint for part (b): Apply the Gram–Schmidt process to $v_1, \ldots, v_n$ and check that the orthonormal set obtained consists of eigenvectors.

ction 7-7:

1. Hint: The Weierstrass approximation theorem holds for continuous functions on closed bounded sets in n-space.

3. Hints: If a and b are finite, there is a polynomial that satisfies (7-14). Why? For the unbounded case, determine a large bounded square S so that the integral of $|k(x, s)|^2$ over the exterior of S is less than 1/2. Approximate $k(x, s)$ by a polynomial in S and by 0 outside of S.

5. Hint: See Prob. 9 in Sec. 7-4.

ction 7-8:

3. Hint: Review the general Fredholm alternative in Sec. 7-7.

7. For $n = 1, 2, 3, \ldots, \lambda_n = (n\pi)^2$, $y_n(x) = c_n \sin n\pi x$.

9. For $n = 0, 1, 2, 3, \ldots, \lambda_n = [(2n + 1)\pi/2]^2$, $y_n(x) = c_n \cos [(2n + 1)\pi/2]x$.

Section 7-9:

8. $y(x) = Gf(x) = \Sigma_{n=1}^{\infty} < f, y_n > y_n(x)$ where $y_n(x) = \sqrt{2} \sin n\pi x$, which (check it) is the Fourier sine series for f on $[0, 1]$.

9. $y(x) = Gf(x) = \Sigma_{n=0}^{\infty} < f, y_n > y_n(x)$ where $y_n(x) = \sqrt{2} \cos (2n + 1)\pi x/2$.

Section 7-10:

1. $\lambda_n = (n\pi)^2$, $y_n(x) = \sqrt{2} \sin n\pi x$ (orthonormalized), for $n = 1, 2, 3, \ldots$. The eigenfunction expansion is the Fourier sine series of f.

4. (a) If $q(x) \not\equiv 0$ is added to the other hypotheses in Th. 10-1, then the conclusions hold for the periodic case (with periodic boundary conditions in (10-1)).

(b) If (10-2) holds and $p(0) = p(1)$, then all eigenvalues are nonnegative and 0 is an eigenvalue if and only if $q = 0$. Moreover, each eigenvalue has multiplicity 1 or 2.

(c) The eigenfunction expansion is the Fourier series expansion of f over the interval $[0, 1]$.

CHAPTER 8

Section 8-1:

1. Let $x_1 = x$, $x_2 = y$, $u(x, y) = u(r \cos \theta, r \sin \theta) = w(r, \theta)$. Then

$$u_x = w_r r_x + w_\theta \theta_x,$$
$$u_{xx} = w_{rr} r_x^2 + 2w_{r\theta} r_x \theta_x + w_{\theta\theta} \theta_x^2 + w_r r_{xx} + w_\theta \theta_{xx}.$$

Similar expressions hold for u_y and u_{yy}. Thus,

$$\Delta w = w_{rr}(r_x^2 + r_y^2) + 2w_{r\theta}(r_x\theta_x + r_y\theta_y) + w_{\theta\theta}(\theta_x^2 + \theta_y^2) + w_r(r_{xx} + r_{yy}) + w_\theta(\theta_{xx} + \theta_{yy}).$$

Since $r = (x^2 + y^2)^{1/2}$, easy calculations give

$$r_x = \frac{x}{r}, \qquad r_y = \frac{y}{r}, \qquad r_{xx} = \frac{1}{r} - \frac{x^2}{r^3}, \qquad r_{yy} = \frac{1}{r} - \frac{y^2}{r^3}.$$

It follows that $r_x^2 + r_y^2 = 1$, $r_{xx} + r_{yy} = 1/r$. Use similar calculations starting with $\theta = \arctan y/x$ to establish that $r_x\theta_x + r_y\theta_y = 0$, $\theta_x^2 + \theta_y^2 = 1/r^2$, and $\theta_{xx} + \theta_{yy} = 0$. It follows that

$$\Delta w = w_{rr} + \frac{1}{r} w_r + \frac{1}{r^2} w_{\theta\theta} = \frac{1}{r}\frac{\partial}{\partial r}\left(r \frac{\partial w}{\partial r}\right) + \frac{1}{r^2}\frac{\partial^2 w}{\partial \theta^2}.$$

4. (a) Let $u = r^n \cos (n\theta)$ and use the result of Prob. 1.

10. (b) From part (a), $\Delta u = r^{1-n}(r^{n-1}u')'$. So, $\Delta u = 0$ implies $r^{n-1}u' = C$ or $u' = C/r^{n-1}$. If $n > 2$, then $u = -C/(n - 2)r^{n-2} + M \equiv (L/r^{n-2}) + M$ with $L = -C/(n - 2)$. If $n = 2$, then similar reasoning leads to $u = L \log r + M$.

ction 8-2:

1. (b)
$$u(x, y) = \sum_{n=1}^{\infty} \alpha_n \sin(\lambda_n x) \sinh(\lambda_n(b - y))$$

where $\lambda_n = n\pi/a$ and

$$\alpha_n = \frac{2}{a \sinh(\lambda_n b)} \int_0^a f(x) \sin(\lambda_n x)\, dx.$$

4. (a)
$$u(r, \theta) = \frac{A_0}{2} + \sum_{n=1}^{\infty} \left(\frac{r}{a}\right)^n [A_n \cos n\theta + B_n \sin n\theta]$$

where $A_0 = \varphi_0/\alpha$, $A_n = a\varphi_n/(n + \alpha a)$, $B_n = a\psi_n/(n + \alpha a)$ and the φ_0, φ_n, ψ_n are the Fourier coefficients of $f(\theta)$ in the expansion

$$f(\theta) = \frac{\varphi_0}{2} + \sum_{n=1}^{\infty} [\varphi_n \cos n\theta + \psi_n \sin n\theta].$$

ction 8-3:

9. Let D be the domain and B its boundary. Let ν denote the exterior unit normal to D and let φ be the given function defined on B. We seek a function $u = u(x)$ such that $Lu = 0$ in D and which satisfies one of the following conditions:

(i) $u = \varphi$ (Dirichlet problem),

(ii) $\dfrac{\partial u}{\partial \nu} = \varphi$ (Neumann problem),

(iii) $\dfrac{\partial u}{\partial \nu} + au = \varphi$ (Robin problem).

In the Robin problem $a = a(x) \geq 0$ is defined on B and is not identically zero. To prove uniqueness, assume there are two solutions, u_1 and u_2, and form their difference $u = u_1 - u_2$. Then u satisfies boundary conditions as above with $\varphi = 0$ in each case. Hence

$$0 = \int_D uLudx = \int_D (u\Delta u - k^2u^2)\, dx = \int_B u \frac{\partial u}{\partial \nu}\, dS - \int_D (|\nabla u|^2 + k^2u^2)\, dx.$$

The boundary integral is zero if u satisfies either homogeneous Dirichlet or homogeneous Neumann boundary conditions. The homogeneous Robin boundary condition implies that

$$\int_B u \frac{\partial u}{\partial \nu}\, dS = -\int_B au^2\, dS \leq 0.$$

In all three cases, use the information above to conclude that $\int_D u^2\, dx = 0$ because $k^2 > 0$; hence, $u(x) \equiv 0$.

Section 8-4:

16. By Poisson's formula (3-22)

$$u(x) = \int_{S_a} \frac{a^2 - |x|^2}{4\pi a|x - y|^3} u(y)\, dS_y \geq \int_{S_a} \frac{(a - |x|)(a + |x|)}{4\pi a(a + |x|)^3} u(y)\, dS_y$$
$$= \frac{a - |x|}{4\pi a(a + |x|)^2} \int_{S_a} u(y)\, dS_y = \frac{(a - |x|)a}{(a + |x|)^2} u(0)$$

by the mean value property.

18. Let x^0 be an arbitrary point in space. Then for any $a > 0$,

$$|u(x^0)| \leq \frac{3}{4\pi a^3} \int_{\mathcal{K}_a(x^0)} |u(y)|\, dy \leq \frac{3}{4\pi a^3} \left(\int_{\mathcal{K}_a(x^0)} |u(y)|^2\, dy\right)^{1/2} \left(\int_{\mathcal{K}_a(x^0)} dy\right)^{1/2}$$
$$\leq \left(\frac{3}{4\pi a^3}\right)^{1/2} \left(\int_{\mathcal{K}_a(x^0)} |u(y)|^2\, dy\right)^{1/2} \leq \left(\frac{3M}{4\pi a^3}\right)^{1/2}.$$

Section 8-5:

12. Construct a ball $\mathcal{K}_\rho(x^0)$, $\rho > 0$, such that $\overline{\mathcal{K}}_\rho(x^0) \subset D$. Let v be the function which is harmonic in $\mathcal{K}_\rho(x^0)$ and is equal to u on $S_\rho(x^0)$, the boundary of $\mathcal{K}_\rho(x^0)$. Then v is defined and continuous on $\overline{\mathcal{K}}_\rho(x^0)$. Show that $v(x) = u(x)$ except perhaps at x^0 by the following argument. (Once the argument is completed, v provides the extension of u to a harmonic function on the entire ball $\mathcal{K}_\rho(x^0)$.) Let $\epsilon > 0$ be arbitrary. It follows from the given limit condition on u and the continuity of v in $\overline{\mathcal{K}}_\rho(x^0)$ that for all $\delta > 0$ sufficiently small, $|u(x)| < \epsilon/2|x - x^0|$ and $|v(x)| < \epsilon/2|x - x^0|$ on $0 < |x - x^0| \leq \delta$. For any such sufficiently small δ, the function $\pm(u - v) + \epsilon/|x - x^0|$ is harmonic in the region $\delta < |x - x^0| < \rho$ and is positive on $|x - x^0| = \delta$ and on $|x - x^0| = \rho$. Conclude that $|u(x) - v(x)| \leq \epsilon/|x - x^0|$ for $\delta \leq |x - x^0| \leq \rho$. Since this inequality holds for all $\delta > 0$ sufficiently small, it follows that $|u(x) - v(x)| \leq \epsilon/|x - x^0|$ for $0 < |x - x^0| \leq \rho$. Finally, fix an x with $0 < |x - x^0| \leq \rho$ and let $\epsilon \to 0$ to conclude that $u(x) = v(x)$.

Section 8-6:

5. (a) Introduce spherical coordinates centered at x. Then $\nabla_y(1/r) = (x - y)/r^3$. On $r = |x - y| = \delta$, $|\nabla_y(1/r)| \leq 1/r^2 \leq 1/\delta^2$, $dS_y = \delta^2\, d\omega$, and $|f(y) - f(x)| \leq M|x - y| = M\delta$. Put these estimates together to obtain the desired conclusion.

Section 8-7:

3.

$$u(x) = \frac{1}{4\pi} \int_{R_2} \frac{\partial}{\partial y_3} \frac{1}{r_{xy}}\bigg|_{y_3=0} \sigma(y_1, y_2)\, dy_1\, dy_2.$$

The resulting "integral equation" is

$$u(x_1, x_2, 0) = f(x_1, x_2) = (1/2)\sigma(x_1, x_2)$$

so that the solution is

$$u(x) = \frac{1}{2\pi} \int_{R_2} \frac{\partial}{\partial y_3} \frac{1}{r_{xy}} \bigg|_{y_3=0} f(y_1, y_2)\, dy_1\, dy_2.$$

12. Of course, you must assume the compatibility condition $\int_{S_a(0)} f(x)\, ds_x = 0$. Assume a solution of the form

$$u(x) = \frac{1}{2\pi} \int_{S_a(0)} \left(\log \frac{1}{r_{xy}}\right) \rho(y)\, ds_y.$$

The integral equation for ρ is

$$f(x) = \frac{1}{2}\rho(x) + \frac{1}{2\pi} \int_{S_a(0)} \left(\frac{\partial}{\partial \nu_x} \log \frac{1}{r_{xy}}\right) \rho(y)\, ds_y$$

$$f(x) = \frac{1}{2}\rho(x) - \frac{1}{2\pi} \int_{S_a(0)} \frac{x-y}{r^2} \cdot \nu_x \rho(y)\, ds_y.$$

Show that $\nu_x = x/a$, $(x - y) \cdot x = a^2 - x \cdot y$, and $r^2 = |x - y|^2 = 2(a^2 - x \cdot y)$. Conclude that

$$f(x) = \frac{1}{2}\rho(x) - \frac{1}{4\pi a} \int_{S_a(0)} \rho(y)\, ds_y.$$

(Incidentally, this equation can hold only if the compatibility condition is satisfied.) Let L be the integral on the right. Then $\rho(x) = 2f(x) + (1/2\pi a)L$. Insert this expression into the formula for $u(x)$ to obtain

$$u(x) = \frac{1}{\pi} \int_{S_a(0)} \left(\log \frac{1}{r_{xy}}\right) f(y)\, ds_y + \frac{L}{4\pi a}.$$

Since the solution is determined only up to an additive constant we can assume the original potential chosen so that $L = 0$.

14. From the equation $\varphi'(z) = \xi_x + i\eta_x = \eta_y - i\xi_y$, conclude that the Cauchy–Riemann equations $\xi_x = \eta_y$ and $\xi_y = -\eta_x$ hold. Use these equations to confirm the given determinant is not zero. Why does the determinant inequality enable you to express u as a function of ξ and η? Now let $u(x, y) = u(\xi, \eta)$ and confirm by chain rule calculations that

$$u_x = u_\xi \xi_x + u_\eta \eta_x, \qquad u_y = u_\xi \xi_y + u_\eta \eta_y,$$

$$u_{xx} = u_{\xi\xi}\xi_x^2 + 2u_{\xi\eta}\xi_x\eta_x + u_{\eta\eta}\eta_x^2 + u_\xi \xi_{xx} + u_\eta \eta_{xx},$$

$$u_{yy} = u_{\xi\xi}\xi_y^2 + 2u_{\xi\eta}\xi_y\eta_y + u_{\eta\eta}\eta_y^2 + u_\xi \xi_{yy} + u_\eta \eta_{yy},$$

$$\Delta_{xy} u = u_{\xi\xi}(\xi_x^2 + \xi_y^2) + 2u_{\xi\eta}(\xi_x\eta_x + \xi_y\eta_y) + u_{\eta\eta}(\eta_x^2 + \eta_y^2) + u_\xi \Delta\xi + u_\eta \Delta\eta.$$

Use the Cauchy–Riemann equations to show that $\xi_x^2 + \xi_y^2 = \eta_x^2 + \eta_y^2 = |\varphi'(z)|^2$, $\xi_x\eta_x + \xi_y\eta_y = 0$, and $\Delta\xi = \Delta\eta = 0$. The desired equality follows.

CHAPTER 9

Section 9-1:

2.

$$u(x, t) = \int_{R_2} k(x - y, at) f(y)\, dy.$$

7. (b) The solution is

$$u(x, t) = \left(\frac{1}{4at}\right)^{3/2} \int_{R_3} e^{-|x-y|^2/4t} f(\rho, y_3)\, dy, \qquad \rho = \sqrt{y_1^2 + y_2^2}.$$

Let $x_1 = r \cos \theta$, $x_2 = r \sin \theta$, $x_3 = z$, $y_1 = \rho \cos \varphi$, $y_2 = \rho \sin \varphi$, $y_3 = \zeta$. Then $(x_1 - y_1)^2 + (x_2 - y_2)^2 = r^2 - 2r\rho \cos (\theta - \varphi) + \rho^2$ and

$$u(x, t) = \left(\frac{1}{4\pi t}\right)^{3/2} \int_{-\infty}^{\infty} \int_0^{\infty} e^{-[r^2+\rho^2+(z-\zeta)^2]/4t} f(\rho, \zeta) \int_0^{2\pi} e^{[2r\rho \cos(\theta-\varphi)]/4t}\, d\varphi \rho d\rho d\zeta$$

$$= \frac{1}{4\sqrt{\pi} t^{3/2}} \int_{-\infty}^{\infty} \int_0^{\infty} e^{-[r^2+\rho^2+(z-\zeta)^2]/4t} I_0\left(\frac{r\rho}{2t}\right) f(\rho, \zeta) \rho d\rho d\zeta.$$

9. The solution in three dimensions is as follows. Let $A(t) = \int_0^t a(\tau)\, d\tau$, $V(t) = \int_0^t v(\tau)\, d\tau$, $C(t) = \int_0^t c(\tau)\, d\tau$, and $k(x, t)$ denote the three-dimensional fundamental solution for the heat equation. Then

$$u(x, t) = e^{C(t)} \int_{R_3} k(x + V(t) - y, A(t)) f(y)\, dy.$$

14. Integrate the expression for $u(x, t)$, interchange the orders of integration and use the fact that the integral of the fundamental solution over R_3 is 1.

Section 9-2:

1.

$$u(r, t) = \sum_{j=1}^{\infty} \alpha_j e^{-a\lambda_j^2 t} J_0(\lambda_j r),$$

where $\lambda_j = \kappa_j / a$ and the κ_j are the positive roots of $J_0(\kappa) = 0$ and

$$\alpha_j = \frac{2}{\lambda_j a^2 J_1^2(\lambda_j a)} \int_0^a f(r) r J_0(\lambda_j r)\, dr.$$

7. Let $\mu_n = (n + 1/2)(\pi/2)$ and $\nu_p = p\pi$ for $n, p = 0, 1, 2, \ldots$ and $\lambda_{np}^2 = \mu_n^2 + \nu_p^2$. Then

$$u(x, y, t) = \sum_{n,p=0}^{\infty} A_{np} e^{-a\lambda_{np}^2 t} \sin (\mu_n x) \cos (\nu_p y).$$

12.

$$c \int_D v^2\, dx = \int_D v \Delta v\, dx = -\int_D |\nabla v|^2\, dx$$

by Green's first identity and the fact that v vanishes on B. If $c \geq 0$, this identity implies that v is identically zero.

ction 9-3:

5.
$$I'(t) = \int_D w(x, t)w_t(x, t)\,dx = a\int_D w(x, t)\Delta w(x, t)\,dx$$
$$= -\int_D |\nabla w(x, t)|^2\,dx \leq 0$$

by the boundary conditions. Thus, $I(t) \leq I(0) = 0$ by the initial condition. Conclude that $w(x, t) \equiv 0$.

17. $G(x, y, t)$ satisfies the heat equation in $x = (x_1, x_2)$ and t for $x_2 > 0$ and $t > 0$ and vanishes when $x_2 = 0$.

ction 9-4:

2.
$$u(x, t) = a\int_0^t k(x, a(t-\tau))\rho(\tau)\,d\tau,$$
$$u_x(x, t) = -a\int_0^t \frac{2x}{4a(t-\tau)}\frac{e^{-x^2/4a(t-\tau)}}{\sqrt{4\pi(t-\tau)}}\rho(\tau)\,d\tau.$$

Now let $x \to 0+$ to conclude that $f(t) = -\rho(t)/2$ and hence
$$u(x, t) = -2a\int_0^t k(x, a(t-\tau))f(\tau)\,d\tau.$$

3. To solve by Laplace transforms, use the convolution theorem and the fact that the Laplace transform of $t^{-1/2}$ is $\sqrt{\pi/s}$. If the problem is done by iteration an infinite series is obtained. To express the series as a function operating on $f(t)$, an interchange of order of integration and summation is necessary. Integrals of the form
$$\int_\tau^t (t-\sigma)^{-1/2}(\sigma-\tau)^r\,d\sigma = (t-\tau)^{r+1/2}\frac{\Gamma(1/2)\Gamma(r+1)}{\Gamma(r+3/2)},$$
where $\Gamma(z)$ is the gamma function, come up.

CHAPTER 10

ction 10-1:

7. Let $\lambda = r/ct$ and $u = f(\lambda)$. The wave equation when u depends only on r and t is $u_{tt} = (c^2/r^2)(r^2u_r)_r$. The function $f(\lambda)$ must satisfy the ordinary differential equation
$$(\lambda^2 - 1)f''(\lambda) + 2\left(\lambda - \frac{1}{\lambda}\right)f'(\lambda) = 0 \quad \text{or} \quad f''(\lambda) + \frac{2}{\lambda}f'(\lambda) = 0.$$

Consequently, $f(\lambda) = (A/\lambda) + B$ for arbitrary constants A and B.

10. Use the vector identity $\nabla \times (\nabla \times \mathbf{E}) = \nabla \operatorname{div} \mathbf{E} - \Delta \mathbf{E}$ and the equation $\operatorname{div} \mathbf{E} = 0$.

Section 10-2:

4. The orthogonality of $v^{(1)}$ and $v^{(2)}$ can be verified by direct calculation. When $\lambda_{nj} \neq \lambda_{pk}$, apply Prob. 3.

9. Let $J_0(\kappa)$ denote the Bessel function of order 0 and let κ_j, $j = 1, 2, \ldots$, denote the positive root of $J_0(\kappa)$. The solution is

$$u(r, t) = \sum_{j=1}^{\infty} J_0(\lambda_j r)[\alpha_j \cos(c\lambda_j t) + \beta_j \sin(c\lambda_j t)]$$

where $\lambda_j = \kappa_j/a$ and the α_j and β_j are chosen so that u satisfies the initial conditions. To do so use the orthogonality relations

$$\int_0^a J_0(\lambda_j r)J_0(\lambda_k r)r\,dr = \begin{cases} 0, & j \neq k \\ (a^2/2)J_1^2(\lambda_j a), & j = k. \end{cases}$$

12.
$$u(r, t) = \frac{1}{r}\sum_{n=1}^{\infty} \beta_n \sin(c\lambda_n t)\sin(\lambda_n r), \qquad \beta_n = \frac{2}{ca\lambda_n}\int_0^a f(r)r\sin(\lambda_n r)\,dr,$$

where $\lambda_n = n\pi/a$.

Section 10-7:

4. Define $\varphi(x_2, x_3, t)$ to be zero for $t \leq 0$. Assume a solution in the form of a double layer potential (7-4). For this let $y = (y_1, y_2)$ and $\nu_y = (-1, 0, 0)$, $r = \sqrt{x_1^2 + (x_2 - y_2)^2 + (x_3 - y_3)^2}$. Then

$$u(x, t) = -\frac{\partial}{\partial x_1}\int_{R_2} \frac{1}{4\pi c^2 r}\,\mu(y, t - r/c)\,dy.$$

The "integral equation" (7-9) becomes $\varphi(x, t) = -(1/2c^2)\mu(x, t)$, $x = (0, x_2, x_3)$ and the solution is

$$u(x, t) = \frac{\partial}{\partial x_1}\int_{R_2} \frac{1}{2\pi r}\,\varphi(y, t - r/c)\,dy.$$

CHAPTER 11

Section 11-1:

1.
$$\delta I = \left(\frac{d}{d\epsilon} I(v + \epsilon\zeta)\right)_{\epsilon=0}$$

$$= \left(\frac{d}{d\epsilon}\int_{t_1}^{t_2}\int_0^L \left[\frac{1}{2}\rho(v_t + \epsilon\zeta_t)^2 - \frac{1}{2}\tau(v_x + \epsilon\zeta_x)^2 + (v + \epsilon\zeta)F\right] dx\,dt\right)_{\epsilon=0}.$$

10.
$$\delta I = \frac{1}{2}\left(\frac{d}{d\epsilon}\int_D e^{-v\cdot x}|\nabla(u+\epsilon\zeta)|^2\,dx\right)_{\epsilon=0} = \int_D e^{-v\cdot x}\nabla u \cdot \nabla\zeta\,dx$$
$$= -\int_D \zeta\,\mathrm{div}\,(e^{-v\cdot x}\nabla u)\,dx = -\int_D \zeta e^{-v\cdot x}[\Delta u - v\cdot\nabla u]\,dx,$$

because ζ was chosen to vanish on the boundary of D.

ection 11-3:

1. $|\langle f_n, h\rangle - \langle f, h\rangle| = |\langle f_n - f, h\rangle| \le \|f_n - f\|\,\|h\|$ by the Schwarz inequality. So $\langle f_n, h\rangle \to \langle f, h\rangle$. Similarly, $\langle h, g_n\rangle \to \langle h, g\rangle$. Then

$$|\langle f_n, g_n\rangle - \langle f, g\rangle| \le |\langle f_n, g_n\rangle - \langle f_n, g\rangle + \langle f_n, g\rangle - \langle f, g\rangle|$$
$$\le \|f_n\|\,\|g_n - g\| + \|f_n - f\|\,\|g\|$$

by the triangle and Schwarz inequalities. Next, $\|f_n\|$ is bounded: there exists N such that $n > N$ implies $\|f_n - f\| < 1$ or $\|f_n\| < \|f\| + 1$. So $\|f_n\|$ is bounded by $M = \max(\|f_1\|, \ldots, \|f_N\|, \|f\| + 1)$.

4.
$$\|f_n - f\|^2 = \langle f_n - f, f_n - f\rangle = \|f_n\|^2 - \langle f, f_n\rangle - \langle f_n, f\rangle + \|f\|^2.$$

Now let $n \to \infty$ and use Prob. 1.

15. Use the Schwarz inequality.

ection 11-5:

2. If $f \in H_1$, then $\|f\|_{H_1}$ is finite and $\|f\|_{H_1}^2 = \|f\|_{L_2}^2 + \|\nabla f\|_{L_2}^2 \ge \|f\|_{L_2}^2$.

5. By Th. 4-2 $\|u\|_{H_1^0}^2 = \|u\|_{L_2}^2 + \|\nabla u\|_{L_2}^2 \le C\|\nabla u\|_{L_2}^2 + \|\nabla u\|_{L_2}^2$.

CHAPTER 12

ection 12-2:

5. $f(\xi)$ satisfies $\beta f'(\xi)x^\alpha t^{\beta-1} = a\alpha^2 f''(\xi)x^{2\alpha-2}t^{2\beta} + af'(\xi)\alpha(\alpha-1)x^{\alpha-2}t^\beta + b\alpha^2 f'^2(\xi)\,x^{2\alpha-2}t^{2\beta}$. Divide by $x^{\alpha-2}t^\beta$, then choose $\alpha = 2$, $\beta = -1$ to obtain $-f'(\xi)\xi = 4a\xi f''(\xi) + 2af'(\xi) + 4b\xi f'^2(\xi)$. Let $y = f'(\xi)$ to obtain the Bernoulli equation $4a\xi y' + (2a + \xi)y + 4b\xi y^2 = 0$, which reduces to the linear equation $4a\xi w' - (2a + \xi)w = 4b\xi$ where $w = 1/y$.

ection 12-3:

3. Multiply by $f''(u)$.

$$f''(u)u_t + f''(u)f(u)_x = 0,$$

$$\frac{\partial}{\partial t} f'(u) + \frac{\partial}{\partial x} \frac{f'^2(u)}{2} = 0.$$

Now, let $w = f'(u)$.

Section 12-5:

1. Assume there are two solutions, u_1 and u_2, and let $u = u_1 - u_2$. Now do Prob. 5 and use the maximum principle.
2. If all the details are checked, this can be a lengthy problem. One mimics the arguments after (5-8).

Suggestions for Further Reading

The field of partial differential equations is vast and there are numerous alternative approaches to it. General treatments of the field containing similar and additional material to that given in this text may be found in:

R. Courant and D. Hilbert: *Methods of mathematical physics*, vols. 1 and 2, Wiley-Interscience, New York, 1953 and 1962. See also volume 2 of the German edition: *Methoden der mathematischen Physik*, Springer-Verlag, Berlin, 1968 (reprint of 1937 edition).

P. Garabedian: *Partial differential equations*, John Wiley & Sons, Inc., New York, 1964.

G. Hellwig: *Partial differential equations*, Blaisdell, Waltham, Mass., 1964.

F. John: *Partial differential equations*, Springer-Verlag, New York, 1982.

O. A. Ladyzhenskaya; *The boundary value problems of mathematical physics*, Springer-Verlag, Berlin, 1985.

A. N. Tychonov and A. A. Samarski: *Partial differential equations of mathematical physics*, Holden-Day San Francisco, 1967.

J. Wloka: *Partielle Differentialgleichungen*, BSB B. G. Teubner Verlagsgesellschaft, Leipzig, 1982.

Considerable additional work has been done on each of the three major types of second-order differential equations. We mention here only a few. For elliptic equations, see

D. Gilbarg and N. S. Trudinger: *Elliptic partial differential equations of second order*, Springer-Verlag, Berlin, 1977.

M. König: *Schauder estimates and boundary value problems for quasilinear partial differential equations*, Les presses de l'Université de Montréal, 1985.

O. A. Ladyzhenskaya and N. N. Ural'tseva: *Linear and quasilinear elliptic equations*, Academic Press, Inc., New York, 1968.

For parabolic equations, see

J. R. Canon, The one-dimensional heat equation, *Encyclopedia of mathematics and its applications*, Addison-Wesley Publishing Co., Inc., Reading, Mass., 1984.

S. D. Eidel'man: *Parabolic systems*, North-Holland Publishing Company, Amsterdam, 1969.

A. Friedman: *Partial differential equations of parabolic type*, Prentice-Hall, Inc., Englewood Cliffs, N.J., 1964.

A. M. Il'm, A. S. Kalashmikov, and O. A. Oleinik: Second order linear equations of parabolic type, *Russian mathematical surveys*, London Mathematical Society, London, 1962.

O. A. Ladyzhenskaya, V. A. Solomikov, N. N. Ural'tseva: *Linear and quasilinear equations of parabolic type*, American Mathematical Society, Providence, R.I., 1968.

For equations of hyperbolic type, see

R. Sakamoto: *Hyperbolic boundary value problems*, Cambridge University Press, Cambridge, 1978.

G. B. Whitham: *Linear and nonlinear waves*, Wiley-Interscience, New York, 1974.

Variational methods have undergone rapid development during recent years. Several sources reporting on these developments are the following.

A. Friedman: *Variational principles and free boundary value problems*, Wiley-Interscience, New York, 1982.

D. Kinderlehrer and G. Stampacchia: *An introduction to variational inequalities and their applications*, Academic Press, Inc., New York, 1980.

C. B. Morrey: *Multiple integrals in the calculus of variations*, Springer-Verlag, New York, 1966.

A deeper understanding of generalized functions, that is, distribution theory, is essential for further study of partial differential equations. The first two sources below give nearly everything one needs to know for applications. The second two sources are less encyclopedic and contain more applications.

I. M. Gelfand and G. E. Shilov: *Generalized functions*, vols. 1–4, Academic Press, Inc., New York, 1964–1968.

L. Hörmander: *The analysis of linear partial differential operators*, vols. 1–4, Springer-Verlag, Berlin, 1985.

J. Mikusiński: *Operational calculus*, vol. 1, Pergamon Press, Inc., Elmsford, N.Y., 1983.

L. Schwartz: *Mathematics for the physical sciences*, Addison-Wesley Publishing Co., Inc., Reading, Mass, 1966.

Books summarizing the state of the theory and applications of free boundary value problems until now are:

A. Bossavit, A. Damlamian, and M. Fremond (eds.): *Free boundary value problems, applications and theory*, vol. 5, Pitman Publishing, Inc., Marshfield, Mass., 1985.

A. Fasano and M. Primicerio (eds.): *Free boundary value problems, theory and applications*, vols. 1 and 2, Pitman Publishing, Inc., Marshfield, Mass., 1983.

L. I. Rubinstein, The Stefan problem, *Translation of Mathematical Monographs*, American Mathematical Society, Providence, R. I., 1971.

I. O. Speck (ed.): *Free boundary value problems, applications and theory*, vol. 3, Pitman Publishing, Inc., Marshfield, Mass., 1985.

The field of applications of partial differential equations is so broad that a complete list of references would be impossible. A good single source is the set of volumes making up the *Handbuch der Physik* (old and new series), published by Springer-Verlag, Berlin, which contains articles in English, French, and German. Less formidable is the book

P. Frank and R. von Mises (eds.): *Die Differentialgleichungen der mathematischen Physik*, vols. 1 and 2, F. Vieweg & Sohn, Braunschweig, 1930–1935.

One can get an idea of how far reaching and deep applications can be by consulting

G. Herglotz: *Vorlesungen über die Mechanik der Kontinua*, BSB B. G. Teubner Verlagsgesellschaft, Leipzig, 1985.

O. A. Ladyzhenskaya: *The mathematical theory of viscous incompressible flow*, Gordon and Breach, Science Publishers, Inc., New York, 1969.

In view of the fact that partial differential equations is such an important area, it should come as no surprise that the literature on numerical solution of both the differential equations themselves and related integral equations is voluminous. Consequently, we list only a few sources which can be used as a starting point for additional study.

K. E. Atkinson: *A survey of numerical methods for the solution of Fredholm integral equations of the second kind*, SIAM, Philadelphia, 1976.

O. Axelsson and V. A. Barker: *Finite element solution of boundary value problems*, Academic Press, Orlando, Fl., 1984.

L. Collatz: *Numerical analysis and functional analysis*, Springer-Verlag, Berlin, 1969.

I. Gladwell and R, Wait (eds.): *A survey of numerical methods for partial differential equations*, Oxford University Press, Oxford, 1979 (also contains a discussion of integral equations' methods).

Index